Brain Science

Series Editors

The "Brain Science Series" provides critical and comprehensive discussions of the most significant areas of neuroscience research, written by scientific leaders and authorities. Each volume in the series presents the latest and most complete information on its respective subject, making it an unrivalled reference source. To analyze brain functions that have been preserved through evolution, a reductive approach based on gene manipulation using models such as genetically modified rodents were mainly used. At the same time, to analyze high-order brain functions that are uniquely acquired only in non-human primates and humans, analyses of complex behaviors using a psychological approach were mainly adopted. The Brain Science Series focuses on clarifying the operating principles of the neural circuitry governing the human mind, as well as the accompanying molecular mechanisms.

Shintaro Funahashi

Dorsolateral Prefrontal Cortex

Working Memory and Executive Functions

Springer

Shintaro Funahashi
Kokoro Research Center
Kyoto University
Kyoto, Kyoto, Japan

ISSN 2570-0197 ISSN 2570-0200 (electronic)
Brain Science
ISBN 978-981-19-7267-6 ISBN 978-981-19-7268-3 (eBook)
https://doi.org/10.1007/978-981-19-7268-3

This Springer imprint is published by the registered company Springer Nature Singapore Pte Ltd.
The registered company address is: 152 Beach Road, #21-01/04 Gateway East, Singapore 189721, Singapore

This book is dedicated to my two great mentors, Profs. Kisou Kubota and Prof. Patricia S. Goldman-Rakic, and my wife, Mihoko Funahashi.

Preface

The prefrontal cortex is known to participate in a variety of higher cognitive functions including appropriate planning, thinking, reasoning, and decision-making. The prefrontal cortex occupies a larger part in the cerebral cortex compared with other association cortices and consists of three main sectors (lateral, medial, and orbital sectors). Among these three sectors, the lateral sector, especially the dorsolateral part of the prefrontal cortex, is very well known to play most important roles in those higher cognitive functions. Since neurophysiological studies using monkeys had been started in 1970s and since neuroimaging studies using human subjects had been started in 1990s, huge amount of information regarding prefrontal functions and their neural mechanisms has been accumulated. Based on accumulated information, several important books focused on prefrontal functions have been published. These include *The Prefrontal Cortex* (fifth edition) by Joaquin M. Fuster and *The Neurobiology of the Prefrontal Cortex* by Richard E. Passingham and Steven P. Wise. The former is the most comprehensive textbook of the prefrontal cortex and the topics include anatomy, pharmacology, animal and human neuropsychology, neurophysiology, human neuroimaging, and a theory of prefrontal functions. I have used this book as the most important reference book from the first edition. In the latter, the authors described functions of each of five sectors (medial, orbital, caudal, dorsal, and ventral) of the prefrontal cortex and summarized the function of the prefrontal cortex as generating goals from current contexts and events. The third book is *Principles of Frontal lobe Function* (second edition) edited by Donald T. Stuss and Robert T. Knight, which is based on the second symposium on the prefrontal cortex held at Toronto in 2010 and consists of a variety of topics including models, social behavior, clinical neuropsychology, psychiatric disorders, and rehabilitation.

These books are important ones to understand functions of the prefrontal cortex. Since 1972, a large amount of neurophysiological results has been accumulated for the functions of the dorsolateral prefrontal cortex. It has been recognized that working memory is a useful concept to understand several important functions of the dorsolateral prefrontal cortex. However, there is no book which focuses on functions of the dorsolateral prefrontal cortex and their relations to working memory.

Therefore, this book focuses mainly on functions of the dorsolateral prefrontal cortex, especially how neural mechanisms related to working memory contribute to these functions. I have investigated dorsolateral prefrontal functions using monkeys by neurophysiological techniques since 1977 at Kyoto University and Yale university. Therefore, I mainly focus on neurophysiological results obtained from monkey dorsolateral prefrontal cortex in this book. Because of this reason, not many studies regarding human neuroimaging studies, pharmacological studies, and theoretical studies are included in this book. However, I think that I can show and summarize important studies to understand dorsolateral prefrontal functions and their neural mechanisms in this book, which include working memory, executive control, cognitive control and memory management, decision-making, metacognition, and top-down control.

Kyoto, Japan
April 27, 2022

Shintaro Funahashi

Acknowledgments

I would like to express my sincere thanks to Prof. Kisou Kubota at Kyoto University for the long-time support and encouragement to my research. I would also like to express my sincere thanks to Late Prof. Patricia S. Goldman-Rakic and Charles J. Bruce at Yale University for their extremely valuable support and kind help during my 7-year stay at Yale University and thereafter. And finally, I would like to express my sincere thanks to the students of my group at Kyoto University, especially to Masato Inoue, Kazuyoshi Takeda, Yumiko Watanabe, Satoe Ichihara-Takeda, Saori Igaki, Kei Watanabe, Jorge Mario Andreau, Kei Mochizuki, and Akio Tanaka. Without working with the students of my group at Kyoto University, this book would not be possible. I would like to thank for the funds (KAKENHI) from the Japanese Ministry of Education, Culture, Sports, Science and Technology (MEXT) and from Japan Society for the Promotion of Science (JSPS) and the funds from Japan Science and technology Agency (JST) for supporting my research activities at Kyoto University. I also express my thanks to Beijing Institute of Technology and Shenzhen Institute of Advanced Technology for supporting my continuation of prefrontal researches in China after my retirement from Kyoto University.

Contents

Chapter 1
Dorsolateral Prefrontal Cortex

1.1 Introduction

The prefrontal cortex is known to play an important role in planning and controlling a variety of cognitive functions and in organizing and controlling various goal-directed behaviors. The prefrontal cortex is an important brain structure related to executive functions (Baddeley 1986; Stuss and Benson 1986; Fuster 2008). Executive control is a neural system necessary for coordinating and supervising the operations of multiple neural systems. To accomplish a specific goal, it is necessary to acquire appropriate information from the environment as well as from memory, find a possible solution, and make a series of decisions. To make correct decisions, apt reasoning and judgment are necessary. Appropriate reasoning, right judgment, and correct decision-making include complex cognitive operations and require simultaneous and coordinated management of multiple functions, such as controlling and focusing attention, inhibiting irrelevant information, managing and planning actions and behavior, monitoring and coding information, and working memory. To coordinately manage these multiple functions, some specific neural mechanism is needed and that is executive control. If the prefrontal cortex plays essential roles in executive control, it is expected that damages to the prefrontal cortex cause an impairment in reasoning, judgment, and decision-making.

In fact, neuropsychological studies show that patients having prefrontal damage exhibit unique characteristics in their behavioral impairment. These patients usually exhibit normal IQ, exhibit comparable performances to normal subjects in most neuropsychological tests, have normal long-term memory, and exhibit normal perceptual, motor, and language skills. However, when the patients need to perform some intellectual activities, such as making a new and adaptive action plan or making the best choice among several alternatives, their performances are profoundly disturbed. These disturbances often cause poor judgment and planning, and inappropriate decision-making. Stuss and Benson (1986) summarized the behavioral characteristics of prefrontal patients as follows: Frontal patients exhibit

S. Funahashi, *Dorsolateral Prefrontal Cortex*, Brain Science,
https://doi.org/10.1007/978-981-19-7268-3_1

a lack of insight and foresight, diminish the capacity for planning, and exhibit decreased initiative. Similar impairments are also observed in animals with prefrontal lesions. The impairments observed in both human prefrontal patients and animals with prefrontal lesions cannot be explained simply by the failure of a single particular function such as perception, motor, or memory. Intellectual activities such as insight, foresight, planning, and decision-making require the coordinated operation of not only perceptual and motor functions but also various cognitive functions including working memory, attention control, and monitoring and managing activities. Therefore, the impairments observed in prefrontal patients can be explained by a failure of coordinated operation of multiple activities. The neural mechanism for achieving coordinated operation of multiple activities is called executive control. Therefore, the syndromes caused by prefrontal damages are called dysexecutive syndromes (Stuss and Benson 1986). In this chapter, general structural and functional features of the prefrontal cortex, especially the dorsolateral prefrontal cortex, are described.

1.2 The Prefrontal Cortex

The prefrontal cortex is a part of the cerebral cortex and is located in the most anterior portion of the cerebral cortex (Fig. 1.1). The cerebral cortex consists of four lobes; the frontal lobe, the parietal lobe, the temporal lobe, and the occipital lobe. The frontal lobe is placed at the anterior portion of the cerebral cortex and is separated from the parietal lobe by the central sulcus and from the temporal lobe by the lateral sulcus. The frontal lobe is divided into two functionally distinct areas

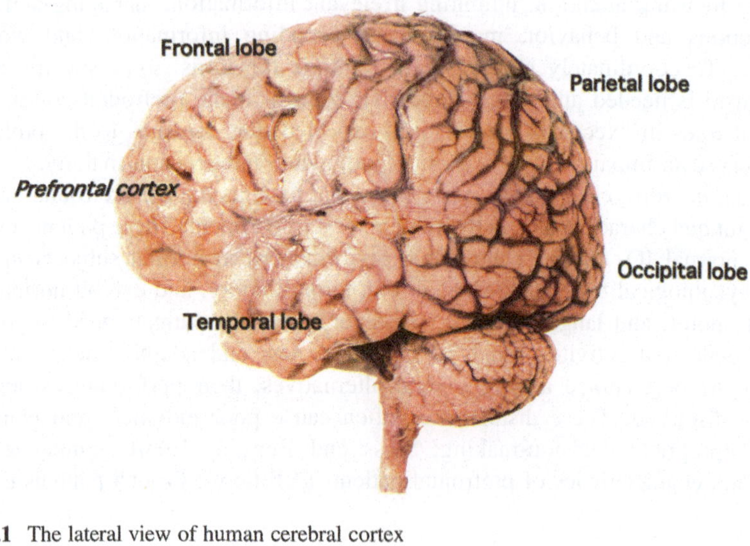

Fig. 1.1 The lateral view of human cerebral cortex

along the anterior-posterior axis. The posterior portion of the frontal lobe is known as the skeletomotor area. The primary motor cortex and the premotor cortex are located in the lateral portion, the supplementary motor area and the pre-supplementary motor area are located in the dorsomedial portion, and the anterior and the posterior cingulate motor areas are located in the medial portion. These motor areas participate in programming, producing, and controlling various voluntary actions of upper and/or lower limbs.

On the other hand, the anterior portion of the frontal lobe is known as the frontal association cortex or the prefrontal cortex. The prefrontal cortex is also divided into three areas: lateral, medial, and orbital areas. Differing from posteriorly located frontal motor areas, the prefrontal cortex does not participate in information processing specifically for a particular sensory modality or for controlling a particular motor behavior. The prefrontal cortex participates in information processing to manage complex behavior to achieve a specific goal as well as controlling higher cognitive functions such as thinking, reasoning, judgment, decision-making, language comprehension, focusing attention, and control of emotion and motivation. Processing multiple sensory modalities and integrating various information are important operations of the prefrontal cortex to achieve complex and higher cognitive functions. Brutkowski (1965) indicated that Hines (1929) was the first person to designate the portion of the frontal cortex locating rostral to the premotor cortex and the frontal eye field as "*the prefrontal cortex.*" Hines (1929) stated about the prefrontal cortex as follows:

> *The frontal lobe, bounded posteriorly by the sulcus centralis (the fissure of Rolando), includes in man very nearly the anterior third of the cortex. The electrically excitable area on the precentral gyrus, and the posterior third of the inferior frontal convolution (Broca's area, field 44) are usually considered separately, yet even then the greater bulk of the enormous subdivision remains. Lying immediately anterior to the precentral convolution is an area (field 6 of Brodmann) of characteristic cyto-architecture, the premotor cortex. This bears a close functional relationship to its posterior neighbor, for certainly it is capable of "facilitating" the activity of the motor cortex and, under certain experimental conditions, itself gives definite movements of the extremities on stimulation. There is, besides, a small area in the center of the second frontal convolution (field 46 of Brodmann; FD_A of Economo and Koskinas) where electrical stimulation in apes (Leyton and Sherrington) and in man (Foerster) is productive, here of conjugate movements of the eyes to the opposite side. The large remainder of the lateral and medial surfaces of the frontal lobe is "silent." This is the prefrontal area.* (Page 534, Hines 1929)

The prefrontal cortex receives a variety of sensory information and emotional and motivational information from other cortical and subcortical areas. The prefrontal cortex integrates various information and generates appropriate outputs to manage operations in other cortical and subcortical areas in order to achieve a specific goal. Therefore, the prefrontal cortex is thought of as the functionally most important area in the cerebral cortex. The prefrontal cortex is thought to be the cortical area that participates in such a higher cognitive function as intelligence.

Not only humans and monkeys but also cats (Markowitsch et al. 1978), rats (Uylings et al. 2003), and mice (Le Merre et al. 2021) have the prefrontal area in the cerebral cortex (Fig. 1.2). The cerebral cortex of cats and rats does not have similar

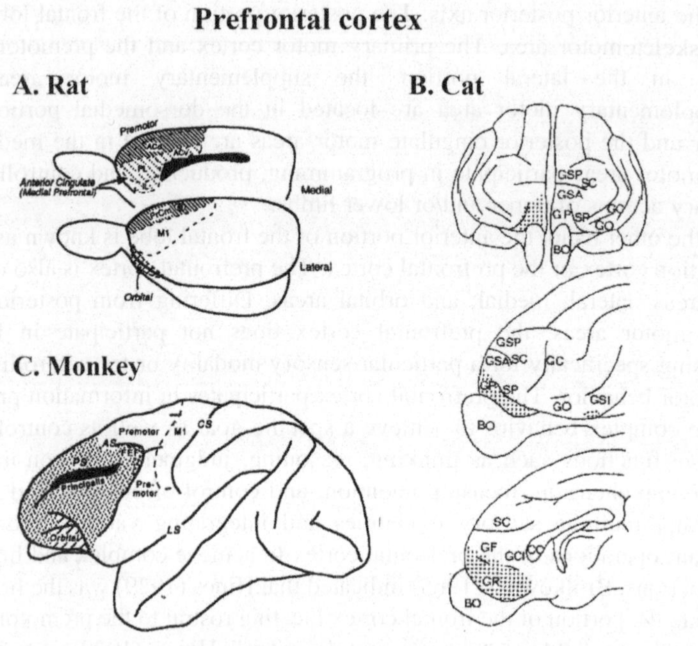

Fig. 1.2 Prefrontal cortex (shaded area) in the brains of rat (a), cat (b), and monkey (c). From Markowitsch et al. (1980) and Preuss (1995) with permission

well-defined patterns of sulci and gyri as the human cerebral cortex has. In addition, patterns of sulci and gyri are different among mammalian species. Therefore, the location and the extent of the prefrontal cortex are not easy to define by surface structures on the cerebral cortex of these animals. The primate brain including the human brain has well-developed granular layer IV in the prefrontal cortex. Therefore, the location and the extent of the prefrontal cortex can be defined as the cortical area which has the granular layer IV in the frontal lobe. In fact, Brodmann (1909) used the presence of the granular layer IV in the cortex anterior to the frontal motor areas as the criterion of the prefrontal cortex. Using this criterion, he determined the prefrontal cortex in the cerebral cortex of marmosets, lemurs, rabbits, ground squirrels, and other animals. However, since mammals other than primates have no granular layer in the frontal lobe, these animals had been considered to have no prefrontal area in the cerebral cortex. On the other hand, since neurons in the mediodorsal nucleus of the thalamus project exclusively to the prefrontal cortex, Rose and Woolsey (1948) proposed that the prefrontal cortex should be defined as the cortical area receiving anatomical projections from the mediodorsal nucleus of the thalamus. Since then, the definition that uses anatomical projections from the mediodorsal nucleus of the thalamus has been used to identify the cortical area of the prefrontal cortex in various animals.

1.3 Features of the Prefrontal Cortex

The prefrontal cortex has several important features. First, the prefrontal cortex has several specific anatomical features. The prefrontal cortex occupies 29% of the total volume of the cerebral cortex in the human brain. The volume of the prefrontal cortex is the largest compared with other areas in the human cerebral cortex (e.g., the motor areas occupy 8%) and the proportion of the human prefrontal cortex is the largest among those of other mammalian animals (Table 1.1). The location and the extent of the prefrontal cortex are defined anatomically as the cortical area which receives projections from the mediodorsal nucleus of the thalamus. Using this definition, the proportion of the prefrontal cortex in the total volume of the cerebral cortex has been estimated in several animals (Table 1.1). These proportions are 29% in human, 17% in the chimpanzee, 11% in the macaque monkey, 8.5% in the lemur, 7% in the dog, and 3.4% in the cat (Elston 2003), indicating that the prefrontal cortex occupies a large volume in the human cerebral cortex and that the prefrontal cortex of the human brain occupies the largest proportion in the cerebral cortex among mammalian species. Bush and Allman (2004) examined the scaling of the frontal cortex using 43 species of mammals including 25 primates and found significant differences in scaling between primates and carnivores. In addition, they also found that the primate frontal cortex is hyper-scaled relative to the rest of the neocortex.

Table 1.1 Comparison of volumes of the granular and agranular areas of the prefrontal cortex and ratios of these areas to the whole volume of the cerebral cortex among various animals. From Elston (2003)

Species	Total cortical volume (mm^3)	Granular FC	Agranular FC	Total FC
Man	108,221	30,254 = **27.9**	6117 = 5.7	36,371 = 33.6
Man	135,470	39,287 = **29.0**	9833 = 7.3	49,120 = 36.3
Chimpanzee	39,572	6719 = **16.9**	5389 = 13.6	12,108 = 30.5
Gibbon	16,301	1839 = **11.3**	1651 = 10.1	3490 = 20.4
Mandrill	31,321	2186 = **10.1**	2194 = 10.3	4362 = 20.4
Baboon	20,594	1967 = **9.5**	2898 = 14.1	4865 = 23.6
Baboon	20,376	2111 = **10.3**	2200 = 10.8	4311 = 21.1
Macaque	15,308	1733 = **11.3**	1817 = 11.9	3550 = 23.2
Guenon	14,641	1625 = **11.1**	1976 = 13.5	3601 = 24.6
Capuchin	13,682	1260 = **9.2**	1822 = 13.3	3082 = 22.5
Marmoset	1649	148 = **8.9**	167 = 10.1	315 = 19.0
Black lemur	4054	337 = **8.3**	453 = 11.2	790 = 19.5
Microcebus	921	70 = **7.2**	94 = 10.2	164 = 17.8
Flying foxa	1097	26 = **2.3**	73 = 6.6	99 = 8.9
Dog	9527	657 = **6.9**	1283 = 13.4	1940 = 20.3
Cat	4474	152 = **3.4**	412 = 9.2	564 = 12.6
Rabbit	1627	36 = **2.2**	148 = 9.1	184 = 11.3
Hedgehog	575	0 = **0**	24 = 4.0	24 = 4.0
Armadillo	2010	0 = **0**	93 = 4.6	93 = 4.6
Opossum	804	0 = **0**	51 = 6.3	51 = 6.3

Bold numbers indicate % of the prefrontal cortical volume against the whole cerebral cortical volume

Donahue et al. (2018) performed an accurate analysis of regional differences in the content of the cortical gray matter and found that the proportion of the cortical gray matter that the prefrontal cortex occupies in human brain is up to 1.9-fold greater compared with macaque's brain and 1.2-fold greater compared with chimpanzee's brain. Further, Barrett et al. (2020) used a novel tractography method and demonstrated that frontal networks including within and beyond the frontal lobe occupied 66% of total brain white matter in humans compared with 48% in macaque species. This difference was also observed in projection, commissural, inter-lobar, and intra-lobar association tracts, indicating that greater communication pathways are present within the frontal lobe and between the frontal lobe and other cortical areas in human brain. Among the association tracts, the greatest difference was found between humans and monkeys for tracts involved in motor planning, auditory memory, top-down control of sensory information, and visuospatial attention. These results indicate that the human frontal lobe significantly involves in higher cognitive functions and language.

Second, it has been known that the prefrontal cortex takes a longer time to mature, compared with other cortical areas. As a method to estimate the degree of maturation in cortical development, the degree of myelination of intrinsic nerve fibers in a given cortical area had often been used (Flechsig 1901, 1920). In the matured cortical areas, most of the intrinsic nerve fibers are myelinated, while in the early phase of cortical development, most of these fibers are not myelinated and the proportion of myelinated fibers is different among cortical areas depending on the level of their development. For example, most of the intrinsic nerve fibers are myelinated at birth in the primary sensory and motor areas (black areas in Fig. 1.3). However, in the sensory and motor association areas, intrinsic nerve fibers become mostly myelinated after birth and the completion of the myelination takes longer time compared with the primary areas (shaded areas in Fig. 1.3). Therefore, the degree of the myelination of intrinsic nerve fibers had been used as an index for estimating the degree of maturation of a particular cortical area. The myelination of intrinsic nerve fibers in the association cortices is shown to continue for a much longer time after birth (white areas in Fig. 1.3). Among association cortices, the prefrontal cortex is the last cortical area that full myelination of intrinsic nerve fibers is completed (see numbers in Fig. 1.3). The anatomical maturation of the prefrontal cortex takes the longest time after the birth and the maturation process of the prefrontal cortex seems to continue until the adulthood. Flechsig (1901, 1920) also proposed that the sequence of the development of cortical functions followed the sequence of the myelination of intrinsic nerve fibers in the development of the cerebral cortex. This suggests that the prefrontal cortex is the last cortical area to complete its functional maturation. Sensory and motor functions are basic and essential functions for surviving and escaping from a variety of dangers in the natural environment and these functions are indispensable to newborns for surviving after birth. Therefore, the cortical areas responsible for sensory and motor functions should be functionally matured at birth. In fact, intrinsic fibers in the primary sensory and motor areas (black areas in Fig. 1.3) are mostly myelinated at birth. However, learning and acquiring such skills as how to adapt to natural and social environment require

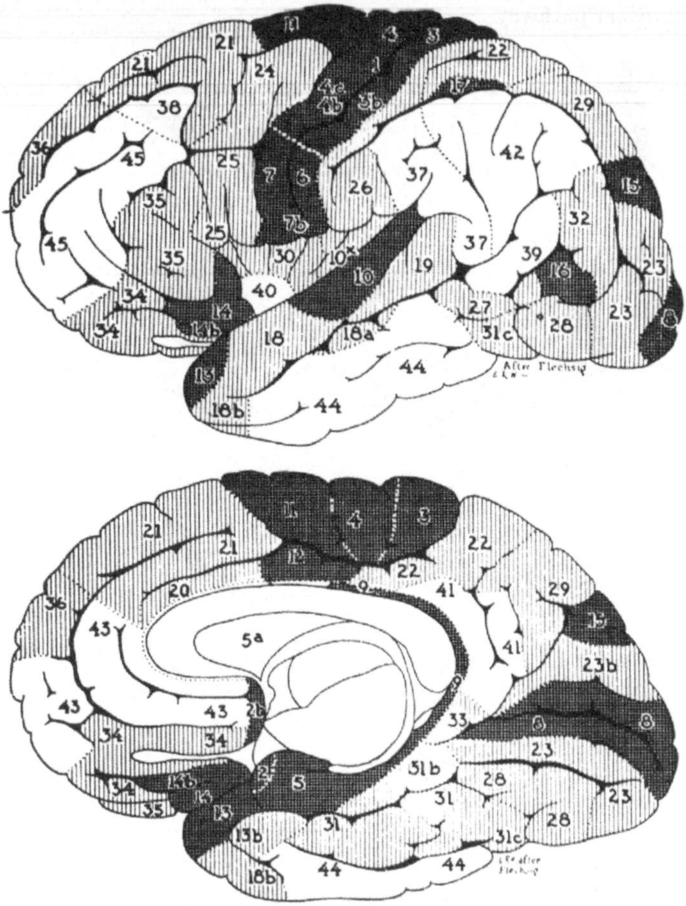

Fig. 1.3 Developmental sequence of the myelination of intrinsic nerve fibers in human cerebral cortex. Numbers indicate the order of the completion of the myelination during cortical development. From Flechsig (1920)

various experiences after birth and most of these skills are associated with the functions of the association cortices. The evidence that structural and functional maturation of the prefrontal cortex takes a longer time after birth suggests that the prefrontal cortex is an essential brain area not for sensory and motor functions to survive in the natural environment but for acquiring and developing skills to well adapt to natural as well as social environment.

Third, the lateral prefrontal cortex has unique anatomical relations with other cortical areas. The lateral prefrontal cortex mainly receives visual, auditory, and somatosensory information, while the medial and the orbital prefrontal areas mainly

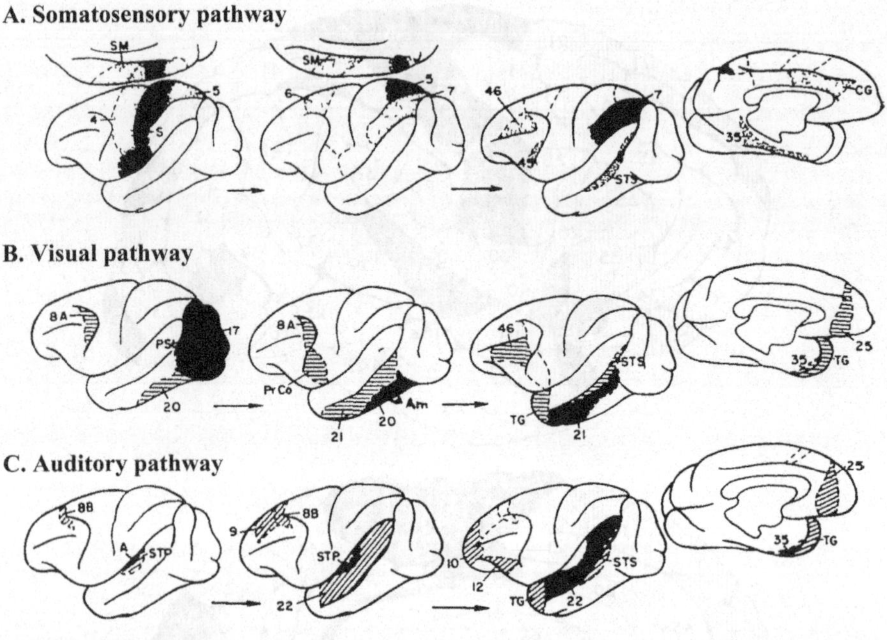

Fig. 1.4 Examples of sensory pathways in the monkey cerebral cortex. (**a**) Somatosensory pathway. (**b**) Visual pathway (ventral pathway). (**c**) Auditory pathways. From Jones and Powell (1970) with permission

receive olfactory, gustatory, and visceral information. An important feature is that the lateral prefrontal cortex does not receive sensory information directly from the primary sensory areas but receives such information from the sensory association cortices in the parietal, temporal, and occipital lobes. For example, visual information first arrives at the primary visual cortex (area 17) in the occipital lobe, then is passed to the inferior temporal cortex (areas 20 and 21) through pre-striate visual areas (PSt), and eventually reaches the lateral prefrontal cortex (area 46) (Fig. 1.4b). Similarly, auditory and somatosensory information first arrives at the primary areas, is passed to the sensory association areas surrounding the primary areas, then to the temporal and the parietal association cortices, and eventually reaches the lateral prefrontal cortex (Fig. 1.4a, c). Thus, the lateral prefrontal cortex is the cortical place where visual, auditory, and somatosensory information is accumulated. Therefore, the prefrontal cortex is the cortical area at the highest hierarchy of sensory information processing. In addition, the sensory information accumulating in the prefrontal cortex is the information that had been processed in the primary areas and in the sensory association areas. Therefore, the prefrontal cortex can monitor information processing in the posterior cortices through bottom-up sensory pathways.

The lateral prefrontal cortex not just receives sensory information from the posterior association cortices through bottom-up pathways but also sends information to these cortices through top-down pathways. The lateral prefrontal cortex has reciprocal connections with the posterior association cortices. Through these reciprocal connections, the prefrontal cortex sends important signals to the posterior association cortices to control information processing in these cortices. This control mechanism by the prefrontal cortex is called top-down control. Top-down control by the prefrontal cortex is an important key concept to understand "executive functions" which are essential functions of the prefrontal cortex.

The lateral prefrontal cortex also sends information to the motor areas in the frontal lobe. The information regarding the voluntary action plan is made in the lateral prefrontal cortex. The lateral prefrontal cortex does not send motor information directly to the primary motor cortex but to the motor association areas such as the pre-supplementary motor area (pre-SMA), the supplementary motor area (SMA), and the premotor area (PM). The pre-SMA sends information to the SMA and the PM, and the SMA and the PM send information to the primary motor cortex. The primary motor cortex is the final output area where motor information is output to the basal ganglia, the brain stem, and the spinal cord. Therefore, the lateral prefrontal cortex is considered the cortical area at the highest hierarchy in the information processing for voluntary actions.

Thus, these features indicate that the prefrontal cortex not only participates in information processing for sensory, perceptual, and motor functions, but also participates in processing and integrating various information to achieve complex cognitive functions such as thinking, reasoning, judging, controlling attention, decision-making, and complex voluntary action planning.

1.4 Anatomical Features of the Prefrontal Cortex

1.4.1 Cytoarchitectonic Features

The prefrontal cortex is divided into several cytoarchitectonically different areas. In the monkey brain, the cytoarchitectonic map of the prefrontal cortex had been proposed by Brodmann (1909), Walker (1940), and Petrides and Pandya (1994). Brodmann (1909) divided the human cerebral cortex into about 50 different areas and numerically designated those areas. He also used the same numerical designation for the monkey's cerebral cortex (Brodmann 1909) (Fig. 1.5a). In the rhesus monkey, the frontal lobe is defined as the cortical area located in the rostral portion from the central sulcus. Brodmann divided the frontal lobe into two regions; *precentral region* and *frontal region*. The *precentral region* corresponded to the frontal motor areas and was defined by the fact that this area had large pyramidal cells (Betz cells) in layer V and no granular layer IV. On the other hand, the *frontal region* corresponded to the prefrontal cortex and was defined by the fact that this area had a well-developed granular layer IV. Thus, Brodmann's criterion to define the *frontal region* was the presence of the granular layer IV in the cortex anterior to the motor areas. He discriminated five areas (areas 8, 9, 10, 11, and 12) in the *frontal region* of the monkey brain (Fig. 1.5a). Area 8 is located in the middle of the lateral

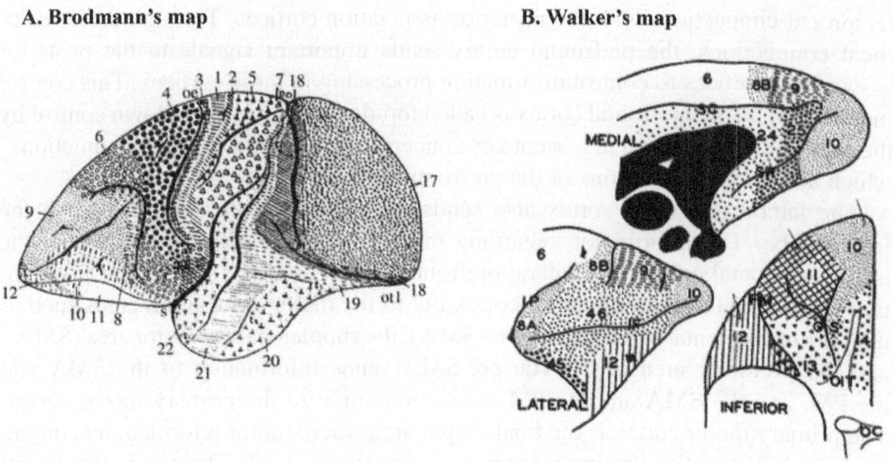

Fig. 1.5 Cytoarchitectonic maps of the monkey's prefrontal cortex. (a) Brodmann's map of the cerebral cortex. (b) Walker's map of the prefrontal cortex. Modified from Garey (1994) and Walker (1940)

surface of the frontal lobe and forms a narrow band stretching along the anterior bank of the arcuate sulcus. Area 9 is the granular frontal area and encompasses a large part of the anterior half of the lateral and medial sides of the frontal lobe. Area 10 is the lateral orbital area and area 11 is the medial orbital area. Area 12 is the frontopolar area including the frontal pole. Regarding the cytoarchitectonic map of the prefrontal area, Brodmann (1909) described:

> *The number of areas is limited to five, while in man there are eight of nine clearly differentiated individual fields. The detailed homologies of these areas are quite unclear; one can only say with certainty that in the monkey no structural equivalents of areas 44, 45, and 46 can be demonstrated. In general, then the inferior frontal gyrus and the central part of the middle frontal gyrus are missing in the lower monkeys.* (Page 134, English edition by Garey (1994))

Brodmann's cytoarchitectonic map of the monkey brain was not always consistent with the map of the human brain. In addition, Brodmann examined brain sections cut horizontally in order to perform cytoarchitectonic classification of the prefrontal cortex. However, the plane of the section often gave a false impression of the cortical architecture. For example, if the plane of the section was not at the right angle to the surface, the cortex might appear thicker than it really was. Therefore, it was important to examine brain sections cut in several different planes. The cortex should be cut at the right angle at least in one plane. Walker (1940) described Brodmann's definition of area 10 as an example of this miss leading.

> *Take area 10, for example. Brodmann (1909) examining sections cut horizontally states that area 10 differentiates itself by the fact that the sixth layer runs parallel to the surface of the cortex. Coronal sections of this cortex, however, show no such arrangement, for the axons of this layer run posteriorly-thus appearing only in horizontally cut sections to lie at right angles to the cortex.* (Page 62–63, Walker 1940)

Therefore, Walker (1940) reexamined the cytoarchitecture of the monkey's frontal lobe.

Walker (1940) performed cytoarchitectonic examinations of the monkey's prefrontal cortex using serial sections of over 40 frontal lobes which were cut in various planes at 20 μm thickness. Walker subdivided Brodmann's *frontal region* into several different areas (e.g., added areas 10, 46, and 45), and constructed a new cytoarchitectonic map (Fig. 1.5b). In the lateral portion of the prefrontal cortex, Walker assigned area 46 to the cortex within and surrounding the principal sulcus, which was the typical frontal granular cortex and was characterized by a well-developed granular layer IV. The dorsal convexity was assigned to area 9, which laid along the midline and on the medial surface of the frontal lobe and was differentiated from area 46 by narrower granular layer IV. The most anterior portion was assigned to area 10, which covered the frontal pole and was a narrow cortex with a well-developed granular layer IV. The cortex anterior of the arcuate sulcus was divided into area 8B, area 8A, and area 45, which had less well-developed granular layer IV and larger pyramidal cells in layers III and V. After Walker's cytoarchitectonic map was published, his map had been used widely to describe the cortical area where lesions were made in the prefrontal areas in behavioral studies, where tracers were injected in anatomical studies, and where neural activity was recorded in the prefrontal cortex in neurophysiological studies.

However, recent quick and vast advances in human neuroimaging studies allow to visualize changes of neural activity in specific regions of the human brain in relation to a particular aspect of the cognitive process. Since a large amount of important anatomical, behavioral, and physiological data has been accumulated by animal studies, especially studies using macaque monkeys, a crosstalk has become necessary between functional neuroimaging studies by human subjects and anatomical, behavioral, and physiological studies by macaque monkeys to correctly interpret the results obtained by human neuroimaging studies and to understand functions and information processing occurred in a given brain area. However, one major problem which made it difficult to achieve an effective crosstalk was that the architectonic map of the human cerebral cortex and that of the macaque monkey's cerebral cortex were not consistent in several important aspects. For example, it was not clear whether the same architectonically designated area is homologous between two species or not (e.g., whether or not area 46 of the monkey brain assigned by Walker is homologous of area 46 in human brain assigned by Talairach's atlas). A large number of anatomical, behavioral, and neurophysiological studies have been performed on the monkey prefrontal cortex. At the same time, a large number of human neuroimaging studies have been performed in the prefrontal cortex since the 1990s and a huge amount of data have been accumulated. Therefore, whether the same architectonically designated area is homologous between two species or not became an important issue for understanding the functions of the prefrontal cortex.

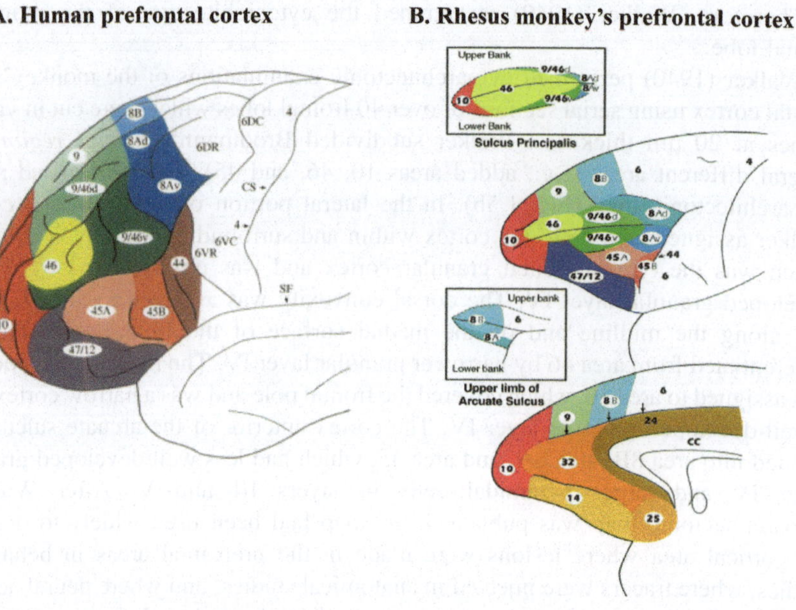

Fig. 1.6 Cytoarchitectonic maps of the human prefrontal cortex (**a**) and the rhesus monkey's prefrontal cortex (**b**). From Petrides and Pandya (1999) with permission

Petrides and Pandya reexamined the cytoarchitecture of the frontal lobe in both human brain and macaque monkey's brain and introduced new cytoarchitectonic maps of the prefrontal cortex for human brain and macaque monkey's brain based on the common cytoarchitectonic and topographic criteria (Pandya and Yeterian 1996; Petrides and Pandya 1999; Petrides 2005; Petrides et al. 2012) (Fig. 1.6). In the new cytoarchitectonic map, the macaque's lateral prefrontal cortex is divided into 12 areas. These include areas 10, 9, 46, 47/12, 9/46d, 9/46v, 8Ad, 8Av, 8B, 45A, 45B, and 44. This cytoarchitectonic map indicates that the basic architecture of the prefrontal cortex is similar between human brain and rhesus monkey's brain and that the area assigned by the same number is cytoarchitectonically homologous between the two species. These maps suggest that function and information processing in a given area of the human prefrontal cortex can be predicted by neurophysiological results obtained from the same area of the monkey's prefrontal cortex. Although this may not always be the case, cytoarchitectonic homology of a particular prefrontal area between human brain and monkey brain is important to interpret functions of that area by the data obtained by either monkey experiments or human neuroimaging studies.

1.4.2 *Cytoarchitectonic Features of the Lateral Prefrontal Cortex*

The lateral prefrontal cortex is generally divided into two sectors based on anatomical and functional differences: dorsolateral and ventrolateral (Barbas and Pandya 1989; Hoshi 2006; Gerbella et al. 2013; Saleem et al. 2014; Borra et al. 2019). The dorsolateral prefrontal cortex corresponds to the dorsal convexity cortex and the cortex within and surrounding the principal sulcus. Based on the cytoarchitectonic map of the macaque brain, the dorsolateral prefrontal cortex includes areas 10, 9, 46, 9/46d, 9/46v, and 8 (8Ad, 8Av, 8B) (Fig. 1.6). The ventrolateral prefrontal cortex corresponds to the ventral convexity cortex and includes areas 47/12, 45A, 45B, and 44 in the cytoarchitectonic map of the macaque brain (Fig. 1.6).

The dorsolateral prefrontal cortex sometimes applies to the cortex within and surrounding the principal sulcus. Petrides and Pandya (1994) cytoarchitectonically divided this cortical area into three areas (areas 46, 9/46d, and 9/46v). Area 46 is located in the rostral half of the principal sulcus, while area 9/46 is located in the caudal half of the principal sulcus. Both areas 46 and 9/46 have well-developed granular layer IV and have the same pattern of anatomical connections with other cortical areas. The difference between the two areas is based on the observation that area 46 does not receive inputs from the lateral parietal cortex while area 9/46 does (Petrides and Pandya 1994). Area 9/46 is further divided into the dorsal portion (area 9/46d) and the ventral portion (area 9/46v) based on small cytoarchitectural differences (Petrides and Pandya 1994). Although Petrides and Pandya (1994) divided the cortex within and surrounding the principal sulcus into three different areas, these three areas are considered as a single area 46 in this book.

Area 8 is usually included in the dorsolateral prefrontal cortex. However, area 8 is a distinct area from other lateral prefrontal areas. Since area 8 has specific functions related to eye movement control, area 8 is called the frontal eye field.

1.4.3 *Relations to the Thalamic Mediodorsal Nucleus*

Another distinct anatomical feature of the prefrontal cortex is that the prefrontal cortex has anatomical connections exclusively with the mediodorsal nucleus of the thalamus (Fig. 1.7). The mediodorsal nucleus of the thalamus has been known as a higher-order thalamic relay nucleus which plays important roles in higher cognitive functions such as learning, memory, and decision-making (Mitchell 2015). Distinct projection patterns between the prefrontal cortex and the thalamic mediodorsal nucleus have been observed not only in primates (Tobias 1975; Tanaka Jr 1976; Goldman-Rakic and Porrino 1985; Barbas et al. 1991; Siwek and Pandya 1991), but also in rats (Krettek and Price 1977, Condé et al. 1990) and cats (Markowitsch et al. 1978; Musil and Olson 1991; Tanibuchi 1992). Therefore, anatomical projections from the thalamic mediodorsal nucleus have been used to define the location and the

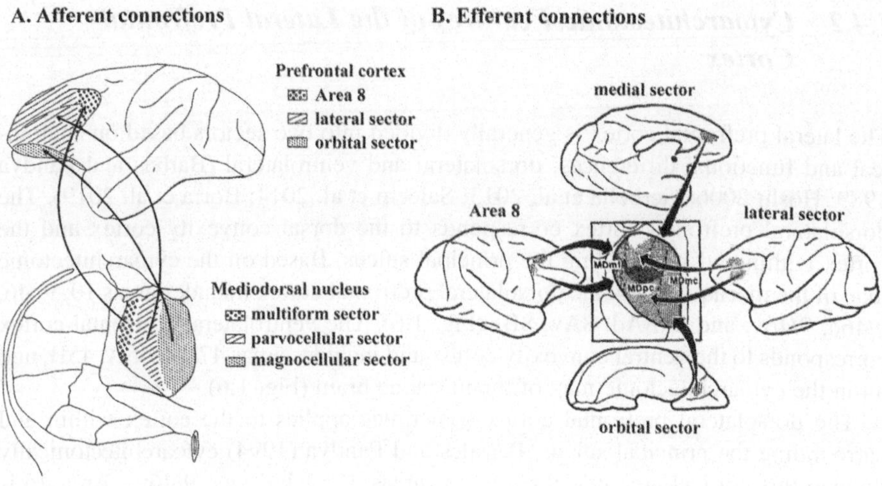

Fig. 1.7 Anatomical connections between the prefrontal cortex and the mediodorsal nucleus of the thalamus. (**a**) Afferent connections. From Fuster (2015) with permission. (**b**) Efferent connections. From Siwek and Pandya (1991) with permission

extent of the prefrontal cortex on the cerebral cortex in various animals. In primates, the mediodorsal nucleus consists of three cytoarchitectonically different sectors; magnocellular, parvocellular, and multiform sectors (Jones 1985) (Fig. 1.7). These three sectors have characteristic and topographic connection patterns with the prefrontal cortex (Tobias 1975; Barbas et al. 1991; Siwek and Pandya 1991). The magnocellular sector is located in the medial part of the nucleus, contains cells of larger size, and has connections mainly with the orbital and the medial parts of the prefrontal cortex. The parvocellular sector is located in the intermediate part of the nucleus, contains mostly small cells, and has connections mainly with the lateral part of the prefrontal cortex. The multiform sector is located in the most lateral part of the nucleus, which is between the parvocellular sector of the mediodorsal nucleus and the central lateral nucleus and has preferential connections with the frontal eye field. The lateral prefrontal cortex can be anatomically defined as the prefrontal area which has anatomical connections with the parvocellular sector of the thalamic mediodorsal nucleus.

Although the location and the extent of the prefrontal cortex have been defined anatomically by projections from the thalamic mediodorsal nucleus, refined anatomical methods revealed that the prefrontal cortex receives projections not only from the thalamic mediodorsal nucleus but also from other thalamic nuclei including the intralaminar nucleus, the midline nuclei, the anterior medial nucleus, and the rostral part of the ventral complex (Kievit and Kuypers 1977; Giguere and Goldman-Rakic 1988; Barbas et al. 1991). In addition, it has been shown that the mediodorsal nucleus projects not only to the prefrontal cortex but also to other cortical areas including the premotor cortex, the temporal cortex, and the parietal cortex (Kievit and Kuypers 1977; Akert and Von Monakov 1980; Giguere and Goldman-Rakic

1988). However, the prefrontal cortex not only receives projections from the thalamic mediodorsal nucleus but also sends outputs to the mediodorsal nucleus and constructs strong reciprocal anatomical connections exclusively with the thalamic mediodorsal nucleus. Therefore, it has been proposed that the prefrontal cortex should be defined as the cortical area which has strong and exclusive reciprocal connections with the thalamic mediodorsal nucleus (Uylings and Van Eden 1990; Uylings et al. 2003).

1.4.4 Granular Layer IV

Another distinct anatomical feature of the lateral prefrontal cortex is that this cortex is a six-layered isocortex and has a well-developed granular layer IV in primates (Fig. 1.8). Therefore, the prefrontal cortex is also called the *frontal granular cortex.*

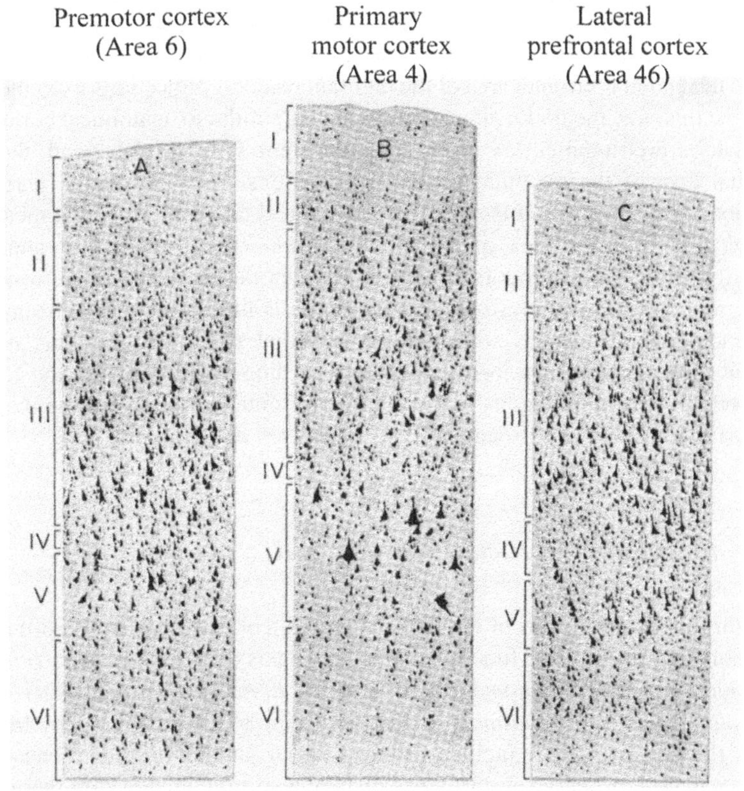

Fig. 1.8 Cellular structures of three frontal areas: area 6, area 4, and area 46. Area 46 has well-developed granular layer IV. Area 6 has layer IV but not a well-developed granular layer. Area 4 has no layer IV

On the other hand, the motor areas in the frontal lobe do not have well-developed or granular layer IV but have large pyramidal cells (e.g., Betz cells) in layer V. Therefore, the presence or absence of well-developed granular layer IV is an important criterion to determine the prefrontal cortex and to anatomically differentiate the lateral prefrontal cortex from the motor areas. Therefore, the lateral prefrontal cortex can be defined as the cortical area that has well-developed granular layer IV and has reciprocal connections with the parvocellular sector of the thalamic mediodorsal nucleus.

Rat's brain is also known to have the prefrontal cortex (Fig. 1.2) and consists of two areas: medial and sulcal prefrontal areas. Whether homologous areas are present in the cerebral cortex between two different species or not depends on the histological similarity in their fine structures and the similarity of afferent connections of these areas between two species. As has been described before, in the monkey brain, the prefrontal cortex is defined as the cortical area which has well-developed granular layer IV and reciprocal connections with the thalamic mediodorsal nucleus. Cytoarchitectonic comparison of the prefrontal cortex between monkeys and rats revealed that the granular layer IV is present in the monkey brain but is absent or greatly reduced in the rat brain. However, the prefrontal cortex of both species has anatomical connections with the thalamic mediodorsal nucleus (for rats, Leonard 1969). The medial prefrontal area of the rat brain receives projections from the lateral part of the thalamic mediodorsal nucleus, which is similar to anatomical connections of the lateral prefrontal cortex of the monkey brain. On the other hand, the sulcal prefrontal area of the rat brain receives projections from the medial part of the mediodorsal nucleus, which is similar to anatomical connections of the medial and the orbital prefrontal cortex of the monkey brain. The similarity of anatomical connections with the mediodorsal nucleus between the rat's medial prefrontal area and the monkey's lateral prefrontal cortex suggests that these two prefrontal areas are homologous. However, since the rat's medial prefrontal area has no clear granular layer IV, there has been a debate regarding whether or not the rat brain has a cortical area homologous to the lateral prefrontal cortex of the monkey brain (Preuss 1995; Brown and Bowman 2002; Uylings et al. 2003).

1.4.5 Mesulam's Cortical Classification

Based on functional aspects of the human cerebral cortex, Mesulam (2000) divided the cerebral cortex into four functionally distinct areas (Fig. 1.9); *primary (idiotypic) areas*, *modality-specific (unimodal) association areas*, *high-order (heteromodal) association areas*, and *paralimbic areas*. According to the definitions by Mesulam (2000), the *primary areas* include primary visual, auditory, somatosensory, and motor areas. The primary sensory areas display a specific architecture representing the highest level of development in granularization and lamination, while the primary motor area displays a unique architecture characterized by the presence of giant pyramidal neurons known as Betz cells. Both *modality-specific* and *high-order*

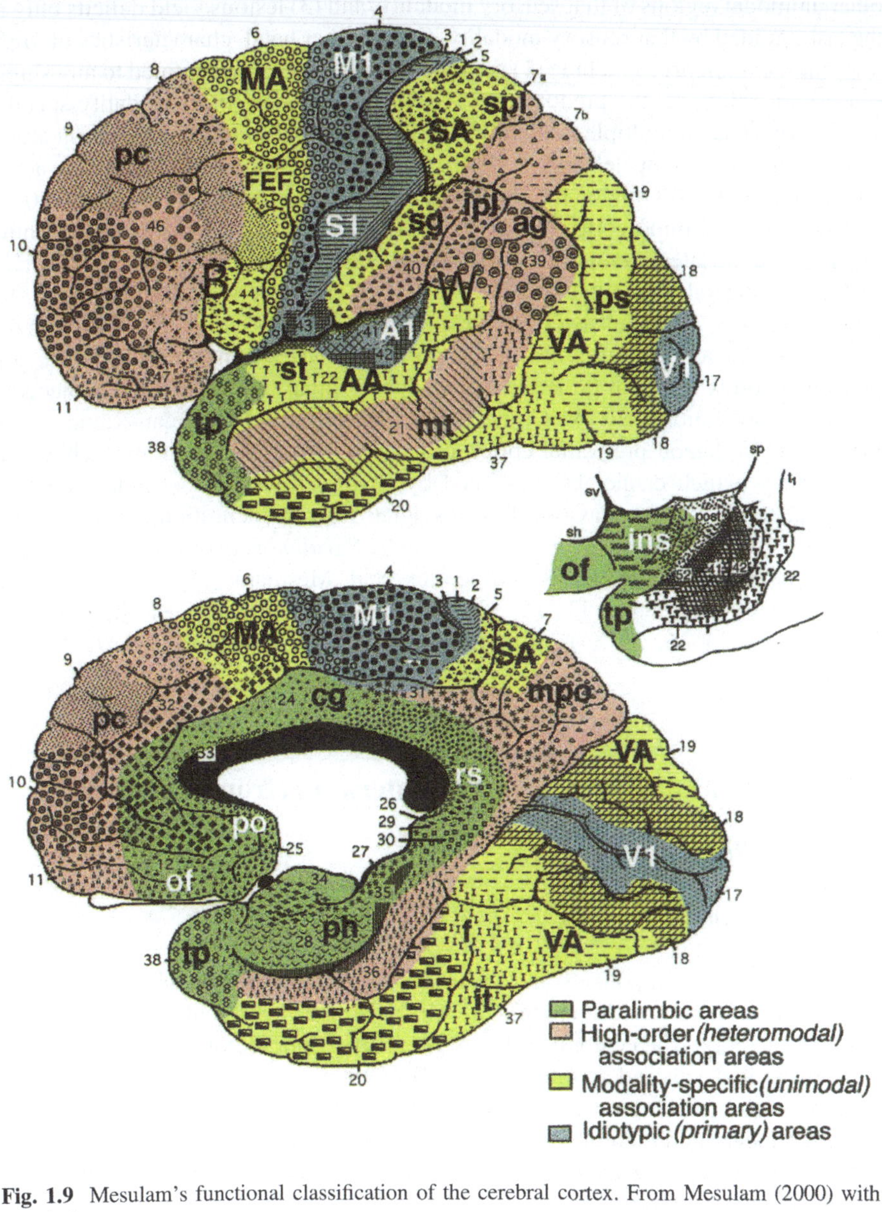

Fig. 1.9 Mesulam's functional classification of the cerebral cortex. From Mesulam (2000) with permission

association areas are characterized by a six-layered homotypical architecture. Essential characteristics of *modality-specific association areas* include (1) constituent neurons predominantly respond to stimuli in only a single sensory modality, (2) the predominant sensory information comes from the primary sensory area and other unimodal regions of that sensory modality, and (3) lesions yield deficits only in the tasks guided by that sensory modality. On the other hand, characteristics of *high-order association areas* include (1) neuronal responses are not confined to any single sensory modality, (2) the predominant sensory inputs come from modality-specific association areas in multiple modalities and from other high-order association areas, and (3) the deficits by lesions in these areas are always multimodal, and never confined to the tasks under the guidance of a single modality. Mesulam (2000) included the prefrontal cortex (areas 8, 9, 10, 45, 46), the inferior parietal lobule (area 7), and the lateral temporal cortex (area 21, 22) in high-order association areas. In fact, the lateral prefrontal cortex does not receive anatomical inputs from the primary sensory areas but receives inputs from modality-specific association areas and high-order association areas in the posterior cortices. In addition, the lateral prefrontal cortex does not project directly to the primary motor area but projects to the motor association areas (McGuire et al. 1991b; Bates and Goldman-Rakic 1993). Lesions in the lateral prefrontal cortex produce complex cognitive and behavioral impairments which cannot be explained by an impairment of a single modality. These anatomical and behavioral features agree with the definition that the lateral prefrontal cortex is a high-order association area. *Paralimbic areas* are characterized by the architecture known as mesocortex and Mesulam (2000) includes the orbitofrontal cortex (areas 11, 12, 13), the temporal pole (area 38), the cingulate cortex (areas 23–26, 29–33), and the parahippocampal areas (areas 27, 28, 35) paralimbic areas.

1.5 Anatomical Features of the Lateral Prefrontal Cortex

1.5.1 Cellular Components

1.5.1.1 Pyramidal Neurons

Knowing cellular components and their circuitry within the dorsolateral prefrontal cortex is crucial for understanding core functions such as executive functions and working memory. The cerebral cortex is generally composed of two kinds of nerve cells: pyramidal cells and non-pyramidal cells. Pyramidal cells are the most common neurons in the cerebral cortex (occupied about 85% of all neural population) and construct local neural circuits with other pyramidal cells and non-pyramidal cells. Pyramidal cells are the major source of intrinsic excitatory synapses in the local neural circuitry and the spines on their dendrites are the main targets of excitatory synapses. Therefore, monosynaptic connections with two or more pyramidal neurons provide the basic circuitry for recurrent excitation, which is the key mechanism underlying persistent activity observed in the prefrontal cortex during working memory performances. In addition, pyramidal cells are important output neurons in the cerebral cortex and construct cortico-cortical and cortico-subcortical connections. Therefore, pyramidal cells are principal components of both structural and functional aspects in intra- and inter-cortical circuitries.

Elston (2003) examined detailed structures of pyramidal neurons in the monkey prefrontal cortex. Pyramidal cells in the prefrontal cortex have twice more branching complexity and four times more dendritic spines than those in the primary visual cortex, suggesting that the complex structure of pyramidal neurons in the prefrontal cortex endow prefrontal cortical circuitry with more complex functional capabilities. Pyramidal cells are mainly distributed in layers II, III, and V. Arnsten et al. (2012) suggested layer-specific contributions to delayed-response performances in the dorsolateral prefrontal cortex, such that neurons in layer II exhibit cue-period activity, neurons in layers III and V exhibit delay-period activity, and neurons in layer V exhibit response-period activity. It has been known that pyramidal cells in layers II and III contribute to cortico-cortical connections, while pyramidal cells in layer V contribute to cortico-subcortical connections. This suggests that pyramidal neurons in the upper layers receive sensory information from the posterior cortices through cortico-cortical connections and represent sensory information such as cue-period activity. Pyramidal cells in the middle layers form recurrent excitatory circuitries with other pyramidal cells and exhibit tonic sustained activation during the delay period in working memory tasks. And, pyramidal cells in the deep layer provide outputs to other cortical and subcortical areas and exhibit response-related activity in working memory tasks. Glausier et al. (2021) showed in the dorsolateral prefrontal cortex that the density of spines receiving inputs from pyramidal neurons is greater in the middle layer than in the deeper layer, that pyramidal neurons have dendritic spines innervated by both pyramidal neurons and interneurons, and that the density of spines dually innervated by pyramidal neurons and interneurons is greater in the middle layer than in the deeper layer. These data support that the middle layer of the dorsolateral prefrontal cortex plays important roles in gating, integrating, and fine-tuning of synaptic functions by recurrent excitatory microcircuits, which also play important roles in selecting appropriate information and for actively maintaining information. In fact, a comparison of pyramidal neurons in layer III between the primary visual cortex and the prefrontal cortex in monkey revealed that layer III pyramidal neurons in the prefrontal cortex are less excitable and had larger dendrites with a higher density of dendritic spines than those in the visual cortex. Similarly, González-Burgos et al. (2019) compared layer III pyramidal neurons between the prefrontal cortex and the posterior parietal cortex in the monkey. They found that pyramidal neurons can be divided into regular-spiking types and bursting types based on the difference in physiological properties, that most of pyramidal neurons are regular-spiking types in the posterior parietal cortex, while bursting types of pyramidal neurons are abundant in the prefrontal cortex, and that layer III pyramidal neurons in the prefrontal cortex have larger and more complex basal dendrites with a higher density of dendritic spines than those in the posterior parietal cortex. These observations suggest that physiological and morphological features of pyramidal neurons, especially those in layer III, contribute significantly to information processing specific in the prefrontal cortex, such as sensorimotor integration and mnemonic processing.

1.5.1.2 Nonpyramidal Neurons

Pyramidal neurons occupy a large proportion (about 85%) in cortical neural population and play important roles in information processing in the cerebral cortex. On the other hand, although nonpyramidal neurons occupy a small proportion (15%) compared with pyramidal neurons, nonpyramidal neurons or local interneurons also play important roles in information processing. Nonpyramidal local interneurons provide inhibitory control over the activity of pyramidal neurons by releasing GABA from the presynaptic terminals. A variety of local interneurons have been identified and known to show various shapes. Morphologically, nonpyramidal interneurons were classified into double-bouquet cells, neuroglia form cells, Martinotti cells, chandelier cells, and basket cells. Since these interneurons contain calcium-binding proteins such as calretinin, calbindin, and parvalbumin, these neurons have been classified based on which calcium-binding protein is included. Condé et al. (1994) examined the distribution and morphology of interneurons containing calcium-binding proteins in the monkey prefrontal cortex (Fig. 1.10). Calretinin was found in double-bouquet cells, calbindin was found in neuroglia form cells and Martinotti cells, and parvalbumin was found in chandelier and basket cells. Calretinin-positive cells were accumulated densely in layers I and II,

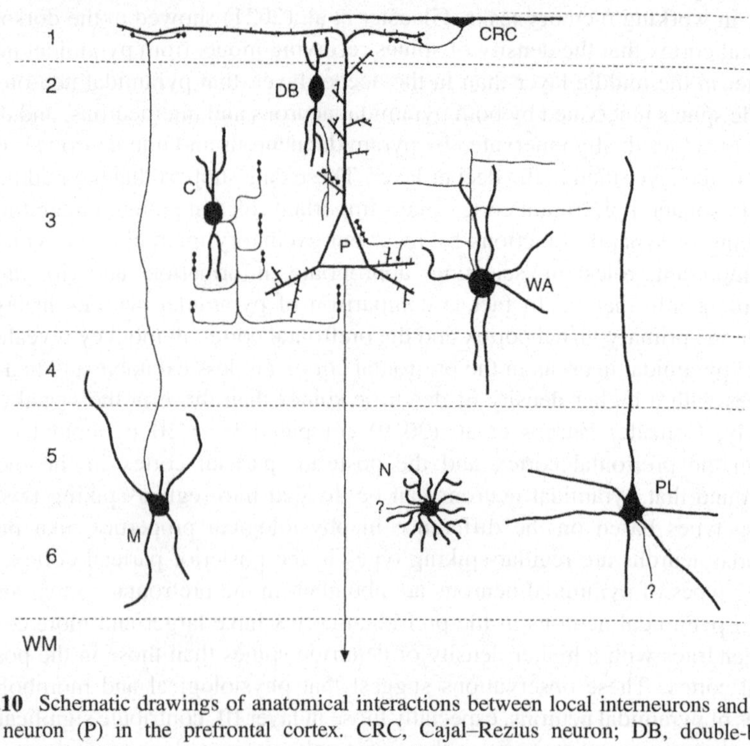

Fig. 1.10 Schematic drawings of anatomical interactions between local interneurons and a pyramidal neuron (P) in the prefrontal cortex. CRC, Cajal–Rezius neuron; DB, double-bouquet neuron; C, chandelier neuron; WA, basket neuron; M, Martinotti cell; N, neurogliaform neuron; PL, pyramidal-like neuron. From Condé et al. (1994) with permission

calbindin-positive cells were accumulated in layer II and superficial layer III, and parvalbumin-positive cells were accumulated in the middle layers. Parvalbumin-positive chandelier cells have been shown to innervate the axon's initial segment of the pyramidal neuron (Lewis and Lund 1990), while basket cells have been shown to innervate soma and/or proximal dendrites of the pyramidal neuron (Fish et al. 2013).

1.5.1.3 Interactions Between Pyramidal and Nonpyramidal Neurons

Interactions between pyramidal neurons and adjacent local interneurons are basic components of local circuits within the cerebral cortex (Fig. 1.11). Pyramidal neurons receive inhibitory GABAergic neurotransmission from local interneurons. In the visual cortex, these inhibitory inputs to pyramidal neurons act as lateral inhibition to produce orientation selectivity and direction selectivity. Similarly, in the prefrontal cortex, inhibitory inputs from local interneurons are thought to act as lateral inhibition on pyramidal neurons and shape spatial tunings of prefrontal neurons that exhibit directional delay-period activity in spatial working memory tasks. For example, Rao et al. (1999) showed that local interneurons such as putative basket and chandelier cells exhibited spatially tuned sensory-, mnemonic-, or response-related activity during spatial working memory performances (oculomotor delayed-response task) and that spatial selectivity of these interneurons was very

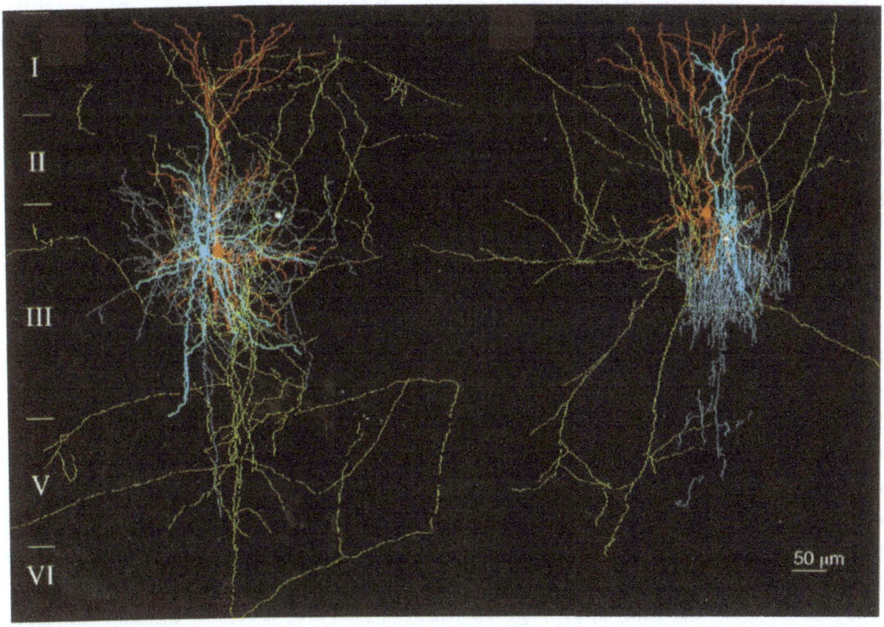

Fig. 1.11 Synaptic interactions between pyramidal neurons (orange and green) and interneurons (blue). *Left*: Interactions between pyramidal neurons and basket cells. *Right*: Interactions between pyramidal neurons and chandelier cells. From Krimer and Goldman-Rakic (2001) with permission

similar to that of pyramidal neurons. In addition, they also showed that directional tunings of local interneurons are very similar as those of neighboring pyramidal neurons. These results suggest that local interneurons provide inhibitory outputs to neighboring pyramidal neurons and that these inhibitory outputs play some role to shape spatially tuned activity of pyramidal neurons (iso-directional inhibition). On the other hand, prefrontal neurons exhibiting directional delay-period activity often showed suppressive responses during the delay period when the visual cues were presented opposite to the best direction, suggesting the presence of cross-directional inhibition between two groups of neurons which exhibit 180° difference in the best direction of delay-period activity. Rao et al. (2000) showed that local iontophoretic application of bicuculine methiodide (GABA receptor antagonist) in the prefrontal cortex produced the loss of spatial tunings in both pyramidal neurons and interneurons and caused the reduction of both iso-directional inhibition and cross-directional inhibition. These results indicate the importance of local interneurons among pyramidal and interneurons through GABAergic synaptic transmission in the production of spatial tunings in both sensorimotor and mnemonic processes.

On the contrary, Rodermund et al. (2020) examined the effects of excitatory glutamatergic functions and inhibitory GABAergic functions on prefrontal single-neuron activity by iontophoretic administration of MK-801 (NMDA receptor antagonist) and bicuculline ($GABA_A$ receptor antagonist) while monkeys performed a go/no-go color or shape discrimination task by holding or releasing a lever. The administration of MK-801 resulted in a slight enhancement of stimulus selectivity and encoding capacity of prefrontal neurons, while the administration of bicuculline resulted in improvement of the selectivity of delay-period activity, an increase of signal-to-noise ratio during the delay period, and the enhancement of encoding capacity of neurons. Although their findings are disparate from the results of Rao et al. (2000), these discrepancies might be caused by the differences in task features (spatial vs. feature base) or cognitive requirement (more motor oriented vs. more memory oriented), or a dose of bicuculine.

1.5.1.4 Modulation of Activity by Catecholamine and Monoamine

Catecholamine and monoamine play important roles in information processing in the prefrontal cortex (Arnsten 1998; Robbins and Arnsten 2009; Arnsten et al. 2012). Prefrontal neurons are very sensitive to changes in neuromodulatory inputs from norepinephrine (NE) and dopamine (DA) systems. This sensitivity leads to marked changes in working memory-related functions of the prefrontal cortex. Brozoski et al. (1979) showed that the depletion of DA and NE in the prefrontal cortex by infusion of 6-hydroxy-dopamine (6-OHDA) caused an impairment of spatial working memory performance, but not of visual discrimination performance in monkeys. They also showed that apomorphine (mixed D1/D2 receptor agonist of DA) improved working memory deficits in catecholamine-depleted monkeys, suggesting the importance of DA for working memory functions in the prefrontal cortex. In fact, the block of D1 receptors by local injection of D1 receptor antagonists in the prefrontal cortex markedly impairs behavioral performance of spatial working memory tasks, but not of the performance of sensory-guided control tasks (Sawaguchi and Goldman-Rakic 1991; Sawaguchi and Goldman-Rakic 1994). However, the behavioral effect of DA was found to be dose-dependent. Williams and Goldman-Rakic (1995) showed that iontophoretic application of D1 antagonist (Scheme 39166) in the prefrontal cortex selectively enhanced spatial selectivity of memory-related activity (delay-period activity) of prefrontal neurons. However, iontophoretic application of D1 agonist (SKF 38393) inhibited enhanced memory-related activity caused by D1 antagonist application. An important finding is that D1 receptors modulate memory-related activity of prefrontal neurons and that the effect of selective D1 antagonist on the activity of prefrontal neurons shows dose-dependent and exhibit an inverted-U shape (Goldman-Rakic et al. 2000; Vijayraghavan et al. 2007). The spatial tuning ability of prefrontal neurons in the spatial working memory performance is enhanced at the moderate and optimal level of D1 occupancy but reduced at higher and lower levels of D1 occupancy. Therefore, higher or lower levels of D1 occupancy in prefrontal neurons cause impairment of the spatial working memory performance because of diminishing spatial selectivity of sustained spatial memory-related activity in prefrontal neurons. Dose dependency and inverted U shape modulation of dopamine effect are the principal features in the prefrontal cortex (Goldman-Rakic et al. 2000; Seamans and Yang 2004).

Dopamine D1 receptors play important roles to regulate the working memory process in the prefrontal cortex. However, prefrontal neurons have not only D1 receptors but also D2 receptors. Wang et al. (2004a, b) showed that D1 and D2 differentially modulate prefrontal task-related activity. D1 receptors selectively modulated persistent memory-related activity (delay-period activity) but showed no effect on memory-guided motor responses (saccades) while the monkey performed an oculomotor delayed-response task. However, D2 receptors selectively modulate memory-related motor responses but showed no modulation on memory-related activity.

Krimer et al. (1997) quantitatively examined dopamine terminals on identified pyramidal and non-pyramidal neurons in the macaque prefrontal cortex (Fig. 1.12). They found about 90 dopamine terminals per pyramidal neuron in layer III with a highly regular distribution on its spiny dendritic trees. On the other hand, non-pyramidal interneurons in layer III had half as dense in the number of dopamine terminals with irregular distribution and large variation from neuron to neuron. The organized distribution of dopamine terminals on a pyramidal neuron suggests that dopamine terminals are placed strategically along the dendrite to modulate excitatory inputs to the neuron (Krimer et al. 1997).

1.5.2 Anatomical Heterogeneity in Area 46

Prefrontal area 46 corresponds to the cortical area within and surrounding the principal sulcus. Rostro-caudal functional heterogeneity has been reported in area

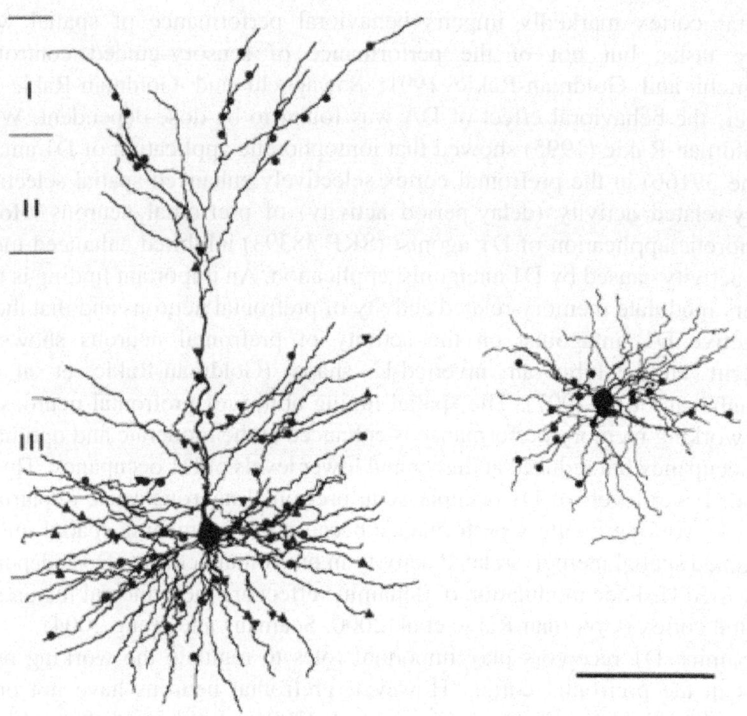

Fig. 1.12 Distribution of dopamine terminals on a pyramidal neuron (*left*) and an interneuron (*right*) in the monkey prefrontal cortex. Scale bar, 100 μm. Modified from Krimer et al. (1997) with permission

46 (Barbas and Pandya 1989; Pandya et al. 1971; Gerbella et al. 2013; Borra et al. 2019). For example, based on the cytoarchitectonic analysis, Barbas and Pandya (1989) suggested rostrocaudal architectonic difference in both dorsal and ventral portions of area 46, although they also indicated wide inter-areal anatomical connections within the prefrontal cortex. Pandya et al. (1971) showed that, although the prefrontal cortex projects widely to the temporal, parietal, and cingulate cortices, different portions of the principal sulcal area (area 46) exhibit different preferences in anatomical connections with specific cortical regions. For example, the rostral area 46 projects to the rostral part of the superior temporal sulcus and the cingulate cortex. The middle area 46 has widespread connections with the rostral and the middle part of the superior temporal sulcus as well as the cingulate cortex. The caudal area 46 projects to the parietal lobe and the rostral and the caudal part of the superior temporal sulcus.

Gerbella et al. (2013) and Borra et al. (2019) divided the principal sulcal cortex (area 46) into two parts (dorsal part, 46d, and ventral part, 46v) and examined cortico-cortical connections of both parts of area 46 (Fig. 1.13). The dorsal part (46d) and the ventral part (46v) correspond to the dorsal bank and the ventral bank of the principal sulcus, respectively. Borra et al. (2019) further divided the dorsal part

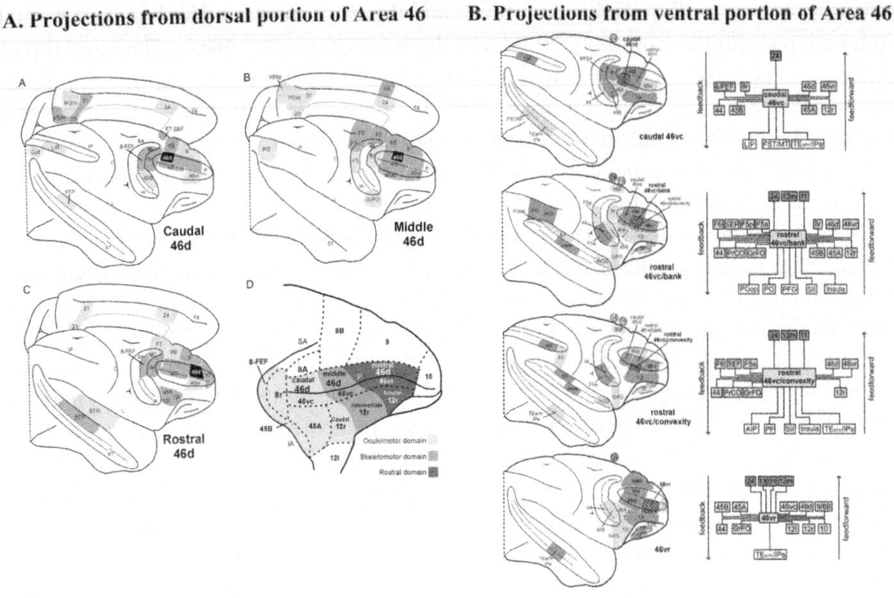

Fig. 1.13 Cortico-cortical connections within the lateral prefrontal cortex projecting from the dorsal portion (**a**) and the ventral portion (**b**) of area 46. (**a**) From Borra et al. (2019) with permission. (**b**) From Gerbella et al. (2013) with permission

of area 46 (46d) into rostral (46dr), middle (46 dm), and caudal (46 dc) portions and examined cortico-cortical connections (Fig. 1.13a). The rostral part of area 46d (46dr) has almost exclusive and extensive connections with the prefrontal areas 9, 8, 10, 45, 46, and 47/12 and has exclusive connections with the superior temporal areas and the cingulate areas (areas 24, 31). The middle part of area 46d (46 dm) has robust and topographical connections with prefrontal areas 9, 46v, and the dorsal part of area 8, the parietal areas (PGm and V6A), the frontal sensorimotor areas (dorsal premotor cortex), and the anterior cingulate area. This part is preferentially connected with arm-related motor areas in the parietal and premotor cortices. The caudal part of the area 46d (46 dc) has robust and topographical connections with prefrontal areas 9 and 46, and the dorsal part of area 8, the parietal areas (V6A, PGm, LIP), and the cingulate area. Since projections from area 46 dc to the dorsal part of area 8 had been reported (Petrides and Pandya 1999), this part of area 46d has a preferential connection with frontal and parietal oculomotor areas.

On the other hand, Gerbella et al. (2013) divided the ventral part of area 46 (46v) into rostral (46vr), rostral/bank (46vc/bank), rostral/convexity (46vc/convexity), and caudal (46vc) portions and examined cortico-cortical connections (Fig. 1.13b). The rostral part of area 46v (46vr) has exclusive and extensive intra-prefrontal connections and has connections limited with the cingulate cortex (area 24) and the inferotemporal cortex. The caudal part of area 46v (46vc) has mostly intra-prefrontal connectivity with ventrolateral areas and robust connections with frontal and parietal sensorimotor areas. Especially, the most caudal part of the area 46v has preferential connections with the frontal eye field and the inferior parietal areas. The caudal part of the ventral bank of the principal sulcus (46vc/bank) has connections with hand/mouth-related ventral premotor areas, area SII, the insula, hand/arm-related parietal areas (PFG and PG) and areas 11 and 24. The caudal part of the ventral convexity (46vc/convexity) has connections with hand/mouth-related ventral premotor areas, area SII, the insula, and the hand-related area AIP.

It has been shown that the dorsolateral prefrontal cortex has heavy intra-prefrontal connections with the lateral portion of the prefrontal cortex (Pandya and Kuypers 1969; Pandya and Yeterian 1996; Passingham et al. 2002) and with the ventromedial (orbital) portion of the prefrontal cortex (Cavada et al. 2000). Gerbella et al. (2013) and Borra et al. (2019) further showed that both dorsal and ventral parts of area 46 have exclusive and extensive intra-prefrontal connections with the dorsal and the ventral convexity of the lateral prefrontal cortex, respectively, and have connections with the dorsal and the ventral part of the frontal eye field (area 8), respectively. In addition, the rostral part of area 46 has exclusive and extensive intra-prefrontal connections in the lateral prefrontal cortex, while the middle and the caudal part of area 46 have connections with other cortical areas including the frontal eye field, the premotor cortex, the supplementary and pre-supplementary motor areas, and the parietal and temporal cortices. Thus, both rostrocaudal as well as dorsal-ventral anatomical heterogeneities have been reported in anatomical connections within the prefrontal area 46 and with other cortical areas. This anatomical heterogeneity suggests the presence of functional heterogeneity in the prefrontal area 46.

1.5.3 Cortico-Cortical Connections

1.5.3.1 Afferent Connections

Jones and Powell (1970) summarized the basic structures of afferent sensory pathways to the prefrontal cortex (Fig. 1.4). Visual, auditory, and somatosensory information first accumulate in the primary area of each modality. Such information is transmitted to adjacent modality-specific association areas in the occipital, temporal, and parietal lobes, and then to the high-order association areas in the temporal and parietal lobes adjacent to the modality-specific association areas. Sensory information received in the temporal and parietal high-order association areas is then transmitted to the lateral prefrontal cortex, which is also classified as the high-order association area (Fig. 1.4). The fact that the lateral prefrontal cortex receives sensory information mainly from the posterior high-order association areas is consistent with other anatomical studies (Yeterian et al. 2012; Barbas 2015). The laminar origin of cortical projections revealed that neurons in the upper (supragranular) layers of the posterior association cortices project to the prefrontal cortex (Barbas 1986), indicating that the sensory information flow from the posterior association areas to the prefrontal cortex is the feed-forward type.

The dorsolateral prefrontal cortex receives projections from the parietal cortex including the lateral posterior parietal cortex (area PG) (Pandya and Seltzer 1982; Cavada and Goldman-Rakic 1989; Petrides 2005) and the medial parietal cortex (area PGm or Pm) (Pandya and Seltzer 1982; Cavada and Goldman-Rakic 1989). Topographic patterns of projections have been observed between separate areas in the parietal cortex and those of the prefrontal cortex (Cavada and Goldman-Rakic 1989). Posterior parietal area 7a projects to the fundus and both banks of the middle portion of the principal sulcus and the frontal eye field. Parietal area 7ip (the posterior bank of the intraparietal sulcus) has connections with the posterior portion of the principal sulcus and the frontal eye field and adjacent areas. Parietal area 7m (the medial parietal surface) connected with the posterior portion of the dorsal bank of the principal sulcus and the frontal eye field.

The dorsolateral prefrontal cortex receives projections from the caudal inferior temporal cortex (area 20) (Chavis and Pandya 1976), which is visual association areas, and from the dorsal bank of the superior temporal sulcus, which is auditory association areas (Hackett et al. 1999; Petrides 2005; Saleem et al. 2014). The middle part of the superior temporal gyrus predominantly projects to the lateral prefrontal cortex (area 9 and 46) and the posterior part of the superior temporal gyrus also projects to the lateral prefrontal cortex (area 46 and 8) (Petrides and Pandya 1988). Especially, the dorsal bank of the principal sulcus and the dorsal pre-arcuate area of the frontal eye field have been shown to have reciprocal connections with the auditory association cortex (Chavis and Pandya 1976; Hackett et al. 1999) and the caudal superior temporal gyrus (Petrides 2005; Saleem et al. 2014).

The dorsolateral prefrontal cortex also receives projections from the limbic areas including the anterior and posterior cingulate cortex (Pandya et al. 1981; Bates and Goldman-Rakic 1993; Pandya and Yeterian 1996; Petrides and Pandya 1999; Morris et al. 1999; Morecraft et al. 2004; Petrides 2005; Saleem et al. 2014), the retrosplenial cortex (Petrides 2005; Saleem et al. 2014), and the rostral temporal cortex (Pandya and Yeterian 1996; Petrides 2005). Regarding the connections between the dorsolateral prefrontal cortex and the hippocampus, Goldman-Rakic et al. (1984) showed that the dorsolateral prefrontal cortex had reciprocal connections with the fundus of the rhinal sulcus and the posterior portion of the parahippocampal gyrus and received projections from the presubiculum through the fronto-occipital fasciculus and the cingulum bundle. However, most hippocampus neurons have been shown to project to the medial and orbital prefrontal cortices (Barbas and Blatt 1995). Similarly, the amygdala is known to project to the medial and the orbital prefrontal cortex and not to project directly to the dorsolateral prefrontal cortex (Porrino et al. 1981; Barbas and De Olmos 1990). Although direct projections were not observed to the dorsolateral prefrontal cortex from the medial temporal lobe, Hirata et al. (2013) showed disynaptic projections from the medial temporal lobe to the dorsal portion of area 46. They used retrograde transsynaptic transport of rabies virus and showed that 3 days after rabies injections into the dorsal area 46, a large number of labeled neurons were found in the hippocampus, the perirhinal cortex, and the parahippocampal cortex. However, rabies injections into the ventral area 46 caused majority of labeled neurons in the inferior temporal cortex.

These results indicate that the main disynaptic target of the medial temporal lobe including the hippocampus is the dorsal area 46 and not the ventral area 46.

On the other hand, the ventrolateral prefrontal cortex receives projections from the inferior temporal cortex, which is visual association areas (Pandya and Kuypers 1969; Ungerleider et al. 1989; Pandya and Yeterian 1990; Webster et al. 1994; Petrides 2005; Saleem et al. 2014). Topographic patterns of projections have been observed in the connections between the ventral prefrontal cortex and the inferior temporal cortex (Pandya and Yeterian 1990; Webster et al. 1994). The anterior portion of the inferior temporal cortex (area TE) projects to the ventral convexity of the lateral prefrontal cortex including areas 45 and 8 and the inferior limb of the arcuate sulcus, while the posterior portion of the inferior temporal cortex (area TEO) projects to areas 8 and 12 and the anterior bank of the arcuate sulcus. In addition, projections from area TE terminate mainly layers I and V/VI of the prefrontal cortex, while projections from area TEO terminate all layers (Webster et al. 1994). The ventrolateral prefrontal cortex receives inputs from the auditory association cortex in the superior temporal cortex (Pandya and Kuypers 1969). The ventrolateral prefrontal cortex (areas 45 and 47/12) has connections with surrounding ventrolateral prefrontal areas (Pandya and Kuypers 1969; Pandya and Yeterian 1996; Passingham et al. 2002) and receives projections from the orbitofrontal cortex (Pandya et al. 1971; Pandya and Yeterian 1996).

1.5.3.2 Efferent Connections

Since cortico-cortical connections between the prefrontal cortex and the posterior association cortices are mainly reciprocal, the cortical areas from that the prefrontal cortex receives anatomical projections receive projections from the prefrontal cortex (Fig. 1.14). The dorsolateral prefrontal cortex projects to the temporal cortex including the superior temporal gyrus (Pandya and Kuypers 1969) and to the intraparietal sulcus (Pandya and Kuypers 1969). The dorsolateral prefrontal cortex also projects to the frontal motor areas including the dorsal premotor cortex (PMd) (Barbas and Pandya 1987; Petrides and Pandya 1999, 2002; Wang et al. 2002), the supplementary eye field (area F7-SEF) (Pandya and Yeterian 1996; Luppino et al. 2003), the rostral portion of the supplementary eye field (area F7-non SEF) (Pandya and Yeterian 1996; Luppino et al. 2003), and the pre-supplementary motor area (pre-SMA or area F6) (Pandya and Yeterian 1996; Luppino et al. 1993).

Similarly, the ventrolateral prefrontal cortex projects to the inferior temporal gyrus (Pandya and Kuypers 1969), the rostral TE, and the dorsomedial portion of the temporal pole (Mohedano-Moriano et al. 2015), the intraparietal sulcus (Pandya and Kuypers 1969; Pandya and Seltzer 1982; Saleem et al. 2014), the frontoparietal operculum (Preuss and Goldman-Rakic 1989; Cipolloni and Pandya 1999; Petrides and Pandya 2002) and ventral limbic areas including the perirhinal cortex (Petrides 2005), the parahippocampal gyrus (Petrides 2005), and the insula (Saleem et al. 2014). This area also projects to the ventral premotor cortex (PMv) (Pandya and

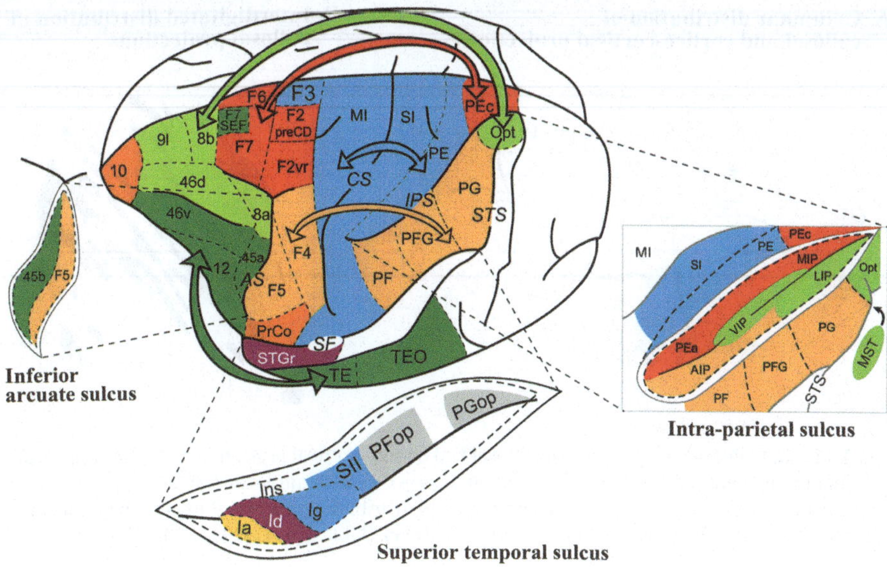

Fig. 1.14 Reciprocal connections between the prefrontal cortex and the parietal and the temporal cortices. Same color indicates dominant reciprocal connections between two areas. From Giarrocco and Averbeck (2021) with permission

Kuypers 1969; Barbas and Pandya 1987; Pandya and Yeterian 1996; Petrides and Pandya 1999, 2002; Wang et al. 2002).

1.5.3.3 Callosal Connections

Interhemispheric connections have been known in the prefrontal cortex since callosal terminals were found in the prefrontal cortex after a complete transection of the corpus callosum. Jacobson and Trojanowski (1977a, b) examined interhemispheric afferents to the dorsolateral prefrontal cortex. Areas 46, 45, 10, 9, and 8a receive extensive interhemispheric projections from the homotopical areas and also receive projections from the heterotopic areas of the prefrontal cortex. However, these areas receive interhemispheric projections less extensively from the cingulate cortex, the superior temporal sulcus, and the insula. Neurons projecting to the prefrontal cortex are primarily located in layer III and distributed in waxed and waned form in cell density in the homotopical cortex and in patch and cluster form in the heterotopic cortex. Cortical neurons projecting to other cortical areas in the ipsilateral hemisphere are classified as association neurons and neurons projecting to cortical areas in the contralateral hemisphere are classified as callosal neurons. Both association and callosal neurons are predominantly located in layer III. Schwartz and Goldman-Rakic (1982) used double-labeling of fluorescent dyes and showed that neurons in the principal sulcus have both callosal and association axons.

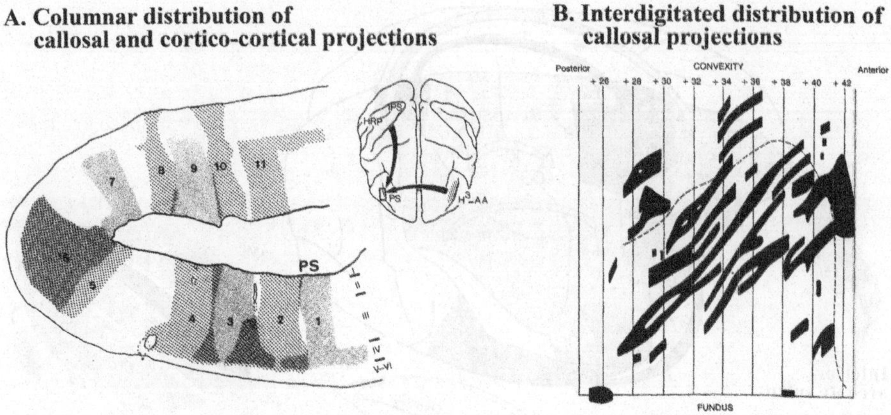

Fig. 1.15 (**a**) Columnar distribution of callosal and cortico-cortical projections. Course stipples and fine stipples indicate callosal projection areas and cortico-cortical projection areas, respectively. (**b**) Two-dimensional reconstruction of the terminal field distribution of callosal fibers. The dashed line indicates the rim of the principal sulcus. From Goldman-Rakic (1984) with permission

These neurons send callosal axons either to a cytoarchitectonic area homotopic to the hemisphere where these neurons are located or to different heterotopic areas in the contralateral hemisphere. Further, Schwartz and Goldman-Rakic (1984) examined features of callosal and association projections of neurons in the principal sulcus. They found that association and callosal neurons are distinct cell groups with similar organization. These neurons have common topography and common morphologies, are distributed with higher density in supra-granular layers and have uneven tangential distribution reminiscent of the inter-digitation of their terminal fields (Fig. 1.15). Based on uneven tangential distributions of callosal and association neurons and inter-digitation-like distributions of terminal fields of these neurons in the principal sulcus, Goldman-Rakic (1984) indicated that the prefrontal cortex is organized in modular units with distinct inputs and outputs (e.g., cortical columns).

1.5.4 Cortico-Subcortical Connections

1.5.4.1 Anatomical Connections with the Thalamus

The prefrontal cortex receives inputs from the thalamus, mainly from the mediodorsal nucleus. At the same time, the prefrontal cortex projects to the mediodorsal nucleus (Russchen et al. 1987). Either hemisphere of the prefrontal cortex reciprocally connects with bilateral mediodorsal nuclei (Preuss and Goldman-Rakic 1987). As described before, the mediodorsal nucleus of the thalamus has been subdivided into three sectors; magnocellular, parvocellular, and multiform sectors. The magnocellular sector, which is located in the medial portion of the nucleus, has

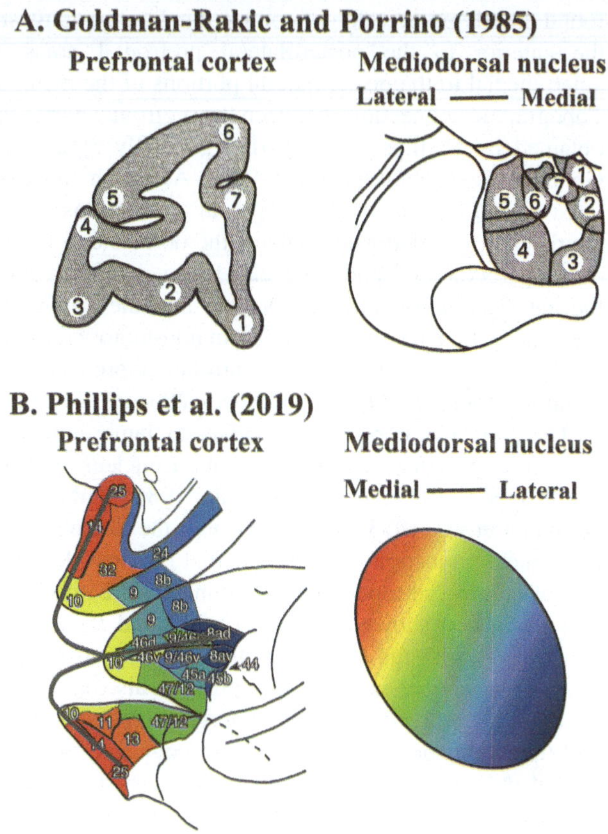

Fig. 1.16 Two examples of topographic connection patterns between the prefrontal cortex and the mediodorsal nucleus of the thalamus. From Goldman-Rakic and Porrino (1985) and Phillips et al. (2019) with permission

preferential connections with the orbitofrontal cortex. The parvocellular sector, which is located in the central portion of the nucleus, has preferential connections with the lateral portion of the prefrontal cortex. And, the multiform sector, which is located in the most lateral portion of the nucleus, has preferential connections with the arcuate area of the prefrontal cortex, which corresponds to the frontal eye field. Projections from the mediodorsal nucleus terminate mostly in the middle layers and sparsely in layer I in the prefrontal cortex. Although connectivity patterns between the different areas of the prefrontal cortex and the different sectors of the thalamic mediodorsal nucleus have been generally agreed, there are some discrepancies in precise area-to-area mapping rules. Goldman-Rakic and Porrino (1985) showed topographic connection patterns between different areas of the prefrontal cortex and the mediodorsal nucleus (Fig. 1.16a). Their projection patterns are slightly rotated, such that the ventromedial prefrontal cortex is connected to the most dorsomedial portion of the nucleus.

Recently, Phillips et al. (2019) performed probabilistic tractography on diffusion MRI data from macaque monkeys and estimated connections between the mediodorsal nucleus and 19 cytoarchitectonic areas of the prefrontal cortex. They found orderly progression of connectional topography, such that the ventromedial to the anterior to the posterolateral prefrontal areas are represented from the anteromedial to the posterolateral portions of the mediodorsal nucleus (Fig. 1.16b). Topographic connections between the prefrontal cortex and the mediodorsal nucleus obtained by Phillips et al. (2019) generally agree with the results obtained by Goldman-Rakic and Porrino (1985). Although the dorsolateral prefrontal cortex (areas 9 and 46) has major reciprocal connections with the parvocellular sector of the thalamic mediodorsal nucleus, the dorsolateral prefrontal cortex, especially the dorsal parts (areas 9 and 46d), also receives inputs from the parvocellular sector of the anterior ventral nucleus (VApc) and the ventrolateral nucleus (VLc) of the thalamus, which receives inputs from non-motor areas of the cerebellum (Middleton and Strick 2001). However, the ventrolateral prefrontal cortex does not receive these thalamic inputs (Middleton and Strick 2001).

The prefrontal cortex also receives thalamic inputs from other nuclei including the medial pulvinar, the anterior ventral nucleus, and intralaminar nucleus (Kievit and Kuypers 1977; Asanuma et al. 1985; Goldman-Rakic and Porrino 1985). Asanuma et al. (1985) had shown that the dorsolateral prefrontal cortex received anatomical projections from the medial pulvinar. Romanski et al. (1997) examined the selectivity and topography of connections between the prefrontal cortex and the pulvinar. They showed that the central and the lateral pulvinar have reciprocal connections with the lateral and the orbital prefrontal cortex (areas 8, 46, 45, 11, 12, 13). The central and the lateral pulvinar also have reciprocal connections with the posterior parietal cortex and the inferior temporal cortex, suggesting that the central and the lateral pulvinar contribute to cognitive functions. On the other hand, the medial pulvinar has connections with the medial and the orbital prefrontal cortex (areas 9, 24, 25, 10, 12) and has reciprocal connections with the temporal pole, the amygdala, and the cingulate cortex, suggesting that the medial pulvinar contributes to emotional and motivational functions. Thus, the topography of anatomical connections between the prefrontal cortex and the pulvinar agrees with functional specificity observed in the prefrontal cortex.

1.5.4.2 Prefrontal Connections with the Basal Ganglia

It had been known that the prefrontal cortex projects to the head of the caudate nucleus. Goldman and Nauta (1977a, b) showed that the middle one-third of the dorsal bank of the principal sulcus projects to the head of the caudate nucleus and that prefronto-caudate projections organize anatomical and functional mosaic. The dorsolateral prefrontal cortex also projects to the striatum, mainly to the head and body of the caudate nucleus (McGuire et al. 1991a; Yeterian and Pandya 1991). The dorsal prefrontal cortex predominantly projects to the dorsal and the central portion of the head and the body of the caudate nucleus, while the ventral prefrontal cortex projects mainly to the ventral and the central portion of the caudate nucleus. The cortex surrounding the principal sulcus projects to the intermediate portion of the head and the body of the caudate nucleus (McGuire et al. 1991b). Arikuni and

Kubota (1986) showed that pyramidal cells in layer V mainly project to the head of the caudate nucleus.

Cortico-striatal connections had been thought to have an anterior-posterior organization, such that the frontal cortex projects to the head of the caudate nucleus, the parietal cortex projects to the body, and the temporal and the occipital cortex project to the ventral and the dorsal region of the tail. However, Selemon and Goldman-Rakic (1985) showed that all these cortical areas project to longitudinal territories. The dorsolateral prefrontal region projects to the central portion of the striatum and the superior arcuate region projects to the dorsolateral portion.

1.5.4.3 Prefrontal Connections with Other Subcortical Structures

Prefrontal connections with the superior colliculus have been known. The dorsal portion of the principal sulcus projects to the intermediate and the deep layers of the superior colliculus ipsilateral to the prefrontal cortex (Goldman and Nauta 1976). The dorsal convexity, the principal sulcus, and the frontal eye field also project to the superior colliculus (Leichnetz et al. 1981). The frontal eye field preferentially projects to the dorsal part of the superior colliculus, while the dorsolateral prefrontal cortex preferentially projects to the deep part of the superior colliculus (Leichnetz et al. 1981). Recently, Borra et al. (2014) reported the ventrolateral prefrontal area (rostral area 46vc and rostral area 12) projecting to the superior colliculus.

Connections between the brainstem and the prefrontal cortex had been examined by retrograde transport of HRP. Porrino and Goldman-Rakic (1982) found that the dorsolateral prefrontal cortex (areas 9 and 46) receives projects from the ventral tegmental area, the substantia nigra, the caudal mesencephalon, the dorsal raphe, the locus coeruleus, and the medial parabrachial nucleus. Since labeled neurons in these brainstem regions correspond to monoamine-containing neurons with regard to their locations and morphology, these neurons are likely to be the sources of dopamine, norepinephrine, and serotonin found in the prefrontal cortex. Further, Ilinsky et al. (1985) showed that the lateral part of the substantia nigra projects to the lateral part of the parvocellular thalamic nucleus, which projects to the posterior portion of the principal sulcus and the frontal eye field, suggesting that this pathway contributes eye movement control.

It has been shown that the cerebellar dentate nucleus projects to the dorsal prefrontal cortex (area 9) in monkeys by stimulation studies using monkeys (Sasaki et al. 1979).

1.5.5 Intracortical Structures: Columnar Organization

Columnar organization has been observed anatomically and functionally in the cerebral cortex. Cortico-cortical fibers have been shown to terminate in distinct vertically oriented columns with a variable width of 300–700 μm. Goldman and

Nauta (1977b) examined terminal distributions of cortico-cortical connections in the prefrontal cortex by local injections of tritiated (^{3}H) amino acids into the principal sulcus area and showed that labeled terminals in the prefrontal cortex were present in distinct vertically oriented columns with 200–500 μm wide, extended across all layers of the cortex, and alternated in regular sequence with columns with comparable width that had no labeled terminals. Bugbee and Goldman-Rakic (1983) compared columnar structures of terminal fields observed in the prefrontal cortex between rhesus monkeys and squirrel monkeys and found that both species had similar size of columnar structures in terminal fields, suggesting that phylogenetic expansion of the cortical surface is accompanied by the increase in the number, not in the width, of columnar structures. In addition, Goldman-Rakic and Schwartz (1982) showed separate and alternating terminal fields in the prefrontal cortex in ipsilateral and contralateral projections to the prefrontal cortex (Fig. 1.15). Terminal fields of callosal projections from the contralateral prefrontal cortex were interdigitated with those of association projections from the parietal cortex ipsilateral to the prefrontal cortex (Fig. 1.15). These terminal fields in the prefrontal cortex formed column-like distributions of 300–700 μm wide. Similar modular patterns of terminal fields have been observed in the projections from the prefrontal cortex in the caudate nucleus and in the hippocampal region (Goldman-Rakic 1984). Further, columnar aggregations of prefrontal neurons have been observed in the neurons that project to the mediodorsal nucleus of the thalamus (Arikuni et al. 1983).

Lund et al. (1993) examined intra-areal and laterally spreading patterns of projections of pyramidal neurons by small injections of biocytin and compared these patterns among several cortical areas of rhesus monkey brain. In the prefrontal cortex (areas 46 and 9) (Fig. 1.17a), axon processes traveled away from the injection site in all directions, mainly in the superficial layers. A small injection of biocytin produced stripe-like connectivity patterns in the prefrontal cortex (Fig. 1.17a). On the contrary, injection of biocytin produced a punctate discontinuous pattern of connections in the visual areas (V1, V2, V4), the somatosensory areas (areas 3b, 1, 2), and the primary motor cortex (area 4) (Fig. 1.17b). Although all cortical areas seem to have a discontinuous stripe or patch-like organization with similar width, stripe-like connectivity patterns, which seem to be specific in the prefrontal cortex, suggest that these laterally spreading patterns of projections in prefrontal pyramidal neurons significantly contribute to integrate various information input from multiple cortical and subcortical areas and operate executive functions.

1.6 Functional Features of the Lateral Prefrontal Cortex

When a particular area in the cerebral cortex is damaged, a unique functional deficit would be usually observed. For example, damages in the cerebral motor areas produce severe impairments in the generation and the control of voluntary movements. Damages in the primary visual cortex cause impairments in the detection and

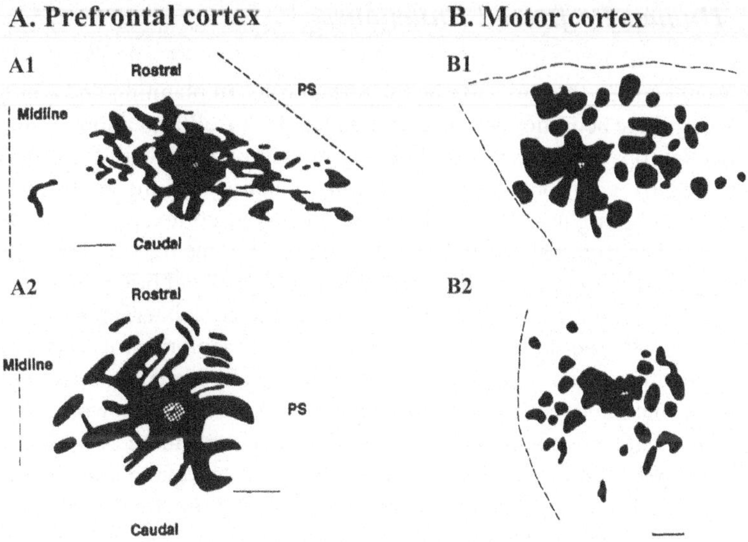

Fig. 1.17 Surface views of the distribution of biocytin-labeled terminals on the prefrontal cortex and the motor cortex. (**a**) Two examples (*a1* and *a2*) of local injections of biocytin in the prefrontal cortex (areas 46 and 9). PS: principal sulcus. (**b**) Two examples (*b1* and *b2*) of local injections of biocytin in the primary motor cortex (area 4). The dashed line at the left indicates the central sulcus. Stippled circles: injection sites of biocytin. Scale bar: 1 mm. From Lund et al. (1993) with permission

perception of visual information. And, damages in the medial temporal lobe including the hippocampus cause severe impairment in memory functions. Similarly, patients having prefrontal damages also exhibit unique impairments but exhibit different and complex features in impairments. As described at the beginning of this chapter, patients having prefrontal damages usually have normal long-term memory and exhibit normal perceptual, motor, and language skills. They even exhibit normal IQ and normal performances in most neuropsychological tests. However, they exhibit difficulties in making a new and adaptive action plan and making the best choice among several alternatives. These difficulties are thought to be caused by poor judgment and planning, and inappropriate decision-making. In addition, patients having prefrontal damages are also shown to exhibit a lack of insight and foresight, diminish their capacity for planning, and exhibit decreased initiative (Stuss and Benson 1986). These impairments cannot be explained simply by the failure of a single particular function such as perception, motor, or memory, but would be explained by a failure of coordinated operation of multiple functions including working memory, attention control, and monitoring and managing activities, which has been called executive control. Therefore, the prefrontal cortex is a unique brain area within the cerebral cortex and is closely linked to human intelligence.

1.6.1 Human Prefrontal Functions

The prefrontal cortex is known to play important roles in planning and controlling a variety of cognitive activities and in organizing and controlling various goal-directed behaviors. Neuropsychological studies using human patients with focal brain damages significantly contribute to understanding functions of the prefrontal cortex. Szczepanski and Knight (2014) divided the prefrontal cortex into five subregions (dorsolateral, ventrolateral, rostral, orbitofrontal, and medial) and reviewed the functions of each subregion based on neuropsychological and brain imaging studies. They listed the following as important functions of the dorsolateral prefrontal cortex: working memory, top-down attention, declarative memory, rule learning, task switching, planning and problem-solving, and novelty detection. Human lesion studies show that the dorsolateral prefrontal cortex is essential for working memory function especially for maintaining information, although there is a view that the dorsolateral prefrontal cortex is important not for simply maintaining information but for monitoring and manipulating working memory contents. Patients having unilateral damage to the dorsolateral prefrontal cortex exhibit decrease in event-related potential when they pay attention to visual stimuli presented in the ipsilateral visual cortex, suggesting that the dorsolateral prefrontal cortex provides top-down facilitatory inputs to sensory cortices. Although damages to the prefrontal cortex do not produce severe amnesic disorder, damages to the dorsolateral prefrontal cortex induce impairments in encoding and retrieving information from episodic memory, especially when information is associated with a particular episode. It is also well known that patients having damages to the dorsolateral prefrontal cortex exhibit impairments in performance of the Wisconsin card sorting test, which requires an ability for rule learning and task switching, and the Tower of London task, which requires an ability for planning and problem-solving. Thus, the dorsolateral prefrontal cortex is shown to participate in such functions as working memory, top-down attention, declarative memory, rule learning, task switching, planning, and problem-solving.

On the other hand, Szczepanski and Knight (2014) listed spatial attention, response inhibition, and language as important functions of the ventrolateral prefrontal cortex. Many studies showed that the ventrolateral prefrontal cortex, especially the right hemisphere, plays an important role in inhibitory control and flexible adjustment of motor plan, because patients having damages to this area exhibit impairments in performance of the go/no-go task and the stop-signal reaction time task. These patients show an increased number of errors on no-go trials in the go/no-go task and significantly slower reaction time to stop the response in the stop-signal task. Further, as functions of the rostral prefrontal cortex, Szczepanski and Knight (2014) listed multitasking control, prospective memory, creativity, mentalizing, and metacognition, although patients having damages in the rostral prefrontal cortex often exhibit no deficit on traditional neuropsychological test batteries and show normal performance on IQ tests.

In animal studies, the delayed-response task and behavioral tasks with the delay period have been widely used to examine functions of the prefrontal cortex. Verin et al. (1993) examined the behavioral effects by damages to the dorsolateral prefrontal cortex on a delayed-response task, a delayed alternation task, and a delayed reversal task in which human patients needed to find the rule by themselves. Patients with limited dorsolateral prefrontal lesions by vascular damages showed deficits in the delayed-response task and a stereotype behavior (perseveration) in the delayed alternation and the delayed reversal task. Based on the patients' behavioral deficits, they concluded that the prefrontal cortex plays a role in behavioral adaptation by inhibiting not only ongoing behavioral programs but also previously established programs.

1.6.2 Functional Heterogeneity of the Prefrontal Cortex

As is described in detail in Chap. 2, since Jacobsen (1936) first reported that rhesus monkeys with large bilateral lesions of the prefrontal cortex exhibited severe and long-lasting impairments of delayed-response performances, the delayed-response task became one of the important behavioral tasks to experimentally examine prefrontal functions in animals. To perform the delayed-response task correctly, the monkey is required to retain the baited position (either right or left) during the delay period. Therefore, delayed-response deficits were thought to be caused by an impairment of the short-term storage of spatial information. Subsequently, not only delayed-response tasks but also other behavioral tasks, which require short-term storage of non-spatial information, were shown to be impaired by bilateral lesions of the prefrontal cortex. Therefore, it was thought that an important function of the prefrontal cortex is the short-term storage of spatial as well as non-spatial information.

Goldman and Rosvold (1970) examined whether the critical factor causing impairments of the delayed-response and the delayed alternation was the delay, the spatial feature, or the combination of both. They used a spatial delayed alternation task (including the spatial feature and the delay), a conditional position response test (including only the spatial feature), and a simultaneous pattern discrimination (not including the spatial feature and the delay). In the conditional position response test, a reward was delivered on the left (right) box when the auditory cue was delivered from the speaker above (below) the cage. Lesions in the principal sulcus region impaired performance of the spatial task with delay (delayed alternation), while lesions in the arcuate sulcus region impaired performance of the spatial task without delay (conditional position response). The simultaneous pattern discrimination was not impaired in both lesion conditions. Thus, these results supported that the critical factors causing impairments of delayed-response tasks are the delay and the spatial feature.

Goldman and Rosvold (1970) showed that lesions in the different areas of the lateral prefrontal cortex produce different behavioral deficits. Stamm (1973) trained successive visual discrimination tasks and an auditory conditional position task and examined the effects of bilateral lesions of different areas of the prefrontal cortex on these behavioral tasks. Lesions of the ventral convexity caused severe impairments in performances of visual discrimination, but lesions of the dorsal convexity and the arcuate region showed no impairment. On the other hand, lesions of the arcuate region caused impairments in performances of the conditional position task, but lesions of the dorsal and the ventral convexity showed no impairment. These results agree with the results obtained by Goldman and Rosvold (1970) and are supported by Petrides (1982, 1985, 1986).

Rosenkilde (1979) summarized the results of prefrontal lesion studies using monkeys. He divided the prefrontal cortex into five regions (sulcus principalis, superior prefrontal convexity, inferior prefrontal convexity, arcuate concavity, and medial orbital cortex) and summarized lesion effects obtained in each of these five regions. Lesions of the sulcus principalis (corresponding to area 46) consistently produced impairments in spatial delayed-response and spatial delayed alternation tasks, but produce slight or no impairment in spatial and object reversal tasks, object alternation tasks, delayed matching-to-sample tasks, visual, auditory, and tactile discrimination tasks, and positional response tasks. On the other hand, lesions of the inferior prefrontal convexity produced impairments in object alternation tasks, delayed matching-to-sample tasks, and go/no-go tasks, but produce no impairment in spatial delayed-responses, spatial delayed alternations, spatial and object reversal tasks, and positional response tasks. Lesions of the dorsal prefrontal convexity produced impairments in the tasks using kinesthetic or proprioceptive cues, such as a go/no-go kinesthetic discrimination task and a go/no-go delayed alternation task. Thus, these results indicate that the lateral prefrontal cortex consists of heterogeneous structures from a functional point of view.

Dorsal-ventral functional heterogeneity has been observed in the lateral prefrontal cortex by lesion studies. Passingham (1975) examined the effects of bilateral lesions of either dorsolateral or ventrolateral prefrontal cortex on a delayed matching-to-color task and a delayed spatial alternation task and found that the dorsolateral lesions did not impair the performance of the delayed matching-to-color task but impaired delayed spatial alternation performance. On the contrary, the ventrolateral prefrontal lesions impaired the performance of the delayed matching-to-color task even at the shortest delay. Thus, the lateral prefrontal cortex is not functionally homogeneous but can be functionally divided into three regions: dorsolateral, ventrolateral, and peri-arcuate areas. The dorsolateral prefrontal cortex participates in spatial aspects of information processing, the ventrolateral prefrontal cortex participates in non-spatial aspects of information processing, and the peri-arcuate area of the prefrontal cortex participates in learning requiring the association of different modalities.

Functional heterogeneity between the dorsolateral and the ventrolateral prefrontal cortex is further supported by several lesion experiments using monkeys. It had been shown that monkeys with lesions of the lateral prefrontal cortex exhibited impairment in performing behavioral tasks including memory of sequential movements or temporal order of stimulus presentation. Pinto-Hamuy and Linck (1965) used monkeys with bilateral lesions of the lateral prefrontal cortex and examined whether these monkeys exhibited impairments in internally ordered sequence condition (push three cued panels with ordered sequence) or externally ordered sequence condition (push panels sequentially guided by light cues). They found that monkeys exhibited impairments in internally ordered condition and not in externally ordered condition, suggesting that bilateral lesions of the lateral prefrontal cortex cause the lack of internal representation of ordered sequence. Brody and Pribram (1978) reported similar results obtained from monkeys with bilateral lesions of the lateral prefrontal cortex when they performed sequential panel pressing behavior in internally and externally ordered conditions. Passingham (1978) examined whether the prefrontal cortex managed information about self-movement or not by training monkeys to perform a counting task. In this task, monkeys were required to first press a panel either 1 or 5 times to turn the light off, and then either immediately or after a 1 s to 5 s delay, they were required to repeat the same number of panel press. Monkeys with lesions in the dorsal convexity of the lateral prefrontal cortex (area 9) were impaired on this task, while monkeys with lesions in the principal sulcus (area 46) showed no impairment. This result suggests that the dorsal convexity of the lateral prefrontal cortex involves in the processing of self-movement information and that information processing of self-motion operating in this region contributes to control sequential actions.

On the other hand, Passingham (1985) used a puzzle box and examined the effects of lateral prefrontal lesions on sequential movements. To open the box and get a reward, the monkey was required to press the button, pull the handle, and then open the door (sequence of two movements) or to press the button, pull the handle, depress the lever, and then open the door (sequence of three movements). He found no impairment in performances of these sequential movements in monkeys with bilateral lesions of the dorsolateral prefrontal cortex but found impairment in monkeys with lesions of arcuate areas (areas 8 and 6). However, Petrides (1991) showed that monkeys with bilateral lesions of the dorsolateral prefrontal cortex exhibited impairment in serial order memory but monkeys with bilateral lesions of areas 8 and 6 did not. He sequentially presented three, four, or five different items, then presented two of these items simultaneously, and asked the monkey which item was presented first. Monkeys with bilateral lesions of areas 46 and 9 were severely impaired, while monkeys with lesions of areas 8 and 6 performed as normal control monkeys. Thus, these studies show that the dorsolateral prefrontal cortex participates in working memory of temporal order of stimulus presentation as well as stimulus itself or sequential movements. Petrides (1995) further examined the effect of bilateral lesions of the dorsolateral prefrontal cortex on non-spatial self-ordered and externally ordered working memory tasks. In the self-ordered task, three objects were presented simultaneously at three predetermined positions but the position of each object was randomly changed across trials. The monkey was required to select any one of these objects in every trial but needed to select the object that had not been selected in previous trials. In the externally ordered task, two objects were presented one by one, and then these two objects and one new object were presented simultaneously, and the monkey was asked to select the new object. Monkeys with lesions of the mid-dorsal part of the lateral prefrontal cortex (the cortex surrounding the principal sulcus) exhibited severe and long-lasting impairments on both self-ordered and externally ordered tasks, while monkeys with lesions of the cortex surrounding the arcuate sulcus showed no impairment on both tasks.

Thus, the lateral prefrontal cortex, especially the cortical area surrounding the principal sulcus (Brodmann's area 46), participates in working memory of temporal order of stimulus presentation and sequential movements, while the posterior part of the lateral prefrontal cortex, especially the cortex surrounding the arcuate sulcus (area 8) does not participate in this function. Lateral prefrontal contribution to working memory of temporal order of stimulus presentation and movement sequence is supported by the results obtained by neurophysiological studies (Barone and Joseph 1989; Funahashi et al. 1993, 1997; Ninokura et al. 2003, 2004; Inoue and Mikami 2006).

1.6.3 *Functional Heterogeneity in Human Prefrontal Cortex*

It has been proposed that the frontal lobe is functionally organized along dorsal-ventral axis as well as the rostro-caudal axis. The dorsal-ventral functional organization has been proposed to explain prefrontal roles for processing different information domains (e.g., spatial vs. non-spatial) (Goldman-Rakic 1996a, b; Courtney et al. 1998). Primate neurophysiological studies showed that visuospatial features are predominantly represented in the posterior portion of the dorsolateral prefrontal cortex, while object features are predominantly represented in the posterior portion of the ventrolateral prefrontal cortex (Constantinidis and Qi 2018). On the other hand, the rostro-caudal functional organization is proposed to explain roles for cognitive control processing, such that the rostral prefrontal area involves more abstract forms of control processing compared with more caudal area (Badre and D'Esposito 2009; Badre and Nee 2018).

O'Reilly (2010) proposed that functional properties are organized along multiple axes on the human prefrontal cortex; ventral *what* vs. dorsal *how*, anterior *abstract* vs. posterior *concrete*, and lateral *cold* vs. medial *hot*. For example, based on the meta-analysis of neuroimaging studies of working memory, Wager and Smith (2003) showed that the ventrolateral prefrontal cortex (areas 45 and 44) exclusively participates in encoding object and verbal information and shows no sensitivity to spatial information, indicating that this area participates in a *what* encoding. On the other hand, the dorsolateral prefrontal area (areas 6, 8, and 9) participates in process-oriented information processing, indicating that these areas participate in a *how* process. Bahlmann et al. (2015) examined by fMRI whether separable dorsal-ventral and rostro-caudal axes exist in humans while subjects performed tasks requiring cognitive control with three levels of abstraction using language and spatial stimuli. They found a single rostro-caudal gradient organized according to the level of abstraction, which was independent of processing of the stimulus domain, in the lateral prefrontal cortex. They also found dorsal-ventral segregation engaged by domain-specific information processing in the most rostral region of the lateral prefrontal cortex. These results suggest that a distinct rostro-caudal gradient for processing abstract information and the dorsal-ventral gradient for processing stimulus domain are present independently in the lateral prefrontal cortex. Blumenfeld et al. (2013) examined intrinsic connectivity among dorso-ventral, rostro-caudal, and medio-lateral axes in the prefrontal cortex using resting fMRI connectivity data. They showed that the dorsomedial prefrontal cortex forms parallel connections with the rostro-caudal axes of the dorsolateral and the ventrolateral prefrontal cortex, indicating that these prefrontal areas operate integratively and not independently. Cieslik et al. (2013) showed that the dorsolateral prefrontal cortex can be divided into two distinct subregions (anterior-ventral and posterior-dorsal) based on whole-brain co-activation patterns across neuroimaging studies. The anterior-ventral subregion showed increased connectivity with the anterior cingulate cortex and was more strongly associated with attention and action inhibition, while the posterior-dorsal subregion showed increased connectivity with the intraparietal sulci and was more strongly related to action execution and working memory. Thus, as O'Reilly (2010) proposed, the prefrontal cortex is organized by multiple axes of different functional properties along anterior-posterior dimension, dorsal-ventral dimension, and medial-lateral dimension. Although anatomy and physiology suggest the presence of distinct subregions in the lateral prefrontal cortex along anterior-posterior and dorsal-ventral axes, an important feature is that these subregions are not operated independently but integratively.

1.6.4 Functional Development of the Prefrontal Cortex

The dorsolateral prefrontal cortex plays an increasingly important role in the development of a variety of cognitive functions during the infant and juvenile period. Ablation of the dorsolateral prefrontal cortex 2 months before birth did not affect the functions that were subserved by the intact prefrontal cortex (Goldman and Galkin 1978). However, many functions specific to the dorsolateral prefrontal cortex are known to develop during early postnatal life. Goldman (1971) examined the developmental effects of lesions of the dorsolateral prefrontal cortex on delayed-response performances using infant monkeys. She found that infant monkeys who received dorsolateral prefrontal lesions performed similar accuracies as normal control monkeys when their performance was tested at 12–18 months of age. However, these monkeys exhibited severe impairment in delayed-response performances when their performance was tested at 24 months of age. On the contrary, infant monkeys who received orbitofrontal lesions exhibited impairments in delayed-response performance when their performance was tested at 12–18 months of age. However, these monkeys exhibited no impairment and showed complete recovery when their performance was tested at 24 months of age. These results indicate that the level of cortical maturation or functional maturation is different among different areas within the prefrontal cortex and that the functional maturation of the orbitofrontal cortex is faster than that of the dorsolateral prefrontal cortex. These results also suggest that, although functions related to delayed-response performance are initially achieved by the orbitofrontal cortex, after the maturation of the dorsolateral prefrontal cortex, these functions are controlled by the dorsolateral prefrontal cortex. Goldman and Alexander (1977) and Alexander and Goldman (1978) further examined the functional development of the dorsolateral prefrontal cortex using cryogenic depression of the prefrontal cortex while juvenile monkeys at different postnatal stages performed the delayed-response task. Monkeys at 34–36-month-old exhibited a 21–25% decrement in delayed-response performance, monkeys at 19–31-month-old exhibited a 7–8% decrement, and monkeys at 9–16-month-old exhibited no decrement in delayed-response performance. These results also indicate that it takes longer time for the dorsolateral prefrontal cortex to functionally mature. As was shown by Flechsig (1901, 1920), the anatomical maturation of the prefrontal cortex takes the longest time after birth based on the degree of myelination of intrinsic nerve fibers and the sequence of the development of cortical functions follows the sequence of the myelination of intrinsic nerve fibers in the development of the cerebral cortex. These results indicate that the prefrontal cortex is the last cortical area to complete its functional maturation and that functional maturation of the prefrontal cortex plays an increasingly important contribution to the development of a variety of cognitive functions.

References

Akert K, Von Monakov KH (1980) Relationships of precentral, premotor, and prefrontal cortex to the mediodorsal and intralaminar nuclei of the monkey thalamus. Acta Neurobiol Exp 40:7–25

Alexander GE, Goldman PS (1978) Functional development of the dorsolateral prefrontal cortex: an analysis utilizing reversible cryogenic depression. Brain Res 143:233–249. https://doi.org/10.1016/0006-8993(78)90566-8

Arikuni T, Kubota K (1986) The organization of prefrontocaudate projections and their laminar origin in the macaque monkey: a retrograde study using HRP-gel. J Comp Neurol 244:492–510. https://doi.org/10.1002/cne.902440407

Arikuni T, Sakai M, Kubota K (1983) Columnar aggregation of prefrontal and anterior cingulate cortical cells projecting to the thalamic mediodorsal nucleus in the monkey. J Comp Neurol 220: 116–125. https://doi.org/10.1002/cne.902200111

Arnsten AFT (1998) Catecholamine modulation of prefrontal cortical cognitive function. Trends Cogn Sci 2:436–447. https://doi.org/10.1016/S1364-6613(98)01240-6

Arnsten AFT, Wang MJ, Paspalas CD (2012) Neuromodulation of thought: flexibilities and vulnerabilities in prefrontal cortical network synapses. Neuron 76:223–239. https://doi.org/10.1016/j.neuron.2012.08.038

Asanuma C, Andersen RA, Cowan WM (1985) The thalamic relations of the caudal inferior parietal lobule and the lateral prefrontal cortex in monkeys: divergent cortical projections from cell clusters in the medial pulvinar nucleus. J Comp Neurol 241:357–381. https://doi.org/10.1002/cne.902410309

Baddeley A (1986) Working memory. Oxford University Press, Oxford, UK

Badre D, D'Esposito M (2009) Is the rostro-caudal axis of the frontal lobe hierarchical? Nat Rev Neurosci 10:659–669. https://doi.org/10.1038/nrn2667

Badre D, Nee DE (2018) Frontal cortex and the hierarchical control of behavior. Trends Cogn Sci 22:170–188. https://doi.org/10.1016/j.tics.2017.11.005)

Bahlmann J, Blumenfeld RS, D'Esposito M (2015) The rostro-caudal axis of frontal cortex is sensitive to the domain of stimulus information. Cereb Cortex 25:1815–1826. https://doi.org/10.1093/cercor/bht419

Barbas H (1986) Pattern in the laminar origin of corticocortical connections. J Comp Neurol 252: 415–422. https://doi.org/10.1002/cne.902520310

Barbas H (2015) General cortical and special prefrontal connections: principles from structure to function. Annu Rev Neurosci 38:269–289. https://doi.org/10.1146/annurev-neuro-071714-033936

Barbas H, Blatt GJ (1995) Topographically specific hippocampal projections target functionally distinct prefrontal areas in the rhesus monkey. Hippocampus 5:511–533. https://doi.org/10.1002/hipo.450050604

Barbas H, De Olmos J (1990) Projections from the amygdala to basoventral and mediodorsal prefrontal regions in the rhesus monkey. J Comp Neurol 300:549–571. https://doi.org/10.1002/cne.903000409

Barbas H, Pandya DN (1987) Architecture and frontal cortical connections of the premotor cortex (area 6) in the rhesus monkey. J Comp Neurol 256:211–228. https://doi.org/10.1002/cne.902560203

Barbas H, Pandya DN (1989) Architecture and intrinsic connections of the prefrontal cortex in the rhesus monkey. J Comp Neurol 286:353–375. https://doi.org/10.1002/cne.902860306

Barbas H, Henion TH, Dermon CR (1991) Diverse thalamic projections to the prefrontal cortex in the rhesus monkey. J Comp Neurol 313:65–94. https://doi.org/10.1002/cne.903130106

Barone P, Joseph JP (1989) Prefrontal cortex and spatial sequencing in macaque monkey. Exp Brain Res 78:447–464. https://doi.org/10.1007/BF00230234

Barrett RLC, Dawson M, Dyrby TB, Krug K, Ptito M, D'Arceuil H, Croxson PL, Johnson PJ, Howells H, Forkel SJ, Dell'Acqua F, Catani M (2020) Differences in frontal network anatomy across primate species. J Neurosci 40:2049–2107. https://doi.org/10.1523/jneurosci.1650-18.2019

Bates JF, Goldman-Rakic PS (1993) Prefrontal connections of medial motor areas in the rhesus monkey. J Comp Neurol 336:211–228. https://doi.org/10.1002/cne.903360206

Blumenfeld RS, Nomura EM, Gratton C, D'Esposito M (2013) Lateral prefrontal cortex is organized into parallel dorsal and ventral streams along the rostro-caudal axis. Cereb Cortex 23:2457–2466. https://doi.org/10.1093/cercor/bhs223

Borra E, Gerbella M, Rozzi S, Tonelli S, Luppino G (2014) Projections to the superior colliculus from inferior parietal, ventral premotor, and ventrolateral prefrontal areas involved in controlling goal-directed hand actions in the macaque. Cereb Cortex 24:1054–1065. https://doi.org/10.1093/cercor/bhs392

Borra E, Ferroni CG, Gerbella M, Giorgetti V, Mangiaracina C, Rozzi S, Luppino G (2019) Rostro-caudal connectional heterogeneity of the dorsal part of the macaque prefrontal area 46. Cereb Cortex 29:485–504. https://doi.org/10.1093/cercor/bhx332

Brodmann K (1909) Vergleichende Localisationslehre der Großhirnrinde in ihren Prinzipien dargestellt auf Grund des Zellenbaues. Johann Ambrosius Barth, Leipzig. English edition "Brodmann's Localization in the Cerebral Cortex" translated and edited by L.J. Garay and published from Smith-Gordon, London, 1994

Brody BA, Pribram KH (1978) The role of frontal and parietal cortex in cognitive processing. Brain 101:607–633. https://doi.org/10.1093/brain/101.4.607

Brown VJ, Bowman EM (2002) Rodent models of prefrontal cortical function. Trends Neurosci 25: 340–343. https://doi.org/10.1016/S0166-2236(02)02164-1

Brozoski TJ, Brown RM, Rosvold HE, Goldman PS (1979) Cognitive deficit caused by regional depletion of dopamine in prefrontal cortex of rhesus monkey. Science 205:929–932. https://doi.org/10.1126/science.112679

Brutkowski S (1965) Functions of prefrontal cortex in animals. Physiol Rev 45:721–746. https://doi.org/10.1152/physrev.1965.45.4.721

Bugbee NM, Goldman-Rakic PS (1983) Columnar organization of corticocortical projections in squirrel and rhesus monkeys: similarity of column width in species differing in cortical volume. J Comp Neurol 220:355–364. https://doi.org/10.1002/cne.902200309

Bush EC, Allman JM (2004) The scaling of frontal cortex in primates and carnivores. Proc Natl Acad Sci USA 101:3962–3966. https://doi.org/10.1073/pnas.0305760101

Cavada C, Goldman-Rakic PS (1989) Posterior parietal cortex in rhesus monkey. II. Evidence for segregated corticocortical networks linking sensory and limbic areas with the frontal lobe. J Comp Neurol 287:422–445. https://doi.org/10.1002/cne.902870403

Cavada C, Compañy T, Tejedor J, Cruz-Rizzolo RJ, Reinoso-Suárez F (2000) The anatomical connections of the macaque monkey orbitofrontal cortex. A review. Cereb Cortex 10:220–242. https://doi.org/10.1093/cercor/10.3.220

Chavis DA, Pandya DN (1976) Further observations on corticofrontal connections in the rhesus monkey. Brain res 117:369–386. https://doi.org/10.1016/0006-8993(76)90089-5

Cieslik EC, Zilles K, Caspers S, Roski C, Kellermann TS, Jakobs O, Langner R, Laird AR, Fox PT, Eickhoff SB (2013) Is there "one" DLPFC in cognitive action control? Evidence for heterogeneity from co-activation based parcellation. Cereb Cortex 23:2677–2689. https://doi.org/10.1093/cercor/bhs256

Cipolloni PB, Pandya DN (1999) Cortical connections of the frontoparietal opercular areas in the rhesus monkey. J Comp Neurol 403:431–458. https://doi.org/10.1002/(SICI)1096-9861(19990125)403:4<431::AID-CNE2>3.0.CO;2-1

Condé F, Audinat E, Maire-Lepoivre E, Crépel F (1990) Afferent connections of the medial frontal cortex of the rat. A study using retrograde transport of fluorescent dyes. I. Thalamic afferents. Brain Res Bull 24:341–354. https://doi.org/10.1016/0361-9230(90)90088-H

Condé F, Lund JS, Jacobowitz DM, Baibridge KG, Lewis DA (1994) Local circuit neurons immunoreactive for calretinin, calbindin D-28k or parvalbumin in monkey prefrontal cortex: distribution and morphology. J Comp Neurol 341:95–116. https://doi.org/10.1002/cne.903410109

Constantinidis C, Qi X-L (2018) Representation of spatial and feature information in the monkey dorsal and ventral prefrontal cortex. Front Integr Neurosci 12:31. https://doi.org/10.3389/fnint.2018.00031

Courtney SM, Petit L, Maisog JM, Ungerleider LG, Haxby JV (1998) An area specialized for spatial working memory in human frontal cortex. Science 279:1347–1351. https://doi.org/10.1126/science.279.5355.1347

Donahue CJ, Glasser MF, Preuss TM, Rilling JK, Van Essen DC (2018) Quantitative assessment of prefrontal cortex in humans relative to nonhuman primates. Proc Natl Acad Sci USA 115: e5183–e5192. https://doi.org/10.1073/pnas.1721653115

Elston GN (2003) Cortex, cognition and the cell: new insights into pyramidal neuron and prefrontal function. Cereb Cortex 13:1124–1138. https://doi.org/10.1093/cercor/bhg093

Fish KN, Hoftman GD, Sheikf W, Kitchens M, Lewis DA (2013) Parvalbumin-containing chandelier and basket cell boutons have distinctive modes of maturation in monkey prefrontal cortex. J Neurosci 33:8352–8358. https://doi.org/10.1523/jneurosci.0306-13.2013

Flechsig P (1901) Developmental (myelogenetic) localization of the cerebral cortex in the human subject. Lancet 158:1027–1030. https://doi.org/10.1016/S0140-6736(01)01429-5

Flechsig P (1920) Anatomie des menschlichen Gehirns und Rückenmarks auf myelogenetischer Grundlage. Verlag von Georg Thieme, Leipzig, pp 7–68

Funahashi S, Inoue M, Kubota K (1993) Delay-related activity in the primate prefrontal cortex during sequential reaching tasks with delay. Neurosci Res 18:171–175. https://doi.org/10.1016/0168-0102(93)90019-m

Funahashi S, Inoue M, Kubota K (1997) Delay-period activity in the primate prefrontal cortex encoding multiple spatial positions and their order of presentation. Behav Brain Res 84:203–223. https://doi.org/10.1016/S0166-4328(96)00151-9

Fuster JM (2008) The prefrontal cortex, 4th edn. Academic Press, New York, NY
Fuster JM (2015) The prefrontal cortex, 5th edn. Academic Press, New York, NY
Garey LJ (1994) Brodmann's 'localization in the cerebral cortex'. Smith-Gordon, London, UK
Gerbella M, Borra E, Tonelli S, Rozzi S, Luppino G (2013) Connectional heterogeneity of the ventral part of the macaque area 46. Cereb Cortex 23:967–987. https://doi.org/10.1093/cercor/bhs096
Giarrocco F, Averbeck BB (2021) Organization of parietoprefrontal and temporoprefrontal networks in the macaque. J Neurophysiol 126:1289–1309. https://doi.org/10.1152/jn.00092.2021
Giguere M, Goldman-Rakic PS (1988) Mediodorsal nucleus: areal, laminar, and tangential distribution of afferents and efferents in the frontal lobe of rhesus monkeys. J Comp Neurol 277:195–213. https://doi.org/10.1002/cne.902770204
Glausier JR, Datta D, Fish KN, Chung DW, Melchitzky DS, Lewis DA (2021) Laminar differences in the targeting of dendritic spines by cortical pyramidal neurons and interneurons in human dorsolateral prefrontal cortex. Neuroscience 452:181–191. https://doi.org/10.1016/j.neuroscience.2020.10.022
Goldman PS (1971) Functional development of the prefrontal cortex in early life and the problem of neuronal plasticity. Exp Neurol 32:366–387. https://doi.org/10.1016/0014-4886(71)90005-7
Goldman PS, Alexander GE (1977) Maturation of prefrontal cortex in the monkey revealed by local reversible cryogenic depression. Nature 267:613–615. https://doi.org/10.1038/267613a0
Goldman PS, Galkin TW (1978) Prenatal removal of frontal association cortex in the fetal rhesus monkey: anatomical and functional censequences in postnatal life. Brain Res 152:451–485. https://doi.org/10.1016/0006-8993(78)91103-4
Goldman PS, Nauta WJH (1976) Autoradiographic demonstration of a projection from prefrontal association cortex to the superior colliculus in the rhesus monkey. Brain Res 116:145–149. https://doi.org/10.1016/0006-8993(76)90256-0
Goldman PS, Nauta WJH (1977a) Columnar distribution of cortico-cortical fibers in the frontal association, limbic, and motor cortex of the developing rhesus monkey. Brain Res 122:393–413. https://doi.org/10.1016/0006-8993(77)90453-X
Goldman PS, Nauta WJH (1977b) An intricately patterned prefronto-caudate projection in the rhesus monkey. J Comp Neurol 171:369–385. https://doi.org/10.1002/cns.901710305
Goldman PS, Rosvold HE (1970) Localization of function within the dorsolateral prefrontal cortex of the rhesus monkey. Exp Neurol 27:291–304. https://doi.org/10.1016/0014-4886(70)90222-0
Goldman-Rakic PS (1984) Modular organization of prefrontal cortex. Trends Neurosci 7:419–424. https://doi.org/10.1016/S0166-2236(84)80146-0
Goldman-Rakic PS (1996a) Regional and cellular fractionation of working memory. Proc Natl Acad Sci USA 93:13473–13480. https://doi.org/10.1073/pnas.93.24.13473
Goldman-Rakic PS (1996b) The prefrontal landscape: implications of functional architecture for understanding human mentation and the central executive. Phil Trans R Soc Lond B 351:1445–1453. https://doi.org/10.1098/rstb.1996.0129
Goldman-Rakic PS, Porrino LJ (1985) The primate mediodorsal (MD) nucleus and its projection to the frontal lobe. J Comp Neurol 242:535–560. https://doi.org/10.1002/cne.902420406
Goldman-Rakic PS, Schwartz ML (1982) Interdigitation of contralateral and ipsilateral columnar projections to frontal association cortex in primates. Science 216:755–757. https://doi.org/10.1126/science.6177037
Goldman-Rakic PS, Selemon LD, Schwartz ML (1984) Dual pathways connecting the dorsolateral prefrontal cortex with the hippocampal formation and parahippocampal cortex in the rhesus monkey. Neuroscience 12:719–743. https://doi.org/10.1016/0306-4522(84)90166-0
Goldman-Rakic PS, Muly EC III, Williams GV (2000) D1 receptors in prefrontal cells and circuits. Brain Res Rev 31:295–301. https://doi.org/10.1016/S0165-0173(99)00045-4
González-Burgos G, Miyamae T, Krimer Y, Gulchina Y, Pafundo DE, Krimer O, Bazmi H, Arion D, Enwright JF, Fish KN, Lewis DA (2019) Distinct properties of layer 3 pyramidal neurons from prefrontal and parietal areas of the monkey neocortex. J Neurosci 39:7277–7290. https://doi.org/10.1523/jneurosci.1210-19.2019

Hackett TA, Stepniewska I, Kaas JH (1999) Prefrontal connections of the parabelt auditory cortex in macaque monkeys. Brain Res 817:45–58. https://doi.org/10.1016/S0006-8993(98)01182-2)
Hines M (1929) On cerebral localization. Physiol Rev 9:462–574. https://doi.org/10.1152/physrev.1929.9.3.462
Hirata Y, Miyachi S, Inoue K, Ninomiya T, Takahara D, Hoshi E, Takada M (2013) Dorsal are 46 is a major target of disynaptic projections from the medial temporal lobe. Cereb Cortex 23:2965–2975. https://doi.org/10.1093/cercor/bhs286
Hoshi E (2006) Functional specialization within the dorsolateral prefrontal cortex: a review of anatomical and physiological studies of non-human primates. Neurosci Res 54:73–84. https://doi.org/10.1016/j.neures.2005.10.013
Ilinsky IA, Jouandet ML, Goldman-Rakic PS (1985) Organization of the nigrothalamocortical system in the rhesus monkey. J Comp Neurol 236:315–330. https://doi.org/10.1002/cne.902360304
Inoue M, Mikami A (2006) Prefrontal activity during serial probe reproduction task: encoding, mnemonic, and retrieval processes. J Neurophysiol 95:1008–1041. https://doi.org/10.1152/jn.00552.2005
Jacobsen CF (1936) Studies of cerebral function in primates. I. The functions of the frontal association areas in monkeys. Comp Psychol Monogr 13:1–60
Jacobson S, Trojanowski JQ (1977a) Prefrontal granular cortex of the rhesus monkey. I. Intrahemispheric cortical afferents. Brain Res 132:209–233. https://doi.org/10.1016/0006-8993(77)90417-6
Jacobson S, Trojanowski JQ (1977b) Prefrontal granular cortex of the rhesus monkey. II. Interhemispheric cortical afferents. Brain Res 132:235–245. https://doi.org/10.1016/0006-8993(77)90418-8
Jones EG (1985) The Thalamus. Plenum Press, New York, USA
Jones EG, Powell TPS (1970) An anatomical study of converging sensory pathways within the cerebral cortex of the monkey. Brain 93:793–820. https://doi.org/10.1093/brain/93.4.793
Kievit J, Kuypers HGJM (1977) Organization of the thalamo-cortical connections to the frontal lobe in the rhesus monkey. Exp Brain Res 29:299–322. https://doi.org/10.1007/BF00236173
Krettek JE, Price JL (1977) The cortical projections of the mediodorsal nucleus and adjacent thalamic nuclei in the rat. J Comp Neurol 171:157–191. https://doi.org/10.1002/cne.901710204
Krimer LS, Goldman-Rakic PS (2001) Prefrontal microcircuits: membrane properties and excitatory input of local, medium, and wide arbor interneurons. J Neurosci 21:3788–3796. https://doi.org/10.1523/jneurosci.21-11-03788.2001
Krimer LS, Jakab RL, Goldman-Rakic PS (1997) Quantitative three-dimensional analysis of the catecholaminergic innervation of identified neurons in the macaque prefrontal cortex. J Neurosci 17:7450–7461. https://doi.org/10.1523/jneurosci.17-19-07450.1997
Le Merre P, Ahrlund-Richter S, Carlen M (2021) The mouse prefrontal cortex: unity in diversity. Neuron 109:1925–1944. https://doi.org/10.1016/j.neuron.2021.03.035
Leichnetz GR, Spencer RF, Hardy SGP, Astruc J (1981) The prefrontal corticotectal projection in the monkey; an anterograde and retrograde horseradish peroxidase study. Neuroscience 6:1023–1041. https://doi.org/10.1016/0306-4522(81)90068-3
Leonard CM (1969) The prefrontal cortex of the rat. I. Cortical projection of the mediodorsal nucleus. II. Efferent connections. Brain Res 12:321–343. https://doi.org/10.1016/0006-8993(69)90003-1
Lewis DA, Lund JS (1990) Heterogeneity of chandelier neurons in monkey neocortex: corticotropin-releasing factor- and parvalbumin-immunoreactive populations. J Comp Neurol 293:599–615. https://doi.org/10.1002/cne.902930406
Lund JS, Yoshioka T, Levitt JB (1993) Comparison of intrinsic connectivity in different areas of macaque monkey cerebral cortex. Cereb Cortex 3:148–162. https://doi.org/10.1093/cercor/3.2.148

Luppino G, Matelli M, Camarda R, Rizzolatti G (1993) Corticocortical connections of area F3 (SMA-proper) and area F6 (pre-SMA) in the macaque monkey. J Comp Neurol 338:114–140. https://doi.org/10.1002/cne.903380109

Luppino G, Rozzi S, Calzavara R, Matelli M (2003) Prefrontal and agranular cingulate projections to the dorsal premotor areas F2 and F7 in the macaque monkey. Eur J Neurosci 17:559–578. https://doi.org/10.1046/j.1460-9568.2003.02476.x

Markowitsch HJ, Pritzel M, Divac I (1978) The prefrontal cortex of the cat: anatomical subdivisions based on retrograde labeling of cells in the mediodorsal thalamic nucleus. Exp Brain Res 32: 335–344. https://doi.org/10.1007/BF00238706

Markowitsch HJ, Pritzel M, Kessler J, Guldin W, Freeman RB Jr (1980) Delayed-alternation performance after selective lesions within the prefrontal cortex of the cat. Behav Brain Res 1:67–91. https://doi.org/10.1016/0166-4328(80)90046-7

McGuire PK, Bates JF, Goldman-Rakic PS (1991a) Interhemispheric integration: I. Symmetry and convergence of the corticocortical connections of the left and the right principal sulcus (PS) and the left and the right supplementary motor area (SMA) in the rhesus monkey. Cereb Cortex 1: 390–407. https://doi.org/10.1093/cercor/1.5.390

McGuire PK, Bates JF, Goldman-Rakic PS (1991b) Interhemispheric integration: II. Symmetry and convergence of the corticostriatal projections of the left and the right principal sulcus (PS) and the left and the right supplementary motor area (SMA) of the rhesus monkey. Cereb Cortex 1: 408–417. https://doi.org/10.1093/cercor/1.5.408

Mesulam M-M (2000) Behavioral neuroanatomy: large-scale networks, association cortex, frontal syndromes, the limbic system, and hemispheric specialization. In: Mesulam M-M (ed) Principles of behavioral and cognitive neurology, 2nd edn. Oxford University Press, New York, NY, pp 1–120

Middleton FA, Strick PL (2001) Cerebellar projections to the prefrontal cortex of the primate. J Neurosci 21:700–712. https://doi.org/10.1523/JNEUROSCI.21-02-00700.2001

Mitchell AS (2015) The mediodorsal thalamus as a higher order thalamic relay nucleus important for learning and decision-making. Neurosci Biobehav Rev 54:76–88. https://doi.org/10.1016/j.neubiorev.2015.03.001

Mohedano-Moriano A, Muñoz-López M, Sanz-Arigita E, Pró-Sistiaga P, Merínez-Marcos A, Legidos-Garcia ME, Insausti AM, Cebada-Sánchez S, Arroyo-Jiménez MDM, Marcos P, Attacho-Pérula E, Insausti R (2015) Prefrontal cortex afferents to the anterior temporal lobe in the *Macaca fascicularis* monkey. J. Comp Neurol 523:2570–2598. https://doi.org/10.1002/cne.23805

Morecraft RJ, Cipolloni PB, Stilwell-Morecraft KS, Gedney MT, Pandya DN (2004) Cytoarchitecture and cortical connections of the posterior cingulate and adjacent somatosensory fields in the rhesus monkey. J Comp Neurol 469:37–69. https://doi.org/10.1002/cne.10980

Morris R, Pandya DN, Petrides M (1999) Fiber system linking the mid-dorsolateral frontal cortex with the retrosplenial/presubicular region in the rhesus monkey. J Comp Neurol 407:183–192. https://doi.org/10.1002/(SICI)1096-9861(19990503)407:2<183::AID-CNE3>3.0.CO;2-N

Musil Y, Olson CR (1991) Cortical areas in the medial frontal lobe of the cat delineated by quantitative analysis of thalamic afferents. J Comp Neurol 308:457–466. https://doi.org/10.1002/cne.903080311

Ninokura Y, Mushiake H, Tanji J (2003) Representation of the temporal order of visual objects in the primate lateral prefrontal cortex. J Neurophysiol 89:2868–2873. https://doi.org/10.1152/jn.00647.2002

Ninokura Y, Mushiake H, Tanji J (2004) Integration of temporal order and object information in the monkey lateral prefrontal cortex. J Neurophysiol 91:555–560. https://doi.org/10.1152/jn.00694.2003

O'Reilly RC (2010) The What and How of prefrontal cortical organization. Trends Neurosci 33: 355–361. https://doi.org/10.1016/j.tins.2010.05.002

Pandya DN, Kuypers HGJM (1969) Cortico-cortical connections in the rhesus monkey. Brain Res 13:13–36. https://doi.org/10.1016/0006-8993(69)90141-3

Pandya DN, Seltzer B (1982) Intrinsic connections and architectonics of posterior parietal cortex in the rhesus monkey. J Comp Neurol 204:196–210. https://doi.org/10.1002/cne.902040208

Pandya DN, Yeterian EH (1990) Prefrontal cortex in relation to other cortical areas in rhesus monkey: architecture and connections. Progr Brain Res 85:63–94. https://doi.org/10.1016/S0079-6123(08)62676-X

Pandya DN, Yeterian EH (1996) Comparison of prefrontal architecture and connections. Phil Trans R Soc Lond B 351:1423–1432. https://doi.org/10.1098/rstb.1996.0127

Pandya DN, Dye P, Butters N (1971) Efferent cortico-cortical projections of the prefrontal cortex in the rhesus monkey. Brain Res 31:35–46. https://doi.org/10.1016/0006-8993(71)90632-9

Pandya DN, van Hoesen GW, Mesulam M-M (1981) Efferent connections of the cingulate gyrus in the rhesus monkey. Exp Brain Res 42:319–330. https://doi.org/10.1007/BF00237497

Passingham R (1975) Delayed matching after selective prefrontal lesions in monkeys (*Macaca mulatta*). Brain Res 92:89–102. https://doi.org/10.1016/0006-8993(75)90529-6

Passingham R (1978) Information about movements in monkeys (*Macaca mulatta*) with lesions of dorsal prefrontal cortex. Brain Res 152:313–328. https://doi.org/10.1016/0006-8993(78)90258-5

Passingham RE (1985) Prefrontal cortex and the sequencing of movement in monkeys (*Macaca mulatta*). Neuropsychologia 23:453–462. https://doi.org/10.1016/1028-3932(85)90001-6

Passingham RE, Stephan KE, Kötter R (2002) The anatomical basis of functional localization in the cortex. Nat Rev Neurosci 3:606–616. https://doi.org/10.1038/nrn893

Petrides M (1982) Motor conditional associative-learning after selective prefrontal lesions in the monkey. Behav Brain Res 5:407–413. https://doi.org/10.1016/0166-4328(82)90044-4

Petrides M (1985) Deficits in non-spatial conditional associative learning after periarcuate lesions in the monkey. Behav Brain Res 16:95–101. https://doi.org/10.1016/0166-4328(85)90085-3

Petrides M (1986) The effect of periarcuate lesions in the monkey on the performance of symmetrically and asymmetrically reinforced visual and auditory go, no-go tasks. J Neurosci 6:2054–2063. https://doi.org/10.1523/jneurosci.06-07-02054.1986

Petrides M (1991) Functional specialization within the dorsolateral frontal cortex for serial order memory. Proc R Soc Lond B 246:299–306. https://doi.org/10.1098/rspb.1991.0158

Petrides M (1995) Impairments on nonspatial self-ordered and externally ordered working memory tasksafter lesions of the mid-dorsal part of the lateral frontal cortex in the monkey. J Neurosci 15:359–376. https://doi.org/10.1523/jneurosci.15-01-00359.1995

Petrides M (2005) Lateral prefrontal cortex: architectonic and functional organization. Philosophical Trans R Soc Lond B 360:781–795. https://doi.org/10.1098/rstb.2005.1631

Petrides M, Pandya DN (1988) Association fiber pathways to the frontal cortex from the superior temporal region in the rhesus monkey. J Comp Neurol 273:52–66. https://doi.org/10.1002/cne.902730106

Petrides M, Pandya DN (1994) Comparative architectonic analysis of the human and the macaque frontal cortex. In: Boller F, Spinnler H, Hendler JA (eds) Handbook of neuropsychology, The frontal lobes, vol 9, Sect. 12. Elsevier, Amsterdam, pp 17–58

Petrides M, Pandya DN (1999) Dorsolateral prefrontal cortex: comparative cytoarchitectonic analysis in the human and the macaque brain and corticocortical connection patterns. Eur J Neurosci 11:1011–1036. https://doi.org/10.1046/j.1460-9568.1999.00518.x

Petrides M, Pandya DN (2002) Association pathways of the prefrontal cortex and functional observations. In: Stuss DF, Knight RT (eds) Principles of frontal lobe function. Oxford University Press, New York, pp 31–50

Petrides M, Tomaiuolo F, Yeterian EH, Pandya DN (2012) The prefrontal cortex: comparative architectonic organization in the human and the macaque monkey brains. Cortex 48:46–57. https://doi.org/10.1016/j.cortex.2011.07.002

Phillips JM, Fish LR, Kambi NA, Redinbaugh MJ, Mohanta S, Kecskemeti SR, Saalmann YB (2019) Topographic organization of connections between prefrontal cortex and mediodorsal thalamus: evidence for a general principle of indirect thalamic pathways between directly connected cortical areas. Neuroimage 189:832–846. https://doi.org/10.1016/j.neuroimage.2019.01.078

Pinto-Hamuy T, Linck P (1965) Effect of frontal lesions on performance of sequential tasks by monkeys. Exp Neurol 12:96–107. https://doi.org/10.1016/0014-4886(65)90101-9

Porrino LJ, Goldman-Rakic PS (1982) Brainstem innervation of prefrontal and anterior cingulate cortex in the rhesus monkey revealed by retrograde transport of HRP. J Comp Neurol 205:63–76. https://doi.org/10.1002/cne.902050107

Porrino LJ, Crane AM, Goldman-Rakic PS (1981) Direct and indirect pathways from the amygdala to the frontal lobe in rhesus monkeys. J Comp Neurol 198:121–136. https://doi.org/10.1002/cne.901980111

Preuss TM (1995) Do rats have prefrontal cortex? The Rose-Woolsey-Akert Program reconsidered. J Cogn Neurosci 7:1–24. https://doi.org/10.1162/jocn.1995.7.1.1

Preuss TM, Goldman-Rakic PS (1987) Crossed corticothalamic and thalamocortical connections of macaque prefrontal cortex. J Comp Neurol 257:269–281. https://doi.org/10.1002/cne.902570211

Preuss TM, Goldman-Rakic PS (1989) Connections of the ventral granular frontal cortex of macaques with perisylvian premotor and somatosensory areas: anatomical evidence for somatic representation in primate frontal association cortex. J Comp Neurol 282:293–316. https://doi.org/10.1002/cne.902820210

Rao SG, Williams GV, Goldman-Rakic PS (1999) Isodirectional tuning of adjacent interneurons and pyramidal cells during working memory: evidence for microcolumnar organization in PFC. J Neurophysiol 81:1903–1916. https://doi.org/10.1152/jn.1999.81.4.1903

Rao SG, Williams GV, Goldman-Rakic PS (2000) Destruction and creation of spatial tuning by disinhibition: GABA(A) blockade of prefrontal cortical neurons engaged by working memory. J Neurosci 20:485–494. https://doi.org/10.1523/jneurosci.20-01-00485.2000

Robbins TW, Arnsten AFT (2009) The neuropsychopharmacology of fronto-executive function: monoaminergic modulation. Annu Rev Neurosci 32:267–287. https://doi.org/10.1146/annurev.neuro.051508.135535

Rodermund P, Westendorff S, Nieder A (2020) Blockage of NMDA- and GABA(A) receptors improves working memory selectivity of primate prefrontal neurons. J Neurosci 40:1527–1537. https://doi.org/10.1523/jneurosci.2009-19.2019

Romanski LM, Giguere M, Bates JF, Goldman-Rakic PS (1997) Topographic organization of medial pulvinar connections with the prefrontal cortex in the rhesus monkey. J Comp Neurol 379:313–332. https://doi.org/10.1002/(SICI)1096-9861(19970317)379-3<313:AID-CNE1>3.0.CO;2-6

Rose JE, Woolsey CN (1948) The orbitofrontal cortex and its connections with the mediodorsal nucleus in rabitt, sheep, and cat. Res Publ Assoc Res Nerv Ment Dis 27:210–232

Rosenkilde CE (1979) Functional heterogeneity of the prefrontal cortex in the monkey: a review. Behav Neural Biol 25:301–345. https://doi.org/10.1016/S0163-1047(79)90404-7

Russchen FT, Amaral DG, Price JL (1987) The afferent input to the magnocellular division of the mediodorsal thalamic nucleus in the monkey, *Macaca fascicularis*. J Comp Neurol 256:175–210. https://doi.org/10.1002/cne.902560202

Saleem KS, Miller B, Price JL (2014) Subdivisions and connectional networks of the lateral prefrontal cortex in the macaque monkey. J Comp Neurol 522:1641–1690. https://doi.org/10.1002/cne.23498

Sasaki K, Jinnai K, Gemba H, Hashimoto S, Mizuno N (1979) Projection of the cerebellar dentate nucleus onto the frontal association cortex in monkeys. Exp Brain Res 37:193–198. https://doi.org/10.1007/BF01474266

Sawaguchi T, Goldman-Rakic PS (1991) D1 dopamine receptors in prefrontal cortex: involvement in working memory. Science 251:947–950. https://doi.org/10.1126/science.1825731

Sawaguchi T, Goldman-Rakic PS (1994) The role of D1-dopamine receptor in working memory: local injections of dopamine antagonists into the prefrontal cortex of rhesus monkeys performing an oculomotor delayed-response task. J Neurophysiol 71:515–528. https://doi.org/10.1152/jn.1994.71.2.5152

Schwartz ML, Goldman-Rakic PS (1982) Single cortical neurons have axon collaterals to ipsilateral and contralateral cortex in fetal and adult primates. Nature 299:154–155. https://doi.org/10.1038/299154a0

Schwartz ML, Goldman-Rakic PS (1984) Callosal and intrahemispheric connectivity of the prefrontal association cortex in rhesus monkey: relation between intraparietal and principal sulcal cortex. J Comp Neurol 226:403–420. https://doi.org/10.1002/cne.902260309

Seamans JK, Yang CR (2004) The principal features and mechanisms of dopamine modulation in the prefrontal cortex. Progr Neurobiol 74:1–58. https://doi.org/10.1016/j.pneurobio.2004.05.006

Selemon LD, Goldman-Rakic PS (1985) Longitudinal topography and interdigitation of corticostriatal projections in the rhesus monkey. J Neurosci 5:776–794. https://doi.org/10.1523/jneurosci.05-03-00776.1985

Siwek DF, Pandya DN (1991) Prefrontal projections to the mediodorsal nucleus of the thalamus in the rhesus monkey. J Comp Neurol 312:509–524. https://doi.org/10.1002/cne.903120403

Stamm JS (1973) Functional dissociation between the inferior and arcuate segments of dorsolateral prefrontal cortex in the monkey. Neuropsychologia 11:181–190. https://doi.org/10.1016/0028-3932(73)90006-7

Stuss DT, Benson DF (1986) The frontal lobe. Raven Press, New York, NY

Szczepanski SM, Knight RT (2014) Insights into human behavior from lesions to the prefrontal cortex. Neuron 83:1002–1018. https://doi.org/10.1016/j.neuron.2014.08.011

Tanaka D Jr (1976) Thalamic projections of the dorsomedial prefrontal cortex in the thesus monkey (Macaca mulatta). Brain Res 110:21–38. https://doi.org/10.1016/0006-8993(76)90206-7

Tanibuchi I (1992) Electrophysiological and anatomical studies on thalamic mediodorsal nucleus projections onto the prefrontal cortex in the cat. Brain Res 580:137–146. https://doi.org/10.1016/0006-8993(92)90937-5

Tobias TJ (1975) Afferents to prefrontal cortex from the thalamic mediodorsal nucleus in the rhesus monkey. Brain Res 83:191–212. https://doi.org/10.1016/0006-8993(75)90930-0

Ungerleider LG, Gaffan D, Pelak VS (1989) Projections from inferior temporal cortex to prefrontal cortex via the uncinate fascicle in rhesus monkeys. Exp Brain Res 76:473–484. https://doi.org/10.1007/BF00248903

Uylings HBM, Van Eden CG (1990) Qualitative and quantitative comparison of the prefrontal cortex in rat and in primates, including humans. In: HBM U, Van Eden CG, JPC DB, Corner MA, MPG F (eds) The prefrontal cortex: its structure, function and pathology. Progress in brain research, vol 85. Elsevier, Amsterdam, pp 31–62

Uylings HB, Groenewegen HJ, Kolb B (2003) Do rats have a prefrontal cortex? Behav Brain Res 146:3–17. https://doi.org/10.1016/j.bbr.2003.09.028

Verin M, Partiot A, Pillon B, Malapani C, Agid Y, Dubois B (1993) Delayed response tasks and prefrontal lesions in man – evidence for self generated patterns of behaviour with poor environmental modulation. Neuropsychologia 31:1379–1369. https://doi.org/10.1016/0028-3932(93)90105-9

Vijayraghavan S, Wang M, Birnbaum S, Williams GV, Arnsten AFT (2007) Inverted-U dopamine D1 receptor actions on prefrontal neurons engaged in working memory. Nat Neurosci 10:376–384. https://doi.org/10.1038/nn1846

Wager TD, Smith EE (2003) Neuroimaging studies of working memory. Cogn Affect Behav Neurosci 3:255–274. https://doi.org/10.3758/CABN.3.4.255

Walker AE (1940) A cytoarchitectural study of the prefrontal area of the macaque monkey. J Comp Neurol 73:59–96. https://doi.org/10.1002/cne.900730106

Wang Y, Shima K, Isoda M, Sawamura H, Tanji J (2002) Spatial distribution and density of prefrontal cortical cells projecting to three sectors of the premotor cortex. Neuroreport 13:1341–1344

Wang M, Vijayraghavan S, Goldman-Rakic PS (2004a) Selective D2 receptor actions on the functional circuitry of working memory. Science 303:853–856. https://doi.org/10.1126/science.1091162

Wang X-J, Tegner J, Constantinidis C, Goldman-Rakic PS (2004b) Division of labor among distinct subtypes of inhibitory neurons in a cortical microcircuit of working memory. Proc Natl Acad Sci USA 101:1368–1373. https://doi.org/10.1073/pnas.0305337101

Webster MJ, Bachevalier J, Ungerleider LG (1994) Connections of inferior temporal area TEO and TE with parietal and frontal cortex in macaque monkeys. Cereb Cortex 4:470–483. https://doi.org/10.1093/cercor/4.5.470

Williams GV, Goldman-Rakic PS (1995) Modulation of memory fields by dopamine D1 receptors in prefrontal cortex. Nature 376:572–575. https://doi.org/10.1038/376572a0

Yeterian EH, Pandya DN (1991) Prefrontostriatal connections in relation to cortical architectonic organization in rhesus monkeys. J Comp Neurol 312:43–67. https://doi.org/10.1002/cne.903120105

Yeterian EH, Pandya DN, Tamaiuolo F, Petrides M (2012) The cortical connectivity of the prefrontal cortex in the monkey brain. Cortex 48:58–81. https://doi.org/10.1016/j.cortex.2011.03.004

Chapter 2
Historical Perspective of Prefrontal Research

2.1 Functional Localization of the Cerebral Cortex

The notion that the cerebral cortex is an essential structure for achieving a variety of cognitive functions is now a common sense. However, this was not the case until the middle of the nineteenth century. Until the eighteenth century, the cerebral cortex was thought to be merely a kind of vascular tissue or a glandular structure. In the early nineteenth century, Franz Joseph Gall established that the cerebral cortex consists of two basic structures, which are gray matter and white matter. He recognized that the gray matter was comprised of a mass of nerve cells while the white matter was comprised of a mass of nerve fibers. Gall also established not only the concept that the brain is an important structure for controlling intelligent functions but also the concept that the cerebral cortex, especially the gray matter, plays essential roles in achieving various complex cognitive behaviors (Zola-Morgan 1995). Thus, Gall contributed significantly to establishing the importance of the cerebral cortex as a central structure for achieving various complex behaviors. This notion was an important step for developing a modern view of cerebral functions, that is, the mental faculties are housed in the cerebral cortex. However, at the same time, he emphasized the idea that had been called organology or phrenology. Gall's idea of organology was based on the following thoughts: (1) Moral and intellectual faculties are innate; (2) Their exercise and manifestation depend on the cerebral structure; (3) The brain is the organ for all faculties, tendencies, and feelings; and (4) The brain is composed of as many particular organs as there are faculties, tendencies, and feelings (Zola-Morgan 1995). These thoughts were based on his observations of normal individuals and neurological and psychiatry patients. Then, Gall attempted to correlate the physical aspects of the skull with prominent psychological characteristics of human behavior and personality. He determined 27 separate faculties (Organs) and assigned each faculty to each area on the skull, which was a crude form of functional localization of the cerebral cortex. Interestingly, Gall had assigned the following functions to the frontal lobe: memory of things, memory of

S. Funahashi, *Dorsolateral Prefrontal Cortex*, Brain Science,
https://doi.org/10.1007/978-981-19-7268-3_2

facts, educability, and perfectibility at Organ 11; comparative sagacity (wisdom) at Organ 20; sense of metaphysics at Organ 21; sense of satire and witticism at Organ 22 (see Zola-Morgan 1995). Functions that Gall assigned to the frontal lobe are amazingly similar to current views of prefrontal functions. However, these are exceptional. He assigned the following functions to the occipital lobe: love of offspring at Organ 2; affection and friendship at Organ 3 (see Zola-Morgan 1995). According to organology, a specific region of the brain was thought to be the center for a specific faculty and the surface of the skull was thought to mirror the development of the underlying brain structure. Therefore, it was expected that highly developed faculty in a particular individual could be estimated by palpating the skull. This idea had been further developed by Johann Gaspard Spurzheim as phrenology movement.

At the same time, Gall's idea of functional localization of the cerebral cortex was strongly criticized. For example, his idea was considered to damage established concepts at that time, such as the unity of the soul, human immortality, free will, and the existence of God, because his idea was thought to indicate that the human soul, free will, and the existence of God were generated by a single organ, the cerebral cortex. Therefore, his idea was strongly denied by philosophical, political, and religious societies.

His idea was also criticized by scientists. His method to determine the locus and extent of each of the 27 Organs was thought to be questionable. In addition, by removing anatomically defined areas of the animal's brain and watching this animal's behavior, Marie-Jean-Pierre Flourens showed no loss of any faculty after the removal of a rather large area of the brain and concluded that, although the cerebral cortex carried out specific functions, responsible areas for these functions were distributed throughout the cortex without any localization, which was the concept of cerebral equipotentiality (see Pearce 2009).

However, localization of particular physiological functions was found in the cerebral cortex in animal studies. For example, by electrical stimulation of the surface of the dog's cerebral cortex, Gustav Theodor Fritch and Eduard Hitzig found that the stimulation around the cruciate sulcus produced muscle contractions of either forelimb or hindlimb contralateral to the stimulated hemisphere and that stimulation at no other areas on the cerebral cortex produced muscle contraction of any body part. These observations indicated that the center for controlling limb movements was localized in the cortical area surrounding the cruciate sulcus in the dog's brain (see Breathnach 1992). Localization of the motor area was also determined in the cortical area surrounding the central sulcus of the monkey brain and crude somatotopic representation in the cortical motor area had also been described (Ferrier 1874, 1875a). Theodor Meynert pointed out that brain areas associated with motor functions were mostly located in the frontal portion of the brain, while brain areas associated with sensory functions were located in the posterior portion (Lanczik and Keil 1991). Regarding sensory representation on the cerebral cortex, David Ferrier described the auditory area in the temporal lobe surrounding the lateral sulcus in the monkey brain (Ferrier 1875b).

In the human brain, Pierre Paul Broca showed that the left posterior inferior frontal region in the human brain was the area responsible for speech control without impairment of language comprehension. Broca met a 51-year-old patient who had been without any productive speech for many years. Every time when Broca attempted to get a response to a question, the patient could only produce a single repetitive syllable "tan" and could not produce any word or phrase. After the patient died, a damage was found on the surface of the left frontal lobe. Subsequently, Broca found similar damages in the frontal lobe in patients with speech disturbance and realized that in most of the cases, damages were located in the posterior inferior frontal region on the left side of the brain (see Dronkers et al. 2007). On the other hand, Carl Wernicke found the area responsible for speech recognition in the left temporal region in the human brain. He recognized the fact that the region described by Broca was not the only speech center and asserted that another speech area was located in the secondary auditory center of the superior temporal gyrus. Wernicke realized that the cortical area described by Broca acted as a motor speech center, while the cortical area that he described was a sensory speech center and involved in the comprehension of spoken language (Lanczik and Keil 1991).

Thus, during the second half of the nineteenth century, the evidences showing localization of physiological functions, such as sensory and motor functions, in the cerebral cortex were growingly accumulated both in animal brain and human brain.

2.2 Historical Perspective of Frontal Lobe Functions

2.2.1 Debate Between Holistic Theory and Localization Theory

Experimental studies regarding prefrontal functions had begun in the late nineteenth century. There was an important debate in the history of neuroscience, which was related to the nature of the functional localization in the cerebral cortex. One group of neuroscientists believed that different portions of the cerebral cortex have distinct functions on mental and physical activities (theory of cerebral localization). On the contrary, another group of neuroscientists believed that these functions are accomplished by the brain acting as a whole unit (holistic theory). It had been shown that muscle contraction of the hand, arm, or leg could be produced by the electrical stimulation at one particular cortical area, while sensation could be produced in other cortical areas. It had also been shown that the ablation of a certain brain area yielded an evident loss of motor functions or sensory functions in animal studies. Therefore, the idea that sensory and motor functions were localized within restricted brain areas on the cerebral cortex became growingly popular.

In 1881, a historical debate between David Ferrier and Friedrich Leopold Goltz took place at the International Medical Congress held in London. David Ferrier was a leading physiologist for cerebral localization (Ferrier 1874, 1875a), while Friedrich

Goltz was a prominent anti-localizationist. The scene of the debate between Ferrier and Goltz was described in an article by Tyler and Malessa (2000).

> "*In the presentation to the Physiology Section, Goltz criticized localizationists for their widely divergent conclusions about the exact anatomic sites of cortical centers and for their failure to adequately explain functional restitution after cortical ablations. He noted that localizationist theories could, like an apple, "look very tempting and still have a worm inside." He described his own studies on massive decerebrations in dogs and noted that despite complete destruction of the cortices of both hemispheres these animals failed to exhibit motor weakness or deficits in primary sensation. Ferrier noted that Goltz's results were irreconcilable with his own experiments in monkeys, in which circumscribed lesions produced clear and reproducible functional deficits. Both investigators exhibited animals with cortical ablations. Ferrier's presentation of a hemiplegic monkey prompted Charcot's famous utterance, "C'est un malade!" ["It's a patient!"]. A distinguished committee examined the brains of the animals, and confirmed that Ferrier had indeed succeeded in producing a circumscribed lesion in the frontoparietal cortex, whereas the cortical ablations in Goltz's dogs were much less widespread than anticipated. Ferrier's dramatic demonstration of the effects produced by localized lesions in macaques triumphed over Goltz's unitary view of brain function, providing a major impetus for the subsequent successful development of neurologic surgery.*" (From Abstract of Tyler and Malessa (2000))

2.2.2 Functions of the Frontal Lobe

Regarding functions of the frontal lobe, electrical stimulation on the frontal lobe in monkeys, especially at the posterior portion of the frontal lobe, was known to produce muscle contractions of the upper or lower limbs (Ferrier and Sanderson 1874, 1875a). The unilateral ablation of this brain area was also shown to yield marked impairments of limb movements on the contralateral side of the ablated cortex. Therefore, the posterior portion of the frontal lobe was thought to be the cortical motor center responsible for generating and controlling movements of upper and lower limbs in monkeys.

On the contrary, the anterior portion of the frontal lobe was not excitable by electrical stimulations. The electrical stimulation at this portion produced no movement and no sensation in monkeys. The ablation of this portion produced only slight or doubtful impairment in movement or sensation. Animals deprived of the frontal lobes became apathetic, stolid, or idiotic. These animals also exhibited ugly, irritable, and restless behavior. In a paper published by Franz (1902), he cited the following sentences from Loeb's *Comparative Physiology of the Brain and Comparative Psychology*:

> "*I have repeatedly removed both frontal lobes in dogs. It was impossible to notice the slightest difference in the mental functions of the dog. There is perhaps no operation which is so harmless for a dog as the removal of the frontal lobes.*" (Page 2 in Franz (1902))

> "*And if the anterior parts of both hemispheres be removed, the dog is no longer normal, but idiotic.*" (Page 2 in Franz (1902))

Ferrier (1876) also described the behavior of monkeys with prefrontal lesions in his book *The Functions of the Brain*:

> "*Removal or destruction by the cautery of the anterior-frontal lobes is not followed by any definite physiological results. The animals retain their appetites and instincts, and are capable of exhibiting emotional feeling. The sensory faculties, sight, hearing, touch, taste and smell, remain unimpaired. The powers of voluntary motion are retained in their integrity, and there is little to indicate the presence of such an extensive lesion, or removal of so large a part of the brain.*" (Page 231 in Ferrier (1876))

But,

> "*Notwithstanding this apparent absence of physiological symptoms, I could perceive a very decided alteration in the animal's character and behavior, though it is difficult to state in precise terms the nature of the change. The animals operated on were selected on account of their intelligent character. After the operation, though they might seem to one who had not compared their present with their past, fairly up to the average of monkey intelligence, they had undergone a considerable psychological alteration. Instead of, as before, being actively interested in their surroundings, and curiously prying into all that came within the field of their observation, they remained apathetic, or dull, or dozed off to sleep, responding only to the sensations or impressions of the moment, or varying their listlessness with restless and purposeless wanderings to and fro. While not actually deprived of intelligence, they had lost, to all appearance, the faculty of attentive and intelligent observation.*" (Page 231–232 in Ferrier (1876))

Further, Ferrier and Yeo (1884) described observations based on their experiments using monkeys with brain lesions:

> "*No physiological defect could be discovered, nor could any very definite alteration in the animal's behavior be determined.. .. Till its death by chloroform, eleven weeks later, it continued in perfect health and exhibited no perceptible deviation from the normal.*" (Page 526 in Ferrier and Yeo (1884))

> "*Apart from a degree of dullness or apathy, there was nothing in the animal's behavior at all remarkable or appreciably abnormal.*" (Page 531 in Ferrier and Yeo (1884))

> "*On the whole there seemed mental deterioration, characterized by general apathetic indifference or purposeless unrest, effects which, in comparison with those of other lesions, appear to have relation with lesions of the frontal lobes as such.*" (Page 532 in Ferrier and Yeo (1884)).

On the other hand, Bianchi (1895) described Hitzig's comments regarding the relationship between prefrontal functions and intelligence. That is, Hitzig had anatomically examined the development of the frontal lobes in several animals including cats, dogs, and monkeys, and he concluded that the frontal lobe was the seat of the highest intellectual faculties, based on the parallel relation observed between the development of the frontal lobes and that of the intellect.

Although sensory and motor functions had been shown to be localized on the cerebral cortex, in general, intellectual activities, such as images, memories, or ideas, were not thought to be localized in any particular areas of the cerebral cortex. Ferrier (1876) also considered that the frontal lobe was not the seat of intelligence. He thought that intellect was not distinctly localized, but was the result of functions of the whole centers, and was correlated to the harmonious development of various portions of the brain.

Ferrier (1876) proposed that the frontal lobe was the center of attention. This conclusion was based on the observations of monkeys with bilateral large frontal lobe lesions. His monkeys exhibited some alteration in the temper and behavior without any alteration in their movements, sensations, instincts, feelings, or emotions. However, these monkeys exhibited less interest in the surroundings, less natural inquisitiveness, apathetic, responses to momentary impressions, and low power of attentive observation. However, according to Fuster (1997), Munk (1890) examined the behavior of dogs and monkeys with frontal lobe ablations and found no existence of any loss of attention. He not only found no disorder of intelligence but also found no disorders of various forms of sensation, whenever the motor area was excluded from the lesion area.

Bianchi (1895) himself used dogs and monkeys with ablation of the frontal lobes and examined whether or not the frontal lobe was the seat of coordination and fusion of incoming and outgoing products in the sensory and motor areas. He concluded that the frontal lobes played a role to integrate the products of the sensory-motor regions with emotional states accompanied by perceptions. The fusion of these products could constitute the psychical tone of the individual. Therefore, he thought that the removal of the frontal lobes disaggregated the personality and disabled to synthesize a group of representations, although it did not interfere so much with the perceptions.

Thus, experimental results using animals (dogs and monkeys) generally agreed with the conclusion that the large bilateral ablations of the anterior portion of the frontal lobe did not exhibit any disturbance of movement and sensation. Disturbance of intellectual activity was not clearly observed in these animals, since it was generally considered that brain functions for intellectual activity were not localized in any particular area of the cerebral cortex. However, it had been known that the animals with ablations in the frontal lobe exhibited some behavioral changes like apathetic, stolid, or idiotic.

2.2.3 *Lashley's Principles: Equipotentiality and Mass Action*

The idea of functional localization in the cerebral cortex agreed fairly well with the results obtained by both sensory and motor experiments. However, clinical literatures, which focused on color vision, reading disability, apraxia, aphasia, judgment, or personality change, failed to confirm any schema of functional localization in the cerebral cortex, especially in the association areas. The difficulty to observe functional localization of the cerebral cortex in clinical literatures might be due to the difficulty to estimate the extent of the damaged brain area and identify its anatomical locus on the cerebral cortex. In addition, attempts to localize the intelligence in any particular brain area, such as the prefrontal areas, had failed in both animal studies and human clinical observations. Since it was thought that intelligence would be a product of the activity of the whole cerebral cortex, the mass of relationships among various faculties was thought to be a more important aspect than the implication by localized faculties to understand neural functions such as intelligence.

Lashley (1926, 1931) used rats as subjects and examined the relative significance of functional localization on the cerebral cortex and total mass of the cerebral cortex during the performance of various types of activity. In this study, he tested the influence of the extent of brain injury on learning and retention of the brightness discrimination (rats needed to avoid a darkened compartment and enter an illuminated compartment). Behaviors of more than 150 rats were examined before and after the cerebral lesions of various extents and loci. He found that lesions to the occipital region had no effect on the ability for brightness discrimination and that the amount of practice necessary to re-establish this ability was proportional to the extent of the lesion and independent of its locus. Lashley (1931) further examined the behavioral effects of cortical ablations for rats that performed visual pattern discrimination, brightness discrimination, and maze learning. He made ablations to various extents and at various loci in the rat cerebral cortex. He found that these animals with cerebral lesions learned the complex maze much more slowly than normal unoperated controls. Importantly, the slowness of maze learning was observed no matter which part of the cerebral cortex was ablated. The degree of slowness was proportional to the amount of the cortical tissue ablated, irrespective of the ablated cortical locus. Lashley described the result as follows:

> "*The correlation between the extent of destruction of the cortex and the amount of training necessary to perfect the habit was 0.84, which was as high as that between any two measures of learning ability or of intelligence that we have. The lesions included for this calculation covered all parts of the neocortex and, within the limits of their statistical reliability, the data indicate that equal amounts of destruction in the motor, somesthetic, auditory or visual areas were attended by equal amounts of retardation in learning.*" (Page 249 in Lashley (1931))

> "*There is evidence of mutual dependence of parts in which the specialization of structures seems less important than the mere mass of functional tissue. There are indications that within the entire cortex, for certain functions, and within specialized areas, for others, the subordinate parts are all equally capable of performing the functions of the whole. Even where the highest degree of specialization exists, as in the visual and motor areas, the facts of equivalence of stimuli or equivalence of motor responses preclude any narrowly localized specialization of intercellular connections.*" (Page 34 in Lashley (1933))

Based on these findings, Lashley proposed the law of mass action and the law of equipotentiality for cerebral control mechanisms of various functions. Lashley (1931) described the reason why he used the rat, not the monkey, as subjects for these studies. At the same time, since he did not perform the same experiment using monkeys, he also described the limitation of his conclusions.

> "*The advantages of this material are the simplicity of the animal's behavior, its steadiness in activity under the motivation of hunger and its availability in large numbers. The last has made possible the use of experimental and statistical controls.*" (Page 246 in Lashley (1931))

"The danger of generalization from the rat to man is obvious. Our program includes the use of this material only as a means of outlining problems and gaining clues which must in every case be retested by experiments with primates and by comparison with clinical evidence. So far as we have been able to carry out such controls, there has been a clear agreement between the results with the rat and with primates. Actually these lower animals seem to show the beginnings of every human mental trait and I have come to doubt that the evolution of mammals has introduced any change in the fundamental organization or mechanism of cerebral activity. The enormous differences are in degree rather than in kind." (Page 246 in Lashley (1931))

2.2.4 The Prefrontal Cortex Has Some Specific Function

Lashley (1931) showed that the degree of impairment in learning of complex mazes and other problems involving the temporal organization of behavior was proportional to the mass of removed cortical tissue (mass action), but was relatively unrelated to the locus of the lesion in the cerebral cortex (equipotentiality). Although Lashley understood limitations in his conclusions, he indicated that various cortical areas of the rat brain were functionally equivalent in maze learning and visual discrimination habits. However, the application of these principles to human neurology did not meet general acceptance, because apparent functional localization in motor and sensory aspects was observed in the human and primate cerebral cortex. For example, Jacobsen (1939) stated as follows:

Development of specialized neural structures mediating the more complicated modes of behavior in the higher mammals seems to be clearly established. In the rat there is little evidence for specialized regions of elaboration other than the sensory and motor spheres, although this lack of demonstration may be due to inadequacies of method. On the other hand a number of studies have consistently indicated that the sensory and motor regions participate in a more general type of integration that is dissociated from their functions as sensory regions per se. *Within limits all parts of the cortex appear to be equally capable of mediating this type of integration. In the primates, however, at least some of the more complicated behavioral adjustments clearly depend on the intactness of restricted areas, especially the parieto-temporal and frontal regions, and injury to these areas results in disturbances not found with damage to any of the sensory or motor projection fields.* (Page 319 in Jacobsen (1939))

Attempt to find some specific function of the prefrontal cortex was further performed using animals with frontal lobe ablations. For example, Franz (1902) trained cats to perform simple sensory motor association tasks, in which cats were required to escape from the box by pulling a string, pressing a button, etc. Memory of these habits was maintained for 7–8 weeks after training in normal controls. He showed that these habits were lost after bilateral lesions in the frontal lobe, although lesions in other brain areas did not produce the loss of these habits. Unilateral lesions of the frontal lobe produced a partial loss or a slowing of the learning of the habits. Importantly, only newly formed habits were lost after the removal of the frontal lobe, while inherited and long-standing habits were retained. The emotional conditions were practically the same after as before the operation. In the subsequent study,

Franz (1910) found no definite motor inabilities but found the distinct loss of associations (memory of habits) following the extirpation of the frontal lobes in cats and monkeys.

Jacobsen (1931) performed a study to answer the following questions regarding prefrontal functions using monkeys. (1) Does injury to the prefrontal cortex result in the loss of previously acquired visuo-kinesthetic-motor habits or in the loss of visual discrimination? (2) If there is an impairment on these habits, can these habits be relearned without the prefrontal cortex? (3) Can new manipulation skills be established without the prefrontal cortex? (4) Is there any permanent change in emotional behavior? (5) Does the removal of the prefrontal cortex exhibit hyperactivity and restlessness? Jacobsen used two series of problem boxes (series A included a pull box, a crank box, a hook and handle box, and a combination box A, while series B included a slide box, a ring box, a hasp box, and a combination box B) for visuo-kinesthetic-motor tasks. He also used the visual discrimination task, in which pairs of simple visual patterns were used (one for rewarded, and the other for non-rewarded). He obtained the following results. (1) Lesions to the prefrontal cortex in monkeys had no effect either on the simple problem-box tests or on the visual discrimination. However, lesions to the prefrontal cortex exhibited the impairment only when the combination box, in which the subjects needed integration of several different actions to solve, was used for the test. (2) Lesions to the prefrontal cortex did not impair the learning ability for solving both simple and combination boxes. (3) No sensory disturbance and no emotional change accompanied by prefrontal lesions. Thus, prefrontal lesioned monkeys exhibited no impairment on solving and relearning problem-box tests and visual discrimination. Jacobsen thought that intelligent behavior was produced by the coordination of various actions. Therefore, based on these results, Jacobsen (1931) concluded that the prefrontal cortex was not "the organ of intelligence."

2.2.5 *Jacobsen's Experiments Using Delayed-Response Task*

In the 1930s, there was a general consensus that the prefrontal cortex did not play an important contribution to movement, sensation, and "intelligence." However, it had been known that animals having lesions in the prefrontal cortex exhibited the loss of some intelligent behaviors, such as integration of multiple actions to solve a problem, difficulty in learning new habits, and idiotic behavior. Therefore, it had been suggested that some important functions could be localized in the prefrontal cortex.

In the mid-1930s, Jacobsen performed a series of experiments and asked the following questions (Jacobsen 1936). (1) Does the prefrontal cortex mediate a type of behavior that the organism is unable to perform in the absence of this region? (2) Is the behavioral deficit observed specific to the prefrontal lesion? Or, does this behavioral deficit also appear after lesion in other cortical areas? (3) Do the principles of equipotentiality and mass action also characterize cortical organization in primates? To answer these questions, Jacobsen (1935, 1936) used two types of

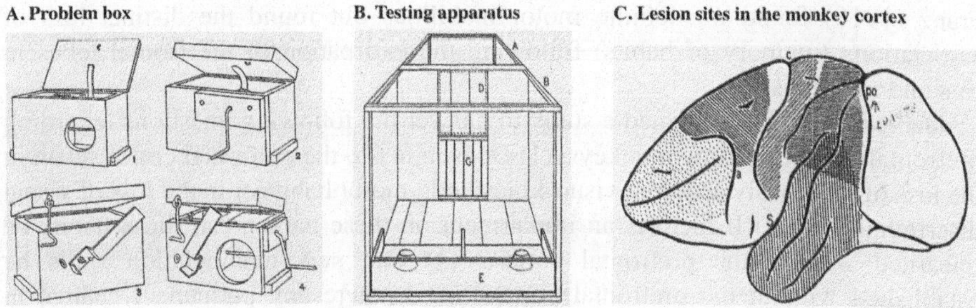

Fig. 2.1 (**a**) Examples of problem boxes. *1.* Pull box. *2.* Crank box. *3.* Hook and handle box. *4.* Combination box. (**b**) A testing apparatus used for delayed response. G is a portcullis grille, D is an opaque door, and E is the tray and cups. (**c**) Schematic summary of the effects of cortical ablations for delayed response. Impairment of delayed response was observed only by ablations at the prefrontal area (dotted area). From Jacobsen (1931, 1936) with permission

behavioral tests. One type of behavioral test was problem box tests and visual discrimination (Fig. 2.1a). In these tests, the cues necessary to solve the tests were available in the subject's environment at the time of response.

> "*The subject is confronted on successive trials by a problem situation which presents the necessary differential cues at the time of response. Thus in comparison and selection between two visual patterns, differential cues impinge upon the animal's sensorium from the test situation itself. Once the subject has learned that food can be obtained from the larger of two paired boxes, the differential cues which determine for a given trial whether the right or left box shall be opened, are inherent in the test situation. In a similar way, each problem box offers to the sophisticated subject differential cues which release motor activities appropriate to the particular task.*" (Page 12 in Jacobsen (1936))

Another type of behavioral test was the delayed-response task (Fig. 2.1b). In this task, the subject needed to recall certain essential cues from recent experience at the time of response.

> "*In testing for delayed response, the differential cues which determine whether the right or left box shall be opened are carefully eliminated from the test situation and must in some manner be supplied by the subject. In the direct method of testing for delayed response, the subject observes the experimenter conceal food under one of two similar test objects. Before there has been opportunity to make a complete response, an opaque door is interposed between subject and test objects, and only after an interval of a few seconds to several minutes is the subject permitted to open one of the two paired test objects, and if its choice has been correct, obtain the food which it earlier saw concealed. In a properly controlled experiment visual, olfactory, or other cues are not inherent in the test situation, and by contrast with discrimination and problem box tasks, the differential cues that determine which particular test object will be selected, must be supplied by the subject either through some sustained activity during the period of delay or by recall from past experience, i.e., observation of the experimenter concealing food at the beginning of the trial.*" (Page 12 in Jacobsen (1936))

In this experiment, Jacobsen (1936) obtained the following results. (1) Unilateral lesion of the prefrontal cortex had no impairment of the delayed-response task as indicated by either the length of the delay or the accuracy of performance. However, bilateral lesions resulted in total and permanent loss of delayed-response ability. This occurred when both hemispheres of the prefrontal cortex were removed either simultaneously or serially. (2) Visual discrimination and simple problem box tests were not impaired by either unilateral or bilateral lesions of the prefrontal cortex. (3) Unilateral and bilateral lesions of the parietal lobe and the temporal lobe caused no impairment on the delayed-response task in either the length of the delay or the accuracy of performance. In the delayed-response task, the subjects need to acquire a necessary information during the cue period, keep holding that information during the delay period, and recall and utilize it at the time of response in order to obtain the reward. Therefore, correct performance of the delayed-response task requires short-term mnemonic process to hold the necessary information during the delay period and recall that information at the time of response. However, visual discrimination and problem box test do not need the short-term mnemonic process that performance of the delayed-response task requires, but require other types of mnemonic processes like long-term memory. Therefore, Jacobsen considered the cause of delayed-response deficit observed in prefrontal lesioned animals as a loss of immediate memory. Jacobsen (1936) stated his conclusions in his paper as follows:

> "*Operated animals are unable to remember after a lapse of a few seconds under which of two cups food is concealed. It looks like "out of sight, out of mind." This deterioration is not generalized. The loss is specific rather than global. Monkeys' behavior is dominated by present external stimuli and the stable habit systems derived from earlier training. The mnemonic influence of immediately past experience is lacking. The animals without the prefrontal cortex learns and retains sensory-motor habits and visual discrimination. But, they are unable to remember for even a few seconds under which of two cups a piece of food had been concealed. The problem boxes and visual discrimination tests favor establishment of stable sensory-motor habits by repeated trials. However, the delayed response test makes quite different demands and is suggestive of the kind of events which in man we characterize as immediate memory.*" (Page 51–53 in Jacobsen (1936), modified and summarized by SF)

Jacobsen and his colleagues provided further support for the thought that the impairment of delayed-response performances was caused by the removal of the prefrontal cortex and not by the removal of other cortical areas. The removal of the temporal lobe or the motor and premotor areas exhibited no evident impairment in delayed-response performances (Fig. 2.1c). Jacobsen and Elder (1936) examined the effects of unilateral and bilateral removal of the temporal lobe on delayed-response performances using one male baboon (*Papio papio*) and found that neither unilateral nor subsequent bilateral removal of the temporal lobe impaired the capacity for the delayed-response. Jacobsen and Haslerud (1936) examined the effects of simultaneous bilateral removal of the motor and premotor areas on delayed-response performances. The animal did not maintain any fixed orientation during the delay period. However, it moved freely in the cage during delayed-response performances. A comparison of the animal's performances between pre- and post-operative periods revealed no impairment on delayed-response performances and exhibited even some improvement in delayed-response performances in the postoperative period.

A series of studies by Jacobsen and his colleagues clearly indicated that the prefrontal cortex was the responsible cortical area for performing the delayed-response task correctly. However, it was not clear whether the delayed-response deficit was caused by the loss of immediate memory as Jacobsen suggested or by other behavioral or instrumental factors. Jacobsen and Nissen (1937) examined performances of both the delayed-response task and the delayed alternation task using two rhesus monkeys in the same behavioral apparatus. The method for acquiring the information needed to maintain during the delay period is different between the two tasks. In the delayed-response task, necessary information (a baited position) is presented visually during the cue period, and after the end of the delay period, the animal is allowed to make a choice based on the visual information presented during the cue period. Therefore, visual information plays an important role to make a correct choice at the time of response. However, in the delayed alternation task, the animal is required to make a behavioral response to the position opposite to where it had made at the preceding trial. Therefore, kinesthetic information plays an important role in making a correct response. After bilateral ablations of the prefrontal cortex, permanent loss of the ability for both delayed-response and delayed alternation performances was observed. However, unilateral ablations caused little disturbance and no permanent deficit in both task performances. Although maintaining the information during the delay interval is essential for performing both tasks, the information needed to maintain during the delay interval is different between two tasks: visual information in the delayed-response task and kinesthetic information in the delayed-alternation task. Therefore, Jacobsen and Nissen (1937) concluded that the loss of the capacity of delayed-response performances after prefrontal lesions is caused by a loss of mnemonic capacity, not by a specific behavioral capacity needed only for delayed-response performances.

2.2.6 Delayed-Response Deficits Caused by Prefrontal Lesions

In primate behavioral studies, Wisconsin general test apparatus (WGTA) had been frequently used (Fig. 2.2). In WGTA, the subject is kept in a small cage. A small table is placed in front of the cage, where necessary stimuli and reward are presented. In the delayed-response task, an opaque screen located at the front of the cage is first raised after several seconds of inter-trial interval. While the subject watches on the table, the experimenter places a reward in one of two food wells on the table (cue period), and then covers both food wells with identical opaque plates, so that the subject cannot see where the reward is hidden. Then, the opaque screen in front of the cage is lowered and a delay period of several seconds is introduced. At the end of the delay period, the opaque screen is raised and the subject is allowed to select either food well to get the reward (response period). If the subject selects a correct food well where the reward is hidden, the reward is given. Since both food wells are covered with identical opaque plates, no clue suggesting where the reward is hidden is provided to the subject during the response period. Therefore, the subject needs to

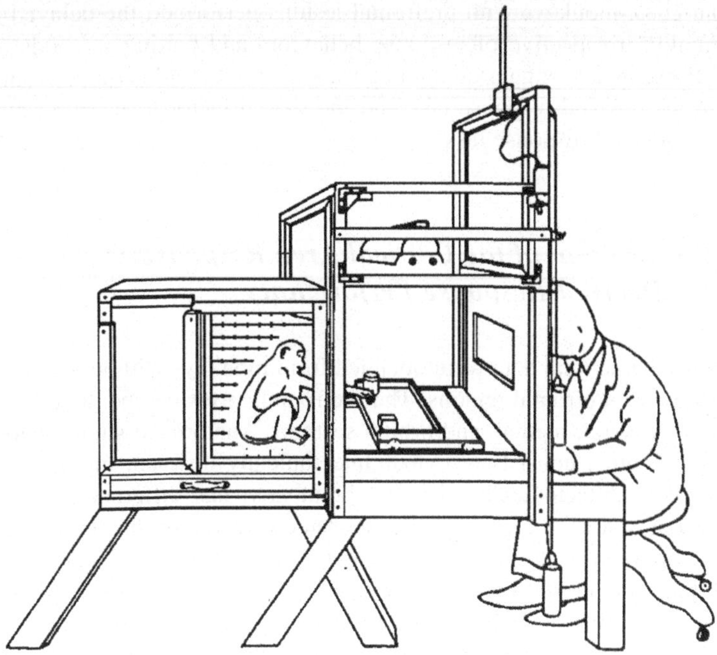

Fig. 2.2 Schematic drawing of Wisconsin general test apparatus

retain the baited position internally during the delay period and utilizes it to make a choice.

Impairment of delayed-response performances had been repeatedly observed in monkeys having bilateral prefrontal lesions after the first publication by Jacobsen (1936). Finan (1939) used mangabey monkeys (*Cercocebus torquatus atys*) and examined the effects of unilateral and bilateral prefrontal lesions on three behavioral tests which included a temporal discrimination maze test, a shuttle-box test, and the delayed-response test. Unilateral lesions caused no impairment in any of these tests. However, bilateral lesions caused total and permanent loss of an ability to perform the delayed-response task but caused no impairment in the maze test and the shuttle-box test. Finan (1942) also found that mangabey monkeys having bilateral ablation of the frontal association area (areas 8, 9, 10, 11, and 12) exhibited a complete loss of the delayed-response over an extended period of testing. Although the importance of the prefrontal cortex in delayed-response performances had been shown, the importance of spatial information processing in delayed-response performances had also been shown. Nissen et al. (1938) examined delayed-response performances with and without spatial cues in chimpanzees and showed that delayed-response performances required a highly developed symbolic mechanism for spatial cues. Mishkin and Pribram (1956) compared the effects of delayed-response performances between the traditional delayed-response task and its variations (e.g., using non-spatial cues, go/no-go responses) and found that, whenever spatial cues were replaced by non-spatial cues, monkeys with prefrontal lesions performed the delayed-response tasks normally, irrespective of response behaviors and baiting methods (direct or indirect). These results suggested the importance of intact bilateral prefrontal cortex to perform the delayed-response task and the spatial factor to interpret the impairment of the delayed-response task.

2.2.7 Localization of the Cortical Area Responsible for Delayed-Response Performance

Most of these behavioral data were obtained from monkeys with lesions of bilateral and almost entire prefrontal regions. The prefrontal cortex can be divided into three major sectors (lateral, medial, and orbital sectors). Anatomical studies showed that the prefrontal cortex consists of cytoarchitectonically different multiple areas and that each cytoarchitectonically different brain area has different functions from others. Lesion studies suggested that the lateral sector of the prefrontal cortex plays an important role in delayed-response performances. Therefore, it was expected that the brain region responsible for delayed-response performances could be localized in cytoarchitectonically defined area in the prefrontal cortex.

Blum (1952) examined the effects of bilateral ablations of the dorsal, mid-dorsal, or ventral region of the lateral prefrontal cortex on delayed-response performances. Six rhesus monkeys were used as subjects: one with complete bilateral lesions of the lateral prefrontal cortex, one with bilateral lesions of the dorsolateral region, three with bilateral lesions of the mid-lateral region, and one with bilateral lesions of the ventrolateral region. The visual delayed-response test with delays of 5–30 s and the auditory delayed-response test (buzzer and doorbell) with delays of 2.5–12.5 s were introduced in a WGTA. The significant impairment of the visual delayed-response test was observed in all animals except one with ventrolateral lesion. Similarly, all subjects, except one with ventrolateral lesion, exhibited impairment in the auditory delayed-response test. However, no subjects exhibited impairment in the auditory discrimination test (the buzzer associated with the right box selection, while the doorbell associated with the left box selection). Thus, the results indicated (1) the impairment of the delayed-response performance was not caused by sensory deficits, (2) animals with bilateral mid-lateral lesions exhibited the most severe impairments in both visual and auditory delayed-response tests, while animals with ventrolateral or dorsolateral lesions exhibited comparatively slight deficits on both tests. These results suggested that the lateral prefrontal cortex is not a unified structure from functional point of views and that the mid-lateral region of the prefrontal cortex is a cortical area responsible for delayed-response performance.

Similarly, Pribram and his colleagues demonstrated that the cortical area focally concerned with delayed-response types of functions was limited at least to the lateral surface of the frontal lobe anterior to the arcuate sulcus. Pribram (1950) showed that baboons with bilateral ablations of the lateral prefrontal cortex exhibited long-lasting impairment of delayed-response performances, although some pharmacological and physiological conditions improved their delayed-response performances. Pribram et al. (1952) compared the effects of delayed-response performances between lesions of the dorsolateral prefrontal cortex and those of the ventrolateral prefrontal cortex of baboons and showed that the dorsolateral prefrontal cortex is especially important for performing the delayed-response task.

Mishkin (1957) examined monkeys' performances of the delayed alternation task in a WGTA and showed the importance of the mid-lateral prefrontal cortex for delayed alternation performances. In the delayed alternation task, monkeys needed to select either right or left box alternately after several seconds of the delay. Among 10 rhesus monkeys used as subjects, two received lesions extending from the ventromedial to the ventrolateral edge of the frontal lobe (control lesion), two received lesions extending from the ventrolateral edge of the frontal lobe to the inferior lip of the principal sulcus (inferior frontal lesion), four received lesions including the lips and banks of the principal sulcus (mid-lateral frontal lesion), and two received lesions extending from the superior lip of the principal sulcus to the longitudinal fissure (superior frontal lesion). As a result, the four monkeys with mid-lateral frontal lesions exhibited the poorest post-operative performance of the delayed alternation. Monkeys with mid-lateral frontal lesions did not show any consistent relationship between the locus or the extent of the cortical damage and the magnitude of behavioral deficit. Mid-lateral frontal lesions produced marked and sustained impairment of delayed alternation performances, while lesions limited to either the superior lateral surface or the ventral surface produced no impairment, suggesting the importance of the cortical area surrounding the principal sulcus for delayed alternation performances.

Gross and Weiskrantz (1962) also showed an importance of the cortical area surrounding the principal sulcus for delayed-response performance. In their study, one group of monkeys received ablations of the principal sulcus, and another group of monkeys received ablations of the lateral frontal cortex excluding the principal sulcus. Acquisition and retention of auditory discrimination, delayed-response, and visual discrimination tasks were compared between the two groups. Monkeys with principal sulcus lesions exhibited greater impairment on the delayed-response task than the monkeys with lesions sparing the principal sulcus. Monkeys with lesions sparing the principal sulcus exhibited greater impairment on the auditory discrimination task. However, both groups of monkeys exhibited no impairment of the visual discrimination task. These results confirmed that the cortical locus crucial for delayed-response performances is the cortical region surrounding the principal sulcus. These results also indicated that the focal regions crucial for delayed-response performances and for auditory discrimination performances can be dissociable by making smaller lesions restricted within the principal sulcal area or not.

The cortical area surrounding the principal sulcus had been shown to be a critical area for both delayed-response and delayed alternation performances. However, this evidence was based on the comparison of the behavioral studies between monkeys with lesions in the dorsal, the mid-dorsal, and the ventral prefrontal cortex (behavioral analysis based on the dorsal-ventral axis). Butters and Pandya (1969) examined the effects of delayed alternation performances by lesions on the anterior-posterior axis of the principal sulcus. Twelve rhesus monkeys learned the delayed alternation task before having lesions in the principal sulcus. Retention of delayed alternation performances was examined in monkeys with bilateral lesions of either anterior, middle, or posterior sectors of the principal sulcus, monkeys with bilateral lesions of the peri-arcuate cortex, and monkeys with the posterior parietal cortex (Fig. 2.3a). As a result, only monkeys with bilateral lesions of the middle one-third of the principal sulcus area exhibited severe impairment of delayed alternation performances and could not relearn this task after the lesion. Lesions of 9 mm in the extent and limited to the middle one-third of the principal sulcus resulted in delayed alternation deficits at least as large as those reported after total ablation of the sulcus. On the other hand, lesions in other sectors of the principal sulcus or in the peri-arcuate cortex exhibited only mild-to-moderate deficits on delayed alternation performances.

Butters et al. (1971) further examined whether the dorsal bank and the ventral bank of the principal sulcus differently contributed to delayed-alternation performances or not. Among 14 rhesus monkeys used as subjects, 7 received lesions in the dorsal bank of the middle principal sulcus (group DB), 4 received lesions in the ventral bank of the middle principal sulcus (group VB), 2 received lesions in both banks of the middle principal sulcus (group CB), and the remaining one received the first lesion in the ventral bank and the second lesion in the dorsal bank of the middle principal sulcus separately (Fig. 2.3b). All monkeys were trained on the delayed alternation task in the WGTA before lesion. As a result, no significant difference was observed in error rates between DB and VB, while a significant difference was observed between DB and CB and between VB and CB, indicating that both banks of the middle principal sulcus are crucial for delayed alternation performances. The lesion was localized in the middle one-third of the principal sulcus and bilateral lesions caused damages to cortical tissue in less than 30% of the total volume of the principal sulcus. This result indicates that not the lesion size but the location of the lesion in the principal sulcus is the critical factor to produce delayed alternation deficits.

Since Jacobsen first reported an impairment of delayed-response performance in monkeys with bilateral prefrontal lesions, the prefrontal cortex had been considered to be a critical brain area to perform the delayed-response task and the delayed alternation task. Initially, large bilateral lesions were made in monkey's prefrontal cortex to examine prefrontal functions. However, in subsequent studies, smaller and localized lesions were made either unilaterally or bilaterally to examine the effects on delayed-response performances. As a result, the middle one-third of the principal sulcus region was found to be the critical area for both delayed-response and delayed alternation performances.

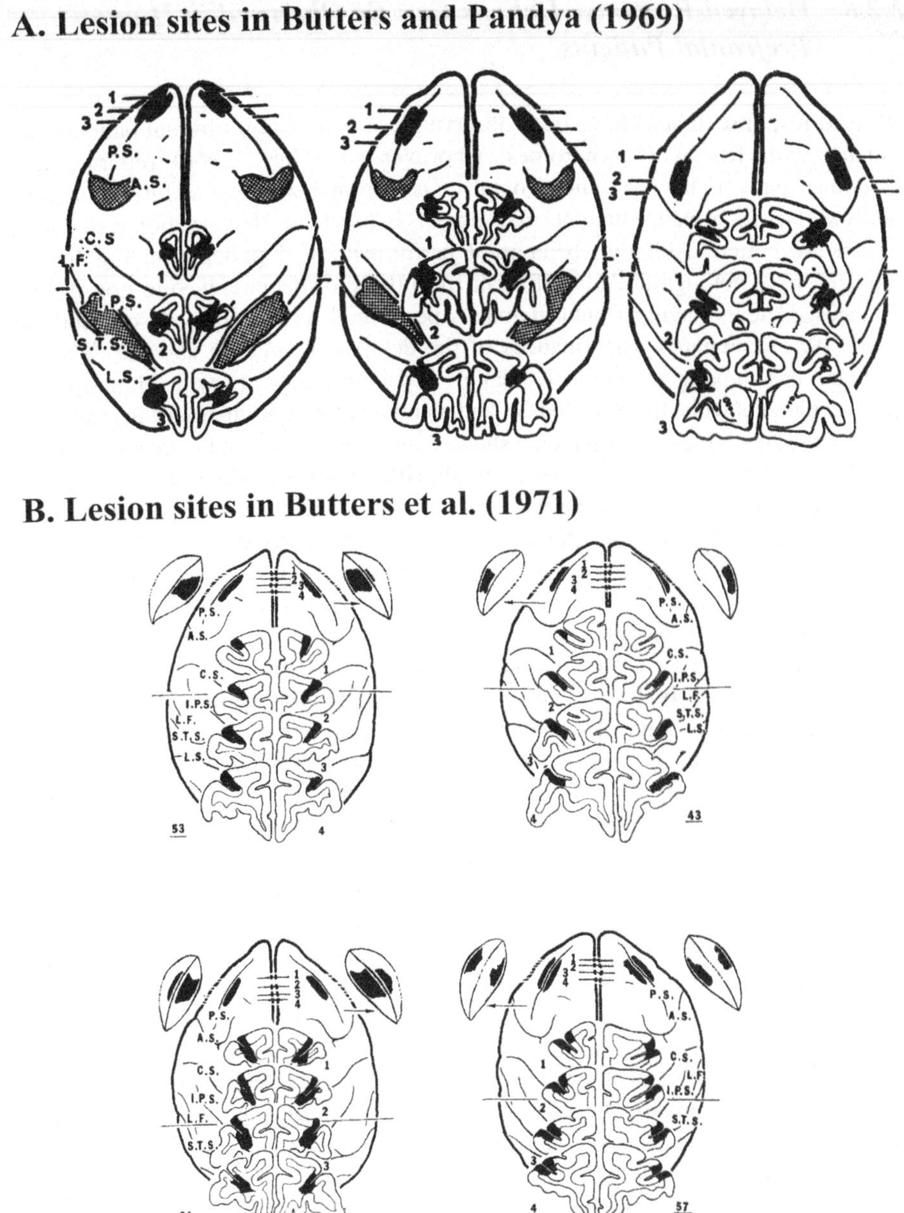

Fig. 2.3 (**a**) Reconstructions of lesion sites of three monkeys' brains. Lesion site of each monkey was the anterior, the mid-, and the posterior portion of the principal sulcus, respectively. Black areas represent the extent of damaged areas of the principal sulcus. From Butters and Pandya (1969) with permission. (**b**) Reconstructions of lesion sites in the principal sulcus. All lesions were localized in the middle portion of the principal sulcus. From Butters et al. (1971) with permission

2.2.8 Delayed-Response Deficits Can Be Observed in Human Prefrontal Patients

Delayed-response deficits have been observed also in human prefrontal patients and patients with Korsakoff syndromes. Chorover and Cole (1966) reported that 33 human patients having either frontal or non-frontal damages exhibited significantly worse performance than normal subjects in the delayed alternation task with a 10-s delay period. Although behavioral performances of patients having frontal damages were not significantly different from those of patients having non-frontal damages, patients having frontal damages performed more poorly. Oscar-Berman et al. (1982) compared performances of the delayed alternation and the delayed-response task among alcoholic Korsakoff's patients, patients with Huntington's disease, patients with Broca's aphasia, and alcoholism without memory impairment. They found that Korsakoff's patients showed impairments in both delayed-response and delayed alternation performances, while Huntington's patients showed impairment only in delayed alternation performances. Freedman and Oscar-Berman (1986a) examined delayed-response deficits in amnesic patients with Parkinson's disease and Alzheimer's disease and found that patients with Parkinson's disease showed impairment only in delayed-response performance, while patients with Alzheimer's disease showed impairment in both delayed-response and delayed alternation performances. Although dysfunction of the frontal lobe was implicated for impairment of both task performances, behavioral effects observed in Alzheimer's patients suggested that impairment observed in Alzheimer's patients was caused by dysfunction of the dorsolateral prefrontal area combined with other regions. Freedman and Oscar-Berman (1986b) further examined human patients' performances of the delayed-response task (0-, 10-, 30-, and 60-s delay) and delayed alternation task (5-s delay) using a modified version of WGTA. Among 35 subjects examined, 6 had bilateral frontal lobe lesions, 5 had amnesia, 12 had alcoholic Korsakoff's syndrome, and 12 were alcoholic controls. The frontal patients were significantly impaired on both delayed-response and delayed alternation performances. Korsakoff's patients were significantly impaired only on delayed alternation performances. Since amnesic patients did not exhibit any impairment on performances of both tasks, they concluded that deficits observed on both delayed-response and delayed alternation performances were not related to memory loss but were caused by impairment of frontal lobe functions. The observation that Korsakoff's patients exhibited delayed-response deficits was also reported by Oscar-Berman et al. (1992). Since alcoholic Korsakoff's syndrome accompanies prefrontal damages, the observations that Korsakoff's patients exhibit delayed-response deficits support the importance of the prefrontal cortex for delayed-response performances. Pascual-Leone and Hallett (1994) examined the contribution of the dorsolateral prefrontal cortex on delayed-response performances using repetitive transcranial magnetic stimulation and showed that unilateral prefrontal stimulation during the delay period caused a significantly greater number of response errors compared with the stimulation at the motor cortex.

D'Esposito and Postle (1999) tried to address whether mnemonic components of memory tasks depend purely on functions of the prefrontal cortex by meta-analysis of human experiments published between 1960 and 1999. Selected reports examining whether the performance of spatial delayed-response tasks was distracted or not by patients with lesions of the dorsolateral prefrontal cortex showed that lesions of the prefrontal cortex disrupt delayed-response performances. In addition, the left ventrolateral prefrontal lesions (Brodmann's areas 44 and 45) disrupted verbal, but not spatial, delayed-response performances, while the dorsolateral prefrontal lesions (areas 9 and 46 and possibly area 8), especially of the right hemisphere, disrupted spatial-delayed response performances. D'Esposito and Postle (1999) also found reports showing that delayed-response performances were impaired more when the delay period was filled with distractors than when it was unfilled, suggesting that prefrontal functions include not only spatial mnemonic function but also inhibition of prepotent responses, gating behaviorally irrelevant stimuli, and maintaining information in a noisy environment. Anyway, the result that dorsolateral prefrontal lesions cause an impairment on spatial delayed-response performances in human subjects is consistent with the results obtained from prefrontal lesion studies using monkeys.

2.2.9 *Delayed-Response Deficits Can Also Be Observed in Rats and Dogs*

Delayed-response deficits had been observed not only in monkeys with prefrontal lesions but also in other animals (e.g., rats) with lesions of the frontal area. Miles and Blomquist (1960) showed that squirrel monkeys having bilateral ablations of the lateral prefrontal cortex exhibited long-lasting impairment in delayed-response performances, but did not exhibit impairment in object discrimination performances. The area of the prefrontal cortex in rats and cats had been defined as the cortical area that receives anatomical projections from the mediodorsal nucleus of the thalamus. Wikmark et al. (1973) reported that rats with lesions in the anteromedial cortex, which receives projections from the mediodorsal nucleus of the thalamus, exhibited an impairment in retention of delayed-alternation performances in a T-maze. Divac (1972) showed an impairment of delayed-alternation performances in cats with lesions of the prefrontal cortex. Markowitsch et al. (1980) also showed that cats having lesions in the presylvian and the proreal sectors of the frontal lobe exhibited the most severe impairment in delayed-alternation performances (10-s delay) in a maze situation. Lawicka et al. (1966) showed that dogs with proreal lesions exhibited an impairment on a three-choice delayed-response task and that the impairment became particularly severe when distraction was inserted during the delay period. Thus, delayed-response deficits have been observed not only in monkeys with prefrontal lesions but also in rats, cats, and dogs with lesions in the cortex corresponding to the primate prefrontal cortex.

2.3 Behavioral Deficits in Other Delay Tasks Caused by Prefrontal Lesions

2.3.1 *Delayed Alternation Task and Importance of Spatial Factor*

In addition to the delayed-response deficit, monkeys with prefrontal lesions exhibit impairment in spatial delayed-alternation performances (Mishkin 1957; Butters and Pandya 1969; Butters et al. 1971). In the spatial delayed-alternation task, monkeys need to select two separate baited positions alternately (e.g., right – left – right – left - . . .) after the delay. Differing from the delayed-response task, the action that the subject performed in the preceding trial (e.g., selection of the right position) becomes the cue to perform the action in the coming trial (e.g., selection of the left position) after the delay period. Pribram (1961) used an object alternation task, in which two different objects were placed at two places randomly selected from six places. After bilateral prefrontal lesions, he examined monkeys' performance in the condition that the positions of the two objects that the subject needed to select alternately were randomized. Monkeys with bilateral prefrontal lesions exhibited a 65% level of performances, while control monkeys exhibited 80% correct performances. Based on the result, Pribram (1961) concluded that the spatial factor did not account for the deficit of alternation performances in prefrontal lesioned monkeys. However, this result only indicated that absolute spatial position was not an important factor to account for the deficit in the object-alternation task and still indicated the importance of the relativity of two spatial positions (e.g., one object is located on the right of the other object) to interprete the deficit of alternation performances.

Pribram and Tubb (1967) considered that the prefrontal cortex is related to an organism's ability to make an effective division in the stream of stimuli and therefore that an impairment of spatial delayed-alternation performances observed in prefrontal lesioned monkeys might be caused by an interference effect between successive inputs due to insufficient temporal resolution. They inserted a 15-s interval between each R-L couplet (R-L. .. R-L. .. R-L.. .) by modifying the classical 5-s delayed-alternation task (R-L-R-L). This modification caused very few errors on the task performed by prefrontal lesions monkeys. They concluded that proper parsing of the stimulus stream is more important variable for correct performance in the delayed-alternation task. Pribram and his colleagues (Pribram and Tubbs 1967; Pribram et al. 1977) suggested that a critical factor in spatial delayed alternation performance was the proper parsing of the stimulus stream exposed to the organism. However, their procedure might alter the task from spatial alternation to time discrimination (e.g., a long delay cueing a leftward response and a short delay cueing a rightward response) and the improvement of animals' task performance might be caused by an intact ability of the time discrimination. In time discrimination, the choice of the correct position depends on the perception of the temporal cue, not on the memory of the spatial cue. Therefore, if the impairment in spatial delayed alternation performances is related to the deficit of spatial memory caused by principal sulcus lesions, monkeys with these lesions should be unimpaired on the time discrimination task. Rosenkilde et al. (1981) examined this issue using 12 rhesus monkeys (6 received lesions of the cortex in the principal sulcus, 4 received lesions of the inferior prefrontal convexity, and 2 received lesions of the cortex in the arcuate sulcus). In the time discrimination task, monkeys were required to select the left plate following the 10-s interval or the right plate following the 30-s interval. Monkeys' performances were also examined by the spatial delayed alternation and the object discrimination reversal task. As a result, the monkeys with the removal of the cortex surrounding the principal sulcus were unaffected on time discrimination performances, even though they failed to learn the spatial delayed alternation. On the other hand, monkeys with inferior prefrontal ablations exhibited marked impairment on all these tasks, while monkeys with the removal of the cortex surrounding the arcuate sulcus were not impaired on any of these tasks. These results showed that the cortex surrounding the principal sulcus participates neither in the temporal discrimination nor in the temporal structuring (parsing). These results also showed that the function of the principal sulcus region is closely related to the spatial factors used in the tasks. These data supported the view that the deficit after dorsolateral prefrontal lesions is caused by a specific disorder for remembering the spatial cue.

An importance of spatial factors had been repeatedly suggested in the interpretation of the deficits observed in monkeys after dorsolateral prefrontal lesions. Goldman and Rosvold (1970) further examined whether the critical factor caused by these impairments was the spatial factor, the delay, or the combination of both. They trained monkeys to perform the spatial delayed alternation task (spatial task with delay) and a conditional position response task (spatial task without delay). They found that the dorsolateral prefrontal lesions caused impairment on the delayed spatial alternation task but not on the conditional position response task. However, removal of the arcuate sulcus impaired only on the conditional position response task. Thus, the results indicated that functions of the dorsolateral prefrontal cortex include both spatial and mnemonic factors.

To examine whether non-spatial factors also play a role in the functions of the dorsolateral prefrontal cortex, performance of the delayed spatial alternation was compared with the performance of the delayed object alternation in prefrontal lesioned monkeys. In the delayed object-alternation, the subject is required to select one of two different objects alternately after the delay, regardless of the positions of these objects. Although the subject needs to choose one of two options alternately after the delay in both spatial-alternation and object-alternation, in the object-alternation task, spatial information of each object at a given trial is not necessary for correct choice, but information regarding which object has been chosen at the preceding trial is essential. Pribram and Mishkin (1956) compared an effect of bilateral large prefrontal cortical lesions between delayed spatial-alternation and delayed object-alternation performances in monkeys and found that these lesions impaired both alternation performances to similar degree. Similar results had been observed by Mishkin and Pribram (1955). However, by making more localized lesions within the prefrontal cortex, Mishkin et al. (1969) showed that rhesus monkeys having bilateral lesions in the dorsolateral prefrontal cortex including the principal sulcus were impaired on spatial-alternation performances but not on object-alternation performances, while monkeys having lesions in the ventrolateral prefrontal cortex were impaired on both spatial- and object-alternation performances. Further, Mishkin and Manning (1978) performed an experiment using delayed object-alternation, delayed object-matching, and delayed color-matching tasks. Half the monkeys received lesions only in the principal sulcus and the other half received lesions in the inferior convexity. Monkeys with principal sulcus lesions exhibited only small and transient impairment on these non-spatial tasks, while monkeys with inferior convexity lesions exhibited severe and long-lasting impairment on all three tasks. Thus, the principal sulcus appears to participate in spatial mnemonic functions, while the inferior convexity of the prefrontal cortex appears to participate in non-spatial memory functions.

Importance of spatial modality in the interpretation of delayed-alternation deficits observed in prefrontal lesioned monkeys is not related to what kind of response behavior is used. Since the delayed-alternation task had been tested in the WGTA, the subject's hand and arm movements were response behaviors. However, Stamm and Weber-Levine (1971) and Stamm et al. (1971) used the delayed alternation task in a maze condition, in which the monkeys come and go between the start and the choice compartments by locomotion. Monkeys received bilateral ablations of either the whole lateral prefrontal cortex including the principal sulcus, only the cortex in the banks and fundus of the principal sulcus, or the lateral prefrontal cortex excluding the principal sulcus. Again, they observed severe impairment in delayed alternation performances on monkeys with bilateral prefrontal lesions including the principal sulcus. Similarly, Pribram (1961) examined the effects of prefrontal lesions on delayed alternation performances, in which monkeys were required to make lever press responses between two reinforcement schedules alternately. In the fixed ratio schedule, monkeys needed to complete 40 consecutive lever presses to obtain a reward, while in the fixed interval schedule, monkeys could obtain a reward at the first lever press only after a 4-min interval. Initially, monkeys were instructed which schedule was in operation by different color cues, but subsequently, no external cue was shown to monkeys. Therefore, monkeys needed to decide which of two behavioral patterns was appropriate at any given time. As a result, unoperated normal monkeys took about 4 weeks to reach 85% correct performance, while bilateral prefrontal lesioned monkeys could not reach the criterion. These results indicated that the deficits of alternation performance caused by prefrontal ablations were not specific to sensory or motor modality.

Thus, delayed-alternation deficits observed in prefrontal lesioned monkeys are independent of response behaviors used. Spatial modality is an important factor to interpret delayed alternation deficits observed in prefrontal lesioned monkeys.

2.3.2 Delayed Matching-to-Sample Task and Other Delay Tasks

Anderson et al. (1976) showed that monkeys with lesions in the frontal lobe significantly impaired in both the delayed-response task and the object permanency test, which was similar to the task used by Piaget to test the development of object permanency in human infants. Both impairments were considered to be attributable to an inability of spatial mnemonic processes. Spatial modality-specific impairment was also observed in the monkeys with lesions of the cortex surrounding the principal sulcus when recognition memory tasks were used. Passingham (1975) examined performances of a delayed matching-to-color task, a visual discrimination task, and a delayed spatial-alternation task using two groups of rhesus monkeys; one group had bilateral lesions in the dorsolateral prefrontal cortex including the principal sulcus and the other group had bilateral lesions in the ventrolateral prefrontal cortex excluding the principal sulcus. Monkeys having bilateral lesions in the dorsolateral prefrontal cortex exhibited impairment on spatial delayed-alternation performances but not on delayed matching-to-color and visual discrimination performances, while monkeys having lesions in the ventrolateral prefrontal cortex exhibited severe impairments on delayed matching-to-color performances even at the shortest delay.

Bachevalier and Mishkin (1986) trained 9 cynomolgus monkeys (*Macaca fascicularis*) on a delayed non-matching-to-sample task, a concurrent object discrimination task with 24-h inter-trial interval, and a spatial delayed-response task in the WGTA. The dorsolateral prefrontal lesion extended from the frontal pole to the fundus of the arcuate sulcus and from the ventral lip of the principal sulcus to the midline, while the ventromedial prefrontal lesion included the mid-orbital and medial orbital cortex and the anterior half of the cingulate cortex. As a result, the monkeys with the dorsolateral prefrontal lesion had no impairment on the delayed non-matching-to-sample task, whereas the monkeys with the ventromedial prefrontal lesions were severely impaired on this task. On the other hand, the monkeys with the dorsolateral prefrontal lesion exhibited severe impairment on the spatial delayed-response task (up to 5 s of the delay), whereas the monkeys with the ventromedial prefrontal lesions exhibited no impairment. Neither group of monkeys was impaired on the object discrimination task. Thus, these results support the view that the delayed-response deficit observed after dorsolateral prefrontal lesions is caused by a mnemonic impairment of spatial modality selectively rather than a memory loss of a more general nature.

Bachevalier and Mishkin (1986) indicated that the dorsolateral prefrontal cortex has little or no contribution to non-spatial memory tasks, whereas the ventromedial prefrontal cortex is critical for visual recognition memory. Kowalska et al. (1991) showed that large bilateral lesions in the ventrolateral prefrontal cortex impaired initial relearning of the delayed non-matching-to-sample task but not impaired on the subsequent learning of this task, and that combined bilateral lesions of the dorsolateral and the ventrolateral prefrontal cortex impaired both initial and subsequent learning of this task. However, these lesions did not impair visual discrimination performances. These results indicate that the lateral prefrontal cortex does not participate in long-term memory functions but participates in short-term memory functions.

Passingham (1972) examined the effects of bilateral lesions of the lateral prefrontal cortex on a visual object discrimination task (triangle vs. square) and a visual simultaneous discrimination task (light vs. no light) and compared these effects with those of bilateral lesions of the orbitofrontal cortex. Monkeys with bilateral orbitofrontal lesions showed impairment on both learning and retention of both discrimination tasks. However, monkeys with bilateral lesions of the lateral prefrontal cortex exhibited no impairment on both discrimination tasks. Similarly, monkeys with bilateral lesions of the lateral prefrontal cortex exhibited no impairment on tactile discrimination tasks, but monkeys with bilateral orbitofrontal lesions exhibited impairments (Passingham and Ettlinger 1972).

Thus, the bilateral lesions of the dorsolateral prefrontal cortex including the principal sulcus impair the performance of the delayed-response task and the delayed spatial-alternation task, but do not impair performances of the delayed object-alternation task and the delayed matching-to-sample tasks, indicating the importance of both delay and spatial factors. On the other hand, the lesions of the ventral prefrontal cortex excluding the principal sulcus impair performances of the delayed object-alternation and the delayed matching-to-sample tasks, but do not impair performances of the delayed-response task and the delayed spatial-alternation task, indicating the importance of both delay and non-spatial factors. Further, the lesions of the lateral prefrontal cortex do not impair performances of the discrimination tasks, indicating that the lateral prefrontal cortex does not participate in long-term memory but could participate in short-term memory.

2.4 Cryogenic Depression of Prefrontal Functions

Cooling at a particular cortical area produces temporary depression of physiological functions of that area without any obvious physical damage to the cortical tissue (Brooks 1983). Cooling of cortical tissue suppresses the generation of action potentials in cortical neurons. Since the generation of action potentials requires energy consumption, this neural process is closely associated with intracellular metabolism. Intracellular metabolism is achieved by enzyme activity. The strength of enzyme activity depends on surrounding brain temperature and becomes maximum at its optimum temperature (e.g., 36 °C). Therefore, if the brain temperature becomes lower than the optimum value, the strength of enzyme activity decreases, and therefore, the generation of action potentials would be suppressed and firing frequency would be reduced. If the brain temperature drops to 10 °C, neurons cannot produce action potential anymore (Brooks 1983). However, if the brain temperature returns to the optimum value, all metabolic functions would be reinstalled and action potentials are generated. Cooling procedure causes temporary suppression of physiological activities of cortical neurons including generation of action potentials with minimum physical damage to the cortical tissue and re-warming reinstalls these physiological functions. Therefore, the reversible cooling method has an advantage over the lesion method which produces permanent unrecovered damages in the cortex.

Fuster and Alexander (1970) chronically implanted cryogenic probes over the surface of the dorsolateral prefrontal cortex (the major part of Brodmann's areas 9 and 10) and examined the behavioral effects of prefrontal cooling on delayed-response performances using monkeys. Bilateral cooling of the dorsolateral prefrontal cortex at the subdural temperature of 25–30 °C impaired delayed-response performances reversibly, while unilateral cooling of the same cortical area impaired delayed-response performances to a lesser extent. Bauer and Fuster (1976) also examined the effects of prefrontal cooling on delayed matching-to-sample and delayed-response performances using monkeys. Monkeys' behavioral performances (errors, reaction times, and motor activities) were examined under three conditions: prefrontal cooling (FC), parietal cooling (PC), and non-cooling (NC) conditions. The proportion of correct responses decreased in the NC condition as a function of the delay length in both tasks. Significant decrease in correct responses was observed in the FC condition in both tasks, but not observed in the PC condition. Although reaction times and motor activities increased as a function of the delay length, these increases were enhanced in the FC condition in both tasks. These results indicate that functional depression of bilateral dorsolateral prefrontal cortices produces severer impairment in behavioral tasks with delay compared with unilateral depression. These results also indicate that delayed-response deficits are specific to the dysfunction of the dorsolateral prefrontal cortex, which is consistent with the results obtained by ablation studies in the prefrontal cortex.

Cooling of the dorsolateral prefrontal cortex produced a significant decrement in the proportion of correct responses in both tasks but produced small changes in reaction times and motor activities. Therefore, Bauer and Fuster (1976) concluded that prefrontal cooling causes impairment of short-term memory of both spatial and non-spatial information. This short-term memory hypothesis was partly supported by the results of other prefrontal cooling studies by Fuster and Bauer (1974). Cooling of the prefrontal cortex caused an impairment on the performance of a delayed matching-to-sample task. The high incidence of errors was directly proportional to the delay length. Importantly, auditory and visual distractors presented during the delay period elicited more errors during the cooling of the prefrontal cortex than in uncooling normal conditions. In this task, short-term memory process needs to keep relevant information during the delay period without disturbance by distractors. Therefore, they suggested that the role of the prefrontal cortex in performing the delayed matching-to-sample task is short-term memory.

Quintana and Fuster (1993) also examined the effects of reversible cooling of the dorsolateral prefrontal cortex and the posterior parietal cortex and showed that the prefrontal cortex participates in broader functions of visuomotor processing and integration of both spatial and non-spatial information, while the parietal cortex participates in spatial processing and memory. They used a delayed matching-to-color task and a conditional position discrimination task. In the delayed matching-to-color task, either red or green stimulus was presented during the cue period and the monkey was required to select the same color stimulus after a 5-s or 12-s delay period. In the delayed conditional position discrimination task, the monkey was required to select either the left or the right window after the delay when red (blue) or green (yellow) stimulus was presented at the center window during the cue period, respectively. Prefrontal cooling caused impairments in the performance of both tasks and the effect of impairment increased with the delay length. However, parietal cooling caused impairments in performance of the delayed conditional position discrimination task and the delay length had no effect on the performance of both tasks. Thus, these results also indicate strong relations between functions of the dorsolateral prefrontal cortex and spatial information processing and short-term memory.

2.5 What Causes Delayed-Response Deficits?

2.5.1 Factors Affecting Delayed-Response Performances

Jacobsen and his colleagues first reported that delayed-response performances were impaired by bilateral ablations of the prefrontal cortex, not by ablations of the parietal, temporal, or premotor cortices. Subsequently, the focal area responsible for delayed-response performance was found within the prefrontal cortex. The bilateral ablations of the middle one-third of the cortex surrounding the principal sulcus were enough to produce severe and permanent impairment of delayed-response performances in monkeys. So far, the general agreement is that the dorsolateral prefrontal cortex is the critical cortical area for delayed-response performances. However, there were some disagreements on the cause of delayed-response deficits and functions of the dorsolateral prefrontal cortex.

To perform the delayed-response task correctly, the subject is required to keep the spatial information (baited position) in mind during the delay period. Since the experimenter randomly selects the baited position, the subject is required to utilize memorized spatial information to select the correct baited position during the response period. Although monkeys with bilateral lesions of the prefrontal cortex exhibit significantly higher error rates in delayed-response performances, basic behavioral habit needed to perform the delayed-response task (e.g., watching the experimenter showing the baiting position and waiting the choice response until the end of the delay period) was maintained normally after the lesion. Therefore, Jacobsen (1936) concluded that the cause of the delayed-response deficits observed in prefrontal lesioned monkeys was "*a loss of immediate memory*." Further, by comparing prefrontal lesioned monkeys' behavior between the delayed-response task and the delayed alternation task, Jacobsen and Nissen (1937) concluded that the delayed-response impairment after prefrontal lesions was caused by a memory loss, not by specific behavior needed to perform the delayed-response task but not to perform the delayed alternation task.

However, subsequent studies brought out some doubt against Jacobsen's conclusion. Nissen and Elder (1935) used chimpanzees and examined the influence of incentive strength by changing the quantity of food in delayed-response performances. They found that the increase of incentive strength improved delayed-response performances. Finan (1942) trained the delayed-response in two monkeys (*Cercocebus torquatus atys*), which had bilateral ablations of the prefrontal cortex (Brodmann's areas 8, 9, 10, 11, and 12). These monkeys exhibited severe impairment of delayed-response performances for a long time. However, when the same reward was given at the correct position during the cue period as would be given during the response period (discrimination between rewarded position and unrewarded position was introduced during the cue period), correct behaviors were partially reinstalled. Therefore, Finan (1942) concluded that the loss of the immediate memory as an interpretation of delayed-response capacity was doubtful because of the demonstration that successful response became possible under the discriminative condition. Meyer and Harlow (1952) used 8 rhesus monkeys and examined the effects of learning, length of delay (5, 10, 15, and 20 s), amount of reward (1, 2, 3, and 4 pieces of food), kind of cue differentiation (different objects of food wells), and opaque screen interpolation on delayed-response performances. Marked facilitation of delayed-response performances was observed when non-spatial cues were presented. Interpolation of an opaque screen as a distractor during the delay period increased errors by 50%. Correct performance ratios varied as a function of the amount of reward. However, the influence of the delay length was moderate in the 5-s to 20-s range. Further, Gross (1963) examined the effect of food deprivation on delayed-response performances. Three monkeys that had mid-lateral frontal lesions and impaired on delayed-response performances were tested on the performance of the delayed-response task following 2, 26, and 50 hours of food deprivation. As a result, food deprivation improved monkey's delayed-response performances. Based on these results, Gross (1963) suggested that the food deprivation procedure enhanced the value of the cue stimulus at the time of baiting. However, he also indicated that the finding that prefrontal lesioned monkeys exhibited improvement in delayed-response performances in hungry conditions did not reveal anything about the cause of their deficits, because the hunger also improved delayed-response performances in unoperated monkeys to a similar extent as prefrontal lesioned monkeys.

Thus, not the delay length but the saliency of the cue, the amount of the reward, and the distraction during the delay period affect delayed-response performances. Although short-term memory of the baited position would be one important factor to understand delayed-response deficits in prefrontal lesioned monkeys, other factors such as the saliency of the cue, the amount and the type of the reward, and presence or absence of distractors would also be the causes of delayed-response deficits.

2.5.2 Increased Susceptibility, Distractibility, and Hyperactivity to Irrelevant Stimuli

Malmo (1942) examined the effect of distractors presented during the delay period on delayed-response performances in the WGTA and showed that the impairment of delayed-response performances in prefrontal lesioned animals was not caused by the loss of immediate memory but by increased susceptibility to interference. He considered that the delayed-response task consisted of a series of learning-retention processes and is a fragile habit. He thought that learning takes place at the time that food is shown in one cup and retention is tested at the end of the delay period. Therefore, the delayed-response task is basically one-trial learning. Since the delayed response is a fragile habit, uncontrolled extraneous visual stimuli may interfere with the retention of this habit.

He used light stimuli as the cue instead of food stimuli and examined delayed-response performances of prefrontal lesioned monkeys under two conditions. In *non-interpolate condition*, first, the cue light was turned on for 2 s either on the right or the left. Then, the animal was kept in the dark for the 10-s delay period. During the delay period, the animal was permitted to select the right or the left panel to obtain a reward. In *interpolate condition*, first, the cue light was turned on for 2 s either on the right or the left. Then, a 10-s delay period was introduced. During the 10-s delay period, the animal was first kept in the dark for 2.5 s, then the room light was turned on for the next 5 s, and the animal was kept in the dark again for 2.5 s. During the response period, the animal was permitted to select the right or the left panel. The room light was placed at the center of the ceiling in the WGTA and was used as a distractor presenting during the delay period. He found that monkeys having bilateral removal of the prefrontal cortex exhibited no impairment in delayed-response performances in the non-interpolate condition, but exhibited impairment in the interpolate condition. The result showed that the difference between normal and prefrontal lesioned monkeys with respect to delayed-response performances was not the difference between the presence or the absence of the capacity for delayed response (memory function), but was the difference in the degree of susceptibility to the interfering effects of extraneous stimuli occurring during the delay period. Therefore, he concluded that the impairment of delayed-response performances in prefrontal animal was not caused by the loss of memory (forgetting the information) but by increased susceptibility to interference.

Malmo (1942) showed that ambient illumination during the delay period acted as a distractor on the performance of the delay tasks in the WGTA. However, Bauer and Fuster (1978) reported no effect of ambient illumination on the performance of these delay tasks. They examined the performances of the delayed matching-to-sample and the delayed-response tasks under the normal lighting condition and the dark condition. The cooling of the lateral prefrontal cortex increased errors on trials with longer delay. They did not observe the effect of the difference of ambient illumination on the performance of these delay tasks during prefrontal cooling conditions. Discrepancy in the results between Malmo (1942) and Bauer and Fuster (1978) might be caused by the difference in the experimental situation. Malmo (1942) performed behavioral experiments using the WGTA, in which the subject can move and behave freely during the delay period. However, in the experiment by Bauer and Fuster (1978), the subject's body was set in the monkey chair and the subject's head was immobilized and directed toward the position where the stimuli were presented during the experiment. Freely movable situation in the WGTA during the delay period might allow stronger susceptibility to interference than the experimental situation of Bauer and Fuster (1978), in which the subject's head was immobilized and directed toward the stimulus position during the delay period.

On the other hand, it had been shown that prefrontal lesioned monkeys exhibit difficulty of discrimination between relevant and irrelevant information for behavioral responses. LeVere and Bartus (1971, 1972) examined the behavioral effect on the discrimination task in which both relevant and irrelevant discriminative stimuli were presented sequentially before the response. The normal monkey performed the task significantly better when the relevant stimulus occurred first and was followed by the irrelevant stimulus. However, when the irrelevant stimulus occurred first and was followed by the relevant stimulus, the normal monkey exhibited substantial difficulty in this task. Based on these results, Bartus and LeVere (1977) further examined the behavioral effects of distractors using rhesus monkeys having ablations of the dorsolateral prefrontal cortex (cortical area centralized on the principal sulcus and the immediate adjacent cortex) in the WGTA. Prior to the experiment, they confirmed that the monkeys having ablations of the prefrontal cortex exhibited an impairment on the 10-s delayed-response task. The monkeys performed a pattern discrimination task. Three different conditions were used: *R-I condition*, relevant stimulus pattern presented first, then irrelevant second; *I-R condition*, irrelevant first, relevant second; *R-R condition*, only relevant stimulus pattern was presented. As a result, the monkeys with lesions of the dorsolateral prefrontal cortex were severely impaired when irrelevant stimuli were presented after the relevant stimuli (R-I condition) just before the choice response. These data demonstrated that the presentation of irrelevant stimuli before the choice response causes a significant negative effect on the correct choice responses in prefrontal lesioned monkeys. Therefore, they concluded that one function of the dorsolateral prefrontal cortex involves the suppression of irrelevant stimuli interposed between the information necessary to guide a behavior and the behavior itself.

Hyperactivity and distractibility have been frequently observed after the removal of the prefrontal cortex. Therefore, it had been considered that hyperactivity and distractibility produced by the prefrontal lesions were the cause of the delayed-response deficits. Lawicka (1969) showed that dogs having lesions limited in the cortex surrounding the gyrus proreus exhibited increased motor responses and became hyper-reactive to external stimuli. This behavioral tendency caused delayed-response deficits in the dogs having lesions limited in the gyrus proreus. Difficulty in the suppression of responses to irrelevant stimuli or distractors can be observed in many animals having ablations of the prefrontal cortex. If the removal of the prefrontal cortex causes hyper-reactivity or stronger distractibility to the external stimuli and if these behavioral tendencies are the primary factor of delayed-response deficits, it could be predicted that the reduction of reactivity or distractibility by some methods (e.g., sedatives) improves attentiveness to the relevant stimulus and recovers the ability to perform the delayed-response task. Wade (1947) used two rhesus monkeys having bilateral dorsolateral prefrontal lesions and examined the effect of intra-peritoneal injections of Nembutal on delayed-response performances in these monkeys. These monkeys exhibited impairment on delayed-response performances with a 5-s delay period in no-sedative conditions. However, after the injection of Nembutal, these monkeys performed the delayed-response task reliably above the chance level at up to a 25-s delay. Based on these results, Wade (1947) concluded that the observed failure of delayed-response performances was caused by an inattention due to increased reactivity and distractibility, not by the loss of immediate memory. Similar improvement of delayed-response deficits was observed in prefrontal lesioned monkeys when Nembutal or insulin was injected intramuscularly (Pribram 1950).

2.5.3 Kinesthetic Effects

Severe behavioral deficits after dorsolateral prefrontal lesions in monkeys have been observed repeatedly on spatial delayed-response and spatial delayed-alternation performances. It had been known that spatial factors play important roles in performing these tasks. However, some researchers argued that prefrontal lesioned animals could solve these spatial tasks using kinesthetic information, such that the monkey turned to the right if the reward was presented at the right food well, while it turned to the left if the reward was presented at the left food well. It had been known that dogs having lesions limited in the cortex surrounding the gyrus proreus exhibited severe impairments on delayed-response performances. But, Lawicka (1969) found that dogs' impairments on delayed-response performances gradually improved during post-operative training and that this improvement was mainly due to the fact that the dogs learned to keep their body orientation toward the baited food well which was signaled at the beginning of the trial. Whenever the dogs changed their body orientation during the delay period, their performance became a chance level. These results suggested that dogs had used the kinesthetic or motoric cues produced by their own movements to solve the task. Similarly, lesion studies using monkeys showed that a failure to use the kinesthetic orientation caused impairment on delayed alternation performances (Mishkin 1964). Lesion experiments using animals had been usually performed in a cage (e.g., WGTA) or a wide field. Animals could freely walk around within the cage and orient to any direction during the delay period. Therefore, animals could learn to use kinesthetic cues to solve the delayed-response task.

Kinesthetic effects on performances of the delay tasks had been examined under more controlled conditions of animals' behaviors. Pribram (1961) examined kinesthetic effects on delayed alternation performances, in which monkeys were required to make lever press responses between two reinforcement schedules alternately. In the fixed ratio (FR) schedule, monkeys needed to complete 40 consecutive lever presses to obtain a reward (FR40), while in the fixed interval (FI) schedule, monkeys could obtain a reward at the first lever press only after a 4-minutes interval (FI40). Initially, monkeys were informed which reward schedule they needed to perform by different external color cues. However, subsequently, no external cue was introduced to the monkeys. Therefore, monkeys needed to decide which of two behavioral patterns was appropriate for any particular trial of delayed alternation performances. As a result, unoperated normal monkeys took about 4 weeks to reach 85% correct performance, while bilateral prefrontal lesioned monkeys could not reach the criterion. Manning (1978) also examined the importance of kinesthetic cues using monkeys. In his experiment, monkeys were required to first perform a fixed number of pressing of a centrally located lever and then select either red or white key or either left or right key. In one condition, the monkey needed to choose a red key after 32 lever presses (FR32) or a white key after 64 lever presses (FR64). In the other condition, the monkey needed to choose the left key press after 64 lever presses (FR64) or the right key press after 32 lever presses (FR32). The positions of the color keys were randomized across trials (fixed ratio discrimination task). Bilateral ablations of the dorsolateral prefrontal cortex including the principal sulcus impaired monkeys' performances in both conditions. Importantly, dorsolateral prefrontal lesions produced no impairment in sensory or spatial processing but produced severe impairments on kinesthetic discrimination performances.

Mishkin et al. (1977) examined the importance of kinesthetic cues on behavioral performances of prefrontal lesioned monkeys using other kinds of behavioral methods. In his study, rhesus monkeys were required to move a sliding knob horizontally with the hand or the mouth. This device permitted either a small or a large displacement. If the displacement was large (5.2 cm), the animal could obtain a reward (Go condition), while if the displacement was small (2.6 cm), the animal could not obtain a reward (No-Go condition). Although the task was a Go/No-Go choice by sliding a knob, this task allowed to examine the animal's ability to discriminate kinesthetic cues (large or small displacement). Monkeys with bilateral lesions in the dorsal prefrontal cortex exhibited only mild impairments on this task. However, monkeys with combined bilateral lesions of the dorsal prefrontal cortex and the premotor cortex exhibited severe impairments on this task.

Although the kinesthetic factor plays a role in the performances of the delay tasks, it had been shown that other factors including the spatial factor and the mnemonic factor are also essential to understanding the behavioral effects of dorsolateral prefrontal lesions. Passingham (1978) examined whether the behavioral deficits caused by the lesion of the cortex surrounding the principal sulcus were related to the loss of spatial information or kinesthetic information or both. To distinguish these two possibilities, he used a delayed counting task, in which the subjects needed to use the cue produced by their own movements. The monkeys first needed to press a key one or five times to turn on the light (cue period). After up to a 5-s delay, they needed to repeat the same number of key press as they did during the cue period. One group of rhesus monkeys received bilateral ablations of the dorsal convexity of the prefrontal cortex including the principal sulcus. The other group of monkeys received bilateral ablations of the cortex only around the principal sulcus. The results showed that the monkeys with principal sulcus lesions showed no impairment in this task, while the monkeys with lesions in the dorsal convexity exhibited impairments of this task. The results also showed that both groups of monkeys exhibited impairments of spatial delayed alternation performances. In the delayed counting task, both cues and responses were the same action and no spatial component was included. However, in the spatial delayed alternation task, the animals needed to remember the spatial position where the reward was obtained and make a response toward the different spatial position. The monkeys with lesions in the principal sulcus area showed impairments in the spatial delayed alternation task but no impairments in the delayed counting task, indicating that the spatial factor is more essential for understanding the behavioral effects of dorsolateral prefrontal lesions than the kinesthetic factor, although the kinesthetic factor also play a role in spatial task performances.

Kojima et al. (1982) used two versions of the two-choice delayed-response task while the monkey was in the cage and examined the effects of the body orientation during the delay period. In one version (DR1), the correct response key was accessible for the monkey throughout a trial. Therefore, the monkey was allowed to orient toward the correct response key and press it during the delay period. In the other version (DR2), the monkey was required to press an unrelated third key during the delay period, which prevented the body orientation toward the correct response key and inhibited pressing the correct key. After bilateral ablations of the dorsolateral prefrontal cortex, monkeys exhibited no impairment on DR1 performances but exhibited marked impairments on DR2 performances, which was caused by strong position preferences. Behavioral examinations revealed that, although training to eliminate the position biases on the DR2 task improved the performance, prefrontal lesioned monkeys tended to exhibit position preference and shift their responses to the incorrect key whenever the correct cue was presented in the unpreferred side. In addition, shifting to the incorrect key always occurred early in the delay period regardless of the length of the delay period. Thus, although kinesthetic cues seem to help solve delayed-response problems, these results also indicate that delayed-response deficits are caused in part by the deficit in encoding the correct spatial information in memory.

2.6 Mnemonic Hypothesis of Prefrontal Function

2.6.1 Problems of Behavior Control During the Delay Period

Since Jacobsen (1936) first reported severe and long-lasting impairment of delayed-response performance observed in monkeys with large bilateral prefrontal lesions, delayed-response deficits in these monkeys have been repeatedly observed in many studies. Monkeys having bilateral lesions of the lateral prefrontal cortex still know behavioral rules of the delayed-response task, such as how to behave during the trial and how to behave at each trial event. Although these monkeys know to select one food well during the response period to obtain a reward, they fail to select a correct food well. To perform the delayed-response task correctly, the monkeys are required to maintain the spatial information where the reward was baited during the delay period. They are also required to utilize maintained spatial information to perform a correct selection. In addition, since the position where the reward is baited changes randomly from trial to trial, they are required to update spatial information that is needed to maintain. Jacobsen (1936) concluded that bilateral lesions of the lateral prefrontal cortex caused impairments of immediate memory (short-term memory). However, there had been disagreements among researchers regarding the psychological processes needed to perform the delayed-response task and the cause of the difficulty that prefrontal lesioned monkeys exhibited in delayed-response performances. Subsequent studies examined the performance of the delayed-response task under various conditions and showed that delayed-response deficits might not be caused by the impairment of short-term memory but could be caused by impairments of other functions, such as susceptibility to interferences (Malmo 1942), inability to suppress interfering events (Bartus and LeVere 1977), and kinesthetic factors (Pribram 1961; Manning 1978). These disagreements might be partly caused by an apparatus used for testing monkey's behavior. Classically, the Wisconsin General Test Apparatus (WGTA) had been used to test delayed-response performances (Fig. 2.2). In the WGTA, a monkey is placed in a small cage and can freely move within the cage. A reward is placed in one of two food wells (right and left) during the cue period, then an opaque screen placed in front of the cage is lowered, and the delay period is introduced for several seconds. Because the opaque screen is lowered during the delay period, the experimenter cannot observe the monkey's behavior and, at the same time, cannot control the monkey's behavior in the cage.

Using the WGTA for behavioral studies has some limitations. For example, the number and positions of the cue are limited. Many studies used two positions (right and left) as baited positions. In addition, since the monkey's behavior is not restrained in the cage, the monkey could freely move around within the cage throughout a trial as well as during an inter-trial interval. The monkey can view the baited position from various angles and sit anywhere in the cage during the delay and the response period. Therefore, the experimenter cannot control the monkey's behavior during the delay period. Thus, unrestrained conditions of the monkeys' behavior in the WGTA make it difficult to interpret psychological processes that are tapped by the delayed-response task and to assess the source of the difficulty observed in prefrontal lesioned monkeys.

2.6.2 An Oculomotor Version of the Delayed-Response Task

To overcome the limitations that experiments using the WGTA had, Funahashi et al. (1993) developed an oculomotor version of the delayed-response task (ODR task), which is based on a memory-guided saccade task originally introduced by Hikosaka

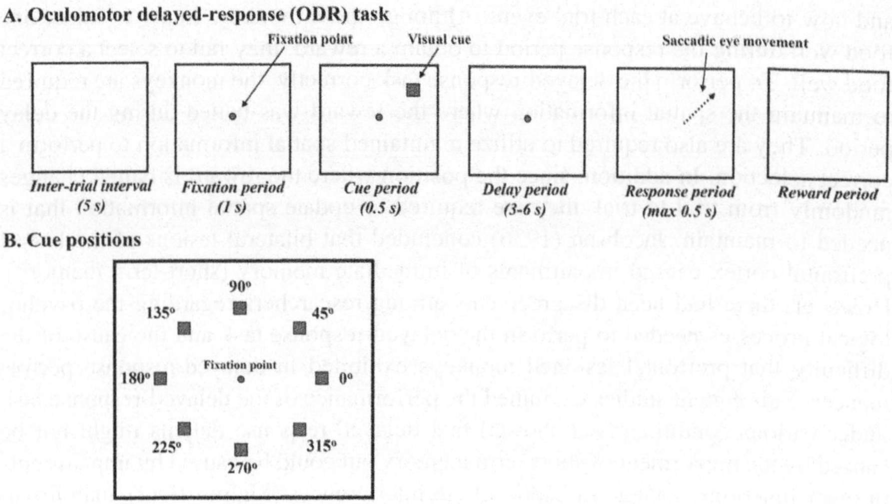

Fig. 2.4 (**a**) Temporal sequence of task events in the oculomotor delayed-response (ODR) task. (**b**) A spatial arrangement of eight visual cue positions

and Wurtz (1983). In the ODR task, the subject is required to maintain fixation at a fixation point during the delay period and make a memory-guided saccade to the position where the visual cue has been presented during the delay period (Fig. 2.4).

In the ODR task, the monkey is set in the primate chair and his head is fixed by a head-restrained instrument. The monkey faces a TV monitor, on which a fixation point and visual cues are presented. The monkey's gaze position and eye movements are continuously monitored during the experiment. The monkey is required to maintain fixation at a fixation point as long as the fixation point is presented on the monitor. As is shown in Fig. 2.4a, after an inter-trial interval of several seconds, a fixation point (a small spot) is presented at the center of the monitor. If the monkey looks at the fixation point for 1 s (fixation period), a visual cue (a white-filled circle) is presented for 0.5 s (cue period). The position of the visual cue is selected randomly from predetermined positions around the fixation point (e.g., Fig. 2.4b). The monkey is required to maintain fixation at the fixation point during the cue period and the subsequent delay period. At the end of the delay period, the fixation point is extinguished, which is the Go signal for the monkey to make a saccade within a 0.5-s response period to the position where the visual cue was presented. Since no visual stimulus is present during the response period, the monkey is required to make a memory-guided saccade to the target position. If the monkey makes a correct saccade, a drop of liquid reward is given.

The ODR task allows to present the visual cue at any position selected from multiple positions in the visual field. Although eight predetermined positions were used in a fixed eccentricity (17°) in Funahashi et al. (1989, 1990, 1991, 1993), it is easy to change the eccentricity and the number of positions of the visual cue. It is impossible for the monkey to use postural and ocular orientations or kinesthetic cues to solve the task, because the monkey needs to maintain fixation at the fixation point during the delay period and because the position of the visual cue is selected randomly not from two positions (left or right) but from multiple positions. The use of the saccadic eye movement as response behaviors allows to analyze the monkey's behavioral responses quantitatively. For example, reaction times, movement times, trajectories of eye movements, and the distribution of endpoints of saccades can be examined quantitatively and compared among different cue conditions and different delay-length conditions.

2.6.3 *Mnemonic Scotoma Observed in Prefrontal Lesioned Monkeys*

Using the ODR task, Funahashi et al. (1993) examined behavioral deficits on ODR performances in four monkeys with unilateral or bilateral lesions of the dorsolateral prefrontal cortex (Fig. 2.5). The lesion sites were restricted within the cortex within and around the principal sulcus. Among four monkeys, two monkeys (monkeys KY and YM) did not participate in any behavioral and neurophysiological studies preoperatively, while two other monkeys (JN and DN) participated in single-neuron recording studies preoperatively. Before having lesions, these monkeys performed the ODR task very well for all 8 cue conditions and exhibited only small error rates even in the longest delay length (6-s delay). After having lesions in the left side of the lateral prefrontal cortex, error rates of ODR performances increased in trials in which visual cues were presented in the right visual field (contralateral to the lesioned hemisphere) (especially 0° and 45° directions). The increase in error rates was obvious in trials with longer delay length (6-s delay) for these monkeys. Two monkeys (JN and YM) further received the second lesions in the right side of the lateral prefrontal cortex 2.5 months (JN) or 1.5 months (YM) after the first lesion. The second lesion caused an increase in error rates in trials in which visual cues were presented in the left visual field (e.g., 180° and 225° directions). Two monkeys (JN and DN) were used for single-neuron recording studies prior to the lesion study and had extensive insertions of metal microelectrodes for recording single-neuron activities from the right prefrontal cortex (DN) or the left prefrontal cortex (JN). A large number of electrode insertions might cause damages to the prefrontal cortex, especially in monkey DN. Preoperatively, monkey DN exhibited higher error rates in trials in which visual cues were presented in the left visual field (e.g., 180° and 225° directions) compared with trials in which visual cues were presented in the right visual field. After the lesion in the left side of the lateral prefrontal cortex, error rates increased in trials in which visual cues were presented in the right visual field (e.g., 315° direction). These results indicate that unilateral lesions in the dorsolateral prefrontal cortex cause significant behavioral deficits in ODR performances. The degree of the behavioral deficit is proportional to the length of the delay period. Importantly, monkeys having unilateral lesions of the dorsolateral prefrontal cortex

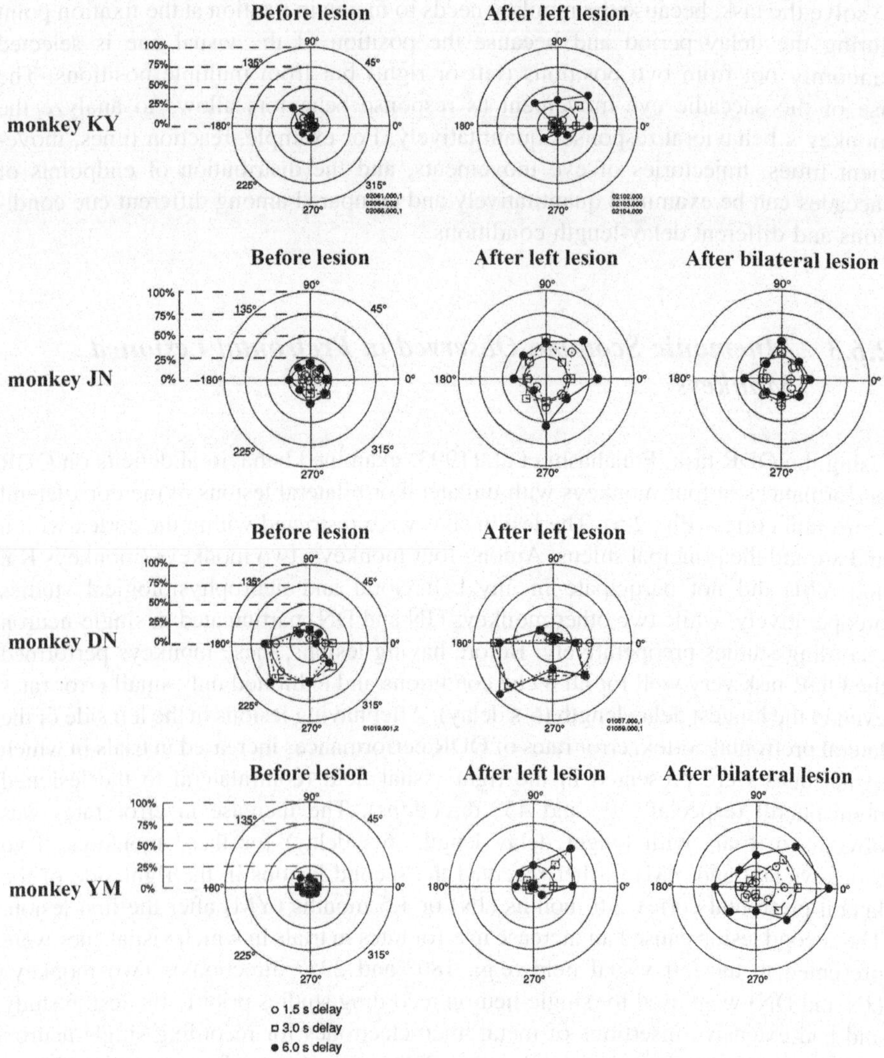

Fig. 2.5 Effects on error rates of memory-guided saccades during ODR performances after unilateral left lesion and bilateral lesions of the principal sulcal cortex for four monkeys. From Funahashi et al. (1993) with permission

exhibited specific impairments of the ODR task, that is, the impairment was observed only in the trials in which visual cues were presented in the visual field contralateral to the lesioned hemisphere.

Figures 2.6 and 2.7 show examples of behavioral results obtained from one monkey (KY). Figure 2.6a shows the lesioned area in the principal sulcal cortex. The lesion area was extended almost the entire length of the principal sulcus and included both the upper and lower banks and fundus of the principal sulcus. But, the

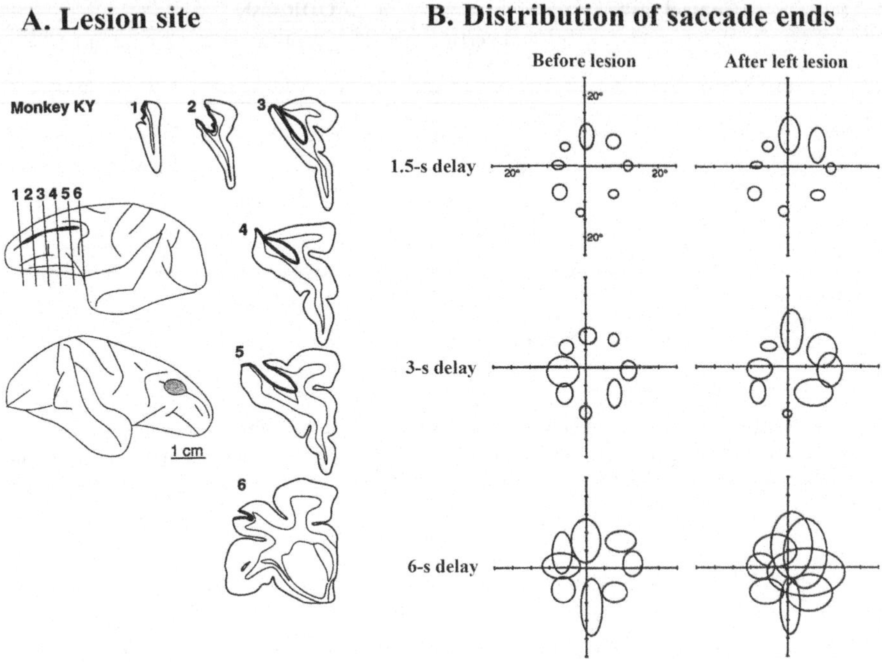

Fig. 2.6 (**a**) Reconstructions of the lesions site (the left principal sulcal area) of monkey KY. Shaded area in the right hemisphere indicates areas recording single-neuron activity before lesion. (**b**) Variations of endpoints of memory-guided saccades during ODR performances under three different delay lengths (1.5 s, 3 s, and 6 s). Each ellipse indicates overall saccade accuracy for each cue position. Each ellipse is centered at the mean x- and y-coordinates of saccade ends for each cue condition, and the horizontal and vertical axes of the ellipse equal the SD of these x- and y-coordinates, respectively. From Funahashi et al. (1993) with permission

lesion was restricted within the principal sulcal area. The arcuate sulcus was not included in the lesion. Figure 2.6b shows monkey's performances of the ODR task before and after the lesion. Preoperatively, monkey KY performed well in the ODR task for all eight cue directions under all three delay lengths, although the monkey showed some difficulty in a 6-s delay when visual cues were presented in the left visual field. This difficulty may be caused by damages produced by single-neuron recordings in the right prefrontal area (see Fig. 2.6a). Following the left principal sulcus lesion, errors in ODR performances substantially increased mostly for trials with longer delay and for trials in which visual cues were presented in the right visual field. Types of errors in the ODR task include an inaccurate saccade (e.g., wrong direction, overshoot, or undershoot) or slow initiation of a saccade (the maximum length of the response period was set at 0.5 s). Monkey KY mostly made saccades within the 0.5-s response period before and after the lesion. The increase in error rates after the lesion was caused by the increase of inaccurate saccades. The comparison of the distribution of endpoints of saccades between before and after the lesion is shown in Fig. 2.7 for the 45° and the 225° cue conditions. Before the

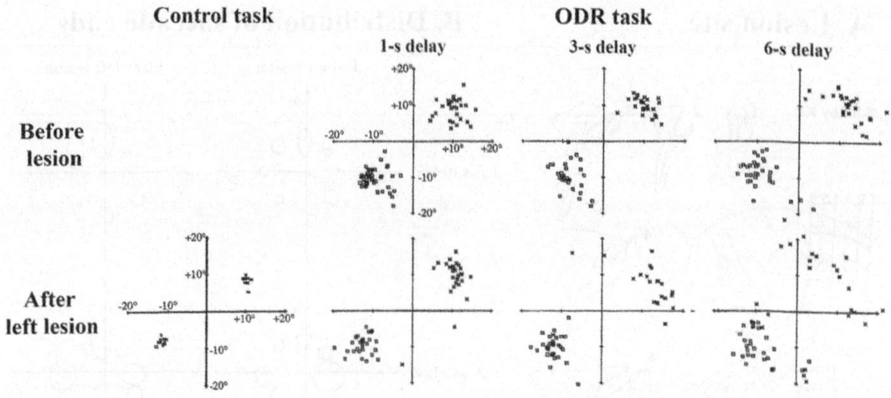

Fig. 2.7 Distributions of saccade end in the 45° trials (ipsilateral to the lesioned hemisphere) and the 225° trials (contralateral to the lesioned hemisphere) during performances of the control task and the ODR task under different lengths of the delay period. From Funahashi et al. (1993) with permission

lesion, the width of saccade distribution was comparable for these two cue conditions, such that nearly all saccades ended within the correct quadrant, although the distribution of saccade endpoints increased fairly with the increase of the delay length for both cue conditions. However, after the lesion was made in the left principal sulcus region, the distribution of saccade endpoints increased, especially in the 6-s delay condition for the 45° cue condition. In contrast, the distribution of saccade endpoints for the 225° cue condition was little affected by the lesion, even in the 6-s delay condition. An increased variability of saccade endpoints in the longer delay condition (the 3-s and 6-s delay conditions) after the lesion was observed for all three rightward (contralateral) (45°, 0°, 315°) and upward (90°) cue conditions (Fig. 2.6b). However, in the 1.5-s delay condition, the monkey made accurate saccades both before and after the lesion. The variability of saccades directed toward visual cues presented in the left (ipsilateral) visual field was similar between before and after the lesion in all delay lengths.

Figure 2.8 shows another monkey's results in this experiment. This monkey (JN) received the first lesion in the left lateral prefrontal cortex. The first lesion site included posterior two-thirds of the principal sulcus and a large portion of the adjacent superior dorsolateral convexity (Fig. 2.8a). The arcuate sulcus was spared. About 2.5 months after the first lesion, the second comparable lesion was made in the right lateral prefrontal cortex (Fig. 2.8a). The arcuate sulcus was again spared. Preoperatively, ODR performances were excellent for all cue conditions and delay conditions. After the first lesion in the left hemisphere, a marked increase of errors was observed in the trials in which visual cues were presented in the right visual field (contralateral to the first lesion), especially for 45°, 0°, and 270° cue conditions. The error rates increased with the length of the delay period. The second lesion in the right hemisphere produced additional impairments in ODR performances, especially in the trials in which visual cues were presented in the left visual field (contralateral

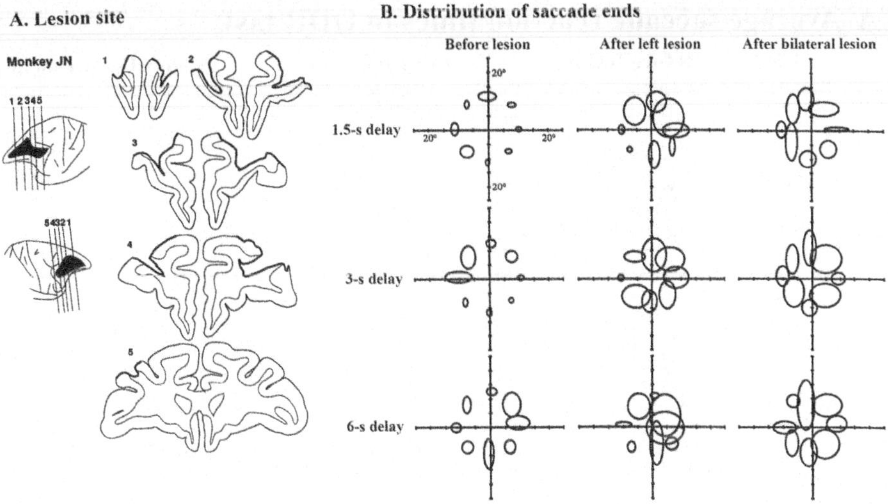

Fig. 2.8 (**a**) Reconstructions of the lesions site (the left principal sulcal area) of monkey JN. Lesions included the principal sulcal area and a large portion of the lateral prefrontal cortex. (**b**) Variations of endpoints of memory-guided saccades during ODR performances under three different delay lengths (1.5, 3, and 6 s). Each ellipse indicates overall saccade accuracy for each cue position. Each ellipse is centered at the mean x- and y-coordinates of saccade ends for each cue condition, and the horizontal and vertical axes of the ellipse equal the SD of these x- and y-coordinates, respectively. From Funahashi et al. (1993) with permission

to the second lesion). Figure 2.8b shows distribution of saccade endpoints and its variability across different cue, delay, and lesion conditions. The distribution of endpoints of saccades toward the right visual field increased after the first lesion in the left lateral prefrontal cortex and the distribution of endpoints of saccades toward the left visual field increased after the second lesion in the right lateral prefrontal cortex. The results shown in Fig. 2.8b emphasize primarily contralateral nature of the lesion effect on memory-guided saccade performances in the ODR task. Like monkey KY, monkey JN consistently made saccades within the 0.5-s response period for all trials in all cue and delay conditions both before and after the lesion. Therefore, again the deficits following the lesions were caused by inaccurate saccades, mainly by saccades directed to wrong directions.

Figure 2.9a shows the reaction times of saccades in the ODR task for two monkeys (KY and JN) before and after the lesions. No marked change was observed in reaction times for saccades directed toward either ipsilateral or contralateral visual field. Reaction times were also consistent among different delay lengths. Unlike behavioral results, no significant change in reaction times was observed in the ODR task after unilateral prefrontal lesions. The lesion of the frontal eye field was excluded in monkeys KY and JN. However, one monkey (YM) received a lesion including the frontal eye field. After monkey YM received a lesion in the left prefrontal cortex including the frontal eye field, reaction times of saccades in the ODR task became markedly longer for saccades directed toward the right visual

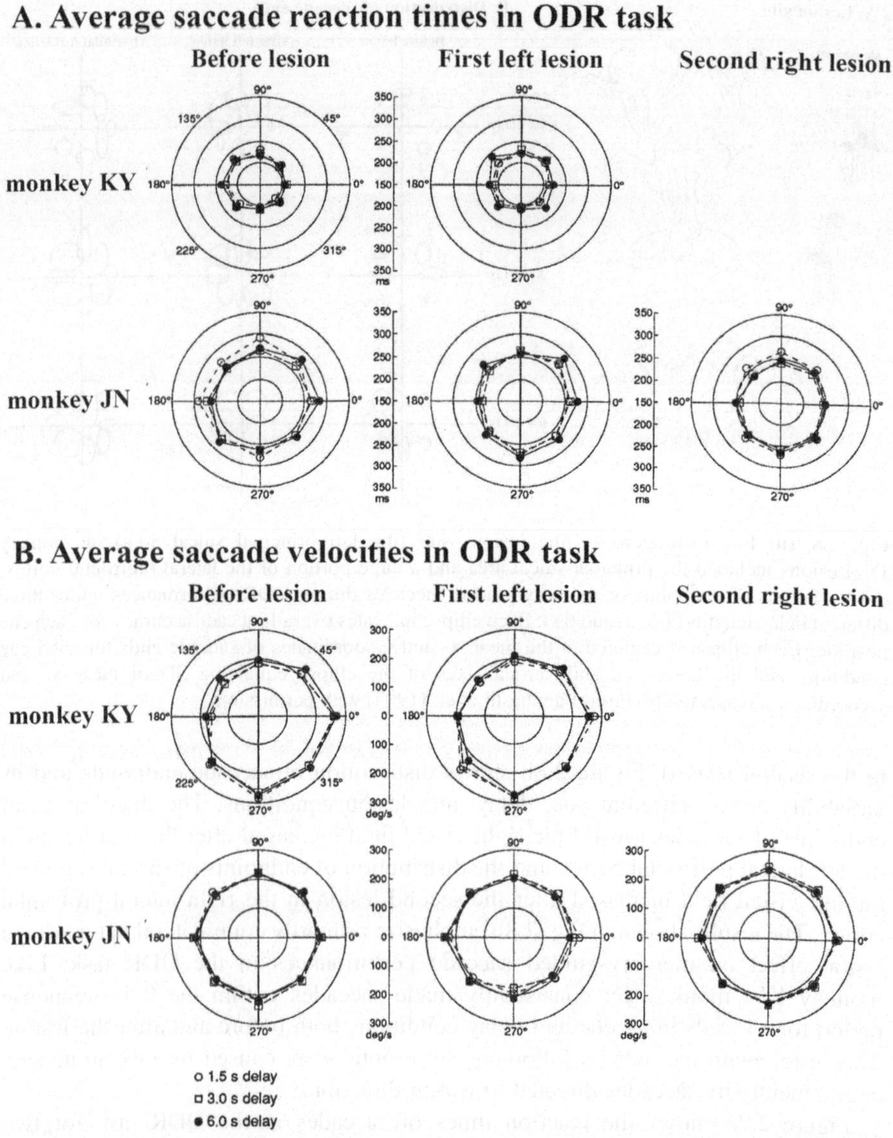

Fig. 2.9 (**a**) Average reaction times of saccades toward 8 cue directions during the response period of the ODR task. Saccade reaction times were largely unaffected by the lesions of the principal sulcus cortex. (**b**) Average saccade velocities of saccades during the response period of the ODR task. Saccade velocities were also unaffected by the lesions of the principal sulcus cortex. From Funahashi et al. (1993) with permission

field. After the second lesion in the right prefrontal cortex excluding the frontal eye field, longer reaction times were maintained for saccades toward the right visual field and no change in reaction times was observed for saccades toward the left visual field.

Figure 2.9b shows plots of saccadic velocity in the ODR task for two monkeys (KY and JN) before and after their lesions. No marked difference was observed in saccade velocities for any direction if the lesion spared the frontal eye field. The mean saccade velocities were consistent across different delay lengths for each saccade direction. Thus, no significant change was observed in average saccade velocity in the ODR task after unilateral prefrontal lesions. However, when the frontal eye field was included in the lesion, a marked decrease in saccade velocities was observed for saccades directed toward the visual field contralateral to the lesion.

Thus, after monkeys received unilateral lesions of the lateral prefrontal cortex, the impairment observed during ODR performance was characterized by eye movements toward an inappropriate direction (see Figs. 2.5, 2.6, 2.7, and 2.8). Unilateral lesions of the lateral prefrontal cortex produced significant impairments on memory-guided saccades to contralateral cues, but only mild impairments on memory-guided saccades to ipsilateral cues. When the second lesion was added in the opposite hemisphere, the impairment of memory-guided saccades was extended to both visual fields. However, unilateral or bilateral lesions of the lateral prefrontal cortex had little or no effect on the performance of visually-guided saccades. The degree of the impairment depended on the length of the delay period. The performance of memory-guided saccades was nearly normal at the shortest (1.5 s) delay condition but became progressively worse as the length of the delay period increased up to 6.0 s. However, saccadic reaction times and saccade velocities showed no difference before and after the lesions.

Based on these observations, Funahashi et al. (1993) concluded as follows:

> *The present results strengthen the evidence that the delayed-response deficits of monkeys with prefrontal lesions are caused by failure to maintain a transient memory "trace" in working memory, and indicate for the first time that working memory mechanisms are lateralized: memories for visuospatial coordinates in each hemifield are processed primarily in the contralateral prefrontal cortex. These findings provide evidence for the concept of mnemonic hemianopias and mnemonic scotomas, that is, memory deficits for particular hemifields or visual field locations, unaccompanied by simple sensory or motor deficits.*
> (Page 1479 in Funahashi et al. (1993))

2.6.4 *Contralateral Bias of Spatial Mnemonic Process in the Prefrontal Cortex*

Contralateral bias observed in the lateral prefrontal cortex for mnemonic processing of visuospatial information is consistent with contralateral bias for processing of visual information in the brain. In the visual system, the information of the visual stimulus presented in either visual hemifield is processed mainly in the hemisphere contralateral to the hemifield. Therefore, the prefrontal cortex in either hemisphere would mainly receive information of visual stimuli presented in the contralateral visual field.

Contralateral bias of spatial mnemonic processing has been observed in other behavioral studies. Sawaguchi and Goldman-Rakic (1991, 1994) used the ODR task and showed that local injections of dopamine receptor D1 antagonists (SCH23390 and SCH39166) into the lateral prefrontal cortex induced an increase in directional errors (decrease of saccade accuracy) and an increase of reaction times in memory-guided saccades only when visual cues were presented in the visual field contralateral to the injected hemisphere. These deficits were dose-dependent (higher dose induced more errors) and sensitive to the length of the delay period (longer delay was associated with the decrease of saccade accuracy). However, the same injection did not affect saccade accuracy and reaction times of visually-guided saccade performances, indicating that the impairment of mnemonic processes causes deficits in ODR performances. In addition, the local injections of a selective D2 antagonist (sulpiride) and a D2/D3 antagonist (raclopride) showed no effect on memory-guided saccades, indicating that dopamine D1 receptors are essential for ODR performances. Sawaguchi and Iba (2001) further showed that local injections of muscimol into the lateral prefrontal cortex impaired memory-guided saccades directed toward a few specific target positions that were located in the contralateral visual field. Again, the impairment was not observed in visually-guided saccades. Pasternak et al. (2015) used a working memory task in which monkeys were required to compare the directions of two moving stimuli separated by the delay period and found that the unilateral lesion of the lateral prefrontal cortex impaired task performances when moving stimuli were presented in the visual field contralateral to the lesioned hemisphere. These findings support the ideas of mnemonic scotomas and contralateral bias of impairments proposed by Funahashi et al. (1993). Monkeys having lesions in the lateral prefrontal cortex exhibit severer behavioral deficits in the delay tasks when visual cues are presented in the visual field contralateral to the lesioned hemisphere than when visual cues are presented in the ipsilateral visual field. Prefrontal lesioned monkeys do not exhibit impairments in sensory and motor functions but exhibit impairments in visuospatial mnemonic functions.

Further, the ideas of mnemonic hemianopias or mnemonic scotomas and the contralateral bias of impairments are also supported by the results obtained in neurophysiological experiments using monkeys. Funahashi et al. (1989) and Takeda and Funahashi (2002) analyzed single-neuron activity recorded from the dorsolateral prefrontal cortex while monkeys performed the ODR task and found that many prefrontal neurons exhibited directional delay-period activity whose preferred direction was directed toward the visual field contralateral to the hemisphere where the neuron was located. These studies also showed that prefrontal neurons exhibiting directional delay-period activity maintain visuospatial information presented within the visual field contralateral to the recording hemisphere.

2.6.5 *Working Memory Is an Important Function of the Prefrontal Cortex*

Mnemonic scotoma is an important concept to understand prefrontal functions. However, Tsujimoto and Postle (2012) claimed that the concept of "mnemonic scotoma" needed to be reconsidered. Their opinion was based on evidence that subtle differences in experimental procedures can lead to different conclusions regarding the cause of delayed-response deficits. Tsujimoto and Postle (2012) cited unpublished results obtained by Wajima and Sawaguchi (2004). They observed that the local injection of a small amount of bicuculline in the lateral prefrontal cortex produced erroneous saccades in the ODR task when the visual cues were presented at a particular area in the visual field, mainly in the contralateral visual field. However, no such deficit was observed under visually guided conditions. These results confirmed the results obtained by Funahashi et al. (1993), Sawaguchi and Goldman-Rakic (1991, 1994), and Sawaguchi and Iba (2001). However, Wajima and Sawaguchi (2004) also observed that monkeys performed two successive saccades quickly in most of the error trials. In these trials, the first saccade was misdirected, but the second saccade often ended at the correct target position, although monkeys could not obtain the reward. In addition, they also observed that the erroneous first saccade tended to be directed toward the position that had been the target on the previous trial. Therefore, they concluded that, although monkeys maintained an intact memory of the cue position, they mistakenly selected a wrong target for the first saccade response, because the information regarding the target for the preceding trial interfered in the memory of the target for the current trial. Based on these observations, Tsujimoto and Postle (2012) argued against the concept of mnemonic scotoma, which is memory deficits for particular hemifields or visual field locations.

Although a local and unilateral injection of bicuculline can disturb the function of only a restricted portion of the prefrontal cortex, most of the prefrontal areas are functionally intact. Therefore, a single injection of bicuculline may not be sufficient to disrupt the working memory function of any single spatial location, because of the lack of a topographic mnemonic map in the prefrontal cortex. In addition, functionally intact areas can hold spatial information necessary for any given cue condition, since each prefrontal neuron has its memory field in the visual field and since neurons having memory fields in different positions in the visual field are intermingled within the prefrontal cortex. Further, other brain areas such as the posterior parietal cortex might compensate for disturbed prefrontal mnemonic functions, since the posterior parietal cortex also participates in visuospatial working memory processes (Chafee and Goldman-Rakic 1998, 2000). Since mnemonic receptive fields of many prefrontal neurons are often extended in bilateral visual fields, neurons in the contralateral hemisphere against the injection site could guide correct saccades during ODR performances. Recently Genovesio et al. (2014) reported that some prefrontal neurons encoded spatial information for the previous trial which was not relevant for a current trial. Therefore, these neurons' activity might generate an erroneous initial saccade toward the target of the preceding trial. However, since many other prefrontal neurons hold correct and relevant spatial information for the current trial, saccades to the correct target could be generated in most of the trials. Thus, the results obtained by Wajima and Sawaguchi (2004) well support the importance of the prefrontal cortex in spatial working memory process.

In order to consider the mnemonic functions of the prefrontal cortex, the importance of delay-period activity should be emphasized. As is described in Chap. 3, the importance of delay-period activity is that this activity can represent a variety of information and, therefore, could participate in a variety of information processing. This feature of delay-period activity enables to support a variety of important operations in the prefrontal cortex, such as executive control. At the same time, the following features of delay-period activity agree with the notion that delay-period activity is a neural correlate of a mechanism for temporarily holding information in working memory. First, delay-period activity persists during the delay period and the duration of delay-period activity depends on the length of the delay period. Second, delay-period activity is not observed or truncated in the trial when the subject makes an error. Third, most delay-period activity exhibit selectivity to the features of sensory cues or response behaviors. Fourth, a great majority of delay-period activity represent retrospective information, while a minority represents prospective information. Fifth, the information represented by delay-period activity can be processed during the delay period. These features indicate that delay-period activity not only serves to maintain information but also participates in manipulating and processing information. These observations support that delay-period activity observed in the prefrontal cortex is a neural correlate of a mechanism for temporarily maintaining information in working memory. Thus, the mnemonic function must be the most important function of the prefrontal cortex and delay-period activity observed in prefrontal neurons plays important roles to achieve mnemonic functions.

References

Anderson RM, Hunt SC, Stoep AV, Pribram KH (1976) Object permanency and delayed response as spatial context in monkeys with frontal lesions. Neuropsychologia 14:481–490. https://doi.org/10.1016/0028-3932(76)90076-2

Bachevalier J, Mishkin M (1986) Visual recognition impairment follows ventromedial but not dorsolateral prefrontal lesions in monkeys. Behav Brain Res 20:249–261. https://doi.org/10.1016/0166-4328(86)90225-1

Bartus RT, Levere TE (1977) Frontal decortication in rhesus monkeys: a test of the interference hypothesis. Brain Res 119:233–248. https://doi.org/10.1016/0006-8993(77)90103-2

Bauer RH, Fuster JM (1976) Delayed-matching and delayed-response deficit from cooling dorsolateral prefrontal cortex in monkeys. J Comp Physiol Psychol 90:293–302. https://doi.org/10.1037/h0087996

Bauer RH, Fuster JM (1978) The effect of ambient illumination on delayed-matching and delayed-response deficits from cooling dorsolateral prefrontal cortex. Behav Biol 22:60–66. https://doi.org/10.1016/S0091-6773(78)92019-9

Bianchi L (1895) The functions of the frontal lobes. Brain 18:497–522. (translation from the original manuscript by A. de Watteville). https://doi.org/10.1093/brain/18.4.473
Blum RA (1952) Effects of subtotal lesions of frontal granular cortex on delayed reaction in monkeys. AMA Arch Neurol Psychiatry 67:375–386. https://doi.org/10.1001/archneurpsyc.1952.02320150108012
Breathnach CS (1992) Eduald Hitzig, neurophysiologist and psychiatrist. Hist Psychiatry 3:329–338. https://doi.org/10.1177/0957154x9200301104
Brooks VB (1983) Study of brain function by local, reversible cooling. Rev Physiol Biochem Pharmacol 95:1–109. https://doi.org/10.1007/bfb0034097
Butters N, Pandya D (1969) Retention of delayed alternation: effect of selective lesions of sulcus principalis. Science 165:1271–1273. https://doi.org/10.1126/science.165.3899.1271
Butters N, Pandya D, Sanders K, Dye P (1971) Behavioral deficits in monkeys after selective lesions within the middle third of sulcus principalis. J Comp Physiol Psychol 76:8–14. https://doi.org/10.1037/h0031037
Chafee MV, Goldman-Rakic PS (1998) Matching patterns of activity in primate prefrontal area 8a and parietal area 7ip neurons during a spatial working memory task. J Neurophysiol 79:2919–2940. https://doi.org/10.1152/jn.1998.79.6.2919
Chafee MV, Goldman-Rakic PS (2000) Inactivation of parietal and prefrontal cortex reveals interdependence of neural activity during memory-guided saccades. J Neurophysiol 83:1550–1566. https://doi.org/10.1152/jn.2000.83.3.1550
Chorover SL, Cole M (1966) Delayed alternation performance in patients with cerebral lesions. Neuropsychologia 4:1–7. https://doi.org/10.1016/0028-3932(66)90016-9
D'Esposito M, Postle BR (1999) The dependence of span and delayed-response performance on prefrontal cortex. Neuropsychologia 37:1303–1315. https://doi.org/10.1016/S0028-3932(99)00021-4
Divac I (1972) Delayed alternation in cats with lesions of the prefrontal cortex and the caudate nucleus. Physiol Behavi 8:519–522. https://doi.org/10.1016/0031-9384(72)90339-3
Dronkers NF, Plaisant O, Iba-Zizen MT, Cabanis EA (2007) Paul Broca's historic cases: high resolution MR imaging of the brains of Leborgne and Lelong. Brain 130:1432–1441. https://doi.org/10.1093/brain/awm042
Ferrier D (1874) On the localization of the functions of the brain. Br Med J 2(729):766–767. https://doi.org/10.1136/bmj.2.729.766
Ferrier D (1875a) Experiments on the brain of monkeys. No. 1. Proc R Soc Lond 23:409–430. https://doi.org/10.1098/rspl.1874.0058
Ferrier D (1875b) The Croonian Lecture. Experiments on the brain of monkeys (second series). Phil Trans R Soc Lond 165:433–488. https://doi.org/10.1098/rstl.1875.0016
Ferrier D (1876) The functions of the brain. Smith Elder & Co., London
Ferrier D, Sanderson JB (1874) The localization of function in the brain. Proc R Soc Lond 22:229–232. https://doi.org/10.1098/rspl.1873.0032
Ferrier D, Yeo GF (1884) A record of experiments on the effects of lesion of different regions of the cerebral hemispheres. Phil Trans R Soc Lond 175:479–564. https://doi.org/10.1098/rstl.1884.0020
Finan JL (1939) Effects of frontal lobe lesions on temporally organized behavior in monkeys. J Neurophysiol 2:208–226. https://doi.org/10.1152/jn.1939.2.3.208
Finan JL (1942) Delayed response with pre-delay reinforcement in monkeys after the removal of the frontal lobes. Am J Psychol 55:202–214. https://doi.org/10.2307/1417079
Franz SI (1902) On the functions of the cerebrum: I. The frontal lobes in relation to the production and retention of simple sensory-motor habits. Am J Physiol 8:1–22. https://doi.org/10.1152/ajplegacy.1902.8.1.1
Franz SI (1910) On the association functions of the cerebrum. J Phil Psychol Sci Methods 7:673–683. https://doi.org/10.2307/2013276

Freedman M, Oscar-Berman M (1986a) Selective delayed response deficits in Parkinson's and Alzheimer's disease. Arch Neurol 43:886–890. https://doi.org/10.1001/archneur.1986.00520090026011

Freedman M, Oscar-Berman M (1986b) Bilateral frontal lobe disease and selective delayed response deficits in humans. Behav Neurosci 100:337–342. https://doi.org/10.1037/0735-7044.100.3.337

Funahashi S, Bruce CJ, Goldman-Rakic PS (1989) Mnemonic coding of visual space in the monkey's dorsolateral prefrontal cortex. J Neurophysiol 61:331–349. https://doi.org/10.1152/jn.1989.61.2.331

Funahashi S, Bruce CJ, Goldman-Rakic PS (1990) Visuospatial coding in primate prefrontal neurons revealed by oculomotor paradigms. J Neurophysio 63:814–831. https://doi.org/10.1152/jn.1990.63.4.814

Funahashi S, Bruce CJ, Goldman-Rakic PS (1991) Neuronal activity related to saccadic eye movements in the monkey's dorsolateral prefrontal cortex. J Neurophysiol 65:1464–1483. https://doi.org/10.1152/jn.1991.65.6.1464

Funahashi S, Bruce CJ, Goldman-Rakic PS (1993) Dorsolateral prefrontal lesions and oculomotor delayed-response performance: evidence for mnemonic "scotomas.". J Neurosci 13:1479–1497. https://doi.org/10.1523/jneurosci.13-04-01479.1993

Fuster JM (1997) The prefrontal cortex: anatomy, physiology, and neuropsychology of the frontal lobe, 3rd edn. Lippincott-Raven, Philadelphia, PA

Fuster JM, Alexander GE (1970) Delayed response deficit by cryogenic depression of frontal cortex. Brain Res 20:85–90. https://doi.org/10.1016/0006-8993(70)90156-3

Fuster JM, Bauer RH (1974) Visual short-term memory deficit from hypothermia of frontal cortex. Brain Res 81:393–400. https://doi.org/10.1016/0006-8993(74)90838-5

Genovesio A, Tsujimoto S, Navarra G, Falcone R, Wise SP (2014) Autonomous encoding of irrelevant foals and outcomes by prefrontal cortex neurons. J Neurosci 34:1970–1978. https://doi.org/10.1523/jneurosci.3228-13.2014

Goldman PS, Rosvold HE (1970) Localization of function within the dorsolateral prefrontal cortex of the rhesus monkey. Exp Neurol 27:291–304. https://doi.org/10.1016/0014-4886(70)90222-0

Gross CG (1963) Effect of deprivation on delayed response and delayed alternation performance by normal and brain operate monkeys. J Comp Physiol Psychol 56:48–51

Gross CG, Weiskrantz L (1962) Evidence for dissociation of impairment on auditory discrimination and delayed response following lateral frontal lesions in monkeys. Exp Neurol 5:453–476. https://doi.org/10.1016/0014-4886(62)90057-2

Hikosaka O, Wurtz RH (1983) Visual and oculomotor functions of monkey substantia nigra pars reticulata. III. Memory-contingent visual and saccade responses. J Neurophysiol 49:1268–1284. https://doi.org/10.1152/jn.1983.49.5.1268

Jacobsen CF (1931) A study of cerebral functions in learning. The frontal lobes. J Comp Neurol 52: 271–340. https://doi.org/10.1002/cne.900520205

Jacobsen CF (1935) Functions of frontal association area in primates. Arch Neurol Psychiatry 33: 558–569. https://doi.org/10.1001/archneurpsyc.1935.02250150108009

Jacobsen CF (1936) Studies of cerebral function in primates. I. The functions of the frontal association areas in monkeys. Comp Psychol Monogr 13:1–60

Jacobsen CF (1939) The effects of extirpations on higher brain processes. Physiol Rev 19:303–322. https://doi.org/10.1152/physrev.1939.19.3.303

Jacobsen CF, Elder JH (1936) Studies of cerebral function in primates. II. The effect of temporal lobe lesions on delayed response in monkeys. Comp Psychol Monogr 13:61–65

Jacobsen CF, Haslerud GM (1936) Studies of cerebral function in primates. III. A note on the effect of motor and premotor area lesions on delayed response in monkeys. Comp Psychol Monogr 13: 66–68

Jacobsen CF, Nissen HW (1937) Studies of cerebral function in primates. IV. The effects of frontal lobe lesions on the delayed alternation habit in monkeys. J Comp Psychol 23:101–112. https://doi.org/10.1037/h0056632

Kojima S, Kojima M, Goldman-Rakic PS (1982) Operant behavioral analysis of memory loss in monkeys with prefrontal lesions. Brain Res 248:51–59. https://doi.org/10.1016/0006-8993(82)91145-3

Kowalska DM, Bachevalier J, Mishkin M (1991) The role of the inferior prefrontal convexity in performance of delayed nonmatching-to-sample. Neuropsychologia 29:583–600. https://doi.org/10.1016/0028-3932(91)90012-W

Lanczik M, Keil G (1991) Darl Wernicke's localization theory and its significance for the development of scientific psychiatry. Hist Psychiatry 2:171–180. https://doi.org/10.1177/0957154x9100200604

Lashley KS (1926) Studies of cerebral function in learning. VII. The relation between cerebral mass, learning, and retention. J Comp Neurol 41:1–58. https://doi.org/10.1002/cne.900410102

Lashley KS (1931) Mass action in cerebral function. Science 73:245–254. https://doi.org/10.1126/science.73.1888.245

Lashley KS (1933) Integrative functions of the cerebral cortex. Physiol Rev 13:1–42. https://doi.org/10.1152/physrev.1933.13.1.1

Lawicka W (1969) A proposed mechanism for delayed response impairment in prefrontal animals. Acta Biol Exp 29:401–414

Lawicka W, Mishkin M, Kreiner J, Brutkowski S (1966) Delayed response deficit in dogs after selective ablations of the proreal gyrus. Acta Biol Exp 26:309–322

LeVere TE, Bartus RT (1971) Stimulus information and primate discrimination learning: pre-response utilization of stimulus information. J Comp Physiol Psychol 77:200–205. https://doi.org/10.1037/h0031662

LeVere TE, Bartus RT (1972) Stimulus information and primate discrimination learning: utilization of pre-response information following acquisition. J Com Physiol Psychol 79:432–437. https://doi.org/10.1037/h0032845

Malmo RB (1942) Interference factors in delayed response in monkeys after removal of frontal lobes. J Neurophysiol 5:295–308. https://doi.org/10.1152/jn.1942.5.4.295

Manning FJ (1978) Dorsolateral prefrontal cortex lesions and discrimination of movement-produced cues by rhesus monkeys. Brain Res 149:77–88. https://doi.org/10.1016/0006-8993(78)90588-7

Markowitsch HJ, Pritzel M, Kessler J, Guldin W, Freeman RB Jr (1980) Delayed-alternation performance after selective lesions within the prefrontal cortex of the cat. Behav Brain Res 1: 67–91. https://doi.org/10.1016/0166-4328(80)90046-7

Meyer DR, Harlow HF (1952) Effects of multiple variables on delayed response performance by monkeys. J Gen Psychol 81:53–61. https://doi.org/10.1080/08856559.1952.10533634

Miles RC, Blomquist AJ (1960) Frontal lesions and behavioral deficits in monkey. J Neurophysiol 23:471–484. https://doi.org/10.1152/jn.1960.23.5.471

Mishkin M (1957) Effects of small frontal lesions on delayed alternation in monkeys. J Neurophysiol 20:616–622. https://doi.org/10.1152/jn.1957.20.6.615

Mishkin M (1964) Preservation of central sets after frontal lesions in monkeys. In: Warren JM, Akert K (eds) The frontal granular cortex and behavior. McGraw-Hill, New York, pp 219–241

Mishkin M, Manning FJ (1978) Non-spatial memory after selective prefrontal lesions in monkeys. Brain Res 143:313–323. https://doi.org/10.1016/0006-8993(78)90571-1

Mishkin M, Pribram KH (1955) Analysis of the effects of frontal lesions in monkeys: I. Variations of delayed alternation. J Comp Physiol Psychol 48:492–495. https://doi.org/10.1037/h0040318

Mishkin M, Pribram KH (1956) Analysis of the effects of frontal lesions in monkey. II. Variations of delayed response. J Comp Physiol Psychol 49:36–40. https://doi.org/10.1037/h0040592

Mishkin M, Vest B, Waxler RM, Rosvold HE (1969) A re-examination of the effects of frontal lesions on object alternation. Neuropsychologia 7:357–363. https://doi.org/10.1016/0028-3932(69)90060-8

Mishkin M, Pohl W, Rosenkilde CE (1977) Kinesthetic discrimination after prefrontal lesions in monkeys. Brain Res 130:163–168. https://doi.org/10.1016/0006-8993(77)90853-8

Munk H (1990) Uber die Funktionen der Grosshirnrinde. Hirschwald, Berlin

Nissen HW, Elder JH (1935) The influence of amount of incentive on delayed response performances of chimpanzees. J Gen Psychol 47:49–72. https://doi.org/10.1080/08856559.1935.9943884

Nissen HW, Riesen AH, Nowlis V (1938) Delayed response and discrimination learning by chimpanzees. J Comp Psychol 26:361–386. https://doi.org/10.1037/h0056387

Oscar-Berman M, Zola-Morgan SM, Oberg RGE, Bonner RT (1982) Comparative neuropsychology and Korsakoff's syndrome. III. Delayed response, delayed alternation and DRL performance. Neuropsychologia 20:187–202. https://doi.org/10.1016/0028-3932(82)90009-4

Oscar-Berman M, Hutner N, Bonner RT (1992) Visual and auditory spatial and nonspatial delayed-response performance by Korsakoff and non-Korsakoff alcoholic and aging individuals. Behav Neurosci 106:613–622. https://doi.org/10.1037/0735-7044.106.4.613

Pascual-Leone A, Hallett M (1994) Induction of errors in a delayed response task by repetitive transcranial magnetic stimulation of the dorsolateral prefrontal cortex. Neuroreport 5:2517–2520

Passingham RE (1972) Visual discrimination learning after selective prefrontal ablations in monkeys (*Macaca mulatta*). Neuropsychologia 10:27–39. https://doi.org/10.1016/0028-3932(72)90040-1

Passingham R (1975) Delayed matching after selective prefrontal lesions in monkeys (*Macaca mulatta*). Brain Res 92:89–102. https://doi.org/10.1016/0006-8993(75)90529-6

Passingham R (1978) Information about movements in monkeys (*Macaca mulatta*) with lesions of dorsal prefrontal cortex. Brain Res 152:313–328. https://doi.org/10.1016/0006-8993(78)90258-5

Passingham RE, Ettlinger G (1972) Tactile discrimination learning after selective prefrontal ablations in monkeys (*Macaca mulatta*). Neuropsychologia 10:17–26. https://doi.org/10.1016/0028-3932(72)90039-5

Pasternak T, Lui LL, Spineli PM (2015) Unilateral prefrontal lesions impair memory-guided comparisons of contralateral visual motion. J Neurosci 36:7095–7105. https://doi.org/10.1523/jneurosci.5265-14.2015

Pearce JMS (2009) Marie-Jean-Pierre Flourens (1794–1867) and cortical localization. Eur Neurol 61:311–314. https://doi.org/10.1159/000206858

Pribram KH (1950) Some physical and pharmacological factors affecting delayed response performance of baboons following frontal lobectomy. J Neurophysiol 13:373–382. https://doi.org/10.1152/jn.1950.13.5.373

Pribram KH (1961) A further experimental analysis of the behavioral deficit that follows injury to the primate frontal cortex. Exp Neurol 3:432–466. https://doi.org/10.1016/0014-4886(61)90021-8

Pribram KH, Tubb WE (1967) Short-term memory, parsing, and the primate frontal cortex. Science 156:1765–1767. https://doi.org/10.1126/science.156.3783.1765

Pribram KH, Mishkin M (1956) Analysis of the effects of frontal lesions in monkey. III. Object alternation. J Comp Physiol Psychol 49:41–45. https://doi.org/10.1037/h0046248

Pribram KH, Mishkin M, Rosvold HE, Kaplan SJ (1952) Effects on delayed-response performance of lesions of dorsolateral and ventrolateral frontal cortex of baboons. J Comp Physiol Psychol 45:565–575. https://doi.org/10.1037/h00611240

Pribram KH, Plotkin HC, Anderson RM, Leong D (1977) Information sources in the delayed alternation task for normal and "frontal" monkeys. Neuropsychologia 15:329–340. https://doi.org/10.1016/0028-3932(77)90042-2

Quintana J, Fuster JM (1993) Spatial and temporal factors in the role of prefrontal and parietal cortex in visuomotor integration. Cereb Cortex 3:122–132. https://doi.org/10.1093/cercor/3.2.122

Rosenkilde CE, Rosvold HE, Mishkin M (1981) Time discrimination with positional responses after selective prefrontal lesions in monkeys. Brain Res 210:129–144. https://doi.org/10.1016/0006-8993(81)90890-8

Sawaguchi T, Goldman-Rakic PS (1991) D1 dopamine receptors in prefrontal cortex: involvement in working memory. Science 251:947–950. https://doi.org/10.1126/science.1825731

Sawaguchi T, Goldman-Rakic PS (1994) The role of D1-dopamine receptor in working memory: local injections of dopamine antagonists into the prefrontal cortex of rhesus monkeys performing an oculomotor delayed-response task. J Neurophysiol 71:515–528. https://doi.org/10.1152/jn.1994.71.2.5152

Sawaguchi T, Iba M (2001) Prefrontal cortical representation of visuospatial working memory in monkeys examined by local inactivation with muscimol. J Neurophysiol 86:2041–2053. https://doi.org/10.1152/jn.2001.86.4.2041

Stamm JS, Weber-Levine ML (1971) Delayed alternation impairments following selective prefrontal cortical ablations in monkeys. Exp Neurol 33:263–278. https://doi.org/10.1016/0014-4886(71)90020-3

Stamm JS, Stepien I, Levine M (1971) Locomotor delayed alternation by prefrontally ablated monkeys. Psychonomic Sci 22:51–52. https://doi.org/10.3758/bf03335934

Takeda K, Funahashi S (2002) Prefrontal task-related activity representing visual cue location or saccade direction in spatial working memory tasks. J Neurophysiol 87:567–588. https://doi.org/10.1152/jn.00249.2001

Tsujimoto S, Postle BR (2012) The prefrontal cortex and oculomotor delayed response: a reconsideration of the "mnemonic scotoma.". J Cogn Neurosci 24:627–635. https://doi.org/10.1162/jocn_a_00171

Tyler KL, Malessa R (2000) The Goltz-Ferrier debates and the triumph of cerebral localizationist theory. Neurol 55:1015–1024. https://doi.org/10.1212/wnl.55.7.1015

Wade M (1947) The effect of sedatives upon delayed response in monkeys following removal of the prefrontal lobes. J Neurophysiol 10:57–61. https://doi.org/10.1152/jn.1947.10.1.57

Wajima K, Sawaguchi T (2004) The role of GABAergic inhibition in suppressing perseverative responses in the monkey prefrontal cortex. Neurosci Res 50(Suppl 1):3–317

Wikmark RGE, Divac I, Weiss R (1973) Retention of spatial delayed alternation in rats with lesions in the frontal lobes. Brain Behav Evol 8:329–339. https://doi.org/10.1159/000124362

Zola-Morgan S (1995) Localization of brain function: the legacy of Franz Joseph Gall (1758–1828). Annu Rev Neurosci 18:359–383. https://doi.org/10.1146/annurev.ne.18.030195.002043

Chapter 3
Working Memory and Prefrontal Functions

3.1 Introduction

The prefrontal cortex plays important roles in a variety of higher cognitive functions, such as thinking, reasoning, planning, and decision-making (Goldman-Rakic 1987; Miller 2000; Miller and Cohen 2001; Fuster 2008; Funahashi 2015). Neural mechanisms for the prefrontal cortex to operate these functions have been extensively examined using human subjects and animals. Especially, studies using nonhuman primates have contributed significantly to understanding prefrontal functions and their neural mechanisms. Since Jacobsen (1936) first reported that monkeys having bilateral lesions in the prefrontal cortex exhibited severe and long-lasting impairments of delayed-response performances, the delayed-response task has been one of the important behavioral tasks to examine prefrontal functions using animal models. Since the monkey is required to retain the baited position during the delay period to correctly perform the delayed-response task, delayed-response deficits had been thought to be caused by an impairment of short-term storage of spatial information. Subsequently, not only the tasks requiring short-term storage of spatial information but also behavioral tasks requiring the storage of non-spatial information were shown to be impaired by bilateral lesions of the prefrontal cortex. Therefore, an important function of the prefrontal cortex was thought to be the short-term storage of various information. Although behavioral studies repeatedly exhibited delayed-response deficits in monkeys having prefrontal lesions, no agreement regarding the cause of delayed-response deficits had been reached among researchers. Some researchers concluded that delayed-response deficits were caused by an impairment of short-term memory. However, other researchers concluded that these deficits were caused by an increase in susceptibility to interference stimuli or difficulty to use kinesthetic information. In addition, the disagreement regarding behavioral deficits after prefrontal lesions was also present between animal studies and human clinical studies. Especially, human patients having damages in the prefrontal cortex did not exhibit delayed-response deficits and exhibit any impairment in memory functions.

S. Funahashi, *Dorsolateral Prefrontal Cortex*, Brain Science,
https://doi.org/10.1007/978-981-19-7268-3_3

To resolve these disagreements among different behavioral results and to understand the functions of both human and monkey prefrontal cortex using a common concept, Goldman-Rakic (1987) proposed an idea of "*working memory*" as a function of the prefrontal cortex. Working memory is described as a mechanism to hold necessary information temporarily "on line" to achieve a task. Working memory is an essential mechanism not only to perform the delayed-response task in monkeys but also to perform a variety of cognitive activities in humans. She proposed that behavioral deficits observed in both monkeys and humans having prefrontal lesions can be explained using the concept of working memory. She described the functions of the prefrontal cortex as follows in her review article:

> *The neural mechanisms resident in the principal sulcal cortex are not themselves necessary for spatial perception, long-term memory, or execution of motor responses; these functions are carried out by sensory analytic cortices of the parietal lobe, limbic structures, and premotor and motor structures, all of which are connected with the principal sulcus. Yet under particular circumstances,* i.e., *when information is not contained in the stimulus, the coordination, facilitation, or negation of these processes may be under prefrontal control. The prefrontal cortex can play this role because of its elemental capacity to access and hold "on line" information relevant to the task at hand. It seems possible that many integrated higher-order functions including language, concept formation, and planning for the future may be built on this functional element* (Page 406 in Goldman-Rakic 1987).

Although initially she used the term "*representational memory* or *knowledge*," she later used the term "*working memory*" to describe the mechanism to hold information online in mind in the service of comprehension, thinking, and planning (Goldman-Rakic 1995, 1996). After this proposal, the idea that working memory is a key concept to understand and explain prefrontal functions, especially functions of the dorsolateral prefrontal cortex, has been supported by numerous human neuroimaging studies and animal studies.

3.2 Short-Term Memory and Working Memory

The term "*working memory*" is described as a unique short-term active storage mechanism that is used to achieve a variety of cognitive activities including thinking, reasoning, judging, decision-making, and language comprehension. It has been known that human memory consists of two distinct systems; *short-term memory* and *long-term memory*. Since working memory is a mechanism to retain information for a short period of time, working memory is classified into short-term memory. However, working memory is usually differentiated from classical psychological definition of short-term memory. Short-term memory often refers to the mechanism for simple temporary storage of information with very limited capacity, whereas working memory refers to the mechanism having both the temporary storage and manipulation processes of information.

3.2.1 Modal Model of Human Memory

The notion that human memory consists of two distinct systems (short- and long-term memory) has been supported by studies using human amnesic patients. For example, the patient HM, who had damages in the medial temporal lobe, exhibited severe impairments in long-term memory, but exhibited normal ability in short-term memory (Milner 1964, 1982). On the other hand, the patient who had damages in the parieto-temporal junction exhibited impairments in short-term memory but exhibited normal ability in long-term memory (Warrington and Shallice 1969). These observations support that human memory consists of two distinct systems and that short-term memory is maintained by a distinct neural mechanism differed from the mechanism for long-term memory in the brain. To explain the relations between two memory systems and their functions in human memory, Atkinson and Shiffrin (1968) proposed a model of human memory system, which has been known as a *modal model* (Fig. 3.1).

The modal model has three basic memory components; sensory memory, short-term memory, and long-term memory. In this model, whole information acquired by sensory receptors is first stored in the sensory memory (sensory registers) for a short period of time (e.g., up to several hundred ms). In the sensory memory, attended information is kept during this period, while unattended information is lost. Therefore, only attended information is then sent to the short-term memory (short-term store, STS). In the short-term memory, each information is evaluated by rehearsing,

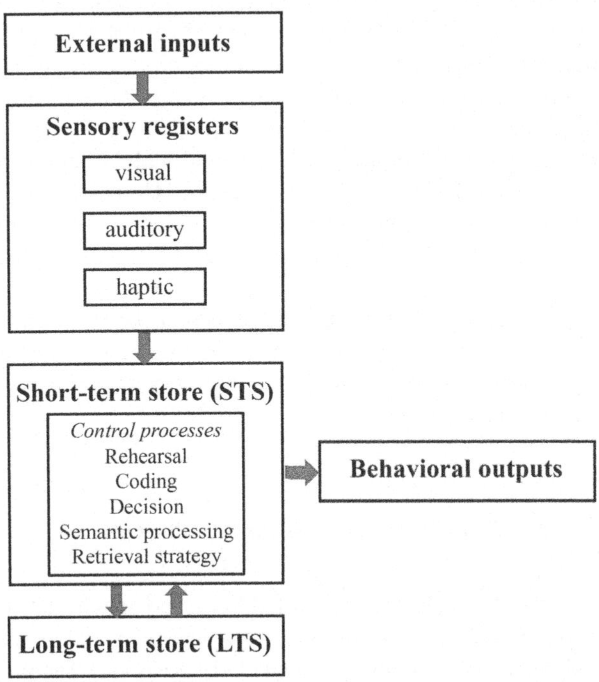

Fig. 3.1 Schematic drawing of the modal model proposed by Atkinson and Shiffrin (1968)

encoding, judging, and semantic processing and only selected information is transferred to the long-term memory (long-term store, LTS). The long-term memory stores those selected information as episodic or semantic memory for a long period of time. Thus, in the modal model, the short-term memory acts as a mechanism to select which information is to be sent to the long-term memory.

On the other hand, when the subject performs cognitive activities including thinking, reasoning, judging, decision-making, and language comprehension, information necessary for these activities is not only obtained from the external environment or the long-term memory or both but also needed to hold temporarily for processing to produce an appropriate output. Therefore, the short-term memory also acts as "working memory" to temporarily hold necessary information for processing. The modal model assumed that the short-term memory includes two important functions; one for selecting appropriate information to send it to the long-term memory and one for temporarily holding information for processing to create necessary information. To perform these two functions, it is expected that the short-term memory has a relatively larger amount of memory capacity. The capacity of the short-term memory had been measured by a memory span test and Miller (1956) showed that the capacity of the short-term memory is 7 ± 2 chunks, indicating that the capacity is very limited. Therefore, a serious question was raised, that is, how the short-term memory achieves these two functions simultaneously in spite of such a small storage capacity.

Further, the modal model encountered several major problems as Baddeley (2012) indicated. (1) Although this model assumed that information holding in the short-term memory guaranteed transfer to the long-term memory, Craik and Lockhart (1972) showed that deeper and more elaborate processing was crucial for this transfer. (2) The patient with defective short-term memory (a digit span of only two items and an absence of the recency effect in free recall) exhibited no impairment in long-term memory (Warrington and Shallice 1969). (3) According to the modal model, the short-term memory was thought to act as working memory and play important roles in cognitive activities. However, the capacity of the short-term memory had been shown to be very limited and was only 7 ± 2 chunks. If patients have defective short-term memory, they would be expected to exhibit severe intellectual and cognitive deficits. However, these patients did not exhibit these deficits. Therefore, questions were raised whether the short-term memory actually involves both selecting appropriate information to send it to the long-term memory and acting as "working memory" for achieving cognitive activities, in spite of such a small storage capacity.

3.2.2 *Baddeley and Hitch's Experiment*

To answer these questions, Baddeley and Hitch (1974) asked human subjects to perform both a serial verbal recall of a sequence of spoken digits and a grammatical judgment of visually presented English sentences simultaneously (dual-task condition) and examined whether the increase of sequence length of digits that the subject needed to recall affected the performance of the grammatical judgment or not. The rationale of their experiment was that, if the sequence length of digits that the subjects needed to recall increased beyond the maximum of the memory span (7 ± 2 chunks), they would exhibit some trouble in the performance of grammatical judgment. Because the serial digit recall task is likely to occupy the limited capacity of the short-term storage to varying degrees, this procedure was expected to produce functional deficits progressively in the grammatical judgment task as the sequence length of digits increased. The result showed that, when the sequence length of digits increased, response times to make judgments increased linearly with the increase of the number of digits. However, the subject's performance of the grammatical judgment task was stable and exhibited a very low error rate even at the heaviest memory load condition. Based on these results and results obtained by other studies, Baddeley and Hitch (1974) abandoned the assumption that "working memory" comprised a single unitary store, such as the short-term store in the modal model, and proposed an independent memory mechanism from the short-term memory and a model of working memory composed by multiple components.

3.3 Working Memory

3.3.1 Working Memory Model Proposed by Baddeley

Baddeley and Hitch (1974) first proposed the current concept of working memory. They defined working memory as a brain system that provides for temporary storage and manipulation of information that is necessary for performing complex cognitive tasks such as language comprehension, learning, and reasoning. Baddeley and Logie (1999) further described working memory as a mechanism that allowed humans to comprehend and mentally represent their immediate environment, retain information about their immediate past experience, support the acquisition of new knowledge, solve problems, and formulate, relate, and act on current goals.

The classical and the most influential model of working memory had been proposed by Baddeley (2000) (Fig. 3.2). Baddeley's original model of working memory consists of three components (Baddeley and Hitch 1974; Baddeley 1986). One master component is called the *central executive*, which is a central processor with a limited capacity and controls two slave components which are called the *phonological loop* and the *visuospatial sketchpad*. The *phonological loop* is a system for speech perception and language comprehension. This component includes mechanisms for temporary storage of speech-based information. The phonological loop is further fractionated into a passive phonological store and an active rehearsal system (Baddeley and Logie 1999). Although subvocal rehearsal is assumed to be necessary for maintaining information, the storage rather than the rehearsal of information is emphasized. On the other hand, the *visuospatial sketchpad* is a system for processing visuospatial information and information that

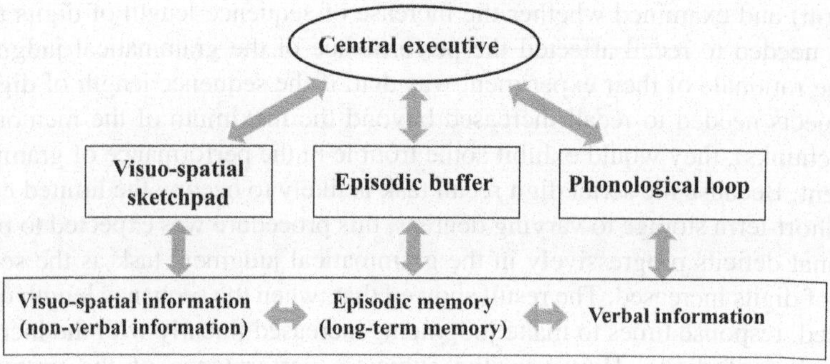

Fig. 3.2 Baddeley's model of working memory. From Baddeley (2000) with permission

cannot be processed by language. The visuospatial sketchpad includes mechanisms for temporary storage of visual images and spatial information. The visuospatial sketchpad is also further fractionated into a passive visual cache and the inner scribe, which is an active and spatially based rehearsal system (Baddeley and Logie 1999). Although the original model of working memory had two slave components (Baddeley 1983, 1986, 1996), Baddeley (2000) proposed a revised model of working memory that includes an additional slave component, that is the *episodic buffer* (Fig. 3.2). The episodic buffer is a temporary storage buffer with limited capacity and has functions to integrate information arose from a variety of sources including long-term memory. The episodic buffer is assumed to hold integrated episodes or chunks and not only acts as a buffer between the phonological loop and the visuospatial sketchpad, but also links working memory with perception and long-term memory. Thus, all three slave components include mechanisms for temporary storage of information. The phonological loop temporarily stores speech-based information, the visuospatial sketchpad temporarily stores visual images and spatial information, and the episodic buffer temporarily stores integrated episodes and chunks entered from perception and long-term memory. On the other hand, the *central executive* is the master component for coordinating and integrating operations of these three slave components to achieve a current goal. The central executive is thought to be a system for allocating a limited capacity of memory resources to each of the slave components based on their demands. If visuospatial processing becomes more demanding than language processing, more memory will be allocated to visuospatial processing. By doing this, an appropriate control process or strategy would eventually be selected to accomplish a current goal. Recently, Baddeley (2012) proposed a revised model of working memory, but its basic structure is the same.

3.3.2 *Other Models and Definitions of Working Memory*

Although the model proposed by Baddeley is well-known, several other models and definitions of working memory have also been proposed (see a book edited by Miyake and Shah 1999). Baddeley's definition of working memory seems to stress more on holding and processing information. The components of his working memory model are generally independent of long- and short-term memory systems. Similarly, Kieras et al. (1999) described working memory as a mechanism for encompassing the entire ensemble of temporary stored codes, knowledge representations, and procedures whereby information is maintained, updated, and applied for performing perceptual-motor and cognitive tasks. Young and Lewis (1999) also referred to working memory as mechanisms that maintain and provide access to information created or retrieved during the performance of a task.

However, there is a definition that working memory is an activated state of long-term memory. Focusing attention is considered to play a significant role in this process. Cowan (1999) described working memory as a temporarily activated subset of long-term memory by focusing attention. He proposed an "embedded-processes model" including five principles. One important principle is that information on working memory comes from hierarchically organized three faculties; long-term memory, the subset of long-term memory that is currently activated, and the subset of activated memory that is in the focus of attention and awareness. Another important principle is that, although activation of memory is time limited, the focus of attention is limited by the capacity, not by time. His definition of working memory is activated memory by attention and he considered the importance of attention (both voluntary and involuntary) and focusing attention in working memory processes. Engle et al. (1999) also considered that working memory is an active state of long-term memory. They described that working memory is a system consisting of units of long-term memory which are activated above the threshold by controlled attention.

On the other hand, there are more physiologically or computationally oriented definitions of working memory. Schneider (1999) defined working memory as short-term activation and connection changes that support maintenance, planning, and the organization of cognitive operations in the performance of short-term tasks. O'Reilly et al. (1999) defined working memory as controlled processing that involves active maintenance and/or rapid learning. In working memory, controlled processing is an emergent property of dynamic interactions of multiple brain systems.

Thus, although working memory is one kind of short-term memory, working memory is distinguished from the psychological definition of short-term memory. The definition and the explanation of working memory are somewhat different among researchers. However, working memory can be defined as the mechanism involving the control, regulation, and active maintenance of task-relevant information in the service of complex cognition including both novel and familiar skilled tasks (Miyake and Shah 1999).

3.3.3 *Working Memory in Animal Studies*

The original concept of working memory was first proposed by Baddeley and Hitch (1974) in human cognitive psychology. The term "*working memory*" has been extensively used to explain prefrontal functions. However, the term "*working memory*" had also been used in animal studies using rats and mice originally for explaining the functions of the hippocampus (Mahut 1971; O'Keefe and Nadel 1978; Olton et al. 1979; Olton and Papas 1979; Nadel and MacDonald 1980; Aggleton et al. 1986; Jarrard and Elmes 1982).

3.3.3.1 Working Memory Versus Reference Memory

The concept of working memory was constructed based on the results obtained from human memory studies. However, the term "*working memory*" had also been used in animal studies together with the term "*reference memory*" (Knowlton et al. 1989; Stokes and Best 1990; Ando and Ohashi 1991; Sakurai 1992; Bushnell and Levin 1993; Prior et al. 1997; Gresack and Frick 2003). In animal studies, "*working memory*" is defined as a memory system that retains specific information necessary only for the current trial, not for subsequent trials (Honig 1978; Olton et al. 1979). On the other hand, "*reference memory*" is defined as a memory system that retains particular information over many trials and even for an entire experiment that would last for several weeks or months (Honig 1978; Olton et al. 1979). The critical difference between working memory tasks and reference memory tasks is that, in the former, different stimuli guide different responses in different trials, while in the latter, the same stimuli guide the same choice responses repeatedly across many trials.

The delayed-response task is a typical example of working memory tasks (Fig. 3.3b, c). In this task, the subject is shown a baited position briefly during the cue period and is required to hold that position during the delay period. Since the baited position changes randomly from trial to trial, the subject is required to hold the spatial information during the delay period to make a correct choice. The delayed matching-to-sample task is another example of working memory tasks. In this task, the subject is first shown a sample stimulus briefly, required to hold the features of that stimulus during the delay period, and needed to select the same stimulus among others. Similar to the delayed-response task, since the sample stimulus varies randomly from trial to trial, the subject is required to hold information regarding the sample stimulus during the delay period. In both delayed-response and delayed matching-to-sample tasks, the subject needs to hold information necessary and specific only for the current trial, not for preceding or subsequent trials. Therefore, these tasks are classified as working memory tasks.

On the other hand, the visual discrimination task is a typical example of reference memory tasks (Fig. 3.3e). In this task, the subject's response to a particular stimulus is always associated with the reward throughout the entire experiment, while

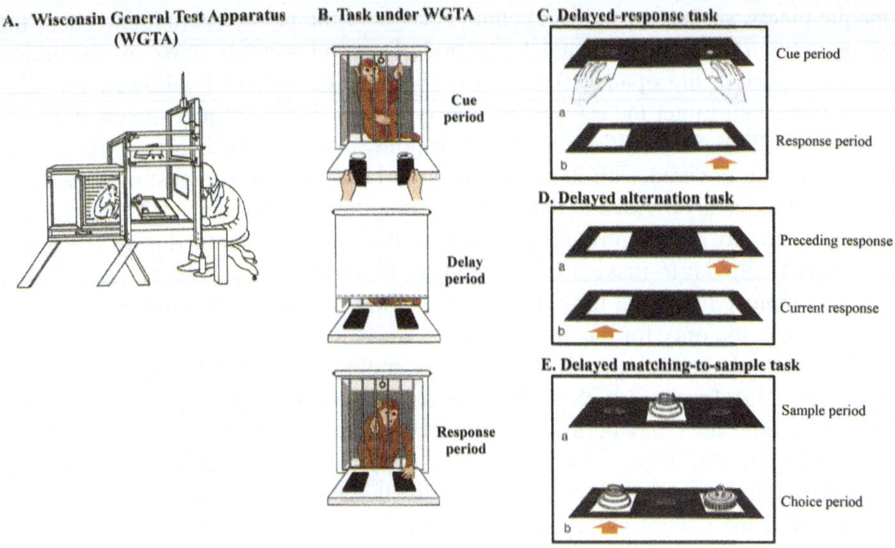

Fig. 3.3 (**a**) Wisconsin general test apparatus (WGTA). (**b**) An example of a monkey's behavioral performance in the WGTA during a trial of the delayed-response task. (**c**–**e**) Stimulus presentations in three examples of behavioral tasks performed using the WGTA. Orange arrows indicate the sides for the correct choice in these behavioral tasks

responses to other stimuli are not associated with the reward. Therefore, the subject is required to establish an association between a particular stimulus and a choice response and needed to maintain this association for entire experimental sessions. Since these features are similar to those of long-term memory in humans, reference memory tasks are often used to examine neural mechanisms of long-term memory in animals (Olton et al. 1979).

3.3.3.2 Spatial Working Memory Tasks

The delayed-response task requires the subject to hold the position where the reward was baited during the delay period and the subject utilizes memorized spatial information to guide a response. Retaining spatial information during the delay period is required for this task. Therefore, the delayed-response task is considered a spatial working memory task. Behavioral performances of the delayed-response task had been examined frequently using Wisconsin general test apparatus (WGTA) (Fig. 3.3a). In the WGTA, the subject is kept in a small cage. A small table is placed in front of the cage, on which necessary stimuli and a reward are presented (Fig. 3.3b, c). In the delayed-response task, an opaque screen is installed in front of the cage and this screen is first raised after several seconds of inter-trial interval. While the subject watches the table, the experimenter places a reward in one of two food wells on the table (cue period), and then covers both food wells with identical opaque plates, so that the subject cannot see where the reward is hidden. Then, the opaque screen is lowered and a delay period of several seconds starts. At the end of the delay period, the opaque screen is raised and the subject is allowed to select either food well to get the reward (response period). If the subject selects the food well where the reward is hidden (correct response), the reward is given. Since both food wells are covered with identical opaque plates, no clue suggesting where the reward is hidden is provided to the subject during the response period. Therefore, the subject needs to hold information regarding the baited position during the delay period and utilize it to make a correct choice. In addition, since the baited position varies randomly from trial to trial, the information needed to hold during the delay period is specific only for the current trial.

The delayed alternation task is another example of the spatial working memory task (Fig. 3.3d). In this task, the subject needs to select the right or the left food well alternately after the delay period. At the very first trial, the reward is hidden in both food wells and the subject can select either food well to get the reward. From the second trial, the reward is hidden in the food well opposite to the one that the subject has obtained the reward in the preceding trial. Therefore, the subject needs to hold which food well the reward was hidden in the preceding trial during the delay period. As a result, the monkey needs not to select either right or left food well but to select right and left food wells alternately.

The delayed matching-to-position task is also a spatial working memory task. In this task, the subject needs to hold the position where the sample stimulus was presented during the sample period and to make a response (e.g., lever release response) if the test stimulus is presented at the same position where the sample stimulus was presented after the delay. The position where the sample stimulus is presented changes randomly from trial to trial and the position where the test stimulus is presented is not always the same as the position where the sample stimulus is presented. The subject is required to judge whether the sample and the test stimulus are presented at the same position or not.

Thus, when the subjects perform these tasks, they need to hold spatial information specific only for the current trial, not for preceding or subsequent trials. Therefore, these tasks are classified as spatial working memory tasks.

3.3.3.3 Non-spatial Working Memory Tasks and Recognition Memory Tasks

The delayed matching-to-sample task and the delayed non-matching-to-sample task are called non-spatial working memory tasks, because visual attributes other than spatial features need to be retained during the delay period (Fig. 3.3e). In these tasks, one item (or an object) is presented as a sample stimulus during the sample period and, after the delay period, the sample item is presented again together with other items (choice period). The subject needs to select the sample item (delayed matching-to-sample task) or other items (delayed non-matching-to-sample task) to get a reward. The information that the subject needs to retain during the delay period is not the spatial information but non-spatial visual features, such as the object itself, its shape, or its color. Therefore, these tasks are called non-spatial working memory tasks (Aggleton et al. 1986; Raffaele and Olton 1988; Ando and Ohashi 1991; Gresack and Frick 2003).

Delayed matching- or non-matching-to-sample tasks are non-spatial working memory tasks. During the choice period of these tasks, both the sample item and a non-sample item are presented simultaneously. Therefore, the subject can select the sample item not only based on the memory of the sample item but also based on other information, such as familiarity or sameness. Classification of items based on "familiar or unfamiliar" or "similar or different" can often lead to the correct choice in these tasks. Therefore, these tasks are also called *recognition memory* tasks (Bachevalier and Mishkin 1986; Miller et al. 1991; Overman et al. 1992; Zola et al. 2000; Brown and Aggleton 2001; Squire et al. 2007; Eichenbaum et al. 2007). Recognition memory tasks have frequently been used to examine mnemonic functions of the hippocampus and the medial temporal lobe (Squire et al. 2007; Eichenbaum et al. 2007), while the delayed-response task has been widely used to examine prefrontal functions (Goldman-Rakic 1987; Fuster 2008; Rosenkilde 1979).

3.3.4 Advantages of Delayed-Response Task in Searching Neural Mechanisms of Working Memory

Neural mechanisms of working memory have been examined while animals performed a variety of spatial and non-spatial working memory tasks and recognition memory tasks. When neural mechanisms of working memory are examined neurophysiologically using animals, the delayed-response task offers several advantages over recognition memory tasks. First, in the delayed-response task, one simple visual stimulus can be used as the spatial cue and presented at various positions in the visual field across trials. Difference in the task parameter across trials is only the position of the visual stimulus. Spatial position of the visual stimulus is the only parameter needed to be considered to interpret neural activity. Secondly, the position of the visual stimulus is easier to change across trials and is described more accurately and quantitatively compared with the physical parameters of items and objects (e.g., color, shape, and size) used in recognition memory tasks. When the subject's head is fixed and when the subject is required to maintain fixation during the trial, the position of the visual stimulus can be accurately expressed in retinotopic coordinates. Thirdly, monkeys are able to report the position of the visual stimulus on the panel or the monitor by pointing by hand or eye movements. Pointing behaviors and eye movements are easy to quantitatively analyze behavioral parameters (e.g., reaction times, response times, and velocities), describe the accuracy of these behaviors, and depict movement trajectories during the response period. Thus, the delayed-response task has several advantages when detailed neural mechanisms of the working memory process are examined in the brain using animals.

3.3.5 *Working Memory As an Important Concept for Prefrontal Functions*

Jacobsen (1936) first shown that monkeys with bilateral prefrontal lesions exhibited a severe and long-lasting impairment of delayed-response performances. Since then, it had been shown that monkeys with bilateral lesions of the lateral prefrontal cortex consistently exhibit behavioral deficits in the tasks with delay, which include the delayed-response task, the delayed alternation task, and the delayed matching-to-sample task. These tasks require short-time storage of spatial or non-spatial information during the delay period. Therefore, the function of the lateral prefrontal cortex was thought to be short-term memory. Subsequent studies indicated that bilateral lesions restricted only in the middle one-third of the principal sulcus area are enough to produce severe and long-lasting impairments of delayed-response performances (Butters and Pandya 1969). It had been also shown that spatial modality is an important factor to interpret delayed-response deficits in animal studies.

However, human patients having damage in the prefrontal cortex exhibit disturbances in a variety of functions and abilities including attention, synthetic reasoning, and planning, an ability to grasp an essence of a situation, the use of past experience, the regulation of the behavior by verbal instructions, the initiative, the spontaneity, verbal and constructional fluency, spatial orientation, and global features of personality (Luria et al. 1964; Stuss and Benson 1986; Goldman-Rakic 1987; Fuster 1997). The reasons why a variety of symptoms are observed in human patients having damages in the prefrontal cortex were thought to be that the human prefrontal cortex might include functionally different multiple systems and that each patient might have a damage in the different areas of the prefrontal cortex with different sizes. In fact, anatomical studies of the prefrontal cortex of monkeys and humans showed that the prefrontal cortex consists of cytoarchitectonically different areas (Petrides and Pandya 1994). Lesion studies using monkeys showed that the prefrontal cortex includes functionally different areas because the damages to different areas produced different behavioral deficits (Rosenkilde 1979). Therefore, to understand prefrontal functions in detail, it is necessary to know what symptoms or behavioral deficits can be observed when the damage is localized within a single cytoarchitectonic area of the prefrontal cortex.

When focusing on the function of the cortex within and surrounding the principal sulcus (Area 46), monkeys with bilateral lesions of this cortical area have consistently exhibited delayed-response deficits. The delayed-response task requires an animal's ability to hold necessary information "on line" during the delay period and to use internalized representation of the visual cue to guide an appropriate behavior during the response period because no external stimulus provides a correct response to the animal during the response period. Therefore, the important function of the dorsolateral prefrontal cortex was thought to be short-term memory. However, there had been inconsistency between the results obtained by animal lesion studies and clinical observations of human frontal patients. For example, delayed-response deficits had been frequently observed in monkey studies but were not observed in human prefrontal patients. In addition, controversies regarding the psychological processes necessary for delayed-response performances and the source of the difficulty that lesioned monkeys exhibited had been present among prefrontal researchers.

To solve these inconsistencies, Goldman-Rakic (1987) proposed the idea that the lateral prefrontal cortex preeminently engages only when internalized or inner models of reality are used to regulate the behavior. She used the term *representational memory* or *knowledge* to describe short-term internalized representation. She explained behavioral deficits observed in human prefrontal patients as well as monkeys having lesions in the cortex within and surrounding the principal sulcus based on her concept of *representational memory*. For example, one feature of behavioral deficits observed in human patients is distractibility, which is thought to be an excessive dependence on or sensitivity to external stimuli, not on internalized representation. However, distractibility can be explained by the deprivation of access to internalized representation or representational memory. The deprivation of these functions forces the patients to rely strongly on external stimuli presented at the response. "Out of sight, out of mind" well describes the behavior of human prefrontal patients as well as monkeys having prefrontal lesions.

The perseveration of incorrect responses is another feature of behavioral deficits. This feature is frequently observed in the Wisconsin card sorting test and the Stroop test in human prefrontal patients. Goldman-Rakic indicated that perseverative behavior can also be explained by the loss of internalized representation and representational memory for guiding behavior. Wisconsin card sorting test uses 128 cards ($4 \times 4 \times 4 \times 2$), each of which consists of 1–4 drawings of one of 4 shapes (circle, triangle, plus, star) with one of 4 colors (red, green, yellow, blue). The subject is required to continuously pick up a card from a stack of 128 cards and sort it based on a particular category (either color or shape or number). The experimenter determines the category to sort the card and sometimes changes the sorting category without notification to the subject. Therefore, the subject is required to flexibly adapt to the change of the sorting category by trial-and-error fashion and to sort the cards based on either current or new categories. Prefrontal patients are known to be able to sort the cards continuously if the sorting category is fixed. However, when a new sorting category is introduced unexpectedly, they exhibit difficulty to change the sorting category (perseveration) and to find a new category by trial and error.

Similarly, in the Stroop test, a name of a color (e.g., BLUE) is written by a colored ink (e.g., red ink) on a card (e.g., **BLUE**) and the subject is required to quickly answer the color of the ink (e.g., red). When the colored word "**BLUE**" has abruptly appeared, answering "blue" is an impulsive and prepotent response. However, the subject is required to inhibit an impulsive and prepotent response by answering "blue" and make a correct answer "red." Prefrontal patients are known to show difficulty to inhibit an impulsive response and exhibit a prepotent response "blue," which corresponds to perseverative behavior.

Similar to distractibility, perseverative behaviors observed in human prefrontal patients in the Wisconsin card sorting test and the Stroop test can also be explained by the deprivation of access to internalized representation or representational memory for guiding appropriate behaviors. The deprivation of these functions forces the patients to rely strongly on external stimuli presented at the response, which is a typical behavioral pattern that prefrontal patients exhibit. Similar behavioral patterns are observed in monkeys having lesions in the lateral prefrontal cortex when they perform the delayed-response task. These monkeys often exhibit the selection of the position not based on the internalized information (memory of the cue position) but guided by external stimuli at the response (random selection). Thus, Goldman-Rakic's proposal was that the lateral prefrontal cortex preeminently engages only when internalized or inner models of reality are used to regulate the behavior. Her proposal would explain behavioral deficits observed in both human prefrontal patients and monkeys having prefrontal lesions.

Goldman-Rakic (1987) introduced "*representational memory* or *knowledge*" as a concept to understand functions of the lateral prefrontal cortex. She proposed that the prefrontal cortex has mechanisms to hold appropriate information "on line" as internalized representation (*representational memory*) for a short period of time and to use this representation to guide appropriate behavior during the response period. Goldman-Rakic initially used the term "*representational memory*" to describe the mechanism to access and internally hold information relevant to the task "on line." However, she subsequently used the term "*working memory*" to describe this mechanism (Goldman-Rakic 1995, 1996), because of the similarity between the concept of working memory and the concept of representational memory.

Since then, the idea that working memory is an important concept to understand and explain the functions of the lateral prefrontal cortex has been widely accepted and supported by numerous human neuroimaging studies (Curtis 2006). For example, Sweeney et al. (1996) used the PET imaging and showed activation of the dorsolateral prefrontal cortex while human subjects performed an oculomotor delayed-response task. Courtney et al. (1996) examined cortical activation by PET using the object and the spatial working memory tasks and showed that regional cerebral blood flow (rCBF) increased in the inferior frontal cortex during the object (face) working memory task, while rCBF increased in the superior frontal sulcus in the spatial working memory task. Courtney et al. (1997) further showed sustained activation in the lateral prefrontal areas (anterior middle frontal gyrus, posterior middle, and inferior frontal gyri) during the delay period by fMRI while human subjects performed the object working memory task. Belger et al. (1998) used fMRI and showed that the dorsolateral prefrontal cortex (right middle frontal gyrus) was activated during spatial working memory performances, while bilateral middle frontal gyrus and the left inferior frontal gyrus were activated during object working memory performances. Postle et al. (2000) also showed the contribution of the dorsolateral prefrontal cortex in working memory processes, especially for the manipulation of the information held in working memory. Thus, the contribution of the lateral prefrontal cortex has been widely reported by human neuroimaging studies, which supports the idea that working memory is an important concept to understand and explain the functions of the lateral prefrontal cortex.

3.4 Working Memory in the Prefrontal Cortex

3.4.1 Search for Neural Correlates of Working Memory in the Prefrontal Cortex

To perform the delayed-response task correctly, the subject is required to hold visuospatial information during the delay period. Therefore, some neural mechanism is required to hold necessary visuospatial information temporarily during the delay period, which corresponds to the working memory. Delayed-response deficits are not caused by sensory impairments, motor impairments, motivational deficits, or global memory loss. Therefore, delayed-response deficits are thought to be caused by impairments of working memory functions. Lesion studies show that the cortex within and surrounding the principal sulcus (Walker's area 46) is a critical brain area for performing the delayed-response task specifically for visuospatial domain. Therefore, it is expected that physiological investigations in the brain area within and surrounding the principal sulcus provide important information for the neural mechanism of working memory.

By electrophysiological studies, Stamm and Rosen (1972) found surface negative sustained cortical potential shifts, which was called contingent negative variation (CNV), on the prefrontal cortex while monkeys performed a delayed-response task. Although surface negative potential shifts were observed before the cue presentation, following the behavioral response, and during the delay period, sustained negative potential shifts during the delay period (CNV) were recorded only from the prefrontal cortex. The magnitude of CNV was significantly correlated with the level of correct performance. Therefore, they interpreted CNV as a neural correlate of short-term spatial memory function. In addition, they applied anodal polarization to the surface of the prefrontal cortex and obtained higher acquisition rates in delayed-response training in monkeys. These results suggested that neurons in the prefrontal cortex change their excitability during delayed-response performances and that the excitability of prefrontal neurons is maintained during the delay period.

3.4.2 Kubota's and Fuster's Studies in the Prefrontal Cortex

Kubota and Niki (1971) first reported single-neuron activity recorded from the dorsolateral prefrontal cortex while wake and behaving monkeys performed a delayed alternation task. In their task, monkeys pressed two keys alternately by hand after a delay period. They found two types of prefrontal neurons (E and D units). E units were activated before the onset of the triceps EMG for key pressing. The magnitude of the activity depended on which key was pressed and the activation continued until the end of the key press. Similar activation was observed during simple alternating key pressing outside the task. On the other hand, D units were activated during the delay period and this activation stopped at the beginning of the response period. This activation was not observed during simple alternating key pressing. Based on these results, they concluded that E units were causally coupled with the initiation of voluntary key pressing, while D units were responsible for maintaining the state of the delay until voluntary key pressing started. Since discharge rates of D units during the delay period were not different between right and left key pressing, they thought that D units were not related to the memory storage or retrieval of the memory for the lever pressing.

> *D units are hardly correlated with the memory storage or retrieval of the memory for the lever pressing. Frequency of D unit during delay phases is not apparently different between right and left lever pressings. However, this interpretation doses not exclude the possibility that the activities of neurons in the prefrontal cortex represent a memory function related to the choice of the correct performance on the basis of immediate past experience,* i.e., *remembering the spatially directed response on the preceding trial.* (Page 346 in Kubota and Niki 1971)

Fuster and Alexander (1971) also reported single-neuron activity in the dorsolateral prefrontal cortex and the mediodorsal nucleus (MD) of the thalamus while monkeys performed a delayed-response task under the condition similar to that in the Wisconsin General Test Apparatus. Food wells were placed at the left and the right on the table and monkeys were required to select the baited food well by hand after the delay period. They found neurons exhibiting activation during the cue period and during the delay period. An important observation was that some prefrontal neurons exhibited and maintained higher discharge rates throughout the delay period compared with discharge rates during the inter-trial interval.

> *The temporal patterns of firing frequency observed in prefrontal and thalamic units during cue and delay periods suggest the participation of these units in the acquisition and temporary storage of sensory information which are implicated in delay response performance. Their function, however, does not seem to be the neural coding of information contained in the test cues, at least according to a frequency code, for we have not found any unit showing differential reactions to the two positions of the reward.* (Page 654 in Fuster and Alexander 1971)

Fuster and Alexander (1971) concluded that excitatory activity of prefrontal and MD neurons was related to the focusing of attention on the information that was placed in temporary memory storage for prospective utilization. They thought that

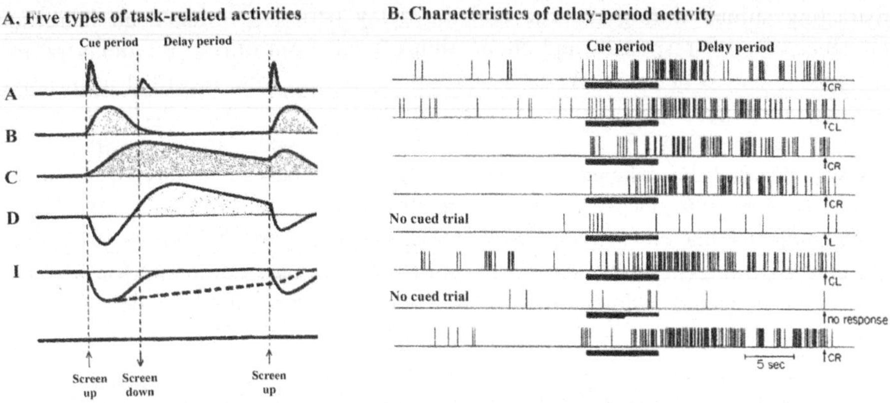

Fig. 3.4 (**a**) Schematic examples of five temporal patterns of delay-period activity observed in the prefrontal cortex during the delayed-response task. (**b**) Delay-period activity in rewarded trials and non-rewarded trials. CR, CL, and L indicate correct right response, correct left response, and left response with no reward, respectively. From Fuster (1973) with permission

excitatory activity was an integral component of basic functions of prefrontal-MD neural circuits in active attention.

Further, Fuster (1973) examined single-neuron activity (328 neurons) recorded from the dorsolateral prefrontal cortex while monkeys performed delayed-response tasks under various conditions (e.g., with reward, without reward, and in action-withholding condition). He observed five types (Type A, B, C, D, and I) of task-related prefrontal activity (Fig. 3.4a). Type A (6%) exhibited transient activations at the onset of the cue period, at the initiation of the delay period, and at the onset of the response period. All these timings corresponded to either up or down of the opaque screen and to either initiation or termination of hand movements. Therefore, Type A was thought to be related to these sensory or motor phenomena. Type B (12%) exhibited activation throughout the cue period and during the response period. Monkeys actively searched the objects and events in the testing apparatus by eye and hand movements during the cue and response periods. Therefore, Type B is thought to be related to these movements. Type C (33%) exhibited activation from the beginning of the cue period and maintained activation throughout the delay period. Some neurons exhibited higher discharge rates during the entire delay period which lasted for longer than 60 s. However, no evidence of particular relations was found between the degree of activation during the delay period and the side of the choice response. Type D (17%) exhibited inhibition during the cue period but excitation during the delay period. Type I (12%) exhibited inhibitory responses during the cue and the response periods. When delayed-response trials were introduced without reward (no reward was placed in either of food wells during the cue period), Types A and B exhibited similar activations as normal trials with reward, while Types C, D, and I exhibited no activation, especially during the delay period. In error trials and trials with no visual cue (Fig. 3.4b), the activation was relatively weak or absent compared with correct trials and normal trials. Presentation of distracting stimuli during the cue and the delay periods produced an increased incidence of error responses and attenuation of activation during the delay period in Types C and D neurons. When the prefrontal activity was recorded from naive untrained monkeys, only neurons exhibiting either no task-related activity or activity assigned to Types A, B, or I categories were observed. However, after these monkeys learned the delayed-response task, all five types of activity ($n = 79$; A, 5%; B, 14%; C, 35%; D, 15%; I, 9%) were observed. Based on these results, Fuster (1973) indicated that elevated discharge during the delay period observed in Types C and D was contingent on the significance of the cue as a memorandum for prospective action. In accord with his previous conclusion (Fuster and Alexander 1971), he concluded as follows:

> *An attractive possibility is that, between cue and response, the prefrontal cortex and the mediodorsal nucleus of the thalamus are bound by reverberating activity essential for retention of information about the cue. However, the absence of a relationship between unit reactions and laterality of the cue suggests that information about the latter is not retained in such a system. It would appear more likely that, during the delay, this frontothalamic system tonically facilitates the function of structures more directly involved in visual discrimination; namely, the visual association cortex and limbic formations, which are not only sources of afferents to the system but targets of some of its efferent. This hypothesis incorporates the function of the frontothalamic system in a simple and plausible view of short-term memory,* i.e., *the view that this form of memory basically consists of the continued facilitation of neural activity in cerebral structures recently engaged in sensory processing.* (Page 75 in Fuster 1973)

3.4.3 Differential Delay-Period Activity.

In the delayed-response task, the baited position changes between right and left randomly from trial to trial. Therefore, the subject needs to actively retain information regarding the baited position during the delay period. If delay-period activity is a neural correlate of the mnemonic process that retains spatial information during the delay period, this activity should exhibit differential patterns of activation depending on the difference in the baited positions. Although Kubota and Niki (1971) found differential activation in prefrontal neurons during the response period of the delayed alternation task, studies by Kubota and Niki (1971), Fuster and Alexander (1971), and Fuster (1973) did not find any differential activity during the delay period depending on the difference of the baited positions.

Niki (1974a) found directionally selective firing in prefrontal neurons in correlation with the left and the right trials of the delayed alternation task. He observed three types of prefrontal activity (Types I, II, and III). Type I, possibly corresponded to E units of Kubota and Niki (1971), exhibited excitation prior to the release of the hold key. Type II, possibly corresponded to D units, exhibited inhibition during the response to the chosen key. Type III exhibited excitation after the initiation of the choice response. He observed directionally selective firings in half of these prefrontal neurons. He indicated that directionally selective Types I and II neurons would involve in the initiation and execution of responses to a given position, while

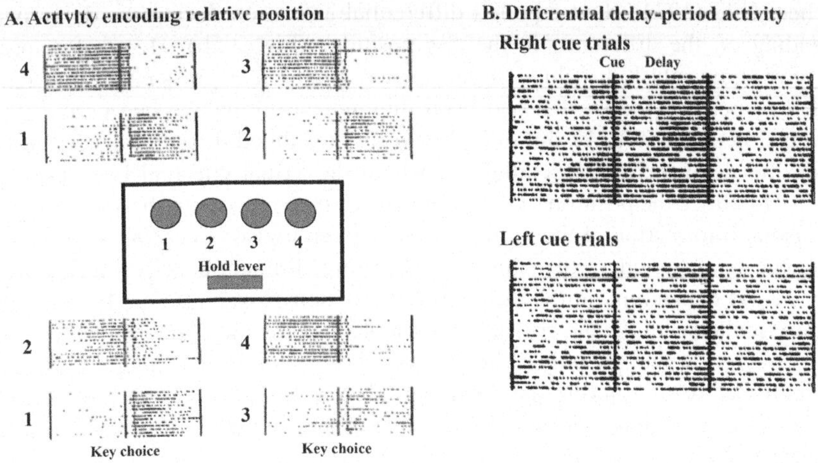

Fig. 3.5 (**a**) Prefrontal activity representing relative positions of the visual cues during delayed alternation performances. From Niki (1974b) with permission. (**b**) Differential delay-period activity observed in prefrontal neurons during delayed-response performances. From Niki (1974c) with permission

directionally selective Type III neurons might involve in feedback from the response. Importantly, he also found few prefrontal neurons which exhibited directionally selective activity during the delay period.

Niki (1974b) further examined whether directionally selective activity of prefrontal neurons encoded relative or absolute direction of the response during delayed alternation performances. To examine this issue, he prepared a panel with four horizontally arranged choice keys and selected two keys as choice keys in any given block of trials (Fig. 3.5a). By systematically selecting two choice keys from the four keys, the same key became either the right or the left choice key depending on which keys were selected as a pair. For example, the key 2 in Fig. 3.5a became the right choice key when the key 1 was selected as a pair, while the same key 2 became the left choice key when the key 3 was selected as a pair. By examining directional selectivity of the same neuron's activity in different combinations of choice keys, he found that, although all directionally selective activities ($n = 111$) were affected by the spatial position of the choice key, most of these neurons ($n = 91$) exhibited directionally selective activity during the response period and encoded the absolute position of the choice key. In these neurons, the key position in space with respect to the subject's body affected directional selectivity and the difference of discharge rates during the subject's reaching to different choice keys. On the contrary, directionally selective activity during the delay period was affected by the relative position of the choice keys, although the number of these neurons were small ($n = 20$).

Then, Niki (1974c) first reported differential activation during the delay period depending on the difference in the cue positions (Fig. 3.5b). He used a manual delayed-response task, in which monkeys were required to press either the right or the left response window that had been illuminated during the cue period. Some prefrontal neurons (6%, 17 out of 273 neurons) exhibited different discharge rates during the delay period depending on where the visual cue had been presented (Fig. 3.5b). Since the position of the visual cue changed randomly between right and left, spatial information of the visual cue on a given trial was necessary to maintain during the delay period to solve this task. Therefore, differential activation during the delay period depending on the different cue positions was considered to be spatial memory-related activity and critically contribute to performing the delayed-response task correctly. Although only a small proportion of prefrontal neurons exhibited differential activation during the delay period, Niki (1974c) was the first to report showing that prefrontal neurons exhibited spatial short-term memory-related activity during delayed-response performances.

3.4.4 Delay-Period Activity Is a Neural Correlate of the Mnemonic Process in the Prefrontal Cortex?

In delayed-response performances, subjects are required to retain not only information regarding where the visual cue was presented but also information regarding where the response behavior would be directed. If differential delay-period activity reflects a neural process related to temporarily holding information necessary for delayed-response performances, this activity would represent information regarding either the location of the visual cue or the direction of the behavioral response.

Niki and Watanabe (1976a, b) examined this issue. They trained monkeys to perform two kinds of the delayed-response task and the conditional position task with delay. They used a panel with four response windows at up, down, left, and right positions (Fig. 3.6a). In the delayed-response task with right-or-left choice, the visual cue is presented at either right or left response window and the subject is required to press the corresponding key during the response period, while in the delayed-response task with up-or-down choice, the visual cue is presented at either up or down response window and the subject is required to press the corresponding key. In the conditional position task, the subject is required to press either right or left response window when the visual cue is presented at up or down response window, respectively. The rationale behind this experiment is that, if delay-period activity represents information regarding the location of the visual cue, significant delay-period activity would be observed in the trials when the visual cue is presented at one response window (e.g., down position) both in the delayed-response task with up-or-down choice and in the conditional position task, regardless of the direction of the behavioral response (Fig. 3.6b). On the other hand, if delay-period activity represents information regarding the direction of the behavioral response, significant

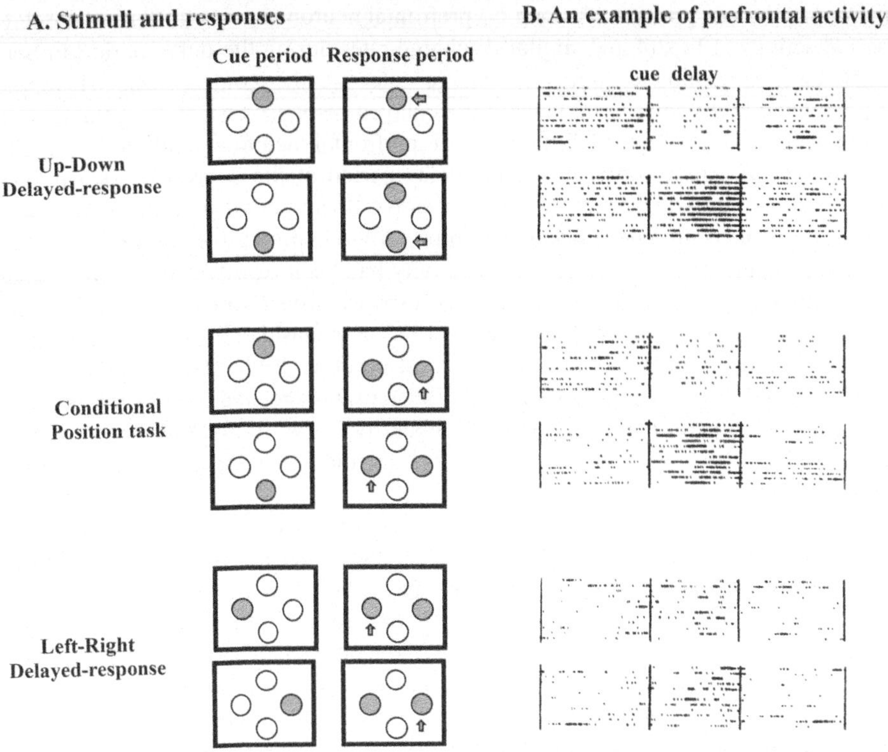

Fig. 3.6 (**a**) Schematic drawings of stimulus presentations during the cue and the response periods for three behavioral tasks used by Niki and Watanabe (1976a). Small arrow indicates the correct response position. (**b**) An example of delay-period activity representing spatial information of the visual cue, not representing response directions. From Niki and Watanabe (1976b) with permission

delay-period activity would be observed in the trials when the behavioral response was directed toward one direction (e.g., leftward direction) both in the delayed-response task with right-or-left choice and in the conditional position task, regardless of where the visual cue is presented. They compared the directional selectivity of delay-period activity in the same neuron among three task conditions and found that 78% (25/32) of differential delay-period activity represented the location of the visual cue (Fig. 3.6b), whereas the remaining 22% (7/32) represented the direction of the behavioral response. These results strongly support that differential delay-period activity is a neural correlated of the mechanism to temporarily retain information necessary for delayed-response performances.

To further confirm that delay-period activity is a neural correlated of the mechanism for temporarily retaining information, Kojima and Goldman-Rakic (1984) compared delay-period activity of the same prefrontal neuron while the monkey performed the delayed-response task and the control task. In the control task, as the cue was presented throughout the delay period, the monkey was not required to memorize the cue position. Among 62 prefrontal neurons having differential delay-period activity, 12 exhibited similar discharge rates during the delay period in both tasks and 32 exhibited higher discharge rates in the control task. However, 12 exhibited higher discharge rates in the delayed-response task than in the control task. Thus, although a small number of prefrontal neurons exhibited spatially discriminative activity only during the delay period of the delayed-response task, this result supports that spatially discriminative delay-period activity would be a neural correlate of the mechanism for temporarily retaining information.

Directionally selective delay-period activity had been reported in several studies while monkeys performed manual delayed-response tasks (Niki 1974c; Kojima and Goldman-Rakic 1982, 1984; Batuev et al. 1985; Funahashi et al. 1997) and manual delayed alternation tasks (Niki 1974a, b; Carlson et al. 1990, 1997). Carlson et al. (1997) found that a large proportion (73%) of task-related prefrontal neurons exhibited delay-period activity and that a majority (79%) of neurons having delay-period activity exhibited spatial selectivity and did not respond to any sensory stimuli unrelated to task performances.

Delay-period activity reported by Fuster (1973), Niki (1974c), and Niki and Watanabe (1976a) showed tonic sustained excitation during the delay period. The length of the delay period is different among studies. For example, Fuster (1973) used rather long delay length (e.g., up to 32 s), while Niki (1974c) and Niki and Watanabe (1976b) used the delay length of 3 s. Kojima and Goldman-Rakic (1982) examined the effects of the delay length on the temporal pattern of delay-period activity. They used delay lengths of 2, 4, 8, or 12 s and found that a majority of prefrontal neurons (62%) exhibited a fixed discharge pattern during the delay period regardless of the delay length, while a minority (33%) altered the discharge pattern during the delay period in relation to the delay length. Similar observations were reported by Batuev et al. (1985). Interestingly, a small number of prefrontal neurons exhibited spatially selective delay-period activity and the most of these neurons exhibited the peak activity in the first few seconds of the delay period (Kojima and Goldman-Rakic 1982). Stamm (1969) had shown that, when electrical stimulation was applied in the principal sulcal area during the first few seconds of the delay period, monkey's performance of the delayed-response task dropped to the chance level. These results support the importance of delay-period activity as a neural correlate of mnemonic processes in the prefrontal cortex.

3.4.5 Differential Delay-Period Activity As a Neural Correlate of Mnemonic Process?

Differential delay-period activity is an important finding to understand the mnemonic functions of the prefrontal cortex. However, this idea had some weaknesses to be fully accepted. For example, the studies by Kubota and Niki (1971), Fuster (1973), Niki (1974a, b, c), and Niki and Watanabe (1976a) used only two positions as the positions of the visual cues and two directions as response directions (usually right and left). To further confirm the importance of delay-period activity in spatial mnemonic processes, directional selectivity of delay-period activity should be examined under the behavioral tasks using multiple cue positions and analyzed spatial or directional selectivity more quantitatively. Further, although directional selectivity of prefrontal activity was affected by absolute or relative position of the response key, the magnitude of prefrontal activity might be affected by the key position in space with respect to the subject's body or head or by the difference of the subject's reaching to different choice keys (Niki 1974b).

Although monkey's hand and arm movements were controlled, monkey's eye movements were not monitored and controlled during behavioral performances. Eye movements and the direction of the gaze have been known to affect the magnitude of prefrontal activity. Many prefrontal neurons exhibit eye movement-related activity and show directionally selective activation or inhibition (Joseph and Barone 1987; Barone and Joseph 1989; Boch and Goldberg 1989; Funahashi et al. 1991; Funahashi 2014). When a visual stimulus is unexpectedly presented somewhere in the visual field, animals tend to exhibit reflexive orienting response to this stimulus and make a saccade to the stimulus. Therefore, excitatory response observed during the cue period might be the response related to the orienting saccade caused by the presentation of the visual cue. In addition, when monkeys and humans perform a reaching behavior toward a target by hand, an eye movement toward the target is first initiated to capture the target, then the hand-reaching movement is initiated (Biguer et al. 1982). The coupling of an eye movement and a hand-reaching movement is usually observed during the reaching behavior toward a target. Therefore, excitatory activity observed during the response period in Niki (1974a, b, c) and Niki and Watanabe (1976) might be activity related to the eye movement. In addition, non-differential activation during the delay period reported by Fuster (1973) might also be eye movement-related activity, since eye movements were not controlled during the experiment and since monkeys could make eye movements freely toward any direction at any moment during the long delay period.

In addition, prefrontal neurons are known to exhibit gaze-related activity, which is tonic sustained activation during gazing at a certain position (Suzuki and Azuma 1977; Suzuki et al. 1979; Boussaoud et al. 1993). The magnitude of gaze-related activity changes depending on the gaze direction (Boussaoud et al. 1993). In addition, it has been shown that the gazing angle affects the magnitude of visual, mnemonic, and motor activity. This effect is called "angle-of-gaze" effect and has been observed in the posterior parietal cortex (Andersen and Mountcastle 1983; Andersen et al. 1985a, b; Squatrito and Maioli 1996; Andersen et al. 1997), the visual cortex (Galletti and Battaglini 1989; Trotter and Celebrini 1999; Rosenbluth and Allman 2002), and the premotor cortex (Boussaoud 1995; Mushiake et al. 1997; Boussaoud et al. 1998). Since posterior parietal neurons project to the prefrontal cortex (Pandya and Seltzer 1982; Cavada and Goldman-Rakic 1989; Petrides 2005), prefrontal neurons could also exhibit angle-of-gaze effects on both visual and mnemonic activity. Since monkey's gaze position was not controlled during the delay period in the previous studies, monkeys might look at the position where the visual cue was presented and maintain gazing at that position during the delay period. Gazing at different spatial positions might produce different angle-of-gaze effects on delay-period activity. Therefore, differential delay-period activity observed in the previous studies might be a product of the angle-of-gaze effect.

Thus, to make a firm conclusion that differential delay-period activity is a neural correlate of spatial mnemonic mechanism, further studies were needed to examine characteristics of delay-period activity under multiple cue conditions and well control of the eye movement during the delay period.

3.5 An oculomotor Delayed-Response Task

3.5.1 The Oculomotor Delayed-Response Task

Funahashi et al. (1989) developed an oculomotor version of the delayed-response task (an oculomotor delayed-response task, ODR task) (Fig. 3.7) to further examine functions and characteristics of delay-period activity under the behavioral task using multiple cue positions and under the control of monkey's eye movements during the delay period. In the ODR task, the subject is required to make a memory-guided saccade to the position where the visual cue was presented after the delay period. The ODR task is a modification of a memory-guided saccade task originally introduced by Hikosaka and Wurtz (1983).

In the ODR task, the monkey is set in a primate chair in a dark, sound-attenuated small room and faces a TV monitor, on which a fixation point and visual cues are

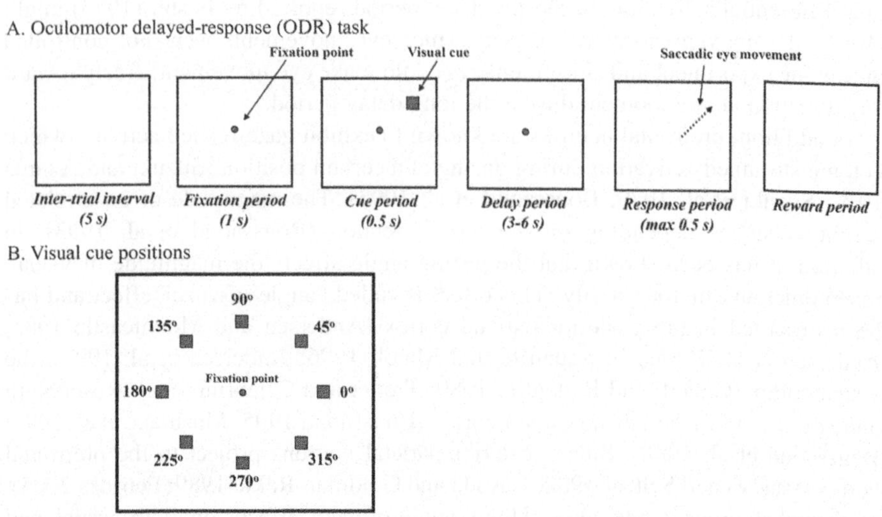

Fig. 3.7 (**a**) Schematic drawing of the temporal sequence of the oculomotor delayed-response (ODR) task. (**b**) Visual cue positions used for the ODR task

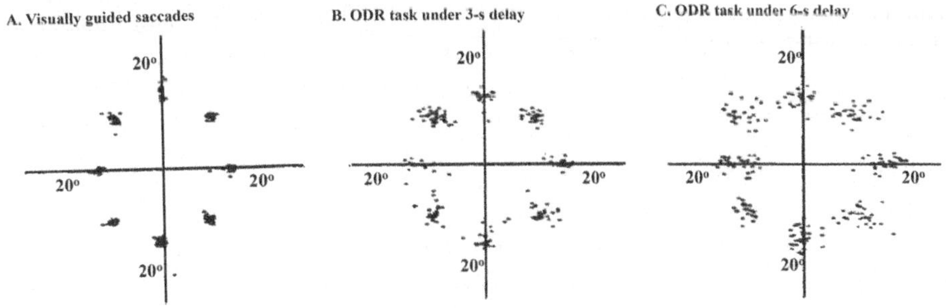

Fig. 3.8 Distributions of saccade ends in different task conditions. (**a**) Visually guided saccade task (control task). (**b**) The ODR task under the 3-s delay condition. (**c**) The ODR task under the 6-s delay condition. Eccentricity of visual cue positions was 13° in visual angle. From Funahashi et al. (1989) with permission

presented. Monkey's eye movements are continuously monitored by a magnetic search coil technique (Robinson 1963) or a technique using an infra-red camera (e.g., EyeLink, ISCAN). The monkey is trained to look at a fixation point as long as the fixation point is presented on the monitor. After an inter-trial interval of several seconds, a fixation point (a small visual stimulus) is presented at the center of the monitor. If the monkey looks at the fixation point for 1 s (fixation period), a visual cue (e.g., a white-filled circle) is presented for 0.5 s (cue period) at one of eight predetermined locations around the fixation point (e.g., the eccentricity was 17°) (Fig. 3.7b). The monkey is required to maintain fixation at the fixation point during the cue period and the subsequent delay period. The length of the delay period is usually 3 s but sometime 6 s or 12 s. At the end of the delay period, the fixation point is extinguished, which is the Go signal for the monkey to make a saccade within 0.4 s (response period) to the position where the visual cue was presented. No visual stimulus is present during the response period. If the monkey makes a correct saccade to the cue position, a drop of a liquid reward is given. Figure 3.8 shows distributions of saccade ends in the visually guided saccade task (A) and in two different delay conditions of the ODR task (B and C). Although distributions of saccade ends became wider in the 6-s delay condition compared with those in the 3-s delay condition, saccade ends in both the 3-s and the 6-s delay conditions converged around the target positions in all cue conditions.

In the ODR task, the monkey's head is immobilized by a head restraining instrument and the monkey is required to maintain gazing at the fixation point during the cue and the delay periods. Differed from the delayed-response task using hand-reaching movements in the previous studies, the ODR task allows to present visual cues at multiple positions in the visual field. Since the monkey maintains gazing at the fixation point during the cue period, the monkey's orienting eye movement toward the visual cue is prevented and the position of the visual cue is described by the retinotopic coordinates. In addition, gazing at the fixation point during the delay period allows to prevent the monkey's postural orientation (e.g., angle-of-gaze effect) during this period. Further, saccadic eye movements as behavioral responses allow us to analyze response behaviors quantitatively using various saccade parameters (e.g., accuracy, latency, direction, amplitude, duration, and trajectory). Thus, the ODR task has several advantages over the task conditions in the previous studies. Therefore, the ODR task and its derivatives have been frequently used for prefrontal mnemonic studies (Sawaguchi and Goldman-Rakic 1991; Wilson et al. 1993; Sawaguchi and Golaman-Rakic 1994; Williams and Goldman-Rakic 1995; Hasegawa et al. 1998a, b, 2000; Chafee and Goldman-Rakic 1998, 2000; Sawaguchi and Yamane 1999; Constantinidis et al. 2001a, b; Sawaguchi and Iba 2001; Takeda and Funahashi 2002; Williams et al. 2002; Tsujimoto and Sawaguchi 2004; Meyer et al. 2007; Qi et al. 2011; Zhou et al. 2012).

3.5.2 Task-Related Prefrontal Activity in the ODR Task

Funahashi et al. (1989, 1990, 1991) reported three typical task-related activities (cue-, delay-, and response-period activity) observed in the dorsolateral prefrontal cortex while monkeys performed the ODR task (Fig. 3.9). Among 434 neurons

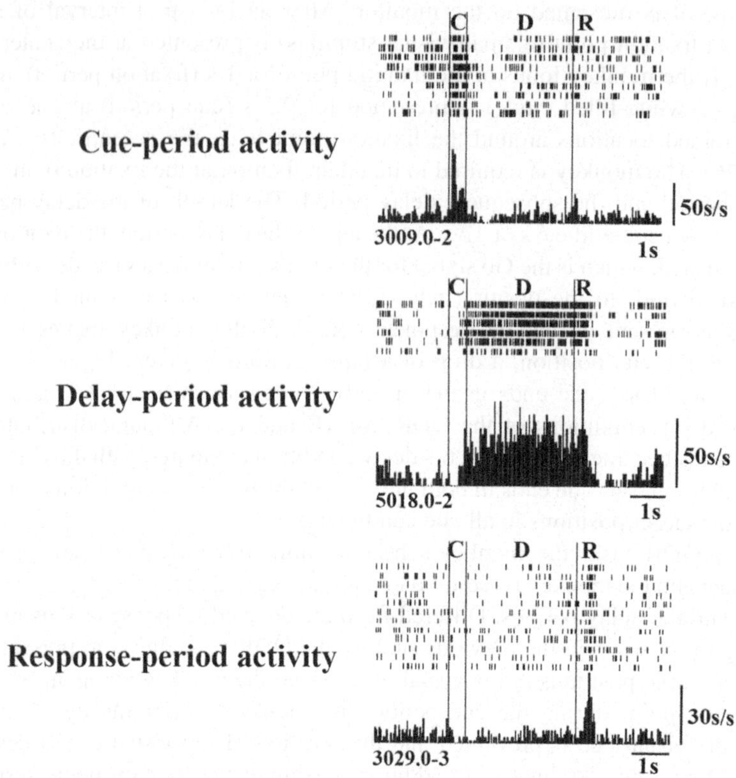

Fig. 3.9 Typical examples of three task-related activities observed during the ODR task in the prefrontal cortex. C, D, and R indicate the cue, the delay, and the response period, respectively

analyzed, 261 exhibited task-related activity, and 74 (28%), 168 (64%), and 147 (56%) exhibited task-related activity during the cue, the delay, and the response periods, respectively. Among them, 21, 73, and 62 exhibited only cue-, delay-, and response-period activity, respectively, and the remaining 105 neurons exhibited task-related activity during two or three task events. Hasegawa et al. (2000) reported that, among 157 neurons examined, 53 (34%) showed visual responses during the cue period. Takeda and Funahashi (2002) reported that 39%, 48%, and 62% of 160 task-related prefrontal neurons exhibited cue-, delay-, and response-period activity during ODR performances, respectively, and that 32%, 51%, and 63% of 174 neurons exhibited cue-, delay-, and response-period activity during the different version of the ODR task (rotatory ODR task), respectively. Iba and Sawaguchi (2002) showed that, among 521 task-related neurons, 97 (19%), 411 (79%), and 13 (2%) exhibited cue-, delay-, and response-period activity, respectively, while monkeys performed an oculomotor delayed visual search task and the ODR task. Thus, while monkeys performed the ODR task or its derivatives, 30–40% of task-related prefrontal neurons exhibited cue-period activity, 50–70% exhibited delay-period activity, and about 60% exhibited response-period activity.

One example of task-related activity is cue-period activity (Fig. 3.9), which is a phasic excitatory or inhibitory response occurring about 120 ms after the visual cue presentation (Funahashi et al. 1990). The studies using the ODR task or similar tasks reveal that 30–40% of task-related prefrontal neurons exhibit cue-period activity. Most cue-period activity exhibits spatial selectivity, such that cue-period activity is observed only when the visual cues are presented within a certain area of the visual field (Funahashi et al. 1990; Hasegawa et al. 2000; Iba and Sawaguchi 2002). Similar excitatory or inhibitory responses are also observed in a visual probe task, in which visual stimuli are presented at the same positions as those in the ODR task but had no behavioral significance (Funahashi et al. 1990). Prefrontal neurons are known to respond to simple visual stimuli (e.g., spot or bar stimuli) and have visual receptive fields (Mikami et al. 1982; Suzuki and Azuma 1983). Therefore, the cue-period activity observed in the ODR task would be visual responses to the presentation of the visual cues. The directional selectivity of cue-period activity would be a result that prefrontal visual neurons have visual receptive fields (Funahashi et al. 1990; Funahashi 2013a). Details of the features of cue-period activity will be described in Chap. 5: Sensory and Motor Processing.

Another example of task-related activities is response-period activity (Fig. 3.9), which is a phasic excitatory or inhibitory activity observed during the response period. Since monkeys perform saccadic eye movements during the response period in the ODR task, response-period activity can be considered to be saccade-related activity. Response-period activity in the ODR task is classified into either pre-saccadic or post-saccadic based on whether response-period activity starts before the initiation of the eye movement or not. Interestingly, the great majority of response-period activities are post-saccadic (Funahashi et al. 1991; Takeda and Funahashi 2002; Funahashi 2014). Majority of post-saccadic response-period activities are initiated about 100 ms after the initiation of the eye movement. Therefore, post-saccadic response-period activity does not participate in the initiation, the execution, or the control of saccadic eye movements. However, almost all post-saccadic response-period activities exhibit directional selectivity, such that the activity is observed only when the monkey performs saccadic eye movements toward a particular direction, mainly contralateral to the hemisphere where the neuron is located (Funahashi et al. 1991; Funahashi 2013b, 2014). In addition, the initiation timing of post-saccadic response-period activity coincided with the termination of delay-period activity (Goldman-Rakic et al. 1990), suggesting that post-saccadic response-period activity is a feedback signal from oculomotor centers and participates in the termination of delay-period activity (Funahashi and Kubota 1994; Funahashi 2014). On the other hand, neurons exhibiting pre-saccadic response-period activity consist of a minority group among prefrontal saccade-related neurons. However, response characteristics of pre-saccadic response-period activity are similar to pre-saccadic activity observed in frontal eye field neurons, which participate in the generation and the control of saccadic eye movements (Bruce and Goldberg 1985). Therefore, pre-saccadic response-period activity observed in dorsolateral prefrontal neurons is also thought to participate in the generation and the control of saccadic eye movements (Funahashi 2013a, 2014). Details of the features of pre- and post-saccadic response-period activities will be described in Chap. 5: Sensory and Motor Processing.

3.6 Delay-Period Activity in the ODR Task

An important example of task-related activity observed during the ODR task is delay-period activity (Fig. 3.9). Delay-period activity has been observed in 50–70% of prefrontal neurons during the ODR task. Funahashi et al. (1989) had initially reported that, among 170 prefrontal neurons having task-related activity, 87 (51%) exhibited delay-period activity. Subsequently, Funahashi et al. (1990) showed that, among 261 task-related prefrontal neurons, 168 (64%) exhibited delay-period activity. Among them, delay-period activity was only task-related activity for 73 neurons (43%), while the remaining 95 neurons (62%) exhibited delay-period activity as well as other task-related activity. Among 87 neurons having delay-period activity, 50 neurons (57%) exhibited excitatory delay-period activity and 37 (43%) exhibited inhibitory delay-period activity (Funahashi et al. 1989) (Fig. 3.10).

3.6.1 Sustained Nature of Delay-Period Activity

Delay-period activity observed in dorsolateral prefrontal neurons has several important features which support the notion that delay-period activity is a neural correlate of the mechanism for temporarily holding information. Prefrontal neurons having delay-period activity play important roles in working memory process. First, the majority of prefrontal neurons having delay-period activity exhibit sustained

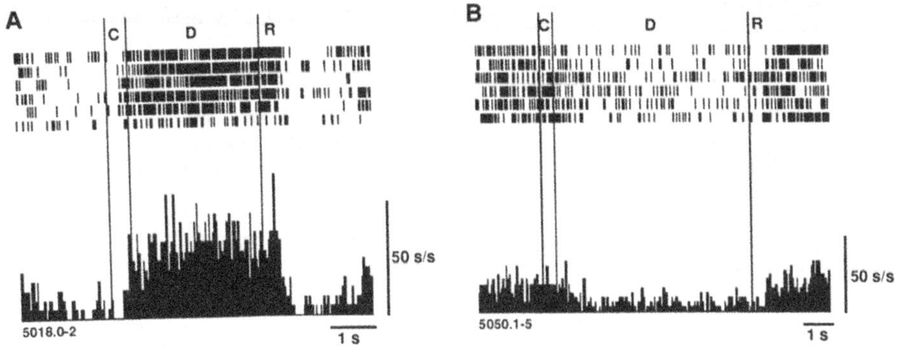

Fig. 3.10 Examples of excitatory (**a**) and inhibitory (**b**) delay-period activity observed in the prefrontal cortex

activation during the delay period (Fig. 3.10). Delay-period activity is often initiated 200–400 ms after the onset of the visual cue, maintains throughout the delay period, and usually terminates just after the initiation of saccadic eye movements during the response period. Prefrontal neurons having sustained delay-period activity often have phasic cue-period activity with the same directional selectivity between two activities, suggesting that cue-period activity triggers the generation of delay-period activity. Both excitatory and inhibitory delay-period activities are observed in the prefrontal cortex, although neurons having excitatory delay-period activity are the major. And, although both excitatory and inhibitory delay-period activity are sustained activity during the delay period (Fig. 3.10), a minority of prefrontal neurons exhibit gradually increasing or decreasing activity during the delay period.

When the length of the delay period is changed, the duration of delay-period activity is prolonged or shortened depending on the length of the delay period in the ODR task (Fig. 3.11a). Even though the length of the delay period is changed, same directional selectivity is maintained during whole delay period. The majority of prefrontal neurons exhibit sustained activity with similar discharge rates along the delay period, while the minority exhibit systematical changes in the temporal pattern of delay-period activity depending on different lengths of the delay period (Funahashi et al. 1989). For example, when three different delay lengths (1.5, 3, and 6 s) were introduced randomly in the ODR task, the temporal pattern of delay-period activity during the first 1.5 s of the delay period was very similar across all three delay lengths, and the temporal pattern of delay-period activity during the next 1.5 s of the delay period was the same in the 3-s and 6-s delay lengths (Funahashi et al. 1989) (Fig. 3.11a). Similar temporal features of delay-period activity had also been observed during classical delayed-response performances using hand-reaching behaviors (Fuster 1973; Kojima and Goldman-Rakic 1982).

Delay-period activity is observed and maintained during the delay period only in correct trials (Fig. 3.11b). However, in error trials, delay-period activity is not observed or is truncated (Funahashi et al. 1989). Population analysis revealed that the magnitude of delay-period activity is significantly greater in correct trials than in

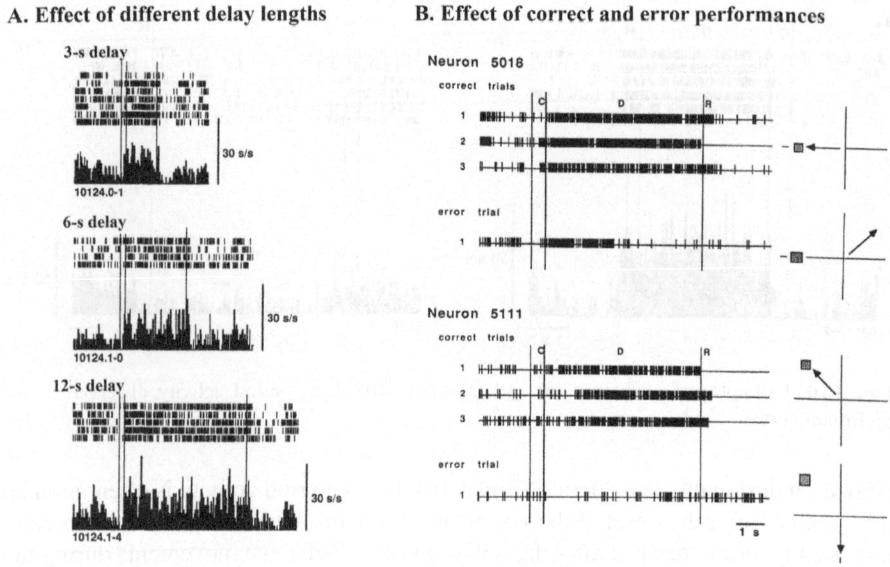

Fig. 3.11 (**a**) Effects of temporal patterns of delay-period activity under different delay lengths. (**b**) Comparisons of delay-period activity between correct trials and error trials. From Funahashi et al. (1989) with permission

error trials (Funahashi et al. 1989). Similar results had been reported by Fuster (1973). Presentation of distractor stimuli during the delay period increases the ratio of incorrect responses and error responses are often associated with lower discharge rates of delay-period activity.

These results indicate the importance of delay-period activity to make a correct response. An important feature of delay-period activity is that this activity is tonically sustained during the delay period. Delay-period activity is maintained along whole delay period in correct trials, but not maintained in error trials. These features support the notion that delay-period activity is a neural correlate of the mechanism for working memory.

3.6.2 Directional Selectivity

Another important feature of delay-period activity is that a great majority of delay-period activity observed in the ODR task shows directional selectivity. Whether delay-period activity shows directional selectivity or not is determined by examining whether a statistically significant difference is present in the mean discharge rates of delay-period activity among different cue conditions. Delay-period activity having directional selectivity has been called *directional* delay-period activity. Directional delay-period activity is observed only when the visual cue is presented at one or a

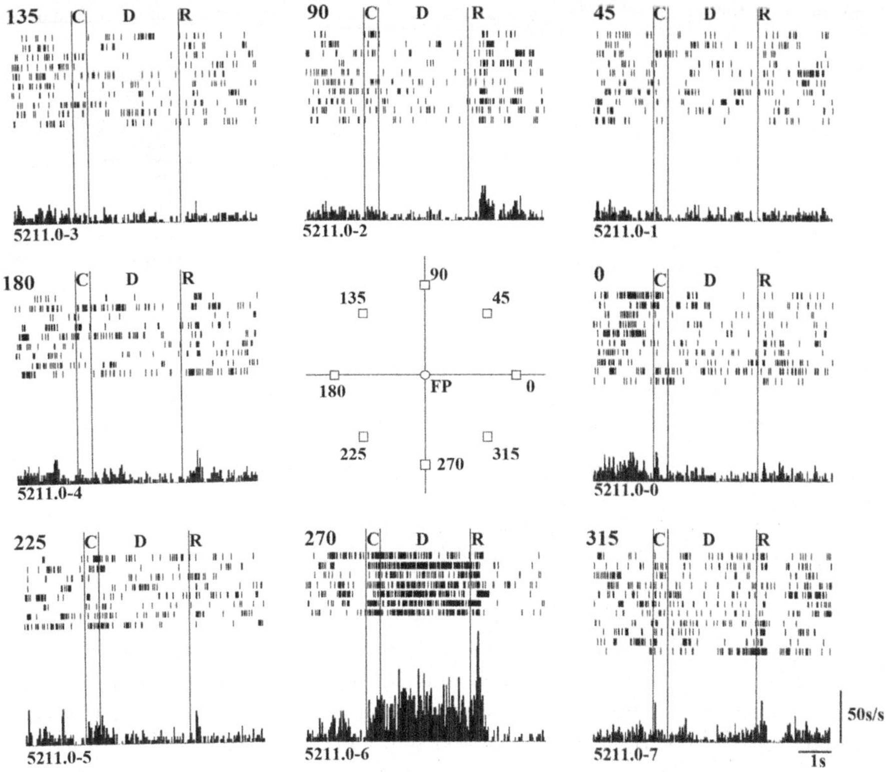

Fig. 3.12 An example of directional delay-period activity. C, D, R, and FP indicate the cue period, the delay period, the response period, and the fixation point, respectively. From Funahashi et al. (1989) with permission

few adjacent spatial positions in the visual field (Fig. 3.12). Directional selectivity of delay-period activity was observed in 79% (69/87, Funahashi et al. (1989)) or 93% (71/76, Takeda and Funahashi (2002)) of prefrontal neurons having this activity. On the other hand, there are prefrontal neurons that exhibit significant delay-period activity across all cue conditions and show no directional selectivity. Delay-period activity exhibiting no directional selectivity is called *omni-directional* delay-period activity. Omni-directional delay-period activity was observed in 21% (18/87, Funahashi et al. (1989)) or 7% (5/76, Takeda and Funahashi (2002)) of prefrontal neurons having delay-period activity.

Excitatory and inhibitory delay-period activities are observed in prefrontal neurons. Funahashi et al. (1989) showed that, for neurons having excitatory delay-period activity ($n = 50$), 92% exhibited directional selectivity and 8% exhibited omni-directional selectivity, while for neurons having inhibitory delay-period activity ($n = 37$), 62% exhibited directional selectivity and 38% exhibited omni-directional selectivity. Similarly, Zhou et al. (2012) reported that 23.5% and 3.5%

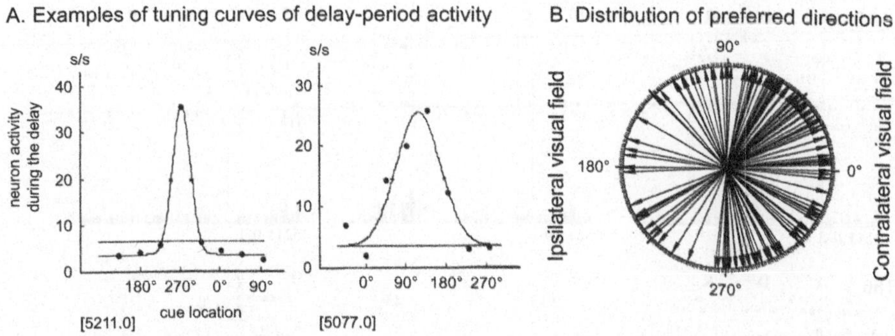

Fig. 3.13 Tuning characteristics of directional delay-period activity. (**a**) Two examples of tuning curves are obtained by delay-period activity. Left, narrower tuning feature. Right, wider tuning feature. (**b**) Distribution of preferred directions of excitatory delay-period activity of prefrontal neurons. Majority of prefrontal neurons have preferred directions in the contralateral visual field. From Funahashi et al. (1989) with permission

of total prefrontal samples ($n = 682$) exhibited excitatory and inhibitory directional delay-period activity, respectively.

Directional delay-period activity has a preferred direction. The preferred direction of delay-period activity can be determined as the direction at which the maximum delay-period activity is observed and is obtained by making a tuning curve by fitting the mean discharge rate during the delay period for each cue direction on a Gaussian function (Fig. 3.13a). Both excitatory and inhibitory delay-period activity exhibit directional tunings. Preferred directions of directional delay-period activity differ from neuron to neuron and distribute across whole directions around the fixation point. For excitatory delay-period activity, the majority of preferred directions are directed toward the visual field contralateral to the hemisphere where neurons are recorded (contralateral bias) (Funahashi et al., 1989; Sawaguchi and Yamane 1999; Takeda and Funahashi 2004) (Fig. 3.13b). However, for inhibitory delay-period activity, contralateral bias of directional selectivity is not obvious (Funahashi et al. 1989). The mean of the tuning width, which is the width of the tuning curve at the level where the magnitude of delay-period activity is 37% ($1/e$, $e = 2.71828$) below the peak discharge rate in the tuning curve, was 26.8° and 43.5° for excitatory and inhibitory directional delay-period activity, respectively.

Wilson et al. (1994) differentiated putative pyramidal neurons (regular-spiking neurons) from putative interneurons (fast-spiking neurons) based on the difference in both the amplitude and the shape of the action potential. They showed that visual receptive fields of both groups of neurons were similar in the size and the location in the visual field (contralateral bias). However, they also showed that the direction of the visual response (excitation or inhibition) to the same visual stimulus in interneurons was often inverted to those of adjacent pyramidal neurons. Similar inverted response characteristics were also observed in delay-period activity between pyramidal neurons and interneurons. Zhou et al. (2012) used the same method to differentiate two groups of prefrontal neurons and examined detailed characteristics of directional tunings of delay-period activity between closely adjacent pyramidal and non-pyramidal interneurons. They found that a majority of both groups of neurons exhibited directional delay-period activity with similar tuning features and that inverted features of the directional tuning of delay-period activity were observed between two groups of neurons. They suggest that these inverted tuning features play a role to maintain or stabilize directional tunings of delay-period activity of pyramidal neurons.

Regarding GABAergic interneurons in the primate prefrontal cortex, three major interneurons (parvalbumin (PV)-, calretinin (CR)-, and calbindin (CB)-containing neurons) are known. PV neurons provide inhibitory outputs to the proximal portions of pyramidal neurons, CR neurons provide inhibitory outputs to other GABAergic neurons, and CB neurons send axon branches mainly to layer I and selectively modulate interactions between inputs coming from different cortical layers (Zaitsev et al. 2005). Regarding functions of these interneurons, Wang et al. (2004a, b) proposed that the tuning of delay-period activity is maintained by the controlled action of widespread inhibition of pyramidal neurons by PV neurons and by localized disinhibition of pyramidal neurons via interneuron-targeting CR neurons. They also proposed that the robustness of the tuning is controlled by CB neurons. Based on this hypothesis, Zhou et al. (2012) proposed that CB neurons would be the interneurons exhibiting inverted tuning features to adjacent pyramidal neurons. While pyramidal neurons are activated, adjacent CR neurons are activated by excitatory inputs from pyramidal neurons and inhibit adjacent CB neurons. In turn, these CB neurons disinhibit pyramidal neurons. These local interactions between pyramidal neurons and adjacent GABAergic CB interneurons would play a role to filter out distracting stimuli and stabilize directional selectivity of delay-period activity of pyramidal neurons. Prefrontal neurons having excitatory directional delay-period activity often exhibit inhibitory delay-period activity at the cue positions opposite to the preferred direction. These tuning features may also be explained by interactions between pyramidal neurons and adjacent CB neurons.

3.6.3 *Robustness of Delay-Period Activity*

The robustness of sustained delay-period activity has been shown using the task in which distractor stimuli are presented during the delay period. Delay-period activity is maintained regardless of the presentation of distractors during the delay period. For example, Qi et al. (2010) compared delay-period activity between dorsolateral prefrontal neurons and posterior parietal neurons while the monkey performed a spatial delayed matching-to-position task, in which the monkey needed to release the lever when the visual stimulus was presented at the same position as the position where the visual cue had been presented. A distractor stimulus was presented several times at different positions during the delay period. Among 148 task-related prefrontal neurons examined, 48 (32%) exhibited delay-period activity and 79% (38/48) maintained sustained activation during the whole delay period even when the distractor was presented at several different positions. On the other hand, among 119 task-related parietal neurons, 36 (30%) exhibited delay-period activity and 47% (17/36) maintained sustained activation while distractors were presented during the delay period. Thus, although sustained delay-period activity was observed in both prefrontal and parietal neurons, delay-period activity of parietal neurons was affected more by the presentations of distractors. Suzuki and Gottlieb (2013) examined responses to the distractor presented during the delay period in dorsolateral prefrontal neurons and lateral intraparietal neurons while the monkey performed a memory-guided saccade task, in which a salient distractor was presented at a variable timing and location during the delay period. In prefrontal neurons, the responses to the distractor presented during the delay period were strongly suppressed, while in parietal neurons, the responses to the distractor were present and not suppressed. The magnitude of the distractor response in prefrontal neurons was positively correlated with error rates in behavioral performances, but this correlation was not observed in parietal neurons. These results suggest that the robustness of sustained delay-period activity is an important factor to perform behavioral responses correctly. Further, di Pellegrino and Wise (1993) used a similar delayed matching-to-position task as Qi et al. (2010) used and compared delay-period activity between prefrontal and premotor neurons. In their task, the monkey was required to memorize the position of the visual cue. After the delay, the monkey needed to move a manipulandum either to the position where the visual cue was presented (congruent condition) or to the predetermined position (incongruent condition). Distractor stimuli were presented during the delay period. Spatial selectivity of delay-period activity observed just after the cue presentation was the same between both conditions in both prefrontal and premotor neurons, suggesting that spatial selectivity of delay-period activity was affected by the visual cue, not by the movement direction. In addition, about one-third of prefrontal neurons exhibited sustained delay-period activity in spite of the distractor presentation during the delay period. These results indicate that delay-period activity is sustained during the delay period in spite of the distractor presentation in prefrontal neurons. Thus, the robustness is an important feature of sustained delay-period activity in the prefrontal cortex.

3.6.4 Learning Effects

There are evidences that delay-period activity is a product of the learning of the behavioral tasks. For example, Fuster (1973) showed that a significant number of prefrontal neurons having delay-period activity was observed in only well-trained monkeys and that no sustained activation or inhibition during the delay period was found in untrained monkeys although visual responses were found. Meyer et al. (2007) analyzed prefrontal activity while naïve monkeys performed a task similar to the ODR task. In this task, while the monkey maintained fixation at a fixation point, the same visual stimulus was presented briefly twice with a 1.5-s inter-stimulus interval at either the same position or the opposite position to the first position. The monkey needed only to maintain fixation during a trial and received a reward 1.5 s after the presentation of the second stimulus without any response. Therefore, visual stimuli had no behavioral meaning. They found that a population of prefrontal neurons (20%, 52/254) exhibited persistent activation during the 1.5-s inter-stimulus interval and that this persistent activation exhibited spatial selectivity. These results suggest that prefrontal neurons have some intrinsic physiological mechanism that generates persistent activation when receiving a particular sensory input. Qi et al. (2011) also compared prefrontal activity between before and after the learning of the delayed matching-to-position task using eye movements and found that, after learning this task, more prefrontal neurons exhibited a larger magnitude of activation during the 1.5-s delay period. Especially, regular spiking neurons (putative pyramidal neurons) exhibited more increased firing rates during the delay period after learning the task. However, the strength of stimulus selectivity of delay-period activity was reduced after learning the task. These results indicate that the development of delay-period activity is learning-dependent because the proportion of neurons having delay-period activity and the magnitude of delay-period activity increase after learning the working memory task. Further, Qi and Constantinidis (2012a, b, 2015) examined the effects of behavioral training of the working memory task on prefrontal task-related activity by comparing noise correlations of pairs of neurons and Fano factors between before and after learning the task. The trial-to-trial correlation of discharge rates between pairs of prefrontal neurons (noise correlation) was decreased after the monkey learned to perform the task and the greatest decrease was observed between neurons that exhibited similar spatial tunings (Qi and Constantinidis 2012a). Variability of neural discharges (Fano factor) decreased after learning the task in prefrontal neurons that exhibited elevated discharge rates during the task events (Qi and Constantinidis 2012b). In addition, Fano factors obtained from prefrontal neurons were consistently lower than those obtained from posterior parietal neurons. Since values of Fano factors were negatively correlated with the performance of the working memory task, the stimulus information was more reliably represented in prefrontal neurons than posterior parietal neurons (Qi and Constantinidis 2015). Further, although working memory capacity is limited, it has been shown that practice improves its capacity. Tang et al. (2019) trained monkeys to perform a working memory task requiring memory of multiple stimuli. Although behavioral performance decreased with an increase in number of stimuli, the practice of the task under an increased number of stimuli improved monkeys' performance. After the practice of the task, they found that more prefrontal neurons became active as behavioral performance improved and that improved working memory capacity was associated with the decrease of variability of firing rates along a trial.

Thus, training of behavioral tasks affects the magnitude and selectivity of task-related activity, the temporal pattern and variability of neural discharges, and the correlations of neural discharges between pairs of neurons in the prefrontal cortex. These results suggest that discharge characteristics and stimulus selectivity of prefrontal neurons change flexibly during training of behavioral tasks and that information regarding the remembered stimulus becomes more effectively represented in prefrontal neurons after learning the task.

3.6.5 Delay-Period Activity Is a Neural Correlate of Working Memory Process

Delay-period activity is a sustained activity observed during the delay period of working memory tasks, exhibits selectivity depending on the position where the visual cue is presented, is robust against external distractions, and is observed only in correct trials. Since information needed to make a correct response differs from trial to trial in working memory tasks, specific neural mechanisms are needed to actively maintain trial-specific information during the delay period and to utilize this information to guide the response behavior. Features of delay-period activity support the idea that tonic sustained delay-period activity is a neural correlate of the mechanism for temporary active maintenance of trial-specific information. Since the delay-period activity of each prefrontal neuron exhibited different directional preferences during ODR performances, different information would be maintained by neurons having different selectivity of delay-period activity. Thus, the directional delay-period activity observed in prefrontal neurons is a neural correlate of the mechanism for temporary active maintenance of trial-specific information in working memory processes.

3.6.6 Mnemonic Receptive Field

Directional delay-period activity has been observed in many prefrontal neurons during ODR performances. During the delay period, a small fixation point is only the visible stimulus, the monkey's eye movement is restricted, and the monkey's body movement is also restricted. Therefore, the stimulus condition and the monkey's behavior are the same during the delay period across all cue conditions. The only difference across trials is the cue position. In an example shown in Fig. 3.12, sustained excitatory delay-period activity is observed only when the visual cue is presented at the lower visual field (e.g., 270° position). Therefore, the directional selectivity of delay-period activity is thought to be caused by the difference in visual cue positions. Analyses of directional characteristics of delay-period activity revealed (1) the distribution of preferred directions of delay-period activity has contralateral bias, although preferred directions cover all possible directions around the fixation point and (2) the mean width of tuning curves is 27° for excitatory delay-period activity, indicating that the most of prefrontal neurons exhibit delay-period activity when visual cues are presented within about one-quarter of the visual field.

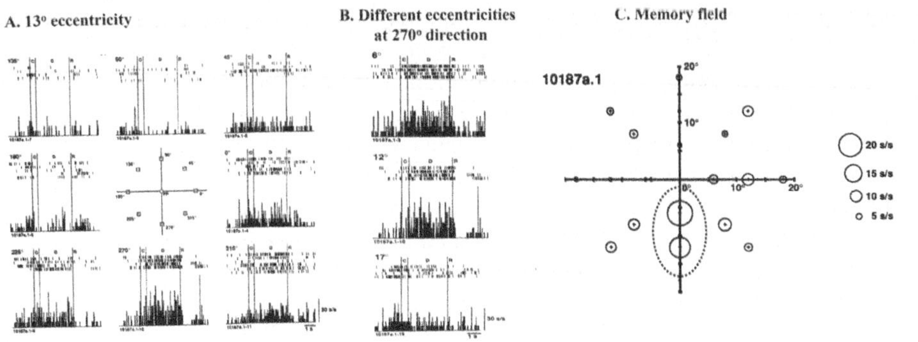

Fig. 3.14 (**a**) An example of delay-period activity observed during the ODR performances when eight visual cues were presented at 13° eccentricity. (**b**) Delay-period activity observed during the ODR performances when visual cues were presented at three different eccentricities at the 270° direction. (**c**) Estimated position of the memory field of neuron 10187a in the visual field. Each circle corresponds to the position of the visual cue examined and the size of each circle indicates the mean discharge rate of delay-period activity when the visual cue was presented at that position

Based on tuning characteristics of delay-period activity, Funahashi et al. (1989) proposed that prefrontal neurons having directional delay-period activity have mnemonic receptive fields (*memory fields*) in the visual field, similar as visual neurons have visual receptive fields in the visual field. To examine the feature of memory fields of prefrontal neurons, a modified version of the ODR task was introduced, in which 13 or 20 visual cues were randomly presented at three different eccentricities (6°, 12°, and 17°) around the fixation point. Fig. 3.14 shows an example of a prefrontal neuron having directional delay-period activity, whose preferred direction was 270° in the 13° eccentricity condition (Fig. 3.14a). The directional preference was maintained, but the mean discharge rate was modulated by the eccentricity of the cue position (Fig. 3.14b). Thus, the mean discharge rate of delay-period activity was modulated by both positions and eccentricities of the visual cue in the visual field. Figure 3.14c shows an estimated memory field of this prefrontal neuron. This result supports the idea that prefrontal neurons having directional delay-period activity have memory fields in the visual field. Although the size of the memory field is different from neuron to neuron, the mean tuning width (27°) of excitatory directional delay-period activity suggests that the size of the memory field is about one-quarter of the visual field.

The evidence that prefrontal neurons exhibiting delay-period activity have memory fields is supported by Rainer et al. (1998b). They used a delayed match-to-object-and-place task, in which the monkey was required to remember both the shape and the position of the sample object during a 1-s delay period. They identified two groups of prefrontal neurons; "what-and-where" neurons (46%, 68/149), whose responses depended on both the sample object and its position, and "where" neurons (49%, 73/149), whose responses depended only on the position of the sample object. Both neurons had memory fields mainly in the visual field contralateral to the hemisphere where neurons were located. "What-and-where" neurons exhibited delay-period activity only when the preferred object was presented within the memory field. Both "what-and-where" neurons and "where" neurons exhibited similar characteristics in the size and position of the memory field.

Neurons in the frontal eye field have been shown to exhibit directional delay-period activity in ODR performances (Funahashi et al. 1989). Leavitt et al. (2018) used the ODR task in which the positions of visual cues were arranged on the 4x4 matrix and analyzed neural activity collected from the frontal eye field (area 8a). They found that, although neurons have spatial mnemonic receptive fields similar as those of dorsolateral prefrontal neurons, mnemonic receptive fields exhibited quadrantic bias, such that spatial mnemonic representations were biased by the vertical and horizontal meridians of the visual field, dividing mnemonic field into quadrants. In addition, memory fields are anatomically organized and clustered in a manner that partially reflects not retinotopic but the geometric properties of visual space.

3.6.7 *Delay-Period Activity Represents Retrospective and Prospective Information*

Response characteristics of delay-period activity suggest that this activity is a neural correlate of the mechanism for temporarily maintaining information in working memory processes. If this is the case, what information does delay-period activity represent while the subject performs a working memory task?

When the subject performs the delayed-response task correctly, the information needed to be maintained during the delay period is either the position of the visual cue (baited position) or the direction of the behavioral response. Does delay-period activity during delayed-response performances represent information regarding the position of the visual cue (retrospective information) or the direction of the behavioral response (prospective information)? As shown in Fig. 3.6, Niki and Watanabe (1976a) first examined this question. They used three manual working memory tasks (a right-left delayed-response task, an up-down delayed-response task, and a conditional position task with delay) and compared directional selectivity of differential delay-period activity of the same neuron among these task conditions (Fig. 3.6). They found that 78% (25/32) of neurons having differential delay-period activity encoded the position of the visual cue, whereas the remaining 22% (7/32) encoded the direction of the behavioral response.

Funahashi et al. (1993b) further examined the same issue using two types of ODR tasks; a standard ODR task, in which the monkey needs to make a saccade to the position where the visual cue was presented after the delay, and an anti-saccade ODR task, in which the monkey needs to make a saccade to the direction opposite to the position where the visual cue was presented (Fig. 3.15). They compared directional selectivity of delayed-period activity recorded from the same prefrontal neuron between these two tasks. Figure 3.15 shows an example of delay-period

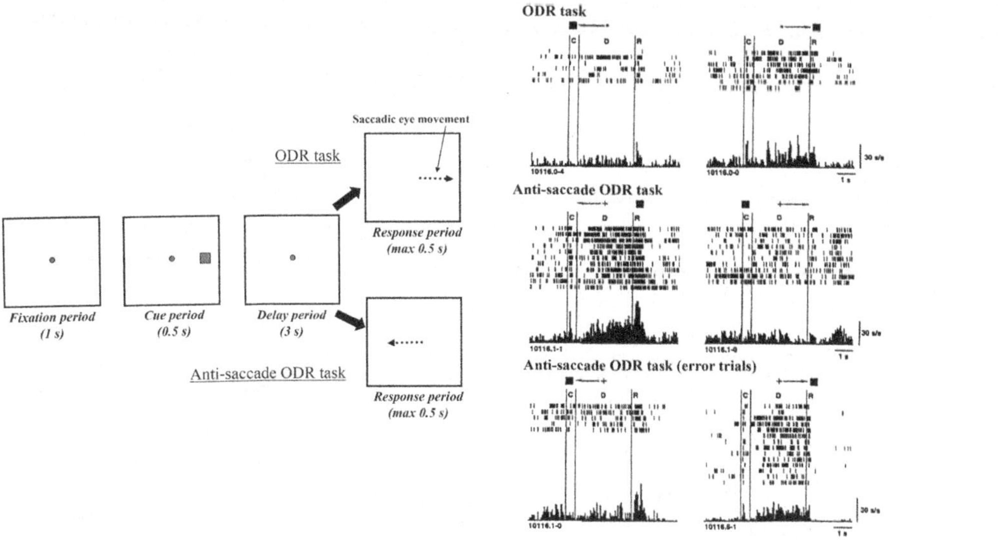

Fig. 3.15 *Left*: Temporal sequence of task events in the standard ODR and the anti-saccade (AS) ODR tasks. *Right*: Examples of delay-period activity during these two tasks were recorded from the same prefrontal neuron. Filled squares indicate the positions of the visual cues and arrows indicate the directions of saccades during the response period. From Funahashi et al. (1993b) with permission

activity representing the position of the visual cue. This neuron exhibits delay-period activity when the visual cue is presented on the right, regardless of the direction of the saccade and correctness of response behavior. They found that 59% (30/51) of prefrontal neurons having directional delay-period activity encoded the position of the visual cue, while 25% (13/51) encoded the direction of the saccade.

These studies indicate that prefrontal neurons having directional delay-period activity can be classified into two groups; one encoding the position of the visual cue (retrospective information) and another encoding the direction of the behavioral response (prospective information). These studies also indicate that neurons encoding retrospective information are the majority in the dorsolateral prefrontal cortex. However, Niki and Watanabe (1976b) and Funahashi et al. (1993a, b, c) used only two cue positions and response directions (right and left) for the visual cues and behavioral responses, respectively. To further confirm the result that the majority of directional delay-period activity represent retrospective information in prefrontal neurons, it may need to examine this issue under the task using multiple positions for the visual cue.

Takeda and Funahashi (2002) compared directional selectivity of delay-period activity between two kinds of ODR tasks; the original ODR task with eight cue positions and a rotatory ODR (R-ODR) task with four cue positions (Fig. 3.16). In the R-ODR task, the monkey is required to make a memory-guided saccade 90° clockwise to the direction where the visual cue was presented.

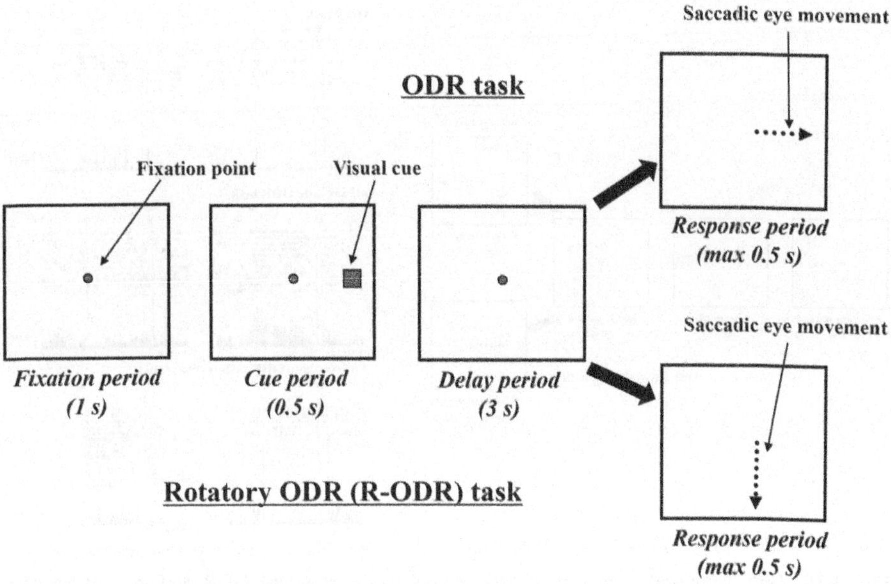

Fig. 3.16 Temporal sequence of task events in the ODR task and the rotatory ODR (R-ODR) task

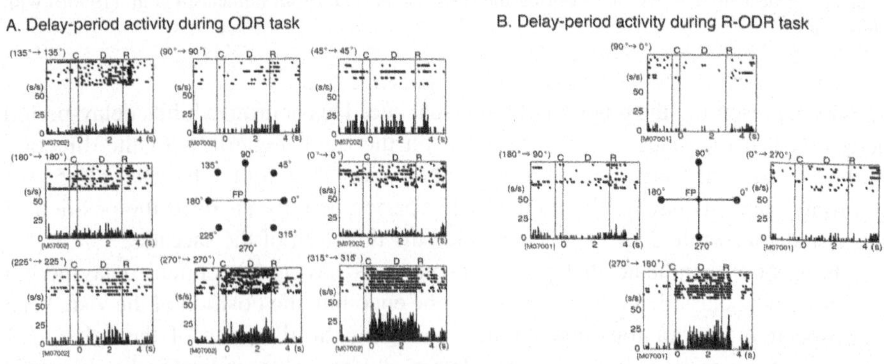

Fig. 3.17 An example of delay-period activity representing retrospective information (the position of the visual cue). (**a**) Delay-period activity during the ODR task. (**b**) Delay-period activity during the R-ODR task. Both delay-period activities were recorded from the same prefrontal neuron. Each figure including rasters and histograms is arranged based on the position where the visual cue was presented. From Takeda and Funahashi (2002) with permission

Figure 3.17 shows an example of delay-period activity recorded from the same prefrontal neuron (M070) while the monkey performed the ODR and the R-ODR tasks independently. This neuron exhibited significant delay-period activity in the ODR task when the visual cues were presented in the lower right visual field (270° and 315° positions). The same neuron also exhibited significant delay-period activity in the R-ODR task when the visual cue was presented at the 270° position. Although

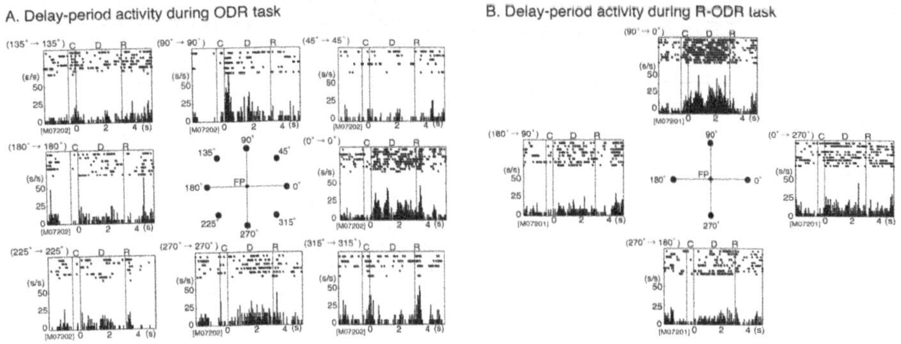

Fig. 3.18 An example of delay-period activity encoding prospective information (the direction of the saccade). (**a**) Delay-period activity during the ODR task. (**b**) Delay-period activity during the R-ODR task. Both delay-period activities were recorded from the same prefrontal neuron. Each figure including rasters and histograms is arranged based on the position where the visual cue was presented. From Takeda and Funahashi (2002) with permission

significant delay-period activity was observed when the visual cue was presented at the same 270° position in both tasks, the saccade directions were the 270° direction in the 270° cue condition of the ODR task and the 180° direction in the 270° cue condition of the R-ODR task. The directions of the saccades during the response period were different between two tasks. However, delay-period activity was observed when the visual cue was presented at the same position in both tasks. Therefore, it can be concluded that delay-period activity of this prefrontal neuron represents the information regarding the position of the visual cue, not the direction of the saccade.

Figure 3.18 shows another example of delay-period activity recorded from the same neuron (M072) while the monkey performed both ODR and R-ODR tasks. This neuron exhibited significant delay-period activity when the visual cue was presented at the 0° cue position in the ODR task and at the 90° cue position in the R-ODR task. The preferred cue positions were different between two tasks. However, the correct saccades were directed toward the 0° direction both in the 0° cue condition of the ODR task and in the 90° cue condition of the R-ODR task. Although delay-period activity was observed when the visual cue was presented in different positions between two tasks, the directions of the saccades performed during the response period were the same in both tasks. Therefore, it is concluded that delay-period activity of this neuron represents the information regarding the direction of the saccade, not the position of the visual cue.

Tuning curves of delay-period activity were constructed for each prefrontal neuron using the mean discharge rates during the delay period across all cue conditions during ODR and R-ODR performances (Fig. 3.19). Preferred directions of tuning curves obtained from two tasks were compared for each neuron. The positions of the visual cues, not the directions of the saccades, were used to construct all tuning curves. Therefore, if the preferred directions were the same between the ODR and the R-ODR tasks for a particular neuron, it could be concluded that

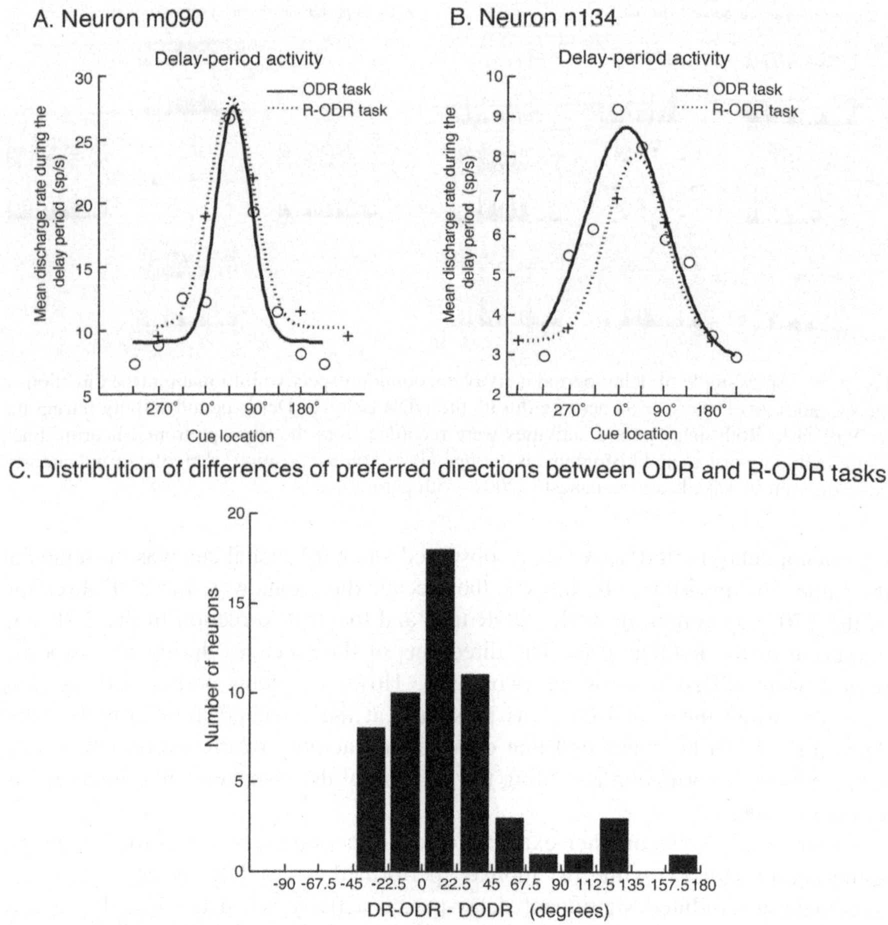

Fig. 3.19 (**a**, **b**) Examples of tuning curves of delay-period activity under ODR and R-ODR tasks. (**c**) Distribution of differences of preferred directions of delay-period activities between ODR and R-ODR tasks. From Takeda and Funahashi (2002) with permission

delay-period activity observed in this neuron represents the information regarding the position of the visual cue. If the preferred direction obtained from the ODR task had a 90° difference from the preferred direction obtained from the R-ODR task, it could be concluded that delay-period activity observed in the neuron represents the information regarding the direction of the saccade. Figure 3.19a, b show examples of tuning curves of delay-period activity obtained from two prefrontal neurons (M090 and N134) during performances of two ODR tasks. Tuning curves obtained from two ODR tasks are overlapped and the peaks of these two tuning curves are close for both neurons (Fig. 3.19a, b), indicating that delay-period activity of both neurons represents the information regarding the position of the visual cue, not the direction of the saccade. Fig. 3.19C shows distributions of the difference of the peak directions of two tuning curves (peak direction in R-ODR minus peak direction in ODR). The comparison of preferred directions between two ODR tasks for each neuron revealed that 86% of directional delay-period activity represent the information regarding the position of the visual cue, whereas 13% represent the information regarding the direction of the saccade (Fig. 3.19c). Thus, in prefrontal neurons, the majority of delay-period activity represent retrospective information (e.g., the position of the visual cue), while the minority represent prospective information (e.g., the direction of the saccade).

The result that the majority of delay-period activity represent retrospective information (sensory cues) has also been shown using other behavioral tasks (Wilson et al. 1993; di Pellegrino and Wise 1993; Sawaguchi and Yamane 1999; Hoshi et al. 2000; Constantinidis et al. 2001a; Fukushima et al. 2004). For example, Wilson et al. (1993) reported that sustained activity during the delay period was generally specific to the cue stimuli while monkeys performed a delay task using visual patterns and visual objects as cue stimuli. di Pellegrino and Wise (1993) also showed that about 60% of delay-period activity observed in prefrontal neurons represent information regarding sensory cues while a monkey performed a delayed matching-to-position task. Similarly, Sawaguchi and Yamane (1999) used a delayed matching-to-position task and found similar results. In their task, the monkey was required to maintain fixation during the sample and the 3-s delay periods and make either a lever holding response (no-go response) or a lever releasing response (go response) depending on whether or not the matching stimulus was presented at the same position as the position where the sample stimulus was presented. They found that 90% (165/184) of prefrontal neurons having sustained delay-period activity showed selectivity to the stimulus position, not to the response behavior (lever releasing or holding response). Constantinidis et al. (2001a) also showed that a majority of delay-period activity did not represent motor attributes (prospective information) but represented sensory attributes (retrospective information). They used the ODR task with two visual cues, which were presented simultaneously during the 0.5-s cue period and one of which was brighter in luminance than the other. Monkeys were required to look at a fixation point during the cue and the 3-s delay periods and then make a saccade to the brighter visual cue. They found that the discharge rate during the delay period was modulated by the luminance intensity of the visual cue when the brighter cue had presented within the neuron's receptive field and that discharge rate during the delay period was directly correlated with the monkey's behavioral performance. Fukushima et al. (2004) used an analog of the ODR task and examined delay-period activity of prefrontal neurons. In their task, the monkey was required to internally shift the saccade target position by one position on every flash of a target-shift cue, which was always the same in shape, color, and size and appeared at the center up to three times within a single trial. They found two kinds of delay-period activity selectively representing the target position. The early dominant target-selective delay-period activity emerged early in the delay period after the presentation of the target-shift cue, while the late dominant target-selective delay-period activity gradually built up toward the end of the delay period. The timing of the occurrence and the temporal pattern of delay period activity suggest that the early dominant target-selective delay-period activity corresponds to delay-period activity representing retrospective sensory informations, whereas the late dominant target-selective delay-period activity corresponds to delay-period activity representing prospective motor information. Further, Hoshi et al. (2000) used a delayed matching-to-sample task performed under two conditions (location-matching and shape-matching conditions) and showed that the most of delay-period activity (88% of 226 delay-related prefrontal neurons) represented the information of the sample stimulus and retain this information until the choice period. Thus, these studies indicate that a large population of prefrontal neurons maintain the information regarding the visual cue throughout the delay period and confirm that delay-period activity of the majority of prefrontal neurons represents not prospective motor information but retrospective sensory information.

3.6.8 Delay-Period Activity Represents Relative or Absolute Spatial Information

When we recognize the position of a visual stimulus, we may use either of two coding mechanisms; absolute coding or relative coding. For example, when we designate the position of a glass placed at the center of the table, we may describe either that a glass is placed at the center of the table or that a glass is on my right side. In the former case, the description of the position of the glass is always the same regardless of where we are in a room and which direction we orient. This is the absolute coding method to designate the spatial information. However, in the latter case, the description of the glass position changes depending on where we are in a room and which direction we orient. For example, if we are on the left side of the table, the glass is on my right side, while if we are on the right side of the table, the glass is on my left side. This is the relative coding method. Thus, there are two distinct coding methods to designate the position of the item. Therefore, when we examine neural mechanisms for spatial working memory, we need to consider which coding method is used for maintaining and representing spatial information by delay-period activity.

When the subject performs the delayed-response task, it is necessary to maintain spatial information of the visual cue during the delay period and, in fact, the majority of delay-period activities have been shown to represent spatial information of the visual cue. An important question to understand working memory mechanisms in the prefrontal cortex is which coding method is used to represent visuospatial information by directional delay-period activity. Niki (1974b) first addressed this question using the delayed alternation task. In his task, two choice keys are selected from four horizontally arranged keys in any given block of trials (see Fig. 3.5a). Therefore, one particular choice key becomes either the right key or the left key depending on which choice key is selected as a pair of that key. He analyzed the same neuron's activity under different pairing of choice keys and found that

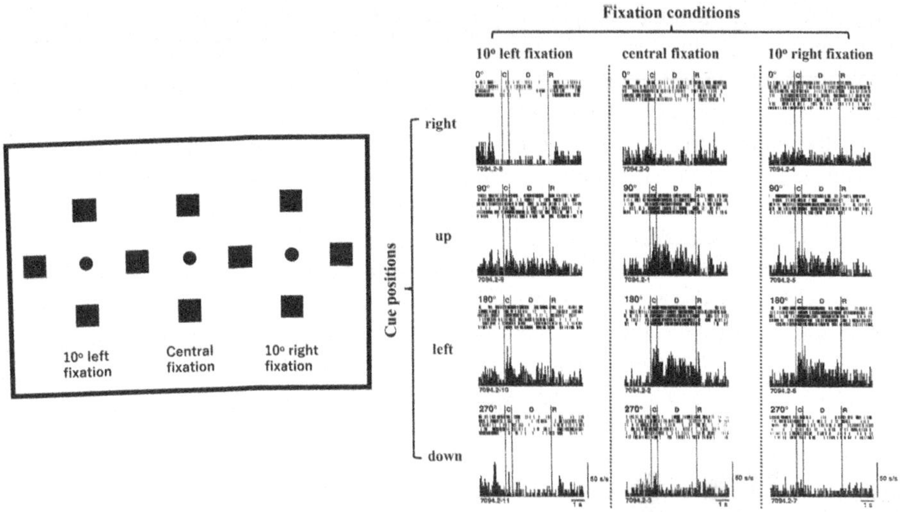

Fig. 3.20 *Left*: Arrangements of fixation points and positions of visual cues. *Right*: An example of delay-period activity when the visual cue was presented at one of four positions (left, right, up, and down) during one of three fixation conditions

directionally selective response-period activity was affected by the absolute position of the chosen key, while directionally selective delay-period activity was affected by the relative position of the chosen key, although the number of neurons exhibited delay activity was small (18%, 20/111).

Whether directional delay-period activity represented the relative or absolute position of the visual cue was further examined using the ODR task (Funahashi and Takeda 2002b). In this experiment, the monkey's head was fixed during the experiment and prefrontal activity of the same prefrontal neuron was examined under three different fixation conditions (Fig. 3.20). Three fixation points were arranged horizontally and separated by 10° in visual angle. Four peripheral cues (right, left, up, and down) were arranged around each fixation point at an eccentricity of 10°. Which fixation point was used and where the visual cue was presented were randomized in any given trial. In this experiment, the right cue at the 10° left fixation condition is presented at the same position as the left cue at the central fixation condition. Similarly, the right cue at the central fixation condition is presented at the same position as the left cue at the 10° right fixation condition. Therefore, if delay-period activity represents the absolute position of the visual cue, significant delay-period activity would be observed when the visual cue is presented at the right position in the 10° left fixation condition and at the left position in the central fixation condition. If delay-period activity represents the relative position of the visual cue, significant delay-period activity would be observed whenever the visual cue is presented at the right or the left position in all fixation conditions. Figure 3.20 shows an example of directional delay-period activity under three fixation conditions. Significant delay-period activity was observed on the left (180°) cue trials under the central and the 10° right fixation conditions. This example clearly shows that directional delay-period activity represents the relative position of the visual cue with respect to the subject's gazing position.

At the same time, the magnitude of delay-period activity is affected by the difference in the gaze directions. The neuron shown in Fig. 3.20 exhibited stronger delay-period activity in the central fixation condition compared with the 10° right and the 10° left fixation conditions. This effect is similar to an "angle-of-gaze" effect observed in other cortical areas. Angle-of-gaze effects on visual responses have been observed in neurons of the posterior parietal cortex (Andersen and Mountcastle 1983; Andersen et al. 1985a, b; Squatrito and Maioli 1996; Andersen et al. 1997), the area V3 (Galletti and Battaglini 1989), and the visual cortex (Trotter and Celebrini 1999; Rosenbluth and Allman 2002). The angle-of-gaze effects on set-related activity (Boussaoud 1995; Boussaoud et al. 1998) and movement-related activity (Mushiake et al. 1997; Boussaoud et al. 1998) have also been observed in neurons of the premotor cortex. Since the prefrontal cortex has anatomical connections with the posterior parietal cortex, it would be possible that delay-period activity of prefrontal neurons is affected by the inputs from the posterior parietal cortex exhibiting the angle-of-gaze effects and therefore that the magnitude of directional delay-period activity is affected by the gazing angles. However, directional selectivity of delay-period activity is maintained across different gazing conditions, confirming that the relative coding method is used to represent spatial information by delay-period activity in the prefrontal cortex.

3.7 Delay-Period Activity Encoding Non-spatial Visual Information

3.7.1 Delay-Period Activity in Non-spatial Visual Working Memory Tasks

Sustained activation during the delay period has been observed in the dorsolateral prefrontal cortex while monkeys performed not only the spatial working memory tasks like the delayed-response task but also a variety of non-spatial working memory tasks like the delayed matching-to-sample task. Although delay-period activity was observed in many prefrontal neurons during the performance of non-spatial working memory tasks, only a small percentage of neurons exhibited differential delay-period activity depending on the difference in non-spatial stimuli (e.g., color differences). Kubota et al. (1980) examined the activity of dorsolateral prefrontal neurons while the monkey performed a delayed matching-to-color task using green and red lights. In this task, the monkey needs to press an "Yes" lever by hand after 1–5 s delay if both the sample and the test stimulus are the same color, while the monkey needs to press a "No" lever if colors of two stimuli are different.

Excitatory delay-period activity was observed in 19 neurons (19% of neurons recorded, 19/99). However, these neurons did not exhibit differential changes in delay-period activity depending on the difference in the colors. Therefore, they concluded that delay-period activity was not related to short-term memory of the sample color. Fuster et al. (1982) obtained similar results as those of Kubota et al. (1980) using similar delayed matching-to-sample tasks with red and green stimuli. They presented either color stimulus during the sample period and after the delay period, presented two color stimuli simultaneously during the choice period. The monkeys were required to select the same color as had been presented during the sample period after the 16- to 20-s delay. They found 76 lateral prefrontal neurons exhibiting delay-period activity and only 7% of these neurons exhibited differential activity depending on different sample colors. They discussed the reason why only a small number of prefrontal neurons exhibited differential delay-period activity in this task and concluded that, although delay-period activity engages in a variety of short-term memory processes, the color is only one of many attributes to memorize in short-term memory by prefrontal neurons. Quintana et al. (1988) also showed a small number of neurons (15%) exhibiting differential delay-period activity during the 18-s delay period depending on the difference in the color cues. They used the delayed matching-to-sample task and a delayed conditional position discrimination task. In the latter task, yellow or blue stimulus is presented during the sample period, and after the delay, the monkey is required to select the right key when the sample cue was yellow or the left key when the sample cue was blue. In the delayed matching-to-sample task, the monkey needs to remember the sample color during the delay period, while in the delayed conditional position discrimination task, the monkey needs not to retain the sample color but to recall the response direction during the delay period. Therefore, the comparison of delay-period activity of the same neuron between two tasks clarifies whether delay-period activity represents information regarding the sample color or the response direction. They found that both delay-period activities representing the sample color and delay-period activity representing the response direction can be observed in prefrontal neurons, although neurons exhibiting differential delay-period activity were small population (15%). Thus, although delay-period activity was observed in many prefrontal neurons, a small percentage of neurons exhibited differential delay-period activity depending on the difference in non-spatial color stimuli.

Wilson et al. (1993) analyzed prefrontal activity under oculomotor types of delayed tasks either using spatial cues presented on the right or the left position or using pattern or object cues presented over the fixation point. They showed that 77% (24/31) of neurons having delay-period activity exhibited selectively to non-spatial memoranda and that delay-period activity exhibited selectivity generally specific to certain patterns or object stimuli. Miller et al. (1996) used the delayed matching-to-sample task and examined delay-period activity in the prefrontal cortex. In their task, the presentation of the sample object (0.5 s) was followed by the presentations (0.5 s) of up to 5 non-matching objects with a 1-s inter-stimulus interval. The task was terminated with the appearance of the sample object and the monkey's bar release response. They found that, among 56% (82/145) of prefrontal neurons exhibited delay-period activity, 28% (40/145) exhibited selectivity to the sample object and that sample selectivity of delay-period activity was maintained throughout the delay period even when non-matching objects were presented as distractors during the delay period. These results indicate that delay-period activity of prefrontal neurons represents non-spatial object information and maintains it during the delay period and that sustained delay-period activity is robust against distractors presented during the delay period. Further, Rainer et al. (1998a) examined lateral prefrontal activity using the delayed matching-to-sample object's place task. In this task, a simple array of three objects is presented during the sample period. The monkey is required to remember the position of the target object during a 1.5-s delay period and to release a lever during the test period if the target object appears at the same position as before. They showed that delay-period activity was affected by the target object itself in 32% (31/97) of prefrontal neurons and by the position of the target object within the array in 46% (45/97) of neurons. Further, Hasegawa et al. (1998a, b) showed that sample cue-specific delay-period activity was observed in a few prefrontal neurons (2/22) while the monkey performed an oculomotor delayed matching-to-sample task. Thus, the delay-period activity observed in prefrontal neurons exhibits selectivity to specific colors or objects while the monkey performs non-spatial working memory tasks, indicating that delay-period activity not only represents spatial information but also represents non-spatial information such as color and object.

It has been shown that delay-period activity represents information regarding visual motion. While monkeys performed a delayed matching-to-motion direction task, Hussar and Pasternak (2012) showed delay-period activity representing remembered direction of visual motion in the dorsolateral prefrontal cortex, although this representation was transient and appeared at different timings during the delay period in different neurons. They also showed that delay-period activity representing the direction of visual motion was mainly observed in putative pyramidal neurons.

Further, it has been shown that delay-period activity represents not only physical attributes of the object but also other features of the task at the same time. Inoue and Mikami (2006) used a serial probe reproduction task (conditional delayed matching-to-sample task) and showed that the magnitude of delay-period activity was affected by the physical features of non-spatial visual cues. They used three object stimuli as the first and the second cues and two colored stimuli (red and green) as color cues. In this task, while the monkey looks at a fixation point, two object stimuli are presented sequentially for 0.5 s (first and second cue period) separated by a 1-s interval (first delay period) and, after a 1-s delay (second delay period), a color cue is presented for 0.5 s (color cue period), which instructs the monkey which object (target object) is needed to select by an eye movement after a 1-s delay (third delay period). Red and green color cues indicate the monkey's selection of the first and the second object as a target object, respectively. During the response period, three object stimuli are simultaneously presented at three predetermined positions randomly and the monkey needs to make a saccade to the target object. Since information regarding the behavioral response is provided only during the response period (only after the end of the third delay period), neural activities observed during the first, the second, and the third delay periods are thought to represent information regarding the object and/or the order of its presentation. Delay-period activity was observed in 183 neurons during the first and the second delay period. Among these, 40 (22%), 18 (10%), and 64 (35%) neurons exhibited object-selective delay-period activity during both delay periods regardless of the order of stimulus presentation, only during the first delay period, and only during the second delay period, respectively. Delay-period activity was observed in 260 neurons during the third delay period. Among these, 102 (39%), 61 (23%), and 38 (15%) neurons exhibited delay-period activity affected by both the color cue and the target object, only by the color cue, and only by the target object, respectively. Thus, the magnitude of delay-period activity is affected by the difference in physical features of visual cues in addition to the order of their presentation, indicating that delay-period activity represents not only non-spatial information such as physical attributes of the object cues but also other features of the task, such as the order of cue presentation.

Thus, although the percentage of prefrontal neurons having delay-period activity representing non-spatial information varies among studies, these studies show that delay-period activity of a population of prefrontal neurons represents information regarding non-spatial visual features. Both spatial and non-spatial visual information are represented and maintained by sustained delay-period activity during the performance of working memory tasks, indicating and supporting that sustained delay-period activity is the neural correlate of the mechanism for working memory.

3.7.2 Delay-Period Activity Representing Object Categories.

Simple visual feature (e.g., color, shape, or direction of motion) has been used as the sample and the test stimuli to examine what information delay-period activity represents during non-spatial working memory tasks. However, not only simple visual features but also various complex physical features are usually needed to be remembered temporarily. Further, abstract and categorical information of objects are also often needed to be remembered. Therefore, it is expected that delay-period activity can represent abstract and categorical information of visual stimuli. Freedman et al. (2001, 2002) examined whether or not prefrontal neurons could encode more complex and abstract visual features such as categorical information of visual stimuli. In their experiment, a large continuous set of photographs generated from prototype photographs of three cats and three dogs were prepared by a three-dimensional morphing system (Fig. 3.21b). Different amounts of morphing from the prototype photographs generate thousands of unique sample images which continuously vary the shape of the cat and the dog, and one or more arbitrary boundaries for categorization can be set to categorize the sample image into "cat" or "dog." The monkey performs the delayed matching-to-sample task and is asked to categorize the sample image morphed from a prototype photograph into either "cat" or "dog" (Fig. 3.21a). In this task, the sample image is first presented for 600 ms, then after a 1-s delay period, the test image is presented for 600 ms. If the test image belongs to the same category as the sample image (e.g., both images are cat photographs), the

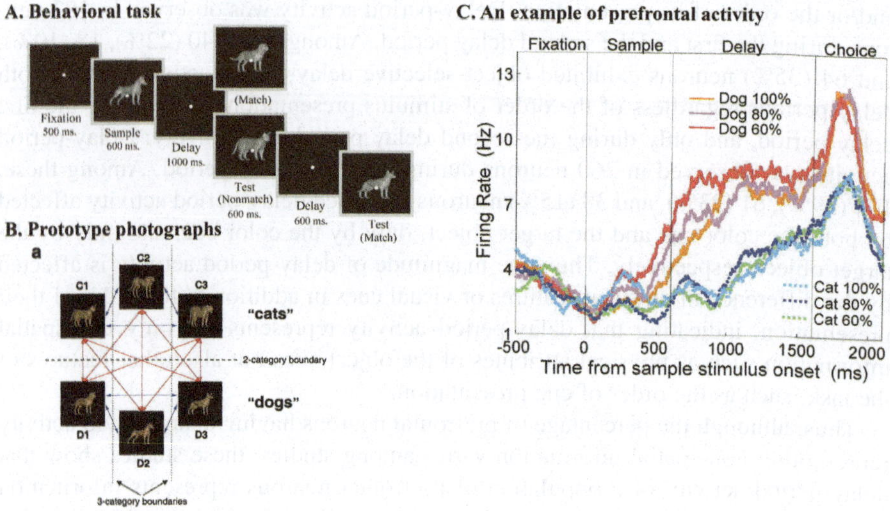

Fig. 3.21 (**a**) Temporal sequence of task events in the delayed matching-to-sample task. (**b**) Prototype photographs and directions for modification by morphing. (**c**) Examples of prefrontal activity. This neuron exhibits significant delay-period activity when the sample image is categorized as dog, regardless of the degree of morphing. From Freedman et al. (2002) with permission

monkey needs to release the lever. They found significantly different magnitude of activity between cat and dog images during the sample period in 19% (74/395; 39 for cat and 35 for dog) of sampled neurons and during the delay period in 13% (51/395; 27 for cat and 24 for dog) of these neurons. Figure 3.21c shows an example of delay-period activity whose magnitude was larger when dog images were presented as the sample image. Differential activation of delay-period activity was observed between two image categories (cat vs. dog), but not observed between two different degrees of morphing images within the same category (e.g., dog 100% vs. dog 60%). Therefore, differential activation of delay-period activity is thought to represent the difference in the perceptual category of sample images and not represent the difference in the perception of sample images (Miller et al. 2002, 2003). They calculated category indices, in which positive values indicate the presence of a greater difference in activity between two categories, and showed that more prefrontal neurons exhibited larger positive values of category indices for delay-period activity than for sample-period activity. Temporal characteristics of category representation by prefrontal neurons revealed that more neurons convey category signals late in the sample period than during the delay period and that the strongest category signal is observed in the late delay period and the early test period. Thus, delay-period activity of prefrontal neurons can represent not only information regarding concrete physical features of the visual stimulus but also information regarding abstract features of the visual stimulus including stimulus category.

Using the same task and the same image sets, Freedman et al. (2003) compared activity between prefrontal neurons and inferior temporal neurons. Inferior temporal neurons exhibited differential activation based on the categorical difference of the sample image preferentially during the sample period, but exhibited no differential activation during the delay period and the test period. On the other hand, prefrontal neurons exhibited differential activation based on the categorical difference between the delay period and the test period. This result indicates that inferior temporal neurons mainly involve in the analysis of currently viewed images, while prefrontal neurons involve more in encoding and representing information which has behavioral significance or meaning for the correct performance of the task. Similar results have been observed by the comparison of population activity between prefrontal and inferior temporal neurons using a population decoding method (Meyers et al. 2008).

3.7.3 Robustness of Delay-Period Activity

Miller et al. (1996) used the delayed matching-to-sample task and compared the robustness of delay-period activity between dorsolateral prefrontal neurons and inferior temporal neurons. In this task, the presentation of the sample object is followed by the presentation of up to five distracting objects during the delay period. They found that, in prefrontal neurons, delay-period activity showed selectivity to the sample object and was maintained throughout the delay period even when distractors were presented during the delay period. In inferior temporal neurons, delay-period activity was observed and exhibited selectivity to the sample object, but was disrupted by intervening distractors presented during the delay period (Miller et al. 1991, 1993; Miller and Desimone 1994). These results indicate that, although delay-period activity observed in both prefrontal and inferior temporal neurons represents and maintains non-spatial object information, the robustness against distractions is a specific feature of delay-period activity in prefrontal neurons. The robustness of delay-period activity against distractions has also been observed in spatial working memory tasks (Qi et al. 2010; di Pellegrino and Wise 1993). Therefore, the robustness of delay-period activity is an important feature of prefrontal neurons for understanding neural mechanisms of working memory processes in the prefrontal cortex.

3.7.4 Retrospective Coding Versus Prospective Coding

Analyses of delay-period activity in spatial working memory tasks revealed that the majority of delay-period activity represents retrospective sensory information while the minority represents prospective motor information. Similar retrospective and prospective representation of information by delay-period activity have been observed while monkeys perform non-spatial working memory tasks. Quintana and Fuster (1999) examined prefrontal contribution to working memory and the preparation for behavioral action using a delayed matching-to-sample task and a delayed conditional position discrimination task. In the delayed matching-to-sample task, either red or green color is presented at the center window during the sample period. After a 12-s delay, two colors are presented at the right and the left windows simultaneously, and the monkey is required to select a window with the same color as the sample color. In the delayed conditional position discrimination task, either red or green color is presented at the center window during the sample period. After a 12-s delay, the monkey is required to select either the left or the right window depending on the sample color. If the sample color is red (green), the monkey needs to select the left (right) window. They found two groups of prefrontal neurons; memory cells and set cells. The activity of memory cells represented the sample color, independent of the response direction, and the discharge rates of memory cells gradually decreased during the delay period, indicating that delay-period activity of memory cells represents retrospective sensory information (sample stimulus). In contrast, the activity of set cells represented the response direction. The discharge rates of set cells gradually increased during the delay period and the increase of discharge rates was in proportion to the predictability of the response direction, indicating that delay-period activity of set cells represents prospective moto information. Thus, delay-period activity representing either retrospective or prospective information can be found in prefrontal neurons during non-spatial working memory tasks. The proportion of neurons encoding retrospective information was similar to those encoding prospective information (Quintana and Fuster 1992, 1999).

On the other hand, Rainer et al. (1999) examined whether the delay-period activity of prefrontal neurons retains information about the sample stimulus or the expected target stimulus while the monkey performs a delayed paired association task. In this task, the monkey needs to remember three arbitrary pairs consisting of different objects by learning before performing the task. During the task, one sample object is first present, then after a delay, one test object is presented. If the test object is the object associated with the sample object during learning, the monkey is required to make a behavioral response (release a lever). Therefore, at a given trial, the monkey needs to remember the sample object (retrospective coding) and, at the same time, recall the correct test object associated with the sample object (prospective coding) during the delay period. Although the monkey made a few errors during this experiment, most errors were caused by the similarity between the target object and non-target object, not by the similarity between the sample object and non-target object, indicating that the monkey relies on the prospective coding strategy. They found two activities during the delay period; prospective activity reflecting information regarding an expected target object and sensory-related (retrospective) activity reflecting information regarding the sample object. The activity related to information processing for the sample object dominated during the sample period. The number of neurons having sensory-related activity was gradually decreased during the time from the sample period to the delay period. However, the number of neurons having prospective activity was increased during the delay period. Thus, this result also indicates the presence of prefrontal neurons encoding either retrospective sensory or prospective motor information while the monkey performs non-spatial working memory tasks.

Rainer and Miller (2002) reported some features of delay-period activity encoding retrospective information. They used an object delayed matching-to-sample task, in which familiar and novel objects for the monkey and noise patterns were used as sample stimuli. They found sample-specific gradually decreasing activity during a 1-s delay period, which corresponds to delay-period activity representing retrospective information. This delay-period activity was enhanced when novel objects and noise patterns were used as sample stimuli, but not enhanced when familiar objects were used as sample stimuli. Therefore, delay-period activity that exhibits sample specificity and gradually decreases during the delay period seems to represents not only retrospective sensory information but also more complex experience-dependent dynamics.

These results indicate that delay-period activity representing either retrospective or prospective information is found not only during non-spatial working memory tasks but also during spatial working memory tasks. These results also indicate that whether delay-period activity represents retrospective information or prospective information is predictable by the temporal pattern of delay-period activity, such that delay-period activity representing retrospective information tends to decrease along the delay period, whereas delay-period activity representing prospective information tends to increase along the delay period. Takeda and Funahashi (2004, 2007) also observed differences in the temporal pattern between delay-period activity representing retrospective information and delay-period activity representing prospective information during the performance of spatial working memory tasks (ODR task). Delay-period activity representing retrospective sensory information exhibited tonic sustained activation along the delay period, while delay-period activity representing prospective motor information exhibited gradually increasing activity along the delay period. Thus, the difference in temporal patterns of delay-period activity predicts whether delay-period activity represents retrospective information or prospective information.

3.7.5 *Interactions Between Spatial and Non-spatial Information*

Delay-period activity represents spatial as well as non-spatial visual features. However, it is not clear whether each prefrontal neuron encodes either spatial or non-spatial information by delay-period activity and whether prefrontal neurons encoding spatial information show different spatial distribution on the prefrontal cortex from neurons encoding non-spatial information. Wilson et al. (1993) examined these issues using the ODR task and the object discrimination task. They showed that prefrontal neurons selectively responding to the objects were found in the ventral convexity of the lateral prefrontal cortex, but not found in the cortex surrounding the principal sulcus. Funahashi et al. (1989) showed that many neurons having spatially selective delay-period activity were found in and surrounding the principal sulcus during the spatial working memory task (the ODR task). Therefore, Wilson et al. (1993) concluded that prefrontal neurons selectively encoding spatial information show separate and discrete distribution on the prefrontal cortex from neurons selectively encoding non-spatial information, such that the formers distribute mainly in the cortex within and surrounding the principal sulcus while the latters distribute mainly in the ventral convexity of the prefrontal cortex.

On the other hand, Rao et al. (1997) examined the activity of lateral prefrontal neurons while the monkey performed a delay task requiring working memory of both the object itself and its spatial position. They showed that about 52% (64/123) of prefrontal neurons exhibited delay-period activity tuning both the object and its position, while 7% (8/123) and 41% (51/123) exhibited delay-period activity tuning either the object or its position, respectively. Although a large number of prefrontal neurons having delay-period activity (92%, 115/123) encoded spatial position of the object, the information regarding the object itself and its spatial position were integrated on many individual prefrontal neurons. These results indicate that neurons encoding spatial information are not separately distributed on the lateral prefrontal cortex from neurons encoding non-spatial information. Rather, neurons encoding spatial information are distributed widely on the lateral prefrontal cortex, and spatial and non-spatial information are encoded integratively on many neurons in the lateral prefrontal cortex. Rainer et al. (1998a) also examined interactions of spatial and non-spatial information on lateral prefrontal neurons using the delayed matching-to-sample task with an array of three objects. In this task, first the target object is shown to the monkey in cued trials before the monkey performs the task. Then, a sample array of three objects is presented to the monkey briefly. The monkey is required to find the target object and remember its position. After a brief delay, a test array is presented and the monkey needs to release a lever if the target object appears at the same position as it was in the sample array. They showed that delay-period activity is affected by the position of the target object within the array in 46% (45/97) of lateral prefrontal neurons and by the target object itself in 32% (31/97) neurons and that no spatially separate distribution is found between neurons encoding spatial information and neurons encoding non-spatial information. Rainer et al. (1998b) examined prefrontal neurons' memory fields by presenting visual objects at various positions and requested monkeys to remember both the object itself and its position. They found that 46% (68/149) of prefrontal neurons having delay-period activity encoded information regarding both the sample object and its spatial position, while 49% encoded information regarding only the spatial position of the sample object. Further, Hoshi et al. (2000) examined prefrontal activity while the monkey performed a delayed matching-to-sample task under two conditions (position-matching and shape-matching conditions). In this task, a sample stimulus (circle or triangle) is presented at one of three positions (right, left, or top) during a 1-s sample period. In the position-matching condition, after a 3-s delay, the sample stimulus is presented at all three positions and the subject is required to reach the position by hand where the sample stimulus was presented during the sample period. In the shape-matching condition, both circle and triangle stimuli are presented at two of three positions, respectively, and the subject is required to reach the position where the sample stimulus is presented. They found that, among 226 prefrontal neurons having delay-period activity (32% of total sample), 54% (121), 5% (11), and 12% (27) exhibit position-selective, shape-selective, and position- and shape-selective delay-period activity, respectively, and that neurons having position-selective delay-period activity and neurons having shape-selective delay-period activity are intermingled in the lateral prefrontal cortex.

If delay-period activity is examined using either spatial or non-spatial working memory task, the information represented by delay-period activity would be inevitably either spatial or non-spatial, respectively. However, examining the same neuron's delay-period activity under both spatial and non-spatial working memory tasks reveals that a population of prefrontal neurons encode both spatial- and non-spatial information, indicating that both spatial and non-spatial information are integrated into individual prefrontal neurons. Although there are prefrontal neurons encoding only non-spatial information, the great majority of prefrontal neurons encode the spatial attribute of the stimulus and these neurons are distributed widely on the lateral prefrontal cortex, suggesting that spatial information processing is an important function of the lateral prefrontal cortex.

3.8 Delay-Period Activity Encoding Somatosensory Information

Delay-period activity representing somatosensory information has been reported in the prefrontal cortex. Romo et al. (1999) used a somatosensory discrimination task (a delayed matching-to-frequency task) and showed that the discharge rate of delay-period activity in prefrontal neurons varied as a monotonic function of the base stimulus frequency. In this task (Hernández et al. 1997), the monkey is required to discriminate the difference of the frequency between two mechanical vibrations applied to the fingertips (Fig. 3.22a). Both the base frequency (sample stimulus) and the comparison frequency (test stimulus) are changed systematically (Fig. 3.22b). The base and the comparison stimuli are sequentially delivered for 1 s with an inter-stimulus interval (delay period) of 3 s. The monkey needs to indicate whether the comparison frequency is lower or higher than the base frequency by pressing a corresponding switch by hand. Therefore, the monkey is required to remember the base frequency during the delay period. Romo et al. (1999) found that the discharge rate of delay-period activity shows either a monotonic increase or a decrease depending on the increase of the base frequency. Brody et al. (2003) further showed that most neurons in the inferior convexity of the lateral prefrontal cortex exhibit delay-period activity whose discharge rate changes monotonically depending on the frequency of the base stimulus which changes linearly from 10 to 34 Hz (Fig. 3.22c). Therefore, they concluded that delay-period activity of prefrontal neurons encodes the frequency of the base stimulus monotonically. Different prefrontal neurons exhibited frequency-encoding activity during different

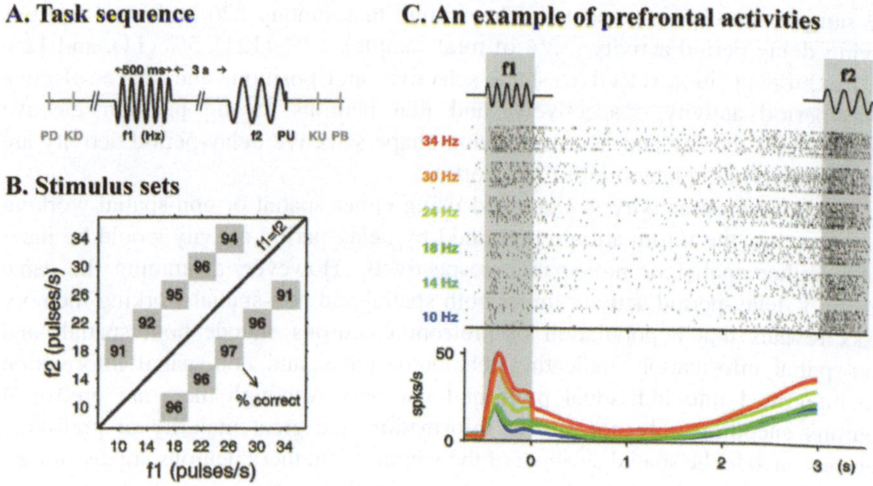

Fig. 3.22 (**a**) Temporal sequence of two somatosensory stimuli (f1 and f2) in the delayed matching-to-frequency task. (**b**) Stimulus sets of two frequency stimuli (f1 and F2). (**c**) Examples of delay-period activity recorded from one prefrontal neuron when a different stimulus set was used. From Brody et al. (2003) with permission

epochs of the delay period. Early neurons (35%, 121/345) exhibited differential frequency-encoding activity during the early epoch of the delay period. Persistent neurons (26%, 90/345) exhibited differential activation throughout the delay period. And, late neurons (29%, 101/345) exhibited differential activation during the late epoch of the delay period. In addition to the prefrontal cortex, neural activity was also examined in the second somatosensory cortex and the medial premotor cortex. The second somatosensory cortex included early neurons, but did not include persistent neurons (Salinas et al. 2000; Romo et al. 2002). The premotor cortex included late neurons predominantly, although other neurons were also observed (Hernández et al. 2002). Frequency-encoding neurons were observed in these cortical areas and exhibited a monotonic or linear change in the discharge rate during the delay period depending on the base frequency. Persistent neurons also encoded information regarding the base frequency (representing retrospective information) and were observed predominantly in the prefrontal cortex.

When the spatial working memory tasks (e.g., the ODR task) are used, prefrontal neurons exhibit delay-period activity selectively representing information regarding particular cue positions and delay-period activity of individual prefrontal neurons represents different spatial information (spatially tuned). Similarly, when the non-spatial working memory tasks (e.g., delayed matching-to-sample task) are used, prefrontal neurons exhibit delay-period activity selectively representing information regarding the sample object itself or some physical features of the sample object. Thus, individual features of both spatial and non-spatial information seem to be encoded by delay-period activity of individual prefrontal neurons. This method for information encoding is similar to the labelled-line code observed in sensory systems. Therefore, each of prefrontal neurons may be labelled by what feature delay-period activity represents.

However, Romo's group (Romo et al. 1999; Brody et al. 2003) showed that the prefrontal neurons can use monotonic one-dimensional representation of stimulus information by linearly increasing or decreasing sustained delay-period activity and that this encoding method is used to represent the quantity of one-dimensional sensory information in working memory. Thus, the nervous system uses not only the labeled line coding method but also a monotonic linear encoding method (one-dimensional representation) to encode linearly altering strength or intensity, or frequency of a sensory stimulus.

3.9 Delay-Period Activity Representing Motor Information

Among prefrontal neurons having directional delay-period activity, 10–25% of these neurons encode information regarding forthcoming behavioral responses in spatial working memory tasks. Niki and Watanabe (1976a) compared the same neuron's delay-period activity among two kinds of spatial delayed-response tasks and a conditional position discrimination task, and found that 78% (25/32) of prefrontal neurons having differential delay-period activity encoded the location of the visual cue, whereas the remaining 22% (7/32) encoded the direction of the behavioral response. Funahashi et al. (1993a) found that 59% (30/51) of prefrontal neurons having directional delay-period activity encoded the position of the visual cue, while 25% (13/51) encoded the direction of the saccade. Takeda and Funahashi (2002) showed that 13% of prefrontal neurons having directional delay-period activity encoded the direction of the saccade. Delay-period activity representing the direction of the saccade tends to exhibit gradually increasing activity along the delay period, although sustained delay-period activity representing the direction of the saccade was also present (Takeda and Funahashi 2004, 2007).

Similarly, Hasegawa et al. (1998a) showed delay-period activity representing information regarding forthcoming saccades. They used an oculomotor delayed matching-to-sample task, in which the monkey is required to first determine the direction of the saccade (left, right, up, or down) based on the combination of the sample and the matching stimuli presented sequentially with a 1-s or 1.5-s inter-stimulus interval (first delay period) and then perform a saccade 1.5 s after the termination of the matching stimulus (second delay period). They found that 80% of prefrontal neurons were activated during the second delay period and that delay-period activity of these neurons represented information regarding the direction of the forthcoming saccade, and not represented the matching stimulus. When the delayed-response task is performed, both information regarding the position of the visual cue (sensory information) and information regarding the direction of the motor response (motor information) are necessary. Therefore, it is expected that neurons encoding sensory information and neurons encoding motor information during the delay period are both present in the prefrontal cortex. However, Hasegawa et al. (1998a) showed that, when motor information is the only information necessary to retain during the delay period, delay-period activity of the most neurons represent motor information.

Niki and Tsutsui (1997) showed prefrontal activity representing and retaining information of self-performed movement. They trained the monkey to perform a conditional discrimination of the monkey's own movement. In this task, the monkey is required to push a handle either 3 or 12 times during the cue period as a motor cue. After the delay, the monkey needs to move the handle either leftward or rightward depending on the number of handle-pushing during the cue period. They found that many prefrontal neurons exhibited differential delay-period activity depending on 3- or 12-push conditions. Then, they examined the same neuron's delay-period activity between the conditional discrimination task of own movement and the visual discrimination task using left-or-right lever movements. They found that some prefrontal neurons exhibited differential delay-period activity in both tasks, indicating that delay-period activity of these neurons represents information regarding forthcoming motor behavior. They also found that other prefrontal neurons exhibited delay-period activity only in the conditional discrimination task of own movement, indicating that delay-period activity of these neurons represents information regarding self-performed movement.

A similar feature of delay-period activity was also observed by Procyk and Goldman-Rakic (2006). They asked monkeys to perform the delayed-response task with four visual cues and a self-organized problem-solving task using eye and hand-reaching movements. In the former task, the monkey needs to first make a saccade and then make a hand-reaching response to the target after the delay. In the latter task, four visual targets are presented simultaneously and the monkey first needs to find the correct target by trial and error by first making a saccade and then making a hand-reaching response (search period). If the monkey finds the correct target, it needs to repeat the same sequential eye and hand-reaching movements to the correct target three times (repetition period). They found that delay-period activity involved in the delayed response task also participates in the problem-solving task with the same preference for the target position, because the preference for the target position of delay-period activity observed in the delayed-response task was also maintained during both the search period and the repetition period of the problem-solving task. Since information regarding self-performed movement is an important factor for the monkey to find the correct target by trial and error in the problem-solving task, delay-period activity representing information regarding self-performed movement plays important role in this task.

3.10 Delay-Period Activity Representing Task Difference and Behavioral Rule

Maintaining the task rule is a fundamental process to control a variety of cognitive functions and achieve a goal. Delay-period activity representing the difference of tasks or behavioral rules has been observed in the lateral prefrontal cortex. White and Wise (1999) asked the monkey to perform a task with two different rules (conditional rule and spatial rule) and examined whether or not the prefrontal cortex plays a role to guide a behavior based on learned rules. In this task, either blue or yellow fixation point appeared at the beginning of the trial indicates that the current trial is under the conditional rule or the spatial rule, respectively. After the presentation of the fixation point, one of four complex-colored visual cues is presented at one of four positions (right, left, up, and down) for 0.25 s. After a variable delay of 0.45–0.75 s (first delay), the same visual cue reappears at the same position for 0.25 s. After another variable delay of 0.45–0.75 s (second delay), four white visual targets are simultaneously presented at four positions. Then, the monkey is required to make a saccade to the visual target assigned by the visual cue, maintain fixation at that visual target for 0.5–1.5 s, and release the bar immediately after the visual target is dimmed. In the conditional rule (blue fixation point), each visual cue is uniquely associated with one of four target positions, and the shape and the color of the visual cue indicate which visual target would dim, regardless of where the visual cue was repeatedly presented. In the spatial rule (yellow fixation point), the position of the visual cue indicates which visual target would dim, regardless of what shape and color the visual cue had. They found that 40% (89/221) of task-related prefrontal neurons exhibited significantly different activities depending on the difference of the rules. Although activity depending on the difference in the rules was observed throughout the task epochs, delay-period activity affected by the difference of the task rules was also observed in 54% (51/95) and 52% (59/114) of task-related neurons during the first and the second delay period, respectively. This rule-dependent effect was not caused by the change in neural excitability across trials, the difference in the color of the fixation point, and the difference in the relation of cue positions and saccade directions between two rules. In addition, neurons exhibiting rule-dependent activity were observed throughout the dorsolateral prefrontal cortex.

Wallis et al. (2001) also reported prefrontal activity representing behavioral rules. They also used two rules. In the matching rule, the monkey needs to release a lever when two successively presented objects (sample and test objects) with 1.5 s intervals are identical. In the nonmatch rule, the monkey needs to release a lever when two successively presented objects are different. Which rule is applied in any given trial is informed by an additional cue (e.g., a drop of juice or green background) that is presented with the sample object. They found that 41% (200/492) of prefrontal neurons exhibited rule-selective activity, which showed stronger excitation during the 1.5-s delay period when the monkey performed trials under either match or nonmatch rule. The activity of these neurons is considered rule-selective, because response features of the activity were not explained by the physical features of the sample object or the cue, by the anticipation of the behavioral response, and by reward expectation. In addition, 34% (167/492) of sampled neurons exhibited interactive effects in delay-period activity between the behavioral rule and the sample object. Rule-selective delay-period activity was observed not only in the dorsolateral prefrontal cortex but also in the ventrolateral and the orbital prefrontal cortex in similar percentages (29%, 24%, and 23%).

Prefrontal neurons having directional delay-period activity are affected by behavioral rules and flexibly change discharge rates of delay-period activity depending on the change of the behavioral rule. Hoshi et al. (2000) showed rule-dependent modulation of activation timing in the lateral prefrontal cortex while the monkey performed a task under two rules; location-matching rule and shape-matching rule. In this task, one sample cue (either triangle or circle) first appears at one of three predetermined positions (top, left, and right) (sample cue period). After a 3-s delay period, the choice cues are presented (choice cue period) and these choice cues indicate which rule is applied for the monkey to make a response after the presentation of the Go signal. In the location-matching rule, either three triangles or three circles are presented at three predetermined positions during the 1.5-s choice cue period, and the monkey is required to make a hand-reaching behavior to the position where the sample cue was presented. In the shape-matching rule, both one triangle and one circle are presented simultaneously at two positions randomly selected from three predetermined positions during the 1.5-s choice cue period, and the monkey is required to reach the position where the same shape as the sample cue is presented. During the 3-s delay period, they found delay-period activity selective to either the position (54%, 121/226) or the shape (5%, 11/226) or both (12%, 27/226) of the sample cue, indicating that delay-period activity represents information regarding spatial and non-spatial features of the sample cue. They classified prefrontal neurons into three groups based on the difference in the activity during the choice cue period. One group of neurons (88% of 226 neurons) exhibited sustained activity during the delay period and the choice cue period, which represented information regarding the sample cue (delay-period activity). Second group of neurons (32%, 117/362) exhibited sustained activity during the choice cue period, which represented information regarding reaching the target position (target-position selective activity). Third group of neurons (46%, 165/362) exhibited excitatory activity during the choice cue period, which represented the configuration of the choice cue stimuli (three circles or three triangles or a circle and a triangle) regardless of the difference of the reaching positions across trials (configuration activity). Among third group of neurons, 23 (14%) and 88 (53%) were significantly more active in the location-matching rule and the shape-matching rule, respectively. The remaining 54 (33%) had no significant difference in activation between two rules. Figure 3.23 shows temporal changes of three types of prefrontal activity along a trial in the location- and the shape-matching rules. Green, blue, and red lines indicate population histograms of delay-period activity, configuration activity, and target-position selective activity, respectively. Before the rule information is provided to the monkey, prefrontal neurons encoded information regarding the sample cue as sample stimulus-specific

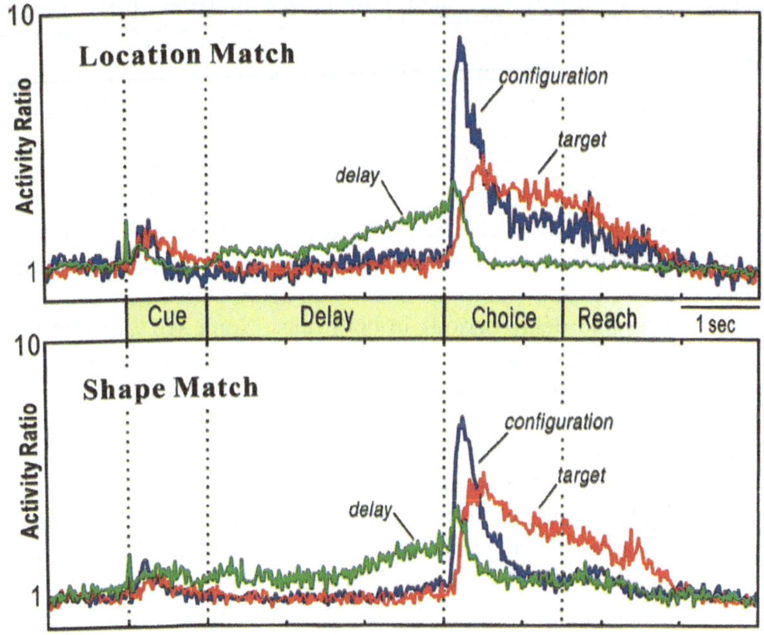

Fig. 3.23 Temporal changes of three types of prefrontal activity along a trial in location-matching and shape-matching rules. Each colored line indicates population histogram of delay-period activity, configuration activity, and target-position selective activity, respectively. From Hoshi et al. 2000) with permission

delay-period activity. When the choice cues were presented, neurons having configuration activity exhibited phasic excitation, which represented the behavioral rule. Since the magnitude of phasic excitation of configuration activity changed depending on the location-matching or the shape-matching rule, the reaching target position was determined by the combination of the sample cue, the choice cue, and the behavioral rule. In fact, target-position selective activity developed after the activation of configuration activity. Figure 3.23 shows an importance of the temporal activation pattern of neural activity representing different information in information processing and an importance of the activation timing of the activity representing behavioral rules in appropriate decision-making for performing correct behavioral responses.

Rule-dependent modulation of activation timing of delay-period activity is also observed in the lateral prefrontal cortex. Amemori and Sawaguchi (2006a) showed that prefrontal neurons having delay-period activity changed representing information depending on the behavioral rule introduced during the delay period. They used an oculomotor version of a two-choice delayed-response task including pro- and anti-saccade conditions. In this task, a triangle is presented as a visual cue for 0.8 s either on the right or on the left of the fixation point and then the first delay period of 2 s is introduced. At the end of the first delay period, either a square or a cross is presented for 0.6 s as the rule cue, which instructs the direction of the saccade to the monkey, and then the second delay period of 2 s is introduced. During the response period, if the rule cue is the square, the monkey needs to make a saccade to the position where the visual cue was presented (pro-saccade rule). If the rule cue is a cross, the monkey needs to make a saccade opposite to the position where the visual cue was presented (anri-saccade rule). Among 301 prefrontal neurons having directional delay-period activity, 33% (98/301) showed directional delay-period activity only during the first delay period in both rules, while 37% (113/301) showed directional delay-period activity only during the second delay period in both rules. The remaining 30% (90/301) exhibited directional delay-period activity during both the first and the second delay periods in both rules. Some (62%; 56/90) of these neurons exhibited sustained delay-period activity throughout both first and second delay periods in the pro-saccade condition when the visual cue was presented at the preferred position. However, these neurons exhibited sustained delay-period activity only during either the first delay period or the second delay period in the anti-saccade condition. Therefore, it is concluded that these neurons exhibit directional delay-period activity only when the preferred cue position and the preferred saccade direction are the same. On the other hand, the remaining neurons (38%; 34/90) exhibited sustained delay-period activity throughout both first and second delay periods in the anti-saccade condition, but exhibited delay-period activity only during either the first or the second delay period in the pro-saccade condition. Therefore, it is concluded that these neurons exhibit delay-period activity when the preferred cue position and the preferred saccade direction are opposite. Although the former neurons seem to represent the visual cue position while the latter neurons seem to represent the saccade direction, these results indicate that prefrontal neurons encoding specific task-related information flexibly modulate their discharge rates depending on the variation of its necessity caused by the change of the behavioral rule.

Delay-period activity observed in studies by Hoshi et al. (2000) and Amemori and Sawaguchi (2006b) did not represent information regarding the behavioral rule, but showed that information represented by prefrontal delay-period activity was flexibly turned on and off by external signals such as rule cues in order to adapt new behavioral situations. Flexible change of information represented by delay-period activity depending on the difference of the task and the rule would be beneficial to perform multiple tasks simultaneously or alternately.

Hussein et al. (2014) further indicated that the dorsolateral prefrontal cortex contributes to the maintenance of the task rules. They asked the monkey to alternately perform either pro-saccade or anti-saccade tasks unpredictably in two rules (cued rule and uncued rule) and applied bilateral cryogenic deactivation to the inferior dorsolateral prefrontal cortex (idlPFC) (the cortex lining the caudal principal sulcus) and the dorsally adjacent superior dorsolateral prefrontal cortex (sdlPFC). Deactivation of the idlPFC caused impairments in both pro- and anti-saccade performances only in the uncued rule condition, while deactivation of the sdlPFC caused no impairment in either task. Combined deactivation of both areas caused impairments only in anti-saccade tasks. Since task rule-dependent impairment was observed only after the cryogenic deactivation of the idlPFC, these results indicate the importance of the dorsolateral prefrontal cortex for maintaining task rules.

Rule-dependent delay-period activation has also been observed in the human prefrontal cortex. Bunge et al. (2003) asked human volunteers to perform a behavioral task requiring association of unfamiliar shapes and nonwords in match and non-match rules and examined activation areas in human prefrontal cortex by event-related fMRI. The magnitude of activation was correlated with the difference of the rule during the cue and the delay periods in the left ventrolateral prefrontal cortex, not in the dorsolateral prefrontal cortex. Greater activation was observed when the subject maintained a particular rule (match or non-match), not a response type (go left or go right). This result indicates that not the dorsolateral but the ventrolateral prefrontal cortex participates in the retrieval and the maintenance of behavioral rule representation.

3.11 Delay-Period Activity Representing Reward Expectation

Prefrontal neurons, especially orbitofrontal neurons, have shown to exhibit reward-related activity responding to the reward delivery during task performance and exhibit activity representing not only the expectation of the reward delivery but also quality and quantity of the reward (Rosenkilde et al. 1981a, b; Watanabe 1996; Leon and Shadlen 1999; Tremblay and Schultz 1999, 2000a, 2000b; Hikosaka and Watanabe 2000). Since the dorsolateral prefrontal cortex receives anatomical inputs from the orbitofrontal cortex (Cavada et al. 2000), dorsolateral prefrontal neurons are also known to exhibit reward-related activity.

Watanabe (1996) showed an effect of reward expectation on delay-period activity using a delayed-response task under three reward conditions (a visible food-reward condition, a cued food-reward condition, and a cued liquid-reward condition). In the visible food-reward condition, the monkey is able to see the food reward (raisin, cabbage, apple, etc.) directly at one of the two response windows (right and left) during the cue period, and after a 5-s delay, the monkey is required to select the response window where the reward was placed during the cue period and get the reward by hand. In the cued food-reward and the cued liquid-reward conditions, one of two windows (right or left) is illuminated during the cue period, and after a 5-s delay, both windows are illuminated and the monkey is required to select the correct light window to obtain either food reward (raisin, cabbage, apple, etc.) or liquid reward (water, orange juice, grape juice, etc.). In this experiment, the same reward is used for several consecutive trials. Therefore, the monkey can expect which food or liquid reward would be given at each trial. Delay-period activity was examined in 42 neurons using more than three kinds of reward. Among them, 21 exhibited different magnitude of delay-period activity depending on the difference of the reward. The magnitude of reward-dependent delay-period activity was affected by the kind of reward that the monkey would receive at the correct response. Since which reward would be given was not indicated to the monkey during the cue period in the cued reward condition, reward-dependent delay-period activity can be considered to reflect an expectation of the reward.

Leon and Shadlen (1999) also showed that delay-period activity is affected by spatial information as well as reward information. They examined the interaction between spatial information processing and reward expectation in lateral prefrontal neurons. They trained monkeys to perform a memory-guided saccade task under two conditions (spatial cue first condition and reward cue first condition). In both conditions, the monkey is required to maintain gazing at a fixation point. The spatial cue indicates a saccade direction and is briefly presented at one of two or four peripheral positions. The reward cue indicates the amount of reward that would be obtained after making a correct saccade and is presented over the fixation point. The amount of the reward (large or small) is indicated by the difference in the color of the reward cue. In the reward cue first condition, the reward cue is presented first, then the spatial cue is followed after 0.4–0.6 s delay. In the spatial cue first condition, the spatial cue is presented first, then the reward cue is followed after 0.8–1.5 s delay. Then, the delay period of 1.0–4.0 s is introduced in both conditions and after the delay period, the monkey is required to make a saccade to the position where the spatial cue was presented. They observed that, although delay-period activity was affected by the direction of the saccade, this activity was enhanced by the expectation of the large reward and that reward-related enhancement of delay-period activity was evident when the spatial cue was presented in the neuron's response field. They also examined reward-related enhancement of delay-period activity in the frontal eye field and found no enhancement in these neurons. Thus, mixed representation of both spatial mnemonic information and information regarding reward expectation can be observed in delay-period activity of dorsolateral prefrontal neurons, but not in frontal eye field neurons. Therefore, mixed representation of spatial mnemonic information and reward expectation must be a distinct feature of delay-period activity in the dorsolateral prefrontal cortex.

Kobayashi et al. (2002) also observed mixed representation of visuospatial processing and reward processing in the lateral prefrontal cortex. They used an oculomotor delayed-response task with asymmetric reward schedule. The position of the visual cue (left or right) indicates the saccade target after the delay period (0.9–2.1 s) and the color of the visual cue (red or yellow) indicates whether a reward will be given or not given during the response period. While the monkey looks at a fixation point, the visual cue is presented briefly (0.2 s) and, after a delay (0.9–2.1 s), the monkey is required to make a saccade to the target position within 0.4 s. In no-reward trials, the monkey will not receive any reward regardless of whether the monkey makes a correct saccade or not. They found delay-period activity representing visuospatial information (only the positions of the visual cue), delay-period activity representing reward expectation (only whether the reward would be given in the current trial or not), and delay-period activity representing both visuospatial information and reward expectation. Delay-period activity representing both visuospatial information and reward expectation was further classified into two groups; one exhibited higher discharge rates in the reward present condition and the other exhibited higher discharge rates in the reward absent condition. Spatial discriminability of delay-period activity was improved by the reward expectation in neurons exhibiting higher discharge rates in the reward present condition. Therefore, the reward information improves mnemonic processing in the prefrontal cortex.

Watanabe et al. (2002) further analyzed the effect of reward expectation on delay-period activity using a delayed reaction task. In this task, an instruction cue indicates either the presence or the absence of the reward in a given trial. In the visible food condition, the monkey can see the presence or the absence of the food reward through a window during a 1-s instruction period. In the cued food or the cued liquid condition, either red or green color signal is presented during the instruction period and indicates the presence or the absence of the reward during the reward period. The monkey is required to keep pressing the lever during the 1-s instruction period and a 5-s delay period and release the lever and press a correct response window after a Go signal. The monkey receives one of several kinds of food or liquid reward repeatedly in a block of many (~50) trials. Among 179 neurons that exhibited discriminative activity between rewarded and unrewarded trials during the instruction period and/or the delay period, 106 and 69 exhibited higher and lower discharge rates in rewarded trials compared with unrewarded trials, respectively, regardless of whether the task was the visible food condition or the cued condition. The monkey's preference was examined using different kinds of food and liquid rewards. Among 72 neurons, 8, 14, and 50 exhibited reward-discriminative activity only in rewarded trials, only in unrewarded trials, and in both rewarded and unrewarded trials, respectively. The majority (76%, 55/72) exhibited significantly greater activation when the preferred reward was expected to be obtained or omitted. Thus, these results showed that delay-period activity represents not only whether the reward would be obtained in a given trial or not but also which reward would be obtained or omitted in a given trial. In addition, some (39%, 28/72) of these neurons exhibited a reward-discriminative sustained change of baseline activity during the 10 or 12 s pre-instruction period or during an entire trial. These neurons are thought to encode contextual information regarding which reward would be obtained in the next trial.

Delay-period activity affected by the reward expectation was further observed by Tsujimoto and Sawaguchi (2005), who showed that delay-period activity of dorso-lateral prefrontal neurons was affected by the length of the predicted delay between the Go signal and the reward delivery (0.5 s of short delay and 2.0 s of long delay) in the ODR task. Amemori and Sawaguchi (2006a) also observed enhancement of directional selectivity of delay-period activity by reward expectation.

Thus, delay-period activity exhibits mixed representation of not only spatial- and non-spatial information but also an expectation of the reward delivery and the reward itself. The magnitude of delay-period activity encoding spatial or non-spatial information is affected by the expectation of the reward. In addition, delay-period activity represents not only whether the reward would be obtained in a given trial or not but also what reward would be obtained or omitted. Delay-period activity also represents contextual information regarding what reward would be obtained in the next trial. Reward-related information is important information represented by delay-period activity in the prefrontal cortex.

3.12 Delay-Period Activity Representing Temporal Order of Stimulus Presentation

It had been shown that monkeys with bilateral lesions in the dorsolateral prefrontal cortex exhibit impairments of behavioral tasks that require memory of sequential movements or temporal order of stimulus presentation (Pinto-Hamuy and Linck 1965; Brody and Pribram 1978; Passingham 1985; Petrides 1991). Similarly in humans, Rowe et al. (2000) asked human subjects to perform a spatial working memory task, in which the subject needed to remember the positions of three visual cues during the delay period and then choose one position based on an instruction. Using fMRI, they found that area 46 was activated in relation to the selection of particular spatial position from memorized positions, while area 8 was activated in relation to the maintenance of the spatial positions in working memory. Rowe and Passingham (2001) further examined the role of the prefrontal cortex in working memory performance using an event-related fMRI method. In the task, human subjects were required to remember spatial positions of the visual cues as well as the temporal order of their presentation. The results showed that the memory for spatial positions of the visual cues and the temporal order of their presentation was associated with activation of area 8, while the selection of one appropriate cue position based on the temporal order of its presentation was associated with activation of the dorsolateral (area 46) and the ventrolateral prefrontal cortex. Thus, both of these studies suggest that the dorsolateral prefrontal cortex (area 46) specifically participates in the selection of the particular item based on the information stored in working memory.

Using monkeys, it has been examined whether or not delay-period activity is able to represent information regarding the temporal order of stimulus presentation during delayed-response performances. In this task, multiple cues are sequentially presented and sequential behavioral responses are required based on the temporal order of cue presentation. The results showed that delay-period activity represents information regarding a particular pair of visual cues, the temporal order of cue presentations, and a particular visual cue presented in a particular temporal order.

Barone and Joseph (1989) used a spatial sequencing task and examined the prefrontal contribution to sequential movement control. In their task, while the monkey looks at a fixation point and presses a key by hand, three light-targets are turned on one after another in random order and remain on at a constant intensity. After the fixation point is turned off, all three light targets are fully illuminated for 0.8–2.0 s (fixation period), during which the monkey is able to look at the first light-target. After all light-targets dim, the monkey is allowed to make a hand-reaching behavior to the first light-target. If the monkey correctly presses the first light-target, the remaining two light-targets are fully illuminated for 0.8–2.0 s, during which the monkey is able to look at the second light-target. If these two light-targets dimmed, the monkey needs to make a reaching behavior to the second light-target. If the monkey correctly presses the second light-target, the remaining light-target is fully illuminated for 0.8–2.0 s. If the light-target is dimmed, the monkey is allowed to make a reaching behavior to the third light-target. If the monkey correctly reaches the three light-targets in the correct temporal order, the reward is given. Barone and Joseph (1989) found two groups of prefrontal neurons (visual tonic cells and context cells) and both groups of neurons exhibited sequence-dependent responses. Visual tonic cells (11.5%, 35/302) showed sustained activation following the onset of the light-target as long as the monkey maintained fixation at the light-target. Among visual tonic cells, 68.5% (24/35) exhibited sustained activation during the fixation period when a particular light-target appeared first in the sequence, while 31.5% (11/35) exhibited sustained activation during the fixation period only when a particular light-target appeared second in a given sequence. On the other hand, context cells (36.5%, 116/302) showed activation during the fixation at a specific light-target when the monkey performed hand-reaching behavior toward that light-target. The activation of context cells depended on which light-target had been pressed previously or was going to be pressed next. Sequence-dependent sustained activation during the fixation period was mainly observed in the superior arcuate area.

Funahashi et al. (1993c, 1997) showed delay-period activity representing information regarding a particular pair of spatial cue positions and the order of their presentation in the dorsolateral prefrontal cortex. They used a delayed reaching task and a delayed sequential reaching task with horizontally arranged three targets (left, middle, and right). In the delayed reaching task, the monkey is required to keep pressing a hold key during the pre-cue (1.5 s), the cue (0.5 s), and the delay (2–5 s) periods. The visual cue is presented at one of three target positions for 0.5 s, and after the delay period, the monkey is required to make a hand-reaching behavior to the target position and press the target. In the delayed sequential reaching task, after the monkey presses the hold key for 1.5 s, visual cues are sequentially presented at two of three target positions one by one for 0.5 s. After 2–5 s of the delay period, the monkey is required to make hand-reaching responses to two target positions with the same temporal order as that during the cue presentation. Among 72 neurons having delay-period activity in the delayed sequential reaching task, 19 (26%) had delay-period activity representing a particular position of the visual cue. Among these neurons, 13 (68%) showed delay-period activity observing only when the visual cue was presented at a particular position in a particular temporal order. Figure 3.24a is an example of this delay-period activity. In this neuron, delay-period activity was observed only when the first cue was presented in the right position (R-M and R-L trials), but not observed when the visual cue was presented on the right position secondly (L-R and M-R trials). In the delayed reaching task, delay-period activity was observed when the visual cue was presented on the right position (R trials). The remaining six neurons exhibited delay-period activity only when the second cue was

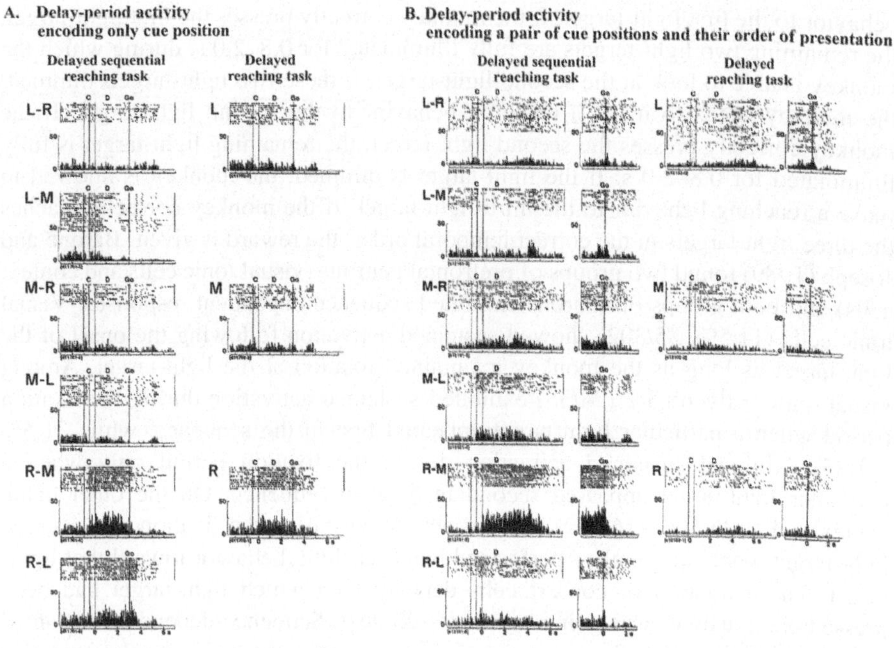

Fig. 3.24 Examples of delay-period activity of the same prefrontal neuron during a delayed sequential reaching task and a delayed reaching task. (**a**) Delay-period activity representing the right cue position. (**b**) Delay-period activity representing a pair of cue positions (right and middle) as well as the order of their presentation (right first and middle second). From Funahashi et al. (1997) with permission

presented at a particular position in the delayed sequential reaching task and exhibited no delay-period activity in the delayed reaching task. On the other hand, 39 neurons (54%) had delay-period activity representing a particular pair of cue positions. Among these, 32 (82%) showed delay-period activity observing only when a particular pair of visual cues were presented in a particular temporal order. Figure 3.24b is an example of this delay-period activity. In these neurons, delay-period activity was observed only when the visual cues were presented on the right and the middle positions as a pair and only when two visual cues were presented in a particular temporal order (first right and second middle; R-M trials). This neuron did not exhibit delay-period activity in the delayed reaching task. Thus, delay-period activity of prefrontal neurons represents information regarding not only a single and particular position of the visual cue but also a particular spatial combination of visual cues. In addition, delay-period activity also represents information regarding the temporal order of visual cue presentations. The presence of delay-period activity representing multiple positions of the visual cues and their temporal order of presentation in the dorsolateral prefrontal cortex indicates that this cortical area plays an important role in planning and preparing sequential behaviors. Similar results have been observed in prefrontal neurons when the monkey performed an oculomotor version of the delayed-response task which included the successive presentation of two visual cues during the cue period and sequential performance of two saccades toward two visual cues during the response period (Inoue and Funahashi 2002).

Further, Qi et al. (2015) reported prefrontal activity representing the temporal order of stimulus positions using oculomotor versions of the spatial delayed-response task. In their task, a spatial cue is presented sequentially at two of eight predetermined positions with an inter-stimulus interval of 1.5 s and, after a 1.5-s delay period, the monkey is required to make a saccade to either the first position or the second position depending on the color of the fixation point. The magnitude of delay-period activity was modulated in prefrontal neurons depending on whether the first cue position or the second cue position was needed to be remembered. Enhanced activation of delay-period activity representing the position of the first cue or the second cue was observed when the position of the first cue or the second cue was needed to be remembered, respectively. Konecky et al. (2017) also observed prefrontal activity representing the serial position of the sample stimulus while the monkey was required to hold three sequentially presented samples (either different positions or different colors) during the sample and the delay periods. Prefrontal neurons exhibited phasic delay-period activity immediately after the sample presentation and this activity was affected by the serial position of the sample stimulus.

Specificity of prefrontal delay-period activity in representation of temporal order of stimulus presentation was shown by comparing features of delay-period activity between prefrontal neurons and other cortical neurons. Qi et al. (2015) showed that enhanced delay-period activity representing the position of the first cue or the second cue was observed in prefrontal neurons when the position of the first cue or the second cue was needed to be remembered, respectively. However, in posterior parietal neurons, enhanced activation was observed in delay-period activity representing the position of the second cue only when the position of the second cue was needed to be remembered. Carpenter et al. (2018) showed neural activity representing serial order of stimulus presentation in the dorsolateral prefrontal cortex and compared its activity with the activity observed in the motor cortex and the premotor cortex. In their task, while the monkey keeps a cursor within the center hold window, up to five visual stimuli are sequentially presented and remained on the screen (list presentation period). At the end of the list presentation period, one of the stimuli (test stimulus) changes its color. The rule of the task is to move the cursor toward the test stimulus immediately followed the presentation of its stimulus. Many prefrontal neurons changed their activity according to the serial position of the list in a relative manner across different list lengths. In addition, these activities were affected by both the spatial position of the visual stimulus and the serial position of the list. In the motor cortex and the premotor cortex, neurons having task-related activity during the list presentation period also exhibited direction-selective activity during the response period, while in many prefrontal neurons, task-related activity was observed only during the list presentation period.

Averbeck and Lee (2007) examined prefrontal activity while the monkey performed a sequential decision task. In this task, the monkey needs to perform a saccade to the correct target three times in succession during a trial. Each time, two visual targets are simultaneously presented to the monkey and the monkey is required to select the correct target by trial and error to continue the trial. Only after the monkey correctly selects the target by a saccade at each of three times in succession, a reward is given. Therefore, the monkey needs to remember the correct target positions throughout a trial. They found that prefrontal activity reflected the sequence of decisions. This activity was initiated in advance of the execution of sequential saccades. This activity predicted where in the sequence the monkey made an error. These results indicate that prefrontal activity is able to retain information regarding behavioral sequence.

Delay-period activity representing the temporal order of stimulus presentation has also been observed in the prefrontal cortex when non-spatial working memory tasks were used. Ninokura et al. (2003, 2004) used a delayed multiple reaching task. In this task, while the monkey looks at a fixation point, three colored object stimuli are presented for 0.5 s over the fixation point sequentially with 1-s inter-stimulus interval (cue period). The temporal order of the object presentation is randomized across trials. After 1.5-s delay period, three object stimuli are presented simultaneously at three predetermined positions for 0.5 s and then the fixation point disappears, indicating the Go signal to make a response. The monkey is required to perform sequential hand-reaching responses to three object positions in the same order as these objects were presented during the cue period. Among 302 prefrontal neurons having excitatory or inhibitory delay-period activity, 129 (43%) exhibited delay-period activity selective to the temporal order of the object presentation. Among these, 40 (31%) exhibited delay-period activity only when three object stimuli were presented in only one specific order during the cue period, while 89 (69%) exhibited delay-period activity when three object stimuli were presented in more than two specific orders. The remaining173 exhibited delay-period activity, but delay-period activity showed no selectivity to the temporal order of object presentation.

Similarly, Inoue and Mikami (2006) used a serial probe reproduction task (conditional delayed matching-to-sample task) and examined effects of the temporal order of stimulus presentation on delay-period activity of prefrontal neurons. They used three object stimuli as sample and test stimuli and two color stimuli (red and green) as condition stimuli. In this task, while the monkey looks at a fixation point, two of three object stimuli are sequentially presented over the fixation point for 0.5 s separated by 1-s delay (first delay). After 1-s delay (second delay), a color stimulus (either red or green) is presented for 0.5 s, which instructs the monkey which object is needed to select by an eye movement after 1-s delay (third delay). Red or green indicates the monkey to select the first or the second object, respectively. They found 131 neurons that exhibited object-selective delay-period activity during the first and/or the second delay period. Among them, 35 (27%) and 30 (23%) neurons

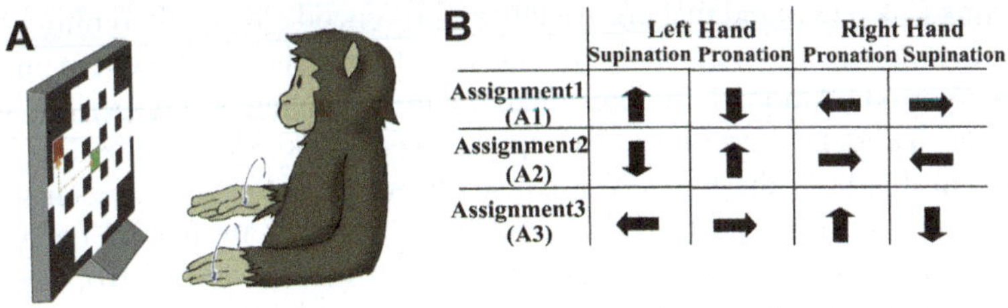

Fig. 3.25 (**a**) Path-planning task on an artificial maze presented on a computer monitor. (**b**) Three combinations between the direction of the cursor movement on the maze (left or right, or up or down) and the direction of the wrist rotation (supination or pronation) of each arm. From Mushiake et al. (2001) with permission

exhibited object-selective delay-period activity only during the second delay period and delay-period activity of these neurons represented information about the first object and the second object, respectively. These results indicate that delay-period activity retains information of both a specific object and the temporal order of its presentation.

Mushiake and others had examined prefrontal contributions to control goal-oriented sequential behavior using a path-planning task on an artificial maze presented on a computer monitor (Mushiake et al. 2001; Saito et al. 2005; Mushiake et al. 2006; Shima et al. 2007). They showed that prefrontal delay-period activity represented information regarding either each of sub-goals to reach the final goal or the final goal. They also showed that delay-period activity represented the direction of the cursor movement on the maze and the temporal order of its move, but not represents the direction of the wrist movement to move the cursor on the maze. In their experiments, the monkey is asked to find an appropriate path to reach the goal on the maze by moving a cursor on the monitor step by step (Fig. 3.25a). To perform this task, the monkey initially needs to learn an association between the direction of the cursor movement on the maze (left or right, or up or down) and the direction of the wrist rotation (supination or pronation) of each arm (e.g., up and down movements of the cursor correspond to supination and pronation of the left wrist, respectively, and left and right movements of the cursor correspond to pronation and supination of the right wrist, respectively) (Fig. 3.25b). First, when the monkey sets the cursor at a start position for 1.0 s, a goal position (final goal) is indicated on the maze for 1.0 s. After 1.0-s delay (first delay), the color of the cursor changes from green to yellow, and the monkey allows to move the cursor to the first position (immediate goal) and holds the cursor at that position for 1.0 s (second delay). The monkey is required to repeat this behavior step by step until the cursor reaches the final goal. Mushiake et al. (2001) found that the monkey always took the shortest path to reach the goal and that the monkey changed the path to the goal flexibly and adaptively when the experimenter put distracters on the way to the goal. Then, Mushiake et al. (2006) examined prefrontal activity while the monkey performed the path-finding task and found that prefrontal neurons encode the forthcoming direction of the cursor movement on the maze, not the direction of the wrist movement, during the delay period. Among 470 prefrontal neurons having delay-period activity, 176 neurons had delay-period activity representing the position of the final goal on the maze. In this task, the monkey was usually required to move the cursor three times to reach the final goal. They found three kinds of prefrontal delay-period activity, each of which represented the direction of the first (111/210), the second (87/210), or the third cursor movement (95/210), respectively, but not the directions of wrist movements for moving a cursor on the maze. They also examined activity of motor cortex neurons and found that all motor cortex neurons encoded the directions of wrist movements for moving a cursor on the maze, not the directions of the cursor movements on the maze. Further, Saito et al. (2005) analyzed prefrontal activity during the first delay period and found three kinds of delay-period activity. The first one (9%, 44/481) represented the position of the immediate goal, the second one (50%, 239/481) represented the position of the final goal, and the third one (41%, 198/481) represented the positions of both the immediate and the final goal. Only 2% (33/1136) of prefrontal neurons encoded the directions of wrist movements. These results indicate that prefrontal delay-period activity is able to retain information regarding behavioral sequence step by step. Retaining step-by-step behavioral sequences by different groups of prefrontal neurons could play significant role to control complex behaviors.

Genovesio et al. (2006) used a conditional eye movement task and observed similar results as Saito et al. (2005) observed. In their task, the monkey needs to select one of three spatial goals (left, right, and up positions) by an eye movement depending on the symbolic visual cues (S or φ) that appeared on the fixation point. The goal for the current trial (current goal) is determined by a combination of the symbolic visual cue that appeared in the preceding trial and the goal that the monkey selected in the preceding trial (preceding goal). During initial fixation period of 1 s, three spatial goals are presented. When the symbolic visual cue presented in the current trial is the same as that in the preceding trial, the monkey is required to select the same goal as the preceding goal in the current trial. When the symbolic visual cue changes, the monkey needs to select one of other goals differed from the preceding goal. They found that prefrontal neurons encode the position of either the previous goal (13%, 108/827) or the future goal (13%, 105/827) during the fixation period and that few neurons encode both positions. These results, together with the results obtained by Saito et al. (2005), indicate that the selectivity of these neurons' activity does not directly link to motor performances but links to visuospatial features and that separate population of prefrontal neurons encodes spatial information of either each sub-goal or the final goal when the subject needs to perform goal-directed sequential behaviors.

3.13 Delay-Period Activity Representing Numerical Quantity

Nieder et al. (2002) used a delayed matching-to-quantity task and showed delay-period activity representing numerical quantity of visual items. In their study, while the monkey grasps a lever and looks at a fixation point, a sample stimulus is presented for 800 ms. The sample stimulus includes one to five dots, which appeared in 24 possible positions on a 5 × 5 matrix. The size of each dot is randomly varied at five different levels and the number of dots is varied from trial to trial. After a 1-s delay period, a test stimulus is presented for 1.2 s. If the test stimulus includes the same number of dots as the sample stimulus had (match condition), the monkey needs to release the lever. If the number of dots in the test stimulus is different from that in the sample stimulus (non-match condition), the monkey is required to keep holding the lever until the test stimulus having the same number of dots as the sample stimulus is presented. Monkey's behavioral performance was affected only by the number of dots that the sample stimulus had, such that the behavioral performance became worse as the number of dots increased. The physical appearance of each numerical quantity (density and arrangement of dots) varied widely. However, such physical appearance was not affected on behavioral performances, indicating that the monkey solves the task based only on the difference in numerical quantity, not on low-level visual features of the sample stimulus. In neurophysiological studies, among 352 prefrontal neurons examined, 37% (131/352) and 32% (111/352) exhibited significantly different activities during the sample and the delay periods, respectively, depending on the number of dots in the sample stimulus. Each neuron exhibited the peak activity at one of the dot quantities (preferred value) and showed a systematic decrease of activity as the number of dots changed from the preferred value. The activity of most quantity-selective neurons was not affected by the difference in the physical appearance of numerical quantity, indicating that the quantity of dots is the critical factor for the activity of these neurons. These results indicate that the activity of the lateral prefrontal neurons represents abstract visual information, such as numerical quantity of visual items.

Nieder and Miller (2003) further examined the representation and scaling of numerical information in prefrontal activity. They used the sample stimulus including one to ten dots and examined whether or not numerical judgment in the delayed matching-to-quantity task follows the psychological low like the Weber-Fechner low. They described behavioral performances as "% same as sample," which corresponds to the ratio that the monkey judges whether the test stimulus contains the same number of dots as the sample stimulus has. When the monkey's behavioral performance is described using "% same as sample" method, the maximum % value is observed when the same number of dots is included in both the sample and the test stimuli and the % value decreases with the increase of the difference of the dot quantity between the sample and the test stimuli. They examined the changes in behavioral performance using % values (behavioral performance curves) across different dot quantities of the sample stimulus and found that these behavioral performance curves are symmetrical when plotted on a logarithmic scale, but become asymmetrical when plotted on a linear scale. This result indicates that the behavioral performance of the monkey's numerical judgment follows the Weber-Fechner law. Similar analysis was performed using activity of 131 and 111 prefrontal neurons which exhibited quantity selectivity during the sample and the delay period, respectively. They found that neurons' response magnitude curves across different dot quantities of the test stimulus are symmetrical when plotted on a logarithmic scale for both groups of prefrontal neurons, indicating that prefrontal activity during the sample and the delay period also follows the Weber-Fechner law for the numerical judgment. The same neural representation was found for the scaling of numerical information in both sample-period activity and delay-period activity.

Thus, Nieder et al. (2002) and Nieder and Miller (2003) showed prefrontal participation in the numerical representation of visual objects. However, the posterior parietal cortex is also known to participate in numerical representation for action (Sawamura et al. 2002). Nieder and Miller (2004) compared numerosity-selective activity between prefrontal neurons and posterior parietal neurons using the delayed matching-to-quantity task and showed that, although both groups of neurons exhibited similar response characteristics in numerical representation of visual objects, posterior parietal neurons responded earlier than prefrontal neurons, suggesting that the numerosity information flows from the posterior parietal cortex to the prefrontal cortex.

On the other hand, numerical information is often associated with quantity and proportional information. Evaluation of quantity and proportional information based on numerical information is an important cognitive function. Bongard and Nieder (2010) examined whether or not the activity of prefrontal neurons represents abstract quantitative rule ("greater than" or "less than" rules) and found that predominant activity observed in prefrontal neurons reflects abstract quantitative rule. Further, Eiselt and Nieder (2013) showed that the majority of rule-selective prefrontal neurons responded only when the quantitative rule was applied to one specific magnitude type, while other neurons represented the quantitative rule related to both magnitudes. Interestingly, pure sensory- and memory-related activity were not observed while the monkey performed the response based on "greater than" or "less than" rule (Bongard and Nieder 2010). Further, Vallentin and Nieder (2008) asked the monkey to discriminate proportions (1:4, 2:4, 3:4, and 4:4) specified by different lengths of bars and showed that the monkey can demonstrate an understanding of proportionality because the monkey discriminated these proportions comparable to those shown by humans. They analyzed the activity of lateral prefrontal neurons while the monkey performed a proportion discrimination task and found that lateral prefrontal neurons selectively activated to preferred proportions during the sample and the delay period, regardless of the physical features of the stimulus.

Thus, the delay-period activity observed in lateral prefrontal neurons can represent not only information regarding numerical quantity but also information regarding quantity and proportionality by an encoding method similar to a labeled-line coding (Nieder and Merten 2007).

3.14 Delay-Period Activity Representing Relative Distance Between Stimuli

Delay-period activity representing temporal and spatial distance has been reported. For mnemonic representation of temporal distance, Onoe et al. (2001) showed a crucial role of the dorsolateral prefrontal cortex in timekeeping and time perception. They measured regional cerebral blood flow (rCBF) using PET while the monkey performed a time discrimination task and observed a significant increase of rCBF in the dorsolateral prefrontal cortex during this task. They also observed selective impairment of this task by a local application of bicuculline. These results indicated an importance of the dorsolateral prefrontal cortex in time perception. Similar results had been obtained in human studies (Rao et al. 2001). To examine how the prefrontal cortex participates in keeping temporal information, Sakurai et al. (2004) examined single-neuron activity of prefrontal neurons while the monkey performed a task that required working memory of stimulus duration (delayed matching-to-sample of stimulus duration). In this task, the monkey needs to press a lever and maintain fixation at a fixation point. After a variable fixation period (0.3–1.5 s), a green square appears over the fixation point for either 0.5 s or 2.0 s (sample period). After the 3-s delay period, the same green square appears over the fixation point for either 0.5 s or 2.0 s (comparison period). After a variable fixation period (0.3–1.5 s), the size of the fixation point shrinks temporary for 1.0 s and then goes back to the original size for 1.0 s. If the presentation time of the green square is the same between the sample period and the comparison period, the monkey needs to release the lever when the size of the fixation point goes back to the original size. If the presentation time of the green square is different between the sample period and the comparison period, the monkey needs to release the lever when the size of the fixation point shrinks. Among 811 neurons recorded in the prefrontal cortex, 83 (10%) exhibited sustained activation during the delay period or during both the sample and the delay period only when the green square was presented for either long or short duration during the sample period. These results indicate that a group of prefrontal neurons exhibit sustained delay-period activity representing specific length of the stimulus presentation, not representing the length of the stimulus presentation for each trial.

Oshio et al. (2006) further examined delay-period activity representing the duration of the stimulus presentation in the prefrontal cortex. They used a duration-discrimination task. While the monkey presses a hold lever and looks at a fixation point, red (R) and blue (B) squares are successively presented one by one for either long (L) or short (S) period with 1.0 s inter-stimulus interval (first delay). Therefore, two different sequences in the stimulus duration (long-short sequence, LS; short-long sequence, SL) and two different sequences in color stimuli (red-blue sequence, RB; blue-red sequence, BR) are present. The long presentation time was selected from 1.2 to 1.6 s and the short presentation time was 0.6–1.0 s shorter than the long presentation time. One second after the second square is presented (second delay), both red and blue squares are simultaneously presented on right and left sides and the monkey is required to press a key associated with the square presented longer time. They defined neural activity "duration-related" if significant difference was observed between LS and SL trials for either the first or the second delay period. Among 562 neurons analyzed, 93 (17%) showed duration-related activity, and among these, 61 and 37 showed duration-related activity during the first and the second delay period, respectively. During the first delay period, duration-related activity was more often observed in LS trials than in SL trials, while during the second delay period, duration-related activity was more often observed in SL trials than in LS trials. The strength of duration-related activity observed during the first delay period was correlated with the category of the duration (longer or shorter) of the first stimulus. These results indicate that the delay-period activity of a group of prefrontal neurons represents the duration of stimulus presentation and that the duration of stimulus presentation is encoded categorically, such as long vs. short.

Genovesio et al. (2009) used a duration-discrimination task and showed that prefrontal activity represents the length of the stimulus presentation and relative length of the stimulus presentation (e.g., longer or shorter than the first stimulus time). In this task, while the monkey looks at a fixation point, the first visual stimulus is presented on the fixation point for 100–3200 ms and followed by the first delay (0.4, 0.8, or 1.2 s). Then, the second visual stimulus is presented for 50–2400 ms and followed by the second delay (0.0, 0.4, or 0.8 s). A blue circle and a red square are used for these two visual stimuli. After the second delay, two visual stimuli reappear simultaneously, one on the left and the other on the right, and the monkey is required to touch the key associated with the visual stimulus that was presented for a longer time. Among 1720 neurons (1286 from the periarcuate area and 434 from the dorsolateral prefrontal area), 25% exhibited delay-period activity during the first delay period and these neurons exhibited a significant linear correlation (either positive or negative) between the magnitude of delay-period activity during the first delay period and the length of the first stimulus presentation. A similar linear correlation was observed between the magnitude of delay-period activity during the second delay period and the length of the second stimulus presentation. Prefrontal neurons having a significant correlation between first delay-period activity and the length of the first stimulus presentation were different from neurons having significant correlation between second-delay-period activity and the length of the second stimulus presentation. Delay-period activity of only 15% of neurons represented both the length of the first stimulus presentation and the length of the second stimulus presentation. Genovesio et al. (2009) also showed that prefrontal activity represents the relative length of the stimulus presentation, such that prefrontal neurons encode whether the first or the second stimulus is presented for a longer time or whether the red or the blue stimulus is presented for a longer time. The activity of prefrontal neurons did not represent information regarding the magnitude of the time difference of stimulus presentation between the first and the second stimulus. This result agrees with the results obtained by Sakurai et al. (2004) and Oshio et al. (2006).

Interestingly, prefrontal neurons are shown to use the same coding method for encoding spatial distance between two stimuli and for encoding the relative difference in the presentation times between two stimuli. Genovesio et al. (2011)

examined how prefrontal neurons encode the relative distance of two stimuli from the fixation point using a task requiring relative-distance discrimination. In this task, while the monkey looks at a fixation point, the first stimulus is presented directly above the fixation point for 1.0 s and followed by the first delay period (0.4 or 0.8 s). Then, the second stimulus is presented directly below the fixation point for 1.0 s and followed by the second delay period (0.0, 0.4, or 0.8 s). A blue circle and a red square are used for two stimuli. After the second delay, two stimuli reappear simultaneously, one on the left and the other on the right of the fixation point, and the monkey is required to touch the key associated with the stimulus that appeared at farther position from the fixation point. They showed that 9% (144/1671) of analyzed prefrontal neurons encoded the relative distance of the stimulus (far or near) from the fixation point during the first delay period. Many of these neurons used abstract and categorical representation (either close to or far from the fixation point) during the first delay period and its representation was independent of the spatial position of the stimulus.

Thus, the delay-period activity of prefrontal neurons can represent the temporal length of the stimulus presentation as well as the spatial distance of the stimulus from the reference point using relative and categorical encoding methods (long or short, or closer or farther), not parametrical encoding method (how much longer, or how much farther).

3.15 Delay-Period Activity Representing Estimation of Behavioral Timing

Niki and Watanabe (1979) examined single-neuron activity recorded from the dorsolateral prefrontal cortex while the monkey performed a task including differential reinforcement of long latency. In their task, the monkey first needs to press a holding lever for 3–5 s to turn on a hit-key. After turning on the hit-key, the monkey needs to keep pressing the hold lever for more than 2 s (delay period) and then press the hit-key to get a reward. To get the reward, the monkey is required to measure the timing of the behavioral response. Among 252 neurons obtained from the prefrontal cortex, 7 (3%) exhibited gradually increasing activity before turning on the hit-key, 73 (29%) exhibited gradually increasing activity during the delay period, and 33 (13%) exhibited sustained activity during the delay period. Based on the temporal pattern of prefrontal activity and the period when the activity is observed, gradually increasing activity before turning on the hit-key is thought to participate in the anticipation of the stimulus presentation, while gradually increasing activity and sustained activity during the delay period are thought to participate in the estimation of the timing of behavioral responses. Kojima et al. (1981) also observed gradually increasing activation in prefrontal neurons during the delay period while the monkey performed a delayed-response task without the Go signal presentation. In this task, after the monkey presses a hold key for 3 s, either the right or the left visual cue turns on for 1 s (cue period), then the delay period is introduced. The monkey is required to keep pressing the hold key for more than 0.9 s (delay period), and then to press the key that is associated with the illuminated cue during the cue period without the Go signal. Therefore, the monkey is required to measure the timing of the behavioral response by itself. Among 177 task-related prefrontal neurons, 66 (37%) exhibited gradually increasing activity before the initiation of behavioral responses. Gradually increasing activity that occurred during the delay period is thought to participate in the estimation of the timing of behavioral response. Thus, these results indicate that delay-period activity is used to estimate the timing of the behavioral response, which is also an important function of working memory.

3.16 Information Processing in Working Memory

3.16.1 Information Represented by Delay-Period Activity Changes

Working memory is defined as neural mechanisms providing both temporary storage and processing of information in order to perform complex cognitive tasks, such as language comprehension, judgment, reasoning, and decision-making (Baddeley and Hitch 1974; Baddeley 1986, 1992, 2012). Neurophysiological studies using monkeys have revealed that delay-period activity observed in prefrontal neurons during performance of working memory tasks is a neural correlate of the mechanism for temporary active storage of information in working memory processes. However, to fully understand the neural mechanisms of working memory, it is important to know not only how information is actively stored but also how information is processed. Delay-period activity of each prefrontal neuron exhibits different selectivity to spatial and non-spatial visual stimuli. Similarly, a population of prefrontal neurons exhibits delay-period activity with similar stimulus selectivity. Therefore, a particular piece of information (e.g., a spatial position of a visual cue) is encoded by a population of prefrontal neurons whose activities exhibit similar stimulus selectivity. Therefore, to understand the neural mechanisms of information processing in working memory, it is necessary to know not only what information each prefrontal neuron encodes and what information a population of prefrontal neurons encodes but also how information encoded by a population of prefrontal neurons alters or transforms with the progress of the task performance.

It has been shown that information represented by delay-period activity alters with the progress of learning the task. Asaad et al. (1998) examined prefrontal delay-period activity and showed that the timing of its appearance and the magnitude of its activity were affected by the learning of the task. They used a stimulus–response association task, in which monkeys need to learn an arbitrary association between visual objects and directions of saccades (e.g., an object A instructs a right-going saccade, whereas an object B instructs a left-going saccade) by trial and error.

Monkeys are also required to perform this task under a reversal condition of the stimulus–response association repeatedly. Delay-period activity of individual neurons represented either the visual object or the saccade direction associated with the object. They found that the magnitude and the timing of delay-period activity representing prospective saccade information were affected by the progress of the learning of the task, such that delay-period activity representing the direction of the impending saccade appeared progressively earlier after the object presentation in parallel with the increase of the number of reversal trainings.

It has been shown that delay-period activity not only represents visual information but also reflects information processing relevant to current behavioral demands, suggesting that information representing in delay-period activity alters during the delay period depending on the current behavioral demands. Rainer et al. (1999) used a delayed paired associate task and a delayed matching-to-sample task and showed that delay-period activity not only held retrospective information but also participated in information processing to produce prospective information. A population analysis of prefrontal activity revealed that, during the sample presentation or the early phase of the delay period, prefrontal activity represented information regarding the sample object. However, toward the end of the delay period, prefrontal activity began to represent information regarding the anticipated target object. In addition, the magnitude of delay-period activity representing the sample object gradually decreased with the progress of the delay period, whereas the magnitude of delay-period activity representing the target object gradually increased during the delay period. These results indicate that information representing in a population of delay-period activity alters gradually from retrospective sample object information to prospective target object information during the delay period. Quintana and Fuster (1999) also showed that prefrontal activity representing the cue color tended to diminish with the progress of the delay period, whereas activity representing the response direction tended to increase during the delay period. Sawaguchi and Yamane (1999) used a spatial delayed matching-to-sample task and showed gradual change of sharpness of spatial tuning and spatial discriminability with the progress of the delay period. In the early phase of the delay period, delay-period activity tuned spatial information broadly, while in the later phase, spatial tuning of delay-period activity became sharper and spatial discriminability increased. Further, Takeda and Funahashi (2002) compared the directional selectivity of delay-period activity between the ODR and the R-ODR tasks in individual prefrontal neurons. While the monkey performed these tasks, cue-period activity represented information regarding only visual cues. Delay-period activity represented information regarding either the positions of the visual cue or the directions of the saccade. However, response-period activity mostly represented information regarding the directions of the saccade. For correct behavioral performances in the R-ODR task, the information for the saccade direction is needed to be produced from the information of the visual cue. Therefore, the transformation from sensory information to motor information is needed during the delay period. These results suggest that the transformation from sensory information to motor information occurs by altering information represented by delay-period activity gradually during the delay period. Thus, the alteration of information representing delay-period activity occurs in the prefrontal cortex during the delay period while working memory tasks are performed.

3.16.2 Population-Vector Analysis

The population-vector analysis had been introduced by Georgopoulos's group to explain how a population of neurons having direction-tuned activity in the primary motor cortex generated trajectories of hand-reaching movements toward various target positions in the two-dimensional space (Georgopoulos et al. 1986, 1988; Georgopoulos 1988; Schwartz et al. 1988) and in the three-dimensional space (Kettner et al. 1988). Each motor cortex neuron exhibits a different magnitude of movement-related activity across different directions of the hand-reaching movement. Response characteristics and directional selectivity of movement-related activity for each neuron can be depicted by a tuning curve constructed by discharge rates during reaching movements across examined directions. Using an obtained tuning curve, a concept of "cell vector" was introduced to describe response magnitude (mean discharge rate) during the reaching movement toward a particular direction. The direction of the cell vector corresponds to the direction at which the maximum discharge rate is observed in the neuron's tuning curve, while the length of the cell vector corresponds to the mean discharge rate obtained from the tuning curve during the reaching movement toward a given direction. Therefore, the direction of the cell vector remains constant regardless of reaching directions, while the length of the cell vector changes depending on the reaching direction (because discharge rate changes across different reaching directions). Then, using cell vectors obtained by activities of all recorded neurons during the reaching movement toward a particular direction, a "population vector" is calculated using a population of cell vectors to estimate whether the direction of the population vector accurately indicates the direction of monkey's reaching behavior or not. The population vector is a weighted sum of all cell vectors obtained when the monkey performed the reaching movement toward a particular direction. Georgopoulos' group showed that the direction of the population vector accurately indicated the direction of the reaching movement that the monkey actually performed. Using this population vector method, Georgopoulos (1988) showed that when the direction of the reaching movement is instructed, the population vector toward the instructed direction is gradually developed during the preparation period before the initiation of the movement. Further, Georgopoulos et al. (1989) showed that, when the monkey was asked to perform a reaching movement 90° counterclockwise from the initial direction instructed by the visual cue, the direction of the population vector was initially directed toward the visual cue, but then gradually rotated from the direction of the visual cue to the actual reaching direction during the reaction time period, and finally directed toward the reaching direction at the initiation of the movement. These results not only indicate how the motor program to perform a reaching movement toward a particular target is developed in the motor cortex during the reaction time period, but also indicate that the population vector method is useful to depict the neural process related to information processing, especially alteration and transformation of information represented by a population of neural activity, in working memory process operating in the prefrontal cortex.

3.16.3 Population-Vector Analysis of Delay-Period Activity

Georgopoulos's group showed the usefulness of population-vector analysis to visualize information processing for preparing arm-reaching movement in the primary motor cortex. In the prefrontal cortex, a large population of neurons exhibit directional selectivity during the delay period and have either cue-coding or response-coding delay-period activity. The transformation from sensory information to motor information seems to be happened during the delay period. Therefore, applying the population vector analysis to delay-period activity is thought to be useful to visualize neural processes related to the transformation from sensory information to motor information during the delay period in the prefrontal cortex.

The same method was used to obtain cell vectors and population vectors by prefrontal activities as was used in the primary motor cortex by Georgopoulos's group. Funahashi and Takeda (2002a) and Takeda and Funahashi (2004) calculated population vectors using a population of task-related activity obtained from prefrontal neurons during the ODR and the R-ODR tasks. To confirm that the population vector method is applicable to prefrontal activity, Takeda and Funahashi (2004) first examined whether or not the best directions of both cue- and response-period activity for all prefrontal neurons uniformly distributed around the fixation point. As seen in Fig. 3.26a and c, although the distribution of the best directions of cue-period activity exhibited weak contralateral bias, the best directions of both cue- and response-period activity were distributed uniformly around the fixation point. Takeda and Funahashi (2004) then examined whether or not the population vector obtained by a population of either cue- or response-period activity correctly indicates the direction of the visual cue or the saccade across different cue conditions, respectively (Fig. 3.26b, d). Comparisons between the directions of population vectors and the directions of visual cues and saccades revealed that the population vector obtained by cue-period activity correctly indicates the direction of the visual cue for each cue condition, and similarly that the population vector obtained by a population of response-period activity correctly indicates the direction of the saccade for each saccade condition (Takeda and Funahashi 2004). These results confirmed that the population vector method is applicable to visualize the neural process for transformation from sensory information to motor information during the delay period in both the ODR and the R-ODR tasks.

Next, Takeda and Funahashi (2004) first determined the best direction of delay-period activity for each neuron using a tuning curve obtained by fitting mean discharge rates of a 3-s delay period across all examined directions in the ODR

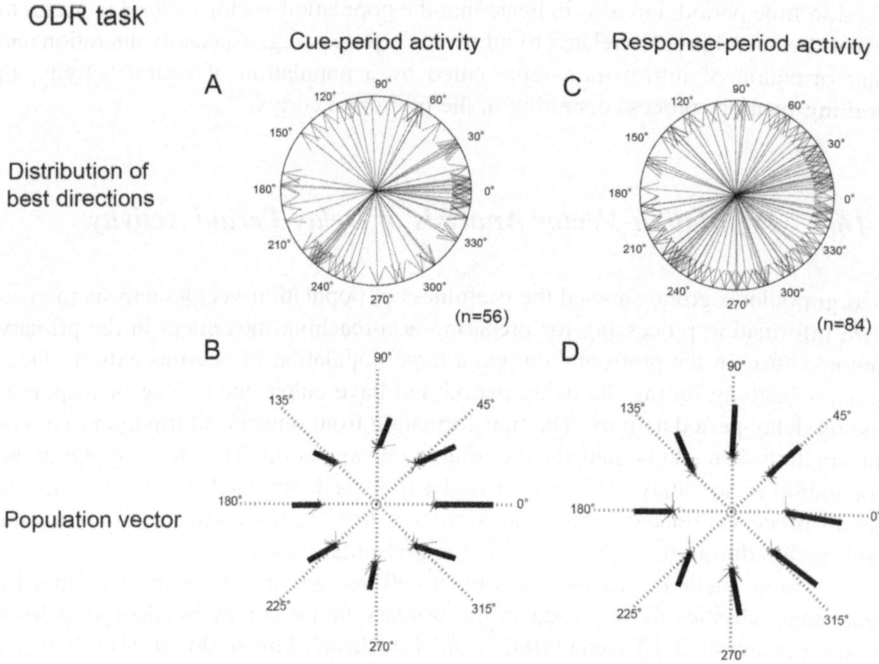

Fig. 3.26 Distributions of best directions of cue- (**a**) and response-period activity (**c**) during the ODR task and population vectors (thick lines) obtained by populations of cue- (**b**) and response-period activity (**d**) during the ODR task. From Takeda and Funahashi (2004) with permission

task on the Gaussian function. Obtained best direction was used as the direction of the cell vector for a given neuron. A 250-ms time window was set along a trial, a cell vector for a particular cue condition (e.g., 180° condition) was obtained by mean discharge rate within each time window and by the best direction for each neuron. The population vector in this cue condition was calculated by cell vectors of all neurons obtained within this time window. Then, a 250-ms time window was moved in a 50-ms time step from the cue onset to the end of the response period (500 ms after the go signal), a population vector was calculated at each time point. Using this method, population vectors for one cue condition were obtained from the cue onset to the end of the response period in a 50-ms time step.

Figure 3.27a shows population vectors along a ODR trial of the 180° cue condition. Since the direction of the visual cue and the direction of the saccade are the same in the ODR task, population vectors were mostly directed toward the 180° direction along the trial, indicating that the same directional preference is maintained by a population of prefrontal activity during the delay period in the ODR task. To confirm this observation across all cue conditions, the temporal change of the mean difference between the direction of the population vector and the direction of the visual cue was examined along the trial for all cue conditions. As is seen in Fig. 3.27C, the mean difference between the direction of the population vector and

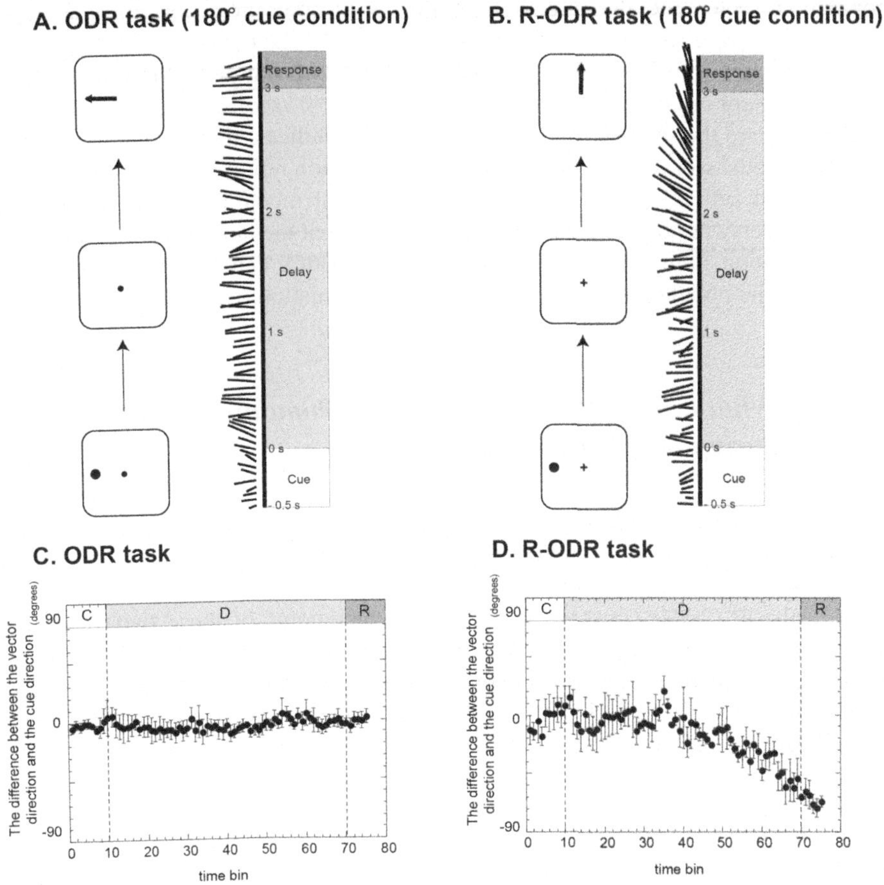

Fig. 3.27 Directions of population vectors along the performance of the 180° cue direction trial of the ODR (**a**) and the R-ODR (**b**) tasks. Temporal changes of directions of population vectors along the performance of the trials of the ODR (**c**) and the R-ODR (**d**) tasks. From Takeda and Funahashi (2004) with permission

the direction of the visual cue was maintained close to 0° along the trial, indicating that the directional information at each cue condition is maintained throughout the delay period by a population of delay-period activity.

Figure 3.27b shows population vectors along a R-ODR trial of the 180° cue condition. In the 180° cue condition of the R-ODR task, the visual cue is presented at the 180° position, and the correct saccade is directed toward the 90° position. As seen in Fig. 3.27b, population vectors were directed toward the 180° direction during the cue period and at the beginning of the delay period. However, population vectors began to rotate slowly from 180° to 90° from the middle of the delay period. Finally, population vectors were directed toward the 90° direction during the response period. As seen in Fig. 3.27d, the mean difference between the direction of the population vector and the direction of the visual cue across all cue conditions gradually changed from 0° to 90° with the progress of the delay period. In general, the population vector began to rotate at approximately 2 s after the start of the delay period and kept rotating at a constant speed (about 90°/sec) until the vectors were directed toward the saccade direction. These results indicate that, while the monkey performs spatial working memory tasks, the information represented by a population of prefrontal activity changes gradually from visual modality to motor modality during the delay period. These results also indicate that the neural process related to the information transformation in the prefrontal cortex can be visualized as the rotation of the population vectors obtained by a population of prefrontal activity.

3.16.4 What Causes the Rotation of the Population Vector during the Delay Period?

Takeda and Funahashi (2002) classified directional delay-period activity into two groups; one representing retrospective sensory information and the other representing prospective motor information. Based on this classification, Takeda and Funahashi (2004) constructed population histograms of these two groups of delay-period activity (Fig. 3.28b). The population histogram constructed by delay-period activity representing retrospective information exhibited sustained activation throughout the delay period (Fig. 3.28b, blue), whereas the population histogram constructed by delay-period activity representing prospective information exhibited gradually increasing activation toward the response period (Fig. 3.28b, red). Comparing the temporal change of the directions of population vectors between the ODR and the R-ODR tasks (Fig. 3.28a) and comparing the temporal change of population histograms between delay-period activity representing retrospective information and the activity representing prospective information (Fig. 3.28b) indicate that an increase of delay-period activity representing prospective information over the activity representing retrospective information is roughly synchronized with the start of the rotation of the population vector in the R-ODR task. Therefore, gradually increasing activation of the neurons encoding prospective information might cause the temporal change of the directions of population vectors, although neurons encoding prospective information are minor group in the prefrontal cortex. Thus, not only the neurons maintaining information by sustained delay-period activity but also the neurons exhibiting temporal change of delay-period activity contribute to information processing for working memory in the prefrontal cortex.

Although many prefrontal neurons exhibit task-related activity during delayed-response performances, some neurons exhibit only one kind of task-related activity (e.g., delay-period activity) but other neurons exhibit two or more kinds of task-related activity (e.g., cue- and delay-period activity). Characteristics of cue-period activity observed in the ODR task are similar to visual responses in the visual probe task (Funahashi et al. 1990; Takeda and Funahashi 2007), indicating

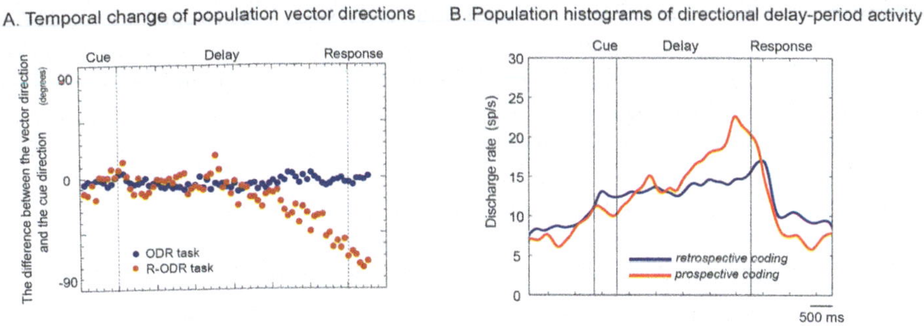

Fig. 3.28 (**a**) The mean difference between the direction of the population vector and the direction of the visual cue across all cue conditions in ODR trials (blue) and R-ODR trials (red). (**b**) Population histograms constructed by directional delay-period activity representing retrospective information (blue) and prospective information (red) in the ODR task. From Takeda and Funahashi (2004) with permission

that cue-period activity corresponds to the visual response of prefrontal neurons. There are prefrontal neurons having both cue- and delay-period activities and the best directions of these two activities are usually similar in these neurons (Takeda and Funahashi 2007), indicating that the visual input plays an important role to generate delay-period activity and determine directional selectivity of this activity in these neurons. Similarly, there are prefrontal neurons having both delay- and response-period activities and the best directions of these two activities are also similar in these neurons (Takeda and Funahashi 2007), indicating that delay-period activity plays an important role to prepare the forthcoming saccade and determine directional selectivity of the saccade in these neurons. However, there are prefrontal neurons having both cue- and response-period activities, but the best direction of cue-period activity is not always similar to that of response-period activity in these neurons (Takeda and Funahashi 2007), indicating that the saccade information is represented independently from the visual cue information in these neurons. In addition, prefrontal neurons having only delay-period activity consist of two independent groups; one representing visual cue information and the other representing saccade information (Takeda and Funahashi 2002, 2004). Therefore, several questions are raised, such as how prospective saccade information is generated in the prefrontal cortex and whether prospective saccade information is generated within the prefrontal cortex by interactions among neurons having different task-related activities or is received from other brain areas where saccade-related activity is observed.

One possibility is that the prefrontal cortex receives prospective saccade information from other brain areas such as the frontal eye field or the posterior parietal cortex (Funahashi 2013b) because these brain areas are known to participate in eye movement control and include neurons exhibiting saccade-related activity. Bruce and Goldberg (1985) reported visuo-movement activity in frontal eye field neurons, which exhibited similar characteristics as delay-period activity representing saccade information in prefrontal neurons. Gnadt and Andersen (1988) reported sustained activation during the delay period of a memory-guided saccade task in posterior parietal neurons and suggested that this activity represents an intention to make an eye movement toward a specific direction with a specific amplitude. Chafee and Goldman-Rakic (1998) showed that delay-period activity with very similar response characteristics is observed in both the frontal eye field and the posterior parietal cortex as those observed in the prefrontal cortex. Both the frontal eye field and the posterior parietal cortex have anatomical connections with the lateral prefrontal cortex. Therefore, delay-period activity representing prospective saccade information in the frontal eye field and the posterior parietal cortex could contribute to producing delay-period activity representing the saccade information in the prefrontal cortex.

Another possibility is that delay-period activity representing prospective saccade information is received from the mediodorsal nucleus of the thalamus because it has been shown that neurons in this nucleus exhibit saccade-related activity (Sommer and Wurtz 2004a, b; Tanibuchi and Goldman-Rakic 2003, 2005; Watanabe and Funahashi 2004a, b) and because this nucleus has anatomical projections to the prefrontal cortex (Goldman-Rakic and Porrino 1985; Barbas et al. 1991; Siwek and Pandya 1991). Watanabe and Funahashi (2004a, b) examined single-neuron activity in the mediodorsal nucleus of the thalamus using the ODR and the R-ODR tasks. They found that delay-period activity representing the cue direction and the saccade direction is present in the thalamus and that more neurons encoding the saccade direction are present in the thalamus compared with the prefrontal cortex. Similar to prefrontal neurons, delay-period activity representing information of the visual cue exhibits sustained activation during the delay period in mediodorsal neurons, while delay-period activity representing information of the saccade direction exhibits gradually increasing activation during the delay period. A larger proportion of mediodorsal neurons exhibit response-period activity compared with prefrontal neurons, and among them, most neurons exhibit pre-saccadic activity (Fig. 3.29a, b). Watanabe et al. (2009) performed the population vector analysis using a population of thalamic activity and examined temporal changes in directions of population vectors along a trial of the ODR and the R-ODR tasks. Although the directions of the population vectors rotated from the cue direction to the saccade direction during the delay period in the R-ODR task, the start of the rotation was earlier in mediodorsal neurons than prefrontal neurons (Fig. 3.29c). Since the mediodorsal nucleus of the thalamus projects to the prefrontal cortex, delay-period activity of thalamic neurons representing the information of the saccade direction could contribute to produce delay-period activity representing saccade information in the prefrontal cortex (Funahashi 2013b).

Thus, delay-period activity representing retrospective sensory information is thought to be generated by the sensory input to the prefrontal cortex and retrospective sensory information is maintained during the delay period by sustained delay-period activity. However, delay-period activity representing prospective motor information might be produced by the inputs from saccade-related brain areas including the frontal eye field, the posterior parietal cortex, and/or the thalamic

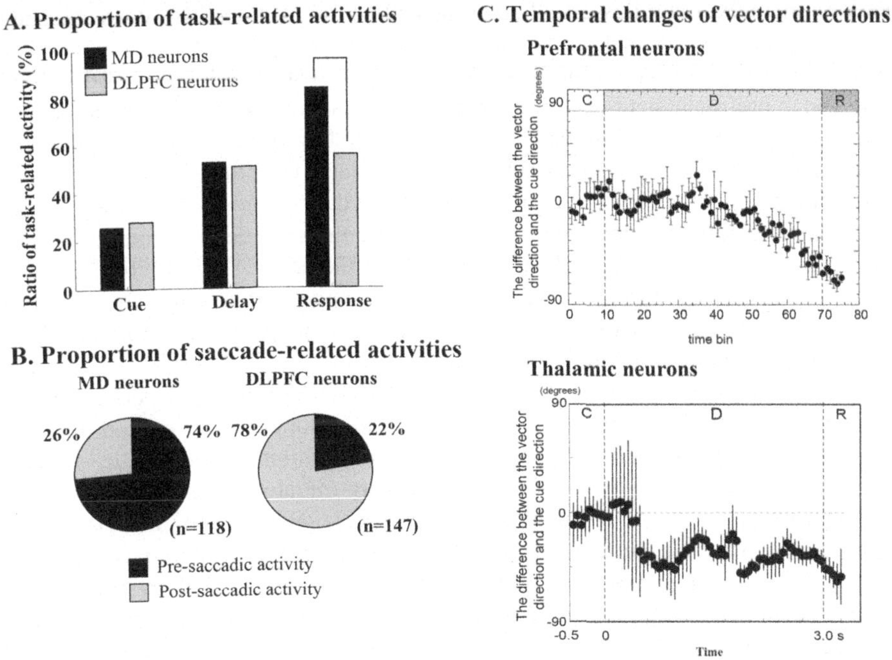

Fig. 3.29 (**a**) Comparison of proportions of task-related activities between prefrontal (DLPFC) and thalamic (MD) neurons. (**b**) Comparison of proportions of pre- and post-saccadic activities between prefrontal (DLPFC) and thalamic (MD) neurons. (**c**) Temporal changes of population vector directions calculated by prefrontal and thalamic activities during the R-ODR task. From Funahashi (2013b) with permission

mediodorsal nucleus. Gradual increase of delay-period activity representing prospective motor information observed in population histograms (see Fig. 3.28b) suggests that prefrontal neurons encoding prospective motor information are gradually recruited by gradually increasing inputs from motor-related brain areas, where population activity representing prospective motor information also exhibits gradual increase during the delay period but its initiation timing is earlier than that in the prefrontal cortex.

3.16.5 Interaction Between Delay-Period Activity and Saccade-Related Activity

Working memory is a neural mechanism that provides temporary active storage of information to accomplish complex cognitive tasks. The information stored in working memory is the one necessary only for a given trial. Therefore, once the behavioral response is done, the information is no longer necessary to store in working memory, regardless of whether the behavioral response was correct or not. Consequently, the information stored in working memory would be eliminated or replaced by new information for the next trial. In fact, sustained delay-period activity observed in prefrontal neurons terminated just after the monkey performed a saccade during the response period in the ODR task (Fig. 3.30a). In the ODR task, since the position of the visual cue changes randomly from trial to trial, the monkey is required to remember a particular position of the visual cue during the delay period for each trial. The termination of delay-period activity after the saccade performance resembles removal of stored information from working memory. A rapid termination of delay-period activity during the response period suggests that some neural signal terminates delay-period activity and that this neural signal plays an important role in controlling the working memory process.

Among a variety of task-related activities observed during ODR performances, post-saccadic activity is a possible neural signal to terminate delay-period activity. The most of saccade-related prefrontal neurons exhibit post-saccadic activity (Funahashi et al. 1991). Post-saccadic activity is initiated after the start of the saccade or often after the termination of the saccade (Fig. 3.30b). Therefore, this activity does not contribute to control the eye movement. However, post-saccadic activity exhibits directional selectivity similar to pre-saccadic activity (Funahashi et al. 1991). Therefore, post-saccadic activity is thought to be a feedback signal, possibly corollary discharge, coming from oculomotor centers. Delay-period activity terminates soon after the monkey makes a saccade during the response period, suggesting that post-saccadic activity acts as the neural signal to actively terminate delay-period activity at the end of each trial. The comparison of temporal patterns between delay-period activity and post-saccadic activity revealed that the termination of delay-period activity coincided with the initiation of post-saccadic activity (Fig. 3.30c). Therefore, it is concluded that post-saccadic activity is an important neural signal to manipulate delay-period activity by terminating the activity at the end of each trial (Goldman-Rakic et al. 1990).

In addition to post-saccadic activity, it has been shown that reward-related activity also plays an important role to terminate delay-period activity at the end of the trial. Ichihara-Takeda et al. (2010) examined whether reward delivery acts as the signal to inhibit delay-period activity or not by unexpectedly delivering a reward during the delay period. Among prefrontal neurons having delay-period activity, unexpected reward delivery in fact inhibited delay-period activity in 34% of these neurons, although 21% and 45% of the remaining neurons exhibited facilitatory response and no response to unexpected reward-delivery, respectively. Since a reward is given at the end of the trial, these results indicate that reward-related response is also a candidate to act as a signal to terminate delay-period activity at the end of the trial.

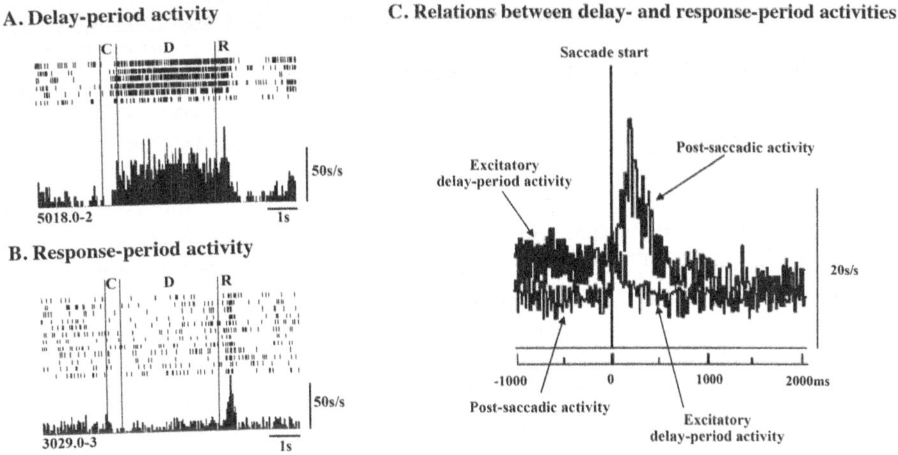

Fig. 3.30 Relations between delay-period activity and post-saccadic activity. (**a**, **b**) Examples of excitatory delay- and response-period activities, respectively. (**c**) Population histograms of delay-period activity and post-saccadic activity aligned at the start of the saccade

3.16.6 Integration of Multiple Information

Single-neuron recording studies showed that the information represented by a population of prefrontal activity changes with the progress of the trial (Asaad et al. 1998; Quintana and Fuster 1999; Rainer et al. 1999; Sawaguchi and Yamane 1999). An analysis of a population of prefrontal activity using the population vector method shows that the information represented by a population of prefrontal activity gradually changes from visual information to motor information during the delay period in the ODR task (Takeda and Funahashi 2004). The inputs from neurons having post-saccadic activity to neurons having delay-period activity act as neural signals to terminate delay-period activity (Goldman-Rakic et al. 1990). These observations indicate that functional interactions among prefrontal neurons encoding various information and exhibiting various task-related activities play key roles in information processing in working memory.

To examine whether or not functional interactions are present among prefrontal neurons and what types of interaction are present among neurons (e.g., excitatory or inhibitory interaction, or serial connection or sharing of common inputs), a cross-correlation analysis has been commonly performed using spiking activity recorded from a pair of simultaneously recorded neurons (Perkel et al. 1967a, b; Aertsen and Gerstein 1985; Gerstein and Aertsen 1985). In this analysis, the relation of the spike timing between two independent neurons (A and B) is examined by constructing temporal distribution of neuron A's spike timings triggered at each of neuron B's spike timing. A temporal correlation of spike timings between two neurons is illustrated in a cross-correlation histogram. If the peak of the cross-correlation histogram is present at time 0 (central peak), which is the timing of neuron B's firing, both neurons A and B tend to fire simultaneously, suggesting that both neurons receive inputs from a common excitatory source (sharing common input). If the peak is displaced from time 0 and if the distribution is asymmetry around the time 0 (displaced peak), either neuron A's firing tends to follow neuron B's firing or neuron A's firing tends to lead neuron B's firing, indicating that neuron A receives excitatory inputs from neuron B or that neuron B receives excitatory inputs from neuron A (serial interaction), respectively. This analysis has been applied in various brain areas including the visual cortex (Toyama et al. 1981a, b; Tanaka 1983; Ts'o et al. 1986; Krüger and Aiple 1988; Hata et al. 1991, 1993), the auditory cortex (Espinosa and Gerstein 1988; Ahissar et al. 1992), the hippocampus (Sakurai 1996,1998), the motor cortex (Murphy et al. 1985a, b; Kwan et al. 1987), and the striatum (Bergman et al. 1998).

Wilson et al. (1994) identified putative pyramidal neurons (regular-spiking, RS) and GABAergic interneurons (fast-spiking, FS) in the prefrontal cortex based on the firing rate and the shape of action potential and compared response characteristics between these two groups of neurons. They found that adjacent RS and FS pairs responded to the same visual stimuli although response direction (excitation or inhibition) was inverted between RS and FS cells and that adjacent RS pairs exhibited similar directional selectivity. Rao et al. (1999) also identified RS and FS cells in the prefrontal cortex and applied the cross-correlation analysis to examine functional interaction between these two cell groups. They found that both FS and RS cells exhibit similar tuning characteristics in delay-period activity and that adjacent neurons, regardless of whether those were FS or RS cells, share similar but not identical directional preferences (iso-directional tuning). Characteristics of iso-directional tuning between FS and RS cells might show a similar function as lateral inhibition, such that tuned activity of adjacent RS cell is suppressed by FS cells exhibiting similar but not identical directional tuning as the RS cell. Wilson et al. (1994) recorded activity of RS-FS pairs located within 400 μm in distance, which corresponds to the size of the functional column in the prefrontal cortex (Goldman and Nauta 1977a; Goldman-Rakic and Schwartz 1982; Schwartz and Goldman-Rakic 1984) and showed that response directions (excitation or inhibition) of FS cells are inverted of those of RS cells. On the other hand, Rao et al. (1999) recorded activity of RS-FS pairs located within 50–60 μm in distance, which corresponds to the size of the microcolumn (Jones 2000) and found iso-directional tuning between FS and RS cells. Based on these observations, Rao et al. (1999) proposed that iso-directional inhibitory interaction is functioning between RS and FS cells within a column, while cross-directional inhibitory interaction functioning between columns in the prefrontal cortex and that these iso- and cross-directional inhibitory interactions play important roles to maintain and stabilize memory fields of prefrontal neurons.

Although functional interactions between RS and FS cells had been examined by cross-correlation analysis, an examination of functional interactions among prefrontal neurons having different task-related activity and different selectivity would also be important to understand information processing in working memory processes. However, these interactions had not been examined in the prefrontal cortex.

Funahashi and Inoue (2000) examined functional interactions among prefrontal neurons having same or different task-related activity using pairs of single neurons simultaneously recorded while the monkey performed the ODR task. Significant peaks were found in cross-correlation histograms in 50% (84/168) of analyzed pairs of prefrontal neurons, and 36% and 45% of these showed excitatory central peaks and displaced peaks, respectively. Neuron pairs both having same task-related activity (56%) exhibited significant functional interactions more than neuron pairs having a different task-related activity (44%). Especially, 71% of neuron pairs both having delay-period activity showed significant functional interactions and these neuron pairs exhibited similar directional selectivity. More neuron pairs having response-period activity exhibited different directional selectivity than neuron pairs having cue- and delay-period activity. In addition, serial interactions (displaced peaks) were observed between neurons having cue-period activity and neurons having delay-period activity and between neurons having delay-period activity and neurons having response-period activity, indicating that the information flow is present in the prefrontal cortex from neurons participating in sensory processing to neurons participating in mnemonic processing and from neurons participating in mnemonic processing to neurons participating in mnemonic processing. These results show that a variety of functional interactions are present not only between pyramidal and non-pyramidal neurons but also among neurons having different task-related activities and having different sensory and motor selectivity in the prefrontal cortex, indicating that functional interactions among prefrontal neurons play important roles for information processing in working memory and participate in managing various functions.

Constantinidis et al. (2001b) collected multiple single-neuron activities using two (1 mm apart) or four (200–300 μm apart) microelectrodes penetrating into the prefrontal cortex and performed cross-correlation analysis using simultaneously recorded activities of prefrontal neuron pairs. They classified cross-correlation histograms into two groups (having either narrow or broader peak) based on whether the width at half peak height was more than 5 ms or less. Cross-correlation histograms having narrow peak suggest mono-synaptic connection between a pair of neurons or that both neurons receive a common input. On the other hand, cross-correlation histograms having broader peak suggest poly-synaptic connection between a pair of neurons. Among 337 pairs of prefrontal neurons having task-related activity, 38 (11%) had narrow peaks and 75 (22%) had broader peaks. Cross-correlation histograms having narrow peaks were frequently observed when two neurons were located less than 300 μm apart. However, the number of the cases in which cross-correlation histograms had narrow peaks decreased dramatically with the increase of the distance between electrodes. Neuron pairs whose positions were 200–300 μm apart exhibited the entire range of spatial tuning preferences, suggesting that the entire visual hemifield is represented by neurons located in 200–300 μm wide region of the prefrontal cortex. In addition, neurons recorded within a distance of 200–300 μm were activated at any task epoch. Intrinsic interconnections within the prefrontal cortex and associational projections to the prefrontal cortex are shown to terminate in a stripe- or columnar-like fashion and the basic size of these stripes or columns is thought to be 200–300 μm (Goldman and Nauta 1977b; Goldman-Rakic and Schwartz 1982; Schwartz and Goldman-Rakic 1984; Levitt et al. 1993; Kritzer and Goldman-Raki 1995; Pucak et al. 1996). Therefore, stripe- or column-like structure is thought to be a basic functional unit for local information processing in the prefrontal cortex similar to ocular dominance columns in the visual cortex (LeVay et al. 1975; Hubel et al. 1977).

Thus, cross-correlation analyses revealed that many pairs of prefrontal neurons both having delay-period activity exhibit significant central peaks and that each neuron consisted of a pair often exhibits different directional selectivity, indicating that mono-synaptic functional interaction between prefrontal neurons having delay-period activity plays an important role for processing and integrating information in working memory. These results show that information processing in working memory is achieved by mono- and multi-synaptic functional interactions among neurons encoding a variety of information.

3.16.7 Dynamic Changes of Functional Interactions Among Prefrontal Neurons

The cross-correlation analysis revealed that many prefrontal neurons have mono- or multi-synaptic functional interactions with other neurons having same or different task-related activity and exhibiting same or different selectivity to stimuli and responses. Cross-correlation histogram is usually constructed using all spiking activity that occurred during one whole recording session, which includes 100 or more trials. In the ODR task, the position of the visual cue is different from trial to trial and each trial includes several different task events. Therefore, spiking activity for constructed cross-correlation histogram reflects various information including sensory, motor, and context information. Each prefrontal neuron often exhibits significant activity in a particular trial condition (e.g., 180° cue condition) and in a particular task epoch (e.g., delay period) with or without particular selectivity (e.g., directional selectivity). Therefore, the strength of the functional interaction between a pair of prefrontal neurons would flexibly change depending on the difference of the trial conditions as well as the difference of the task epochs.

To examine whether or not functional interaction between a pair of prefrontal neurons changes flexibly and dynamically depending on the difference of the trial conditions, a cross-correlation histogram was constructed for each cue condition of the ODR task (Funahashi 2001). Figure 3.31a shows the cross-correlation histogram constructed from spiking activity of two simultaneously recorded neurons during a whole recording session. The cross-correlation histogram has a symmetrical central peak, indicating that both neurons receive a common input. Figure 3.31b shows cross-correlation histograms each of which was constructed from spiking activity during all trials of each of eight cue conditions. Although the central peak is observed in all cross-correlation histograms, the peak height is different across

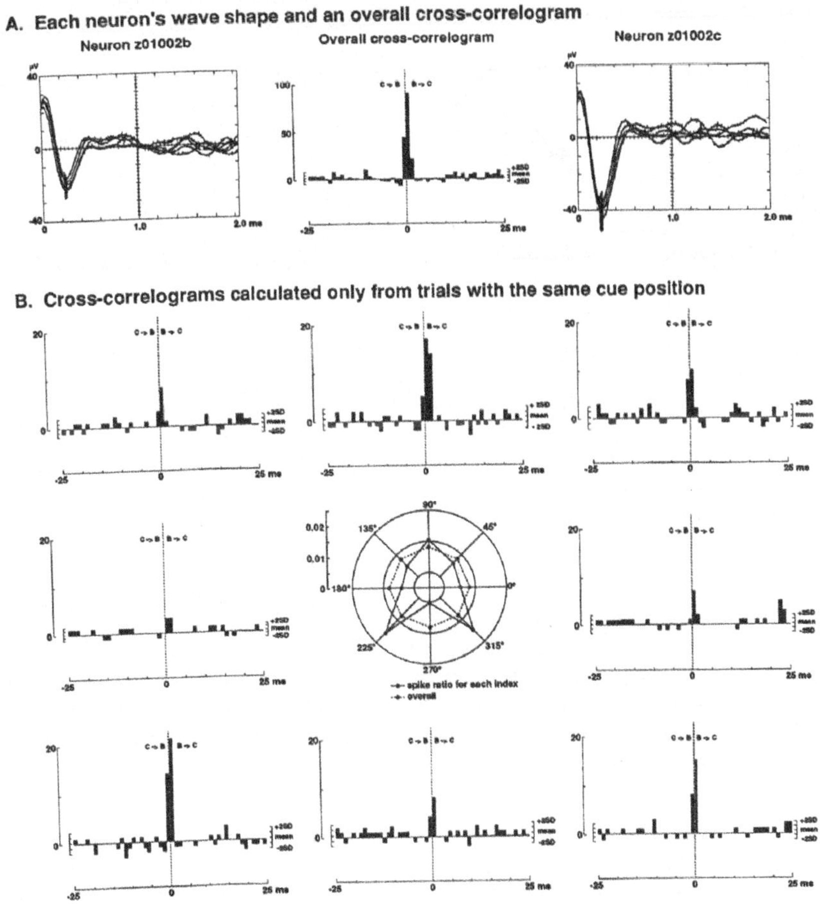

Fig. 3.31 (**a**) Shapes of action potentials of simultaneously recorded two prefrontal neurons and the cross-correlogram obtained by whole spike activities of these two neurons. (**b**) Cross-correlograms obtained by spike activities of these two neurons for each cue condition. The task was the ODR task with 8 cue positions. The center figure indicates polar plots depicting the height of the central peak of the cross-correlogram for each cue condition. From Funahashi (2001) with permission

eight cross-correlation histograms and higher central peak is observed in 90°, 225°, and 315° cue conditions. Similar modulation of the peak height depending on the difference of cue conditions was observed in most of neuron pairs examined. These results indicate that the strength of functional interaction between a pair of prefrontal neurons changes dynamically depending on the difference of trial conditions.

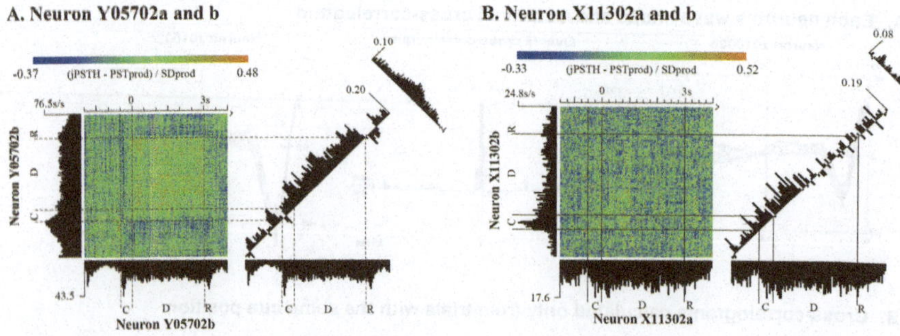

Fig. 3.32 Two examples of joint PSTH analysis. (**a**) The result of joint PSTH analysis using activities simultaneously isolated two single neurons Y05702a and Y05702b. (**b**) The result of joint PSTH analysis using activities simultaneously isolated two single neurons X11302a and X11302b. Color difference indicates the difference in the functional interactions between two neurons. Horizontal and vertical histograms indicate activity for each of the two neurons along trials of the ODR task, respectively. Diagonal histograms on the right are PST coincidence histograms obtained from jPSTH. Diagonal histograms at up-right corner are cross-correlograms obtained by activities of two neurons

Working memory tasks not only include different trial conditions but also include different task events within each trial. Since different task event is introduced one after another with the progress of a trial, the magnitude of activity of each prefrontal neuron would change depending on the relation to the events. Therefore, the strength of functional interaction between a pair of prefrontal neurons would also change with the progress of a trial. A joint peri-stimulus time histogram (jPSTH) analysis, a kind of correlation analysis, is a method to investigate dynamic changes in the strength of the functional interaction (effective connectivity) between two simultaneously recorded neurons with the progress of a trial (Fig. 3.32). This method was developed by Gerstein and Perkel (1969, 1972) and Aertsen et al. (1989). In this analysis, one histogram called the PST coincidence histogram (diagonally oriented histograms in Fig. 3.32) draws the temporal change of the strength of effective connectivity between two simultaneously recorded neurons along a trial. The change in the height of the PST coincidence histogram is independent of task-related modulation of each neuron's firing rate (e.g., visual responses) during a trial (Aertsen et al. 1989). The usefulness of the jPSTH method and the importance of the temporal change of effective connectivity to understand cortical functions had been demonstrated by Vaadia et al. (1995). They performed a simultaneous recording of 6–16 frontal single-neuron activities while monkeys performed a delayed-response task and a Go/No-Go task. They showed that correlated firing between two single-neurons evolved within a fraction of a second and exhibited that correlated firing has systematic relation to the behavioral event but not to the modulation of the firing rate. Thus, the jPSTH method would be a useful method to examine the temporal change of the strength of functional connectivity between simultaneously recorded prefrontal neurons with the progress of a trial, although some caution is needed to interpret the result (Ito and Tsuji 2000).

To examine whether or not the temporal change of the strength of functional connectivity is present between a pair of prefrontal neurons with the progress of a working memory trial, jPSTH was constructed using spiking activity of pairs of prefrontal neurons while monkeys performed the ODR task (Funahashi 2001). Figure 3.32 shows two examples of the results by jPSTH analysis. Both prefrontal neurons shown in Fig. 3.32a exhibited delay-period activity. The cross-correlation histogram shown in the up-right corner of Fig. 3.32a shows a broad symmetrical peak around time 0, suggesting that these two neurons tend to fire synchronously. PST coincidence histogram (diagonal histogram in Fig. 3.32a) indicates that high probability of correlated firing was maintained throughout the delay period, but not observed during the cue period and the response period. On the other hand, in neurons shown in Fig. 3.32b, one neuron (x11302a) exhibited delay-period activity and the other neuron (x11302b) exhibited cue-period activity. The cross-correlation histogram shown in the up-right corner of Fig. 3.32b shows an asymmetrical displaced peak from time 0, suggesting that these two neurons have mono-synaptic interaction and that neuron x11302b sends excitatory outputs to neuron x11302a. PST coincidence histogram indicates that the probability of correlated firings was modulated with the progress of the trial. Mono-synaptic functional connectivity was stronger during the early half of the delay period, but became weaker during the late half of the delay period. The temporal pattern of PST coincidence histogram varied from neuron pair to neuron pair. Some neuron pairs showed stronger correlated firing during either the cue period or the response period, during the entire delay period, or only during the early or late half of the delay period.

Tsujimoto et al. (2008) also examined the temporal change of functional interaction of prefrontal neurons using the jPSTH analysis. They used the task previously used by Genovesio et al. (2005). In this task, the monkey is required to make a saccade to one of three visual targets based on the visual cue presented over the fixation point (one cue indicates the monkey to select the same target as the one selected at the preceding trial, while the other cue indicates the monkey to select one of the other targets). Genovesio et al. (2005) found two groups of prefrontal neurons; one encoded previously selected target (P cells) and the other encoded the target which would be selected for the current trial (F cells). Tsujimoto et al. (2008) examined temporal change of functional interaction in F-F, P-P, and F-P pairs and showed that the strength of the interaction of these pairs changed dynamically along the progress of a trial.

Thus, cross-correlation analysis of spiking activity of a pair of prefrontal neurons revealed that many pairs of adjacent prefrontal neurons have excitatory mono-synaptic functional interactions and exhibit correlated firing. Pairs of prefrontal neurons exhibiting correlated firing often show different task-related activity and/or different selectivity in task-related activity. In addition, dynamic and flexible modulation of the strength of functional interaction is also observed between a pair of adjacent prefrontal neurons depending on the difference in the trial conditions. Further, temporal modulation of the strength of functional interaction is observed with the progress of a trial. Thus, the neural mechanism of information processing in working memory could be achieved by functional interaction among neurons encoding different information and dynamic and flexible modulation of the strength of functional interaction depending on the difference between behavioral context and trial events and the progress of a trial.

3.17 Delay-Period Activity Observed in Other Brain Areas

When the delay period is included in a behavioral task between the presentation of the sensory cue and the behavioral response, sustained excitation, or gradually increasing or decreasing activity are often observed during the delay period. Delay-period activity has been observed not only in prefrontal neurons but also in neurons of other cortical and subcortical areas and not only in association cortices but also in sensory systems (Pasternak and Greenlee 2005). In fact, delay-period activity has been observed in the parietal cortex (Gnadt and Andersen 1988; Crammond and Kalaska 1989; Koch and Fuster 1989; Constantinidis and Steinmetz 1996; Snyder et al. 1997; Chafee and Goldman-Rakic 1998; Quintana and Fuster 1999; Calton et al. 2002; Pesaran et al. 2002; Huk and Shadlen 2005; Nieder et al. 2006; Tudusciuc and Nieder 2007; Katsuki and Constantinidis 2012a, b), the temporal cortex (Fuster and Jervey 1982; Miyashita 1988; Miyashita and Chang 1988; Sakai and Miyashita 1991; Miller et al. 1993; Chelazzi et al. 1998; Yakovlev et al. 1998), the somatosensory cortex (Zhou and Fuster 1996, 1997; Romo et al. 2002; Romo and Salinas 2003), and the premotor cortex (Weinrich and Wise 1982; Kurata and Wise 1988; Crammond and Kalaska 2000; Ohbayashi et al. 2003; Romo and Salinas 2003; Romo et al. 2004). Delay-period activity has also been observed in the visual cortex (Gibson and Maunsell 1997; Supér et al. 2001; Lee et al. 2005), the superior colliculus (Hikosaka and Wurtz 1983; Basso and Wurtz 1998), the basal ganglia (Niki et al. 1972; Soltysik et al. 1975; Hikosaka et al. 1989; Apicella et al. 1992), the hippocampus (Watanabe and Niki 1985; Riches et al. 1991), the thalamus (Watanabe and Funahashi 2004a, b), and even in the spinal cord (Prut and Fetz 1999).

3.17.1 Frontal Eye Field

In the frontal eye field (FEF), Bruce and Goldberg (1985) reported that visuomovement cells exhibited sustained excitatory activity during the delay when the saccade was delayed relative to the target onset. Subsequently, Funahashi et al. (1989) showed that neurons in the FEF exhibited directional delay-period activity during ODR performances similar to dorsolateral prefrontal neurons. Further, Chafee and Goldman-Rakic (1998) found sustained delay-period activity with spatial tuning in FEF (area 8a) neurons while monkeys performed the ODR task. Features of sustained delay-period activity were very similar in temporal patterns and directional tunings to directional delay-period activity observed in the prefrontal cortex. They also compared characteristics of task-related activity observed during the ODR task between neurons in the FEF and neurons in the posterior parietal cortex and found striking similarity in response characteristics including temporal patterns and directional tuning of delay-period activity between these two areas. Sommer and Wurtz (2001) analyzed characteristics of delay-period activity of FEF neurons and examined whether these neurons projected to the superior colliculus or not. In their experiment, the monkey performed a Go/No-Go task either in memory-guided or visually-guided eye movement conditions. They found that delay-period activity of FEF neurons is modulated by Go or No-Go instruction in both conditions (movement-related), only in memory-guided condition (working memory-related), or only in visually-guided condition (visual attention-related) and that all these FEF neurons send information to the superior colliculus, suggesting that delay-period activity in FEF neurons affects saccade-related activity of neurons in the superior colliculus. Clark et al. (2012) used a delayed matching-to-sample task and found that delay-period activity in the FEF represents the information of the sample position independent of its behavioral relevance. Delay-period activity represented the information of the sample position even in the trials in which the matching stimulus was not presented in the neuron's response field. These results indicate that sustained delay-period activity is observed in the FEF with similar response characteristics as the activity observed in the dorsolateral prefrontal cortex. However, these results also show that, although delay-period activity representing retrospective information (e.g., sample position) is observed in the FEF, the ratio of neurons encoding prospective information (e.g., saccade direction) is larger than that of neurons encoding retrospective information in the FEF.

3.17.2 Orbitofrontal Cortex

Functions of the orbitofrontal cortex have been examined mainly with regard to emotional and motivational aspects as well as decision-making. For example, Tremblay and Schultz (1999) examined how orbitofrontal neurons processed reward information during spatial delayed-response performances and showed that differential activity was observed in most orbitofrontal neurons and that this activity represented the monkey's relative preference among available rewards rather than physical properties of the reward. Roesch and Olson (2005) examined the effect of the delay length (shorter delay or longer delay) on reward-related activity in orbitofrontal neurons while monkeys performed a memory-guided saccade task. They showed that orbitofrontal neurons tended to fire more strongly in response to the visual cue that predicted the shorter delay. Therefore, they concluded that orbitofrontal activity represents the time-discounted value of the expected reward. Further, Padoa-Schioppa and Assad (2006) examined how and where the economic values of reward are represented in the brain and showed that orbitofrontal neurons encode the values of an offered item and a chosen item independent of visuospatial factors of items and behavioral responses. Thus, orbitofrontal neurons encode the motivational value of the expected reward and this value is used for an assignment in the economic choice.

However, the orbitofrontal cortex has strong reciprocal connections with the dorsolateral prefrontal cortex (Barbas and Pandya 1989; Carmichael and Price 1995; Cavada et al. 2000), suggesting that the orbitofrontal cortex also contributes to cognitive functions such as working memory. Ichihara-Takeda and Funahashi (2006, 2007) recorded single-neuron activity from the dorsolateral prefrontal cortex and the orbitofrontal cortex while a monkey performed the ODR task and compared characteristics of task-related activity between these two areas. Although all kinds of task-related activity observed in dorsolateral prefrontal neurons were also observed in orbitofrontal neurons, the proportion and response characteristics of task-related activity were different between these two areas. For example, in the dorsolateral prefrontal cortex, the majority of delay-period activity was directionally selective and showed tonic sustained activation. However, in the orbitofrontal cortex, the most delay-period activity exhibited omnidirectional selectivity and showed gradual increase with the progress of the delay period. In addition, reward-period activity is predominant among task-related activities in the orbitofrontal cortex (Ichihara-Takeda and Funahashi 2006). The proportion of neurons having reward-period activity was significantly higher than neurons having other kinds of task-related activity in the orbitofrontal cortex. Since orbitofrontal neurons have been known to encode the motivational value of the expected reward, Ichihara-Takeda and Funahashi (2008) compared motivational effects on delay-period activity between the orbitofrontal cortex and the dorsolateral prefrontal cortex while the monkey performed the ODR task under different reward schedules. In the ODR-1 schedule, the reward was given only after the successful completion of four consecutive trials, whereas in the ODR-2 schedule, the reward was given after every correct trial. Delay-period activity in both areas showed spatial selectivity and spatial characteristics of delay-period activity remained constant in both schedules. Delay-period activity was affected by the reward schedule and the magnitude of this activity gradually increased depending on the proximity of the rewarded trial in the ODR-1 schedule (the max activity was observed at the fourth trial). In addition, more delay-period activity was affected by the reward schedule in the orbitofrontal cortex, whereas more delay-period activity was affected by both the spatial factor and the reward schedule in the dorsolateral prefrontal cortex. These results indicate that, although neurons having sustained delay-period activity are present in the orbitofrontal cortex, this cortical area plays a role in monitoring and detecting reward delivery, whereas the dorsolateral prefrontal cortex plays a role to integrate cognitive and motivational information. Thus, although the orbitofrontal cortex participates in spatial working memory processes, this cortical area is related more to the expectation and the detection of reward delivery.

3.17.3 Parietal Cortex

In the parietal cortex, delay-period activity has been observed in many studies (Gnadt and Andersen 1988; Crammond and Kalaska 1989; Koch and Fuster 1989; Barash et al. 1991a, b; Constantinidis and Steinmetz 1996; Snyder et al. 1997; Quintana and Fuster 1999; Calton et al. 2002; Pesaran et al. 2002; Huk and Shadlen 2005; Nieder et al. 2006; Tudusciuc and Nieder 2007; Katsuki and Constantinidis

2012b). The dorsolateral prefrontal cortex and the posterior parietal cortex share a number of functional properties and often are co-activated in relation to a variety of cognitive activities with similar temporal patterns of activation (Katsuki and Constantinidis 2012b).

In the intraparietal cortex, Gnadt and Andersen (1988) found neurons exhibiting sustained activity during the delay period in the lateral bank of the intraparietal cortex. These neurons remained active during the time period while monkeys looked at a fixation point and remembered the target position for the saccade. Since the activity during the delay period was tuned for specific direction and amplitude of the eye movement, they concluded that this activity represents motor-planning signal. Barash et al. (1991a, b) further showed that neurons in the intraparietal cortex exhibited sustained activity during the delay period in delayed saccade performances. Sustained activity exhibited directional selectivity and had the preferred direction mainly toward the contralateral visual field, which is similar characteristics to prefrontal delay-period activity. Constantinidis and Steinmetz (1996) also showed that sustained activity was observed in area 7a during the delay period in a delayed matching-to-position task and that the discharge rate of sustained delay-period activity was greater when the visual cue was presented at the position where a strong visual response was generated. In addition, they observed anticipatory activity which was characterized by accelerating discharge rates and often associated with facilitatory responses to the matching stimulus. Snyder et al. (1997) examined whether sustained delay-period activity observed in the posterior parietal cortex represented either visual information or the intention to make a movement. They used the task in which the monkey was required to either perform a hand-reaching movement or make a saccade to a single remembered position. Sustained delay-period activity of most posterior parietal neurons was related to the type of the movement being planned, indicating that delay-period activity observed in the posterior parietal cortex represents the intention to make a movement, not the information of the visual signal. Calton et al. (2002) supported the results of Snyder et al. (1997), although they also showed delay-period activity representing not the type of movement to be performed but the target position. Similar result had also been reported by Quintana and Fuster (1999) showing that delay-period activity in the parietal cortex was attuned to response direction and tended to accelerate its firing in anticipation of the response and in proportion to the predictability of response direction. In addition, Nieder et al. (2006) and Tudusciuc and Nieder (2007) had shown the presence of delay-period activity representing numerical information in the parietal cortex.

In the parietal area 5, Crammond and Kalaska (1989) used the task in which the monkey was required to perform an arm-reaching movement toward the direction shown by a visual cue after the delay period and showed sustained activity during the instructed delay period which varied with the direction of the impending movement. This activity is very similar to set-related activity observed in the premotor cortex (Weinrich and Wise 1982; Kurata and Wise 1988). Further, Koch and Fuster (1989) examined parietal activity using a haptic delayed matching-to-sample task and showed that many neurons in area 5 exhibited sustained activation during the delay period, some of which encoded which sample object was palpated.

Thus, these results indicate that, although neurons having persistent delay-period activity are present in the parietal cortex, delay-period activity in this cortical area represents information mostly for preparing and controlling motor performances. Thus, although the persistent delay-period activity is observed in both the prefrontal cortex and the parietal cortex, the information represented by this activity is not always the same between these two cortical areas.

3.17.4 Inferior Temporal Cortex

Sustained delay-period activity has been observed in neurons of the inferior temporal (IT) cortex while monkeys performed memory tasks (Fuster and Jervey 1982; Miyashita 1988; Miyashita and Chang 1988; Miller et al. 1993; Gibson and Maunsell 1997; Chelazzi et al. 1998; Yakovlev et al. 1998). Neurons in the inferior temporal cortex are shown to exhibit stimulus-selective delay-period activity. Mikami and Kubota (1980) used a delayed matching-to-color task and examined task-related activity in the IT cortex. They found only a few IT neurons that exhibited activation during the delay period but did not exhibit differential responses to different color stimuli, although many IT neurons responded to the sample and the matching color stimuli and exhibited differential responses to different color stimuli. Fuster and Jervey (1981, 1982) analyzed single-neuron activity recorded in the IT cortex while monkeys performed the delayed matching-to-sample task. In this task, one color stimulus is presented during the sample period and, after the delay period, two or four color stimuli are presented simultaneously and the monkey is required to select the same color stimulus as the sample stimulus. Although many IT neurons responded to color stimuli and exhibited color-selectivity, they observed a substantial number of IT neurons exhibiting sustained color-selective activation during the delay period. Fuster (1990) also examined the activity of IT neurons during delayed matching-to-sample tasks with color and pattern stimuli and found that 14% of all IT neurons showed delay-period activity with selectivity to a particular feature of the sample stimulus (a particular color or pattern). Miyashita (1988) used the delayed matching-to-sample task and showed that neurons in the anterior ventral temporal cortex exhibited sustained activity during the delay period which exhibited highly selective responses to a few of the 100 colored fractal patterns. Further, Miyashita and Chang (1988) showed that sustained delay-period activity in anterior ventral temporal neurons was highly selective for the object information to be remembered and independent of the physical attributes including size, orientation, color, or position of the object. Nakamura and Kubota (1995) found that 22% of neurons ($n = 311$) recorded in the anterior portion of the inferior temporal cortex including the temporal pole exhibited stimulus-selective delay-period activity while monkeys performed the visual recognition task, in which monkeys were required to release the lever when a new stimulus was presented after the repetitive presentation of the same stimulus. Thus, neurons in the inferior temporal cortex exhibit stimulus-selective delay-period activity.

Delay-period activity had been shown to exhibit different features between prefrontal neurons and IT neurons. Miller et al. (1993) and Miller and Desimone (1994) used the delayed matching-to-sample task to examine neural mechanisms for short-term memory in the IT cortex. In this task, the monkey is required to hold a sample picture (e.g., A) in mind and make a behavioral response when it appears in a sequence of other pictures (e.g., ABBA) which the monkey needs to ignore during the delay period (ABBA task). They found that the activity during the delay period was suppressed in some IT neurons when any picture was presented repeatedly regardless of whether it was relevant or not. However, in other IT neurons, the activity was enhanced only when the picture matched the sample. Miller et al. (1996) further compared characteristics of task-related activity between IT neurons and prefrontal neurons while the monkey performed the ABBA task. Prefrontal neurons more often exhibited sample-selective delay-period activity and enhancement of the activity when the picture matched the sample than IT neurons. Further, in prefrontal neurons, sample-selective delay-period activity survives intervening distractor stimuli presented during the delay period. However, although IT neurons also exhibited sample-selective delay-period activity, this activity was not maintained continuously throughout the delay period. Delay-period activity of IT neurons was temporarily suppressed when the presented picture was irrelevant. These results indicate that the information of the sample stimulus is robustly maintained by sustained delay-period activity throughout the delay period in the prefrontal cortex, whereas it is not in the IT cortex although delay-period activity is observed.

3.17.5 Other Cortical Areas

Delay-period activity has been observed in the visual cortex. Supér et al. (2001) trained the monkey to perform an oculomotor type of the delayed-response task and examined multiple neuron activities in the primary visual cortex. In this task, while the monkey maintains fixation at the fixation point, a motion-defined figure appears briefly at one of three positions on the monitor together with background motion stimulus. After a delay, the fixation point is removed, indicating to the monkey to make a saccade to the position where the figure has been presented. A stronger response to the figure over the response to the background was observed and sustained during the delay period. This figure-background modulation was observed in correct trials and disappeared in incorrect trials. Thus, persistent activation representing the figure-background difference in the primary visual cortex during the delay period is similar to features of directional delay-period activity observed in the prefrontal cortex. Lee et al. (2005) recorded local field potentials and single-neuron activities from the extrastriate visual cortex while the monkey performed the delayed matching-to-sample task and found significant enhancement of theta band oscillation during the delay period, which was correlated with the enhancement of sample-selective delay-period activity. Serences et al. (2009) used a multiboxel pattern analysis of fMRI data while human subjects performed a delayed matching-to-sample task and showed that sustained activation during the delay period is observed in the primary visual cortex and that delay-period activity represents only intentionally stored feature (either orientation or color) of a colored Gabor stimulus.

Delay-period activity has also been observed in the somatosensory cortex (Zhou and Fuster 1996, 1997). Zhou and Fuster (1996) recorded single-neuron activity from the hand area of the somatosensory cortex (area 3a, 3b, 1, and 2) while monkeys performed a haptic delayed matching-to-sample task and found that many neurons exhibit sustained delay-period activity. Delay-period activity observed in the somatosensory cortex was considered to be a neural correlate of short-term memory of tactile stimulus, because this activity showed object selectivity, had no relation to preparation for movement, and observed not in the visuo-haptic matching task but in the haptic-haptic matching task. However, a subsequent study showed that some neurons in the somatosensory cortex exhibited delay-period activity in the visuo-haptic matching task (Zhou and Fuster 1997).

In the premotor cortex, set-related activity has been observed during the preparation period before the initiation of the hand-arm movement (Weinrich and Wise 1982; Kurata and Wise 1988; Crammond and Kalaska 2000; Ohbayashi et al. 2003). Set-related activity is persistent activation that appeared after the presentation of the imperative visual cue and was maintained until the initiation of the movement. Set-related activity also exhibited selectivity to the movement direction. Thus, set-related activity has similar characteristics as those of delay-period activity in the prefrontal cortex. However, set-related activity is a preparatory activity for impending movement and represent movement information.

In the anterior portion of the cingulate cortex, neurons have been shown to exhibit delay-period activity. Niki and Watanabe (1976b) used three kinds of spatial delay tasks (delayed-response tasks with left-right and up-down positions and conditional position discrimination tasks) and showed that 14% (31/222) of recorded neurons exhibited differential delay-period activity depending on the difference of task conditions. Among these, 80% of delay-period activity exhibited differential changes depended on the difference of the visual cue positions, while 20% exhibited differential changes depended on the difference of the response directions. Proportions of delay-period activity representing retrospective and prospective information are similar between cingulate neurons and prefrontal neurons (Niki and Watanabe 1976a).

Thus, persistent delay-period activity has been observed in many areas in the cerebral cortex during working memory tasks and this activity is shown to play important role in information processing in these areas. Information representing delay-period activity is different across different brain areas since information necessary for processing in one brain area is not always the same as that for other areas.

3.17.6 Thalamus

In the thalamus, delay-period activity has been observed in many neurons, especially in the mediodorsal nucleus (MD) of the thalamus. Fuster and Alexander (1971, 1973) first reported activity of MD neurons while monkeys performed the delayed-response task and showed that MD neurons exhibited short-term memory-related activity similar to what was observed in the prefrontal cortex (Fuster 1973). Tanibuchi and Goldman-Rakic (2003) examined the activity of MD neurons while the monkey performed the spatial ODR task and non-spatial tasks using objects and faces and showed that MD neurons responded selectively either to the spatial position of the visual cue in the spatial task or to faces and objects in the non-spatial task. They also showed that the distributions of spatially tuned neurons and non-spatially tuned neurons were segregated within the MD, such that spatially tuned neurons were presented in the area where the MD connected to the dorsolateral prefrontal cortex, whereas non-spatially tuned neurons were present in the area where the MD connected with the ventrolateral prefrontal cortex. The presence of visually selective neurons in the MD was shown by Sommer and Wurtz (2004a). In addition, Sommer and Wurtz (2004a) and Tanibuchi and Goldman-Rakic (2005) showed a high proportion of pre-saccadic neurons in the MD. Wyder et al. (2003, 2004) examined single-neuron activity collected from the central thalamus in which the MD was excluded while monkeys performed a visually-guided delayed saccade task and a memory-guided saccade task. Although visually selective neurons and pre-saccadic neurons were presented in this thalamic area, some neurons exhibited delay-period activity representing the position of visible stimuli regardless of behavioral relevance, some neurons exhibited delay-period activity preferentially representing the position of the current saccade goal in the absence of the stimulus, and the other neurons exhibited delay-period activity representing only the position of the current visible saccade goal.

Characteristics of delay-period activity observed in the MD were fully analyzed by Watanabe and Funahashi (2004a, b). They used the ODR task, examined characteristics of task-related activity observed in MD neurons, and compared these characteristics with those observed in prefrontal neurons (Watanabe and Funahashi 2004a; Funahashi et al. 2004). Similar to the prefrontal cortex, a large proportion (53%) of task-related neurons exhibited tonic sustained delay-period activity, most of which exhibited directional selectivity. However, differed from prefrontal neurons most of which exhibited contralateral bias in the directional selectivity, significant contralateral bias was not observed in directional selectivity of delay-period activity in MD neurons, although many MD neurons had best directions toward the contralateral visual field. Further, information represented by delay-period activity was examined in MD neurons by comparing directional selectivity of delay-period activity between ODR and R-ODR performances (Watanabe and Funahashi 2004b; Funahashi et al. 2004). In the MD, nearly half (56%) of delay-period activity represented information regarding the position of the visual cue, whereas the remaining half (41%) represented information regarding the direction of the saccade. This result indicates that larger proportion of delay-period activity represents the direction of the saccade in the MD than in the prefrontal cortex, suggesting that MD neurons participate more in prospective information processing than prefrontal neurons (Funahashi et al. 2004). MD neurons' participation in prospective information processing had been supported by an analysis using the population vector method. Watanabe et al. (2009) showed that in the R-ODR task, the direction of the population vector started rotation earlier in MD neurons than in prefrontal neurons (see Fig. 3.29c), indicating that the prospective motor information is provided from the MD to the prefrontal cortex. The evidence that the MD participates relatively more in motor aspects of information processing than the prefrontal cortex is supported by DeNicola et al. (2020) which found different contributions for decision-making and state representation between two areas, such that MD neurons specifically contribute to decision-making and response selection, whereas prefrontal neurons contribute to encoding the environmental state for decision-making.

3.17.7 Basal Ganglia

In the basal ganglia, Hikosaka and Wurtz (1983) first introduced the memory-guided saccade task for examining activity of neurons in the substantia nigra pars reticulata and found that a small number of neurons (16%) exhibited memory-contingent sustained activity. In the caudate nucleus, Hikosaka et al. (1989) reported delay-period activity similar to that was observed in the prefrontal cortex while monkeys performed an ODR-like task. Since sustained delay-period activity observed in the caudate nucleus started after the cue presentation, they suggested that delay-period activity represents retrospective spatial information. Further, Apicella et al. (1992) examined neurons in the caudate nucleus and the putamen while monkeys performed a delayed Go/No-Go task using arm reaching movements to the target. They found that some of these neurons exhibited sustained activation between the presentation of the instruction stimulus and the trigger stimulus to make a response. They also found that sustained activity of these neurons represented the expectation of the trigger stimulus independent of arm movements or the preparation or expectation of specific movement, suggesting that sustained activity of these neurons represents prospective sensory information.

3.17.8 Other Subcortical Areas

Delay-period activity has been observed in subcortical structures. In the hippocampus, Watanabe and Niki (1985) showed delay-period activity while monkeys performed the manual delayed-response task. Neurons having delay-period activity exhibited sustained excitation or inhibition or gradually increasing activation and some of them exhibited differential firing patterns during the delay period depending on the difference in the cue positions. Riches et al. (1991) collected neuron activity from the hippocampus, the rhinal cortex, and the parahippocampal gyrus while monkeys performed the delayed matching-to-sample task and reported activation during the delay period (0.5 s) just after the presentation of either the sample or the test stimulus.

In the superior colliculus, it has been known that neurons in the intermediate layer exhibit delay-period activity when monkeys perform a delayed saccade task and a memory-guided saccade task. For example, neurons referred to as prelude burst neurons exhibit long lead activity representing the selection of the saccade during the delay period (Glimcher and Sparks 1992), suggesting that this activity is similar to set-related activity in the premotor cortex. Neurons referred to as buildup neurons have long-leading excitatory activity after the target is identified, suggesting that this activity reflects saccade preparation (Munoz and Wurtz 1995; Kojima et al. 1996; Basso and Wurtz 1998).

In the nucleus basalis of Meynert, Richardson and DeLong (1986) reported that 14% of task-related neurons ($n = 183$) exhibited delay-period activity while monkeys performed a two-choice manual delayed-response task, although none of these neurons exhibited differential delay-period activity.

In the caudal dentate nucleus of the cerebellum, Ashmore and Sommer (2013) examined the activity of saccade-related neurons. The caudal dentate nucleus of the macaque cerebellum has anatomical connections with the frontal eye field and the lateral intraparietal area. They showed that neurons in this area exhibited delay-period activity while monkeys performed sensory-guided self-timed saccade tasks and a memory-guided saccade tasks. Delay-period activity exhibited tuning for saccade directions and directional tuning was maintained throughout the delay period. A minority of neurons exhibited sensory-related delay activity.

Even in the spinal cord, delayed activity had been observed. Prut and Fetz (1999) found pre-movement delay activity in spinal interneurons during the instructed delay period and that this delay activity resembled set-related activity observed in the premotor cortex, indicating that this activity reflects preparatory changes of neural activity before the execution of a movement.

3.18 Delay-Period Activity Is a Neural Mechanism for Temporarily Maintaining Information

When the monkey performs a behavioral task including the delay period between the presentation of a sensory cue and a behavioral response, sustained activation can be observed in single-neuron activity or multiple neuron activities during the delay period. Persistent delay-period activity exhibits several important features. First, delay-period activity is usually maintained actively throughout the delay period. When the length of the delay period changes, the duration of delay-period activity is prolonged or shortened depending on the length of the delay period. However, once the behavioral response is executed, delay-period activity is terminated. Second, persistent delay-period activity can be observed only when the subject performs correct responses. In a trial that the subject makes an error, delay-period activity is not observed at all or terminated sometime during the delay period. Third, delay-period activity exhibits selectivity to a variety of physical features of the stimulus or behavioral contexts. For example, delay-period activity can be observed only when a specific stimulus is presented as a behavioral cue or relevant for a behavioral response. When the subject performs a task with delay, it is required to remember trial-specific information necessary to make a correct response in a given trial. Trial-specific information varies trial by trial. Therefore, the features of delay-period activity indicate that delay-period activity is a neural correlate of the mechanism to hold trial-specific information temporarily during the delay period.

In order to perform a variety of complex cognitive tasks with delay, the subject needs to hold not only sensory information but also other kinds of information necessary to perform the tasks. In fact, delay-period activity observed in the prefrontal cortex has been shown to represent a variety of information including not only sensory information but also task rules, expected reward, numerical quantity, temporal orders of stimulus presentation. These evidences strongly support that delay-period activity is a neural correlate of the mechanism for temporarily holding trial-specific information.

Delay-period activity was first found in the prefrontal cortex. However, this activity can be observed not only in the prefrontal cortex but also in many other cortical and subcortical areas and persistent excitation throughout the delay period is a common and typical feature of delay-period activity across various brain areas. However, each brain region participates in specific functions and performs specific information processing. Therefore, although the temporal pattern of delay-period activity (e.g., sustained activation) is similar across different brain regions, information represented by delay-period activity would be different across different brain regions. In fact, in the prefrontal cortex, the majority of delay-period activity represents retrospective sensory information, while the minority represents prospective motor information. On the other hand, in the parietal cortex, the majority of delay-period activity represents prospective motor information. Similarly, in the thalamic mediodorsal nucleus, the majority of delay-period activity represents prospective motor information. Thus, these results further support the notion that delay-period activity is a neural correlate of the mechanism for temporarily maintaining information necessary to perform a specific behavioral task and that delay-period activity plays an important role in specific information processing in each brain area. Thus, delay-period activity is an essential and important component of working memory processes in many brain areas.

References

Aertsen AMHJ, Gerstein GL (1985) Evaluation of neuronal connectivity: sensitivity of cross-correlation. Brain Res 340:341–354. https://doi.org/10.1016/0006-8993(85)90931-x

Aertsen AMHJ, Gerstein GL, Habib MK, Palm G (1989) Dynamics of neuronal firing correlation: modulation of "effective connectivity". J Neurophysiol 61:900–917. https://doi.org/10.1152/jn.1989.61.5.900

Aggleton JP, Hunt PR, Rawlins JNP (1986) The effects of hippocampal lesions upon spatial and non-spatial tests of working memory. Behav Brain Res 19:133–146. https://doi.org/10.1016/0166-4328(86)90011-2

Ahissar M, Ahissar E, Bergman H, Vaadia E (1992) Encoding of sound-source location and movement: activity of single neurons and interactions between adjacent neurons in the monkey auditory cortex. J Neurophysiol 67:203–215. https://doi.org/10.1152/jn.1992.67.1.203

Amemori K, Sawaguchi T (2006a) Rule-dependent shifting of sensorimotor representation in the primate prefrontal cortex. Eur J Neurosci 23:1895–1909. https://doi.org/10.1111/j.1460-9568.2006.04702.x

Amemori K, Sawaguchi T (2006b) Contrasting effects of reward expectation on sensory and motor memories in primate prefrontal neurons. Cereb Cortex 16:1002–1016. https://doi.org/10.1093/cercor/bhj042

Andersen RA, Mountcastle VB (1983) The influence of the angle of gaze upon the excitability of the light-sensitive neurons of the posterior parietal cortex. J Neurosci 3:532–548. https://doi.org/10.1523/jneurosci.03-03-00532.1983

Andersen RA, Asanuma C, Cowan WM (1985a) Callosal and prefrontal associational projecting cell populations in area 7A of the macaque monkey: a study using retrogradely transported fluorescent dyes. J Comp Neurol 232:443–455. https://doi.org/10.1002/cne.902320403

Andersen RA, Essick GK, Siegel RM (1985b) Encoding of spatial location by posterior parietal neurons. Science 230:456–458. https://doi.org/10.1126/sicnece.4048942

Andersen RA, Snyder LH, Bradley DC, Xing J (1997) Multimodal representation of space in the posterior parietal cortex and its use in planning movements. Annu Rev Neurosci 20:303–330. https://doi.org/10.1146/annurev.neuro.20.1.303

Ando S, Ohashi Y (1991) Longitudinal study on age-related changes of working and reference memory in the rat. Neurosci Lett 128:17–20. https://doi.org/10.1016/0304-3940(91)90750-N

Apicella P, Scarnati E, Ljungberg T, Schultz W (1992) Neuronal activity in monkey striatum related to the expectation of predictable environmental events. J Neurophysiol 68:945–960. https://doi.org/10.1152/jn.1992.68.3.945

Asaad WF, Rainer G, Miller EK (1998) Neural activity in the primate prefrontal cortex during associative learning. Neuron 21:1399–1407. https://doi.org/10.1016/s0896-6273(00)80658-3

Ashmore RC, Sommer MA (2013) Delay activity of saccade-related neurons in the caudal dentate nucleus of the macaque cerebellum. J Neurophysiol 109:2129–2144. https://doi.org/10.1152/jn.00906.2011

Atkinson RC, Shiffrin RM (1968) Human memory: a proposed system and its control processes. In: Spence KW, Spence JT (eds) The psychology of learning and motivation: advances in research and theory. Academic Press, New York, pp 89–195

Averbeck BB, Lee D (2007) Prefrontal neural correlates of memory for sequences. J Neurosci 27:2204–2211. https://doi.org/10.1523/jneurosci.4483-06.2007

Bachevalier J, Mishkin M (1986) Visual recognition impairment follows ventromedial but not dorsolateral prefrontal lesions in monkeys. Behav Brain Res 20:249–261. https://doi.org/10.1016/0166-4328(86)90225-1

Baddeley A (1983) Working memory. Phil Trans R Soc Lond B 302:311–324. https://doi.org/10.1098/rstb.1983.0057

Baddeley A (1986) Working memory. Oxford University Press, Oxford, UK

Baddeley A (1992) Working memory. Science 255:556–559. https://doi.org/10.1126/science.1736359

Baddeley A (1996) The fractionation of working memory. Proc Natl Acad Sci USA 93:13468–13472. https://doi.org/10.1073/pnas.93.24.13468

Baddeley A (2000) The episodic buffer: a new component of working memory? Trends Cogn Sci 4:417–423. https://doi.org/10.1016/S1364-6613(00)01538-2

Baddeley A (2012) Working memory: theories, models, and controversies. Annu Rev Psychol 63:1–29. https://doi.org/10.1146/annurev-psych-120710-100422

Baddeley A, Hitch GJ (1974) Working memory. In: Bower GH (ed) The psychology of learning and motivation, vol 8. Academic Press, New York, pp 47–89

Baddeley A, Logie RH (1999) Working memory: the multiple-component model. In: Miyake A, Shah P (eds) Models of working memory: mechanisms of active maintenance and executive control. Cambridge University Press, Cambridge, UK, pp 28–61

Barash S, Bracewell RM, Fogassi L, Gnadt JW, Andersen RA (1991a) Saccade-related activity in the lateral intraparietal area. I. Temporal properties; comparison with area 7a. J Neurophysiol 66:1095–1108. https://doi.org/10.1152/jn.1991.66.3.1095

Barash S, Bracewell RM, Fogassi L, Gnadt JW, Andersen RA (1991b) Saccade-related activity in the lateral intraparietal area. II. Spatial properties. J Neurophysiol 66:1109–1124. https://doi.org/10.1152/jn.1991.66.3.1109

Barbas H, Pandya DN (1989) Architecture and intrinsic connections of the prefrontal cortex in the rhesus monkey. J Comp Neurol 286:353–375. https://doi.org/10.1002/cne.902860306

Barbas H, Henion TH, Dermon CR (1991) Diverse thalamic projections to the prefrontal cortex in the rhesus monkey. J Comp Neurol 313:65–94. https://doi.org/10.1002/cne.903130106

Barone P, Joseph JP (1989) Prefrontal cortex and spatial sequencing in macaque monkey. Exp Brain Res 78:447–464. https://doi.org/10.1007/BF00230234

Basso MA, Wurtz RH (1998) Modulation of neuronal activity in superior colliculus by changes in target probability. J Neurosci 18:7519–7534. https://doi.org/10.1523/jneurosci.18-18-07519.1998

Batuev AS, Shaefer VI, Orlov AA (1985) Comparative characteristics of unit activity in the prefrontal and parietal areas during delayed performance in monkeys. Behav Brain Res 16:57–70. https://doi.org/10.1016/0166-4328(85)90082-8

Belger A, Puce A, Krystal JH, Gore JC, Goldman-Rakic P, McCarthy G (1998) Dissociation of mnemonic and perceptual processes during spatial and nonspatial working memory using fMRI. Hum Brain Mapping 6:14–32. https://doi.org/10.1002/(SICI)1097-0193(1998)6:1<14::AID-HBM2>3.0.CO;2-O

Bergman H, Feingold A, Nini A, Ras A, Slovin H, Abeles M, Vaadia E (1998) Physiological aspects of information processing in the basal ganglia of normal and parkinsonian primates. Trends Neurosci 21:32–38. https://doi.org/10.1016/s0166-2236(97)01151-x

Biguer B, Jeannerod M, Prablanc C (1982) The coordination of eye, head, and arm movements during reaching at a single visual target. Exp Brain Res 46:301–304. https://doi.org/10.1007/bf00237188

Boch RA, Goldberg ME (1989) Participation of prefrontal neurons in the preparation of visually guided eye movements in the rhesus monkey. J Neurophysiol 61:1064–1084. https://doi.org/10.1152/jn.1989.61.5.1064

Bongard S, Nieder A (2010) Basic mathematical rules are encoded by primate prefrontal cortex neurons. Proc Natl Acad Sci USA 107:2277–2282. https://doi.org/10.1073/pnas.0909180107

Boussaoud D (1995) Primate premotor cortex: modulation of preparatory neuronal activity by gaze angle. J Neurophysiol 73:886–890. https://doi.org/10.1152/jn.1995.73.2.886

Boussaoud D, Barth TM, Wise SP (1993) Effects of gaze on apparent visual responses of frontal cortex neurons. Exp Brain Res 93:423–434. https://doi.org/10.1007/BF00229358

Boussaoud D, Jouffrais C, Bremmer F (1998) Eye position effects on the neuronal activity of dorsal premotor cortex in the macaque monkey. J Neurophysiol 80:1132–1150. https://doi.org/10.1152/jn.1998.80.3.1132

Brody BA, Pribram KH (1978) The role of frontal and parietal cortex in cognitive processing. Brain 101:607–633. https://doi.org/10.1093/brain/101.4.607

Brody CD, Hernandez A, Zainos A, Romo R (2003) Timing and neural encoding of somatosensory parametric working memory in macaque prefrontal cortex. Cereb Cortex 13:1196–1207. https://doi.org/10.1093/cercor/bhg100

Brown MW, Aggleton JP (2001) Recognition memory: what are the roles of the perirhinal cortex and hippocampus? Nat Rev Neurosci 2:51–61. https://doi.org/10.1038/35049064

Bruce CJ, Goldberg ME (1985) Primate frontal eye fields. I. Single neurons discharging before saccades. J Neurophysiol 53:603–635. https://doi.org/10.1152/jn.1985.53.3.603

Bunge SA, Kahn I, Wallis JD, Miller EK, Wagner AD (2003) Neural circuits subserving the retrieval and maintenance of abstract rules. J Neurophysiol 90:3419–3428. https://doi.org/10.1152/jn.00910.2002

Bushnell PJ, Levin ED (1993) Effects of dopaminergic drugs on working and reference memory in rats. Pharmacol Biochem Behav 45:765–776. https://doi.org/10.1016/0091-3057(93)90119-e

Butters N, Pandya D (1969) Retention of delayed alternation: effect of selective lesions of sulcus principalis. Science 165:1271–1273. https://doi.org/10.1126/science.165.3899.1271

Calton JL, Dickinson AR, Snyder LH (2002) Non-spatial, motor-specific activation in posterior parietal cortex. Nat Neurosci 5:80–588. https://doi.org/10.1038/nn0602-862

Carlson S, Tanila H, Pertovaara A, Lähteenmäki A (1990) Vertical and horizontal coding of space in the monkey dorsolateral prefrontal cortex. Brain Res 527:45–149. https://doi.org/10.1016/0006-8993(90)91073-P

Carlson S, Rämä P, Tanila H, Linnankoski I, Mansikka H (1997) Dissociation of mnemonic coding and other functional neuronal processing in the monkey prefrontal cortex. J Neurophysiol 77:761–774. https://doi.org/10.1152/jn.1997.77.2.761

Carmichael ST, Price JL (1995) Limbic connections of the orbital and medial prefrontal cortex in macaque monkeys. J Comp Neurol 363:615–641. https://doi.org/10.1002/cne.903630408

Carpenter AF, Baud-Bovy G, Georgopoulos AP, Pellizzer G (2018) Encoding of serial order in working memory: neuronal activity in motor, premotor, and prefrontal cortex during a memory scanning task. J Neurosci 38:4912–4933. https://doi.org/10.1523/jneurosci.3294-17.2018

Cavada C, Goldman-Rakic PS (1989) Posterior parietal cortex in rhesus monkey. II. Evidence for segregated corticocortical networks linking sensory and limbic areas with the frontal lobe. J Comp Neurol 287:422–445. https://doi.org/10.1002/cne.902870403

Cavada C, Compañy T, Tejedor J, Cruz-Rizzolo RJ, Reinoso-Suárez F (2000) The anatomical connections of the macaque monkey orbitofrontal cortex. A review. Cereb Cortex 10:220–242. https://doi.org/10.1093/cercor/10.3.220

Chafee MV, Goldman-Rakic PS (1998) Matching patterns of activity in primate prefrontal area 8a and parietal area 7ip neurons during a spatial working memory task. J Neurophysiol 79:2919–2940. https://doi.org/10.1152/jn.1998.79.6.2919

Chafee MV, Goldman-Rakic PS (2000) Inactivation of parietal and prefrontal cortex reveals interdependence of neural activity during memory-guided saccades. J Neurophysiol 83:1550–1566. https://doi.org/10.1152/jn.2000.83.3.1550

Chelazzi L, Duncan J, Miller EK, Desimone R (1998) Responses of neurons in inferior temporal cortex during memory-guided visual search. J Neurophysiol 80:2918–2940. https://doi.org/10.1152/jn.1998.80.6.2918

Clark KL, Noudoost B, Moore T (2012) Persistent spatial information in the frontal eye field during object-based short-term memory. J Neurosci 32:10907–10914. https://doi.org/10.1523/jneurosci.1450-12.2012

Constantinidis C, Steinmetz MA (1996) Neuronal activity in posterior parietal area 7a during the delay periods of a spatial memory task. J Neurophysiol 76:1352–1355. https://doi.org/10.1152/jn.1996.76.2.1352

Constantinidis C, Franowicz MN, Goldman-Rakic S (2001a) The sensory nature of mnemonic representation in the primate prefrontal cortex. Nat Neurosci 4:311–316. https://doi.org/10.1038/85179

Constantinidis C, Franowicz MN, Goldman-Rakic PS (2001b) Coding specificity in cortical microcircuits: a multiple-electrode analysis of primate prefrontal cortex. Neurosci 21:3646–3655. https://doi.org/10.1523/jneurosci.21-10-03646.2001

Courtney SM, Ungerleider LG, Keil K, Haxby JV (1996) Object and spatial visual working memory activate separate neural systems in human cortex. Cereb Cortex 6:39–49. https://doi.org/10.1093/cercor/6.1.39

Courtney SM, Ungerleider LG, Keil K, Haxby JV (1997) Transient and sustained activity in a distributed neural system for human working memory. Nature 386:608–611. https://doi.org/10.1038/386608a0

Cowan N (1999) An embedded-processes model of working memory. In: Miyake A, Shah P (eds) Models of working memory: mechanisms of active maintenance and executive control. Cambridge University Press, Cambridge, UK, pp 62–101

Craik FIM, Lockhart RS (1972) Levels of processing: a framework for memory research. J Verbal Learn Verbal Behav 11:671–684. https://doi.org/10.1016/s0022-5371(72)80001-x

Crammond DJ, Kalaska F (1989) Neuronal activity in primate parietal cortex area 5 varies with intended movement direction during an instructed-delay period. Exp Brain Res 76:458–462. https://doi.org/10.1007/BF00247902

Crammond DJ, Kalaska JF (2000) Prior information in motor and premotor cortex: activity during the delay period and effect on pre-movement activity. J Neurophysiol 84:986–1005. https://doi.org/10.1152/jn.2000.84.2.986

Curtis CE (2006) Prefrontal and parietal contributions to spatial working memory. Neuroscience 139:173–180. https://doi.org/10.1016/j.neuroscience.2005.04.070

DeNicola AL, Park M-Y, Crowe DA, MacDonald AW III, Chafee MV (2020) Differential roles of mediodorsal nucleus of the thalamus and prefrontal cortex in decision-making and state representation in a cognitive control task measuring deficits in schizophrenia. J Neurosci 40:1650–1667. https://doi.org/10.1523/jneurosci.1703-19.2020

di Pellegrino G, Wise SP (1993) Visuospatial versus visuomotor activity in the premotor and prefrontal cortex of a primate. J Neurosci 13:1227–1243. https://doi.org/10.1523/jneurosci.13-03-01227.1993

Eichenbaum H, Yonelinas AP, Ranganath C (2007) The medial temporal lobe and recognition memory. Annu Rev Neurosci 30:123–152. https://doi.org/10.1146/annurev.neuro.30.051606.094328

Eiselt A-K, Nieder A (2013) Representation of abstract quantitative rules applied to spatial and numerical magnitudes in primate prefrontal cortex. J Neurosci 33:7526–7534. https://doi.org/10.1523/jneurosci.5827-12.2013

Engle RW, Kane MJ, Tuholski SW (1999) Individual differences in working memory capacity and what they tell us about controlled attention, general fluid intelligence, and functions of the prefrontal cortex. In: Miyake A, Shah P (eds) Models of working memory: mechanisms of active maintenance and executive control. Cambridge University Press, Cambridge, UK, pp 102–134

Espinosa IE, Gerstein GL (1988) Cortical auditory neuron interactions during presentation of 3-tone sequences: effective connectivity. Brain Res 450:39–50. https://doi.org/10.1016/0006-8993(88)91542-9

Freedman DJ, Riesenhuber M, Poggio T, Miller EK (2001) Categorical representation of visual stimuli in the primate prefrontal cortex. Science 291:312–316. https://doi.org/10.1126/science.291.5502.312

Freedman DJ, Riesenhuber M, Poggio T, Miller EK (2002) Visual categorization and the primate prefrontal cortex: neurophysiology and behavior. J Neurophysiol 88:929–941. https://doi.org/10.1152/jn.00040.2002

Freedman DJ, Riesenhuber M, Poggio T, Mille EK (2003) A comparison of primate prefrontal and inferior temporal cortices during visual categorization. J Neurosci 23:5235–5246. https://doi.org/10.1523/jneurosci.23-12-05235.2003

Fukushima T, Hasegawa I, Miyashita Y (2004) Prefrontal neuronal activity encodes spatial target representations sequentially updated after nonspatial target-shift cues. J Neurophysiol 91:1367–1380. https://doi.org/10.1152/jn.00306.2003

Funahashi S (2001) Neuronal mechanisms of executive control by the prefrontal cortex. Neurosci Res 39:147–165. https://doi.org/10.1016/S0168-0102(00)00224-8

Funahashi S (2013a) Space representation in the prefrontal cortex. Prog Neurobiol 103:131–155. https://doi.org/10.1016/j.pneurobio.2012.04.002

Funahashi S (2013b) Thalamic mediodorsal nucleus and its participation in spatial working memory processes: comparison with the prefrontal cortex. Front Syst Neurosci 7:36. https://doi.org/10.3389/fnsys.2013.00036

Funahashi S (2014) Saccade-related activity in the prefrontal cortex: its role in eye movement control and cognitive functions. Front Integr Neurosci 8:54. https://doi.org/10.3389/fnint.2014.00054

Funahashi S (2015) Functions of delay-period activity in the prefrontal cortex and mnemonic scotomas revisited. Front Syst Neurosci 9:2. https://doi.org/10.3389/fnsys.2015.00002

Funahashi S, Inoue M (2000) Neuronal interactions related to working memory processes in the primate prefrontal cortex revealed by cross-correlation analysis. Cereb Cortex 10:535–551. https://doi.org/10.1093/cercor/10.6.535

Funahashi S, Kubota K (1994) Working memory and prefrontal cortex. Neurosci Res 21:1–11. https://doi.org/10.1016/0168-0102(94)90063-9

Funahashi S, Takeda K (2002a) Population vector analysis by primate prefrontal neuron activities. J Biol Physics 28:527–537. https://doi.org/10.1023/A:1020309916014

Funahashi S, Takeda K (2002b) Information processes in the primate prefrontal cortex in relation to working memory processes. Rev Neurosci 13:313–346. https://doi.org/10.1515/revneuro.2002.13.4.313

Funahashi S, Bruce CJ, Goldman-Rakic PS (1989) Mnemonic coding of visual space in the monkey's dorsolateral prefrontal cortex. J Neurophysiol 61:331–349. https://doi.org/10.1152/jn.1989.61.2.331

Funahashi S, Bruce CJ, Goldman-Rakic PS (1990) Visuospatial coding in primate prefrontal neurons revealed by oculomotor paradigms. J Neurophysio 63:814–831. https://doi.org/10.1152/jn.1990.63.4.814

Funahashi S, Bruce CJ, Goldman-Rakic PS (1991) Neuronal activity related to saccadic eye movements in the monkey's dorsolateral prefrontal cortex. J Neurophysiol 65:1464–1483. https://doi.org/10.1152/jn.1991.65.6.1464

Funahashi S, Bruce CJ, Goldman-Rakic PS (1993a) Dorsolateral prefrontal lesions and oculomotor delayed-response performance: evidence for mnemonic "scotomas.". J Neurosci 13:1479–1497. https://doi.org/10.1523/jneurosci.13-04-01479.1993

Funahashi S, Goldman-Rakic CMV, PS. (1993b) Prefrontal neuronal activity in rhesus monkeys performing a delayed anti-saccade task. Nature 365:753–756. https://doi.org/10.1038/365753a0

Funahashi S, Inoue M, Kubota K (1993c) Delay-related activity in the primate prefrontal cortex during sequential reaching tasks with delay. Neurosci Res 18:171–175. https://doi.org/10.1016/0168-0102(93)90019-m

Funahashi S, Inoue M, Kubota K (1997) Delay-period activity in the primate prefrontal cortex encoding multiple spatial positions and their order of presentation. Behav Brain Res 84:203–223. https://doi.org/10.1016/S0166-4328(96)00151-9

Funahashi S, Takeda K, Watanabe Y (2004) Neural mechanisms of spatial working memory: contributions of the dorsolateral prefrontal cortex and the thalamic mediodorsal nucleus. Cogn Affect Behav Neurosci 4:409–420. https://doi.org/10.3758/cabn.4.4.409

Fuster JM (1973) Unit activity in prefrontal cortex during delayed-response performance: neuronal correlates of transient memory. J Neurophysiol 36:61–78. https://doi.org/10.1152/jn.1973.36.1.61

Fuster JM (1990) Inferotemporal units in selective visual attention and short-term memory. J Neurophysiol 64:681–697. https://doi.org/10.1152/jn.1990.64.3.681

Fuster JM (1997) The prefrontal cortex: anatomy, physiology, and neuropsychology of the frontal lobe, 3rd edn. Lippincott-Raven, Philadelphia, PA
Fuster JM (2008) The prefrontal cortex, 4th edn. Academic Press, New York, NY
Fuster JM, Alexander GE (1971) Neuron activity related to short-term memory. Science 173:652–654. https://doi.org/10.1126/science.173.3997.652
Fuster JM, Alexander GE (1973) Firing changes in cells of the nucleus medialis dorsalis associated with delayed response behavior. Brain Res 61:79–91. https://doi.org/10.1016/0006-9883(73)90517-9
Fuster JM, Jervey JP (1981) Inferotemporal neurons distinguish and retain behaviorally relevant features of visual stimuli. Science 212:952–955. https://doi.org/10.1126/science.7233192
Fuster JM, Jervey JP (1982) Neuronal firing in the inferotemporal cortex of the monkey in a visual memory task. J Neurosci 2:361–375. https://doi.org/10.1523/jneurosci.02-03-00361.1982
Fuster JM, Bauer RH, Jervey JP (1982) Cellular discharge in the dorsolateral prefrontal cortex of the monkey in cognitive tasks. Exp Neurol 77:679–694. https://doi.org/10.1016/0014-4886(82)90238-2
Galletti C, Battaglini P (1989) Gaze-dependent visual neurons in area V3A of monkey prestriate cortex. J Neurosci 9:1112–1125. https://doi.org/10.1523/jneurosci.02-03-00361.1982
Genovesio A, Brasted PJ, Mitz AR, Wise SP (2005) Prefrontal cortex activity related to abstract response strategies. Neuron 47:307–320. https://doi.org/10.1016/j.neuron.2005.06.006
Genovesio A, Brasted PJ, Wise SP (2006) Representation of future and previous spatial goals by separate neural populations in prefrontal cortex. J Neurosci 26:7305–7316. https://doi.org/10.1523/jneurosci.0699-06.2006
Genovesio A, Tsujimoto S, Wise SP (2009) Feature- and order-based timing representations in the frontal cortex. Neuron 63:254–266. https://doi.org/10.1016/j.neuron.2009.06.018
Genovesio A, Tsujimoto S, Wise SP (2011) Prefrontal cortex activity during the discrimination of relative distance. J Neurosci 31:3968–3980. https://doi.org/10.1523/JNEUROSCI.5373-10.2011
Georgopoulos AP (1988) Neural integration of movement: role of motor cortex in reaching. FASEB J 2:2849–2857. https://doi.org/10.1096/fasebj.2.13.3139485
Georgopoulos AP, Schwartz AB, Kettner RE (1986) Neuronal population coding of movement direction. Science 233:1416–1419. https://doi.org/10.1126/science.3749885
Georgopoulos AP, Kettner RE, Schwartz AB (1988) Primate motor cortex and free arm movements to visual targets in three-dimensional space. II. Coding of the direction of movement by a neuronal population. J Neurosci 8:2928–2937. https://doi.org/10.1523/jneurosci.08-08-02928.1988
Georgopoulos AP, Lurito JT, Petrides M, Schwartz AB, Massey JT (1989) Mental rotation of the neuronal population vector. Science 243:234–236. https://doi.org/10.1126/science.2911737
Gerstein GL, Aertsen AM (1985) Representation of cooperative firing activity among simultaneously recorded neurons. J Neurophysiol 54:1513–1528. https://doi.org/10.1152/jn.1985.54.6.1513
Gerstein GL, Perkel DH (1969) Simultaneously recorded trains of action potentials: analysis and functional interpretation. Science 164:828–830. https://doi.org/10.1126/science.164.3881.828
Gerstein GL, Perkel DH (1972) Mutual temporal relationships among neuronal spike trains. Statistical techniques for display and analysis. Biophys J 12:453–473. https://doi.org/10.1016/S0006-3495(72)86097-1
Gibson JR, Maunsell JHR (1997) Sensory modality specificity of neural activity related to memory in visual cortex. J Neurophysiol 78:1263–1275. https://doi.org/10.1152/jn.1997.78.3.1263
Glimcher PW, Sparks DL (1992) Movement selection in advance of action in the superior colliculus. Nature 355:542–545. https://doi.org/10.1038/355542a0
Gnadt JW, Andersen RA (1988) Memory related motor planning activity in posterior parietal cortex of macaque. Exp Brain Res 70:216–220. https://doi.org/10.1007/BF00271862

Goldman PS, Nauta WJH (1977a) Columnar distribution of cortico-cortical fibers in the frontal association, limbic, and motor cortex of the developing rhesus monkey. Brain Res 122:393–413. https://doi.org/10.1016/0006-8993(77)90453-X

Goldman PS, Nauta WJH (1977b) An intricately patterned prefronto-caudate projection in the rhesus monkey. J Comp Neurol 171:369–385. https://doi.org/10.1002/cns.901710305

Goldman-Rakic PS (1987) Circuitry of primate prefrontal cortex and regulation of behavior by representational memory. In: Plum F (ed) Handbook of physiology, section 1: the nervous system, Part 1: higher functions of the brain, vol V. American Physiological Society, Bethesda, MD, pp 373–417. https://doi.org/10.1002/cphy.cp010509

Goldman-Rakic PS (1995) Cellular basis of working memory. Neuron 14:477–485. https://doi.org/10.1016/0869-6273(95)90304-6

Goldman-Rakic PS (1996) Regional and cellular fractionation of working memory. Proc Natl Acad Sci USA 93:13473–13480. https://doi.org/10.1073/pnas.93.24.13473

Goldman-Rakic PS, Porrino LJ (1985) The primate mediodorsal (MD) nucleus and its projection to the frontal lobe. J Comp Neurol 242:535–560. https://doi.org/10.1002/cne.902420406

Goldman-Rakic PS, Schwartz ML (1982) Interdigitation of contralateral and ipsilateral columnar projections to frontal association cortex in primates. Science 216:755–757. https://doi.org/10.1126/science.6177037

Goldman-Rakic PS, Funahashi S, Bruce CJ (1990) Neocortical memory circuits. Cold Spring Harbor Symp Quant Biol 55:1025–1038. https://doi.org/10.1101/SQB.1990.055.01.097

Gresack JE, Frick KM (2003) Male mice exhibit better spatial working and reference memory than females in a water-escape radial arm maze task. Brain Res 982:98–107. https://doi.org/10.1016/S0006-8993(03)03000-2

Hasegawa I, Fukushima T, Ihara T, Miyashita Y (1998a) Callosal window between prefrontal cortices: cognitive interaction to retrieve long-term memory. Science 281:814–818. https://doi.org/10.1126/science.281.5378.814

Hasegawa R, Sawaguchi T, Kubota K (1998b) Monkey prefrontal neuronal activity coding the forthcoming saccade in an oculomotor delayed matching-to-sample task. J Neurophysiol 79:322–333. https://doi.org/10.1152/jn.1998.79.1.322

Hasegawa R, Matsumoto M, Mikami A (2000) Search target selection in monkey prefrontal cortex. J Neurophysiol 84:1692–1696. https://doi.org/10.1152/jn.2000.84.3.1692

Hata Y, Tsumoto T, Sato H, Tamura H (1991) Horizontal interactions between visual cortical neurones studied by cross-correlation analysis in the cat. J Physiol 441:593–614. https://doi.org/10.1113/jphysiol.1991.sp018769

Hata Y, Tsumoto T, Sato H, Hagihara K, Tamura H (1993) Development of local horizontal interactions in cat visual cortex studied by cross-correlation analysis. J Neurophysiol 69:40–56. https://doi.org/10.1152/jn.1993.69.1.40

Hernández A, Salinas E, Garcia R, Romo R (1997) Discrimination in the sense of flutter: new psychophysical measurements in monkeys. J Neurosci 17:6391–6400. https://doi.org/10.1523/jneurosci.17-16-06391.1997

Hernández A, Zainos A, Romo R (2002) Temporal evolution of a decision-making process in medial premotor cortex. Neuron 33:959–972. https://doi.org/10.1016/s0896-6273(02)00613-x

Hikosaka K, Watanabe M (2000) Delay activity of orbital and lateral prefrontal neurons of the monkey varying with different rewards. Cereb Cortex 10:263–271. https://doi.org/10.1093/cercor/10.3.263

Hikosaka O, Wurtz RH (1983) Visual and oculomotor functions of monkey substantia nigra pars reticulata. III. Memory-contingent visual and saccade responses. J Neurophysiol 49:1268–1284. https://doi.org/10.1152/jn.1983.49.5.1268

Hikosaka O, Sakamoto M, Usui S (1989) Functional properties of monkey caudate neurons. III. Activities related to expectation of target and reward. J Neurophysiol 61:814–832. https://doi.org/10.1152/jn.1989.61.4.814

Honig WK (1978) Studies of working memory in the pigeon. In: Hulse SH, Fowler H, Honig WK (eds) Cognitive processes in animal behavior. Lawrence Erlbaum, Hillsdale, NJ, pp 211–248

Hoshi E, Shima K, Tanji J (2000) Neuronal activity in the primate prefrontal cortex in the process of motor selection based on two behavioral rules. J Neurophysiol 83:2355–2373. https://doi.org/10.1152/jn.2000.83.4.2355

Hubel DH, Wiesel TN, LeVay S (1977) Plasticity of ocular dominance columns in monkey striate cortex. Phil Trans R Soc Lond 278:377–409. https://doi.org/10.2307/2417674

Huk AC, Shadlen MN (2005) Neural activity in macaque parietal cortex reflects temporal integration of visual motion signals during perceptual decision making. J Neurosci 25:10420–10436. https://doi.org/10.1523/JNEUROSCI.4684-04.2005

Hussar CR, Pasternak T (2012) memory-guided sensory comparisons in the prefrontal cortex: contribution of putative pyramidal cells and interneurons. J Neurosci 32:2747–2761. https://doi.org/10.1523/jneurosci.5135-11.2012

Hussein S, Johnston K, Belbeck B, Lomber SG, Everling S (2014) Functional specialization within macaque dorsolateral prefrontal cortex for the maintenance of task rules and cognitive control. J Cogn Neurosci 26:1918–1927. https://doi.org/10.1162/jocn_a_00608

Iba M, Sawaguchi T (2002) Neuronal activity representing visuospatial mnemonic processes associated with target selection in the monkey dorsolateral prefrontal cortex. Neurosci Res 43:9–22. https://doi.org/10.1016/S0168-0102(02)00013-5

Ichihara-Takeda S, Funahashi S (2006) Reward-period activity in primate dorsolateral prefrontal and orbitofrontal neurons is affected by reward schedules. J Cogn Neurosci 18:212–226. https://doi.org/10.1162/jocn.2006.18.2.212

Ichihara-Takeda S, Funahashi S (2007) Activity of primate orbitofrontal and dorsolateral prefrontal neurons: task-related activity during an oculomotor delayed-response task. Exp Brain Res 181:409–425. https://doi.org/10.1007/s00221-007-0941-0

Ichihara-Takeda S, Funahashi S (2008) Activity of primate orbitofrontal and dorsolateral prefrontal neurons: effects of reward schedule on task-related activity. J Cogn Neurosci 20:563–479. https://doi.org/10.1162/jocn.2008.20047

Ichihara-Takeda S, Takeda K, Funahashi S (2010) Reward acts as a signal to control delay-period activity in delayed-response tasks. Neuroreport 21:367–370

Inoue M, Funahashi S (2002) Prefrontal delay-period activity is affected by visual cues presented outside the memory field. NeuroReport 13:2097–2101

Inoue M, Mikami A (2006) Prefrontal activity during serial probe reproduction task: encoding, mnemonic, and retrieval processes. J Neurophysiol 95:1008–1041. https://doi.org/10.1152/jn.00552.2005

Ito H, Tsuji S (2000) Model dependence in quantification of spike interdependence by joint peri-stimulus time histogram. Neural Comput 12:195–217. https://doi.org/10.1162/089976600300015952

Jacobsen CF (1936) Studies of cerebral function in primates. I. The functions of the frontal association areas in monkeys. Comp Psychol Monogr 13:1–60

Jarrard LE, Elmes DG (1982) Role of retroactive interference in the spatial memory of normal rats and rats with hippocampal lesions. J Comp Physiol Psychol 96:699–711. https://doi.org/10.1037/h0077926

Jones EG (2000) Microcolumns in the cerebral cortex. Proc Natl Aca Sci USA 97:5019–5021. https://doi.org/10.1073/pnas.97.10.5019

Joseph JP, Barone P (1987) Prefrontal unit activity during a delayed oculomotor task in the monkey. Exp Brain Res 67:460–468. https://doi.org/10.1007/BF00247279

Katsuki F, Constantinidis C (2012a) Early involvement of prefrontal cortex in visual bottom-up attention. Nat Neurosci 15:1160–1168. https://doi.org/10.1038/nn.3164

Katsuki F, Constantinidis C (2012b) Unique and shared roles of the posterior parietal and dorsolateral prefrontal cortex in cognitive functions. Front Integr Neurosci 6:17. https://doi.org/10.3389/fnint.2012.00017

Kettner RE, Schwartz AB, Georgopoulos AP (1988) Primate motor cortex and free arm movements to visual targets in three-dimensional space. III. Positional gradients and population coding of

movement direction from various movement origins. J Neurosci 8:2938–2947. https://doi.org/10.1523/jneurosci.08-08-02938.1988
Kieras DE, Meyer DE, Mueller S, Seymour T (1999) Insights into working memory from the perspective of the EPIC architecture for modeling skilled perceptual-motor and cognitive human performance. In: Miyake A, Shah P (eds) Models of working memory: mechanisms of active maintenance and executive control. Cambridge University Press, Cambridge, UK, pp 183–223
Knowlton BJ, Shapiro M, Olton DS (1989) Hippocampal seizures disrupt working memory performance but not reference memory acquisition. Behav Neurosci 103:144–1147. https://doi.org/10.1037/0735-7044.103.5.1144
Kobayashi S, Lauwereyns J, Koizumi M, Sakagami M, Hikosaka O (2002) Influence of reward expectation on visuospatial processing in macaque lateral prefrontal cortex. J Neurophysiol 87:1488–1498. https://doi.org/10.1152/jn.00472.2001
Koch KW, Fuster JM (1989) Unit activity in monkey parietal cortex related to haptic perception and temporary memory. Exp Brain Res 76:292–306. https://doi.org/10.1007/BF00247889
Kojima S, Goldman-Rakic PS (1982) Delay-related activity of prefrontal neurons in rhesus monkeys performing delayed response. Brain Res 248:43–49. https://doi.org/10.1016/0006-8993(82)91145-3
Kojima S, Goldman-Rakic PS (1984) Functional analysis of spatially discriminative neurons in prefrontal cortex of rhesus monkey. Brain Res 291:229–240. https://doi.org/10.1016/0006-8993(84)91255-1
Kojima S, Matsumura M, Kubota K (1981) Prefrontal neuron activity during delayed-response performance without imperative GO signals in the monkey. Exp Neurol 74:396–407. https://doi.org/10.1016/0014-4886(81)90178-3
Kojima J, Matsumura M, Togawa M, Hikosaka O (1996) Tonic activity during visuo-oculomotor behavior in the monkey superior colliculus. Neurosci Res 26:17–28. https://doi.org/10.1016/0168-0102(96)01067-X
Konecky RO, Smith MA, Olson CR (2017) Monkey prefrontal neurons during Sternberg task performance: full contents of working memory or most recent item? J Neurophysiol 117:2269–2281. https://doi.org/10.1152/jn.00541.2016
Kritzer MF, Goldman-Raki PS (1995) Intrinsic circuit organization of the major layers and subleyers of the dorsolateral prefrontal cortex in the rhesus monkey. J Comp Neurol 359:131–143. https://doi.org/10.1002/cne.903590109
Krüger J, Aiple F (1988) Multimicroelectrode investigation of monkey striate cortex: spike train correlations in the infragranular layers. J Neurophysiol 60:798–828. https://doi.org/10.1152/jn.1988.60.2.798
Kubota K, Niki H (1971) Prefrontal cortical unit activity and delayed alternation performance in monkeys. J Neurophysiol 34:337–347. https://doi.org/10.1152/jn.1971.34.3.337
Kubota K, Tonoike M, Mikam A (1980) Neuronal activity in the monkey dorsolateral prefrontal cortex during a discrimination task with delay. Brain Res 183:29–42. https://doi.org/10.1016/0006-8993(80)90117-1
Kurata K, Wise SP (1988) Premotor cortex of rhesus monkeys: set-related activity during two conditional motor tasks. Exp Brain Res 69:327–343. https://doi.org/10.1007/BF00247578
Kwan HC, Murphy JT, Wong YC (1987) Interaction between neurons in precentral cortical zones controlling different joints. Brain Res 400:259–269. https://doi.org/10.1016/0006-8993(87)90625-1
Leavitt ML, Pieper F, Sachs AJ, Matinez-Trujillo JC (2018) A quadrantic bias in prefrontal representation of visual-mnemonic space. Cereb Cortex 28:2405–2421. https://doi.org/10.1093/cercor/bhx142
Lee H, Simpson GV, Logothetis NK, Rainer G (2005) Phase locking of single neuron activity to theta oscillations during working memory in monkey extrastriate visual cortex. Neuron 45:147–156. https://doi.org/10.1016/j.neuron.2004.12.025

Leon MI, Shadlen MN (1999) Effects of expected reward magnitude on the response of neurons in the dorsolateral prefrontal cortex of the macaque. Neuron 24:415–425. https://doi.org/10.1016/S0896-6273(00)80854-5

LeVay S, Hubel DH, Wiesel TN (1975) The pattern of ocular dominance columns in macaque visual cortex revealed by a reduced silver stain. J Comp Neurol 159:559–575. https://doi.org/10.1002/cne.901590408

Levitt JB, Lewis DA, Yoshioka T, Lund JS (1993) Topography of pyramidal neuron intrinsic connections in macaque monkey prefrontal cortex (areas 9 and 46). J Comp Neurol 338:360–376. https://doi.org/10.1002/cne.903380304

Luria A, Pribram KH, Homskaya ED (1964) An experimental analysis of the behavioural disturbance produced by a left frontal arachnoidal endothelioma (meningioma). Neuropsychologia 2:257–280. https://doi.org/10.1016/0028-3932(64)90034-X

Mahut H (1971) Spatial and object reversal learning in monkeys with partial temporal lobe ablations. Neuropsychologia 9:409–424. https://doi.org/10.1016/0028-3932(71)90005-4

Meyer T, Qi X-L, Constantinidis C (2007) Persistent discharges in the prefrontal cortex of monkeys naïve to working memory tasks. Cereb Cortex 17:i70–i76. https://doi.org/10.1093/cercor/bmh063

Meyers EM, Freedman DJ, Kreiman G, Miller EK, Poggio T (2008) Dynamic population coding of category information in inferior temporal and prefrontal cortex. J Neurophysiol 100:1407–1419. https://doi.org/10.1152/jn.90248.2008

Mikami A, Kubota K (1980) Inferotemporal neuron activities and color discrimination with delay. Brain Res 182:65–78. https://doi.org/10.1016/0006-8993(80)90830-6

Mikami A, Ito S, Kubota K (1982) Visual response properties of dorsolateral prefrontal neurons during visual fixation task. J Neurophysiol 47:593–605. https://doi.org/10.1152/jn.1982.47.4.593

Miller GA (1956) The magical number seven, plus or minus two: some limits on our capacity for processing information. Psychol Rev 63:81–97. (101: 343–352, 1994). https://doi.org/10.1037/0033-295x.101.2.343

Miller EK (2000) The prefrontal cortex and cognitive control. Nat Rev Neurosci 1:59–65. https://doi.org/10.1038/35036228

Miller EK, Cohen JD (2001) An integrative theory of prefrontal cortex function. Annu Rev Neurosci 24:167–202. https://doi.org/10.1146/annurev.neuro.24.1.167

Miller EK, Desimone R (1994) Parallel neurons mechanisms for short-term memory. Science 263:520–522. https://doi.org/10.1126/science.8290960

Miller EK, Li L, Desimone R (1991) A neural mechanism for working and recognition memory in inferior temporal cortex. Science 254:1377–1379. https://doi.org/10.1126/science.1962197

Miller EK, Li L, Desimone R (1993) Activity of neurons in anterior inferior temporal cortex during a short-term memory task. J Neurosci 13:1460–1478. https://doi.org/10.1523/jneurosci.13-04-01460.1993

Miller EK, Erickson CA, Desimone R (1996) Neural mechanisms of visual working memory in prefrontal cortex of the macaque. J Neurosci 16:5154–5167. https://doi.org/10.1523/jneurosci.16-16-05154.1996

Miller EK, Freedman DJ, Wallis JD (2002) The prefrontal cortex: categories, concepts and cognition. Phil Trans R Soc Lond B 357:1123–1136. https://doi.org/10.1098/rstb.2002.1099

Miller EK, Nieder A, Freedman DJ, Wallis JD (2003) Neural correlates of categories and concepts. Curr Opin Neurobiol 13:198–203. https://doi.org/10.1016/S0959-4388(03)00037-0

Milner B (1964) Some effects of frontal lobectomy in man. In: Warren JM, Akert K (eds) The frontal granular cortex and behavior. McGraw-Hill, New York, pp 313–334

Milner B (1982) Some cognitive effects of frontal-lobe lesions in man. Phil Trans R Soci Lond B 298:211–226. https://doi.org/10.1098/rstb.1982.0083

Miyake A, Shah P (1999) Toward unified theories of working memory: emerging general consensus, unresolved theoretical issues, and future research directions. In: Miyake A, Shah P (eds)

Models of working memory: mechanisms of active maintenance and executive control. Cambridge University Press, Cambridge, UK, pp 442–481
Miyashita Y (1988) Neuronal correlate of visual associative long-term memory in the primate temporal cortex. Nature 335:817–820. https://doi.org/10.1038/335817a0
Miyashita Y, Chang HS (1988) Neuronal correlate of pictorial short-term memory in the primate temporal cortex. Nature 331:68–70. https://doi.org/10.1038/331068a0
Munoz DP, Wurtz RH (1995) Saccade-related activity in monkey superior colliculus. I. Characteristics of burst and buildup cells. J Neurophysiol 73:2313–2333. https://doi.org/10.1152/jn.1995.73.6.2313
Murphy JT, Kwan HC, Wong YC (1985a) Cross correlation studies in primate motor cortex: synaptic interaction and shared input. Can J Neurol Sci 12:11–23. https://doi.org/10.1017/S0317167100046527
Murphy JT, Kwan HC, Wong YC (1985b) Cross correlation studies in primate motor cortex: event-related correlation. Can J Neurol Sci 12:24–30. https://doi.org/10.1017/S0317167100046539
Mushiake H, Tanatsugu Y, Tanji J (1997) Neuronal activity in the ventral part of premotor cortex during target-rearch movement is modulated by direction of gaze. J Neurophysiol 78:567–571. https://doi.org/10.1152/jn.1997.78.1.567
Mushiake H, Saito N, Sakamoto K, Sato Y, Tanji J (2001) Visually based path-planning by Japanese monkeys. Cogn Brain Res 11:165–169. https://doi.org/10.1016/S0926-6410(00)00067-7
Mushiake H, Saito N, Sakamoto K, Itoyama Y, Tanji J (2006) Activity in the lateral prefrontal cortex reflects multiple steps of future events in action plans. Neuron 50:631–641. https://doi.org/10.1016/j.neuron.2006.03.045
Nadel L, MacDonald L (1980) Hippocampus: cognitive map or working memory? Behav Neural Biol 29:405–409. https://doi.org/10.1016/S0163-1047(80)90430-6
Nakamura K, Kubota K (1995) Mnemonic firing of neurons in the monkey temporal pole during a visual recognition memory task. J Neurophysiol 74:162–178. https://doi.org/10.1152/jn.1995.74.1.162
Nieder A, Merten K (2007) A labeled-line code for small and large numerosities in the monkey prefrontal cortex. J Neurosci 27:5986–5993. https://doi.org/10.1523/jneurosci.1056-07.2007
Nieder A, Miller EK (2003) Coding of cognitive magnitude: compressed scaling of numerical information in the primate prefrontal cortex. Neuron 37:149–157. https://doi.org/10.1016/S0869-6273(02)01144-3
Nieder A, Miller EK (2004) A parieto-frontal network for visual numerical information in the monkey. Proc Natl Acad Sci USA 101:7457–7462. https://doi.org/10.1073/pnas.0402239101
Nieder A, Freedman DJ, Miller EK (2002) Representation of the quantity of visual items in the primate prefrontal cortex. Science 297:1708–1711. https://doi.org/10.1126/science.1072493
Nieder A, Diester I, Tudusciuc O (2006) Temporal and spatial enumeration processes in the primate parietal cortex. Science 313:1431–1435. https://doi.org/10.1126/science.1130308
Niki H (1974a) Prefrontal unit activity during delayed alternation in the monkey. I. Relation to direction of response. Brain Res 68:185–196. https://doi.org/10.1016/0006-8993(74)90388-6
Niki H (1974b) Prefrontal unit activity during delayed alternation in the monkey. II. Relation to absolute versus relative direction of response. Brain Res 68:197–204. https://doi.org/10.1016/0006-8993(74)90389-8
Niki H (1974c) Differential activity of prefrontal units during right and left delayed response trials. Brain Res 70:346–349. https://doi.org/10.1016/0006-8993(74)90324-2
Niki H, Tsutsui K (1997) Prefrontal cortex and behavior: encoding of self-movement in dorsal prefrontal neurons. Epilepsia 38(Suppl. 6):35–39. https://doi.org/10.1111/j.1528-1157.1997.tb00104.x
Niki H, Watanabe M (1976a) Prefrontal unit activity and delayed response: relation to cue location versus direction of response. Brain Res 105:79–88. https://doi.org/10.1016/0006-8993(76)90924-0

Niki H, Watanabe M (1976b) Cingulate unit activity and delayed response. Brain Res 110:381–386. https://doi.org/10.1016/0006-8993(76)90412-1

Niki H, Watanabe M (1979) Prefrontal and cingulate unit activity during timing behavior in the monkey. Brain Res 171:213–224. https://doi.org/10.1016/0006-8993(79)90328-7

Niki H, Sakai M, Kubota K (1972) Delayed alternation performance and unit activity of the caudate head and medial orbitofrontal gyrus in the monkey. Brain Res 38:343–353. https://doi.org/10.1016/0006-8993(72)90717-2

Ninokura Y, Mushiake H, Tanji J (2003) Representation of the temporal order of visual objects in the primate lateral prefrontal cortex. J Neurophysiol 89:2868–2873. https://doi.org/10.1152/jn.00647.2002

Ninokura Y, Mushiake H, Tanji J (2004) Integration of temporal order and object information in the monkey lateral prefrontal cortex. J Neurophysiol 91:555–560. https://doi.org/10.1152/jn.00694.2003

O'Keefe J, Nadel L (1978) The hippocampus as a cognitive map. Oxford University Press, Oxford, UK

O'Reilly RC, Braver TS, Cohen JD (1999) A biologically based computational model of working memory. In: Miyake A, Shah P (eds) Models of working memory: mechanisms of active maintenance and executive control. Cambridge University Press, Cambridge, UK, pp 375–411

Ohbayashi M, Ohki K, Miyashita Y (2003) Conversion of working memory to motor sequence in the monkey premotor cortex. Science 301:233–236. https://doi.org/10.1126/science.1084884

Olton DS, Papas B (1979) Spatial memory and hippocampal function. Neuropsychologia 17:669–682. https://doi.org/10.1016/0028-3932(79)90042-3

Olton DS, Becker JT, Handelmann GE (1979) Hippocampus, space, and memory. Behav Brain Sci 2:313–365. https://doi.org/10.1017/S0140525X00062713

Onoe H, Komori M, Onoe K, Takechi H, Tsukada H, Watanabe Y (2001) Cortical networks recruited for time perception: a monkey positron emission tomography (PST) study. Neuroimage 13:37–45. https://doi.org/10.1006/nimg.2000.0670

Oshio K, Chiba A, Inase M (2006) Delay period activity of monkey prefrontal neurons during duration-discrimination task. Eur J Neurosci 23:2779–2790. https://doi.org/10.1111/j.1460-9568.2006.04781.x

Overman W, Bachevalier J, Turner M, Peuster A (1992) Object recognition versus object discrimination: comparison between human infants and infant monkeys. Behav Neurosci 106:15–29. https://doi.org/10.1037/0735-7044.106.1.15

Padoa-Schioppa C, Assad JA (2006) Neurons in the orbitofrontal cortex encode economic value. Nature 441:223–226. https://doi.org/10.1038/nature04676

Pandya DN, Seltzer B (1982) Intrinsic connections and architectonics of posterior parietal cortex in the rhesus monkey. J Comp Neurol 204:196–210. https://doi.org/10.1002/cne.902040208

Passingham RE (1985) Prefrontal cortex and the sequencing of movement in monkeys (*Macaca mulatta*). Neuropsychologia 23:453–462. https://doi.org/10.1016/1028-3932(85)90001-6

Pasternak T, Greenlee MW (2005) Working memory in primate sensory systems. Nat Rev Neurosci 6:97–107. https://doi.org/10.1038/nrn1603

Perkel DH, Gerstein GL, Moore GP (1967a) Neuronal spike trains and stochastic point processes. I. The single spike train. Biophysical J 7:391–418. https://doi.org/10.1016/S0006-3495(67)86596-2

Perkel DH, Gerstein GL, Moore GP (1967b) Neuronal spike trains and stochastic point processes. II. Simultaneous spike trains. Biophysical J 7:419–440. https://doi.org/10.1016/S0006-3495(67)86597-4

Pesaran B, Pezaris JS, Sahani M, Mitra PP, Andersen RA (2002) Temporal structure in neuronal activity during working memory in macaque parietal cortex. Nat Neurosci 5:805–811. https://doi.org/10.1038/nn890

Petrides M (1991) Functional specialization within the dorsolateral frontal cortex for serial order memory. Proc R Soc Lond B 246:299–306. https://doi.org/10.1098/rspb.1991.0158

Petrides M (2005) Lateral prefrontal cortex: architectonic and functional organization. Philosophical Trans R Soc Lond B 360:781–795. https://doi.org/10.1098/rstb.2005.1631

Petrides M, Pandya DN (1994) Comparative architectonic analysis of the human and the macaque frontal cortex. In: Boller F, Spinnler H, Hendler JA (eds) Handbook of neuropsychology, The frontal lobes, vol 9, Sect. 12. Elsevier, Amsterdam, pp 17–58

Pinto-Hamuy T, Linck P (1965) Effect of frontal lesions on performance of sequential tasks by monkeys. Exp Neurol 12:96–107. https://doi.org/10.1016/0014-4886(65)90101-9

Postle BR, Berger JS, Taich AM, D'Esposito M (2000) Activity in human frontal cortex associated with spatial working memory and saccadic behavior. J Cogn Neurosci 12(Suppl. 2):2–14. https://doi.org/10.1162/089892900564028

Prior H, Schwegler H, Dücker G (1997) Dissociation of spatial reference memory, spatial working memory, and hippocampal mossy fiber distribution in two rat strains differing in emotionality. Beha Brain Res 87:183–194. https://doi.org/10.1016/S0166-4328(97)02282-1

Procyk E, Goldman-Rakic PS (2006) Modulation of dorsolateral prefrontal delay activity during self-organized behavior. J Neurosci 26:11313–11323. https://doi.org/10.1523/jneurosci.2157-06.2006

Prut Y, Fetz EE (1999) Primate spinal interneurons show pre-movement instructed delay activity. Nature 401:590–594. https://doi.org/10.1038/44145

Pucak ML, Levitt JB, Lund JS, Lewis DA (1996) Patterns of intrinsic and associational circuitry in monkey prefrontal cortex. J Comp Neurol 376:614–630. https://doi.org/10.1002/(sici)1096-9861(19961223)376:4<614::aid-cne9>3.0.co;2-4

Qi X-L, Constantinidis C (2012a) Correlated discharges in the primate prefrontal cortex before and after working memory training. Eur J Neurosci 36:3538–3548. https://doi.org/10.1111/j.1460-9568.2012.08267.x

Qi X-L, Constantinidis C (2012b) Variability of prefrontal neuronal discharges before and after training in a working memory task. PLoS One 7:e41053. https://doi.org/10.1371/journal.pone.0041053

Qi X-L, Constantinidis C (2015) Lower neuronal variability in the monkey dorsolateral prefrontal than posterior parietal cortex. J Neurophysiol 114:2194–2203. https://doi.org/10.1152/jn.00454.2015

Qi X-L, Katsuki F, Meyer T, Rawley B, Zhou X, Douglas KL, Constantinidis C (2010) Comparison of neural activity related to working memory in primate dorsolateral prefrontal and posterior parietal cortex. Front Syst Neurosci 4:12. https://doi.org/10.3389/fnsys.2010.00012

Qi X-L, Meyer T, Stanford TR, Constantinidis C (2011) Changes in prefrontal neuronal activity after learning to perform a spatial working memory task. Cereb Cortex 21:2722–2732. https://doi.org/10.1093/cercor/bhr058

Qi X-L, Elworthy AC, Lambert BC, Constantinidis C (2015) Representation of remembered stimuli and task information in the monkey dorsolateral prefrontal and posterior parietal cortex. J Neurophysiol 113:44–57. https://doi.org/10.1152/jn.00413.2014

Quintana J, Fuster JM (1992) Mnemonic and predictive functions of cortical neurons in a memory task. Neuroreport 3:721–724. https://doi.org/10.1097/00001756-199208000-00018

Quintana J, Fuster JM (1999) From perception to action: temporal integrative functions of prefrontal and parietal neurons. Cereb Cortex 9:213–221. https://doi.org/10.1093/cercor/9.3.213

Quintana J, Yajeya J, Fuster JM (1988) Prefrontal representation of stimulus attributes during delay tasks. I. Unit activity in cross-temporal integration of sensory and sensory-motor information. Brain Res 474:211–221. https://doi.org/10.1016/0006-8993(88)90436-2

Raffaele KC, Olton DS (1988) Hippocampal and amygdaloid involvement in working memory for nonspatial stimuli. Behav Neurosci 102:349–355. https://doi.org/10.1037/0735-7044.102.3.349

Rainer G, Miller EK (2002) Time course of object-related neural activity in the primate prefrontal cortex during a short-term memory task. Eur J Neurosci 15:1244–1254. https://doi.org/10.1046/j.1460-9568.2002.01958.x

Rainer G, Asaad WF, Miller EK (1998a) Selective representation of relevant information by neurons in the primate prefrontal cortex. Nature 393:577–579. https://doi.org/10.1038/31235

Rainer G, Asaad WF, Miller EK (1998b) Memory fields of neurons in the primate prefrontal cortex. Proc Natl Acad Sci USA 95:15008–15013

Rainer G, Rao SC, Miller EK (1999) Prospective coding for objects in primate prefrontal cortex. J Neurosci 19:5493–5505. https://doi.org/10.1523/jneurosci.19-13-05493.1999

Rao SC, Rainer G, Miller EK (1997) Integration of what and where in the primate prefrontal cortex. Science 276:821–824. https://doi.org/10.1126/science.276.5313.821

Rao SG, Williams GV, Goldman-Rakic PS (1999) Isodirectional tuning of adjacent interneurons and pyramidal cells during working memory: evidence for microcolumnar organization in PFC. J Neurophysiol 81:1903–1916. https://doi.org/10.1152/jn.1999.81.4.1903

Rao SM, Mayer AR, Harrington DL (2001) The evolution of brain activation during temporal processing. Nat Neurosci 4:317–323. https://doi.org/10.1038/85191

Richardson RT, DeLong MR (1986) Nucleus basalis of Meynert neuronal activity during a delayed response task in monkey. Brain Res 399:364–368. https://doi.org/10.1016/0006-8993(86)91529-5

Riches IP, Wilson FAW, Brown MW (1991) The effects of visual stimulation and memory on neurons of the hippocampal formation and the neighboring parahippocampal gyrus and inferior temporal cortex of the primate. J Neurosci 11:1763–1779. https://doi.org/10.1523/jneurosci.11-06-01763.1991

Robinson DA (1963) A method of measuring eye movement using a scleral search coil in a magnetic field. IEEE Transact Biomed Engineer 10:137–145

Roesch MR, Olson CR (2005) Neuronal activity dependent on anticipated and elapsed delay in macaque prefrontal cortex, frontal and supplementary eye fields, and premotor cortex. J Neurophysiol 94:1469–1497. https://doi.org/10.1152/jn.00064-2005

Romo R, Salinas E (2003) Flutter discrimination: neural codes, perception, memory and decision making. Nat Rev Neurosci 4:203–218. https://doi.org/10.1038/nrn1058

Romo R, Brody CD, Hernández A, Lemus L (1999) Neuronal correlates of parametric working memory in the prefrontal cortex. Nature 399:470–473. https://doi.org/10.1038/20939

Romo R, Hernández A, Zainos A, Lemus L, Brody CD (2002) Neuronal correlates of decision-making in secondary somatosensory cortex. Nat Neurosci 6:1217–1225. https://doi.org/10.1038/nn950

Romo R, Hernández A, Zainos A (2004) Neuronal correlates of a perceptual decision in ventral premotor cortex. Neuron 41:165–173. https://doi.org/10.1016/S0896-6273(03)00817-1

Rosenbluth D, Allman JM (2002) The effect of gaze angle and fixation distance on the responses of neurons in V1, V2, and V4. Neuron 33:143–148. https://doi.org/10.1016/s0896-6273(01)00559-1

Rosenkilde CE (1979) Functional heterogeneity of the prefrontal cortex in the monkey: a review. Behav Neural Biol 25:301–345. https://doi.org/10.1016/S0163-1047(79)90404-7

Rosenkilde CE, Bauer RH, Fuster JM (1981a) Single cell activity in ventral prefrontal cortex of behaving monkeys. Brain Res 209:375–394. https://doi.org/10.1016/0006-8993(81)90160-8

Rosenkilde CE, Rosvold HE, Mishkin M (1981b) Time discrimination with positional responses after selective prefrontal lesions in monkeys. Brain Res 210:129–144. https://doi.org/10.1016/0006-8993(81)90890-8

Rowe JB, Passingham RE (2001) Working memory for location and time: activity in prefrontal area 46 relates to selection rather than maintenance in memory. Neurimage 14:77–86. https://doi.org/10.1006/nimg.2001.0784

Rowe JB, Toni I, Josephs O, Frackowiak RSJ, Passingham RE (2000) The prefrontal cortex: response selection or maintenance within working memory? Science 288:1656–1660. https://doi.org/10.1126/science.288.5471.1656

Saito N, Mushiake H, Sakamoto K, Itoyama Y, Tanji J (2005) Representation of immediate and final behavioral goals in the monkey prefrontal cortex during an instructed delay period. Cereb Cortex 15:1535–1546. https://doi.org/10.1093/cercor/bhi032

Sakai K, Miyashita Y (1991) Neural organization for the long-term memory of paired associates. Nature 354:152–155. https://doi.org/10.1038/354152a0

Sakurai Y (1992) Auditory working and reference memory can be tested in a single situation of stimuli for the rat. Beha Brain Res 50:193–195. https://doi.org/10.1016/S0166-4328(05)80301-8

Sakurai Y (1996) Hippocampal and neocortical cell assemblies encode memory processes for different types of stimuli in the rat. J Neurosci 16:2809–2819. https://doi.org/10.1523/jneurosci.16-08-02809.1996

Sakurai Y (1998) The search for cell assemblies in the working brain. Behav Brain Res 91:1–13. https://doi.org/10.1016/s0166-4328(97)00106-x

Sakurai Y, Takahashi S, Inoue M (2004) Stimulus duration in working memory is represented by neuronal activity in the monkey prefrontal cortex. Eur J Neurosci 20:1069–1080. https://doi.org/10.1111/j.1460-9568.2004.03525

Salinas E, Hernández A, Zainos A, Romo R (2000) Periodicity and firing rate as candidate neural codes for the frequency of vibrotactile stimuli. J Neurosci 20:5503–5515. https://doi.org/10.1523/jneurosci.20-14-05503.2000

Sawaguchi T, Goldman-Rakic PS (1991) D1 dopamine receptors in prefrontal cortex: involvement in working memory. Science 251:947–950. https://doi.org/10.1126/science.1825731

Sawaguchi T, Goldman-Rakic PS (1994) The role of D1-dopamine receptor in working memory: local injections of dopamine antagonists into the prefrontal cortex of rhesus monkeys performing an oculomotor delayed-response task. J Neurophysiol 71:515–528. https://doi.org/10.1152/jn.1994.71.2.5152

Sawaguchi T, Iba M (2001) Prefrontal cortical representation of visuospatial working memory in monkeys examined by local inactivation with muscimol. J Neurophysiol 86:2041–2053. https://doi.org/10.1152/jn.2001.86.4.2041

Sawaguchi T, Yamane I (1999) Properties of delay-period neuronal activity in the monkey dorsolateral prefrontal cortex during a spatial delayed matching-to-sample task. J Neurophysiol 82:2070–2080. https://doi.org/10.1152/1999.82.5.2070

Sawamura H, Shima K, Tanji J (2002) Numerical representation for action in the parietal cortex of the monkey. Nature 415:918–922. https://doi.org/10.1038/415918a

Schneider W (1999) Working memory in a multilevel hybrid connectionist control architecture (CAP2). In: Miyake A, Shah P (eds) Models of working memory: mechanisms of active maintenance and executive control. Cambridge University Press, Cambridge, UK, pp 340–374

Schwartz ML, Goldman-Rakic PS (1984) Callosal and intrahemispheric connectivity of the prefrontal association cortex in rhesus monkey: relation between intraparietal and principal sulcal cortex. J Comp Neurol 226:403–420. https://doi.org/10.1002/cne.902260309

Schwartz AB, Kettner RE, Georgopoulos AP (1988) Primate motor cortex and free arm movements to visual targets in three-dimensional space. I. Relations between single cell discharge and direction of movement. J Neurosci 8:2913–2927. https://doi.org/10.1523/jneurosci.08-08-02913.1988

Serences JT, Ester EF, Vogel EK, Awh E (2009) Stimulus-specific delay activity in human primary visual cortex. Psychol Sci 20:207–214. https://doi.org/10.1111/j.1467-9280.2009.02276.x

Shima K, Isoda M, Mushiake H, Tanji J (2007) Categorization of behavioural sequences in the prefrontal cortex. Nature 445:315–318. https://doi.org/10.1038/nature05470

Siwek DF, Pandya DN (1991) Prefrontal projections to the mediodorsal nucleus of the thalamus in the rhesus monkey. J Comp Neurol 312:509–524. https://doi.org/10.1002/cne.903120403

Snyder LH, Batista AP, Andersen RA (1997) Coding of intention in the posterior parietal cortex. Nature 386:167–170. https://doi.org/10.1038/386167a0

Soltysik S, Hull CD, Buchwald NA, Fekete T (1975) Single unit activity in basal ganglia of monkeys during performance of a delayed response task. Electroencephalogr Clin Neurophysiol 39:65–78. https://doi.org/10.1016/0013-4694(75)90128-5

Sommer MA, Wurtz RH (2001) Frontal eye field sends delay activity related to movement, memory, and vision to the superior colliculus. J Neurophysiol 85:1673–1685. https://doi.org/10.1152/jn.2001.85.4.1673

Sommer MA, Wurtz RH (2004a) What the brain stem tells the frontal cortex. I. Oculomotor signals sent from superior colliculus to frontal eye field via mediodorsal thalamus. J Neurophysiol 91: 1381–1402. https://doi.org/10.1152/jn.00738.2003

Sommer MA, Wurtz RH (2004b) What the brain stem tells the frontal cortex. II. Role of the SC-MD-FEF pathway in corollary discharge. J Neurophysiol 91:1403–1423. https://doi.org/10.1152/jn.00740.2003

Squatrito S, Maioli MG (1996) Gaze field properties of eye position neurons in area MST and 7a of the macaque monkey. Visual Neurosci 13:385–398. https://doi.org/10.1017/S0952523800007628

Squire LR, Wixted JT, Clar RE (2007) Recognition memory and the medial temporal lobe: a new perspective. Nat Rev Neurosci 8:872–883. https://doi.org/10.1038/nrn2154

Stamm JS (1969) Electrical stimulation of monkeys' prefrontal cortex during delayed-response performance. J Comp Physiol Psychol 67:535–546. https://doi.org/10.1037/h0027294

Stamm JS, Rosen SC (1972) Cortical steady potential shifts and anodal polarization during delayed response performance. Acta Neurobiol Exp 32:193–209

Stokes KA, Best PJ (1990) Mediodorsal thalamic lesions impair "reference" and "working" memory in rats. Physiol Behav 47:471–476. https://doi.org/10.1016/0031-9384(90)90111-G

Stuss DT, Benson DF (1986) The frontal lobe. Raven Press, New York, NY

Supér H, Spekreijse H, Lamme VAF (2001) A neural correlate of working memory in the monkey primary visual cortex. Science 293:120–124. https://doi.org/10.1126/science.1060496

Suzuki H, Azuma M (1977) Prefrontal neuronal activity during gazing at a light spot in the monkey. Brain Res 126:497–508. https://doi.org/10.1016/0006-8993(77)90600-X

Suzuki H, Azuma M (1983) Topographic studies on visual neurons in the dorsolateral prefrontal cortex of the monkey. Exp Brain Res 53:47–58. https://doi.org/10.1007/BF00239397

Suzuki M, Gottlieb J (2013) Distinct neural mechanisms of distractor suppression in the frontal and parietal lobe. Nat Neurosci 16:98–104. https://doi.org/10.1038/nn.3282

Suzuki H, Azuma M, Yumiya H (1979) Stimulus and behavioral factors contributing to the activation of monkey prefrontal neurons during gazing. Jpn J Physiol 29:471–489. https://doi.org/10.2170/jjphysiol.29.471)

Sweeney JA, Mintun MA, Kwee S, Wiseman MB, Brown DL, Rosenberg DR, Carl JR (1996) Positron Emission tomography study of voluntary saccadic eye movements and spatial working memory. J Neurophysiol 75:454–468. https://doi.org/10.1152/jn.1996.75.1.454

Takeda K, Funahashi S (2002) Prefrontal task-related activity representing visual cue location or saccade direction in spatial working memory tasks. J Neurophysiol 87:567–588. https://doi.org/10.1152/jn.00249.2001

Takeda K, Funahashi S (2004) Population vector analysis of primate prefrontal activity during spatial working memory. Cereb Cortex 14:1328–1339. https://doi.org/10.1093/cercor/bhh093

Takeda K, Funahashi S (2007) Relationship between prefrontal task-related activity and information flow during spatial working memory performances. Cortex 43:38–52. https://doi.org/10.1016/S0010-9452(08)70444-1

Tanaka K (1983) Cross-correlation analysis of geniculostriate neuronal relationships in cat. J Neurophysiol 49:1303–1318. https://doi.org/10.1152/jn.1983.49.6.1303

Tang H, Qi X-L, Riley MR, Constantinidis C (2019) Working memory capacity is enhanced by distributed prefrontal activation and invariant temporal dynamics. Proc Natl Acad Sci USA 116: 7095–7100. https://doi.org/10.1073/pnas.1817278116

Tanibuchi I, Goldman-Rakic PS (2003) Dissociation of spatial-, object-, and sound-coding neurons in the mediodorsal nucleus of the primate thalamus. J Neurophysiol 89:1067–1077. https://doi.org/10.1152/jn.00207.2002

Tanibuchi I, Goldman-Rakic PS (2005) Comparison of oculomotor neuronal activity in paralaminar and mediodorsal thalamus in the rhesus monkey. J Neurophysiol 93:614–619. https://doi.org/10.1152/jn.00969.2003

Toyama K, Kimura M, Tanaka K (1981a) Cross-correlation analysis of interneuronal connectivity in cat visual cortex. J Neurophysiol 46:191–201. https://doi.org/10.1152/jn.1981.46.2.191

Toyama K, Kimura M, Tanaka K (1981b) Organization of cat visual cortex as investigated by cross-correlation technique. J Neurophysiol 46:202–214. https://doi.org/10.1152/jn.1981.46.2.202

Tremblay L, Schultz W (1999) Relative reward preference in primate orbitofrontal cortex. Nature 398:704–708. https://doi.org/10.1038/19525

Tremblay L, Schultz W (2000a) Reward-related neuronal activity during go-nogo task performance in primate orbitofrontal cortex. J Neurophysiol 83:1864–1876. https://doi.org/10.1152/jn.2000.83.4.1864

Tremblay L, Schultz W (2000b) Modifications of reward expectation-related neuronal activity during learning in primate orbitofrontal cortex. J Neurophysiol 83:1877–1885. https://doi.org/10.1152/jn.2000.83.4.1877

Trotter Y, Celebrini S (1999) Gaze direction controls response gain in primary visual-cortex neurons. Nature 398:239–242. https://doi.org/10.1038/18444

Ts'o DY, Gilbert CD, Wiesel TN (1986) Relationship between horizontal interactions and functional architecture in cat striate cortex as revealed by cross-correlation analysis. J Neurosci 6: 1160–1170. https://doi.org/10.1523/jneurosci.06-04-01160.1986

Tsujimoto S, Sawaguchi T (2004) Properties of delay-period neuronal activity in the primate prefrontal cortex during memory- and sensory-guided saccade tasks. Eur J Neurosci 19:447–457

Tsujimoto S, Sawaguchi T (2005) Neuronal activity representing temporal prediction of reward in the primate prefrontal cortex. J Neurophysiol 93:3678–3692. https://doi.org/10.1152/jn.01149.2004

Tsujimoto S, Genovesio A, Wise SP (2008) Transient neuronal correlations underlying foal selection and maintenance in prefrontal cortex. Cereb Cortex 18:2748–2761. https://doi.org/10.1093/cercor/bhn033

Tudusciuc O, Nieder A (2007) Neuronal population coding of continuous and discrete quantity in the primate posterior parietal cortex. Proc Natl Acad Sci USA 104:14513–14518. https://doi.org/10.1073/pnas.0705495104

Vaadia E, Haalman I, Abeles M, Bergman H, Prut Y, Slovin H, Aertsen A (1995) Dynamics of neuronal interactions in monkey cortex in relation to behavioral events. Nature 373:515–518. https://doi.org/10.1038/373515a0

Vallentin D, Nieder A (2008) Behavioral and prefrontal representation of spatial proportions in the monkey. Curr Biol 18:1420–1425. https://doi.org/10.1016/j.cub.2008.08.042

Wallis JD, Anderson KC, Miller EK (2001) Single neurons in prefrontal cortex encode abstract rules. Nature 411:953–956. https://doi.org/10.1038/35082081

Wang M, Vijayraghavan S, Goldman-Rakic PS (2004a) Selective D2 receptor actions on the functional circuitry of working memory. Science 303:853–856. https://doi.org/10.1126/science.1091162

Wang X-J, Tegner J, Constantinidis C, Goldman-Rakic PS (2004b) Division of labor among distinct subtypes of inhibitory neurons in a cortical microcircuit of working memory. Proc Natl Acad Sci USA 101:1368–1373. https://doi.org/10.1073/pnas.0305337101

Warrington EK, Shallice T (1969) The selective impairment of auditory verbal short-term memory. Brain 92:885–896. https://doi.org/10.1093/brain/92.4.885

Watanabe M (1996) Reward expectancy in primate prefrontal neurons. Nature 382:629–632. https://doi.org/10.1038/382629a0

Watanabe Y, Funahashi S (2004a) Neuronal activity throughout the primate mediodorsal nucleus of the thalamus during oculomotor delayed-response. I. Cue-, delay-, and response-period activity. J Neurophysiol 92:1738–1755. https://doi.org/10.1152/jn.00994.2003

Watanabe Y, Funahashi S (2004b) Neuronal activity throughout the primate mediodorsal nucleus of the thalamus during oculomotor delayed-response. II. Activity encoding visual versus motor signal. J Neurophysiol 92:1756–1769. https://doi.org/10.1152/jn.00995.2003

Watanabe T, Niki H (1985) Hippocampal unit activity and delayed response in the monkey. Brain Res 325:241–254. https://doi.org/10.1016/0006-8993(85)90320-8

Watanabe M, Hikosaka K, Sakagami M, Shirakawa S (2002) Coding and monitoring of motivational context in the primate prefrontal cortex. J Neurosci 22:2391–2400. https://doi.org/10.1523/jneurosci.22-06-02391.2002

Watanabe Y, Takeda K, Funahashi S (2009) Population vector analysis of primate mediodorsal thalamic activity during oculomotor delayed-response performance. Cereb Cortex 19:1313–1321. https://doi.org/10.1093/cercor/bhn170

Weinrich M, Wise SP (1982) The premotor cortex of the monkey. J Neurosci 2:1329–1345. https://doi.org/10.1523/jneurosci.02-09-01329.1982

White IM, Wise SP (1999) Rule-dependent neuronal activity in the prefrontal cortex. Exp Brain Res 126:315–335. https://doi.org/10.1007/s002210050740

Williams GV, Goldman-Rakic PS (1995) Modulation of memory fields by dopamine D1 receptors in prefrontal cortex. Nature 376:572–575. https://doi.org/10.1038/376572a0

Williams GV, Rao SG, Goldman-Rakic PS (2002) The physiological role of 5-HT_{2A} receptors in working memory. J Neurosci 22:2843–2854. https://doi.org/10.1523/jneurosci.22-07-02843.2002

Wilson FAW, Scalaidhe O, SP, Goldman-Rakic PS. (1993) Dissociation of object and spatial processing domains in primate prefrontal cortex. Science 260:1955–1958. https://doi.org/10.1126/science.8316836

Wilson FAW, O'Scalaidhe SP, Goldman-Rakic PS (1994) Functional synergism between putative gamma-aminobutyrate-containing neurons and pyramidal neurons in prefrontal cortex. Proc Natl Acad Sci USA 91:4009–4013. https://doi.org/10.1073/pnas.91.9.4009

Wyder MT, Massoglia DP, Stanford TR (2003) Quantitative assessment of the timing and tuning of visual-related, saccade-related, and delay period activity in primate central thalamus. J Neurophysiol 90:2029–2052. https://doi.org/10.1152/jn.00064.2003

Wyder MT, Massoglia DP, Stanford TR (2004) Contextual modulation of central thalamic delay-period activity: representation of visual and saccade doals. J Neurophysiol 91:2628–2648. https://doi.org/10.1152/jn.01221.2003

Yakovlev V, Fisi S, Berman E, Zohary E (1998) Inter-trial neuronal activity in inferior temporal cortex: a putative vehicle to generate long-term visual associations. Nat Neurosci 1:310–317. https://doi.org/10.1038/1131

Young RM, Lewis RL (1999) The Soar cognitive architecture and human wirking memory. In: Miyake A, Shah P (eds) Models of working memory: mechanisms of active maintenance and executive control. Cambridge University Press, Cambridge, UK, pp 224–256

Zaitsev AV, Gonzalez-Burgos G, Povysheva NV, Kröner S, Lewis DA, Krimer LS (2005) Localization of calcium-binding proteins in physiologically and morphologically characterized interneurons of monkey dorsolateral prefrontal cortex. Cereb Cortex 15:1178–1186. https://doi.org/10.1093/cercor/bhh218

Zhou Y-D, Fuster JM (1996) Mnemonic neuronal activity in somatosensory cortex. Proc Natl Acad Sci USA 93:10533–10537. https://doi.org/10.1073/pnas.93.19.10533

Zhou Y-D, Fuster JM (1997) Neuronal activity of somatosensory cortex in a cross-modal (visuo-haptic) memory task. Exp Brain Res 116:551–555. https://doi.org/10.1007/PL00005783

Zhou X, Katsuki F, Qi X-L, Constantinidis C (2012) Neuron with inverted tuning during the delay periods of working memory tasks in the dorsal prefrontal and posterior parietal cortex. J Neurophysiol 108:31–38. https://doi.org/10.1152/jn.01151.2011

Zola SM, Squire LR, Teng E, Stefanacci L, Buffalo EA, Clark RE (2000) Impaired recognition memory in monkeys after damage limited to the hippocampal region. J Neurosci 20:451–463. https://doi.org/10.1523/jneurosci.20-01-00451.2000

Chapter 4
Executive Control

4.1 Introduction

When we try to accomplish a specific goal in our daily life, we often face various kinds of problems. To solve each problem, we try to acquire necessary information from the environment as well as our memory, find a possible solution, and make a series of decisions to accomplish the solution. To make a correct decision, appropriate reasoning and right judgment are necessary. Reasoning, judgment, and decision-making include complex cognitive operations and require simultaneous and coordinated management of multiple functions. To achieve coordinated management of multiple functions, specific neural function is needed. The neural function that makes it possible to managing multiple neural functions coordinately is called *executive control* (Roberts 1998; Miyake and Shah 1999; Shah and Miyake 1999). Executive functions are functions achieved by the executive control.

It has been known that the prefrontal cortex is an important brain structure to achieve executive functions (Baddeley 1986; Stuss and Benson 1986; Fuster 2008; Funahashi and Andreau 2013). Neuropsychological studies have shown that patients having prefrontal damages exhibit unique characteristics in their behavioral impairments. These patients usually exhibit comparable performances to normal subjects in most neuropsychological tests, have normal long-term memory, and exhibit normal perceptual, motor, and language skills. However, when they need to make a new and adaptive action plan or make the best choice among several alternatives, their intellectual activities are profoundly disturbed. These disturbances often cause poor judgment and planning, and inappropriate decision-making. Stuss and Benson (1986) summarized behavioral characteristics of prefrontal patients as such frontal patients exhibit a lack of insight and foresight, diminished capacity for planning, and decreased initiative. Similar impairments have also been observed in animals with prefrontal lesions. These impairments cannot be explained simply by the failure of a single particular function such as perception, motor, or memory. Intellectual activities such as insight, foresight, planning, and decision-making require coordinated

S. Funahashi, *Dorsolateral Prefrontal Cortex*, Brain Science,
https://doi.org/10.1007/978-981-19-7268-3_4

operation of not only perceptual and motor functions but also various cognitive functions including memory, attention, and controlling and monitoring various functions. Therefore, the impairments observed in frontal patients can be explained by a failure of coordinated operation of multiple functions. Since the neural mechanism for achieving coordinated operation of multiple functions is called executive control, the syndromes caused by prefrontal damage are often called *dysexecutive syndromes* (Stuss and Benson 1986).

4.2 What Is Executive Control?

A variety of explanations regarding executive function, executive process, and executive control are present. Perner and Lang (1999), for example, defined executive processes as "*processes responsible for higher-level action control that are necessary in particular for maintaining a mentally specified goal and for bringing it to fruition against distracting alternatives*." Pineda et al. (1998) included the following processes in executive processes: self-regulation, control of cognition, temporal organization of response to immediate stimuli, planning behavior, and control of attention. Baddeley and Della Salla (1998) explained that executive control is the coordinated operation of multiple control systems operating simultaneously. They indicated that executive control is the main function of the central executive in their working memory model (see Fig. 2.2). Defective central executive cannot oversee the simultaneous and coordinated operation of slave systems (visuospatial sketchpad, phonological loop, and episodic buffer). Therefore, they proposed that functions of the executive control can be examined not under single-task conditions but under dual-task conditions. In dual-task conditions, the subject is required to coordinately manage operations of these slave systems to perform two different tasks simultaneously and independently. In addition to coordinated management of the slave systems, they also indicated that functions of the central executive include selective attention, the capacity to switch attention from one source to another, and the capacity to access and manipulate information in long-term memory (Baddeley and Della Salla 1998). Shallice and Burgess (1991) hypothesized a supervisory executive system, which is similar to the central executive as Baddeley's working memory model, and listed 5 situations in which the supervisory executive system is required. These are (1) planning or decision-making, (2) error correction or troubleshooting, (3) performing responses that have not been well-learned or responses that contain novel sequences of action, (4) judging whether something is dangerous or technically difficult, and (5) overcoming a strong habitual response or resisting temptation. Burgess (1997) listed five functions as executive functions: problem-solving, planning, initiation of activity, cognitive estimation, and prospective memory. Rabbitt (1997) described seven features of the executive control: (1) Executive control is necessary to deal with novel tasks. (2) Executive control extends beyond the current internal or external environment to restructure an interpretation of the past as well as to attempt active control of the future.

(3) Executive control is necessary to initiate new sequences of behavior and also to interrupt other ongoing sequences of responses. (4) Executive control is necessary to prevent inappropriate responses. (5) Executive control can achieve rapid switching from one task to another. (6) Executive control is necessary to monitor performance to correct errors, to alter plans, or recognize new opportunities. (7) Executive control enables attention to be sustained continuously over long periods. Further, Smith and Jonides (1999) summarized five executive processes: (1) focusing attention on relevant information or processes and inhibiting irrelevant ones (attention and inhibition); (2) scheduling processes in complex tasks (task management), which requires the switching of focused attention between tasks; (3) planning a sequence of subtasks to accomplish a goal (planning); (4) updating and checking the contents of working memory to determine the next step in a sequential task (monitoring); (5) coding representations in working memory for time and place of appearance (coding).

Definitions of executive process and functions including executive process are somewhat different from researcher to researcher. However, executive control seems to include the following functions; (1) the control of attention (switching attention from one source to another or focusing attention on one source), (2) the temporal organization of behavior, (3) the planning or scheduling of complex tasks to accomplish a future goal, (4) the capacity to access and manipulate information stored in long-term memory, and (5) the monitoring of current internal and external states.

4.3 Executive Control and the Prefrontal Cortex

Executive control is an important function of the prefrontal cortex. Executive control is a neural system for coordinating and supervising the operations of multiple neural systems. Therefore, if the prefrontal cortex plays an essential role in executive control, damages to the prefrontal cortex would cause impairments mostly in complex cognitive functions. In fact, patients having prefrontal damages show complex cognitive impairments.

> "*Old information and well-established processes appear to be unaffected by frontal lobe damage. However, when intellectual operations demand the creation of a program of action and a choice between several equally probable alternatives, the intellectual activity of patients with a marked 'frontal syndrome' is profoundly disturbed*" (Luria 1969; From Stuss and Benson 1986, p. 202).

Patients with prefrontal damages usually exhibit normal IQs, have normal long-term memory, and exhibit normal perceptual, motor, and language skills. However, neuropsychological studies have shown that prefrontal patients exhibit unique impairments in behavioral performances (Stuss and Benson 1986). The abilities and the disabilities of these functions have been examined in frontal patients by various behavioral tests which require coordinate operation of multiple functions and multiple information processing simultaneously. These behavioral tests include the Wisconsin card sorting test (Milner 1963; Stuss et al. 2000), the Tower of London puzzle (Shallice 1982; Owen et al. 1990), the self-ordered task (Petride and Milner 1982), word or item fluency tests (Milner 1964; Jones-Gotman and Milner 1977; Crowe 1992; Jahanshahi et al. 1998; Jahanshahi and Dirnberger 1999), and various other tests (Milner 1964, 1982; Freedman and Oscar-Berman 1986; Milner et al. 1991). The disturbances of these intellectual activities are caused by poor judgment, planning, and decision-making (Stuss and Benson 1986), by poor temporal organization of behavior (Fuster 1997), or by poor working memory (Goldman-Rakic 1987). Prefrontal patients have been shown to exhibit specific impairments in these neuropsychological tests.

4.3.1 Wisconsin Card Sorting Test

The Wisconsin card sorting test is one of the important tests to examine prefrontal executive functions. Figure 4.1a shows examples of test materials that are presented to the subject. Four stimulus cards and a stack of 128 response cards are used. Stimulus cards are placed in front of the subject. Each of these cards differs in color, form, and number of stimuli on the card (e.g., one red triangle, two green stars, three yellow crosses, and four blue circles). A stack of 128 response cards is also placed in front of the subject. Each response card has a different number of stimuli with different forms and color on the card. The response card shown in Fig. 4.1a has two (number) red (color) circles (form). The subject's task is to sort each response card by a predetermined sorting category (either form, color, or number). The subject sorts each response card by the current sorting category and places it in front of one of the stimulus cards. For example, since the response card shown in Fig. 4.1a has two red circles, this card would be classified into the group of stimulus card 1 if the current sorting category is the color, or into the group of stimulus card 4 if the current sorting category is the form, or into the group of stimulus card 2 if the current sorting category is the number. However, the sorting category is determined by the experimenter before the subject stars a series of sorting and is not informed to the subject. The experimenter informs only whether the subject's classification of the response card is right or wrong to the subject. Therefore, the subject needs to find the correct sorting category by trial and error based on the information from the experimenter (right or wrong) and maintain sorting of the remaining response cards using the same sorting category. However, once the subject achieves 10 consecutive correct sorting by the same category, the experimenter changes the sorting category without any notification to the subject. Unexpected changes in the sorting category will happen several times until the subject completes the classification of all response cards. To accomplish this test, the subject needs to find the correct sorting category by trial and error and to keep the correct soring category in mind for consecutive trials.

Milner (1963) showed that patients having dorsolateral prefrontal lesions exhibit inability to perform this test, whereas patients having inferior and orbital prefrontal

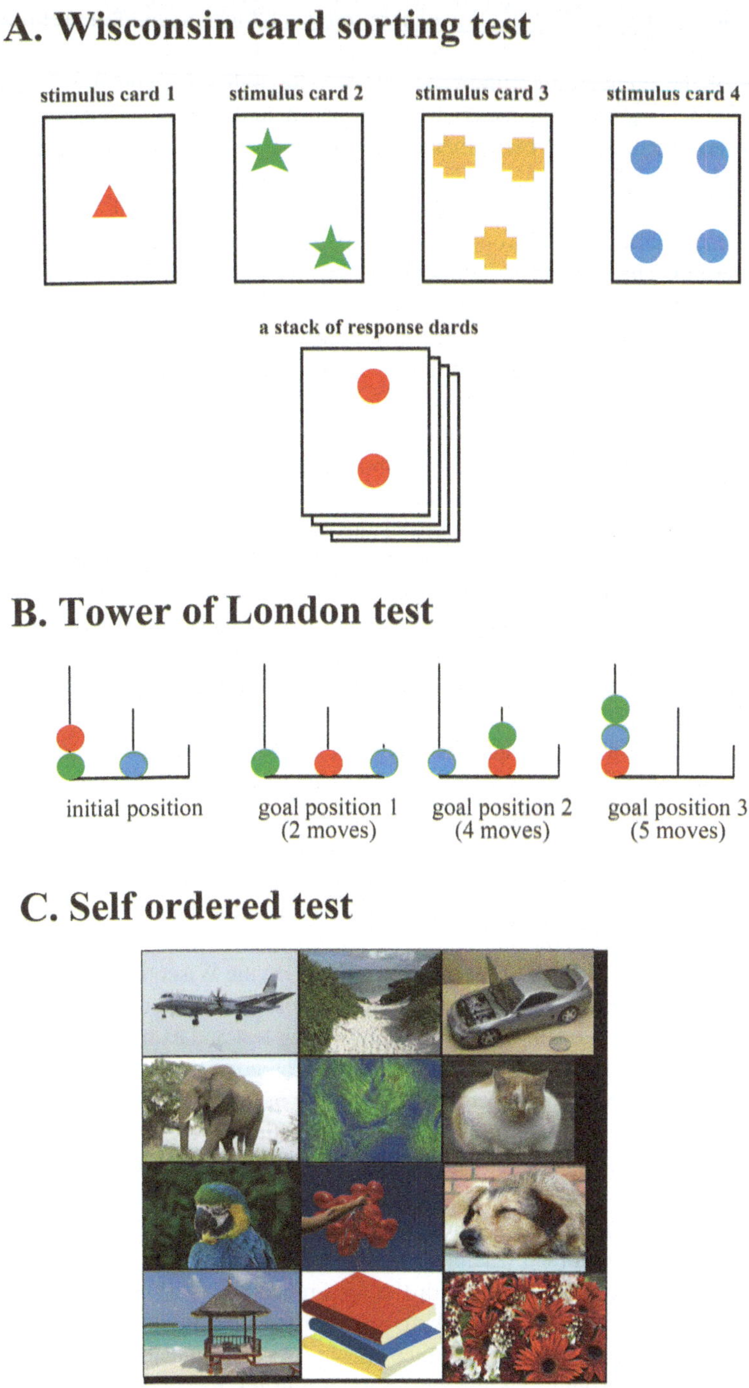

Fig. 4.1 Three examples of neuropsychological tests to examine prefrontal functions

lesions do not. Patients having dorsolateral prefrontal lesions exhibited difficulty to shift from one sorting category to another, possibly because of perseverative interference of the preceding sorting category or inflexibility of behavior. Stuss et al. (2000) further examined characteristics of impairment that patients having prefrontal lesions exhibited during performance of the Wisconsin card sorting test. They compared the performance of this test among four groups of prefrontal patients having single focal lesions (right and left dorsolateral frontal lesions, superior frontal lesions, and inferior medial frontal lesions), patients having non-frontal lesions, and control group. The Wisconsin card sorting test was administered under three conditions. In the first condition, all 128 response cards were administered regardless of the subject's performance. In the second condition, additional 64 cards were administered after informing the subject that there were three ways to sort the cards, by color or number of forms, or form itself. In the third condition, the subject was reminded of the sorting method and informed of the first sorting category, then, additional 64 cards were administered, and after 10 correct sorts, the experimenter informed the subject that the sorting category was changed. Prefrontal patients having the right and left dorsolateral lesions and superior medial lesions attained significantly fewer categories than other groups in all three conditions. Prefrontal patients exhibited stronger perseverating effects caused by preceding sorting category. Patients having non-frontal lesions were not impaired in all three conditions and did not exhibit perseverating effects. Thus, these results show that the normal function of the prefrontal cortex is required to perform the Wisconsin card sorting test correctly.

Performing the Wisconsin card sorting test correctly requires multiple functions including keeping the sorting category in working memory, flexibly switching the sorting category depending on the condition, finding a new sorting category by trial and error, and inhibiting inappropriate responses to distractors, all of which are important components of the executive control. The results obtained by the Wisconsin card sorting test indicate that the dorsolateral prefrontal cortex is an important structure for executive control.

The importance of the prefrontal cortex to perform the Wisconsin card sorting test has been shown by human neuroimaging studies. Konishi et al. (1998) used a computerized version of the Wisconsin card sorting test and examined event-related fMRI signals that occurred at the timing of the category shift. They also prepared three test conditions which differed in the number of the sorting category (one, two, and three categories). They found transient activation in the posterior part of the bilateral inferior frontal sulci and that this activation became larger as the number of the sorting category increased. Based on the result, they suggested that the inferior frontal area plays an essential role in flexible shifting of the sorting category. Konishi et al. (1999a) also examined whether the prefrontal area related to flexible shifting of the sorting category overlapped with the area related to working memory or not. They used an n-back task as a working memory task and found that the prefrontal area activated during the performance of the Wisconsin card sorting test was spatially overlapped with the area activated during the performance of the n-back task. Konishi et al. (1999b) further examined spatial relations of prefrontal areas activated by the Wisconsin card sorting test and a Go/No-Go test by event-related fMRI method. The prefrontal areas exhibiting transient No-Go dominant activation at the Go/No-Go test were spatially overlapped with the prefrontal areas exhibiting transient activation at the timing of the category shift in the card sorting test. These results suggest that the prefrontal cortex plays an essential role in the performance of the Wisconsin card sorting test and that working memory plays an important role to perform the card soring test.

Contribution of the dorsolateral prefrontal cortex to perform the Wisconsin card sorting test has also been shown in the experiments using monkeys. Monkeys with lesions in the dorsolateral prefrontal cortex exhibited impairment in the conflict-induced behavioral adjustment of the Wisconsin card sorting task (Mansouri et al. 2007). The card sorting test in monkey studies used two categories (color and shape categories) and was similar as matching-to-sample tasks, which was different from the Wisconsin card sorting test in human studies. However, Boschin et al. (2017a, b) used an analog of the Wisconsin card sorting test used in monkey studies by Mansouri et al. (2007) and obtained similar behavioral results in human patients having damages in the dorsolateral prefrontal cortex as had been obtained in monkey studies. In the analog of the Wisconsin card sorting task used in monkey studies, the monkey needs to shift visual dimensions between shape and color based on the sorting category. In this condition, monkeys usually exhibit behavioral advantage in the shape dimension compared to the color dimension. Mansouri et al. (2020) showed that monkeys having bilateral lesions of the dorsolateral prefrontal cortex exhibited no impairment in behavioral advantage to the shape dimension but exhibited impairment in monkeys' ability to shift visual dimensions between shape and color depending on the sorting category. Buckley et al. (2009) compared the effects of lesion made at several local areas of the monkey prefrontal cortex and found that only the lesion made at the principal sulcus area impaired maintenance of abstract rules in working memory while monkeys performed a Wisconsin card sorting analog. Further, Mansouri et al. (2006) showed that neurons in the dorsolateral prefrontal cortex exhibited modulation of their activity by the difference of the rules across trials while monkeys performed a Wisconsin card sorting analog. These results further support that the dorsolateral prefrontal cortex contributes to the performance of the Wisconsin card sorting test and that the dorsolateral prefrontal area plays role in maintaining relevant rules across trials which is an important component of executive control.

4.3.2 Tower of London Test

The Tower of London test (or Tower of Hanoi test) is also used to examine prefrontal executive functions (Fig. 4.1b). This is a look-ahead test, in which the subject needs to construct a final configuration of colored blocks from an initial configuration by a series of moves of individual blocks. To solve this task in a predetermined number of moves, the subject needs to determine an appropriate order of moves of individual blocks. Although an appropriate planning of whole moves of individual blocks is necessary, planning of appropriate subgoals is also necessary to reach the final goal.

Therefore, this task can be called a planning task. Shallice (1982) tested this task in 61 patients having unilateral localized lesions in various cortical regions and 20 control subjects. Patients were further grouped into anterior and posterior groups based on whether lesions were located in the anterior or the posterior cerebral cortex, respectively. He compared the mean number of problems solved within 60 s at the first attempt among groups and found a specific deficit on this task in the anterior group having lesions in the left hemisphere. This lesion group exhibited a significantly low percentage of correct moves in all three conditions (2, 4, and 5 move conditions). Since the anterior group is mainly composed of patients having lesions in the frontal lobe, this result suggests that the frontal lobe participates in planning or foresight, which is an important component of the executive function. Owen et al. (1990) also showed that patients having damage in the frontal lobe required significantly more moves to solve the Tower of London task than the controls and solved fewer problems in the minimum number of moves.

Significant relations between the performance of the Tower of London task and the function of the prefrontal cortex have been shown by human neuroimaging studies. Baker et al. (1996) performed a PET study using the Tower of London task and showed activation of a distributed network of cortical areas including the prefrontal cortex and the parietal cortex, which are known to participate in visuospatial working memory. They also observed activation in the rostral prefrontal cortex, which is thought to participate in the executive component of planning such as response selection and evaluation. The importance of the rostral prefrontal cortex for planning has also been reported by Wagner et al. (2006). They indicated that the rostral prefrontal cortex participates in specific processes for planning including simultaneous monitoring of both internally and externally presented information. Focal activation was further observed during performance of the Tower of London task in the fMRI studies (Newman et al. 2003; van den Heuvel et al. 2003; Kaller et al. 2011). Thus, the experiments using the Tower of London task indicate that the prefrontal cortex plays an important role for planning and foresight.

4.3.3 Self-Ordered Test

Self-ordered test is another test to examine prefrontal executive functions. Figure 4.1c shows an example of the stimuli used for the self-ordered test. In this test, the subject is first shown 12 pictures in a particular 3 × 4 arrangement and asked to select any one picture by pointing. After the first selection, the same 12 pictures are rearranged in the same 3 × 4 arrangement and the subject is asked to select one new picture which is not selected before. The subject is required to select one new picture at every selection and this procedure is repeated 11 times. Therefore, the subject needs to remember which pictures have been selected and inhibit to select already-selected pictures. Petride and Milner (1982) introduced four kinds of self-ordered tests using four stimulus materials (abstract designs, representational drawings, high-imagery words, and low-imagery words) to human subjects. The performance of each self-ordered test was examined by 14 patients having the left frontal lesions, 9 having the right frontal lesions, and 56 having the temporal lesions. They found that patients having lesions in the left frontal lobe exhibited significant impairments in all four kinds of self-ordered tests, while patients having lesions in the right frontal lobe exhibited impairments only in two kinds of non-verbal self-ordered tests. Patients having lesions in the temporal lobe exhibited no impairment in all four kinds of self-ordered tests. Thus, these results showed that the left frontal lobe participates in remembering past selections and monitoring responses, and suppressing responses to irrelevant stimuli in both verbal and non-verbal conditions, while the right frontal lobe participates in similar functions but only in non-verbal conditions.

Petrides (1995) used monkeys having lateral prefrontal lesions and showed that these monkeys exhibited impairments in the performance of the self-ordered task. In this task, three containers that differed in color and shape are presented in a row and the positions of these containers are randomly determined during each trial of the three-trial set. The monkey is asked to select one of them in each trial but is required to select the new container that has not been selected in the preceding trial. Petrides (1995) also added the externally ordered task. In this task, two of three containers are randomly selected, each of two containers is presented to the monkey one by one at the first and the second trial, and the monkey selects the presented container at each trial. In the third trial, all three containers are presented in a row and the monkey is asked to select the container that has not been selected. He found that monkeys having lesions in the dorsolateral prefrontal cortex (the lateral cortex extended above the principal sulcus as far as the midline) exhibit severe and long-lasting impairment in both the self-ordered task and the externally ordered task. Interestingly, the critical factor causing this impairment was found to be the size of the stimulus set that the monkey needed to monitor. Thus, these results also support the notion that the dorsolateral prefrontal cortex participates in remembering and monitoring the stimulus set in working memory.

4.3.4 Other Tests Sensitive to Prefrontal Functions

Word or design fluency tests have also been frequently used to examine prefrontal executive functions. In the word fluency test (Milner 1964), the subject needs to write down as many words beginning with the letter "S" as possible for 5 min, and after this performance, the subject needs to write four-letter words beginning with "C" for another 4 min. In the design fluency test (Jones-Gotman and Milner 1977), the subject is allowed for 5 min to create drawings that neither represent actual objects nor are derived from actual objects as many as possible, and then for 4 min to create drawings that consist of exactly four lines as many as possible. It has been shown that patients having lesions in the left prefrontal cortex exhibited impairment in the verbal fluency tests (Milner 1964; Crowe 1992), while patients having lesions in the right prefrontal cortex exhibited impairment in the design fluency test (Jones-Gotman and Milner 1977).

Random number generation test is also used to examine prefrontal functions (Jahanshahi et al. 1998; Jahanshahi and Dirnberger 1999). In this test, the subject is asked to say the number 1–9 in random order for 100 times synchronized with a pacing LED light presented at the rate of once every 1.2 s. It has been shown that patients having frontal lesions performed poorly in random number generation and said the same number or the same sequence of numbers repetitively. PET activation studies using normal subjects showed activation in the dorsolateral prefrontal cortex (areas 46 and 9) during random number generation.

Features that prefrontal patients exhibit in these tests are stereotyped repetition of responses, inability to retrieve information from memory, inability to produce original responses, and inability to monitor their responses. These features would be due to the difficulty in using feedback from previous responses to monitor their actions. Performance of these tests requires working memory to maintain information online, suppression of habitual responses, generation of internally driven responses, and monitoring of responses. Since patients having lateral prefrontal lesions exhibit impairments in these tests, the lateral prefrontal cortex is thought to play important roles in these functions.

Further, patients having frontal lobectomy had been shown to exhibit impairment in a maze learning test (Milner 1964). In this test, the subject needs to learn the correct path from the start to the end on a stylus maze, on which nail heads provide visible stepping stones. The correct path is invisible to the subject. When the subject makes an error, a click sounds. Therefore, the subject must find the correct path by trial and error using an error click. Performance of the maze learning test requires working memory for previous actions, flexible switching of actions depending on the result, and monitoring of the action. Therefore, the frontal lobe, especially the prefrontal cortex, is thought to play roles in these functions.

4.3.5 Prefrontal Contribution to Executive Control

It has been well-known that patients having prefrontal lesions exhibit normal functions in sensation, perception, movement, and long-term memory. However, damages to the prefrontal cortex cause severe impairment in the Wisconsin card sorting test, the Tower of London test, and the self-ordered test. Impairments of these tests are specific in the patients who have damages in the prefrontal cortex and not observed in the patients who have damages in other cortical areas. Impairments of these tests cannot be explained simply by the failure of a particular function such as perception, motor, or long-term memory, but can be explained by the failure of coordinated operation of multiple functions. For example, the Wisconsin card sorting test requires working memory to maintain a sorting category for the necessary time, attention to the specific category of stimuli, flexible response to find new sorting categories by trial and error, and inhibition of responses to irrelevant stimuli. The Tower of London test is a planning task and requires planning of appropriate subgoals to reach the final goal and planning and remembering of an appropriate order of moves of each block. The self-ordered test requires working memory of already selected items, monitoring of self-selection, and inhibition to select inappropriate items. The fluency tests and the random number generation require working memory to hold items, words, or numbers already listed up online, monitoring of own responses, and suppression of habitual behavior to repeat the same response. Correct performance of these behavioral tests requires functions such as maintaining, switching, and focusing attention, planning the temporal organization of behavior, accessing and sorting information stored in long-term memory, and monitoring past and current states. In general, these functions are called executive functions. The fact that patients having focal damages in the prefrontal cortex exhibit severe impairments in these tests indicates that the prefrontal cortex plays important roles to achieve executive functions.

Gläscher et al. (2012) tested 344 individuals who had focal lesions in the cerebral cortex (165 prefrontal patients) using a comprehensive battery of neuropsychological tasks and applied voxel-based lesion symptom mapping. They found two distinct functional–anatomical networks within the prefrontal cortex. One prefrontal network which included the dorsolateral prefrontal cortex and the anterior cingulate cortex associated with executive control such as response inhibition, conflict monitoring, and switching. Executive control shared a common behavioral factor related to set shifting which is linked to the function of the anterior cingulate cortex. The other prefrontal network which included the orbitofrontal cortex, the ventromedial cortex, and the frontopolar cortex associated with value-based decision-making. Their results indicated that the prefrontal network including the dorsolateral prefrontal cortex associates with executive control, whereas the network including the ventral prefrontal cortex associates with decision-making. Thus, the dorsolateral prefrontal cortex is shown to play important roles in executive control.

Therefore, it can be concluded that "*disorders of the executive functions, if present after focal brain damage, appear to be maximal after frontal lobe damage, suggesting that the frontal lobe is the executive center*" (Page 215–216 in Stuss and Benson 1986).

4.4 Executive Control by the Prefrontal Cortex

4.4.1 Control of Attention

Attention can be described as a cognitive process that act to optimize stimulus detection, discrimination, and processing (Sarter and Lustig 2009). By attention, the subject can choose appropriate information among equally potent sources or appropriate action among equally valid actions. Different forms of attention have been known. *Sustained attention* is the subject's state of readiness to detect rarely and unpredictably occurring changes in the stimulus situation over an extended period of time. Sometimes, sustained attention is considered to have the same meaning as working memory. However, sustained attention is different from working memory. Sustained attention emphasizes the sensory aspect of information processing, such as the state of readiness to detect changes that occurred in the stimulus situation, whereas working memory emphasizes holding detected sensory information online for a certain length of time. *Selective attention* emphasizes the subject's consciousness of a targeted information or signal and inhibits the detection and processing of non-target information and signals. *Switching attention* emphasizes the subject's consciousness from one target to another intentionally or automatically by the unexpected appearance of the stimulus. *Divided attention* requires the allocation and the management of limited capacity of attention resources to attend multiple stimuli or perform multiple tasks.

Kowler (2009) described four characteristics of attention. (1) Attention can lower the detection threshold and increase perceived contrast, which produces attention-modulated changes in cell firings (e.g., enhancement of visual response). (2) Improvement of perceptual threshold can be produced by paying attention either to a specific location (space-based attention), to a specific physical feature of the stimulus (e.g., color, shape, size, and orientation) (feature-based attention), or to a specific object itself (object-based attention). (3) Attention to a specific object makes it possible to overcome potential perceptual interference from other visual objects nearby because the amount of resources for attention is limited. (4) Attention controls access to working memory.

In fact, an enhancement of the visual response and the improvement of the discriminative ability to a peripheral visual stimulus has been shown in extra-striate visual neurons when the monkey is required to attend to the peripheral visual stimulus while maintaining fixation at the central fixation point (Moran and Desimone 1985; Spitzer et al. 1988; Motter 1993). The increase of the strength of visual selectivity or the modulation of the extent and the position of visual receptive field have been reported in these extra-striate neurons in association with covert attention without eye movement or overt attention with eye movement to the relevant stimulus (McAdams and Mauncell 1999; Tolias et al. 2001). Enhancement of the visual response has also been observed in neurons of the prefrontal cortex and the frontal eye field in overt attention conditions, in which a visual stimulus is the target of the eye movement (Goldberg and Bushnell 1981; Bruce and Goldberg 1985; Boch and Goldberg 1989), or in covert attention condition, in which the monkey needed to attend a peripheral visual stimulus while fixation at a fixation point (Mikami et al. 1982; Funahashi et al. 1990; Kodaka et al. 1997).

4.4.1.1 Attention-Related Activity Observed in the Prefrontal Cortex

It has been known that the enhancement of the visual response is observed when the visual stimulus has behavioral significance, such that it becomes the target of the saccadic eye movement. Attention-related enhancement of visual response associated with eye movement has been observed in neurons in posterior parietal neurons (Lynch et al. 1977; Robinson et al. 1978; Bushnell et al. 1981). Robinson et al. (1978) found that 47% (27/58) of saccade-related parietal neurons showed an enhancement of visual response when the stimulus presented in the receptive field was the target of the saccadic eye movement. The enhancement was selective for the eye movement directed toward the visual receptive field and was not found for eye movements directed toward the outside of the visual receptive field.

The same enhancement effect was observed for visual responses in prefrontal neurons. Boch and Goldberg (1989) collected single-neuron activity from the banks of the posterior third of the principal sulcus in the prefrontal cortex and showed that 36% (29/79) of visually responsive prefrontal neurons exhibited enhanced responses to the visual stimulus. This enhancement effect was spatially selective and observed only when the visual stimulus was presented within the visual receptive field and was the target of the saccadic eye movement. The enhancement did not observe when the monkey made a saccade to a visual stimulus outside the receptive field even if another visual stimulus was presented within the receptive field.

Attention-related enhancement of visual response was also observed in relation to the effect of the behavioral context. Bushnell et al. (1981) examined the activity of posterior parietal neurons when the monkey performed a peripheral attention task. In this task, the monkey was required to maintain fixation at the central fixation point and to release a key as soon as possible when the peripheral target dimmed. An enhancement of the visual response was observed only when the monkey performed the attention task, not during a simple fixation task, in which the same visual stimulus was presented in the visual receptive field without any key response. In addition, the enhancement effect was spatially selective and observed only when the monkey was required to attend the visual stimulus presented in the neuron's visual receptive field. An enhancement index was calculated using mean firing rates of visual responses in both the attention task and the simple fixation task for 96 neurons and 40% (38/96) of these neurons had an enhancement index of 1.5 or greater. This behavioral context-dependent enhancement of visual responses was independent of saccadic eye movements.

A similar context-dependent enhancement effect of visual responses had been observed in the prefrontal cortex. Mikami et al. (1982) examined an effect of attention on the visual response of prefrontal neurons by comparing the responses to the same visual stimulus between attended condition and non-attended condition. They used a visual fixation task, in which the monkey was required to release a lever when the fixation point dimmed, and a visual detection task, in which the monkey was required to press either the right or the left key depending on whether the visual stimulus was presented or not during the fixation period. They found that a small number of prefrontal neurons (13%, 14/105) exhibited an enhancement of the visual response in the attended condition compared with the visual fixation condition.

Funahashi et al. (1990) compared the same prefrontal neuron's responses to the same visual stimulus presented at the same position in the visual field between two task conditions. In one condition [a visual probe (VP) task], a visual stimulus was presented for 0.5 s while the monkey maintained fixation at the central fixation point. In the other condition [an oculomotor delayed-response (ODR) task], the monkey was required to remember the position of the visual stimulus for a 3-s delay period and make a saccade to that position during the response period. The enhancement index was calculated for 34 prefrontal neurons using the method of Bushnell et al. (1981). Among prefrontal neurons having visual responses, only 7 neurons (21%) exhibited significant enhancement of visual responses (over 2.0) in the ODR condition compared with the VP condition, while 27 had indices between 0.5 and 2.0 (Fig. 4.2), indicating that the magnitude of the visual response is affected only weakly by the difference of the behavioral context in prefrontal neurons.

Further, di Pellegrino and Wise (1993) also shown attention-related enhancement of visual responses in prefrontal neurons. They used a spatial attention task (spatial matching task). In this task, the monkey is required to pay attention to the position where the visual cue was presented in the preceding trial, and needs to move the forelimb to the cued position when the visual cue is presented at the same position as in the preceding trial. They found that spatial attention to the visual stimulus enhanced visual responses in prefrontal neurons as well as premotor neurons.

Thus, an enhancement of visual responses can be observed in prefrontal neurons when the visual stimulus has behavioral significance, such that the visual stimulus is the target of the eye movement or is needed to be remembered. The enhancement of neural activity is considered as a neural correlate of the attention effect observed in single-neuron activity. Although a smaller population of prefrontal neurons is shown to exhibit behavioral context-dependent enhancement of visual responses compared with posterior parietal neurons, this result indicates that prefrontal neurons also participate in behavioral context-dependent attention functions.

4.4.1.2 Attention-Related Activity in the Frontal Eye Field

Eye movements are closely related to *overt attention*. If an attractive object is unexpectedly found or if a salient stimulus is presented in the peripheral portion of the visual field, an eye movement is generated toward the object or the stimulus to look at it. An oculomotor behavior toward a particular item to selectively attend to it over many others is called overt orienting. An eye movement for overt orienting is triggered by the presentation of a particular sensory stimulus. Therefore, attention related to an overt-orienting behavior is often called bottom-up attention. Oculomotor behaviors are often used to examine neural mechanisms of overt attention or bottom-up attention. On the other hand, we can pay attention to a particular item located in the peripheral visual field without any eye movement to shift gazing toward it. This type of attention is called *covert attention* and is described metaphorically as a mental spotlight that illuminates a selected item or a specific space and facilitates information processing for that item or the stimulus within that space. Covert attention is generated not by an externally driven sensory process but by an internally driven voluntary process. Therefore, covert attention is also called top-down attention. Since covert attention is closely linked to overt eye movements, it is proposed that covert attention arises from a latent command of the eye movement although no eye movement is generated.

The presence of an enhancement of the visual response is one important sign to determine whether a particular brain area participates in overt attention or not. It has

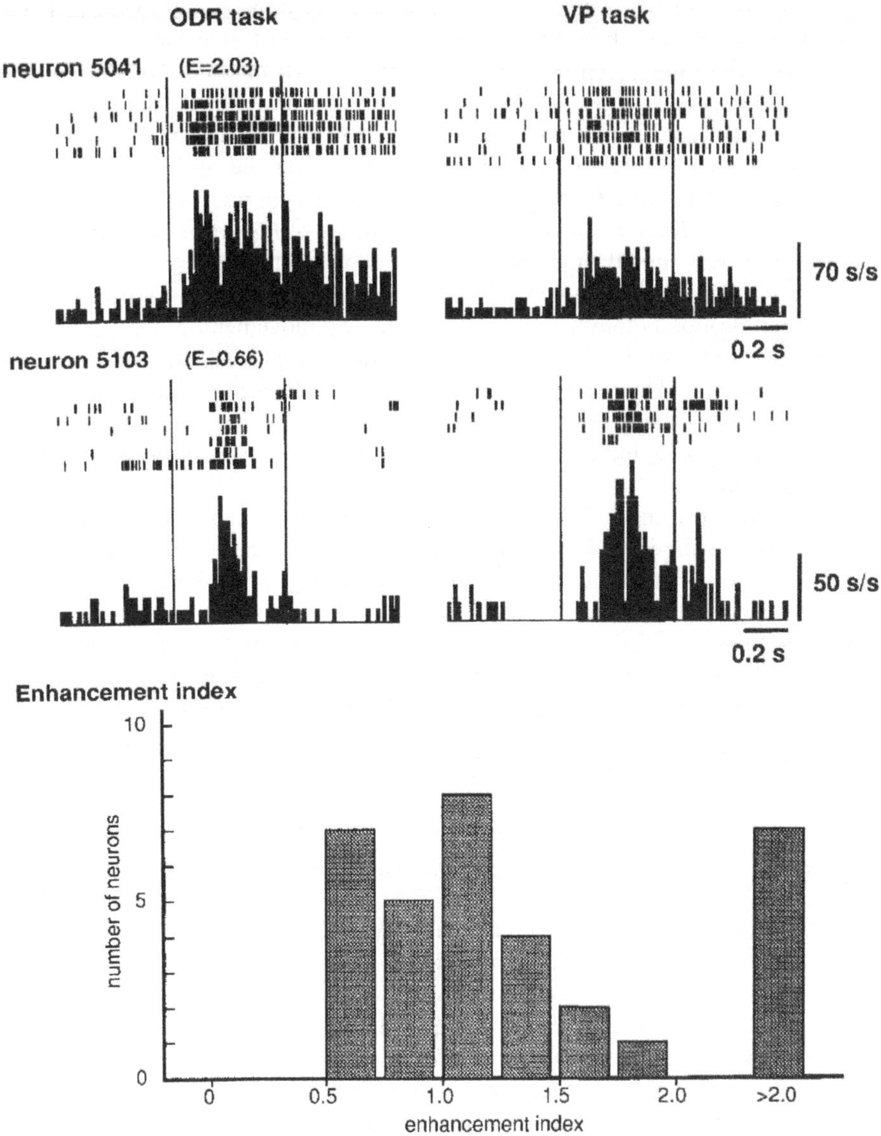

Fig. 4.2 *Top*: Visual responses of two prefrontal neurons in the ODR task and the visual probe (VP) task are shown. Enhancement of visual responses is observed in the ODR task in neuron 5041, but not in neuron 5103. *Bottom*: The distribution of enhancement indices in prefrontal neurons. Few neurons exhibited enhancement indices over 2.0. From Funahashi et al. (1990) with permission

been shown that FEF neurons exhibit an enhancement of visual responses in overt attention conditions, in which a visual stimulus is the target of eye movement (Goldberg and Bushnell 1981; Bruce and Goldberg 1985; Boch and Goldberg

1989). The visual search task has also been used to examine neural mechanisms of overt visual attention (Schall 2002). The visual search task that Schall and Hanes (1993) used, for example, was that, after fixation at a fixation point for 300–500 ms, one visual target and 7 distractors (search stimulus array) were simultaneously presented around the fixation point at the same eccentricity. The visual target was distinguished from distractors by color (e.g., a red target and green distractors). A change in the color of the fixation point was the signal for the monkey to shift the gaze to the target. The magnitude of visual responses of FEF neurons to the search stimulus array was initially the same regardless of whether the visual target or a distractor was in the receptive field. However, neural activity evolved to specify the visual target location before the execution of the eye movement (Schall and Hanes 1993; Schall et al. 1995). Although this target selection process did not depend on saccade production (Thompson et al. 1997), the saccade was an important component of this task. Thus, neurons in the FEF exhibit an enhancement of the visual response in overt orienting eye movement toward the visual target presented in the peripheral visual field.

At the same time, the FEF had been suggested to play a causal role in directing covert spatial attention (Awh et al. 2006). The results obtained by the visual search task have shown that the activity of FEF neurons was not sensitive to specific physical features of the visual stimulus, but exhibited selectivity to the target stimulus in the search array. Selective response to the target reflects the bottom-up saliency of the target itself and the top-down knowledge about the target. Although FEF activity appears to have relations to the saccade performance toward the target, this activity is not directly related to the production of an eye movement (motor command), because the magnitude of activation reflects the behavioral significance of the stimulus itself (e.g., target or distractor) and because significant activation occurs without saccade performance. These results suggest that the FEF not only involved in overt attention but also in covert spatial attention (Thompson and Bichot 2005).

Involvement of the FEF in covert attention has been shown. Kodaka et al. (1997) showed that FEF neurons exhibit modulation of responses to peripheral visual stimuli when the monkey is asked to pay attention to them, irrespective of eye movements. In their task, the monkey is required to release a lever when a peripheral stimulus presented in the visual receptive field dims while the monkey maintains fixation. They found that visual response was either enhanced or suppressed in the attention condition but not in a simple fixation condition. Thompson et al. (2005) used a covert visual search task without eye movement and examined the visual responses of FEF neurons (Fig. 4.3). In this task, the monkey is required to grasp a vertical lever and maintain fixation on a central fixation point throughout a trial. After the monkey looks at the fixation point for 400–800 ms, a target and distractors (search array) are presented at 6 or 8 iso-eccentric peripheral positions. If the monkey turns the lever toward the direction (left or right) where the target is presented, the reward is given. They found that the majority of visually responsive neurons (77%) exhibited an enhancement of visual response before turning the lever when the singleton target was presented in the receptive field, but no enhancement

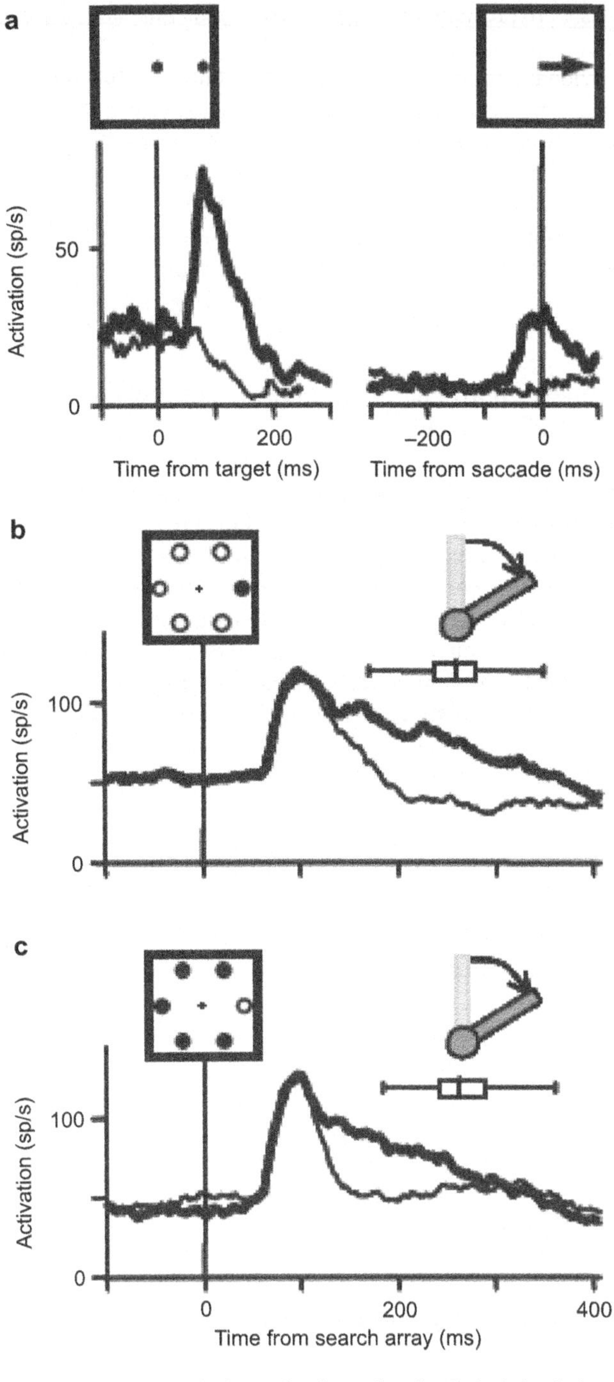

Fig. 4.3 Responses of FEF neuron during a visual search task. (**a**) Activity during memory-guided saccade performance. (**b** and **c**) Activity during visual search performances. Thick lines and thin lines indicate activities when the target was presented within the receptive field and outside of the receptive field, respectively. From Thompson et al. (2005) with permission

when distractors were presented in the receptive field (Fig. 4.3). This result indicates that the enhancement of visual response can be observed without behavioral and physiological evidences of saccade planning. In addition, when attention is allocated to the target of the visual search, the activity of saccade-related neurons is inhibited, suggesting that spatial attention signals are independent of explicit saccade command signals. Covert attention has been thought to link with overt eye movement and arise from the latent command of eye movement. However, Thompson et al. (2005) proposed that spatially selective attention-related visual responses in FEF neurons correspond to the mental spotlight of attention via modulation of ongoing visual processing.

4.4.1.3 Top-Down Control of Visual Information Processing by the Frontal Eye Field

Covert spatial attention modulates the strength of visual response in various cortical areas. Responses to a peripheral visual stimulus are enhanced in extra-striate neurons when covert attention is directed toward this stimulus (Moran and Desimone 1985; Motter 1993). The modulations of visual selectivity and the size of the visual receptive field have also been reported in these extra-striate neurons in association with covert attention (McAdams and Mauncell 1999; Tolias et al. 2001). Although it has been known that attention modulates visual responses in striate and extra-striate neurons, the source of this modulation remained unknown. However, the modulation of visual responses has also been observed in both overt and covert attention conditions in FEF neurons. Neurons in the FEF have reciprocal connections with neurons in the extra-striate cortices (Stanton et al. 1995). Therefore, it can be thought that the outputs from the FEF cause either enhancement or suppression of the activity of neurons in the extra-striate cortices and modulation of information processing in these cortices.

Moore and Fallah (2004) investigated the effect of micro-stimulation applied to the FEF to examine whether or not saccade-related mechanism of the FEF provides spatial top-down attention signals to the extra-striate cortices. In their task, the monkey is required to detect the luminance change of the target while ignoring a flashing distractor. They found that the monkey could detect smaller change in the target luminance if the luminance change is preceded by the focal electrical stimulation at the FEF, whose current intensity was below the level that evoked saccades. Thus, the stimulation of the FEF could increase the sensitivity to detect the visual information and produce transient improvement of an ability of covert spatial attention to the visual target. Similarly, Wardak et al. (2006) reversibly inactivated the FEF by the focal microinjection of muscimol while the monkey performed a covert visual search task. Inactivation of the FEF caused a spatially selective deficit in the execution of the visual search task, such that inactivation of the FEF altered the ability to detect the visual target among distractors. Thus, focal activation or inactivation of the FEF produces spatially selective improvement or deficit in the execution of covert attention tasks, respectively. Further, Moore and Armstrong

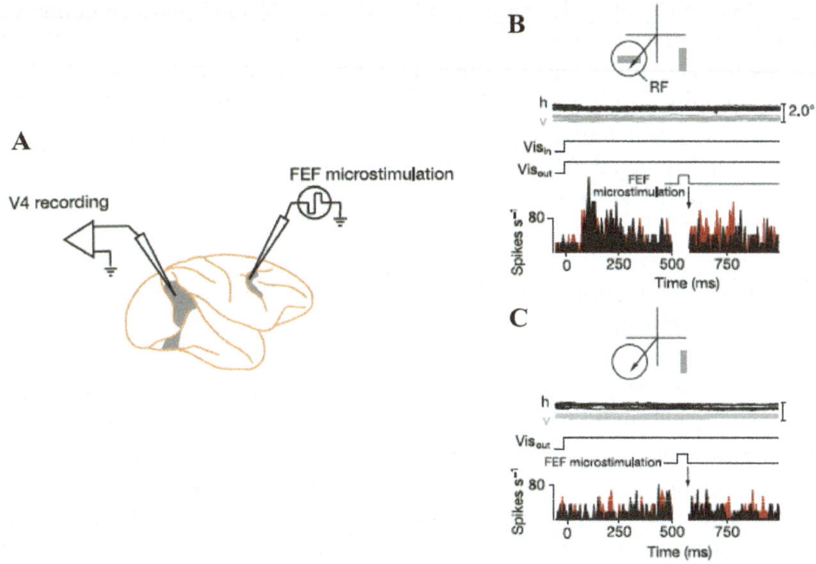

Fig. 4.4 Enhanced visual responses were observed in a V4 neuron by micro-stimulations at the FEF. (**a**) Experimental situation. (**b**) Enhanced response (red) was observed when a visual stimulus was presented within the receptive field of the V4 neuron and the response field of the stimulation site of the FEF. **C**. Enhanced response was not observed when a visual stimulus was presented within the response field of the stimulation site of the FEF but outside of the receptive field of the V4 neuron. From Moore and Armstrong (2003) with permission

(2003) showed that visual responses in the visual area V4 are enhanced by micro-stimulation at the FEF when the position of the visual receptive field of the V4 neuron matches the region of the visual field represented at the stimulation site of the FEF (Fig. 4.4). In addition, it has been known that iontophoretic application of selective dopamine D1 receptor antagonist (SCH23390) increases persistent delay-period activity in the prefrontal cortex (Williams and Goldman-Rakic 1995; Robbins and Arnsten 2009). Noudoost and Moore (2011) locally applied D1 antagonist to the FEF and found that the application of D1 antagonist enhanced the magnitude, the orientation selectivity, and the reliability of V4 visual responses. The enhancement of V4 visual responses was restricted to neurons whose visual receptive fields overlapped with the visual space affected by the D1 receptor manipulation of the FEF. These results confirmed the results obtained by micro-stimulation experiments (Moore and Armstrong 2003). The observation that the focal electrical stimulation of the FEF produces an enhancement of visual response in the extra-striate areas would show a neural mechanism for focusing attention on a specific stimulus presented in a particular visual field. Thompson et al. (2005) showed that the source of the attention signal is the enhanced activity of visually responsive neurons in the FEF. They proposed that spatially selective activity of visually responsive neurons in the FEF corresponds to the "mental spotlight" of attention and modulates ongoing visual information processing in the posterior cortices. Thus, the activity of the FEF modulates the activity of other cortical or subcortical structures via efferent projections (Ninomiya et al. 2012). As was shown by Thompson et al. (2005), the spatially selective visual responses in the FEF would be a neural source of top-down signals for controlling attention-related modulation of activity observed in the other cortical and subcortical structures.

4.4.1.4 Top-Down Control of Visual Information Processing by the Prefrontal Cortex

Although the FEF plays an important role in the top-down control of attention, neuropsychological studies show that the prefrontal cortex also plays important roles in the top-down control of attention. Rossi et al. (2007) made unilateral lesions in the primate prefrontal cortex in combination with the transection of the corpus callosum and the anterior commissure and asked monkeys to discriminate the visual target from the distractor presented in either visual hemifield. Although the monkey exhibited no impairment when the visual target was constantly presented in the affected hemifield for many trials, they were severely impaired when the position of the visual target was switched frequently across trials. They concluded that the prefrontal cortex plays an essential role in the top-down control of attention to allocate attention flexibly on the basis of the task demand.

Neurophysiological studies have also shown prefrontal participation in the top-down control of attention. For example, Everling et al. (2002) used a focusing attention task and examined prefrontal activity (Fig. 4.5). In this task, while the monkey looks at a central fixation point, a white square (cue) appears on the left or the right, indicating a cued position, then 1–4 visual objects or 1–4 pairs of visual objects are presented for 300 ms either unilaterally (same side as the cue position) or bilaterally with 800 ms inter-stimulus interval (Fig. 4.5a). The monkey is required to monitor a stream of visual objects and, if the target object appears at a cued location, the monkey needs to make a saccade to the target object. Therefore, the monkey is required to maintain attention to the cued position and monitor whether the target object appears or not. They found that prefrontal neurons exhibited stronger responses to the target object than non-target objects and discriminated the target object from non-target objects. However, when the target object was presented in the unattended position, the discriminability of prefrontal activity between the target object and non-target objects was diminished (Fig. 4.5b, c). Based on these results, they concluded that the prefrontal cortex participates in the active filtering of visual stimuli presented in the unattended position and that this filtering is achieved by the top-down signal which is generated in the prefrontal cortex and modulates activity in the brain areas that provide visual information to the prefrontal cortex.

Buschman and Miller (2007) compared activities between prefrontal and posterior parietal neurons while the monkey performed visual attention tasks (Fig. 4.6). In these tasks, the monkey is required to find a target among four visual stimuli under

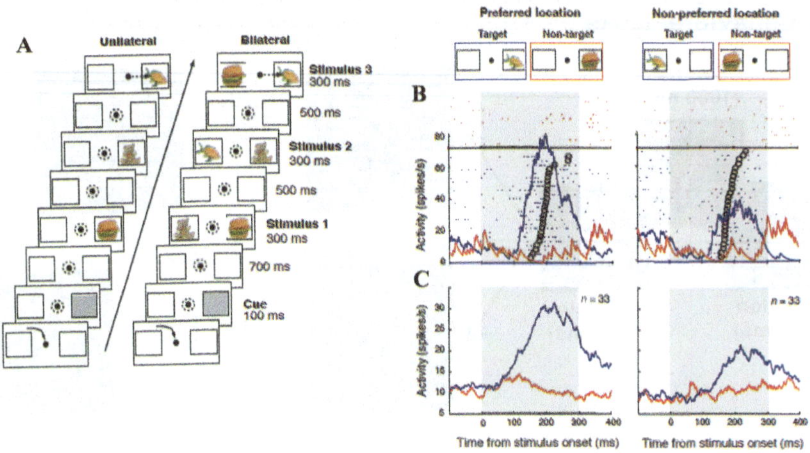

Fig. 4.5 (**a**) Schematic drawing of task events in the focusing attention task. (**b**) Prefrontal activity when the target object was presented at the preferred location (left) and at the non-preferred location (right). From Everling et al. (2002) with permission

two conditions: a visual "pop-out" (bottom-up) condition and a visual search (top-down) condition (Fig. 4.6a). While the monkey looks at a central fixation point, a sample stimulus is presented for 1 s over the fixation point. After a delay period of 500 ms, four stimuli are presented around the fixation point and the monkey is required to make a saccade to the stimulus (target) same as the sample stimulus. The target stimulus and distractors are bars with different colors and orientations. In the visual pop-out condition, the target stimulus and distractors are identical shapes, but distractors are different from the target stimulus in dimensions of color and orientation. Therefore, the salience of the stimulus automatically captures the attention of the target stimulus. In the visual search condition, each distractor independently differs from the target stimulus in dimensions of color and orientation. Since the target stimulus matches some distractors in either color or orientation, the monkey needs to use remembered information regarding the sample stimulus and intentionally select the correct target. They examined response timings of neurons that exhibited selectivity for the target location in two conditions. They found that prefrontal activity reflected the location of the target stimulus first during the top-down condition, whereas parietal activity reflected the location of the target stimulus first during the bottom-up condition (Fig. 4.6b, c). Thus, although both the prefrontal cortex and the posterior parietal cortex participate in the visual attention task, Buschman and Miller (2007) observed that the bottom-up signal appears first in the parietal cortex whereas the top-down signal appears first in the prefrontal cortex (Fig. 4.6b, c). Therefore, they concluded that the top-down signal arises from the prefrontal cortex when the subject performs tasks that require the top-down control of attention. The prefrontal cortex participates in the top-down control of visual

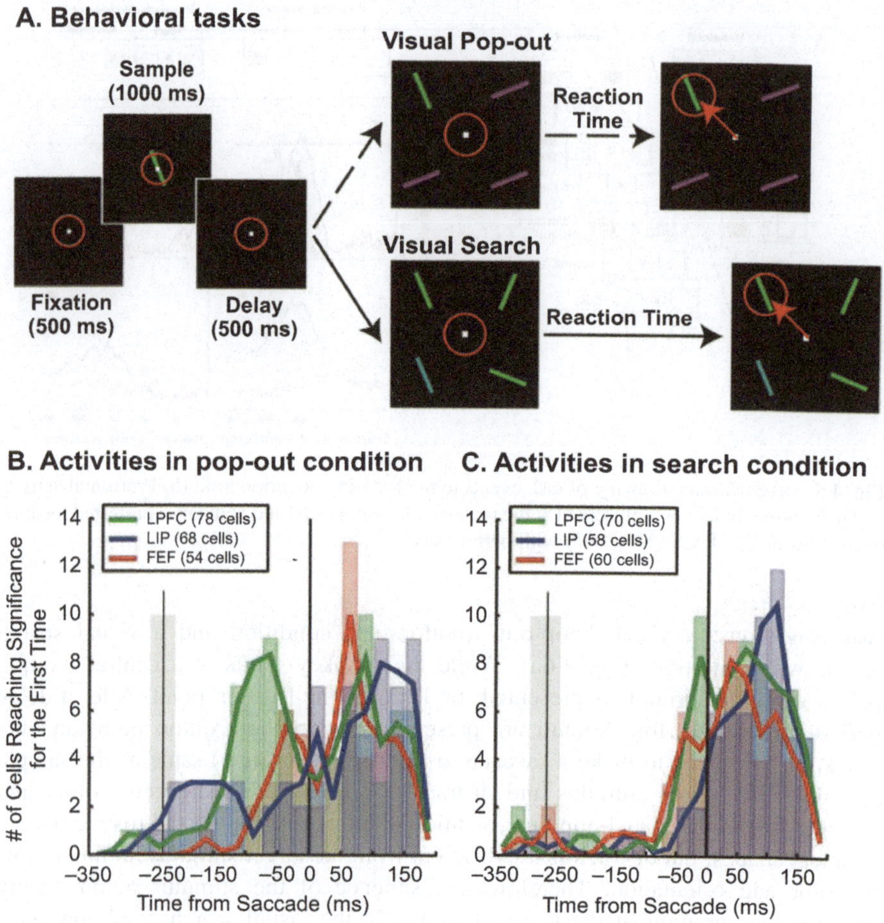

Fig. 4.6 (**a**) Schematic drawing of the visual attention task under two conditions. In the visual pop-out condition, bottom-up attention works, while in the visual search condition, top-down attention is required. (**b** and **c**) Comparisons of response timings of neuron activities in three brain areas between two conditions. In the pop-out condition, activation of LIP neurons was faster than LPFC neurons (**b**), while in the visual search condition, activation of LPFC neurons was faster than LIP neurons (**c**). From Buschman and Miller (2007) with permission

attention by sending top-down signals to the brain areas which provide visual information to the prefrontal cortex in order to filter necessary information from distractors.

In addition, Buschman and Miller (2007) analyzed the degree of synchrony of local field potentials recorded from the prefrontal cortex and the posterior parietal cortex. They found an increase in coherence between these two areas in the middle (22–34 Hz) and upper (35–55 Hz) frequency bands. A greater increase of coherence in the middle-frequency band occurred during the top-down condition, while a greater increase of coherence in the upper-frequency band occurred during the bottom-up condition. This result indicates that the synchrony of neural activity enhances the effectiveness of representation of attended stimulus and that the top-down control system uses coherence of different frequency bands from the bottom-up control system. Buschman and Miller (2009) further compared neural activity between the dorsolateral prefrontal cortex and the FEF using visual attention tasks. They found that the FEF is more involved in shifting attention to search for the target stimulus, while the dorsolateral prefrontal cortex is more involved in identifying the target stimulus once it is searched. They also found an increase in coherence of local field potentials in the middle frequency band (18–34 Hz) between two areas. Based on these results, Miller and Buschman (2013) concluded that top-down and bottom-up attentions are derived from frontal and parietal networks, respectively, and that the interactions of both networks are regulated through rhythmic oscillations.

Buschman and Miller (2007) showed that the prefrontal cortex participates in the top-down control of visual attention while the parietal cortex participates in the bottom-up control of attention by comparing neurons' response latencies to the target stimulus. However, Katsuki and Constantinidis (2012a, b) indicated that the dorsolateral prefrontal cortex involves in bottom-up visual attention as well. They used a delayed matching-to-sample task. In this task, nine square-shape colored stimuli were presented in a 3×3 grid during the sample period and the monkey was required to recognize the salient stimulus differed from other stimuli in color as a target stimulus and remember its position. After the delay, the monkey was required to make a response when the choice stimulus was presented at the same position as the target stimulus was presented.

Although attention-related effects have been known as an enhancement and shortening latency of neural responses to the target stimulus, these changes are caused by the change in sensitivity of neurons to the stimulus. Luo and Maunsell (2018) thought that attention-related changes are also caused by the subject's criterion, which corresponds to how readily the subject can report the presence of the target or withhold from such a report. They examined whether or not attention-related changes in neural activity in the lateral prefrontal cortex exhibits neural signal associated with the subject's criterion. The results showed that the response of prefrontal visual neurons was robustly modulated in association with the change of the subject's criterion, and that in prefrontal neurons, firing rates, spike count correlations between simultaneously recorded neuron pairs, and Fano factors were modulated in relation to the change of the subject's criterion, but not in visual cortex (area V4) neurons. Thus, the activity in lateral prefrontal neurons is affected by the subject's criterion for attention, which is closely related to top-down control of attention. This result supports the notion that the prefrontal cortex participates in top-down attention process.

Prefrontal participation in the top-down control of attention has also been demonstrated by human brain imaging studies. For example, Hopfinger et al. (2000) used a cued spatial attention task, in which the human subject is required to discriminate black and white checkerboards from those containing some gray checks at the cued location. Two checkerboards are presented at both sides of the central fixation point and the position where the subject pays attention (the cue) is indicated by yellow and blue arrows presented at the center of the screen. The subject is told which colored arrow he needs to attend for the session. They showed that the attention-directing cue (colored arrow) selectively activated the superior frontal, the inferior parietal, and the temporal cortices. Since the subject needs to attend to the particular color of the arrow throughout the session, they concluded that these brain areas consist of the network for top-down attention control. Egner et al. (2008) examined brain areas representing spatial- and feature-based search information and neural mechanisms for integrating top-down spatial- and feature-based information during visual search performance by fMRI. In their task, the subject is shown a cue stimulus followed by a search array, which consists of four diamond stimuli peripherally placed (one blue and one red diamond are placed on both sides of the central fixation point). The subject needs to identify a target diamond, which is missing either its upper or lower corner, and indicates which corner is missing by pressing a button. The cue stimulus informs the subject which side he needs to attend (spatial-based search information) or which color he needs to attend (feature-based search information). They found that spatial- and feature-based cue information were represented additively in the frontal and the parietal regions during the preparation period for the visual search. They suggested that this anticipatory integration observed in these regions is related to the generation of a "top-down salience map" (search template for primed target location and feature). Bressler et al. (2008) used Granger causality effects measured by BOLD signals while the subject performed a visual spatial attention task and showed that activity of the FEF and the intraparietal sulcus (IPS) area modulates the activity of the occipital visual areas. They showed that the top-down level of Granger causality from the FEF and the IPS to the occipital visual areas was significantly greater than the bottom-up level and the cortex-wide level. In addition, they showed that Granger causality was significantly greater from the FEF to the IPS than from the IPS to the FEF. These results indicate that the FEF and the IPS modulate activity of the occipital visual areas and the FEF modulates activity of the IPS in relation to visual attention. Thus, the prefrontal cortex including the FEF plays an essential role in selective visual attention and participates in the control of attention using top-down control signals.

4.4.2 *Inhibitory Control*

4.4.2.1 Anti-Saccade Task

The dorsolateral prefrontal cortex has been shown to play an important role to suppress inappropriate reflexive behavior. The anti-saccade task is often used to investigate the neural mechanism for flexible control of behavior and the ability to inhibit behavioral responses to inappropriate stimuli. In the anti-saccade task, the subject needs to suppress reflexive saccade to look at a visual stimulus that unexpectedly appeared in the peripheral visual field and make a saccade away from that stimulus. Therefore, the anti-saccade task has been thought to be the behavioral task that allows to examine neural mechanisms of inhibitory control as well as flexible control of behavior (Munoz and Everling 2004). In addition, since the subject actively suppresses prepotent reflexive saccade to the visual stimulus and must look away from the stimulus, the top-down inhibition of prepotent reflexive saccade is crucial to perform the anti-saccade task. Evidences have been shown that the dorsolateral prefrontal cortex plays important roles to perform the anti-saccade task and to achieve the top-down inhibition of inappropriate reflexive saccades.

In human studies, Guitton et al. (1985) compared performances of pro- and anti-saccade tasks between prefrontal patients and normal subjects. The visual cue is presented randomly at right or left of the central fixation point and includes a symbol with a gap. In the pro-saccade task, the subject is first required to make a saccade to the position where the visual cue is presented and then required to show the direction of the gap in the visual cue (left, right, up, or down) by the thumb. In the anti-saccade task, the subject is required to make a saccade opposite to the direction of the visual cue and indicate the direction of the gap when the visual cue appears at the correct target position after 200 ms or more later. They found that most prefrontal patients having lesions in the dorsolateral and mesial cortex exhibited long-term difficulty in suppressing a saccade to the visual cue in the anti-saccade task. Although these prefrontal patients eventually could make saccades toward correct positions in the anti-saccade task, the saccades toward the correct target were mostly triggered by the appearance of the target in both tasks. Saccade latency was short (80–140 ms) in the pro-saccade task compared to the latency of the correct saccade (170–200 ms) in the anti-saccade task. These results indicate that the dorsolateral prefrontal cortex plays important roles to suppress inappropriate reflex-like oculomotor behavior and to control and generate appropriate voluntary eye movements through complex computational processing operating in the prefrontal cortex. Similar results had been obtained by Pierrot-Deseilligny et al. (1991). They also used anti- and pro-saccade tasks and compared performances of these tasks among patients having limited unilateral lesions in several brain regions including the dorsolateral prefrontal cortex, the FEF, and the posterior parietal cortex. They found that the posterior parietal cortex is the main brain area which facilitates and triggers reflexive visually-guided pro-saccades. However, the error ratio in the anti-saccade task was significantly high in patients having lesions in the dorsolateral prefrontal cortex and the FEF. Therefore, they suggested that the prefrontal cortex is the responsible area which inhibits reflexive visually-guide saccade performances.

Funahashi et al. (1993) examined prefrontal single-neuron activity while the monkey performed a delayed pro-saccade task and a delayed anti-saccade task. In the delayed pro-saccade task, the monkey makes a memory-guided saccade after the delay to the position where the visual cue was briefly presented, while in the delayed anti-saccade task, the monkey makes a memory-guided saccade after the delay to the position opposite to the visual cue position. They found two groups of prefrontal neurons having delay-period activity. Delay-period activity of one group of prefrontal neurons represented the direction of the visual cue (stimulus-dependent), while delay-period activity of the other group of neurons represented the direction of the

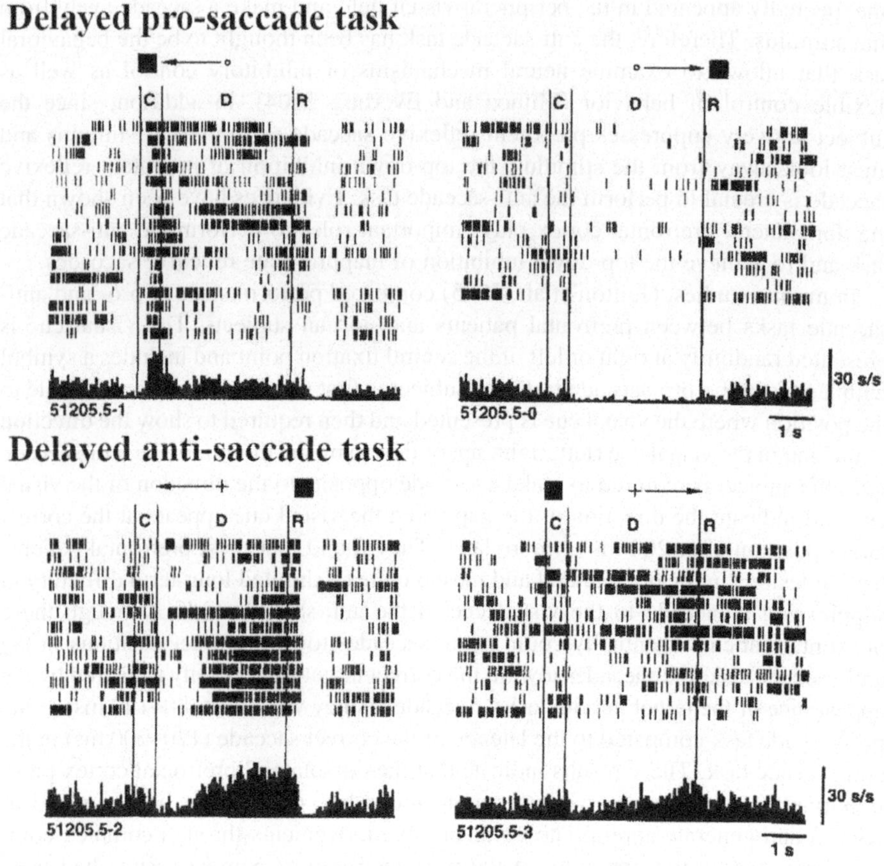

Fig. 4.7 The same prefrontal neuron's activities during delayed pro- and anti-saccade tasks. C, D, and R indicate cue, delay, and response periods. The length of the cue and the delay period were 0.5 and 3.0 s, respectively. Black squares and arrows indicate the position of the visual cue and the direction of the saccade. From Funahashi et al. (1993) with permission

saccade (saccade-dependent). They also found evidence that stimulus-dependent prefrontal neurons have a role to suppress inappropriate saccade responses. The neuron shown in Fig. 4.7 exhibited a complex firing pattern which was affected both by the location of the visual cue and the direction of the saccade. Delay-period activity was increased before the leftward saccades in both tasks, but was suppressed after the presentation of the right visual cue strongly in the pro-saccade task and transiently in the anti-saccade task. The suppression of delay-period activity was associated with the presentation of the visual cue on the right position and not with the rightward saccade. Therefore, the suppression observed in saccade-dependent prefrontal neurons might be driven by the inputs from stimulus-dependent prefrontal neurons. The inhibitory interactions may be present between stimulus- and saccade-dependent prefrontal neurons. Since the monkey was required to maintain fixation at the fixation point and suppress saccades during the delay period, these interactions may contribute to suppressing inappropriate behavioral responses during the delay period.

Zhou et al. (2016) examined prefrontal activity under three conditions of the anti-saccade task (overlap condition, zero-gap condition, and 100-ms-gap condition). In the overlap condition, the onset of the visual cue is 100 ms before the offset of the fixation point. In the zero-gap condition, the onset of the visual cue coincides with the offset of the fixation point. In the 100-ms-gap condition, the onset of the visual cue is 100 ms after the offset of the fixation point, which is the most difficult condition for the monkey because the monkey is required to suppress the eye movement until the onset of the visual cue without the fixation point. Prefrontal neurons exhibited visual responses when the visual cue was presented in the neuron's visual receptive field and the magnitude of visual response became maximum in the most difficult 100-ms-gap condition. Visually responsive prefrontal neurons were activated during anti-saccade trials, in which the visual cue was presented outside of the receptive field and the monkey needed to make a saccade toward the receptive field. And, the level of activation of these visually responsive neurons predicted the level of behavioral performance in the anti-saccade task. Response characteristics observed in these prefrontal neurons were not observed in posterior parietal neurons. Therefore, these results suggest that the prefrontal cortex participates in anti-saccade performances and that visually responsive prefrontal neurons play a role to suppress inappropriate reflexive saccade responses toward the visual stimulus in the anti-saccade task.

Express saccades (saccades with short latency between 70 and 120 ms) are also reflexive eye movements to salient visual stimuli that appeared in the peripheral visual field and are used to examine neural mechanisms of response inhibition. Müri et al. (1999) used a saccade gap task, in which the peripheral visual target appears 200 ms after the offset of the fixation point. They reported that transcranial magnetic stimulation (TMS) at the dorsolateral prefrontal cortex was able to shorten the latency of contralateral visually guided saccades but had no effect on the latency of ipsilateral visually guided saccades. This shortening of the saccade latency was caused by a significant increase in the ratio of contralateral express saccades only when TMS was applied at the end of the 200-ms gap in the saccade gap task. However, earlier TMS applied at the dorsolateral prefrontal cortex had no effect on visually-guided saccades. In addition, TMS applied to the posterior parietal cortex had no effect on the saccade latency during the performance of the saccade gap task, suggesting that TMS interfered with specific functions of the dorsolateral prefrontal cortex by inhibiting the activity of this region. Express saccades are reflexive responses to salient visual stimuli, and the superior colliculus has been known to be a responsible region for the generation of express saccades. A brief suppression of the prefrontal activity by an application of TMS at the onset of the visual stimulus may disinhibit the activity of the superior colliculus, causing an increase of reflexive express saccades. Therefore, these results suggest that the dorsolateral prefrontal cortex controls activity of the superior colliculus and suppresses generation of inappropriate reflexive saccades. Nyffeler et al. (2007) further examined the effect of single-pulse TMS applied at the dorsolateral prefrontal cortex on the inhibition of reflexive saccades during the anti-saccade performance. TMS was applied either 100 ms before, at the same time, or 100 ms after the target onset. As a result, TMS applied 100 ms before the target onset significantly increased reflexive saccades toward the target in the anti-saccade task, suggesting that the critical period that the prefrontal cortex suppresses reflexive saccades in the anti-saccade task is just before target onset and that correct anti-saccade performance is not directly controlled by the prefrontal cortex but is controlled by suppressing inappropriate activity in other brain areas, such as the superior colliculus, by prefrontal activity.

4.4.2.2 Stop-Signal Paradigm

The stop-signal paradigm is used to directly investigate the ability to voluntarily inhibit behavioral responses. In this paradigm, the subject is asked to make a response as quickly as possible when an imperative Go stimulus is presented. In some trials, a "stop signal" is presented just after Go stimulus presentation and instructs the subject to stop and inhibit the ongoing response. The main interest is whether the subject is able to stop the response or not after the presentation of the stop signal. It has been shown that the probability to stop the ongoing response depends on "stop-signal delay (SSD)," which is the time available to interrupt response preparation. The SSD controls the amount of time to detect the stop signal and countermands the response before the response is executed. Human neuroimaging studies showed that an ability to stop the response in the stop-signal paradigm is associated with the activation of the fronto-basal-ganglia circuit, which includes the ventrolateral prefrontal cortex (inferior frontal gyrus), the dorsolateral prefrontal cortex (middle frontal gyrus), and the pre-supplementary motor area (Verbruggen and Logan 2008). It has been shown that the right inferior frontal gyrus is involved in stopping and that this region increases activation when stopping is successful. In monkey studies, it has been shown that the monkey is able to perform a saccade version of the stop-signal task and that neurons in the frontal eye field, the basal ganglia, and the superior colliculus participate in stopping the response in the performance of the saccade stop-signal task (Schall et al. 2017).

4.4.2.3 No-Go Dominant Activity

A go/no-go discrimination task with hand or arm movements has been used to examine neural mechanisms of response inhibition. Sasaki and Gemba (1986) analyzed field potentials recorded from the frontal lobe while the monkey performed a simple reaction-time task with discrimination between Go and No-Go light signals. Significant field potential was observed in the principal sulcus area specifically during No-Go trials (Fig. 4.8). This No-Go field potential appeared 110–150 ms after the cue presentation and 150–210 ms earlier than the initiation of the Go response. No-Go field potential appeared only during No-Go trials irrespective of

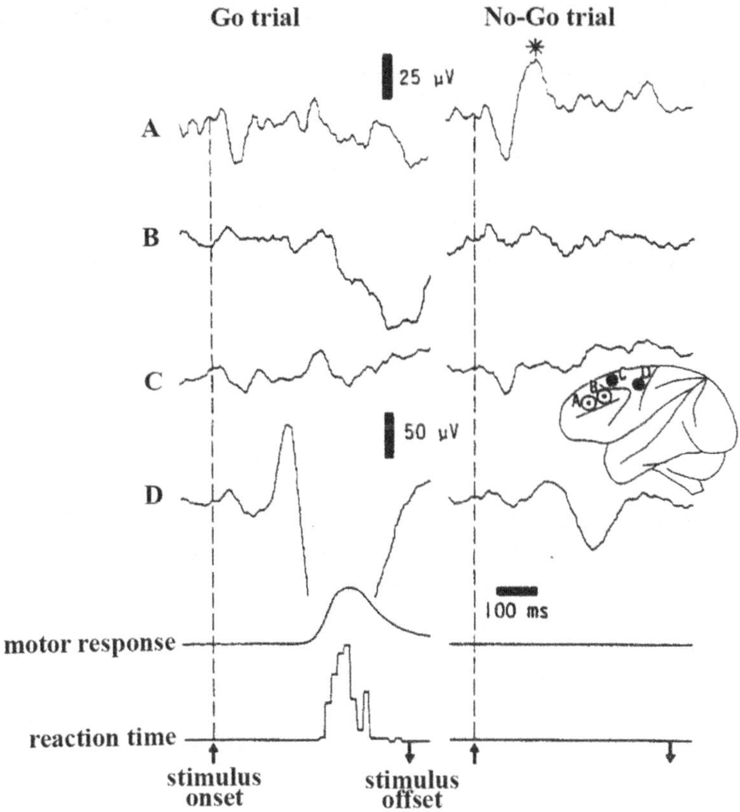

Fig. 4.8 No-go field potentials (shown with an asterisk) observed in no-go trials recorded at the principal sulcus cortex. A, B, C, and D indicate field potentials recorded at the rostral and the caudal parts of the principal sulcus, the premotor cortex, and the motor cortex, respectively. These potentials were the result of a summation of 100 trials. From Gemba and Sasaki (1990) with permission

different colors (Sasaki and Gemba 1986) or different frequencies of auditory stimuli (Gemba and Sasaki 1990) used as Go/No-Go signals. Similar No-Go field potential was observed in the frontal lobe of human subjects (Gemba and Sasaki 1989). The flow pattern of magnetic fields similar to No-Go field potential was observed in the human prefrontal area by MEG measurement while the human subject performed Go/No-Go reaction time tasks (Sasaki et al. 1993). In addition, Konishi et al. (1999b) examined No-Go specific activation in the brain using an event-related fMRI method while the human subject performed a Go/No-Go discrimination task and found transient No-Go dominant activation in the posterior part of the inferior frontal sulcus in the right hemisphere. Observed No-Go dominant activation was irrespective of whether the subject used the right or the left hand for making responses. Further, Sasaki et al. (1989) applied brief electrical pulses to the locus where No-Go field potential was recorded in the monkey prefrontal area and found that the electrical stimulation suppressed Go movements by cancelling the movement or delaying the initiation of the movement. These results indicate that No-Go field potentials represent voluntarily inhibitory processes, which revoke the motor command initiated by external stimuli, and show that the dorsolateral prefrontal cortex participates in active inhibition of movement initiation when the subject is required to actively suppress a particular response behavior.

4.4.2.4 Conflict Conditions

The attention-inhibition process (Smith and Jonides 1999) and switching attention from one source to another (Baddeley and Della Salla 1998) are typical situations in which inhibitory control is necessary. Inhibitory control can be best observed when two processes or stimuli are in conflict condition. Under the conflict condition, one needs to exert active inhibition or "executive inhibition" (Perner and Lang 1999) to an inappropriate response. The Stroop test is a typical test to examine this capacity in human subjects. In the Stroop test, the subject is shown a card on which the name of a color is printed in a different color (e.g., the word "green" is printed in red, "**green**"). The subject is required to report the name of the printing color (e.g., red). To perform this test successfully, the subject must pay attention to the relevant stimulus (the painting color) and inhibit the irrelevant but prepotent response (answering the printed word). Therefore, the subject takes more time to make a correct answer "red" and often makes errors when the word "green" is printed in red color (**green**) than when the word "red" is printed in red color (**red**). Human patients having lesions in the right dorsolateral prefrontal cortex are known to have more errors (Vandrell et al. 1995) compared with normal subjects. Human functional imaging studies have shown that, although numerous brain areas are activated while the subject performed the Stroop test, the area commonly activated among these studies is the anterior cingulate cortex (Peterson et al. 1999). The anterior cingulate cortex is thought to monitor conflicting processes during task performances (Carter et al. 1998; Botvinick et al. 1999).

Another example to test the active inhibitory process is a version of Luria's hand game (Luria 1969). In this task, the subject is required to perform a different behavior from what the experimenter does. For example, if the experimenter presents the pointed finger, the subject is required to make a fist, or if the experimenter taps once, the subject needs to tap twice, and vice versa. The subject's prepotent behavioral response would be an imitation of the experimenter's behavior. However, to perform this game correctly, the subject needs to pay attention to the experimenter's behavior, suppress to generate an irrelevant prepotent behavior, and produce a relevant behavior. These processes are similar to those for correct performances of the Stroop test. Luria (1969) reported that a patient having damage in the left frontal lobe could not perform this task correctly even though she could repeat the experimenter's verbal instruction for this task.

A modified version of the rock-paper-scissors (RPS) game has also been used to examine the inhibitory control process in conflict situation. In the modified RPS game, similar to Luria's hand game, the subject is asked to keep either winning or losing the game. In the losing condition, if the experimenter presents "rock," the subject needs to present "scissors," for example. When the patient having a lesion in the prefrontal cortex was asked to keep losing the RPS game, he could not inhibit his stereotype behavior, such that he showed difficulty to keep losing the game and exhibited a strong tendency to win the game. Kadota et al. (2010) used a modified version of the RPS game and examined fMRI activation in two groups (win group and lose group) of normal human subjects. The loss group exhibited greater activation in the left dorsolateral prefrontal cortex. Application of TMS on the left dorsolateral prefrontal cortex before the stimulus presentation produced erroneous win responses when the subject was asked to keep losing the game. These results suggest that the dorsolateral prefrontal cortex plays a role to suppress the generation of stereotype behaviors.

Matsushima and Tanaka (2012) showed important functions of the prefrontal cortex in behavioral control under conflict condition. They used a covert tracking of two moving objects during fixation of a fixation point and examined single-neuron activity in the monkey lateral prefrontal cortex. In this task, while the monkey looks at the fixation point, two white circles are presented simultaneously and, after a brief delay, one of them changes its color (red) for 300 ms and is back to white, indicating that this circle is the target for a saccade. Then, two circles simultaneously move around the fixation point while the monkey is maintaining fixation at the fixation point. After two circles stop moving, the monkey is required to make a saccade to the target circle. Therefore, the monkey needs to covertly track the target circle while maintaining fixation. In the covert tracking of multiple moving stimuli, the subject needs to monitor movements of both the target and distractors and continuously discriminate the target from distractors. Analyses of single-neuron activity revealed that, although many neurons responded to the target circle and were insensitive to the distractor circle which is visually identical to the target circle, many other neurons responded selectively to the distractor circle. On the other hand, in the overt tracking of a moving stimulus, prefrontal neurons are generally insensitive to the distractors. These results suggest that the lateral prefrontal cortex provides at least two top-down signals. One population of prefrontal neurons that were sensitive to the target circle but insensitive to the distractor would participate in enhancing visual processing of the target by providing top-down signals to the posterior cortical areas, whereas the other population of prefrontal neurons which were sensitive to the distractor would participate in selectively suppressing visual processing of the distractor by providing top-down signals to the posterior cortices. The covert tracking of multiple moving stimuli seems to be a condition similar to the conflict condition. The results of Matsushima and Tanaka (2012) indicate that in the conflict condition, the prefrontal cortex not only provides top-down excitatory signals to enhance information processing for the target stimulus in the posterior cortices but also provides top-down signals to suppress information processing of distractors.

4.4.3 Task Management

When the subject performs multiple tasks simultaneously, task management is necessary to determine which task has higher priority and which task performs first. Task management is an important function under executive control. Since the amount of neural resources available for information processing to achieve cognitive tasks is limited, it is difficult to perform multiple cognitive tasks simultaneously without any interference. Therefore, it often happens that one task is completed without any trouble while the other task is often performed slowly and unsatisfactorily. Thus, when multiple tasks are performed simultaneously without any interference, a mechanism for task management is necessary.

The dual-task paradigm has been used to investigate how task management is achieved. In the dual-task paradigm, the subject is asked to perform two different cognitive tasks simultaneously. The dual-task paradigm has been used to examine functions of the central executive, which is an important component of the working memory model proposed by Baddeley and Hitch (1974). Baddeley and Della Salla (1998) and Logie (1995) defined that executive control is the mechanism to achieve coordinated operation of multiple control systems which work simultaneously. Since executive control is thought to be the main function of the central executive of the working memory model, Baddeley (1986) proposed that the task management mechanism is best analyzed under the dual-task condition.

D'Esposito et al. (1995) investigated activation areas in the cerebral cortex during dual-task conditions by fMRI. In one task, the subject is required to determine whether the word presented is a vegetable or not. In the other task, the subject has to decide whether two visual stimuli differ only with regard to rotation. In the dual-task condition, the subject is required to perform these two tasks concurrently. They found that only the frontal area, especially the dorsolateral prefrontal area (mainly area 46), was activated specifically in the dual-task condition, suggesting that the dorsolateral prefrontal cortex plays a specific role in task management in the dual-task condition. Activation in the dorsolateral prefrontal cortex specifically during the dual-task condition had also been observed by Collette et al. (2005). The prefrontal contribution to the dual-task performance and in the mechanism of task management is further discussed in the Chap. 6 "*Cognitive Control.*"

Another important function of task management is switching from one task to another at appropriate timing and condition to achieve the currently relevant goal. When initiate one particular task and when switch from one task to another are important processes for task management. Genovesio et al. (2005) examined how prefrontal neurons switch from one strategy to another depending on the difference in task conditions. If the same alphabetical symbol as that in the preceding correct trial is presented in the current trial, the monkey is required to select the same target as the one selected in the preceding trial (repeat-stay strategy). However, if a different alphabetical symbol is presented, the monkey is required to select a different target from the one selected in the preceding trial (change-shift strategy). They found strategy effect on task-related activity in 16% of prefrontal neurons

(145/907). In these neurons, a stronger response was observed after the presentation of the alphabetical symbol in the one strategy condition (e.g., repeat-stay strategy) compared with the other condition (e.g., change-shift strategy), regardless of the difference in alphabetical symbols. The strategy effect was not the result of the detection of the difference of the stimuli or repetitive presentation of the same stimulus. Therefore, the strategy effect on task-related prefrontal activity may contribute to manage the selection of the behavioral response depending on the difference in the strategy. Tsujimoto et al. (2011, 2012) used a similar behavioral task as Genovesio et al. (2005) used and examined strategy-related prefrontal activity. In their task, the monkey is required to make a saccade toward either right or left direction based on the visual cue presented over the fixation point. The saccade direction is either the same as the direction rewarded in the preceding trial (stay trial) or the opposite to the direction in the preceding trial (shift trial). The difference in the color or the orientation of the visual cue indicates which strategy the monkey needs to take in the current trial. They found that neurons in the dorsolateral prefrontal cortex encoded either strategies (stay or shift) or response directions (right or left) or both during the delay period. In dorsolateral prefrontal neurons, strategy-related activity was modality specific, while response direction-related activity was modality general. On the other hand, neurons in the orbitofrontal cortex encoded the strategy. However, neurons in the frontal pole encoded neither strategy nor response direction. These results further confirm that the dorsolateral prefrontal cortex participates in task management by representing the behavioral strategy to select the correct response and that the strategy effect of task-related activity contributes to manage the selection of the behavioral response depending on the difference of the strategies.

Tsutsui et al. (2016a, b) also examined how prefrontal neurons participate in task switching and behavioral selection. They prepare three sets of visual stimuli, each of which includes 8 complex abstract figures. Eight visual stimuli included in each set are further divided into two groups (A and B). In each trial, one visual stimulus is selected from either group A or B and presented to the monkey. The monkey is required to perform an appropriate response (licking or avoiding licking) after the delay period based on the presented visual stimulus. Stimuli in group A are associated with liquid reward delivery (licking; Go response) and stimuli in group B are associated with saline delivery (avoiding licking; No-Go response) in one condition. However, the contingency between stimulus groups and delivered reward is reversed in other conditions. Therefore, the monkey needs to categorize each visual stimulus into two groups (making either Go or No-Go response; functional categorization) based on the rule but not based on physical differences of visual stimuli within each group. In fact, they found that single neurons in the dorsolateral prefrontal cortex encoded not the physical feature but the functional category of visual stimuli and that information of functional category was encoded independently from the rule and the contingency between stimulus groups and delivered reward. In addition, contingency information was derived by the integration of category information and rule information. These results suggest that neural processes encoding category information, rule information, and contingency between stimulus and reward are needed to

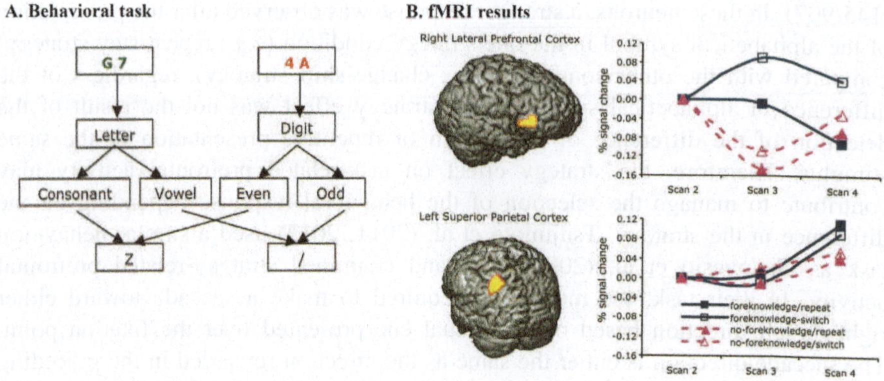

Fig. 4.9 (**a**) Behavioral task. The subjects needed to perform either a letter task or a digit task depending on the color under either foreknowledge condition or no foreknowledge condition. (**b**) Activation patterns observed in the right lateral prefrontal cortex and the left posterior parietal cortex during task performances. From Sohn et al. (2000) with permission

control necessary behavior flexibly based on the task requirement and indicate that the dorsolateral prefrontal cortex participates in this neural processing.

Participation of the lateral prefrontal cortex in task switching has also been shown by human neuroimaging studies. Sohn et al. (2000) asked human subjects to perform a switching task and investigated activated brain areas by fMRI. In their task, a stimulus consists of one letter and one digit with either red or green color (e.g., "4 A") (Fig. 4.9). If the color is red, the task is a digit task and the subject needs to answer whether the digit is even or odd by pressing a corresponding button. If the color is green, the task is a letter task and the subject needs to answer whether the letter is a consonant or a vowel by pressing a button. The subject performs two tasks successively in either foreknowledge condition (the subject knows whether the task switch occurs or not in advance) or no foreknowledge condition. They found that the lateral prefrontal cortex (areas 46 and 45) was activated during the repetition of the same task and exhibited stronger activation in the foreknowledge condition. In addition, this prefrontal area was also activated in the task switching condition and exhibited stronger activation in the no foreknowledge condition. Stronger activation of the lateral prefrontal cortex in the task switching condition was also observed in an event-related fMRI study (Dove et al. 2000).

Task management is necessary not only when multiple tasks are needed to perform simultaneously but also when serial decision-making is needed within a task (e.g., preparing a particular dish for dinner). Although the neurobiological basis of single decision-making has been examined widely, how the brain can link multiple decision-making is an important issue to understand how the brain manages a complex sequential behavior coherently. Abzug and Sommer (2018) used monkeys and examined neural mechanisms for serial decision-making. They used a two-stage sequential decision task, in which each trial consists of the rule selection stage followed by the rule implementation stage. In the rule selection stage, two colored stimuli are presented on the right and the left of the fixation point, which indicate the rules. In the self-selected trials, two different colored stimuli are presented and indicate two different rules, respectively (e.g., a green stimulus corresponds to the size rule, while a red stimulus corresponds to the brightness rule). The monkey itself needs to select either the size rule (pick the smaller target) or the brightness rule (pick the darker target) by making a saccade to either colored stimulus. In the instructed trials, either two green stimuli or two red stimuli are presented at both sides of the fixation point and the monkey needs to select the stimulus at either side of the fixation point by a saccade. If two green stimuli are presented, the selection in the trial is based on the size rule. If two red stimuli are presented, the selection is based on the brightness rule. In the rule implementation stage, two square targets are presented on the left and the right of the fixation point and these two squares vary in size and brightness. The monkeys are required to make a saccade to either target based on the rule which was selected in the rule selection stage. They analyzed single-neuron activity in the supplementary eye field (SEF). The reasons why they examined activity in the SEF are (1) SEF activity anticipates an expected value of the reward when reward probability and reward size are provided explicitly, (2) visual responses of SEF neurons encode forthcoming saccade directions well before the saccade initiation, and (3) SEF activity predicts the performance of upcoming saccade. They found (1) SEF neurons encoded the spatial information of the saccade during the rule selection stage (decision of the saccade direction), but not during the rule implementation stage (discrimination of visual stimuli), suggesting the flexibility of movement-relating tuning in SEF activity. (2) SEF activity linked the sequential decisions of the rule selection and the discrimination of visual stimuli, but not continuously. (3) The rule-coding activity appeared in a just-in-time manner before the rule implementation stage. (4) SEF neurons encoded trial outcomes prospectively before decisions within a trial and retrospectively across multiple trials. These results clarified neural correlates of rule selection and its application and transient signals linking sequential decisions in the SEF and suggested that the SEF participates in serial decision-making as part of a broad network. Although they examined only activities of SEF neurons while the monkey performed a serial decision-making task, they suggested that "As one of the foundations of continuous behavior, serial decision-making merits further study and it will be important to investigate the contributions of other candidate brain areas and how they work with the SEF to create the internal representations that link cognitive operations over time." The prefrontal cortex must be an important candidate brain area. Further studies are needed in the prefrontal cortex.

4.4.4 Shifting Stimulus or Action Categories

Task management in the performance of complex tasks includes a process for flexibly shifting stimulus categories depending on the current goal. The Wisconsin card sorting test is a typical example (see Fig. 4.1a). Flexible shifting of sorting category to find the correct sorting category and maintaining the correct category across trials are necessary for this test. Konishi et al. (1998, 1999b) examined brain areas specifically related to the shift of sorting category during the performance of the Wisconsin card sorting test by event-related fMRI methods (Fig. 4.10). They found that the transient activation at the bilateral dorsolateral prefrontal cortex was associated with the timing of shifting sorting category in the Wisconsin card sorting test (Konishi et al. 1998). They also found that the same prefrontal areas were activated even when the subject was informed of the new correct category of stimulus attributes (i.e., working memory is needed, but not category-shifting) (Konishi et al. 1999a). Therefore, they concluded that category-shifting and working memory could act cooperatively in the same area in the dorsolateral prefrontal cortex (Konishi et al. 1999a). Konishi et al. (1999b) also found transient no-go dominant activation in the posterior part of the right inferior frontal sulcus by transient fMRI analysis. This area spatially overlapped the area related to category-shifting in the Wisconsin card sorting test (Konishi et al. 1998). Therefore, shifting category from one stimulus attribute to another may be closely related to the inhibition of behavioral responses to inappropriate stimuli. These results indicate that the same area within the dorsolateral prefrontal cortex participates in task management as well as attention-inhibition processes.

Nakahara et al. (2002) used a modified Wisconsin card sorting task and examined task-related brain areas, especially related to functions for shifting the category of the stimulus for the right selection, in both humans and monkeys by fMRI. They found transient activation related to shifting the stimulus category in the posterior portion of the ventrolateral prefrontal cortex, which was the cytoarchitectonically equivalent region in both humans and monkeys, indicating that the posterior portion of the ventrolateral prefrontal cortex participates in flexible shifting of the stimulus category for the right selection in the Wisconsin card sorting test. Kamigaki et al. (2012) further examined neural mechanisms for shifting the stimulus category in the prefrontal cortex by recording single-neuron activity while the monkey performed a modified Wisconsin card sorting test. They found a group of neurons that exhibited activity selectively when the monkey was required to shift the stimulus category. These neurons were localized in the posterior bank of the inferior arcuate sulcus. Further, the inactivation of this region by the local injection of muscimol selectively impaired the performance of shifting the category from one stimulus category to another. These results indicate that a localized region in the prefrontal cortex participates in the performance which requires flexible shifting of the category for the right selection. Thus, the experiments using the Wisconsin card sorting test indicate that the prefrontal cortex, especially the posterior portion of this cortex, contributes to behavioral performances which require flexible shift between stimulus categories and between action categories for the right selection and that these functions are also closely related to the top-down control of attention and the task management.

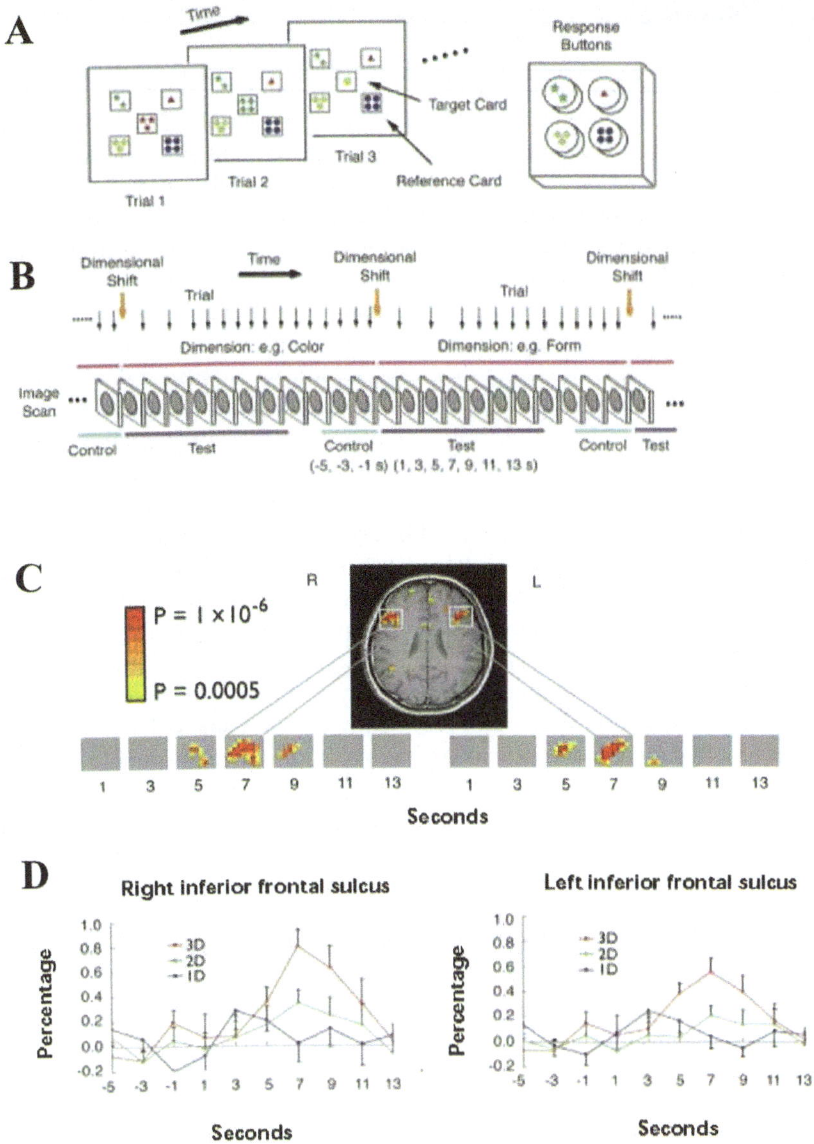

Fig. 4.10 (**a**) Schematic drawing of a computerized Wisconsin card sorting test. (**b**) Data processing strategy under the event-related fMRI method. Each arrow indicates each trial. The time zero is the time immediately after the shift of the sorting category. (**c**) Shift-related activation observed in the bilateral inferior frontal sulci. (**d**) The temporal change of the sift-related signal in the right and left inferior frontal sulci. From Konishi et al. (1998) with permission

4.4.5 Planning

The essence of planning is to maintain a goal throughout a series of intermediate steps. Unterrainer and Owen (2006) described two basic requirements for successful planning; (1) creating a mental representation of the current situation and the goal and (2) establishing which actions are needed to transform the current situation into the goal state. Therefore, to understand the neural mechanism of planning, we need to know how the initial situation and the goal are recognized, how the goal is maintained throughout the process, how a series of actions going through intermediate steps are created, and how the current state is transformed into the next state.

In order to examine neural mechanisms of planning, the Tower of London task has often been used (Fig. 4.1b). In this task, the subject is presented with both the initial configuration and the goal configuration constructed by a set of three colored beads. The subject is asked to rearrange the beads in the initial configuration and match the configuration to the goal configuration in the minimal number of beads' moves (Shallice 1982). To solve this task in the minimal number of moves, the subject first needs to plan the sequence of beads' moves before starting the response behavior. Early incorrect moves often cause unsolvable problems, because the subject needs to retrace and reverse all previous moves to correct incorrect moves. Therefore, the Tower of London task is a typical task requiring planning. The difficulty of the task can be manipulated by varying the initial configuration with respect to the goal configuration.

Shallice (1982) and Owen et al. (1990) showed that patients having prefrontal damages required more moves and more time to complete the Tower of London task. Owen et al. (1990) showed that the amount of time spent for thinking before the first move was not different between patients having prefrontal damages and normal controls. However, patients having prefrontal damages spent significantly more time to think during the rest of the moves. Behavioral impairment of the Tower of London task has been observed in patients having frontal lobe lesions and frontal lobe dysfunctions and in patients having schizophrenia (Unterrainer and Owen 2006). This observation indicates a significant contribution to the prefrontal cortex in planning for making actions, especially for establishing a step of future actions.

Functional neuroimaging studies also show the focal activation of the lateral prefrontal cortex while human subjects performed the Tower of London task. Baker et al. (1996) examined neural systems related to planning by a PET while human subjects performed the Tower of London task. They showed significant activation of the rostral prefrontal cortex in addition to the cortical areas related to visuospatial working memory processes. In addition, enhanced activation was observed in these areas in association with the increase in the task difficulty. Owen et al. (1996a, b) reported activation of the mid-dorsolateral frontal cortex by PET when the subjects performed the Tower of London test. Dagher et al. (1999) also showed the contribution of the dorsolateral prefrontal cortex (areas 9 and 46) bilaterally to the tower of London task by PET. Brain regions where the level of the regional cerebral blood flow was correlated with the complexity of the task were the premotor cortex (area

6), the anterior cingulate cortex (areas 32 and 24), in addition to the dorsolateral prefrontal cortex. They suggest that these brain areas form a network for motor planning that requires both visual processing and execution of a series of movements. Focal activation of the lateral prefrontal cortex during the tower of London performance has also been confirmed by fMRI studies (van den Heuvel et al. 2003; Newman et al. 2003; Unterrainer and Owen 2006; Wagner et al. 2006; Kaller et al. 2011).

On the other hand, Mushiake et al. (2001) developed a maze-based planning task (path finding task) for monkey studies and examined in the prefrontal cortex how the initial situation and the goal are recognized, how the goal is maintained throughout the process, how series of actions going through intermediate steps are created, and how the current state is transformed into the next state (Fig. 4.11). Similar to the Tower of London task, the monkey is shown the initial position by the cursor and the goal position by the visual stimulus on the maze. The monkey is required to move the cursor on the maze by rotating two handles by bilateral wrist rotations, to find an appropriate path to reach the goal, and to acquire the goal in the minimal number of cursor moves. To perform this task, the monkey first needs to learn the association between the direction of the cursor movement and the direction of the wrist rotation (supination or pronation) (Fig. 4.11a). After the monkey learned this association, both the initial cursor position and the goal position are informed on the maze. The monkey is required to find the path and acquire the goal by moving the cursor by rotating both wrists appropriately (Fig. 4.11a). They found that the monkey always took the shortest path to acquire the goal. They also found that the monkey changed the path to the goal flexibly and adaptively when the distracters are set on the way to the goal. Then, Mushiake et al. (2006) examined prefrontal activity while the monkey performed the path-finding task and found that prefrontal neurons encoded each of forthcoming directions of the cursor movement, but not the directions of the wrist rotation, during the preparation period (before the initiation of the first cursor movement) (Fig. 4.11b). In this task, the monkey usually performed three cursor movements to acquire the goal. Mushiake et al. (2006) found three types of prefrontal activity during the preparation period, which represented the direction of either the first, the second, or the third forthcoming cursor movement (Fig. 4.11b). They also examined the activity of neurons in the primary motor cortex during the performance of the path-finding task and found that the activity of all motor cortex neurons represented the directions of the wrist movement (either supination or pronation). Further, Saito et al. (2005) found two kinds of delay-period activity that were observed between the presentation of the goal and the initiation of the cursor movement. One type of delay-period activity represented the position of the immediate goal on the maze that the monkey needed to move the cursor in the first step. The other type of delay-period activity represented the position of the final goal on the maze. In addition, Mushiake et al. (2002) observed prefrontal activation in an event-related fMRI study while human subjects performed the same path-finding task as used in the monkey studies. Thus, these results indicate that prefrontal activity represents the spatial information to acquire the forthcoming intermediate subgoal in a series of behavioral responses as well as maintains the spatial

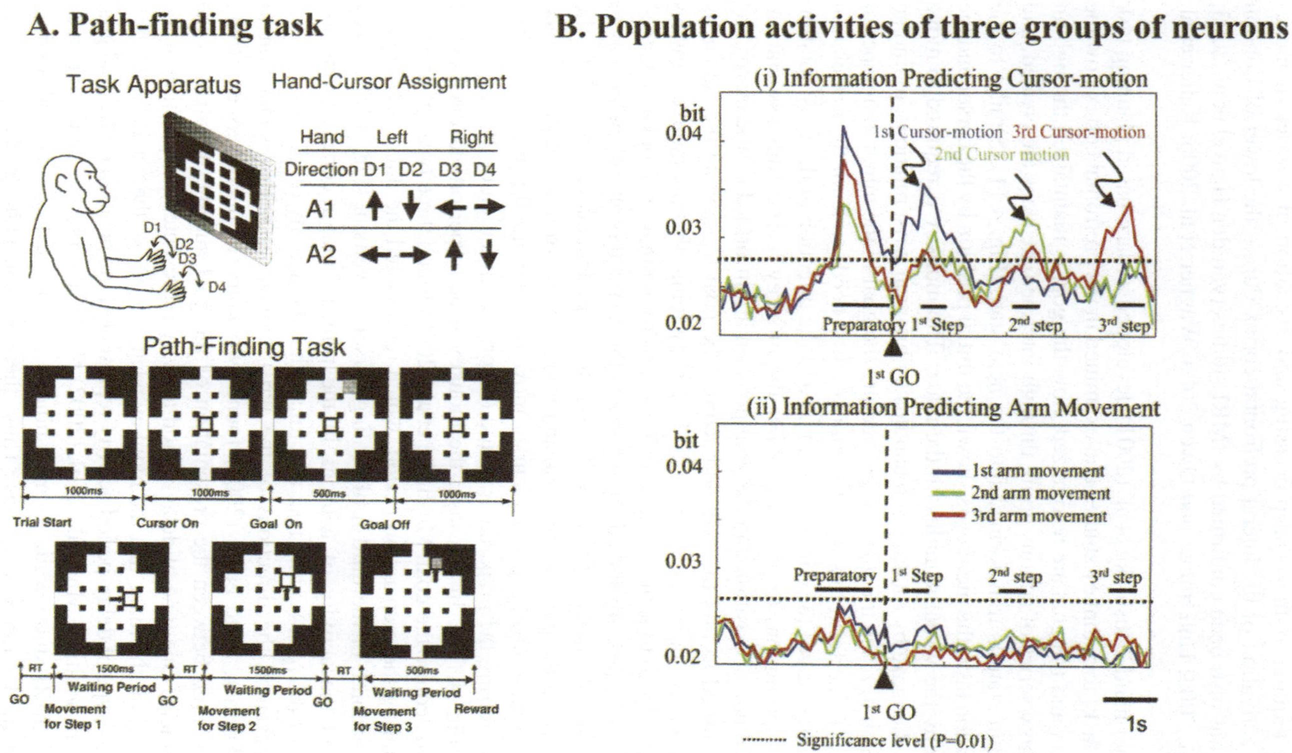

Fig. 4.11 (**a**) Schematic drawings of task apparatus, hand-cursor assignments, and step-by-step cursor movements in the path-finding task. The monkey moved a cursor on the maze by rotating two handles. (**b**) Examples of three groups of prefrontal neurons whose activities encoded different phase of the cursor-motion sequence. Each group of neurons encoded information regarding the direction of predicting cursor-motion at either the first, second, or third phase (*i*) but did not encode information regarding predicting arm movement to move the cursor at each phase (*ii*). (**a**) From Mushiake et al. (2001) with permission. (**b**) From Mushiake et al. (2006) with permission

information of the final goal, that different groups of prefrontal neurons encode different spatial information necessary at different timing (e.g., rightward cursor direction at the second step), and that the combination of activation in different groups of prefrontal neurons contributes to preparing the direction of the forthcoming behavior. The neural mechanism of planning would be a process produced by the combination of activation in different groups of prefrontal neurons to prepare the forthcoming series of actions.

Genovesio et al. (2006) also observed results similar to those of Saito et al. (2005), although their task was not a navigation task. In their task, three visual stimuli (saccade targets) are presented in the peripheral positions and one of two different symbolic visual cues is presented over the fixation point. The monkey needs to select one of three visual targets (future goal) by a saccade based on the combination of the symbolic visual cue that appeared in the preceding trial and the saccade target selected in the preceding trial. When the symbolic visual cue in a given trial is the same as that in the preceding trial, the monkey is required to make a saccade to the same target as in the preceding trial. When a different visual cue is presented, the monkey needs to select one of the two remaining targets in the current trial. They found that prefrontal neurons encoded the position of either the saccade target of the preceding trial or the saccade target of the current trial and that a few neurons encoded both target positions. These results suggest that separate prefrontal networks encode spatial information either for previous goals or for future goals. This segregation might contribute to monitoring which goal has been accomplished and which goal remains pending. This monitoring is important for planning.

Further, Shima et al. (2007) asked monkeys to perform a series of four movements, and examined the activity of lateral prefrontal neurons. Each movement was push, pull, rightward, and leftward turns of a handle by one hand. Three categories of the sequence of four movements were constructed; paired sequence consists of two movements (e.g., turn-turn-push-push), the alternate sequence consists of the alternation of two movements (e.g., pull-turn-pull-turn), and four-repeat sequence requires repetition of the same movement four times (e.g., pull-pull-pull-pull). They found that lateral prefrontal neurons exhibited sustained activation during the preparation period only when a specific category of sequential movements was requested and that this activation was not specific to a particular combination of hand movements but depended only on the difference in the category of sequential movements. Further, this activation was observed during the preparation period before the initiation of the sequential movements and was not observed in error trials. Therefore, the activation observed during the preparation period in lateral prefrontal neurons is considered to represent the preparation of a specific forthcoming action sequence, which corresponds to a neural correlate for planning and preparing a complex sequence of actions, not for execution of these actions.

As indicated previously, successful planning requires creating a mental representation of the current situation and the goal and establishing what actions are needed to transform the current situation into the goal state. A series of studies show that the activity of individual lateral prefrontal neurons represents information regarding either the initial situation, the current situation, or the goal state, and also represents information regarding actions for each intermediate step to reach the goal, indicating that the lateral prefrontal cortex plays an important role for planning.

4.5 How Does the Prefrontal Cortex Achieve Executive Control?

Executive control is the control mechanism to achieve a variety of functions such as focusing and selective attention, inhibition of inappropriate responses, task management, and planning a series of actions to achieve a goal. Not only multiple neural systems are needed to accomplish any of these functions but also a proper mechanism is also needed to manage these neural systems coordinately and flexibly to accomplish a particular goal. The system managing and controlling multiple systems is called executive control. To achieve executive control, a coordinated management of various neural systems operating in various brain regions is necessary. To manage a variety of functions that are operated in multiple brain areas, monitoring and controlling operations of these brain areas are essential by bottom-up signals and top-down signals, respectively. Therefore, anatomical connections between the cortical area responsible for executive control and a variety of brain areas are necessary. It has been shown that the prefrontal cortex plays an important role in executive control. In fact, the prefrontal cortex has wide anatomical connections with various cortical and subcortical structures and most of these connections are reciprocal. Through these reciprocal anatomical connections, the prefrontal cortex not only receives bottom-up signals to monitor operations in other cortical and subcortical structures but also sends top-down signals to control their operations. Reversible inactivation of the prefrontal cortex by cooling modulates task-related as well as spontaneous activity of posterior parietal and temporal neurons, indicating that prefrontal activity in fact affects and modulates information processing in the posterior cortices and acts as top-down control signals. At the same time, the prefrontal cortex receives sensory information from the posterior cortices, feedback information from motor centers, and motivational information, suggesting that the prefrontal cortex receives bottom-up signals and uses these signals for monitoring operations in other brain areas.

In order to consider the neural mechanism for executive control and its relation to prefrontal functions, working memory is an important mechanism to be considered because working memory is a neural mechanism not only for temporary active storage of task-relevant information but also for manipulating and processing stored information and because working memory is an important mechanism to understand prefrontal functions. The specific feature of working memory mechanism operating in the prefrontal cortex is that this system is not for information processing for a specific purpose, such as visual information processing, control of limb movements, or speech comprehension, but for the general purpose to achieve processing a variety of information. Therefore, the working memory system in the prefrontal cortex can be considered as general-purpose working memory system. On the other hand, each brain area except the prefrontal cortex has specific functions, processes particular information for a specific purpose, and uses a working memory system for a specific purpose. Therefore, the working memory system in these brain areas can be considered as purpose-specific working memory system.

Purpose-specific working memory system participates in information processing for a specific purpose, such as processing sensory information, processing linguistic information, or processing skeletomotor or oculomotor information. The purpose-specific working memory system is an important neural component in a cortical area where a particular sensory or motor information processing is performed. The working memory system for processing visual information is an important neural component in the visual cortex, while the working memory system for processing skeletomotor information is an important neural component in the motor cortex.

On the other hand, executive control is the control mechanism to achieve a variety of functions and to manage multiple neural systems coordinately and flexibly to accomplish a particular goal. Therefore, in order to achieve executive control, working memory system for general purpose is required to maintain and process a variety of information which includes not only information of a particular or multiple modalities but also information for controlling and monitoring various activities occurring in other brain areas. General-purpose working memory system is an important neural component in the prefrontal cortex to produce top-down signals and bottom-up signals for controlling and monitoring operations of purpose-specific working memory systems. Outputs from the general-purpose working memory system (top-down control signals) regulate information processing in purpose-specific working memory systems as well as information flows between purpose-specific working memory systems. By controlling and monitoring through top-down signals and bottom-up signals, general-purpose working memory system leads information processing in purpose-specific working memory systems to accomplish a specific goal. Thus, general-purpose working memory system is an important neural component to understand functions of the prefrontal cortex.

4.6 Interactions between the Prefrontal Cortex and Other Cortical Areas

To understand how the prefrontal cortex acts as the general-purpose working memory system, it is important to consider dynamic and flexible functional interactions between the prefrontal cortex and other cortical and subcortical areas. The prefrontal cortex has reciprocal connections with the sensory areas including posterior parietal cortex, the inferior temporal cortex, and the superior temporal polysensory areas. In addition, the prefrontal cortex has connections with the skeletomotor and oculomotor areas including the frontal eye field, the pre-supplementary motor area, the premotor cortex, and the caudate nucleus. The mid-dorsolateral prefrontal cortex has been shown to project to the retrosplenial area 30 and the posterior presubiculum (Morris et al. 1999), suggesting the anatomical substrate of functional interaction between the prefrontal cortex and the hippocampus. These observations indicate that the prefrontal cortex can receive various information including sensory, motor, and emotional information as well as information stored in long-term memory and also can send various information and signals to various brain areas to coordinate information processing operated in various brain regions. Strong reciprocal connections between the prefrontal cortex and various brain areas are the anatomical basis of dynamic and flexible interactions between the general-purpose working memory system and the purpose-specific working memory systems.

Dynamic cortical networks including the prefrontal cortex have been shown directly by correlation analyses in functional brain imaging studies as well as by covariance and correlation analyses of event-related potentials. Using a working memory task of faces, McIntosh et al. (1996) examined functional interactions among different cortical regions by an anatomy-based covariance structural equation model. As a result, they found that different functional interactions among different cortical areas were observed in different task conditions (e.g., different delay lengths) and that a top-down interaction from the prefrontal cortex to the temporal and occipital cortices was observed in a long delay condition. McIntosh et al. (1999) also showed that only aware subjects exhibited learning-related modulations of brain activation in the left prefrontal cortex during an associative learning task and that a correlation analysis of this modulation revealed significant functional connectivity among bilateral prefrontal cortices, the sensory association cortices, and the cerebellum. Gazzaley et al. (2004) used a multivariate analysis method of fMRI signals to examine functional connectivity among different brain areas during a face recognition task with delay. They used activity of the fusiform face area (FFA) as a seed and generated whole-brain correlation maps during the delay period. Brain areas that exhibited significant correlations with the FFA included the dorsolateral and ventrolateral prefrontal cortex, the premotor cortex, the intraparietal sulcus, and the occipitotemporal cortex. These results indicate that coordinated functional interaction among widely distributed brain regions is necessary to actively maintain perceptual information as sustained activation in working memory.

Evidences of homotopic interaction or coactivation of the bilateral prefrontal cortices have been reported (McGuire et al. 1991a, b; Klingberg et al. 1997). Klingberg et al. (1997) reported the increase in bilateral co-activation of the prefrontal cortex and the inferior parietal cortex in association with the increase in task difficulty. They also found that the inferior parietal cortex was often coactivated with the prefrontal cortex in working memory tasks and tasks involving planning. Further, Huettel et al. (2005) showed that the degree of uncertainty in decision-making influenced the magnitude of activation of both prefrontal and parietal cortices. Collette et al. (2005) showed activation of both the prefrontal and the inferior parietal cortex in dual-task performances. Thus, all of these results indicate the importance of functional interactions as well as co-activation between the prefrontal cortex and other cortical and subcortical areas when subjects perform cognitive tasks. These results also indicate that functional interactions and co-activation between the prefrontal cortex and other cortical and subcortical areas are dynamically and flexibly modulated depending on the context of the cognitive task.

Evidences of top-down modulatory effects by the prefrontal cortex have also been shown. Using visual associative memory tasks, Hasegawa et al. (1998a) showed that retrieval of visual images from long-term memory was under the executive control of the prefrontal cortex by the experiment using partial split-brain monkeys. In addition, by neurophysiological experiments using monkeys having resection of the corpus callosum, Tomita et al. (1999) showed that top-down signals from the prefrontal cortex regulated retrieval of associative memory in inferior temporal neurons. These observations indicate that the prefrontal cortex can control information processing in other cortical areas by sending top-down signals (Miyashita 2004). Lumer and Rees (1999) showed by fMRI that covariation of activation was observed in multiple extra-striate areas, the parietal cortex, and the prefrontal cortex in bistable viewing conditions. Coordinated activation of these areas was not linked to either sensory or motor events, but rather reflected perceptual events or internal changes of perception. Therefore, they concluded that functional interactions between visual areas and the prefrontal cortex are important for conscious vision. On the other hand, either enhancement or suppression of the visual response was observed in inferior temporal neurons when two visual stimuli (one preferred and the other non-preferred) were presented simultaneously in the neuron's receptive field. Chelazzi et al. (1998) proposed that this phenomenon can be explained by a "biased competition" model of attention. Based on this model, simultaneously presented objects in the visual receptive field would compete with others for representation within the cortex. They suggested that this competition could be biased by "top-down" feedback signals from brain structures involved in working memory. Based on the results by Hasegawa et al. (1998b), Tomita et al. (1999), and Lumer and Rees (1999), it would be reasonable to think that top-down signals from the prefrontal cortex bias the magnitude of the activity in the visual system. These results indicate the presence and importance of top-down modulatory effects by the prefrontal cortex to perform cognitive tasks. These results also suggest that top-down modulatory effects by the prefrontal cortex affect a wide variety of cognitive activities including sensory perception, directed attention, episodic memory encoding, and retrieval through dynamic and flexible functional interactions. This could be the typical function of the prefrontal cortex as the general-purpose working memory system.

4.7 Conclusions

Executive function is considered to be a product of the coordinated operation of various processes to accomplish a particular goal in a flexible manner. The mechanism or system responsible for the coordinated operation of various processes is called executive control. Since impairments caused by prefrontal damages have often been called dysexecutive syndromes, the prefrontal cortex is considered to play a significant role in executive control. Working memory processes observed in the prefrontal cortex, especially neuronal mechanisms for the temporary storage of information and dynamic and flexible interactions among them, can explain how the prefrontal cortex exerts executive control. The presence of extensive functional interactions among temporal information-storage processes and dynamic and flexible modulation of these interactions have been shown. Evidences for the dynamic and flexible modulation of neuronal interaction are observed in prefrontal neurons by neurophysiological experiments. In addition, recent imaging studies have shown dynamic modulations of functional connectivity between the prefrontal cortex and other cortical and subcortical structures depending upon the characteristics of the task or the temporal context of the task.

Information processing in general can be considered a modification or transformation from one type of representation to another type of representation. The representation of particular information in the prefrontal cortex might be considered a spatial pattern of both active and inactive areas within the prefrontal cortex, or a differential temporal pattern of activity at a particular region within the prefrontal cortex, or a combination of both. Prefrontal neurons that exhibit one type of task-related activity are usually distributed in a wide area of the prefrontal cortex. There is a large overlap of spatial distributions among neurons that exhibit different types of task-related activities (Carlson et al. 1997; Quintana and Fuster 1999; Rainer et al. 1999; Sakagami and Tsutsui 1999). However, some topographic patterns in the distribution of task-related neurons within the prefrontal cortex have been suggested. For example, an experiment using local impairment through the injection of bicuculline revealed a topographic deficit in spatial working memory, such that impairment of the anterior part of the principal sulcal area produced spatial working memory deficits when the visual cues were presented in the lower visual field, whereas impairment of the posterior part produced deficits when the visual cues were presented in the upper visual field (Sawaguchi and Goldman-Rakic 1994). In addition, each type of task-related activity also exhibits some specificities, such as directional selectivity observed in cue-, delay-, and response-period activities (Funahashi et al. 1989, 1990, 1991). Therefore, we might observe flexible and dynamic features of information processing by showing spatio-temporal change in the distribution of active and inactive areas within the prefrontal cortex along the temporal context of the task.

References

Abzug ZM, Sommer MA (2018) Neuronal correlates of serial decision-making in the supplementary eye field. J Neurosci 38:7280–7292. https://doi.org/10.1523/jneurosci.3643-17.2018

Awh E, Armstrong KM, Moore T (2006) Visual and oculomotor selection: links, causes and implications for spatial attention. Trends Cogn Sci 10:124–130. https://doi.org/10.1016/j.tics.2006.01.001

Baddeley A (1986) Working memory. Oxford University Press, Oxford, UK

Baddeley A, Della Salla S (1998) Working memory and executive control. In: Roberts AC, Robbins TW, Weiskrantz L (eds) The prefrontal cortex: executive and cognitive functions. Oxford University Press, Oxford, pp 9–21

Baddeley A, Hitch GJ (1974) Working memory. In: Bower GH (ed) The psychology of learning and motivation, vol 8. Academic Press, New York, pp 47–89

Baker SC, Rogers RD, Owen AM, Frith CD, Dolan RJ, Frackowiak RSJ, Robbins TW (1996) Neural systems engaged by planning: a PET study of the Tower of London Task. Neuropsychologia 34:515–526. https://doi.org/10.1016/0028-3932(95)00133-6

Boch RA, Goldberg ME (1989) Participation of prefrontal neurons in the preparation of visually guided eye movements in the rhesus monkey. J Neurophysiol 61:1064–1084. https://doi.org/10.1152/jn.1989.61.5.1064

Boschin EA, Brkic MM, Simons JS, Buckley MJ (2017a) Distinct roles for the anterior cingulate and dorsolateral prefrontal cortices during conflict between abstract rules. Cereb Cortex 27:34–45. https://doi.org/10.1093/cercor/bhw350

Boschin EA, Mars RB, Buckley MJ (2017b) Transcranial magnetic stimulation to dorsolateral prefrontal cortex affects conflict-induced behavioural adaptation in a Wisconsin Card Sorting Test analogue. Neuropsychologia 94:36–43. https://doi.org/10.1016/j.neuropsychologia.2016.11.015

Botvinick M, Nystrom LE, Fissell K, Carter CS, Cohen JD (1999) Conflict monitoring versus selection-for-action in anterior cingulated cortex. Nature 402:179–181. https://doi.org/10.1038/46035

Bressler SL, Tang W, Sylvester CM, Shulman GL, Corbetta M (2008) Top-down control of human visual cortex by frontal and parietal cortex in anticipatory visual spatial attention. J Neurosci 28:10056–10061. https://doi.org/10.1523/jneurosci.1776-08.2008

Bruce CJ, Goldberg ME (1985) Primate frontal eye fields. I. Single neurons discharging before saccades. J Neurophysiol 53:603–635. https://doi.org/10.1152/jn.1985.53.3.603

Buckley MJ, Mansouri FA, Hoda H, Mahboubi M, Browning PGF, Kwok SC, Phillips A, Tanaka K (2009) Dissociable components of rule-guided behavior depend on distinct medial and prefrontal regions. Science 325:52–58. https://doi.org/10.1126/science.1172377

Burgess PW (1997) Theory and methodology in executive function research. In: Rabbitt P (ed) Methodology of frontal and executive function. Psychology Press, East Sussex, UK, pp 81–116

Buschman TJ, Miller EK (2007) Top-down versus bottom-up control of attention in the prefrontal and posterior parietal cortices. Science 315:1860–1862. https://doi.org/10.1126/science.1138071

Buschman TJ, Miller EK (2009) Serial, covert shifts of attention during visual search are reflected by the frontal eye fields and correlated with population oscillations. Neuron 63:386–396. https://doi.org/10.1016/j.neuron.2009.06.020

Bushnell MC, Goldberg ME, Robinson DL (1981) Behavioral enhancement of visual responses in monkey cerebal cortex. I. Modulation in posterior parietal cortex related to selective visual attention. J Neurophysiol 46:755–772. https://doi.org/10.1152/jn.1981.46.4.755

Carlson S, Rämä P, Tanila H, Linnankoski I, Mansikka H (1997) Dissociation of mnemonic coding and other functional neuronal processing in the monkey prefrontal cortex. J Neurophysiol 77:761–774. https://doi.org/10.1152/jn.1997.77.2.761

Carter CS, Braver TS, Barch DM, Botvinick MM, Noll D, Cohen JD (1998) Anterior cingulate cortex, error detection, and the online monitoring of performance. Science 280:747–749. https://doi.org/10.1126/science.280.5364.747

Chelazzi L, Duncan J, Miller EK, Desimone R (1998) Responses of neurons in inferior temporal cortex during memory-guided visual search. J Neurophysiol 80:2918–2940. https://doi.org/10.1152/jn.1998.80.6.2918

Collette F, Olivier L, Van der Linden M, Laureys S, Delfiore G, Luxen A, Salmon E (2005) Involvement of both prefrontal and inferior parietal cortex in dual-task performance. Cogn Brain Res 24:237–251. https://doi.org/10.1016/j.cogbrainres.2005.01.023

Crowe SF (1992) Dissociation of two frontal lobe syndromes by a test of verbal fluency. J Clin Exp Neuropsychol 14:327–339. https://doi.org/10.1080/01688639208402832
Dagher A, Owen AM, Boecker H, Brooks DJ (1999) Mapping the network for planning: a correlational PET activation study with the Tower of London task. Brain 122:1973–1987. https://doi.org/10.1093/brain/122.10.1973
D'Esposito M, Detre JA, Alsop DC, Shin RK, Atlas S, Grossman M (1995) The neural basis of the central executive system of working memory. Nature 378:279–281. https://doi.org/10.1038/378279a0
di Pellegrino G, Wise SP (1993) Effects of attention on visuomotor activity in the premotor and prefrontal cortex of a primate. Somatosens Motor Res 10:246–262. https://doi.org/10.3109/08990229309028835
Dove A, Pollmann S, Schubert T, Wiggins CJ, von Cramon DY (2000) Prefrontal cortex activation in task switching: an event-related fMRI study. Cogn Brain Res 9:103–109. https://doi.org/10.1016/S0926-6410(99)00029-4
Egner T, Monti JMP, Trittschuh EH, Wieneke CA, Hirsch J, Mesulam M-M (2008) Neural integration of top-down spatial and feature-based information in visual search. J Neurosci 28:6141–6151. https://doi.org/10.1523/jneurosci.1262-08.2008
Everling S, Tinsley CJ, Gaffan D, Duncan J (2002) Filtering of neural signals by focused attention in the monkey prefrontal cortex. Nat Neurosci 5:671–676. https://doi.org/10.1038/nn874
Freedman M, Oscar-Berman M (1986) Bilateral frontal lobe disease and selective delayed response deficits in humans. Behav Neurosci 100:337–342. https://doi.org/10.1037/0735-7044.100.3.337
Funahashi S, Andreau JM (2013) Prefrontal cortex and neural mechanisms of executive function. J Physiol Paris 107:471–482. https://doi.org/10.1016/j.jphysparis.2013.05.001
Funahashi S, Bruce CJ, Goldman-Rakic PS (1989) Mnemonic coding of visual space in the monkey's dorsolateral prefrontal cortex. J Neurophysiol 61:331–349. https://doi.org/10.1152/jn.1989.61.2.331
Funahashi S, Bruce CJ, Goldman-Rakic PS (1990) Visuospatial coding in primate prefrontal neurons revealed by oculomotor paradigms. J Neurophysio 63:814–831. https://doi.org/10.1152/jn.1990.63.4.814
Funahashi S, Bruce CJ, Goldman-Rakic PS (1991) Neuronal activity related to saccadic eye movements in the monkey's dorsolateral prefrontal cortex. J Neurophysiol 65:1464–1483. https://doi.org/10.1152/jn.1991.65.6.1464
Funahashi S, Goldman-Rakic CMV, PS. (1993) Prefrontal neuronal activity in rhesus monkeys performing a delayed anti-saccade task. Nature 365:753–756. https://doi.org/10.1038/365753a0
Fuster JM (1997) The prefrontal cortex: anatomy, physiology, and neuropsychology of the frontal lobe, 3rd edn. Lippincott-Raven, Philadelphia, PA
Fuster JM (2008) The prefrontal cortex, 4th edn. Academic Press, New York, NY
Gazzaley A, Rissman J, D'Esposito M (2004) Functional connectivity during working memory maintenance. Cogn Affect Behav Neurosci 4:580–599. https://doi.org/10.3758/CABN.4.4.580
Gemba H, Sasaki K (1989) Potential related to no-go reaction of go/no-go hand movement task with color discrimination in human. Neurosci Lett 101:263–268. https://doi.org/10.1016/0304-3940(89)90543-0
Gemba H, Sasaki K (1990) Potential related to no-go reaction in go/no-go hand movement with discrimination between tone stimuli of different frequencies in the monkey. Brain Res 537:340–344. https://doi.org/10.1016/0006-8993(90)90382-L
Genovesio A, Brasted PJ, Mitz AR, Wise SP (2005) Prefrontal cortex activity related to abstract response strategies. Neuron 47:307–320. https://doi.org/10.1016/j.neuron.2005.06.006
Genovesio A, Brasted PJ, Wise SP (2006) Representation of future and previous spatial goals by separate neural populations in prefrontal cortex. J Neurosci 26:7305–7316. https://doi.org/10.1523/jneurosci.0699-06.2006

Gläscher J, Adolphs R, Damasio H, Bechara A, Rudrauf D, Calamia M, Paul LK, Tranel D (2012) Lesion mapping of cognitive control and value-based decision making in the prefrontal cortex. Proc Natl Acad Sci USA 109:14681–14686. https://doi.org/10.1073/pnas.1206608109

Goldberg ME, Bushnell MC (1981) Behavioral enhancement of visual responses in monkey cerebral cortex. II. Modulation in frontal eye fields specifically related to saccades. J Neurophysiol 46:773–787. https://doi.org/10.1152/jn.1981.46.4.773

Goldman-Rakic PS (1987) Circuitry of primate prefrontal cortex and regulation of behavior by representational memory. In: Plum F (ed) Handbook of physiology, section 1: the nervous system, Part 1: higher functions of the brain, vol V. American Physiological Society, Bethesda, MD, pp 373–417. https://doi.org/10.1002/cphy.cp010509

Guitton D, Buchtel HA, Douglas RM (1985) Frontal lobe lesions in man cause difficulties in suppressing reflexive glances and in generating goal-directed saccades. Exp Brain Res 58:455–472. https://doi.org/10.1007/BF00235863

Hasegawa I, Fukushima T, Ihara T, Miyashita Y (1998a) Callosal window between prefrontal cortices: cognitive interaction to retrieve long-term memory. Science 281:814–818. https://doi.org/10.1126/science.281.5378.814

Hasegawa R, Sawaguchi T, Kubota K (1998b) Monkey prefrontal neuronal activity coding the forthcoming saccade in an oculomotor delayed matching-to-sample task. J Neurophysiol 79:322–333. https://doi.org/10.1152/jn.1998.79.1.322

Hopfinger JB, Buonocore MH, Mangum GR (2000) The neural mechanisms of top-down attentional control. Nat Neurosci 3:284–291. https://doi.org/10.1038/72999

Huettel SA, Song AW, McCarthy G (2005) Decisions under uncertainty: probabilistic context influences activation of prefrontal and parietal cortices. J Neurosci 25:3304–3311. https://doi.org/10.1523/jneurosci.5070-04.2005

Jahanshahi M, Dirnberger G (1999) The left dorsolateral prefrontal cortex and random generation of responses: studies with transcranial magnetic stimulation. Neuropsychologia 37:181–190. https://doi.org/10.1016/S0028-3932(98)00092-X

Jahanshahi M, Profice P, Brown RG, Ridding MC, Dirnberger G, Rothwell JC (1998) The effects of transcranial magnetic stimulation over the dorsolateral prefrontal cortex on suppression of habitual counting during random number generation. Brain 121:1533–1544. https://doi.org/10.1093/brain/121.8.1533

Jones-Gotman M, Milner B (1977) Design fluency: the invention of nonsense drawings after focal cortical lesions. Neuropsychologia 15:653–674. https://doi.org/10.1016/0028-3932(77)90070-7

Kadota H, Sekiguchi H, Takeuchi S, Miyazaki M, Kohno Y, Nakajima Y (2010) The role of the dorsolateral prefrontal cortex in the inhibition of stereotyped responses. Exp Brain Res 203:593–600. https://doi.org/10.1007/s00221-010-2269-4

Kaller CP, Rahm B, Spreer J, Weiller C, Unterrainer JM (2011) Dissociable contributions of left and right dorsolateral prefrontal cortex in planning. Cereb Cortex 21:307–317. https://doi.org/10.1093/cercor/bhq096

Kamigaki T, Fukushima T, Tamura K, Miyashita Y (2012) Neurodynamics of cognitive set shifting in monkey frontal cortex and its causal impact on behavioral flexibility. J Cogn Neurosci 24:2171–2185. https://doi.org/10.1162/jocn_a_00277

Katsuki F, Constantinidis C (2012a) Early involvement of prefrontal cortex in visual bottom-up attention. Nat Neurosci 15:1160–1168. https://doi.org/10.1038/nn.3164

Katsuki F, Constantinidis C (2012b) Unique and shared roles of the posterior parietal and dorsolateral prefrontal cortex in cognitive functions. Front Integr Neurosci 6:17. https://doi.org/10.3389/fnint.2012.00017

Klingberg T, O'Sullivan BT, Roland PE (1997) Bilateral activation of fronto-parietal networks by incrementing demand in a working memory task. Cereb Cortex 7:465–471. https://doi.org/10.1093/cercor/7.5.465

Kodaka Y, Mikami A, Kubota K (1997) Neuronal activity in the frontal eye field of the monkey is modulated while attention is focused on to a stimulus in the peripheral visual field, irrespective of eye movement. Neurosci Res 28:291–298. https://doi.org/10.1016/S0168-0102(97)00055-2

Konishi S, Nakajima K, Uchida I, Kameyama M, Nakahara K, Sekihara K, Miyashita Y (1998) Transient activation of inferior prefrontal cortex during cognitive set shifting. Nat Neurosci 1: 80–84. https://doi.org/10.1038/283

Konishi S, Kawazu M, Uchida I, Kikyo H, Asakura I, Miyashita Y (1999a) Contribution of working memory to transient activation in human inferior prefrontal cortex during performance of the Wisconsin card sorting test. Cereb Cortex 9:745–753. https://doi.org/10.1093/cercor/9.7.745

Konishi S, Nakajima K, Uchida I, Kikyo H, Kameyama M, Miyashita Y (1999b) Common inhibitory mechanism inhuman inferior prefrontal cortex revealed by event-related functional MRI. Brain 122:981–991. https://doi.org/10.1093/brain/122.5.981

Kowler E (2009) Attention and eye movements. In: Squire LR (ed) Encyclopedia of neuroscience. Elsevier, Amsterdam, pp 605–616

Logie RH (1995) Visuo-spatial working memory. Lawrence Erlbaum Associates, Hove, UK

Lumer ED, Rees G (1999) Covariation of activity in visual and prefrontal cortex associated with subjective visual perception. Proc Natl Acad Sci USA 96:1669–1673. https://doi.org/10.1073/pnas.96.4.1669

Luo TZ, Maunsell JHR (2018) Attentional changes in either criterion or sensitivity are associated with robust modulations in lateral prefrontal cortex. Neuron 97:1382–1393.e7. https://doi.org/10.1016/j.neuron.2018.02.007

Luria AR (1969) Frontal lobe syndromes. In: Vinken PJ, Bruyn GW (eds) Handbook of clinical neurology, vol 2. North Holland, Amsterdam, pp 725–757

Lynch JC, Mountcastle VB, Talbot WH, Yin TCT (1977) Parietal lobe mechanisms for directed visual attention. J Neurophysiol 40:362–389. https://doi.org/10.1152/jn.1977.40.2.362

Mansouri FA, Matsumoto K, Tanaka K (2006) Prefrontal cell activities related to monkeys' success and failure in adapting to rule changes in a Wisconsin card sorting test analog. J Neurosci 26: 2745–2756. https://doi.org/10.1523/jneurosci.5238-05.2006

Mansouri FA, Buckley MJ, Tanaka K (2007) Mnemonic function of the dorsolateral prefrontal cortex in conflict-induced behavioral adjustment. Science 318:987–990. https://doi.org/10.1126/science.1146384

Mansouri FA, Buckley MJ, Fehring DJ, Tanaka K (2020) The role of primate prefrontal cortex in bias and shift between visual dimensions. Cereb Cortex 30:85–99. https://doi.org/10.1093/cercor/bhz072

Matsushima A, Tanaka M (2012) Neuronal correlates of multiple top-down signals during covert tracking of moving objects in macaque prefrontal cortex. J Cogn Neurosci 24:2043–2056. https://doi.org/10.1162/jocn_a_00265

McAdams CJ, Mauncell JH (1999) Effects of attention on orientation-tuinng functions of single neurons in macaque cortical area V4. J Neurosci 19:431–441. https://doi.org/10.1523/JNEUROSCI.19-01-00431.1999

McGuire PK, Bates JF, Goldman-Rakic PS (1991a) Interhemispheric integration: I. Symmetry and convergence of the corticocortical connections of the left and the right principal sulcus (PS) and the left and the right supplementary motor area (SMA) in the rhesus monkey. Cereb Cortex 1: 390–407. https://doi.org/10.1093/cercor/1.5.390

McGuire PK, Bates JF, Goldman-Rakic PS (1991b) Interhemispheric integration: II. Symmetry and convergence of the corticostriatal projections of the left and the right principal sulcus (PS) and the left and the right supplementary motor area (SMA) of the rhesus monkey. Cereb Cortex 1: 408–417. https://doi.org/10.1093/cercor/1.5.408

McIntosh AR, Grady CL, Haxby JV, Ungerleider LG, Horwitz B (1996) Changes in limbic and prefrontal functional interactions in a working memory task for faces. Cereb Cortex 6:571–584. https://doi.org/10.1093/cercor/6.4.571

McIntosh AR, Rajah MN, Lobaugh NJ (1999) Interactions of prefrontal cortex in relation to awareness in sensory learning. Science 284:1531–1533. https://doi.org/10.1126/science.284.5419.1531

Mikami A, Ito S, Kubota K (1982) Modifications of neuron activities of the dorsolateral prefrontal cortex during extrafoveal attention. Behav Brain Res 5:219–223. https://doi.org/10.1016/0166-4328(82)90056-0

Miller EK, Buschman TJ (2013) Cortical circuits for the control of attention. Curr Opin Neurobiol 23:216–222. https://doi.org/10.1016/j.conb.2012.11.011

Miller EK, Li L, Desimone R (1991) A neural mechanism for working and recognition memory in inferior temporal cortex. Science 254:1377–1379. https://doi.org/10.1126/science.1962197

Milner B (1963) Effects of different brain lesions on card sorting: the role of the frontal lobes. Arch Neurol 9:90–100. https://doi.org/10.1001/archneur.1963.00460070100010

Milner B (1964) Some effects of frontal lobectomy in man. In: Warren JM, Akert K (eds) The frontal granular cortex and behavior. McGraw-Hill, New York, pp 313–334

Milner B (1982) Some cognitive effects of frontal-lobe lesions in man. Phil Trans R Soci Lond B 298:211–226. https://doi.org/10.1098/rstb.1982.0083

Miyake A, Shah P (1999) Toward unified theories of working memory: emerging general consensus, unresolved theoretical issues, and future research directions. In: Miyake A, Shah P (eds) Models of working memory: mechanisms of active maintenance and executive control. Cambridge University Press, Cambridge, UK, pp 442–481

Miyashita Y (2004) Cognitive memory: cellular and network machineries and their top-down control. Science 306:435–440. https://doi.org/10.1126/science.1101864

Moore T, Armstrong KM (2003) Selective gating of visual signals by microstimulation of frontal cortex. Nature 421:370–373. https://doi.org/10.1038/nature01341

Moore T, Fallah M (2004) Microstimulation of the frontal eye field and its effects on covert spatial attention. J Neurophysiol 91:152–162. https://doi.org/10.1152/jn.00741.2002

Moran J, Desimone R (1985) Selective attention gates visual processing in the extrastriate cortex. Science 229:782–784. https://doi.org/10.1126/science.4023713

Morris R, Pandya DN, Petrides M (1999) Fiber system linking the mid-dorsolateral frontal cortex with the retrosplenial/presubicular region in the rhesus monkey. J Comp Neurol 407:183–192. https://doi.org/10.1002/(SICI)1096-9861(19990503)407:2<183::AID-CNE3>3.0.CO;2-N

Motter BC (1993) Focal attention produces spatially selective processing in visual cortical areas V1, V2, and V4 in the presence of competing stimuli. J Neurophysiol 70:909–919. https://doi.org/10.1152/jn.1993.70.3.909

Munoz DP, Everling S (2004) Look away: the anti-saccade task and the voluntary control of eye movement. Nat Rev Neurosci 5:218–228. https://doi.org/10.1038/nrn1345

Müri RM, Rivaud S, Gaymard B, Ploner CJ, Vermersch AI, Hess CW, Pierrot-Deseilligny C (1999) Role of the prefrontal cortex in the control of express saccades. A transcranial magnetic stimulations study. Neuropsychologia 37:199–206

Mushiake H, Saito N, Sakamoto K, Sato Y, Tanji J (2001) Visually based path-planning by Japanese monkeys. Cogn Brain Res 11:165–169. https://doi.org/10.1016/S0926-6410(00)00067-7

Mushiake H, Saito N, Furusawa Y, Izumiyama M, Sakamoto K, Shamoto H, Shimizu H, Yoshimoto T (2002) Orderly activations of human cortical areas during path-planning task. Neuroreport 13:423–426

Mushiake H, Saito N, Sakamoto K, Itoyama Y, Tanji J (2006) Activity in the lateral prefrontal cortex reflects multiple steps of future events in action plans. Neuron 50:631–641. https://doi.org/10.1016/j.neuron.2006.03.045

Nakahara K, Hayashi T, Konishi S, Miyashita Y (2002) Functional MRI of macaque monkeys performing a cognitive set-shifting task. Science 295:1532–1538. https://doi.org/10.1126/science.1067653

Newman SD, Carpenter PA, Varma S, Just MA (2003) Frontal and parietal participation in problem solving in the Tower of London: fMRI and computational modeling of planning and high-level perception. Neuropsychologia 41:1668–1682. https://doi.org/10.1016/S0028-3932(03)00091-5

Ninomiya T, Sawamura H, Inoue K-I, Takada M (2012) Segregated pathways carrying frontally derived top-down signals to visual areas MT and V4 in macaques. J Neurosci 32:6851–6858. https://doi.org/10.1523/jneurosci.6295-11.2012

Noudoost B, Moore T (2011) Control of visual cortical signals by prefrontal dopamine. Nature 474: 372–375. https://doi.org/10.1038/nature09995

Nyffeler T, Müri RM, Bucher-Ottiger Y, Pierrot-Deseilligny C, Gaymard B, Rivaud-Pechoux S (2007) Inhibitory control of the human dorsolateral prefrontal cortex during the anti-saccade paradigm – a transcranial magnetic stimulation study. Eur J Neurosci 26:1381–1385. https://doi.org/10.1111/j.1460-0598.2007.05758.x

Owen AM, Downes JJ, Sahakian BJ, Polkey CE, Robbins TW (1990) Planning and spatial working memory following frontal lobe lesions in man. Neuropsychologia 28:1021–1034. https://doi.org/10.1016/0028-3932(90)90137-D

Owen AM, Doyon J, Petrides M, Evans AC (1996a) Planning and spatial working memory: a positron emission tomography study in humans. Eur J Neurosci 8:353–364. https://doi.org/10.1111/j.1460-9568.1996.tb01219.x

Owen AM, Evans AC, Petrides M (1996b) Evidence for a two-stage model of spatial working memory processing within the lateral frontal cortex: a positron emission tomography study. Cereb Cortex 6:31–38. https://doi.org/10.1093/cercor/6.1.31

Perner J, Lang B (1999) Development of theory of mind and executive control. Trends Cogn Sci 3: 337–344. https://doi.org/10.1016/S1364-6613(99)01362-5

Peterson BS, Skudlarski JP, Gatenby C, Zhang H, Anderson AW, Gore JC (1999) An fMRI study of Stroop word-color interface: evidence for cingulate subregions subserving multiple distributed attentional systems. Biol Psychiatry 45:1237–1258. https://doi.org/10.1016/S0006-3223(99)00056-6

Petride M, Milner B (1982) Deficits on subject-ordered tasks after frontal- and temporal-lobe lesions in man. Neuropsychologia 20:249–262. https://doi.org/10.1016/0028-3932(82)90100-2

Petrides M (1995) Impairments on nonspatial self-ordered and externally ordered working memory tasksafter lesions of the mid-dorsal part of the lateral frontal cortex in the monkey. J Neurosci 15:359–376. https://doi.org/10.1523/jneurosci.15-01-00359.1995

Pierrot-Deseilligny CH, Rivaud S, Gaymard B, Agid Y (1991) Cortical control of reflexive visually-guided saccades. Brain 114:1473–1485. https://doi.org/10.1093/brain/114.3.1473

Pineda D, Ardila A, Rosselli M, Cadavid C, Mancheno S, Mejia S (1998) Executive dysfunctions in children with attention deficit hyperactivity disorder. Intern J Neurosci 96:177–196. https://doi.org/10.3109/00207459808986466

Quintana J, Fuster JM (1999) From perception to action: temporal integrative functions of prefrontal and parietal neurons. Cereb Cortex 9:213–221. https://doi.org/10.1093/cercor/9.3.213

Rabbitt P (1997) Introduction: methodologies and models in the study of executive function. In: Rabbitt P (ed) Methodology of frontal and executive function. Psychology Press, Hove, East Sussex, UK, pp 1–38

Rainer G, Rao C, Miller EK (1999) Prospective coding for objects in primate prefrontal cortex. J Neurosci 19:5493–5505. https://doi.org/10.1523/JNEUROSCI.19-13-05493.1999

Robbins TW, Arnsten AFT (2009) The neuropsychopharmacology of fronto-executive function: monoaminergic modulation. Annu Rev Neurosci 32:267–287. https://doi.org/10.1146/annurev.neuro.051508.135535

Roberts AC (1998) Introduction. In: Roberts AC, Robbins TW, Weiskrantz L (eds) The prefrontal cortex: executive and cognitive functions. Oxford University Press, Oxford, UK, pp 1–8

Robinson DL, Goldberg ME, Stanton GB (1978) Parietal association cortex in the primate: sensory mechanisms and behavioral modulation. J Neurophysiol 41:910–932. https://doi.org/10.1152/jn.1978.41.4.910

Rossi AF, Bichot NP, Desimone R, Ungerleider LG (2007) Top-down attentional deficits in macaques with lesions of lateral prefrontal cortex. J Neurosci 27:11306–11314. https://doi.org/10.1523/JNEUROSCI.2939-07.2007

Saito N, Mushiake H, Sakamoto K, Itoyama Y, Tanji J (2005) Representation of immediate and final behavioral goals in the monkey prefrontal cortex during an instructed delay period. Cereb Cortex 15:1535–1546. https://doi.org/10.1093/cercor/bhi032

Sakagami M, Tsutsui K (1999) The hierarchical organization of decision making in the primate prefrontal cortex. Neurosci Res 34:79–89. https://doi.org/10.1016/S0168-0102(99)00038-3

Sarter M, Lustig C (2009) Attentional functions in learning and memory. In: Squire LR (ed) Encyclopedia of neuroscience. Elsevier, Amsterdam, pp 639–645

Sasaki K, Gemba H (1986) Electrical activity in the prefrontal cortex specific to no-go reaction of conditioned hand movement with colour discrimination in the monkey. Exp Brain Res 64:603–606. https://doi.org/10.1007/BF00340499

Sasaki K, Gemba H, Tsujimoto T (1989) Suppression of visually initiated hand movement by stimulation of the prefrontal cortex in the monkey. Brain Res 495:100–107. https://doi.org/10.1016/0006-8993(89)91222-5

Sasaki K, Gemba H, Nambu A, Matsuzaki R (1993) No-go activity in the frontal association cortex of human subjects. Neurosci Res 18:249–252. https://doi.org/10.1016/0168-0102(93)90062-U

Sawaguchi T, Goldman-Rakic PS (1994) The role of D1-dopamine receptor in working memory: local injections of dopamine antagonists into the prefrontal cortex of rhesus monkeys performing an oculomotor delayed-response task. J Neurophysiol 71:515–528. https://doi.org/10.1152/jn.1994.71.2.5152

Schall JD (2002) The neural selection and control of saccades by the frontal eye field. Phil Trans R Soc Lond B 357:1073–1082. https://doi.org/10.1098/rstb.2002.1098

Schall JD, Hanes DP (1993) Neural basis of saccade target selection in frontal eye field during visual search. Nature 366:467–469. https://doi.org/10.1038/366467a0

Schall JD, Hanes DP, Thompson KG, King DJ (1995) Saccade target selection in frontal eye field of macaque. I. Visual and premovement activation. J Neurosci 15:6905–6918. https://doi.org/10.1523/jneurosci.15-10-06905.1995

Schall JD, Palmeri TJ, Logan GD (2017) Models of inhibitory control. Phil Trans R Soc B 372: 20160193. https://doi.org/10.1098/rstb.2016.0193

Shah P, Miyake A (1999) Models of working memory: an introduction. In: Miyake A, Shah P (eds) Models of working memory: mechanisms of active maintenance and executive control. Cambridge University Press, Cambridge, UK, pp 1–27

Shallice T (1982) Specific impairments of planning. Phil Trans R Soc Lond B 298:199–209. https://doi.org/10.1098/rstb.1982.0082

Shallice T, Burgess P (1991) Higher-order cognitive impairments and frontal lobe lesions in man. In: Levin HS, Eisenberg HM, Benton AL (eds) Frontal lobe function and dysfunction. Oxford University Press, New York, NY, pp 125–138

Shima K, Isoda M, Mushiake H, Tanji J (2007) Categorization of behavioural sequences in the prefrontal cortex. Nature 445:315–318. https://doi.org/10.1038/nature05470

Smith EE, Jonides J (1999) Storage and executive processes in the frontal lobes. Science 283:1657–1661. https://doi.org/10.1126/science.283.5408.1657

Sohn M-H, Ursu S, Anderson JR, Stenger VA, Carter CS (2000) The role of prefrontal cortex and posterior parietal cortex in task switching. Proc Natl Acad Sci USA 97:13448–13453. https://doi.org/10.1073/pnas.240460497

Spitzer H, Desimone R, Moran J (1988) Increased attention enhances both behavioral and neuronal performance. Science 240:338–340. https://doi.org/10.1126/science.3353728

Stanton GB, Bruce CJ, Goldberg ME (1995) Topography of projections to posterior cortical areas from the macaque frontal eye fields. J Comp Neurol 353:291–305. https://doi.org/10.1002/cne.903530210

Stuss DT, Benson DF (1986) The frontal lobe. Raven Press, New York, NY

Stuss DT, Levine B, Alexander MP, Hong J, Palumbo C, Hamer L, Murphy KJ, Izukawa D (2000) Wisconsin card sorting test performance in patients with focal frontal and posterior brain damage: effects of lesion location and test structure on separable cognitive processes. Neuropsychologia 38:388–402. https://doi.org/10.1016/S0028-3932(99)00093-7

Thompson KG, Bichot NP (2005) A visual salience map in the primate frontal eye field. Progr Brain Res 147:249–262. https://doi.org/10.1016/S0079-6123(04)47019-8

Thompson KG, Bichot NP, Schall JD (1997) Dissociation of visual discrimination from saccade programming in macaque frontal eye field. J Neurophysiol 77:1046–1050. https://doi.org/10.1152/jn.1997.77.2.1046

Thompson KG, Biscoe KL, Sato TR (2005) Neuronal basis of covert spatial attention in the frontal eye field. J Neurosci 25:9479–9487. https://doi.org/10.1523/jneurosci.0741-05.2005

Tolias AS, Moore T, Smirnakis SM, Tehovnik EJ, Siapas AG, Schiller PH (2001) Eye movements modulate visual receptive fields of V4 neurons. Neuron 29:757–767. https://doi.org/10.1016/S0896-6273(01)00250-1

Tomita H, Ohbayashi M, Nakahara K, Hasegawa I, Miyashita Y (1999) Top-down signal from prefrontal cortex in executive control of memory retrieval. Nature 401:699–703. https://doi.org/10.1038/44372

Tsujimoto S, Genovesio A, Wise SP (2011) Comparison of strategy signals in the dorsolateral and orbitofrontal cortex. J Neurosci 31:4583–4592. https://doi.org/10.1523/jneurosci.5816-10.2011

Tsujimoto S, Genovesio A, Wise SP (2012) Neuronal activity during a cued strategy task: comparison of dorsolateral, orbital, and polar prefrontal cortex. J Neurosci 32:11017–11031. https://doi.org/10.1523/1230-12.2012

Tsutsui K-I, Hosokawa T, Yamada M, Iijima T (2016a) Representation of functional category in the monkey prefrontal cortex and its rule-dependent use for behavioral selection. J Neurosci 36: 3038–3048. https://doi.org/10.1523/jneurosci.2063-15.2016

Tsutsui K-I, Oyama K, Nakamura S, Iijima T (2016b) Comparative overview of visuospatial working memory in monkeys and rats. Front Syst Neurosci 10:99. https://doi.org/10.3389/fnsys.2016.00099

Unterrainer JF, Owen AM (2006) Planning and problem solving: from neuropsychology to functional neuroimaging. J Physiol Paris 99:308–317. https://doi.org/10.1016/j.jphysparis.2006.03.014

van den Heuvel OA, Groenewegen HJ, Barkhof F, Lazeron RHC, van Dyck R, Veltman DJ (2003) Frontostriatal system in planning complexity: a parametric functional magnetic resonance version of Tower of London task. Neuroimage 18:367–374. https://doi.org/10.1016/S1053-8119(02)00010-1

Vandrell P, Junqué C, Pujol J, Jurado MA, Molet J, Grafman J (1995) The role of prefrontal regions in the Stroop task. Neuropsychologia 33:341–352. https://doi.org/10.1016/0028-3932(94)00116-7

Verbruggen F, Logan GD (2008) Response inhibition in the stop-signal paradigm. Trends Cogn Sci 12:418–424. https://doi.org/10.1016/j.tics.2008.07.005

Wagner G, Koch K, Reichenbach JR, Sauer H, Schlösser RGM (2006) The special involvement of the rostrolateral prefrontal cortex in planning abilities: an event-related fMRI study with the Tower of London paradigm. Neuropsychologia 44:2337–2347. https://doi.org/10.1016/j.neuropsychologia.2006.05.014

Wardak C, Ibos G, Duhamel J-R, Olivier E (2006) Contribution of the monkey frontal eye field to covert visual attention. J Neurosci 26:4228–4235. https://doi.org/10.1523/jneurosci.3336-05.2006

Williams GV, Goldman-Rakic PS (1995) Modulation of memory fields by dopamine D1 receptors in prefrontal cortex. Nature 376:572–575. https://doi.org/10.1038/376572a0

Zhou X, Qi X-L, Constantinidis C (2016) Distinct roles of the prefrontal and posterior parietal cortices in response inhibition. Cell Rep 14:2765–2773. https://doi.org/10.1016/j.celrep.2016.02.072

Chapter 5
Sensory and Motor Processing

5.1 Sensory Processing

The prefrontal cortex receives a variety of sensory information and plays critical roles in representing, processing, and integrating sensory information to accomplish various cognitive activities. The lateral prefrontal cortex receives visual, auditory, and somatosensory information from the parietal cortex (Jones and Powell 1970; Jacobson and Trojanowski 1977a; Kawamura and Naito 1984; Petrides and Pandya 1984; Schwartz and Goldman-Rakic 1984; Barbas 1988; Cavada and Goldman-Rakic 1989; Andersen et al. 1990a, b) and the temporal cortex (Jones and Powell 1970; Jacobson and Trojanowski 1977b; Kawamura and Naito 1984; Barbas 1988; Seltzer and Pandya 1989; Ungerleider et al. 1989; Petrides and Pandya 1994) (Fig. 5.1a). The parietal and the temporal association cortices receive sensory information mainly from sensory association areas. The prefrontal cortex does not receive sensory information directly from the primary sensory cortices, indicating that the prefrontal cortex receives information that has been selected and processed in the posterior association cortices.

The dorsolateral prefrontal cortex receives visual, auditory, and somatosensory information from the posterior association cortices and individual sensory information exhibits distinct spatial representation on the lateral prefrontal cortex (Barbas et al. 2011) (Fig. 5.1b). Anatomical projections from the visual association cortices terminate most heavily in the frontal eye field (area 8) and the caudal portion of the principal sulcus area (area 46). Projections from the auditory cortex terminate most robustly in the frontopolar area (area 10), the middle portion of the principal sulcus (area 46), and the anterior portion of the dorsal arcuate sulcus (area 8). And, projections from the somatosensory cortex terminate most heavily in the middle portion of the ventral area 46 and the adjacent area 12. Thus, anatomical results suggest the topographic representation of different sensory modalities in the lateral prefrontal cortex. However, the lateral prefrontal cortex receives not only unimodal information but also multimodal information from the posterior association areas.

S. Funahashi, *Dorsolateral Prefrontal Cortex*, Brain Science,
https://doi.org/10.1007/978-981-19-7268-3_5

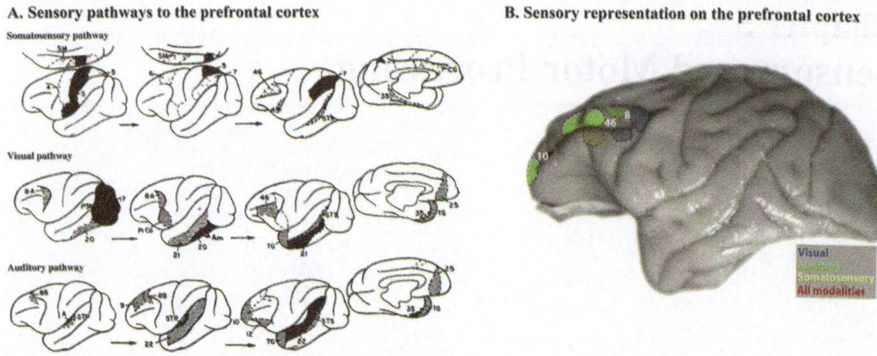

Fig. 5.1 (**a**) Cortical sensory pathways from primary sensory cortices to the prefrontal cortex. From Jones and Powell (1970) with permission. (**b**) Visual, auditory, and somatosensory representations on the dorsolateral prefrontal cortex. Form Barbas et al. (2011) with permission

Neighboring areas in the lateral prefrontal cortex are anatomically interconnected each other and exchange information through these anatomical connections. Therefore, anatomically observed topographic representation may not indicate exclusive processing of individual sensory information within a localized prefrontal area.

However, neurophysiological studies suggest that the dorsolateral prefrontal cortex is functionally compartmentalized with respect to the different sensory modalities. Lateral prefrontal neurons have been known to respond to sensory stimuli having complex physical features. Using complex visual (e.g., objects, food, and pictures in various shapes, colors, and sizes), auditory (e.g., clicks, rustles, steps, and human voice in various pitch and intensity), and somatosensory (e.g., blowing, light touching, pinching, and stroking of the skin) stimuli, Tanila et al. (1992) examined which sensory modality activated each prefrontal neuron while the monkey sit quietly in a monkey chair. Among 134 dorsolateral prefrontal neurons examined, 29% responded to visual stimuli, 2% responded to auditory stimuli, 7% responded to somatosensory stimuli, and only 9% responded to two or more sensory modalities. Tanila et al. (1993) also examined cortical distributions of single neurons exhibiting either sensory or oculomotor responses on the lateral prefrontal cortex. Visual neurons were distributed widely in areas 46 and 9. Somatosensory neurons were concentrated in the middle third of the inferior bank of the principal sulcus. Oculomotor neurons were found in the caudal portion of the principal sulcus. They concluded that, although the dorsolateral prefrontal cortex receives sensory information from the posterior association cortices, this cortical area is functionally compartmentalized with respect to the nature of sensory inputs and response characteristics to sensory stimuli. Further, Saga et al. (2011) used visual, auditory, and tactile stimuli and examined prefrontal responses to these stimuli while monkeys performed a behavioral task. In this task, the monkey receives the same modality stimulus successively with variable intervals and needs to release a key after the fourth stimulus is presented. They found phasic and short-latency responses to each modality of the sensory stimulus either at the first presentation or at the later

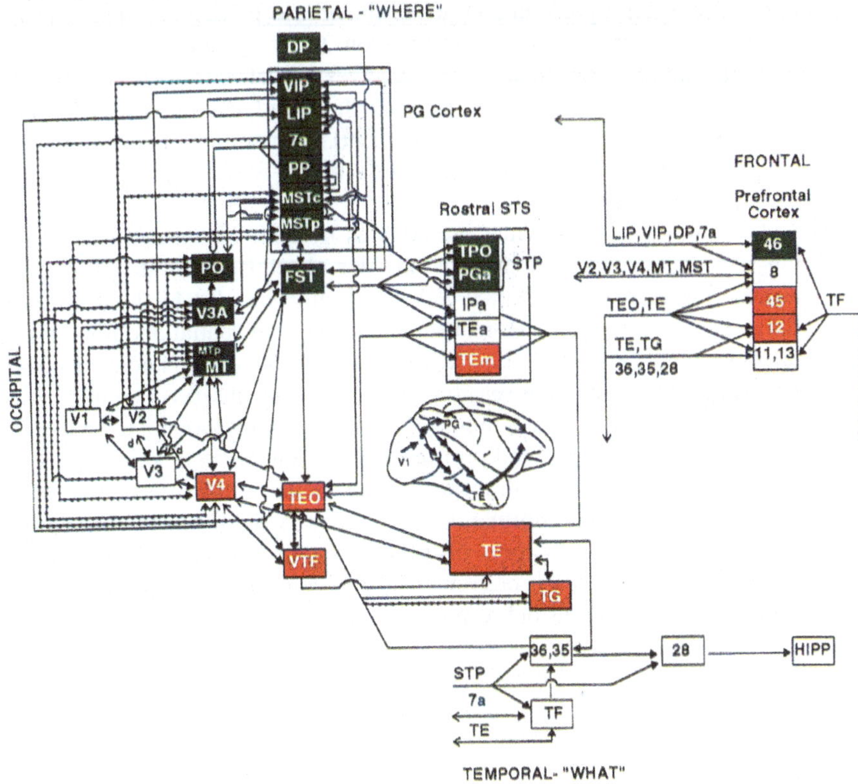

Fig. 5.2 Two visual pathways: parietal "where" pathway and temporal "what" pathway

presentation. Neurons responded to the visual stimulus were accumulated at the posterior-ventral portion of the lateral prefrontal cortex, neurons responded to the auditory stimulus were accumulated at the posterior-dorsal portion, and neurons responded to the tactile stimulus were accumulated at the anterior-ventral portion of the posterior lateral prefrontal cortex. Thus, the dorsolateral prefrontal cortex may be functionally compartmentalized with respect to the different sensory modalities.

Regional differences in spatial and non-spatial visual information processing have also been proposed in the lateral prefrontal cortex. Ungerleider and Mishkin (1982) proposed two distinct visual pathways in the cerebral cortex, which are the dorsal pathway and the ventral pathway (Fig. 5.2). The dorsal visual pathway (parietal pathway) contributes mainly to the processing of spatial aspects of visual information (position and motion direction), while the ventral visual pathway (temporal pathway) contributes mainly to the processing of non-spatial aspects of visual information (color, shape, texture, and object itself). In the lateral prefrontal cortex, the cortex within and surrounding the principal sulcus mainly receives visual information from the posterior parietal cortex through the dorsal pathway (Andersen et al. 1985; Cavada and Goldman-Rakic 1989; Felleman and Van Essen 1991).

Therefore, the dorsolateral prefrontal cortex is thought to participate mainly in the processing of spatial aspects of visual information. On the other hand, the ventral convexity of the lateral prefrontal cortex receives visual information from the temporal cortex through the ventral pathway (Seltzer and Pandya 1989; Petrides and Pandya 1994; Felleman and Van Essen 1991). Therefore, the ventrolateral prefrontal cortex is thought to participate mainly in the processing of non-spatial aspects of visual information.

However, neurons whose activities are affected by both spatial and non-spatial information have been observed in both mid-dorsolateral and mid-ventrolateral prefrontal areas (Quintana et al. 1988; Sakagami and Niki 1994a, b; Rao et al. 1997). Many neurons exhibited task-related activity in both spatial and non-spatial working memory tasks (Rao et al. 1997). In human neuroimaging studies, Rizzuto et al. (2005) showed that spatially selective activation was observed in the human ventrolateral prefrontal cortex where non-spatial information is thought to be mainly processed. Thus, although the lateral prefrontal cortex might be functionally compartmentalized into two sectors with respect to the spatial or the non-spatial information processing, sensory information processing would be performed by extensive interactions among neurons distributing in the wide area of the lateral prefrontal cortex. Therefore, each prefrontal neuron could not respond just to one modality of the sensory stimulus but respond to a range of modalities of the sensory stimulus.

5.1.1 Visual Processing

5.1.1.1 Characteristics of Visual Responses

Responses to visual stimuli are observed in dorsolateral prefrontal neurons. Pigarev et al. (1979) showed that neurons in the posterior portion of the prefrontal cortex (arcuate and pre-arcuate areas) exhibited responses to conventional stationary or moving visual stimuli while the monkey was under a curarized condition. These neurons exhibited no orientation selectivity and no direction selectivity except few neurons. Characteristics of visual responses of prefrontal neurons have also been examined under unanesthetized conditions while the monkey performed conventional fixation task. Ito (1982) examined visual responses of lateral prefrontal neurons using a sensory reaction time task, in which the monkey is required to release the lever as soon as possible when the intensity of either visual or auditory stimulus changes. He found that many prefrontal neurons exhibited either phasic or tonic responses to either visual or auditory stimulus, whereas only few neurons responded to both stimuli. Mikami et al. (1982) examined characteristics of prefrontal visual responses. They presented moving and stationary slit stimuli in the peripheral visual field while the monkey looked at a central fixation point. They found that most of examined neurons exhibited transient excitation to the presentation of the visual stimulus and had visual receptive fields. In addition, some neurons

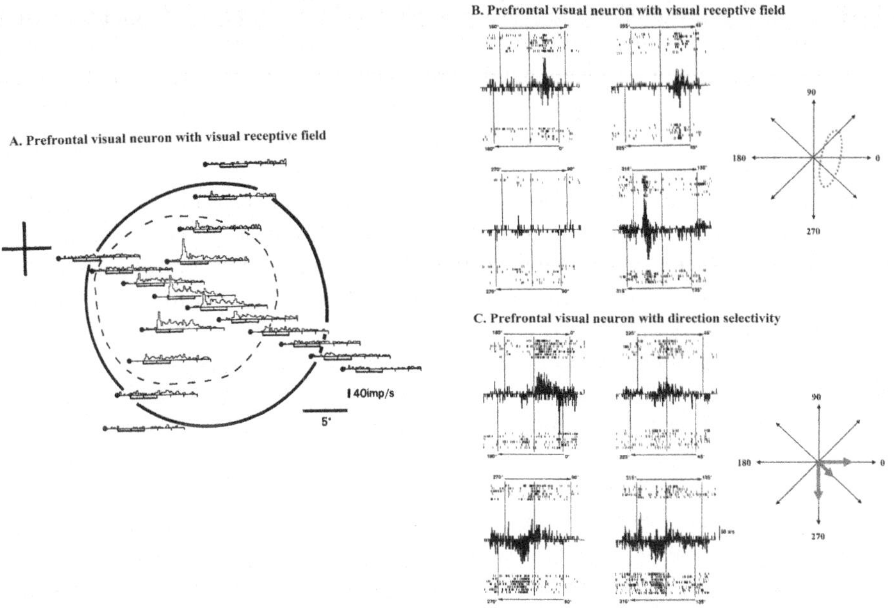

Fig. 5.3 (**a**) Examples of prefrontal visual responses and an estimated position of the visual receptive field. From Suzuki and Azuma (1983) with permission. (**b** and **c**) Two examples of prefrontal visual responses to a moving spot stimulus. (**b**). Visual responses were observed when the spot stimulus passed in the receptive field (dotted circle) regardless of its moving directions. (**c**). Visual responses with direction selectivity. Visual responses were observed when the spot stimulus moved toward particular directions (showed by arrows) in a particular visual field. Directions of the spot movements are indicated by arrows

responded to moving slit stimulus and showed direction selectivity. However, no neuron showed orientation selectivity to stationary slit stimulus. Suzuki and Azuma (1983) also examined features of the visual receptive field of lateral prefrontal neurons (Fig. 5.3a) and found a unique relation between the cortical position of prefrontal visual neuron and the position and the width of the visual receptive field in the visual field.

Direction selectivity had been observed in prefrontal visual neurons when moving visual stimulus was used. Fig. 5.3b, c are examples of directional selectivity of the visual response of prefrontal neurons to a moving spot stimulus. In this study, a small light spot moves between two peripheral positions through the central fixation point (e.g., alternately from the 0° position to the 180° position and from the 180° position to the 0° position, and so on). This method allowed to examine responses of the light spot moving toward and away from the central fixation point in eight directions. Most prefrontal neurons examined showed responses similar as those shown in Fig. 5.3b. Excitatory responses were observed whenever the spot stimulus moved within the neuron's visual receptive field, regardless of moving directions. However, a small number of neurons exhibited excitatory responses only when the spot stimulus moved toward particular directions within the neuron's visual receptive field. An example of a prefrontal neuron shown in Fig. 5.3c had the visual receptive field in the lower right visual field and showed excitatory responses only when the spot stimulus moved within the receptive field in particular directions (rightward, downward, and lower right directions). Thus, prefrontal visual neurons have visual receptive fields in the visual field and some prefrontal visual neurons exhibit direction selectivity to the stimulus moving within the visual receptive field. Hussar and Pasternak (2009, 2013) showed that prefrontal neurons responded to visual motion stimuli as accurately as those in the MT and MST. They also showed that direction selectivity was high during direction discrimination conditions but became low during speed discrimination and simple fixation conditions, suggesting that, although prefrontal neurons exhibit direction selectivity to moving stimuli, the strength of directional selectivity to visual motion stimuli is affected by task conditions.

Some prefrontal neurons exhibit responses to visual stimuli with rather complex visual features, such as natural objects or human and monkey faces (Pigarev et al. 1979; Rizzolatti et al. 1981; Wilson et al. 1993; Scalaidhe et al. 1997). Pigarev et al. (1979) showed that neurons in the pre-arcuate area tended to respond only to complex visual stimuli such as human faces, pieces of white cotton, strips of paper, or wires. Wilson et al. (1993) used a picture fixation task and found that neurons in the ventrolateral prefrontal cortex responded to complex visual stimuli including objects and human or monkey faces. They also showed that responses of these neurons exhibited selectivity to the stimulus, such that one neuron responded to objects but not to faces, while the other neuron responded to faces but not to objects. Ó Scalaidhe et al. (1997) examined whether the area selective for information processing of the object vision was present in the prefrontal cortex or not. They focused on face-selective neurons and found that these face-selective neurons selectively processed information related to the identity of faces and were localized within a restricted area in the ventrolateral prefrontal cortex. Romanski and Diehl (2011) also showed neurons responsive to human and monkey faces in the ventrolateral prefrontal cortex. They indicated that ventrolateral prefrontal neurons responded selectively either to the identity of the face, to the specific view of the face, or to both. Haile et al. (2019) performed the mapping of relative cortical locations where responses to faces, bodies, scenes, disparity, color, and eccentricity were observed in the lateral prefrontal cortex by fMRI while the monkey was maintained fixation. As a result, the face area was localized in the ventrolateral prefrontal area. The color area was located near but distinct area from the face area. The scene area was located in the dorsolateral prefrontal area and overlapped with the disparity area (Fig. 5.4). In the inferior temporal cortex, face-selective neurons had been found in the ventral bank of the superior temporal sulcus and the adjacent cortex in monkeys (Perret et al. 1982; Schwartz et al. 1983; Rolls and Baylis 1986; Baylis et al. 1987). Neurons in the inferior temporal cortex project to the inferior convexity of the lateral prefrontal cortex (Jacobson and Trojanowski 1977a; Kawamura and Naito 1984; Barbas 1988; Rodman 1994; Bullier et al. 1996). Therefore, specific anatomical connections between the inferior temporal cortex

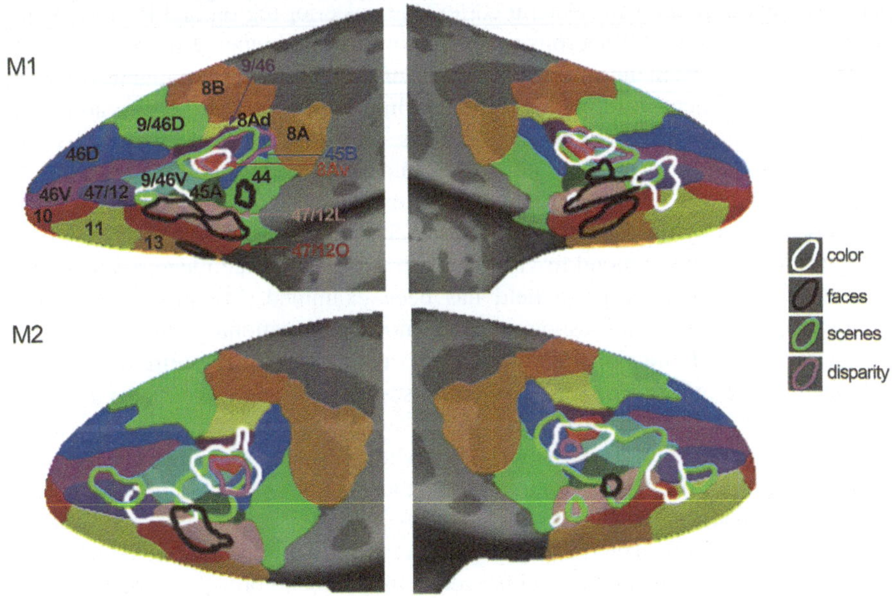

Fig. 5.4 Spatial mappings of relative locations of responses to faces, scenes, disparity, color on the lateral prefrontal cortex. From Haile et al. (2019) with permission

and the ventrolateral prefrontal cortex would cause an accumulation of face-selective neurons in the ventral convexity of the prefrontal cortex.

5.1.1.2 Topographic Representation of Visual Response Features

Anatomical heterogeneity has been shown in the lateral prefrontal cortex. For example, rostro-caudal anatomical heterogeneity has been reported in the prefrontal area 46 (Barbas and Pandya 1989; Pandya et al. 1971; Gerbella et al. 2013; Borra et al. 2019). Gerbella et al. (2013) and Borra et al. (2019) divided the principal sulcal cortex (area 46) into two parts (dorsal part, 46d, and ventral part, 46v) and Borra et al. (2019) further divided the dorsal part of area 46 (46d) into rostral (46dr), middle (46 dm), and caudal (46 dc) portions and showed difference in corticocortical connections among these divided regions. However, it was not known whether anatomical heterogeneity observed in the lateral prefrontal cortex caused functional heterogeneity. Riley et al. (2017) divided the lateral prefrontal cortex into 5 regions (the posterior-dorsal, the mid-dorsal, the anterior-dorsal, the posterior-ventral, and the anterior-ventral regions) and compared characteristics of visual responses recorded from each region based on the selectivity for spatial positions, shapes, and colors of visual stimuli while monkeys viewed visual stimuli passively. In the dorsolateral cortex, neurons in the anterior region showed larger receptive fields, longer response latencies, and lower stimulus selectivity compared to neurons in the posterior region. In the ventrolateral cortex, the posterior region had lower percentage of visual neurons, but neurons in this region showed higher stimulus selectivity compared to neurons in the anterior region. Neurons representing spatial information were more prominent in the dorsal region, while neurons sensitive to colors were more prominent in the ventral region. Thus, functional heterogeneity in visual response properties is present along the anterior-posterior axis as well as the dorsal-ventral axis in the lateral prefrontal cortex. And, this functional heterogeneity would be caused by the anatomical heterogeneity.

Prefrontal neurons respond to visual stimuli and have visual receptive field. The feature of the visual receptive field has been examined. Mikami et al. (1982) examined prefrontal visual responses using moving or stationary slit presented on a wide screen while monkeys performed a fixation task. Many prefrontal neurons exhibited transient excitatory visual responses. Visual receptive fields of these neurons varied in size from 10° × 10° to 60° × 60° and were located mainly in the visual field contralateral to the hemisphere where neurons were located, although some neurons had visual receptive fields in the ipsilateral visual field or in both sides of the visual field.

Suzuki and Azuma (1983) examined features of visual receptive fields of lateral prefrontal neurons and showed unique relations between cortical positions of prefrontal visual neurons and features of their visual receptive fields in the visual field (Fig. 5.5a). Visual neurons were distributed widely in the pre-arcuate and inferior lateral portions of the prefrontal cortex. Activities of visual neurons in each hemisphere mainly represented visual information presented in the contralateral visual field. The width of the receptive field and the eccentricity of the receptive field in the retinal position varied depending on the position of the neuron on the cortical surface. Neurons located in the inferior portion of the inferior lateral prefrontal cortex and close to the inferior arcuate sulcus had smaller size of visual receptive fields and represented visual information presented in the foveal and parafoveal region. On the other hand, neurons located in the superior portion of the inferior lateral prefrontal cortex had larger size of receptive fields and represented visual information presented in the more peripheral region (Fig. 5.5, A2 and A3). The size of the receptive field gradually increased as neurons located more anteriorly without change of the eccentricity of the position of the receptive field (Fig. 5.5, A1 and A4). Based on these observations, they concluded that the inferior lateral part of the prefrontal cortex is functionally differentiated regarding visual properties, such that the inferior portion is mainly related to the information processing for the central vision, while the superior part is related to the information processing for the peripheral vision.

Topographic representation of the receptive fields of visual neurons on the lateral prefrontal cortex suggests the presence of a visuospatial map in this cortical area. By human fMRI studies, visuospatial maps representing the visual field have been identified not only in the striate and prestriate cortices (Arcaro et al. 2009; Engel et al. 1997; Gardner et al. 2008; Kolster et al. 2010; Reppas et al. 1997) but also in the posterior parietal cortex (Arcaro et al. 2011; Sereno et al. 2001; Silver et al. 2005). The visuospatial map representing the visual field has also been demonstrated

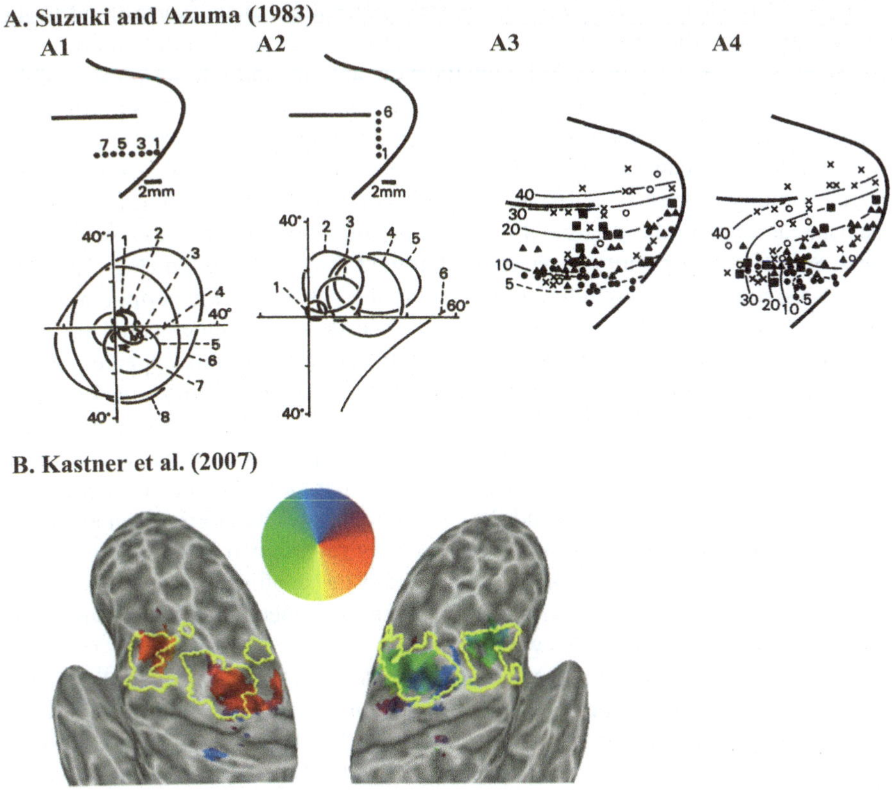

Fig. 5.5 (**a**) Topographic representations of the size and the center eccentricity of visual receptive fields of prefrontal neurons. The size of the visual receptive field gradually increases for anteriorly located neurons without changing its center eccentricity (*A1* and *A4*). More dorsally located neurons have a larger size of the receptive field which represents more eccentric region from the fovea (*A2* and *A3*). From Suzuki and Azuma (1983) with permission. (**b**) Two topographic spatial maps are found in the frontal cortex; one located in the superior branch of the precentral sulcus and caudal portion of the superior frontal sulcus (correspond to the frontal eye field) and the other located in the inferior branch of the superior precentral sulcus and the caudal part of the inferior frontal sulcus. From Kastner et al. (2007) with permission

in the lateral prefrontal cortex. Hagler Jr and Sereno (2006) used an n-back task using human face pictures and asked the subject to judge whether the picture presented at 2-back before matches to the present picture with respect to either identity or location. In the identity condition, the subject needs to monitor whether a particular face is presented repeatedly or not, regardless of whether pictures are presented at the same position or not. In the location condition, the subject needs to monitor whether a face is presented repeatedly at the same position or not, regardless of whether the same face is presented or not. The subject fixates a fixation point and needs to press either button to indicate 2-back match or no match. In the location condition, two visuospatial maps were found; one was found in the inferior frontal sulcus and the other was found in the superior precentral sulcus (correspond to the frontal eye field). Both maps preferentially represented visual information presented in the contralateral visual field. Although the more anterior and inferior portion of the lateral prefrontal cortex was activated during the identity condition, this portion had no visuospatial map. These results indicate that the human lateral prefrontal cortex has multiple visuospatial maps and that these topographic maps represent the visual space in orderly and reproducible manner. Kastner et al. (2007) also showed similar visuospatial maps in the prefrontal cortex using memory-guided response tasks in which 8 or 12 target positions were placed around the fixation point (Fig. 5.5b). In these tasks, the subject is required to remember the position of the peripheral target for several seconds and then either makes a saccade to the remembered position (memory-guided saccade task) or judges whether a test stimulus appeared at the same position or not by pressing a button (spatial working memory task). They found two topographic visuospatial maps in the frontal cortex; one located in the superior branch of the precentral sulcus and the caudal portion of the superior frontal sulcus (correspond to the frontal eye field) and the other located in the inferior branch of the superior precentral sulcus and the caudal portion of the inferior frontal sulcus (Fig. 5.5b). Both visuospatial maps predominantly represented information regarding the contralateral direction of the saccade and the contralateral position of the target. In addition, particular saccade directions and memorized positions were represented in multiple locations within the map.

Thus, both humans and monkeys have multiple visuospatial maps in the lateral prefrontal cortex. These maps are thought to be useful for attending task-relevant visual stimuli present at various spatial positions (Hagler Jr and Sereno 2006) or integrating or interacting spatially separate objects (Suzuki and Azuma 1983). Recently, O'Rawe and Leung (2020) performed connective field mapping to estimate corticocortical (region-to-region) resting-state functional connectivity using resting-state fMRI data. Although topographic, convergent, and divergent connectivity have been postulated to perform different information processing in a given neural network (e.g., convergent connectivity in the visual network is necessary for object information processing), they showed that functional connectivity based on the topographic relationship is maintained between early visual areas, that this topographic organization is a fundamental component of functional connectivity across entire cortex, and that stronger topographic connectivity is found within a cortical network for a particular function than among networks for various functions. The topographic functional connectivity was found in the lateral prefrontal cortex and organized in two axes. One axis consists of the posterior superior frontal gyrus - > the anterior inferior frontal gyrus, which corresponds to the dorsal-ventral functional gradient, and is related to the difference of information-domain specific processing (spatial vs object). The other axis consists of the posterior middle frontal gyrus - > the anterior inferior frontal gyrus and the frontopolar cortex, which corresponds to the posterior-anterior gradient, and is related to the gradient from more concrete information processing such as sensory-motor processing to more abstract information processing such as rule. Topographic functional connectivity contributes to transferring visuospatial information from the posterior cortices to the prefrontal cortex and processing information within the lateral prefrontal cortex and would influence topographic representation of the visuospatial information in the lateral prefrontal cortex.

5.1.1.3 Cue-Period Activity Observed During Performance of Cognitive Tasks

Visual response is also observed in prefrontal neuron while the monkey performs a variety of behavioral tasks. Simple visual stimulus such as a square or circle has been usually used in the delayed-response tasks, while colors, complex shapes, or objects have been used in the delayed matching-to-sample task and the discrimination task. Prefrontal neurons are known to exhibit selective responses to spatial or physical features of the cue, sample, or test stimulus while the monkey performs these behavioral tasks.

5.1.1.3.1 Visual Responses Observed During Delayed-Response Performances

Transient excitation or inhibition to the presentation of the visual cue has been widely observed in prefrontal neurons when the monkey performs the delayed-response task. Responses to the visual cue had been observed during the performance of classical manual delayed-response tasks, in which the visual cue was presented randomly at either right or left position (two-choice delayed-response task). For example, Fuster (1973) classified observed task-related prefrontal activity into five types (A, B, C, D, and I). Type A exhibits phasic excitatory responses at the onset of the cue period, type B exhibits increased discharge throughout the cue period, and type D and I exhibit inhibition during the cue period. Thus, prefrontal neurons are shown to exhibit either excitatory or inhibitory responses to the visual stimulus during delayed-response performance. Kubota et al. (1974) observed neurons exhibiting visual responses to the visual cue in the delayed-response task and called these neurons "visuokinetic neurons" because these neurons not only responded to the visual cue but also were active during the response behavior. Niki (1974) reported phasic excitatory responses to the onset of either the right or the left visual cue. Kojima and Goldman-Rakic (1984) showed spatially discriminative cue-period activity during the delayed-response task.

Although spatially discriminative cue-period activity had been observed during delayed-response performances, it was not clear whether spatially discriminative feature of cue-period activity was related to the feature of the visual receptive field of prefrontal visual neurons, because only two spatial positions (left and right) were used and because the monkey's eye movements were not controlled and monitored. Kojima (1980) examined prefrontal visual responses under the condition in which the monkey's eye movements were controlled during two-choice delayed-response performances, and showed that the majority of prefrontal neurons fired vigorously when the visual cue was presented in the visual field contralateral to the recording hemisphere. Contralateral bias of cue-period activity agrees with the features of visual responses of prefrontal neurons obtained by Mikami et al. (1982) and Suzuki and Azuma (1983). Therefore, spatially discriminative feature of cue-period activity is thought to reflect the feature of visual receptive field of prefrontal visual neurons. However, to further examine spatially discriminative features of cue-period activity and compare with the spatial feature of the visual receptive field of prefrontal neurons, cue-period activity is needed to be examined using multiple cue positions distributed in the whole visual field under controlling eye movement.

5.1.1.3.2 Visual Responses in the ODR Task

An oculomotor version of the delayed-response (ODR) task has been used to examine prefrontal activity. In this task, a spatial visual cue is presented randomly at one of multiple peripheral positions while the monkey is maintained fixation at the central fixation point. Since the same visual cue is presented at multiple positions around the fixation point, cue-period activity can be examined at multiple cue positions distributed in the whole visual field. Further, since the monkey needs to maintain fixation at the central fixation point during the presentation of the visual cue, spatial feature of cue-period activity can be depicted by the similar way as to depict spatial feature of the visual response and compared with the feature of the visual receptive field of prefrontal visual neurons. In addition, since the monkey is required to make a saccade to the position where the visual cue was presented, detecting the visual cue and remembering the position of the visual cue are crucial for correct performance of this task. Therefore, it would be possible to examine whether cue-period activity is a simple response to the visual cue or includes additional cognitive factors such as attention or behavioral context.

Response to the visual cue (cue-period activity) was observed in 28% (74/261) of task-related prefrontal neurons during ODR performances (Fig. 5.6a) (Funahashi et al. 1990). Most (93%) of these responses were transient excitation, while the remaining (7%) were transient inhibition. Response latencies of excitatory cue-period activity were distributed between 37 and 309 ms and the median was 116 ms after the onset of the visual cue. Most (96%) of cue-period activity showed directional selectivity, such that significant cue-period activity was observed only when the visual cue was presented within a particular visual field, while the remaining showed omnidirectional selectivity, such that significant cue-period activity was equally observed in all cue positions tested. Characteristic of directional selectivity of cue-period activity was examined by making a tuning curve which was constructed by implementing the mean discharge rate at each cue position in Gaussian function (Fig. 5.6b). Using obtained tuning curve, the best direction was determined as the direction at which the maximum cue-period activity was observed in the tuning curve. The majority of prefrontal neurons (71%) had best directions toward the visual field contralateral to the hemisphere where the neurons were

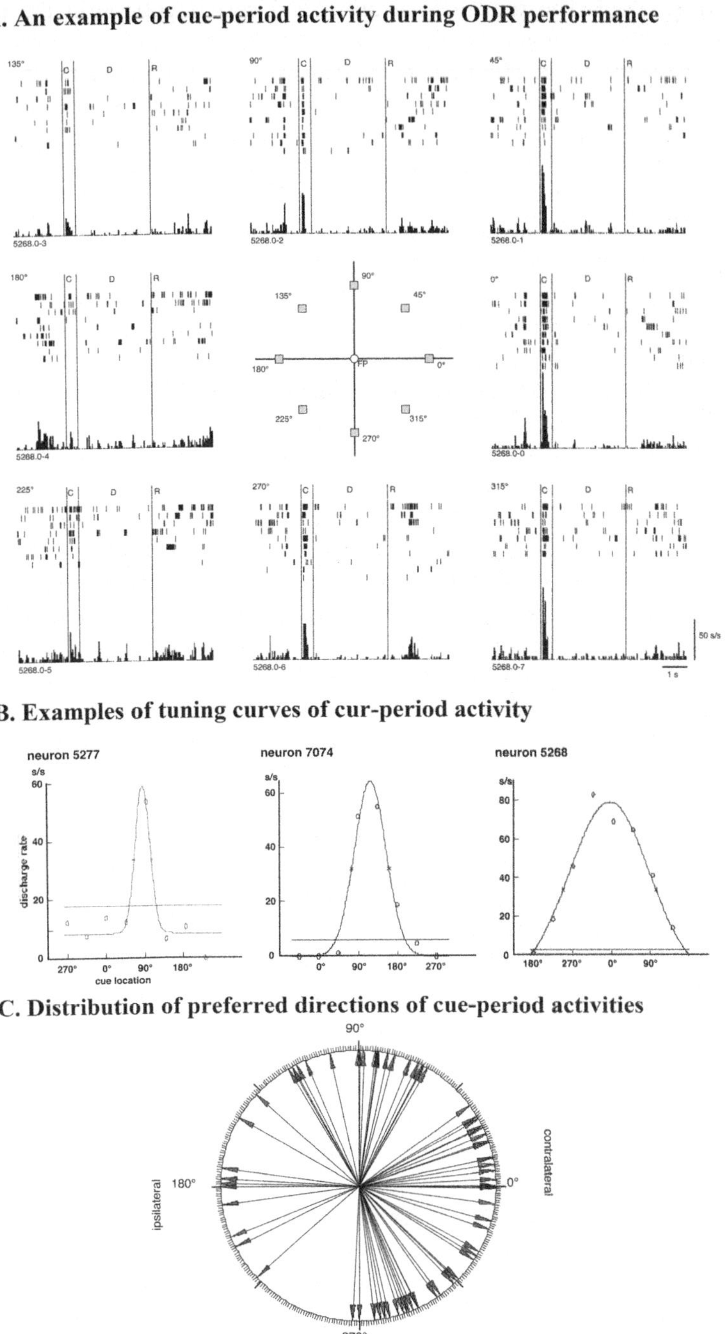

Fig. 5.6 (**a**) An example of cue-period activity observed during the ODR performance. (**b**) Three examples of tuning characteristics of cue-period activities. (**c**) Distribution of preferred directions of cue-period activities. From Funahashi et al. (1990) with permission

located (Fig. 5.6c). The remaining neurons had best directions in the ipsilateral visual field (18%) or along the vertical meridian (11%) (Fig. 5.6c). The width of the directional tuning of cue-period activity was also estimated by the standard deviation parameter of the Gaussian curve. The median value of the tuning width was 37°. These results generally agreed with the features of visual responses (e.g., response latency, receptive field size, and contralateral bias) of prefrontal visual neurons.

It has been shown that the magnitude of the visual response is significantly enhanced when the same visual stimulus is used as a target for a saccadic eye movement. The enhancement of the visual response has been considered to be an effect of attention and widely observed in posterior parietal neurons (Robinson et al. 1978; Bushnell et al. 1981) and frontal eye field neurons (Wurtz and Mohler 1976; Goldberg and Bushnell 1981; Bruce and Goldberg 1985). The visual cue in the ODR task is a behaviorally significant signal, because its position needs to be remembered during the delay period and utilized it as the saccade target during the response period. Therefore, the enhancement may be observed in the response to the visual cue. To examine this issue, the response magnitude and the spatial tuning characteristic were compared between cue-period activity observed in the ODR task and the visual response observed in the visual probe task for the same prefrontal neuron (Funahashi et al. 1990). As a result, strong positive correlations were found for both best directions and tuning widths of tuning curves between cue-period activity in the ODR task and the visual response in the visual probe task. To examine how much behavioral significance of the visual stimulus affected the magnitude of the visual response, an enhancement index (the response to the visual cue in the ODR task relative to the response to the same stimulus in the visual probe task) was calculated for neurons having cue-period activity. The majority (79%) of prefrontal neurons had enhancement indexes between 0.5 and 2.0 and the median was 1.12, indicating that no or little enhancement effect was observed when the magnitude of cue-period activity was compared with the magnitude of the visual response in the visual probe task for individual prefrontal neurons.

These results suggest that the same visual process via receptive field activation elicit cue-period activity in the ODR task and visual response in the visual probe task in prefrontal neurons. In the ODR task, the position of the visual cue is an important information to perform a forthcoming saccade. The spatial information of the visual cue is maintained as directional delay-period activity, which is utilized for saccade performance during the response period. Prefrontal neurons having both cue- and delay-period activity showed similar best directions in both activities. Therefore, cue-period activity in the ODR task may represent visual input received from the posterior cortices and act as a trigger signal to generate directional delay-period activity in prefrontal neurons.

5.1.1.3.3 Visual Responses in Matching-to-Sample Tasks and Discrimination Tasks

In the delayed matching-to-sample task and the visual discrimination task, complex shapes and multiple colors are often used as visual stimuli. Many prefrontal neurons have been shown to respond these stimuli. For example, prefrontal neurons responded to color stimuli presented as sample or matching stimuli. Kubota et al. (1980) reported that prefrontal neurons showed excitatory (83%) or inhibitory (17%) responses to color stimuli (green and red) and that some neurons (15%) exhibited color selectivity. Fuster et al. (1982) also showed that prefrontal neurons exhibited differential (13%) or nondifferential (41%) responses to red and green stimuli during the delayed matching-to-sample task. Quintana et al. (1988) used color stimuli in a delayed matching-to-sample task (red and green) and a delayed conditional position task (yellow and blue) and showed that 55% of sampled prefrontal neurons exhibited responses to color stimuli and that 23% of these responded to only one color while 32% responded to more than two color stimuli. In discrimination tasks, Komatsu (1982) used a Go/No-Go color-discrimination task and showed that 32% of task-related prefrontal neurons had phasic responses to the start light (yellow) or the cue light (red or green). Watanabe (1981, 1986a) showed that 30–40% of prefrontal neurons differentially responded to the color cues (red and green) and the pattern cues while the monkey performed a delayed conditional discrimination task.

Miller et al. (1996) used visual stimuli with complex patterns as sample stimuli in the delayed matching-to-sample task. They found 76% of recorded neurons (110/145) responded to visual stimuli and, among them, 68% and 32% exhibited excitatory and inhibitory responses, respectively. They also found that 37% of visually responded neurons (41/110) exhibited stimulus selectivity. In their task, the same stimulus was presented as a match stimulus during the choice period in one trial or as a nonmatch stimulus (distractor) during the delay period in other trials. Among 75 neurons having excitatory visual responses, 68% (51/75) varied their responses depending on whether the same visual stimulus was used as a matching stimulus or as a distractor, and 63% (23/51) exhibited enhancement of the visual response and 37% (19/51) exhibited suppression of the response when the visual stimulus was used as a matching stimulus.

Asaad et al. (2000) reported task-specific prefrontal activity using three types of working memory tasks (delayed-response task, delayed matching-to-sample task, and delayed conditional task) (Fig. 5.7). In the delayed matching-to-sample task (object task), one sample object is briefly presented over the fixation point and, after a brief delay (1.0 s), the monkey needs to select the sample object by a saccade. In the delayed-response task (spatial task), a small light spot (visual cue) is briefly presented at either the right or the left and, after the delay (1.0 s), the monkey needs to make a saccade to the position where the visual cue was presented. In the delayed conditional task (associative task), a sample object is briefly presented over the fixation point and, after a 1.0-s delay, the monkey needs to make a saccade to the position (right or left) associated with the sample object. They found that many prefrontal neurons exhibited task-dependent difference of activity during the pre-cue

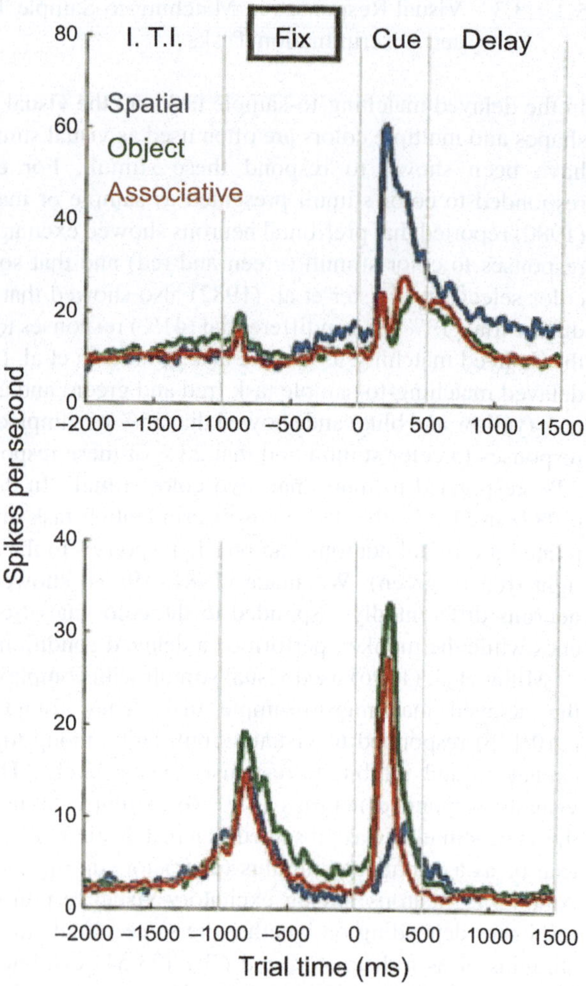

Fig. 5.7 Examples of task-dependent modulation of visual responses for two prefrontal neurons. The activity of the neuron shown in the upper figure diverged after the presentation of the spatial cue in the delayed-response task, while the activity of the neuron shown in the lower figure diverged after the presentations of the spatial cue in the delayed-response task and the associative cue in the delayed conditional task. From Asaad et al. (2000) with permission

fixation period, the cue period, or the response period. Although many neurons (53%, 163/305) exhibited sensory-related activity, this activity was modulated by the difference of the task in significant number of neurons (27%, 84/305). Task-dependent difference of the activity included the modulation of overall firing rate, temporal pattern of firing rates, and stimulus selectivity. Left and right saccades used across all tasks were the same, but stimulus conditions were completely different among tasks. Therefore, task-dependent difference of cue-period activity was mostly caused by the difference of stimulus conditions, although some task-dependent difference of cue-period activity was affected by the difference of the behavioral rules.

In the delayed matching-to-sample task and the discrimination task, physical features of the visual stimulus (e.g., color, shape, position, size, brightness) are

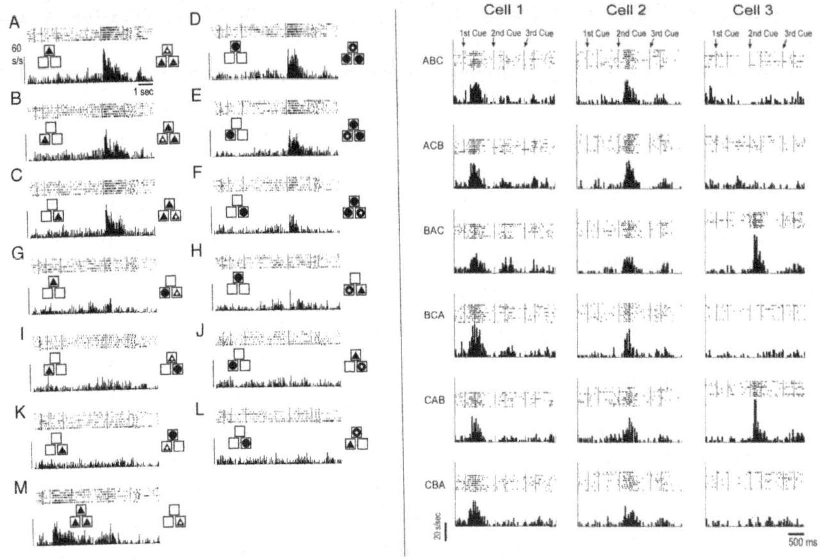

Fig. 5.8 *Left*: Visual responses affected by both position and shape of the stimulus. Visual response was observed only when the same visual stimulus was presented at all three positions. From Hoshi et al. (2000) with permission. *Right*: Visual responses affected only by a rank order of stimulus presentation, regardless of the stimulus itself. From Ninokura et al. (2004) with permission

important factors for the subject to pay attention and maintain during the delay period. Prefrontal neurons have been shown to exhibit selectivity to a particular physical attribute of the visual stimulus. However, in these tasks, each physical attribute of the visual stimulus often has different behavioral meaning or significance. If the subject needs to pay attention to different attribute of the same visual stimulus to perform different response in the same task, the response of prefrontal neurons would become more complex. Hoshi et al. (2000) examined prefrontal activity while the monkey performed a delayed matching-to-sample task under two conditions (location-matching and shape-matching conditions). During the sample period, either a circle or a triangle is presented as a sample cue at one of three positions (right, left, and top). They found that, among 279 prefrontal neurons responded to the sample stimulus, the response to the sample stimulus was affected by the position of the stimulus in 126 (42%) neurons, by the shape of the stimulus in 32 (11%) neurons, and by both position and shape of the stimulus in 63 (21%) neurons (Fig. 5.8, left). These results indicate that prefrontal visual response represents not only one physical attribute of the visual stimulus (e.g., shape or position) but also a combination of multiple attributes (e.g., both shape and position) depending on the requirement of the task.

Ninokura et al. (2004) used a delayed multiple reaching task, in which, while the monkey looks at a fixation point, three object stimuli are sequentially presented as sample cues over the fixation point for 0.5 s with a 1-s inter-stimulus interval. The monkey needs to make sequential reaching responses toward the positions where sample cues were presented after a delay. Therefore, the monkey needs to remember the temporal order of the sample cue presentation. Among 539 neurons responded to sample cues, 141 (26%) were selective for physical attributes (color and/or shape) of the sample cues, 235 (44%) were selective for the rank order (e.g., 94, 63, and 39 were selective for the first, the second, and the third rank, respectively) and irrespective of physical attributes of the sample cues (Fig. 5.8, right), and 163 (30%) were selective for both the physical attributes of the sample cues and their rank order.

Further, Inoue and Mikami (2006) used a serial probe reproduction task (conditional delayed matching-to-sample task) and examined prefrontal responses to three object stimuli (sample and test stimuli) and two colored-stimuli (condition stimuli). In this task, two object stimuli are sequentially presented for 0.5 s with a 1-s inter-stimulus interval and, after a 1-s delay, a color stimulus instructs the monkey which object it needs to select by an eye movement. Among 81 neurons responded to object stimuli, the response magnitude to the same stimulus was not affected by the order of its presentation in 22 neurons (27%). However, in 59 neurons (73%), the response magnitude was affected by the order of the presentation and 33 (56%) showed larger response at the first presentation and 26 (44%) showed larger response at the second presentation.

Kadohisa et al. (2014) found specific difference in temporal development of the visual response between two prefrontal hemispheres. During 50–250 ms after the stimulus onset, the visual response observed in one hemisphere of the prefrontal cortex was restricted to the stimulus presented in the contralateral visual field and exhibited selectivity for both object and category. However, during 300–500 ms after the stimuli onset, many prefrontal neurons responded more strongly to the ipsilateral stimulus than the contralateral stimulus and exhibited selectivity for category. Thus, prefrontal activity encodes a variety of behaviorally relevant visual information including objects, their categories, and their spatial locations.

5.1.1.4 Discrimination of Relevant Visual Stimuli from Irrelevant Ones

The dorsolateral prefrontal cortex has reciprocal connections with the posterior parietal cortex and the inferior temporal cortex and receives visual information from these cortical regions. Therefore, characteristics of visual responses observed in prefrontal neurons are expected to be similar to those observed in posterior parietal neurons and inferior temporal neurons. However, characteristics of visual responses are not always the same between the prefrontal cortex and the posterior cortices.

Prefrontal neurons discriminate behaviorally relevant visual stimuli from irrelevant stimuli and respond only to the relevant stimuli. Yajeya et al. (1988) occasionally inserted behaviorally irrelevant and inconsequential color stimulus (violet) as a sample stimulus while the monkey performed the delayed matching-to-sample task using red and green sample stimuli and the delayed conditional position discrimination task using yellow and blue sample stimuli. In these tasks, the monkey is required to keep pressing a lever by hand during the sample period and the subsequent delay period. When the violet is presented as a sample stimulus, the monkey needs to perform the same behavior as in ordinary trials, but the trial ends without reward. They compared visual responses of prefrontal neurons between behaviorally relevant color stimuli and an irrelevant stimulus and found that the most of prefrontal neurons not only exhibited differential responses to behaviorally relevant sample stimuli but also discriminated irrelevant violet stimulus from relevant stimuli, indicating that prefrontal neurons differentiate visual stimuli not only by the difference of their physical attributes but also by the difference of their behavioral significance.

Prefrontal neurons respond only to the behaviorally relevant visual stimulus, even though irrelevant visual stimuli are presented simultaneously with the relevant stimulus. Rainer et al. (1998a) used a delayed matching-to-sample and location task and examined whether prefrontal visual responses encoded only behaviorally relevant information or not when irrelevant stimuli were presented simultaneously. In this task, three objects are presented simultaneously and one of them is the target during each block of trials. The monkey is taught which object is the target before starting each block. During the sample period, three objects are presented at three fixed positions simultaneously. After the 1.5-s delay period, the same three objects are presented at the same three positions but in different spatial arrangement. The monkey can get a reward if it releases the lever when the target object is presented at the same position as it was presented during the sample period. Rainer et al. (1998a) found that the magnitude of prefrontal sample-period activity varied depending on whether the target object was presented at the neuron's preferred position or not, regardless of which object was a target. However, the presentation of the non-target object at the neuron's preferred position had little or no effect on sample-period activity. Information regarding whether the target object was presented at the neuron's preferred position or not was appeared in prefrontal activity as early as 140 ms after the presentation of the target object. In addition, only relevant information (the target object was presented at the neuron's preferred position) was represented in delay-period activity. These results indicate that prefrontal activity represents only behaviorally relevant information, suggesting that prefrontal neurons participate in the selection of behaviorally relevant information from various information.

Similar observation was reported by Hasegawa et al. (2000). They used a visual search task with a delay. Gratings with different spatial frequencies are used as target and non-target stimuli. The monkey is required to find the target grating whose spatial frequency differs from non-target gratings, all of which have the same spatial frequency. In this task, the target grating is presented simultaneously with 5 non-target gratings during the cue period and, after a brief delay (0.8–1.5 s), the monkey is required to make a memory-guided saccade to the position where the target grating was presented. They observed stronger responses during the cue period only when the target stimulus was presented within the neuron's visual receptive field and no or little response when non-target stimulus was presented

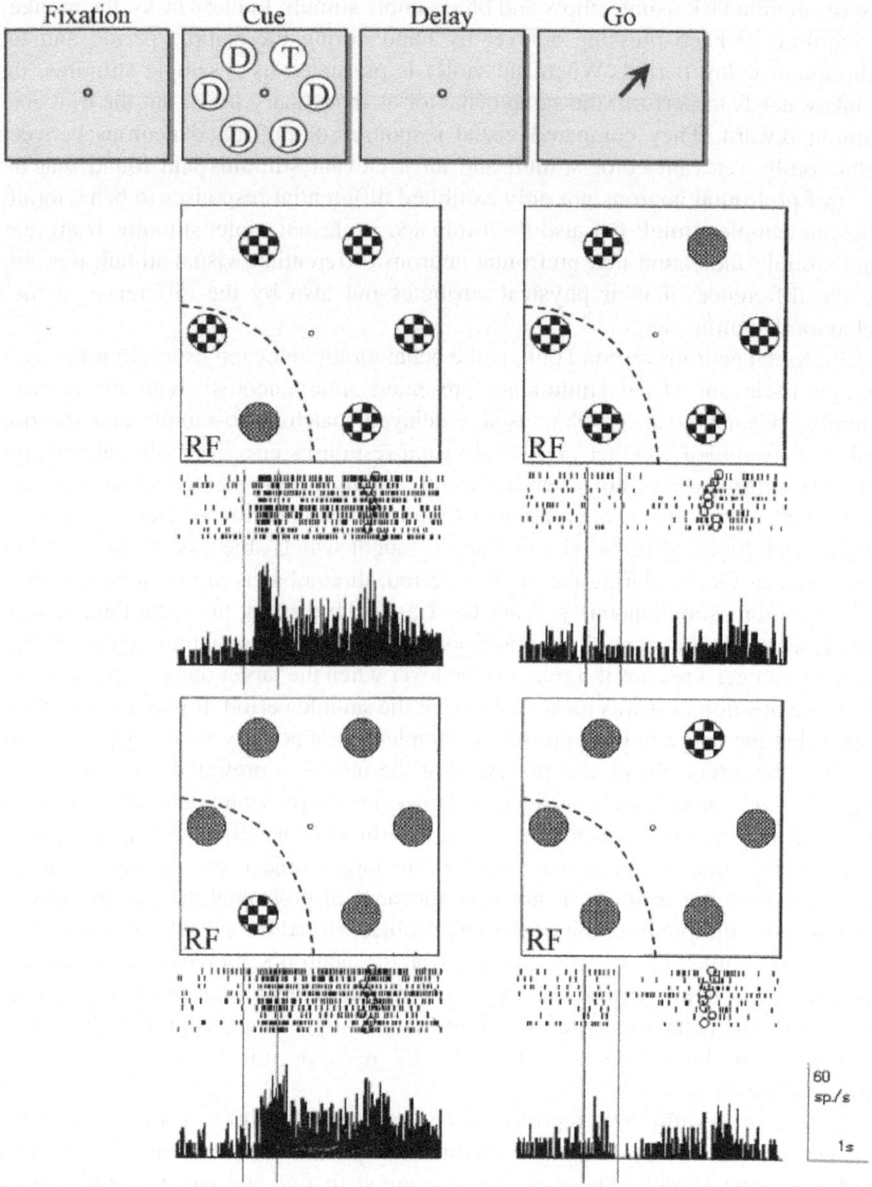

Fig. 5.9 An example of prefrontal visual responses which identify behaviorally relevant stimulus and discriminate whether the relevant stimulus is presented at a particular visual area or not. From Hasegawa et al. (2000) with permission

within the neuron's visual receptive field (Fig. 5.9). Prefrontal visual neurons have visual receptive fields and exhibit significant responses only when the behaviorally relevant stimulus is presented within neurons' visual receptive fields.

However, Lauwereyns et al. (2001) showed that prefrontal neurons also responded to task-irrelevant visual features. They trained the monkey to perform a multi-dimensional Go/No-Go discrimination task, in which the monkey is required to discriminate either color or motion direction of the visual target to perform a correct hand movement. The essence of this task is that the monkey needs to extract the relevant feature (color in the color-discrimination condition and motion direction in the motion-discrimination condition) and ignore irrelevant feature (motion direction in the color-discrimination condition and color in the motion-discrimination condition). They found that around 40% of prefrontal activity were affected by task-irrelevant visual feature and that the response to the task-irrelevant feature was positively correlated with the response to the same feature when it was relevant.

Moran and Desimone (1985) showed that, when both relevant (attended) and irrelevant (unattended) visual stimuli were presented within a neuron's visual receptive field, the response to the irrelevant stimulus was dramatically attenuated in area V4 neurons and inferior temporal neurons, but was not attenuated in prestriate neurons. This result suggests that the filtering relevant information from irrelevant information by attention occurs in the extrastriate and the inferior temporal cortex. Prefrontal neurons receive visual information from the posterior association cortices. Therefore, prefrontal neurons might receive only relevant visual information that has been filtered in the inferior temporal cortex. However, there is also a possibility that the filtering relevant visual information from irrelevant information is performed in the prefrontal cortex. Further examination is needed to understand how behaviorally relevant visual information is selected from various information and whether prefrontal neurons participate in this selection process or just receive relevant visual information which has been filtered in the posterior cortices.

5.1.1.5 Behavioral Context-Dependent Visual Responses

Prefrontal neurons exhibit selectivity to various physical features of the visual stimuli used in behavioral tasks. In these tasks, visual stimuli play an important role to instruct the subject which behavioral response is needed at each given trial. Stimulus selectivity and response magnitude of prefrontal neurons are affected by whether the stimulus is behaviorally relevant or not. The response to a visual stimulus is affected by the difference of its behavioral context. In fact, studies show that, although prefrontal neurons exhibit differential responses to different visual cues, these differential responses represent not only the difference of physical properties of the visual cues but also the difference of behavioral meanings of these cues.

Watanabe (1981) examined prefrontal activity using conditional discrimination tasks with delay. In these tasks, two horizontally arranged response panels (right and left) are placed in front of the monkey. First, two panels are illuminated with either red or green as an instruction cue (IC) and, after a first brief delay, two pattern stimuli (stripe and circle or cross and square) are presented on response panels as discriminative cues (DC). After a second brief delay, both panels are illuminated with white, then the monkey is required to press either right or left response panel by hand. Selecting the response panel having either a circle or a cross is a correct response if IC was red, while selecting the response panel having either a stripe or a square is a correct response if IC was green. Watanabe (1981) found that prefrontal neurons exhibited differential responses to different color or pattern stimuli during the IC or DC period. However, differential visual responses in some prefrontal neurons were not related to the physical difference of pattern stimuli but related to the difference of the response positions where the monkey needed to press. These results show that, although differential responses of many prefrontal neurons are caused by the different physical features of the visual cues, differential responses to different visual cues in some prefrontal neurons are caused by the difference of the behavioral meanings (respond to the right or the left) of the visual cues independent of their physical features (stripe, circle, cross, or square).

Watanabe (1986a) also examined prefrontal visual responses representing not the physical features but the behavioral meaning of the visual stimulus. He used a delayed conditional Go/No-Go discrimination task (Fig. 5.10) and showed that prefrontal visual responses not represented physical features of the visual stimulus but represented information regarding which response (Go or No-Go) was needed to perform. In this task, an instructive color cue (red or green) is first presented, then a discriminative pattern cue (circle or stripe or red or green) is followed after a brief delay. Depending on the combination of two cues, the monkey is required to perform either Go or No-Go response. Among prefrontal neurons having visual responses, one-third of these neurons exhibited differential responses depending on the difference of physical features of either instructive or discriminative cues. The other one-third of neurons exhibited differential responses to different discriminative cues, but their responses were affected not by the difference of the physical features of the discriminative cues but by the difference of behavioral responses (Go or No-Go response). The remaining one-third of neurons exhibited differential responses to different discriminative cues. Their responses were affected by both the difference of the physical features of the discriminative cues and the difference of behavioral responses. These results indicate that prefrontal visual activity not only represents physical features of the visual stimulus but also represents other aspects that the visual stimulus represents including its behavioral relevance or its behavioral meaning (respond to the right or the left, or make Go response or No-Go response).

Watanabe (1990) further showed context-dependent prefrontal visual responses using a stimulus-reward association task, which is similar as those previously used (Watanabe 1981, 1986a). In this task, a visual cue (either circle or stripe, or either cross or square) is first presented, and, after a 1-s delay, an imperative cue (either red or green) is presented. Either pattern is selected randomly and used as the visual cue at any given trial, while either color is used as the imperative cue throughout a block of trials. Specific combination of a pattern and a color (circle and red, stripe and green, plus and red, and square and green) is associated with reward delivery when the monkey correctly releases a lever. However, other combinations of stimuli are associated with no reward delivery even when the monkey correctly releases a lever. Thus, in this task, the monkey is able to predict whether the reward would be given

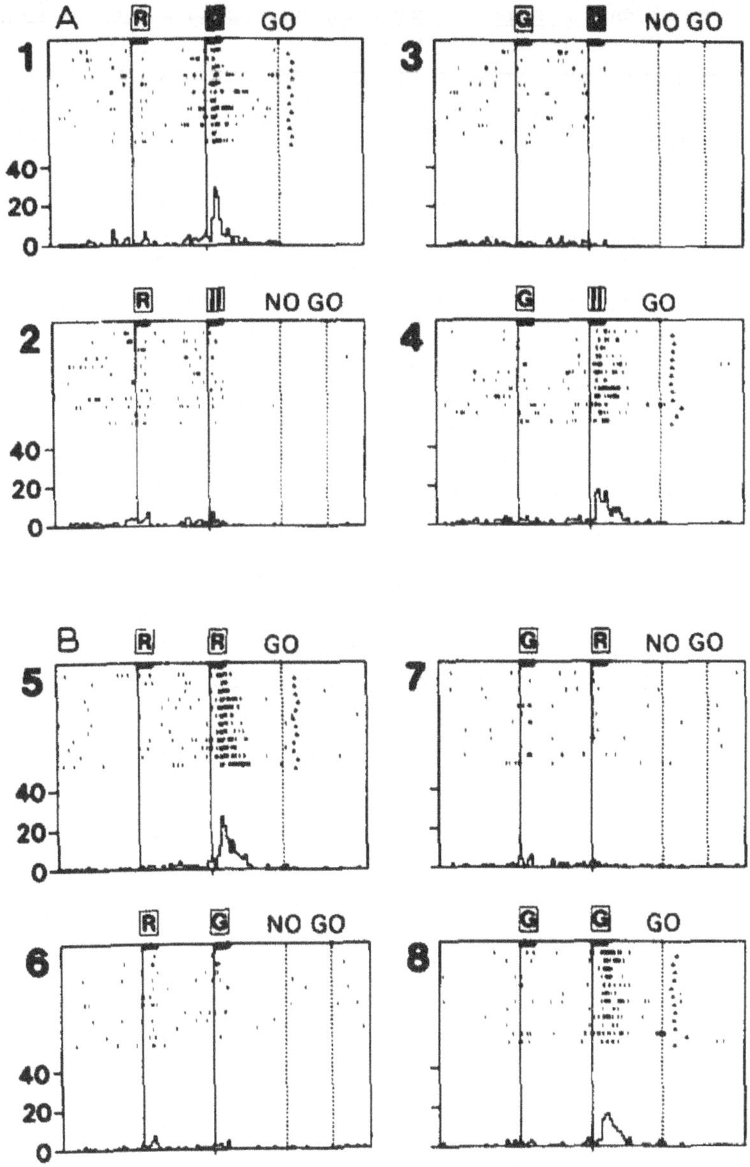

Fig. 5.10 An example of prefrontal visual responses not representing physical features of the visual stimulus but representing information regarding whether Go or No-Go response was needed to perform. From Watanabe (1986a) with permission

or not when the monkey sees the visual cue. Among 179 prefrontal neurons exhibiting differential responses to pattern stimuli, only 7% showed differential responses depending on the difference of physical features of the stimuli, while

67% showed differential responses depending on whether the pattern stimulus is associated with reward delivery or not, regardless of the difference of the physical features. Watanabe (1992) further examined features of the stimulus-reward association observed in prefrontal activity using visual and auditory association tasks and found that some prefrontal neurons exhibited differential responses in relation to the stimulus-reward association independent of the difference of the modalities of the cues. These results further indicate that prefrontal visual responses not only represent physical features of the stimulus, but also represent other aspects of the visual stimulus independent from its physical features, such as information regarding the stimulus-reward association, behavioral relevance, or behavioral meanings of the stimulus.

Similar observations have also been reported by Sakagami and Niki (1994a) (Fig. 5.11). They trained the monkey to perform symmetrically rewarded Go/No-Go discrimination tasks under three conditions (color-matching, shape-matching, and position-matching). The monkey is required to maintain pressing the lever by hand and fixate at a fixation point during the task. The difference of the task condition is informed by the color of the fixation point. During the cue period, one colored shape stimulus is presented as a visual cue at one of four positions (left, right, up, and down) briefly. Sixteen cues [combinations of 4 colors (red, yellow, green, purple) × 4 shapes (circle, stripe, cross, diamond)] were prepared. After a variable delay (1–2 s), the color of the fixation point changes depending on the condition, indicating the signal to perform Go (immediate release the lever) or No-Go responses (delayed release the lever) to obtain the reward. In the color condition, red and yellow signal Go responses, while green and purple signal No-Go responses, regardless of their shapes and positions. In the shape condition, circle and stripe signal Go responses, while cross and diamond signal No-Go responses, regardless of color and position. In the position condition, the left and upper positions signal Go responses, while the right and down positions signal No-Go responses, regardless of color and shape. Among prefrontal neurons responded to the visual cues, 76% exhibited differential responses depending only on the difference of response behaviors (Go or No-Go response), not on the difference of physical features (color, shape, or position) of the cues (Fig. 5.11). Only 2% exhibited differential responses depending on the difference of physical features of the cues.

Sakagami and Tsutsui (1999) further showed behavioral context-dependent prefrontal visual activity using a Go/No-Go selective attention task. In their task, the monkey is required to perform either a lever-pressing (Go) or a lever-holding (No-Go) response depending on the color or the motion direction of the moving colored-dots. In the color condition, the monkey needs to attend the color and ignore the direction of the moving dots, whereas in the motion condition, the monkey needs to attend the moving direction and ignore the color of the dots. Prefrontal neurons showed clear phasic excitatory responses to the cue presentation and exhibited differential activation depending on the difference of the responses (Go or No-Go) in both color and motion conditions (37%), only in the color condition (32%), and only in the motion condition (15%). These differential responses to the visual cues represented the difference of the behavioral context (Go or No-Go responses) and

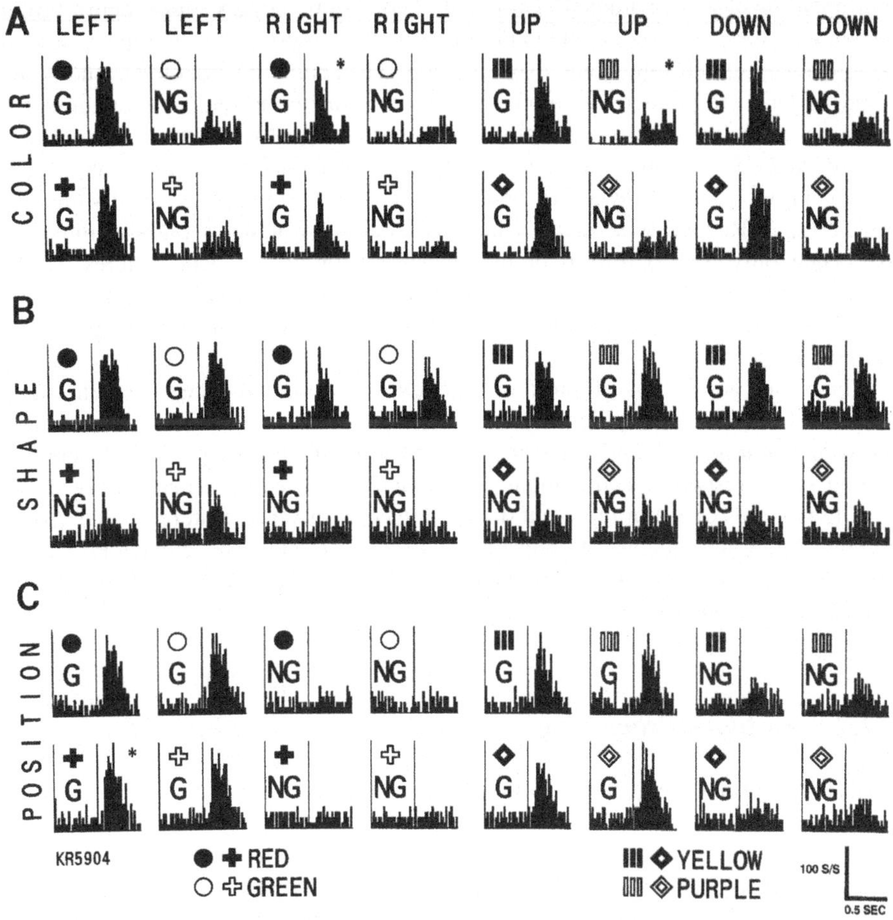

Fig. 5.11 Prefrontal visual responses representing Go responses independent of the color and shape of the visual stimulus. From Sakagami and Niki (1994a) with permission

did not reflect the difference of the physical features of the visual cues in both color and motion conditions.

Hoshi et al. (2000) also showed behavioral context-dependent visual responses in prefrontal neurons. They used a delayed matching-to-sample task under the position-matching condition or the shape-matching condition. During the sample period, a sample cue (triangle or circle) is presented at one of three positions (top, right, or left). After a 3-s delay period, the matching period is introduced. In the position-matching condition, either triangles or circles are presented at all three positions and the monkey is required to select the position where the sample cue was presented by reaching hand movements. In the shape-matching condition, a triangle and a circle are presented at two of three positions and the monkey is required to select the position where the shape same as the sample cue is presented. Sixty-five percent of prefrontal neurons examined exhibited behavioral context-dependent visual responses. These neurons exhibited phasic excitatory responses during the matching period, but exhibited significant responses only when either the position or the shape of the target corresponded to the position of the forthcoming reaching movement.

Pan et al. (2008) examined features of association between visual stimuli and reward prediction in prefrontal neurons using a sequential paired-association task and found two types of reward-related visual responses (reward type and stimulus-reward type). Reward-type visual responses predicted whether a visual stimulus is associated only with a reward or not, regardless of its physical features, while stimulus-reward type visual responses predicted whether a visual stimulus is associated with a reward or not based on whether the stimulus is included in the preferred category or not.

Thus, one population of prefrontal neurons have visual receptive fields and exhibit stimulus selectivity, such that neurons exhibit different responses to different physical features of the visual stimulus. Other population of prefrontal neurons exhibit visual responses and stimulus selectivity. However, the selectivity of these visual responses does not represent the difference of physical features of the visual stimulus but represents the difference of the behavioral relevance, behavioral meaning, or behavioral context of the visual stimulus. These must be distinguished features of prefrontal visual responses.

5.1.2 Auditory Processing

5.1.2.1 Auditory Projections to the Prefrontal Cortex

The prefrontal cortex has been shown to receive auditory inputs from the auditory cortex (Romanski et al. 1999a, 1999b; Romanski and Averbeck 2009; Medalla and Barbas 2014; Plakke and Romanski 2014). The cortical auditory system of the monkey consists of the core region, the belt region, and the parabelt auditory cortex. The centrally located core region contains three subdivisions including the primary auditory area (AI), the rostral field (R), and the rostral temporal field (RT). The belt region of the secondary auditory area is divided into the medial belt and the lateral belt, and the lateral belt is further divided into the antero-lateral field (AL), the middle-lateral field (ML), and the caudal-lateral field (CL). Information stream leaving from the core region makes a series of parallel and topographically organized streams with the lateral belt region, such that AI sends information to CL and ML, while R sends information to AL. Parallel and topographically organized auditory streams go beyond the belt and the parabelt regions and are observed in the superior temporal cortex. The topographic feature of auditory connections to caudal and rostral regions of the superior temporal cortex is thought to be similar to the dorsal and ventral streams of the visual system.

Five regions of the prefrontal cortex receive auditory inputs from the lateral belt (Romanski et al. 1999b; Romanski 2007). These regions include the frontal pole

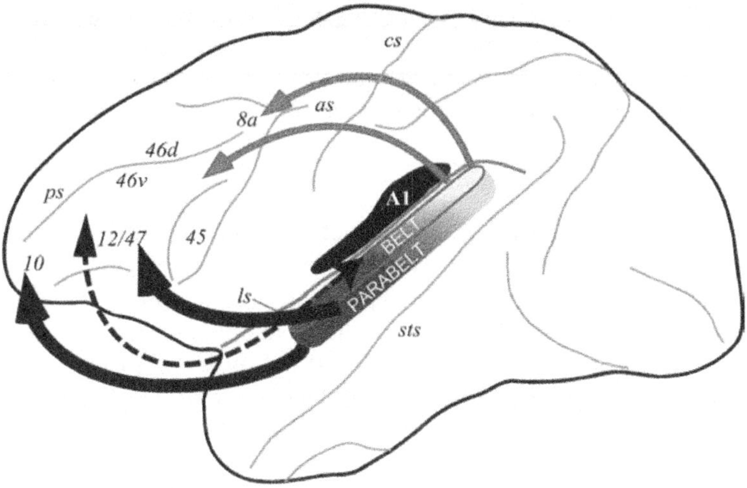

Fig. 5.12 Auditory pathways to the prefrontal cortex. From Romanski (2007) with permission

(area 10), the principal sulcus (area 46), the ventrolateral prefrontal cortex (areas 47 and 45), the lateral orbital area (areas 11 and 12), and the dorsal arcuate area (area 8a). These connections are topographically organized (Fig. 5.12). Projections from the antero-lateral field (AL) of the belt region terminate in the frontal pole, the rostral principal sulcus, the anterior ventrolateral prefrontal cortex, and the lateral orbital area. In contrast, projections from the caudal-lateral field (CL) terminate in the dorsal arcuate area, the caudal principal sulcus, and the caudal ventrolateral prefrontal cortex. Anatomical studies suggest that auditory pathways originating from the rostral and the caudal belt and parabelt regions are analogous to the "what" and "where" pathways in the visual system, respectively. The rostral auditory pathway terminated in the ventrolateral prefrontal cortex may involve in processing "what" features of auditory stimuli, while the caudal auditory pathway terminated in the caudal prefrontal cortex may involve in processing "where" features of auditory stimuli (Romanski et al. 1999b; Romanski and Averbeck 2009).

Although the auditory association cortices project distinct areas in the lateral, orbital, and medial surfaces of the prefrontal cortex in rhesus monkeys, Medalla and Barbas (2014) indicated that the frontopolar area 10 has the densest interconnections with the auditory association cortices. The frontopolar area 10 is the major recipient of auditory inputs in the prefrontal cortex and the major source of outputs from the prefrontal cortex to the auditory cortices. Therefore, the frontopolar area 10 may be an important auditory field for processing auditory information in various cognitive processing.

Similar as monkeys, anatomical connections are present in the human brain between the auditory cortex and the ventrolateral prefrontal cortex. Rocchi et al. (2021) performed electrical stimulations at the auditory cortex in humans and monkeys and examined effective connectivity between the auditory cortex and the prefrontal cortex by fMRI. They found comparable effective connectivity between the caudal auditory cortex and the ventrolateral prefrontal cortex (area 44) in both species, although humans exhibited stronger hemispheric lateralization effects and stronger preference for speech sounds to evoke responses in the ventrolateral prefrontal cortex.

5.1.2.2 Auditory Responses in the Prefrontal Cortex

Several studies using macaque monkeys have shown that prefrontal neurons respond to auditory stimuli and that prefrontal auditory responses exhibit similar spatial selectivity as prefrontal visual responses (Ito 1982; Azuma and Suzuki 1984; Suzuki 1985; Vaadia et al. 1986; Romanski and Goldman-Rakic 2002). Using pure tones, Ito (1982) examined prefrontal responses to visual and auditory stimuli while the monkey performed a simple reaction time task. The auditory responses occurred with 30–100 ms latencies and showed no frequency selectivity and no selectivity to natural sounds such as whistle, hand-clapping, and metal chattering. Prefrontal neurons exhibited phasic responses to either auditory or visual stimuli and only a few neurons responded to both modalities. Vaadia et al. (1986) examined prefrontal activity under four task conditions: auditory localization task (AL; monkey made behavioral responses directed toward the sound source), auditory detection task (AD; monkey made responses when it detected the sound, regardless of the position of the sound source), visual localization task (VL; no sound), and passive listening task (NP). They used a noise burst as an auditory stimulus. Only few prefrontal neurons exhibited auditory responses with similar characteristics as those of neurons in the auditory cortex. Most prefrontal neurons exhibited strong phasic auditory responses with short-latencies during the auditory localization task (Fig. 5.13). However, a variety of response types were observed including early, late, tonic sustained, and inhibitory responses. Auditory responses of these neurons were stronger in the auditory discrimination task compared with the auditory detection task and the passive listening task. Most auditory neurons exhibited a contralateral bias in spatial tuning during the auditory localization task (Fig. 5.13). Bimodal neurons were present in the prefrontal cortex and these bimodal neurons exhibited similar spatial tunings in the auditory and visual localization tasks.

Azuma and Suzuki (1984) had also examined the characteristics of spatial representation of prefrontal auditory neurons while the monkey is sitting in the monkey chair quietly with their head fixed. They examined the spatial tunings of prefrontal auditory neurons using natural sounds, tone pips, or clicks elicited around the monkey's head. They found a contralateral bias in spatial tunings of auditory responses. In addition, the most of auditory neurons exhibited the maximum response to the sound elicited from the lateral or the posterolateral directions. Auditory neurons were found mainly in the dorsal part of the prefrontal cortex.

Suzuki and Azuma (1983) had shown that visual neurons located in the ventral part of the prefrontal cortex represented information of visual stimuli presented in the central region of the contralateral visual field, while neurons located in the dorsal

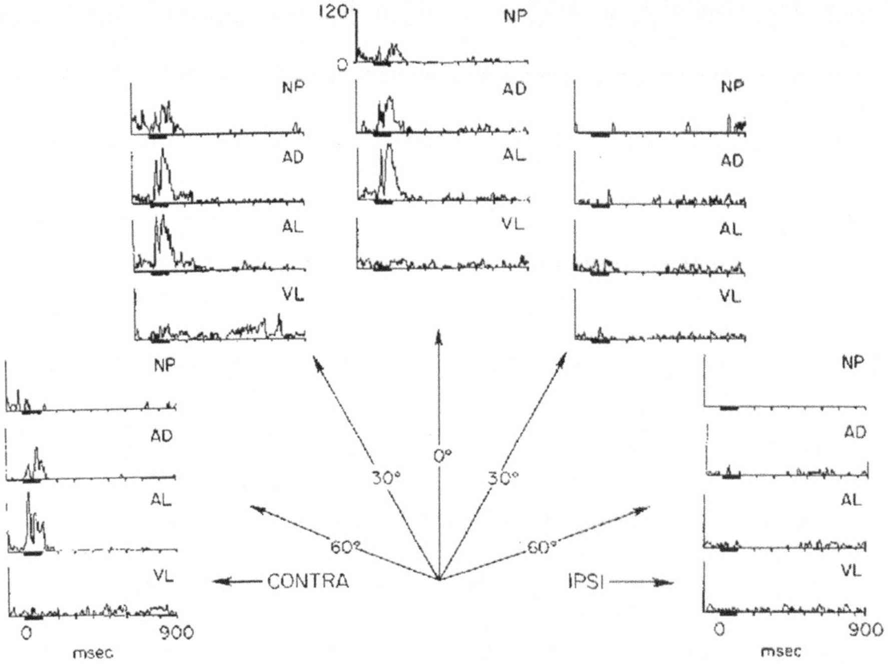

Fig. 5.13 An example of auditory responses observed in a prefrontal neuron under different behavioral conditions. NP, no performance; AD, auditory detection task; AL, auditory localization task; VL, visual localization task. Arrows indicate the directions of auditory stimuli. From Vaadia et al. (1986) with permission

part of the prefrontal cortex represented information of visual stimuli presented in the peripheral region of the visual field. Azuma and Suzuki (1984) showed that auditory neurons in the dorsal part of the prefrontal cortex had spatial tunings directed toward lateral or posterolateral directions. By comparing topographic distributions of visual receptive fields in prefrontal visual neurons with spatial tunings of prefrontal auditory neurons, Suzuki (1985) concluded that prefrontal visual and auditory neurons function together to obtain information raised in the entire contralateral space, such that the receptive fields of prefrontal auditory neurons would cover the spatial area that could not be covered by the receptive fields of prefrontal visual neurons.

The modality specificity of prefrontal activity was also examined using visual and auditory cues. Kikuchi-Yorioka and Sawaguchi (2000) examined the modality specificity of delay-period activity in the prefrontal cortex using an oculomotor version of the delayed-response task (ODR task) under visual and auditory conditions. Among neurons that exhibited directional delay-period activity, 53% were classified as visual-specific neurons, 17% were auditory-specific neurons, and 17% were bimodal neurons. The fact that a majority of prefrontal neurons exhibited modality-specific delay-period activity suggests the presence of parallel processing streams for visual and auditory information in the lateral prefrontal cortex. The spatial tuning of directional delay-period activity was not compared between visual-specific neurons and auditory-specific neurons. However, no significant difference was present in the discriminability of spatial locations between these two groups of neurons, suggesting that the spatial tuning characteristics are similar between visual and auditory neurons in the prefrontal cortex.

5.1.2.3 Selectivity of Auditory Responses and Participation in Vocal Communication

Romanski and her group used a variety of vocalization and non-vocalization stimuli and examined prefrontal responses to these stimuli (Romanski and Goldman-Rakic 2002; Romanski et al. 2004; Averbeck and Romanski 2006; Romanski and Averbeck 2009; Plakke and Romanski 2014) (Fig. 5.14). They found that auditory responsive neurons were located mainly in the ventrolateral part of the prefrontal cortex. Romanski and Goldman-Rakic (2002) recorded activity of 400 prefrontal neurons during auditory stimulus presentation in the fixation task and found 70 neurons responded to auditory stimuli with excitation ($n = 63$) or inhibition ($n = 7$). Vocalization was the most effective auditory stimulus in most of these neurons (52/70). These neurons were responsive to complex sounds including species-specific vocalizations, human speech sounds, and environmental sounds (Fig. 5.14). However, only a few neurons responded to pure tones or noise stimuli.

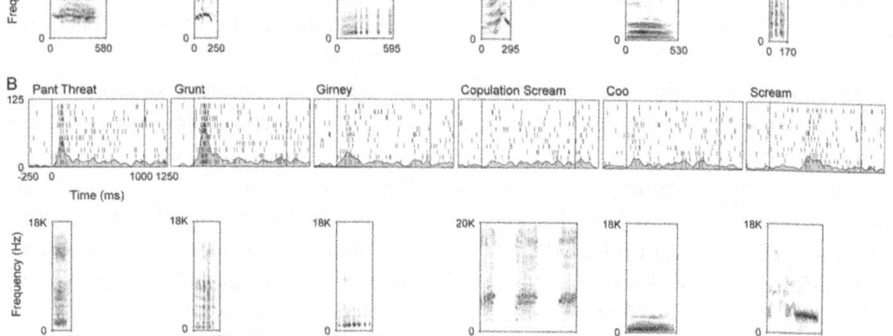

Fig. 5.14 Auditory responses of two prefrontal neurons to several different vocalizations of monkeys. From Romanski (2007) with permission

Romanski et al. (2004) further examined responses of ventrolateral prefrontal neurons to species-specific vocalizations. Most of these auditory neurons responded to a variety of species-specific vocalizations including aggressive calls, coo sound, copulation screaming, and others. However, their information theoretic analysis revealed that these neurons encoded information regarding one or two specific vocalizations. Plakke et al. (2013) and Plakke and Romanski (2014) further reported vocalization-related activity in the ventrolateral prefrontal cortex. Cohen et al. (2004) showed modulations of ventrolateral prefrontal activity by both spatial and non-spatial (vocalization) attributes of an auditory stimulus. They compared responses to spatial and non-spatial attributes of auditory stimuli between ventrolateral prefrontal neurons and neurons in the antero-lateral belt region of the auditory cortex. Non-spatial sensitivity was poorer but spatial selectivity was better in ventrolateral prefrontal neurons. Since most of these neurons responded to complex sounds such as vocalization, they concluded that prefrontal auditory neurons might participate in vocal communication, and not in discriminating between different auditory stimuli. Averbeck and Romanski (2006) reported similar results. They found that static estimates of the spectral and temporal contrasts of specific calls were not effective to discriminate among calls, while a hidden Markov model was more effective. They also found that responses of prefrontal auditory neurons could be predicted more effectively as linear functions of the probabilistic output of the hidden Markov model than as linear functions of the spectral features of the calls. Based on these analyses, they concluded that prefrontal auditory neurons carry signals related to the output of vocalization processing, while the auditory system performs dynamic processing of vocalizations.

For vocal communications, an integration of auditory and visual stimuli is crucial. Although ventrolateral prefrontal neurons had been shown to respond to auditory stimuli including species-specific vocalizations, it had also been shown that only few auditory neurons responded to visual stimuli (Romanski and Goldman-Rakic 2002). However, Sugihara et al. (2006) showed integration of audiovisual communication stimuli in ventrolateral prefrontal neurons. They first examined neural responses to vocalization and non-vocalization sounds, and audiovisual stimuli with face and non-face object, and then tested these responses under five conditions: vocalization alone, static face/object alone, vocalization + static face/object, silent video clip, and video clip with corresponding vocalizations. Neurons having predominantly auditory responses or visual responses exhibited enhancement or suppression when multisensory stimulus conditions were introduced, although a majority showed suppression. Thus, ventrolateral prefrontal neurons exhibited face-selective or object-selective visual responses and responses to complex sounds including human and monkey vocalizations. Some neurons exhibited multisensory responses to the stimuli combining with face and vocalization. Since ventrolateral prefrontal neurons are multisensory and responsive to auditory and visual communication stimuli, this process would help to establish an animal model to investigate the circuit and cellular basis of human communication (Romanski 2007). In addition, the fact that vocalization responsive cells which also respond to faces are found in the ventrolateral prefrontal cortex suggests that this brain area is specialized to process and integrate auditory and visual information for social communication (Romanski 2012). Further, Hwang and Romanski (2015) showed that the ventrolateral prefrontal cortex not only integrates auditory and visual information but also maintains integrated information in working memory.

The Broca's area is located in the ventrolateral prefrontal cortex in the human brain. Vocalization responsive cells have been recorded in the ventrolateral prefrontal cortex in monkey brain. Hage and Nieder (2013, 2015) examined single-neuron activity recorded from the ventrolateral prefrontal cortex while the monkey vocalized in response to visual cues. They found that vocalization-related neurons predicted the preparation of instructed vocalizations. Many vocalization-related neurons became active before vocalization and their activity correlated with parameters of instructed vocalizations. Thus, the ventrolateral prefrontal cortex also participates in planning and volitional initiation of specific vocalization.

5.1.3 Somatosensory Processing

The orbitofrontal cortex and the ventrolateral prefrontal cortex receive somatosensory information. The orbitofrontal cortex receives somatosensory information from the primary and the secondary somatosensory cortex, the insula, and area 7b in the parietal lobe (Carmichael and Price 1995; Cavada and Goldman-Rakic 1989). Similarly, the ventrolateral prefrontal cortex receives somatosensory information from the secondary somatosensory cortex and area 7b in the parietal lobe (Jones and Powell 1970; Petrides and Pandya 1984; Cavada and Goldman-Rakic 1989; Preuss and Goldman-Rakic 1989; Cipolloni and Pandya 1999) (Fig. 5.15). Preuss and Goldman-Rakic (1989) injected tritiated amino acids or horseradish peroxidase in the ventrolateral prefrontal cortex and observed both anterograde and retrograde extensive labeling in the somatosensory cortex. Labeled areas included the secondary somatosensory area and the central portion of the insula. Injections of these tracers in the dorsolateral prefrontal cortex showed no labeling in the somatosensory

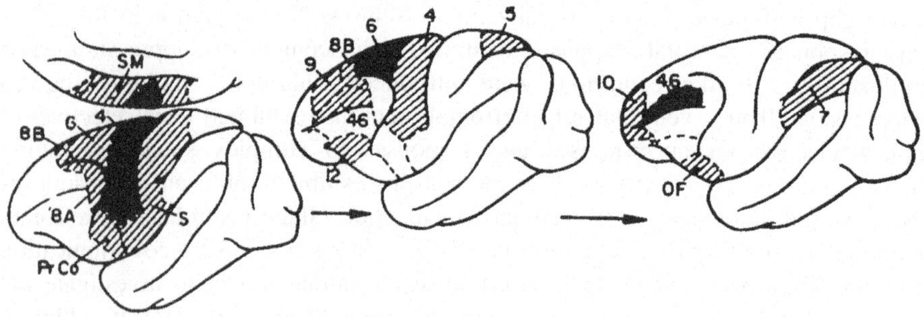

Fig. 5.15 Somatosensory pathways to the prefrontal cortex. From Jones and Powell (1970) with permission

cortex. Based on the result that the ventrolateral prefrontal cortex has strong reciprocal connections with the somatosensory cortex, Preuss and Goldman-Rakic (1989) suggested that the ventrolateral prefrontal cortex contains a working memory mechanism to remember tactile characteristics of recently encountered objects.

Deuel and Mishkin (1977) examined effects of lesions of the dorsal and the ventral prefrontal cortex on tactile discrimination learning and reversal and showed that the dorsal prefrontal lesion yielded deficits in tactile discrimination learning, possibly caused by the contralateral sensory neglect or the spatial disorder, while the ventral prefrontal lesion yielded deficits in both tactile discrimination learning and reversal.

Romo et al. (1999) and Brody et al. (2003) showed that the ventrolateral prefrontal area participated in working memory for tactile stimuli. Romo et al. (1999) used a somatosensory discrimination task (a delayed matching-to-frequency task) (Hernández et al. 1997) and showed that the discharge rate of delay-period activity in prefrontal neurons varied as a monotonic function of the base stimulus frequency. In this task, the monkey is required to discriminate the difference of the frequency between two mechanical vibrations applied to the fingertips. Both the base frequency (sample stimulus) and the comparison frequency (test stimulus) are changed systematically. The monkey needs to indicate whether the comparison frequency is lower or higher than the base frequency by pressing a corresponding switch by hand. They found that the discharge rate of delay-period activity shows either monotonic increase or decrease depending on the increase of the base frequency. Brody et al. (2003) further showed that most neurons in the inferior convexity of the lateral prefrontal cortex exhibit delay-period activity whose discharge rate changes monotonically depending on the frequency of the base stimulus. Therefore, they concluded that delay-period activity of ventrolateral prefrontal neurons represents the frequency of the base stimulus monotonically. Thus, the ventrolateral prefrontal cortex participates in working memory for somatosensory information processing.

In human study, Hagen et al. (2002) examined somatosensory processing in the inferior prefrontal cortex using PET. In the first study, they applied a repetitive tap by von Frey hair at one of four stimulation sites (right or left index finger, right or left great toe) and the subject was required to count the number of pauses in the tactile stimulation. Significant increase of rCBF was observed in the ventral frontal lobe including the posterior-most position of the inferior frontal gyrus and the frontal operculum. In the second study, to examine whether observed activation reflected a passive somatosensory representation, a general attention process, or a modality-specific sensory process, two attention conditions were introduced, in which the subject needed to attend either somatosensory or visual stimulus. In attending somatosensory stimulus, the subject needed to count the number of pauses in the tactile stimulation, while in attending visual stimulus, the subject needed to count the number of times of the luminance change of the fixation point. Significant increase of rCBF in the ventral frontal lobe was observed when attending to the somatosensory stimulus compared with when attending to the visual stimulus. Thus, these results indicate that the posterior inferior prefrontal cortex participates in somatosensory information processing and may play a role to attend to the tactile stimulation.

5.2 Skeletomotor Control

5.2.1 Anatomical Connections Between the Prefrontal Cortex and Cortical Motor Areas

The frontal lobe consists of functionally distinct two sectors. The posterior portion of the frontal lobe is classified as the motor area and participates in the control of skeletomotor actions. The frontal motor area is further divided into two portions: the primary motor cortex and the motor association areas which include the premotor cortex (PM), the supplementary motor area (SMA), the pre-supplementary motor area (pre-SMA), and the cingulate motor area (CMA). The anterior portion of the frontal lobe includes the prefrontal cortex and the frontal eye field (FEF) which participates in eye movement control.

Anatomical connections are present between the prefrontal cortex and the frontal motor areas. The prefrontal cortex does not have direct connections with the primary motor cortex, but has connections with the motor association areas. The dorsolateral prefrontal cortex projects to the pre-SMA (Bates and Goldman-Rakic 1993; Lu et al. 1994; Rizzolatti and Luppino 2001), the PM (Arikuni et al. 1980; Barbas and Pandya 1987; Arikuni et al. 1988; Lu et al. 1994; Ghosh and Gattera 1995; Rizzolatti and Luppino 2001; Luppino et al. 2003), and the CMA (Lu et al. 1994; Picard and Strick 1996) (Fig. 5.16).

For anatomical connections between the lateral prefrontal cortex and the pre-SMA and the CMA, Bates and Goldman-Rakic (1993) showed that multiple prefrontal regions project to the SMA, the CMA, or both. Areas 46, 8a, 9, 11, and 12 are connected reciprocally with the rostral portion of the SMA (the pre-SMA), while areas 46, 8a, and 11 are reciprocally connected with the CMA (Fig. 5.16a). Similarly, Rizzolatti and Luppino (2001) indicated that area 46 and 8 project to the pre-SMA (Fig. 5.16b). Lu et al. (1994) examined connections between the prefrontal area 46 and the SMA and the CMA and showed that area 46 is interconnected with only a portion of the arm representation of the SMA and the caudal CMA and with the entire arm representation of the rostral CMA. Takada et al. (2004) also examined specific connections between the prefrontal area 46 and the pre-SMA and the CMA by tract-tracings. They found that only a few neurons in area 46 project to the pre-SMA and the CMA and that terminals from area 46 are sparsely distributed in both areas. A dual labeling method revealed that the terminal distribution from area 46 and the distribution of neurons projecting to the pre-SMA and the CMA are overlapped in the rostral PM. A tracer injection into the rostral PM revealed that labeling cells are found in area 46 and that labeling terminals are found in the pre-SMA and the CMA. These results indicate that the majority of prefrontal

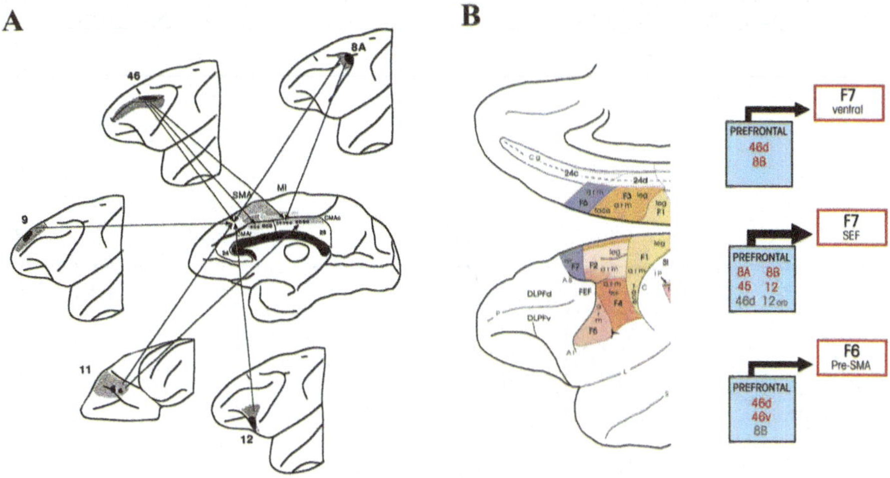

Fig. 5.16 Anatomical connections between the prefrontal cortex and the pre-SMA and the SMA. From Bates and Goldman-Rakic (1993) (**a**) and Rizzolatti and Luppino (2001) (**b**) with permission

neurons in area 46 project to the rostral PM and not directly to the pre-SMA and the CMA and that the pre-SMA and the CMA receive prefrontal outputs through the rostral PM.

For connections between the prefrontal cortex and the PM, Arikuni et al. (1980) injected horseradish peroxidase into the post-arcuate PM and reported that the dorsal post-arcuate PM receives projections from the dorsal pre-arcuate prefrontal area, while the ventral post-arcuate PM receives projections from the ventral pre-arcuate area. The same patterns of connections between the pre- and post-arcuate areas have been observed by Arikuni et al. (1988) and Ghosh and Gattera (1995). Similarly, Rizzolatti and Luppino (2001) and Luppino et al. (2003) indicated that the dorsal portion of the prefrontal area 46 projects to the dorsal PM except the supplementary eye field, while the ventral portion of the prefrontal area 46 projects to the anterior-ventral PM. Barbas and Pandya (1987) showed that, although the dorsal PM receives inputs from relatively restricted neighboring of the dorsal prefrontal cortex, the ventral PM has relatively widespread connections with the ventral prefrontal cortex. Lu et al. (1994) showed substantial interconnections between the prefrontal area 46 and the ventral PM, indicating that the major site of interconnections between the prefrontal cortex and the PM is the ventral PM. Prefrontal area 46 is interconnected with the entire arm representation of the ventral PM.

Miyachi et al. (2005) used a retrograde trans-neuronal transport method of rabies virus and examined anatomical connections between the prefrontal cortex and the frontal motor areas (Fig. 5.17). The virus was injected into the forelimb region and the hindlimb region of the primary motor cortex and the time-dependent distribution pattern of trans-synaptic labeling was analyzed from the second day to the fourth day after injection. As a result, labeling was observed in the PM on the second day and it extended into the rostral PM and area 8 on the third day. On the fourth day, labeling

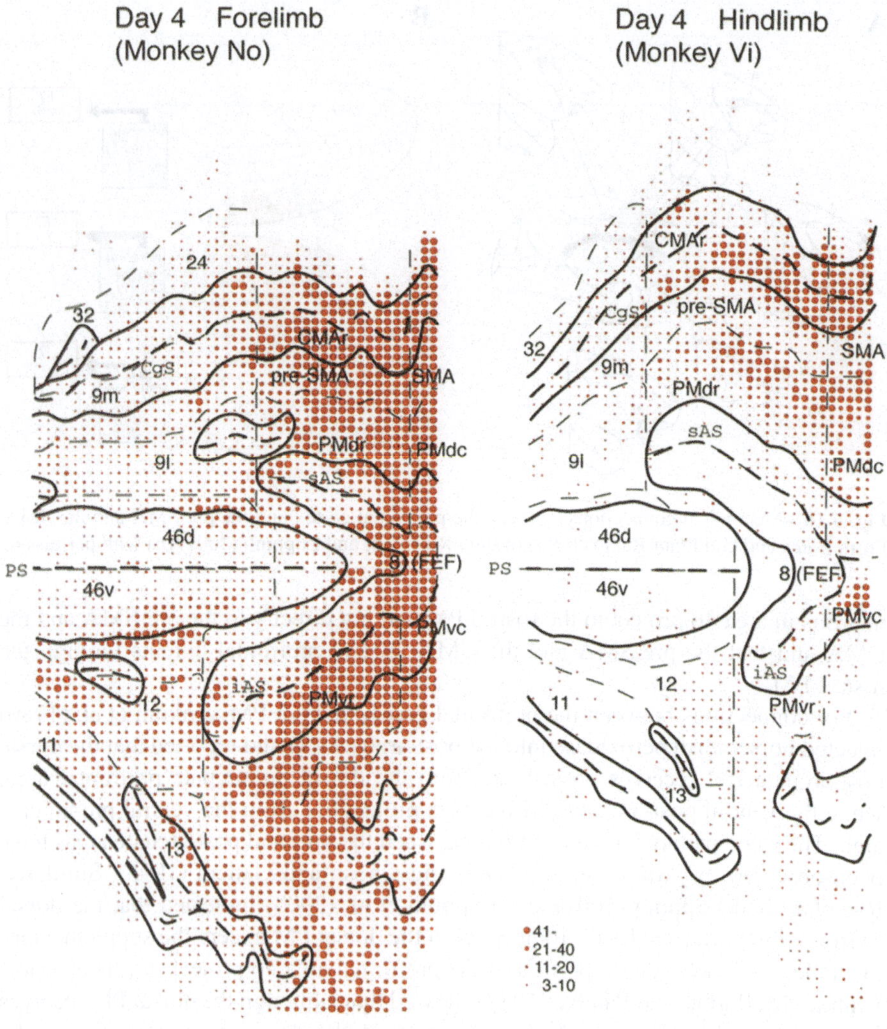

Fig. 5.17 The time-dependent distribution patterns of transsynaptic labeling of rabies virus in the prefrontal cortex. Widespread labeling is observed in the ventrolateral prefrontal cortex when the virus was injected in the forelimb region of the primary motor cortex. From Miyachi et al. (2005) with permission

was observed widely in the ventrolateral prefrontal cortex in the case that virus was injected in the forelimb region (Fig. 5.17), while few labeled neurons were observed in the case that virus was injected in the hindlimb region (Fig. 5.17). These results indicate that information of the ventrolateral prefrontal cortex is mainly transported to the forelimb region of the primary motor cortex through the PM and suggest that the lateral prefrontal cortex mainly participates in the control of forelimb movements and not of hindlimb movements. Takahara et al. (2012) further examined anatomical connections between the lateral prefrontal cortex and the PM and found that the dorsolateral prefrontal cortex directly projects to the dorsal PM and that the ventrolateral prefrontal cortex indirectly projects to the dorsal PM through the pre-SMA.

Thus, the lateral prefrontal cortex (area 46, 8, and 9) projects to the pre-SMA, the PM, and the CMA. When only projections from area 46 to frontal motor areas are considered, the main target of area 46 is the PM, especially the ventral PM. The major site of interconnections between the lateral prefrontal cortex and the PM is the ventral PM. Prefrontal area 46 has anatomical connections extensively with the ventral PM which participates in the control of forelimb movements.

5.2.2 Movement-Related Activity in the Prefrontal Cortex

Although movement-related single-neuron activity was first reported in the frontal eye field while the monkey performed eye movements (Bizzi 1967, 1968), subsequently hand/arm movement-related activity was found in the dorsolateral prefrontal cortex. Kubota and Niki (1971) first reported a transient excitation of prefrontal neurons when the monkey pressed a lever with the hand during delayed alternation performances. This excitatory activity (E units) appeared after the presentation of the Go signal, started 220–300 ms before the initiation of EMG activity recorded from the upper arm, and continued until the end of lever-pressing behavior. This transient activation was also observed during simple alternations of right and left lever-pressing. Therefore, this activation was considered to be a specific activity related to simple lever-pressing behaviors. Fuster (1973) also observed transient activation or suppression of prefrontal neurons during the response period of a manual delayed-response task. In this task, the monkey is required to perform a hand-reaching behavior toward the correct food well to get a food reward. The activation or suppression of prefrontal activity usually began before the initiation of the hand-reaching behavior and was also observed during incorrect choices of the food well. Therefore, he considered that this activity was related to the execution of hand-reaching behaviors. Further, Kubota et al. (1974) found that visually responding prefrontal neurons were often active during motor performances (lever-pressing by hand) in the delayed-response task. Therefore, they termed these neurons "*visuokinetic*." The majority of visuokinetic neurons exhibited excitatory activity during the delay period and the activity during the delay period exhibited statistically significant difference in the discharge rate between left and right cue trials. Therefore, they suggested that these neurons participate in the preparation for forthcoming motor responses.

Activity related to behavioral responses has been widely observed while the monkey performed hand or arm movements during delayed-response tasks (Niki and Watanabe 1976; Rosenkilde et al. 1981a, b; Kojima and Goldman-Rakic 1984), visual discrimination tasks with delay (Kubota et al. 1980; Kojima et al. 1981; Kubota and Komatsu 1985; Yajeya et al. 1988), Go/No-Go discrimination tasks (Watanabe 1986b), and delayed matching-to-sample tasks (Rosenkilde et al. 1981b;

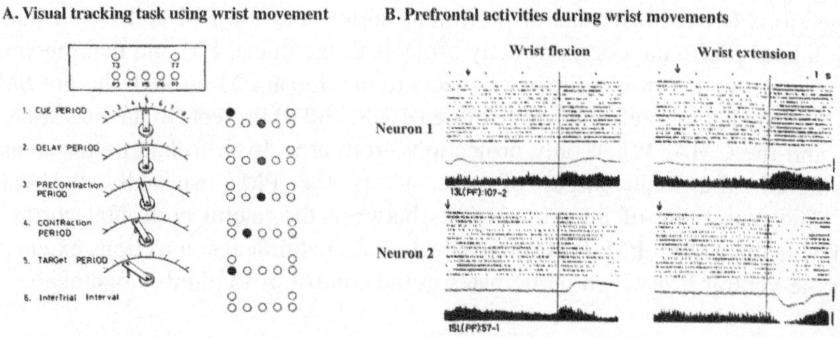

Fig. 5.18 (**a**) Schematic drawing of the visual-tracking task. (**b**) Movement-related activities of prefrontal neurons while the monkey performed wrist extension and flexion movements in the visual-tracking task. From Kubota and Funahashi (1982) with permission

Fuster et al. 1982; Rainer et al. 1998a, b). Most movement-related activities exhibit selectivity, such that movement-related activity is observed only when the monkey performs hand or arm movements toward a particular direction (right or left) or a certain type of response behavior (e.g., Go response). Movement-related activity has been shown to start well before the initiation of the behavioral response. Therefore, movement-related activity observed in prefrontal neurons has been considered to participate in the initiation and the execution of behavioral responses in the tasks.

Movement-related activity has also been observed while the monkey performs simple motor tasks such as visual-tracking movements by hand (Kubota and Funahashi 1982; Funahashi 1983) (Fig. 5.18), visually guided lever-pressing (Sakai 1974), visual reaction time tasks with lever releasing behavior (Sawaguchi 1987; Sawaguchi et al. 1989a, b), isometric force generation task (Fukushi and Sawaguchi 2005), and fixed ratio bar press behavior (Aou et al. 1983). Kubota and Funahashi (1982) compared activity of neurons between the lateral prefrontal cortex and the primary motor cortex while the monkey performed visual-tracking hand movements (left or right direction, Fig. 5.18a) under the delayed-response condition, the visually guided condition, and the control condition. In the delayed-response condition, both prefrontal and motor cortex neurons exhibited either bidirectional (59 for prefrontal and 18 for motor) (Fig. 5.18b, neuron 1) or unidirectional (43 for prefrontal and 32 for motor) activity (Fig. 5.18b, neuron 2). Both prefrontal and motor cortex neurons showed gradual increase of activation in the late phase of the delay period. In the visually guided and the control condition, movement-related activity was less frequently observed in both the prefrontal and the motor cortex. Thus, a majority of movement-related activity observed in both cortical areas showed selectivity to the movement direction, suggesting that prefrontal neurons having movement-related activity involve in movement execution as motor cortex neurons do.

Selectivity of movement-related activity has been observed depending on the difference of the types of movement (Kubota and Komatsu 1985; Watanabe 1986b).

Movement-related activity usually starts firing well before the initiation of the behavioral response (Kubota and Niki 1971; Kubota 1978). These response characteristics of movement-related activity observed in prefrontal neurons are similar to those of movement-related activity observed in primary motor cortex neurons (Evarts 1966; Schwartz et al. 1988). Therefore, movement-related activity observed in prefrontal neurons has been thought to participate in the control of behavioral responses (Kubota 1978). However, some differences are present in response characteristics between prefrontal neurons and motor cortex neurons. For example, in prefrontal neurons, movement-related activity is observed during the response period only when behavioral response is linked with the reward delivery (Kubota and Niki 1971; Kubota et al. 1974; Kubota and Funahashi 1982). This suggests that prefrontal movement-related activity is context dependent and participates not in simple movement execution but in voluntary control of movement linked with a specific cognitive function.

Recently, Bruni et al. (2015) found that ventrolateral prefrontal neurons participate in selection and execution of goal-directed manipulative actions. Borra et al. (2011) showed that the ventrolateral prefrontal cortex, especially the rostral area 12, has dense connections with the caudal portion of the ventral area 46 and has connections with the ventral premotor cortex and the anterior intraparietal area where primarily involve in visuomotor transformation for grasping something by hand. Bruni et al. (2015) examined neural activity in the ventrolateral prefrontal cortex while the monkey performed a variety of tasks including grasp-to-eat or grasp-to-place tasks triggered by visual or auditory cues. They observed unimodal sensory neurons, motor-related neurons, unimodal sensory-and-motor neurons, and multisensory neurons. Activity of some neurons reflected the decision to act or refrain from acting, while activity of others encoded the decision to perform or withhold a specific action. Simone et al. (2015) recorded activity of ventrolateral prefrontal neurons while the monkey performed tasks requiring reaching-grasping actions in different conditions (in light or dark condition, memory-guided condition, and absence of learned rules). This study showed that ventrolateral prefrontal neurons were active during execution of this action, that most movement-related neurons were active during action execution with and without visual feedback, and that these neurons' activity did not change in association with mnemonic conditions and rule conditions. Further, Raos and Savaki (2017) examined prefrontal roles in action perception by ^{14}C-deoxyglucose methods. The monkey performed reaching-to-grasp execution in the light or in the dark conditions or observed the same action executed by an external agent. Observation of action exclusively activated the dorsolateral region (area 9 and dorsal 46). Execution in the dark activated the ventral bank of the principal sulcus and the ventrolateral convexity, while execution in the light activated both banks of the principal sulcus and both dorsal and ventral convexities. These results suggest that the prefrontal cortex contributes to both action generation and action perception or action monitoring.

5.2.3 Response-Period Activity Is Context Dependent

Response-period activity has been observed while the monkey performs a variety of cognitive tasks and considered to be related to behavioral control for executing a particular response. However, it has also been shown that response-period activity is not always related to simple motor control function. Watanabe (1990) examined prefrontal activity using a stimulus-reward association task. In this task, two visual stimuli are presented successively with a 1-s inter-stimulus interval and the combination of these two stimuli informs the monkey whether it will receive a reward or not after lever release response. He showed that a number of prefrontal neurons (75/313, 24%) showed differential responses during the lever release period and that these responses were related to whether a reward was given or not. The monkey made the same lever release behaviors during the lever release period regardless of whether a reward was given or not. The response was time-locked with the lever release responses. These results indicate that these differential responses are movement-related activity and that some movement-related activity represents information regarding not only the motor aspect of the behavioral response but also other aspects of the behavioral task such as reward expectation. Hoshi et al. (1998) also examined prefrontal movement-related activity using two motor tasks with delay (Fig. 5.19). During the sample cue period, either a circle or a triangle is presented at one of three positions. After a 3-s delay period, choice cues are presented. In the location-matching condition, either circle or triangle is presented at all three positions and the monkey is required to touch the position where the sample cue was presented. In the shape matching condition, a circle and a triangle are presented at two of three positions and the monkey is required to touch the position where the same shape as the sample cue is presented. They found that, although 46% of movement-related activity did not show selectivity to the difference of the shapes of the target or the task rules (location matching or shape matching), 36% and 34% of movement-related activity showed selectivity to the shape of the target (Fig. 5.19) and the task rule, respectively. The similar results have also been reported by Hoshi et al. (2000). These results indicate that movement-related activity observed in the prefrontal cortex is not only related to the initiation or the execution of the response movement but also affected by various task-related information including behavioral context or reward expectation or task rules.

To further examine what information is included in response-period activity observed in the prefrontal cortex, Takeda and Funahashi (2002) analyzed prefrontal activity under two task conditions: the oculomotor version of the delayed-response (ODR) task and the rotatory ODR (R-ODR) task. Both tasks used memory-guided saccadic eye movements as response behaviors. They compared saccade-related activity observed during the response period in the same prefrontal neurons between two task conditions. In the ODR task, the direction of the saccade is the direction of the visual cue, while in the R-ODR task, the saccade direction is 90° clockwise from the direction of the visual cue. Therefore, if a prefrontal neuron exhibits the maximum activity when the visual cue is presented at the same position in both tasks, it

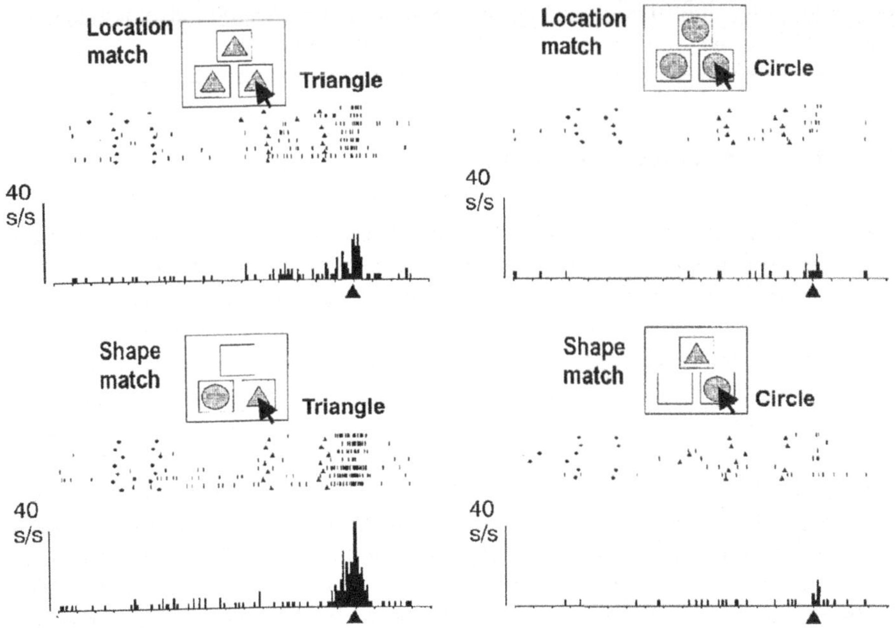

Fig. 5.19 An example of behavioral context-dependent movement-period activity. This neuron was active during reaching toward the lower right key only when a triangle was presented at the key window regardless of task difference. From Hoshi et al. (1998) with permission

can be concluded that this neuron's activity represents information of the position of the visual cue. If a prefrontal neuron exhibits the maximum activity when saccade directions are the same in both tasks, it can be concluded that this neuron's activity represents information of the saccade direction. Among 57 prefrontal neurons exhibiting directionally selective activity during the response period, 58% of response-period activity represented information of the saccade direction, while the remaining 35% represented information of the direction of the visual cue.

These results indicate that, although movement-related activity is observed in lateral prefrontal neurons while the monkey performs a variety of behavioral tasks, not all movement-related activities are directly related to the initiation or the execution of response behaviors. Movement-related activity in lateral prefrontal neurons is not only related to the information processing for the initiation or the execution of the response movements but also related to the processing for various task-related information including behavioral context or reward expectation. Further, it has been shown that movement-related activity participates in regulating or controlling some neuronal processes operating in the prefrontal cortex. For example, it has been proposed that one group of saccade-related activity (post-saccadic activity) acts as a signal to terminate delay-period activity during the ODR performance (Funahashi et al. 1993; Funahashi and Kubota 1994; Goldman-Rakic et al. 1990). Thus, movement-related activity observed in the lateral prefrontal cortex not only represents information regarding response behaviors but also represents a variety of information that are required for task performances.

5.3 Eye Movement Control

It is well known that two eye movement-related brain areas are present in the frontal lobe—the frontal eye field and the supplementary eye field. The frontal eye field is located in the pre-arcuate area in the prefrontal cortex, which corresponds to area 8 based on the cytoarchitectonic classification. The supplementary eye field is located in the dorsomedial portion of the premotor cortex. Electrical stimulations in these two eye fields evoke saccadic eye movements (e.g., Bruce et al. 1985; Schlag and Schlag-Rey 1987; Huerta and Kaas 1990; Russo and Bruce 1993). Lesions of these areas cause deficits in voluntary eye movements (e.g., Latto and Cowey 1971b). Eye movement-related activity is observed in neurons of these eye fields (e.g., Bizzi 1967, 1968; Bizzi and Schiller 1970; Bruce and Goldberg 1985).

Neurophysiological studies using non-human primates reveal that saccade-related activity can be observed not only in the frontal eye field but also in the dorsolateral prefrontal cortex (Boch and Goldberg 1989; Funahashi et al. 1991; Takeda and Funahashi 2002). However, a great majority of saccade-related activity in the dorsolateral prefrontal cortex are post-saccadic, such that saccade-related activity is initiated after the initiation of the saccadic eye movement or after the termination of the eye movement. Most of post-saccadic activity have directional selectivity, such that this activity can be observed only when the monkey makes a saccade toward a particular direction. Further, post-saccadic activity is context dependent, such that this activity can be observed only when the monkey performs a goal-directed saccade within the task and not observed during spontaneous saccades outside the task. Thus, a large number of prefrontal neurons exhibit directional and context-dependent post-saccadic activity, although post-saccadic activity apparently has no functional significance in eye movement control. This suggests that post-saccadic activity observed in the prefrontal cortex has significant functions other than eye movement controls when the subject performs cognitive tasks. Therefore, understanding the right function of saccade-related activity of prefrontal neurons is an important issue to examine.

5.3.1 Historical Perspective on the Prefrontal Contribution to Eye Movement Control

5.3.1.1 Effects of Electrical Stimulation in the Frontal Lobe

Using monkeys, Ferrier (1875a, b) showed that the frontal lobe participates in eye movement control. He applied electrical stimulation to a part of the monkey's frontal lobe and observed eye movements that were directed toward the opposite side of the stimulated hemisphere. He described his observations as follows:

> *"It has already been stated that the antero-frontal regions of the hemispheres give no response to electrical stimulation. Only one exception to this statement is to be made, viz. that in one case irritation of these regions caused the eyes to be tuned to one or other side, according as the electrodes were placed on the opposite hemisphere."* (Page 433 in Ferrier (1875b))

Subsequently, evidences that electrical stimulation of the frontal lobe, especially of the frontal eye field, produced eye movements had been observed in other species including cats, dogs, and apes (chimpanzee and orangutan) (see Crosby et al. (1952) for details). In humans, Holmes (1938) reported that a patient with damages to the frontal lobe was unable to move his eyes in response to a command or to look at an object in any direction. He concluded as follows.

> *"The frontal centres make possible the tuning of gaze in any desired direction and the exploration of space, but they also keep under control, or inhibit, reflexes that are not appropriate to our conduct or our reactions to the world around us."* (Page 112 in Holmes (1938))

Electrical stimulation in the precentral sulcus produced hand or arm movements. Therefore, it had been recognized that the frontal lobe includes the cortical area which supports eye movement control separately from the area which is responsible for skeletomotor control. This notion was supported by both animal and human studies. However, the size of the cortical area where electrical stimulations effectively evoke eye movements and the types of eye movements evoked by electrical stimulations were different among animal species and among investigators (Crosby et al. 1952). These differences were due to several experimental factors. For example, since animals were anesthetized in classic experiments, the animal's state of alertness or degree of anesthesia affected eye movements evoked by electrical stimulations. In addition, the stimulus intensity used in these experiments was not always reported. Therefore, the stimulus intensity might be different among these studies. Further, eye movements were not recorded and analyzed in detail. Therefore, differences in experimental conditions and methods among different experiments might cause inconsistent results regarding the characteristics of eye movements evoked by electrical stimulations in the frontal lobe.

Krieger et al. (1958) examined the effects of anesthesia on electrically evoked eye movements and found that the state of consciousness is important for determining the features of electrically evoked eye movements, as are the location stimulated and the parameters of the stimulation. Therefore, to exclude the effects of anesthesia, Wagman et al. (1961) used rhesus monkeys having high cervical transection and examined eye movements evoked by electrical stimulation at the frontal lobe. As shown in Fig. 5.20a, they found that electrical stimulations evoked eye movements over a rather wide area of the frontal lobe. Eye movements were evoked by electrical stimulations at the frontal cortex near the arcuate sulcus as well as the principal sulcus. The most commonly observed eye movements they found were horizontal conjugate, contralateral saccades. They also frequently observed oblique, either

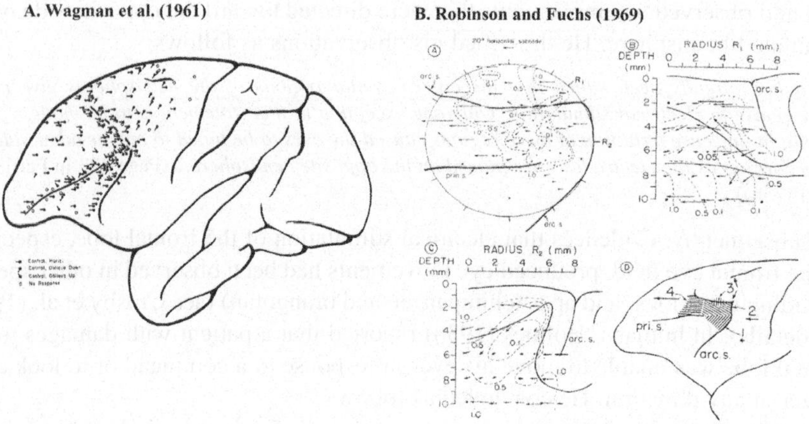

Fig. 5.20 Features of evoked saccades by prefrontal stimulations. (**a**) Results from Wagman et al. (1961). (**b**) Results from Robinson and Fuchs (1969)

upward or downward, contralateral saccades. They examined characteristics of evoked eye movements by stimulating not only the surface of the frontal lobe but also within the cortex. They observed contralateral conjugate saccades by stimulation of the subsurface cortex. Based on these observations, they concluded that the frontal eye field extended anteriorly to the rostral limit of the principal sulcus, medially to the medial surface, and laterally to the lower limit of the arcuate sulcus.

Although Wagman et al. (1961) examined electrically evoked eye movements in the frontal lobe, they neither stated parameters of electrical stimulations, nor recorded evoked eye movements. Robinson and Fuchs (1969) developed a method to accurately record eye movements of awake monkeys and found that brief electrical stimulations (30-ms pulse train, 1-ms pulse at 200 Hz, 0.1–0.5 mA) at the frontal eye field (Brodmann's area 8) produced a single contralateral saccadic eye movement with a typical latency of 25 ms after the stimulation onset (Fig. 5.20b). Most of the evoked saccades were single contralateral saccades. They found that the electrical stimulations at the frontal eye field produced only saccadic eye movements and did not produce smooth pursuit, vergence, or centering eye movements, or nystagmus, although these eye movements were often observed in anesthetized animals and were later found in electrophysiological studies.

Electrical stimulations at the frontal eye field consistently produced saccadic eye movements, indicating that the frontal eye field acts as a premotor region for producing saccadic eye movements. However, the effects of frontal eye field lesions in monkeys did not directly support this notion. For example, Latto and Cowey (1971a, b) showed that, although lesions of the frontal eye field in monkeys produced temporary visual neglect in the contralateral visual field, recovery from the neglect was rapid, and that there was little effect on saccadic eye movements. Schiller et al. (1980) also showed that both unilateral and bilateral lesions of the frontal eye field produced only temporary deficits in eye movements. Monkeys with these lesions exhibited neglect of the contralateral visual field and made fewer saccades toward the targets in that visual field. These results suggested that the frontal eye field participated in the initiation of visually guided saccadic eye movements. However, they showed that paired lesions of both the frontal eye field and the superior colliculus produced severe deficits in eye movements. These monkeys could no longer perform saccadic eye movements toward visual targets and the deficits showed little recovery with time. Thus, although both electrical stimulation studies and lesion studies suggested that the frontal eye field apparently participated in the initiation of visually guided saccadic eye movements, further neurophysiological studies were needed to confirm that neurons in the frontal eye field in fact participate in the execution of saccadic eye movements by showing discharge patterns of neurons during saccadic eye movements.

5.3.1.2 Saccade-Related Single-Neuron Activity in the Frontal Eye Field

Evarts (1966, 1968) developed a new method to record single-neuron activity from the motor cortex of awake behaving monkeys. Using this method, Bizzi (1967) recorded single-neuron activity in the frontal eye field (area 8) of awake monkeys and found that the majority of frontal eye field neurons discharge in relation to the eye position and that these neurons discharge only after the initiation of the eye movement. Subsequently, Bizzi (1968) found two types of neural activity (Type I and Type II) in the frontal eye field of monkeys. Type I neurons exhibited burst activity only after the initiation of the saccadic eye movement and these neurons were classified as having post-saccadic activity. This burst activity lasted for a short time and was observed in complete darkness. This activity was also observed in the fast phase of nystagmus. On the other hand, Type II neurons exhibited a steady discharge when the eyes were oriented in a specific direction and were silent during saccadic eye movements. These neurons were also active during the smooth pursuit eye movement and during the slow phase of nystagmus. Although these experiments were performed with the monkey's head immobilized, Bizzi and Schiller (1970) found that characteristic discharge patterns of both Type I and Type II neurons were maintained regardless of whether the monkey's head movements were restricted or not.

Neurophysiological studies using awake behaving monkeys revealed that neurons in the frontal eye field (area 8) exhibit activation in relation to eye movements. Distinct groups of frontal eye field neurons participate in either saccadic eye movements or smooth pursuit eye movements. Eye movement-related activity is observed during eye movements in complete darkness. Thus, these results support the notion that the frontal eye field participates in eye movement control. However, none of these neurons exhibit activation before the initiation of the eye movement (pre-saccadic activation). Therefore, it was thought that these neurons participate in some complex coordinate mechanism for orienting, since lesions of the prefrontal cortex caused characteristic difficulties of visuomotor performance, such as difficulty in orienting vertical direction under tilted body conditions (Teuber 1964).

Although none of neurons in the frontal eye field exhibited pre-saccadic activity, neurons having visual responses had been observed in the frontal eye field (Mohler et al. 1973) and it had been shown that visual response is enhanced when the visual stimulus is used as the target for the saccadic eye movement (Wurtz and Mohler 1976). Therefore, visual response was the only activity occurred before the initiation of visually guided eye movement and the enhancement of the visual response was the only phenomenon related to the eye movement in the frontal eye field. This enhancement was considered to be a pre-eye movement signal and exclusively related to the eye movement. Goldberg and Bushnell (1981) further examined pre-saccadic enhancement of the visual response under a variety of task conditions and found that the enhancement of the visual response is specifically observed only when the presentation of the visual stimulus preceded eye movement. The enhancement of the visual response is not observed when the visual stimulus is presented within the visual receptive field but is not the target of the eye movement. In addition, the enhancement of the visual response is not observed when the visual stimulus is the target of the reaching behavior without eye movements. Based on these observations, they concluded that pre-saccadic enhancement of the visual response is a cortical component of the neural event that precedes purposeful visually guided saccades. Although pre-saccadic enhancement of the visual response is observed in the frontal eye field, this activity is the response to the presentation of the visual stimulus and is not directly related to the initiation and the control of the eye movement.

Bruce and Goldberg (1984, 1985) first reported pre-saccadic activity in the frontal eye field. They strictly defined the location of the frontal eye field by intracortical microstimulation and found that more than half of sampled neurons exhibited pre-saccadic activity, which was initiated before the start of the saccadic eye movement. The neurons having pre-saccadic activity were classified into three groups (*visual*, *movement*, and *visuomovement* cells). Among these, movement and visuomovement cells exhibited saccade-related activity while the monkey performed eye movements in the dark (Fig. 5.21). Therefore, these cells were considered to be directly related to the initiation and the control of eye movements. Interestingly, movement cells exhibited significant pre-saccadic activity only during performance in learned saccade tasks, but exhibited either no activity or significantly weaker and less consistent activity during spontaneous saccades in the dark. In addition, each cell having pre-saccadic activity exhibited broad tuning with respect to saccade directions and had an optimal direction in which the magnitude of pre-saccadic activity was maximal. Cells having pre-saccadic activity were mostly found within the cortical area where the stimulations of low (< 50 μA) electrical current evoked saccadic eye movements. This cortical area corresponded to the frontal eye field defined by intracortical microstimulation by Bruce et al. (1985). The directions of saccades evoked by intracortical microstimulation were usually the same as the optimal directions of pre-saccadic activities at a given recording site.

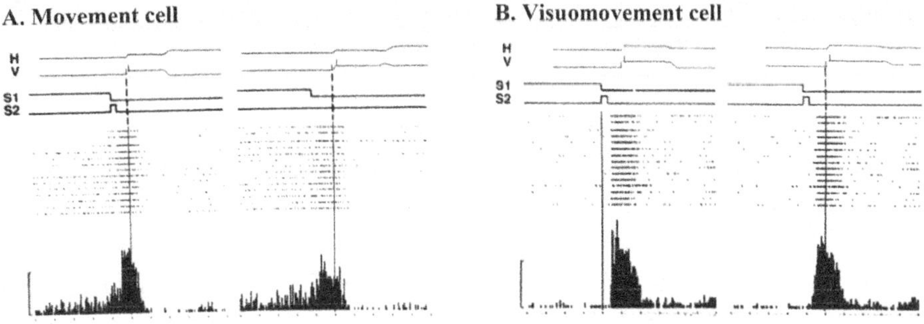

Fig. 5.21 Examples of pre-saccadic activities found in the frontal eye field. (**a**) Activities of a movement cell triggered at the start of the saccade. (**b**) Activities of a visuomovement cell. Left: triggered at the visual stimulus onset. Right: triggered at the start of the saccade. H and V: horizontal and vertical eye traces, respectively. S1 and S2: signals indicating on and off of the fixation point and the target stimulus, respectively. From Bruce and Goldberg (1985) with permission

On the other hand, Bruce and Goldberg (1985) also observed neurons having post-saccadic activity in the frontal eye field. These neurons exhibited only post-saccadic activity during saccades made in conjunction with the task as well as during spontaneous saccades made outside the task. Neurons having only post-saccadic activity were observed in the cortical area where the stimulations of higher (>100 μA) threshold current were required to evoke saccades. Directions of evoked saccades were different from the optimal directions of post-saccadic activities at a given recording site.

Some neurons in the frontal eye field exhibited both pre- and post-saccadic activity (Bruce and Goldberg 1985). Although both pre- and post-saccadic activity showed directional tunings in these neurons, the optimal direction of pre-saccadic activity was usually opposite of the optimal direction of post-saccadic activity. In visually guided or memory-guided saccade tasks, the monkey usually made a saccade back to the central fixation point just after making a saccade to the peripheral target. Therefore, post-saccadic activity observed in these neurons might not be post-saccadic activity but might be pre-saccadic activity related to the saccade toward the central fixation point. Therefore, neurons that exhibited both pre- and post-saccadic activity might be different from neurons that exhibit only post-saccadic activity.

Thus, although Bizzi (1967, 1968) did not find neurons having pre-saccadic activity in the frontal eye field, Bruce and Goldberg (1985) showed the evidence that neurons having pre-saccadic activity are found in the frontal eye field where low-threshold intracortical microstimulations evoke saccadic eye movements. Pre-saccadic activity is directionally tuned and is observed only during task-related saccades and not during spontaneous saccades. Therefore, pre-saccadic activity is thought to play important roles for the initiation and the control of purposeful saccades. These results support the importance of the frontal eye field in generating and controlling voluntary and purposeful saccades. On the other hand, post-saccadic activity is also present in the frontal eye field, but this activity is neither directionally tuned nor context dependent. Therefore, neurons having pre-saccadic activity are distinct from neurons having only post-saccadic activity.

5.3.2 Saccade-Related Activity in the Dorsolateral Prefrontal Cortex

Eye movement-related activity had also been observed in the dorsolateral prefrontal cortex. Kojima (1980) used a delayed saccade task and reported saccade-related activity in the lateral prefrontal cortex when eye movements were directed contralaterally to the recording hemisphere. Joseph and Barone (1987) examined single-neuron activity in the dorsolateral prefrontal cortex while the monkey performed a delayed oculomotor task, in which the monkey was required to perform a saccadic eye movement and an arm reaching movement separately following visual and auditory stimuli. They observed several types of task-related activity (e.g., signal-related pre-saccadic tonic cells, post-saccadic tonic cells, and signal-related phasic cells). Among these, few saccade-related neurons were present, most of which exhibited post-saccadic activation. Subsequently, Barone and Joseph (1989) examined prefrontal activity using a spatial sequencing task, in which the monkey was required to press three targets in the order of their illumination under either a visually guided condition or a memory-guided condition. Although they analyzed visual tonic cells and context cells in detail, they found that saccade-related neurons comprised 11% of their total sample ($n = 302$). Among these neurons, 85% exhibited post-saccadic activation and the remaining exhibited some evidence of pre-saccadic activation. Although Joseph and Barone (1987) and Barone and Joseph (1989) used oculomotor tasks, the main focus of their experiments was on neural activity related to cognitive functions. These studies showed that the dorsolateral prefrontal cortex contained saccade-related neurons. Although the proportion of saccade-related neurons was small, they showed that most of these neurons exhibited post-saccadic activation.

Boch and Goldberg (1989) directly examined characteristics of saccade-related activity in the lateral prefrontal cortex using several oculomotor tasks (e.g., visually guided saccade task, delayed saccade task, and memory-guided saccade task). They analyzed single-neuron activity collected from the posterior one-third of the principal sulcus area. Similar as frontal eye field neurons, visual response was enhanced when the visual stimulus presented within the visual receptive field was the target of the eye movement. No evoked eye movement was observed in the recording area by electrical stimulations using a current as high as 150 μA. They observed post-saccadic activation in 11% of neurons. Post-saccadic activation had no selectivity with respect to saccade direction or amplitude and was observed both during purposive saccades and during spontaneous saccades in total darkness.

Funahashi et al. (1991) also examined saccade-related activity of dorsolateral prefrontal neurons while the monkey performed memory-guided saccades in the oculomotor delayed-response (ODR) task and found that a large number of prefrontal neurons exhibited saccade-related activity when the monkey performed saccades in the dark. Among 434 neurons recorded in the principal sulcal area, one-third ($n = 147$) exhibited saccade-related activity during the response period in the ODR task (Figs. 5.22 and 5.23). Response latencies of saccade-related activity relative to the saccade initiation were distributed between −192 and 460 ms. Great majority (78%) of saccade-related activity were post-saccadic and the remaining 22% were pre-saccadic.

Figure 5.22 shows an example of post-saccadic activity observed in the ODR task. This neuron exhibited phasic excitatory activity only when the saccades were directed toward the left visual field (Fig. 5.22a). No activation was observed when the saccades were directed toward the right visual field and the vertical directions. The response latency at the 180° cue condition was 110 ms after the initiation of the saccadic eye movements (Fig. 5.22b). This neuron exhibited saccade-related activity only during rewarded saccades toward the 180° cue direction (Fig. 5.22b) and did not exhibit significant activity during spontaneous saccades toward the 180° cue direction in the intertrial intervals (Fig. 5.22c).

On the other hand, Fig. 5.23a shows an example of pre-saccadic activity observed in the prefrontal cortex. This neuron exhibited excitatory pre-saccadic activity when the saccades were directed toward the upper left visual field and the response latency at the 135° cue condition was −60 ms (Fig. 5.23b). This neuron also exhibited excitatory post-saccadic activity when the saccades were directed toward the lower right visual field (e.g., 315° cue condition) (Fig. 5.23b). Since the monkey tended to make return saccades directed toward the fixation point after finishing response saccades, post-saccadic activity observed in the 0°, 270°, and 315° cue conditions might correspond to pre-saccadic activation for return saccades towards the fixation point. Both pre- and post-saccadic activities were observed only during rewarded saccades within the task and were not observed during spontaneous saccades in the intertrial interval (Fig. 5.23c). As is seen in Figs. 5.22c and 5.23c, both pre- and post-saccadic activities were observed only in conjunction with task-related saccades and comparable activity was not observed in association with saccades during the intertrial interval in dark. The context dependency of both pre- and post-saccadic activities is an important feature of prefrontal saccade-related activity.

Nearly all neurons that exhibited either pre- or post-saccadic activity showed directional selectivity. Fig. 5.24a shows examples of tuning curves of pre-saccadic activity (left) and post-saccadic activity (right). Tuning curves were constructed by adapting mean firing rates during the response period for each cue condition to Gaussian function. Using obtained tuning curves, the best direction was determined as the peak direction of obtained tuning curve. For the neurons having pre-saccadic activity, the majority (62%) had the best directions toward the visual field contralateral to the recording hemisphere and the remaining had the best directions toward the ipsilateral field (23%) or along the vertical meridian (15%) (Fig. 5.24b). On the other hand, for the neurons having post-saccadic activity, 48% had the best directions toward the contralateral visual field and the remaining had the best directions toward the ipsilateral field (36%) or along the vertical meridian (16%) (Fig. 5.24b).

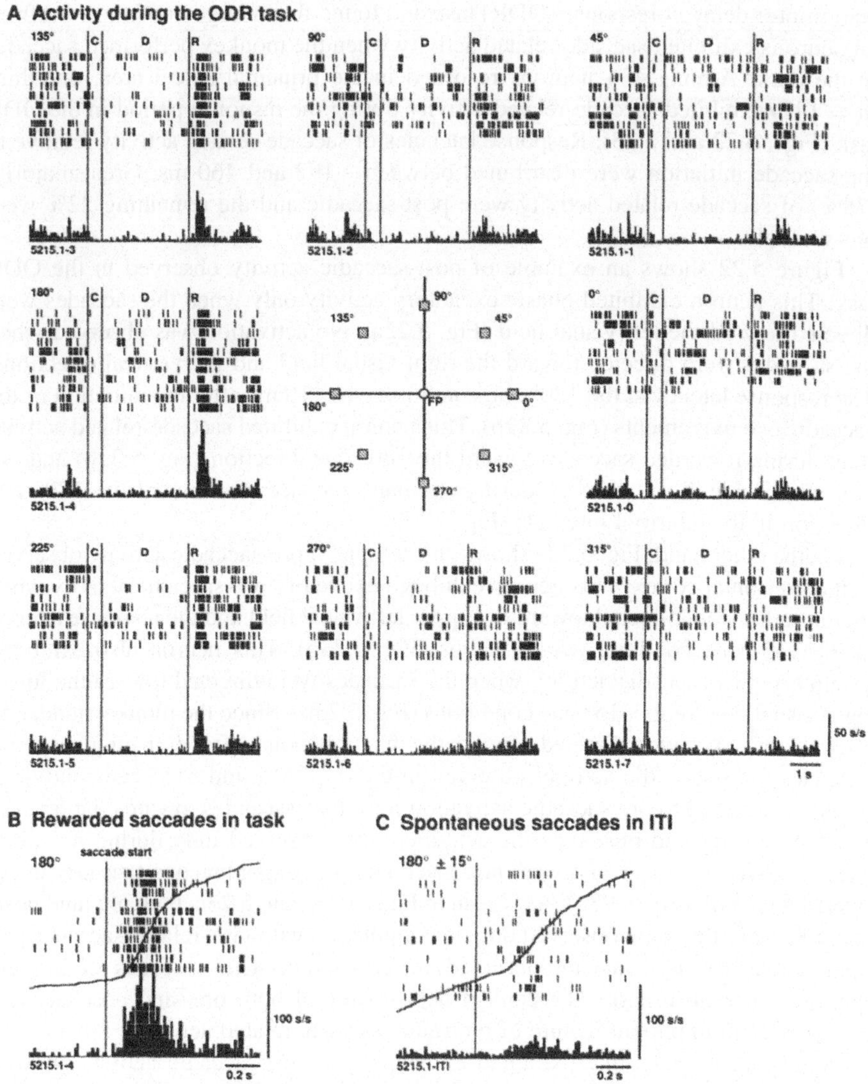

Fig. 5.22 (**a**) An example of post-saccadic activity during ODR performances. Delay length was 3 s. C, D, and R indicate the cue, the delay, and the response period, respectively. (**b**) Response-period activities triggered at the start of the saccade. (**c**) Activities during spontaneous saccades during intertrial intervals. Curved lines indicate time course of cumulative activities, not eye movement traces. From Funahashi et al. (1991) with permission

Significant contralateral bias of the best directions was observed in pre-saccadic activity, but was not observed in post-saccadic activity.

Saccade-related activity was compared between memory-guided conditions in the ODR task and visually guided conditions in the control task in the same prefrontal

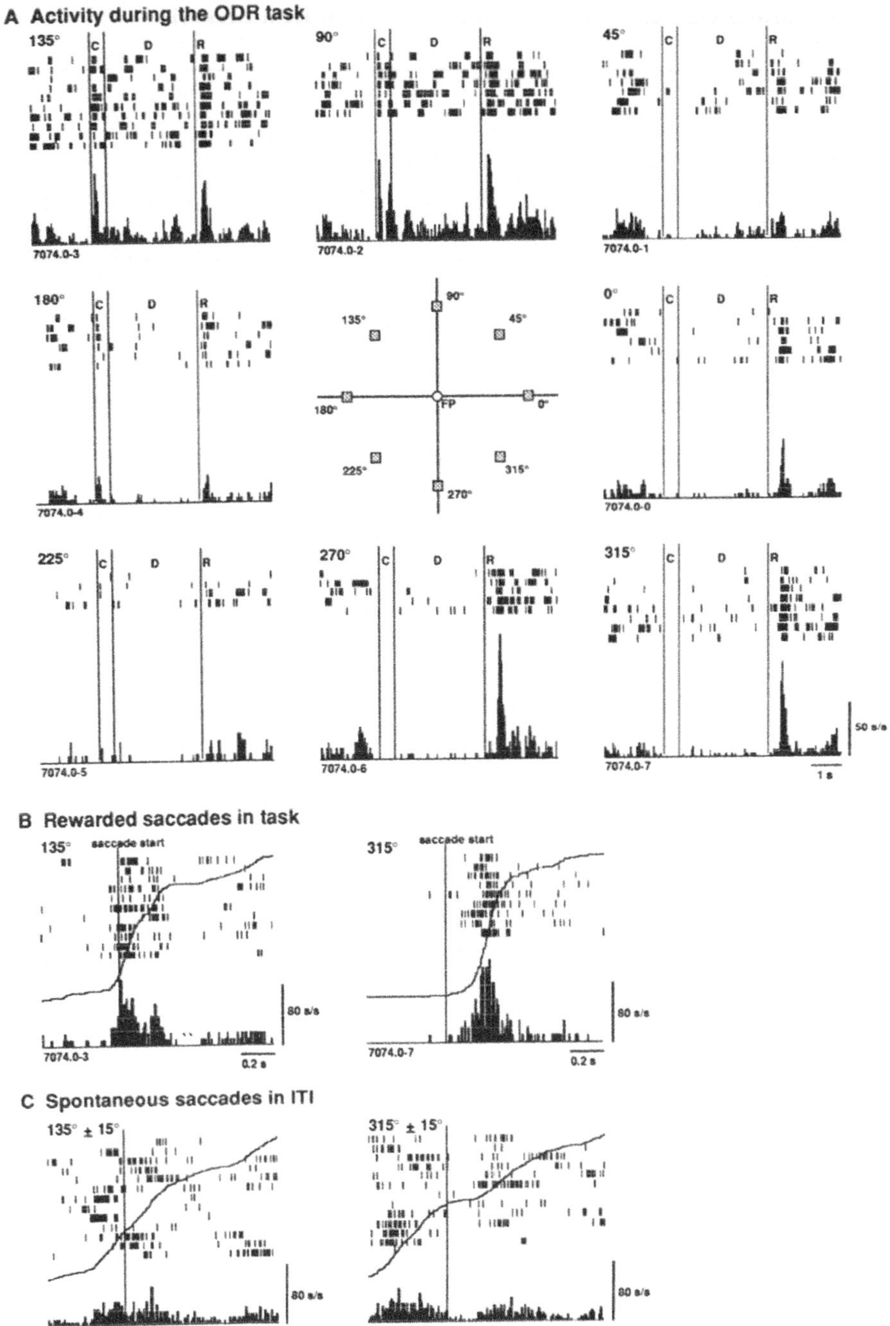

Fig. 5.23 (**a**) An example of pre-saccadic activity during ODR performances. Delay length was 3 s. C, D, and R indicate the cue, the delay, and the response period, respectively. (**b**). Response-period activities triggered at the start of the saccade. Pre-saccadic activation is observed in 135° trials, while post-saccadic activity is observed in 315° trials. (**c**). Activities during spontaneous

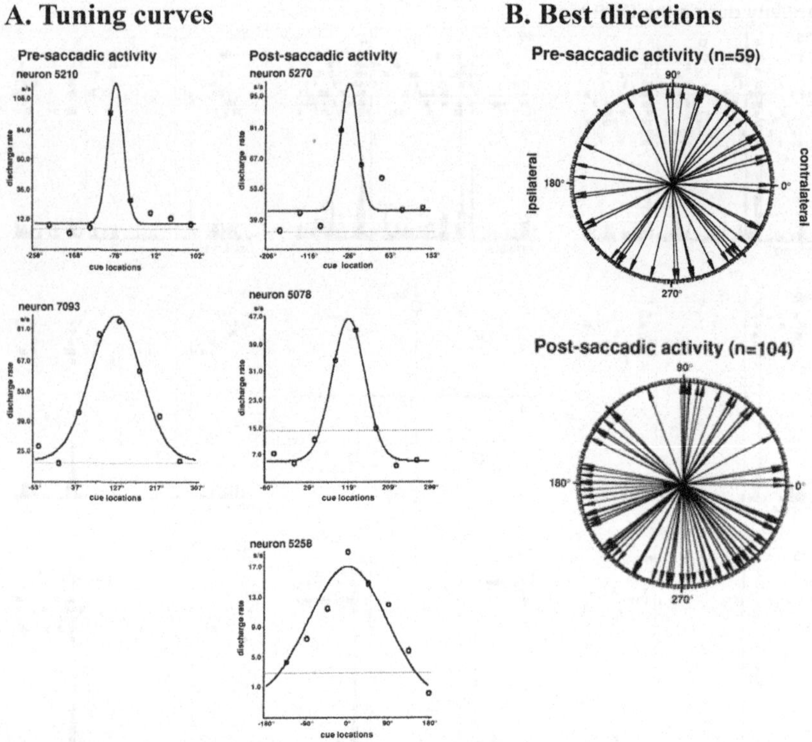

Fig. 5.24 (**a**) Examples of tuning curves based on pre- and post-saccadic activities. (**b**) Distribution of best directions obtained from tuning curves of pre- and post-saccadic activities. From Funahashi et al. (1991) with permission

neuron. Most neurons (40/60) exhibited highly similar directional selectivity and response magnitude in these two conditions, although some neurons exhibited saccade-related activity only during memory-guided conditions.

Takeda and Funahashi (2002) examined information represented by response-period activity. They used the ODR task and the rotatory ODR (R-ODR) task. In the ODR task, the direction of the saccade is the same as the direction of the visual cue, while in the R-ODR task, the direction of the saccade is 90° clockwise from the direction of the visual cue. Therefore, by comparing the directional tunings of response-period activity between the ODR and R-ODR conditions in the same neuron, it could determine whether response-period activity represents the direction

---

Fig. 5.23 (continued) saccades during intertrial intervals. Curved lines indicate time course of cumulative activities, not eye movement traces. From Funahashi et al. (1991) with permission.

of the visual cue or the direction of the saccade. The directions of visual cues and the directions of saccades were all described based on the direction of the visual cue in their experiment. Therefore, if the best directions of response-period activity were the same between two tasks, it could be concluded that this activity represented the direction of the visual cue. On the other hand, if the best direction of response-period activity in the R-ODR task was 90° clockwise from the best direction in the ODR task, it could be concluded that this activity represented the direction of the saccade. Takeda and Funahashi (2002) showed that, among 57 neurons having directional response-period activity, 58% represented the direction of the saccade, while 35% represented the direction of the visual cue. Both pre- and post-saccadic neurons were included in the latter group. Interestingly, most neurons having response-period activity representing the direction of the visual cue also exhibited directional delay-period activity representing the direction of the visual cue. The best direction of response-period activity was almost identical with the best direction of delay-period activity for these neurons. These results indicate that not all saccade-related activities in prefrontal neurons are related to the execution or control of the eye movement. The activity representing the cue direction is observed in relation to the saccade performance but independent from the saccade direction. Therefore, some saccade-related activities do not participate in eye movements but participate in functions other than eye movements.

Thus, the dorsolateral prefrontal cortex includes a large number of saccade-related neurons. Although most of saccade-related activity are post-saccadic, a significant proportion of prefrontal neurons exhibit pre-saccadic activity. Directional selectivity and context dependency are observed in both pre- and post-saccadic activity. These results indicate that prefrontal neurons participate in the control of purposive eye movements similar to frontal eye field neurons. However, since most of saccade-related activity are post-saccadic in the prefrontal cortex and since some saccade-related activities participate in functions other than eye movement control, saccade-related activity in the prefrontal cortex may have different functions from those of saccade-related activity in the frontal eye field.

5.3.3 Comparison of Saccade-Related Activity Between the Prefrontal Cortex and the Frontal Eye Field

An important feature of saccade-related activity in the prefrontal cortex is that both pre- and post-saccadic activity are context dependent. Prefrontal neurons exhibit saccade-related activity only when the monkey performs purposive saccades within the task. The magnitude of both pre- and post-saccadic activity is greatly reduced or both activities are absent during spontaneous saccades performed outside the task. On the other hand, in the frontal eye field, context dependency is observed only in pre-saccadic activity and not in post-saccadic activity (Bruce and Goldberg 1985). Context-dependent pre-saccadic activity is also observed in the posterior parietal cortex (area 7a) (Lynch et al. 1977), the caudate nucleus (Hikosaka et al. 1989), and the substantia nigra pars reticulata (Hikosaka and Wurtz 1983). In these brain areas, most saccade-related neurons exhibit pre-saccadic activity while the monkey performs goal-directed saccades within the task, but these neurons exhibit only weak or no pre-saccadic activity during spontaneous saccades with the same direction and magnitude performed in the dark.

Another important feature of saccade-related activity is that most prefrontal neurons exhibit only post-saccadic activity. Post-saccadic activity begins at the time of the initiation of the saccade, after the initiation of the saccade, or after the termination of the saccade. The median latency of post-saccadic activity is 122 ms after the initiation of the saccade. The mean duration of the saccade during the response period of the ODR task is 45 ms (Funahashi et al. 1989). Therefore, most of post-saccadic activity in the prefrontal cortex begin more than 50 ms after the termination of the saccade.

Post-saccadic activity is usually neglected in many studies, because this activity does not have any clear role in the initiation, the execution, and the control of the eye movements. Post-saccadic activity has usually been considered as a corollary discharge provided from oculomotor structures. If post-saccadic activity is a corollary discharge from oculomotor structures, this activity should be observed with any saccade regardless of whether it is purposive or not. In fact, in the brainstem premotor nuclei (Hepp et al. 1989) and the superior colliculus (Sparks and Mays 1980), neurons are active during every saccade regardless of whether it is performed as a part of the task or spontaneously. Further, in the inferior and the lateral pulvinar (Robinson et al. 1986) and the frontal eye field (Bizzi 1968; Bizzi and Schiller 1970; Bruce and Goldberg 1985; Bruce et al. 1985), post-saccadic activity is observed during saccades regardless of whether they are performed as a part of the task or spontaneously. Therefore, post-saccadic activity observed in these brain areas is likely to be a corollary discharge. On the other hand, post-saccadic activity observed in the prefrontal cortex is context dependent, is observed only during rewarded saccades in the task, and becomes significantly weak or is not observed during spontaneous saccades performed in the intertrial interval. Therefore, post-saccadic activity observed in the prefrontal cortex has different features from the corollary discharge.

In addition, in the frontal eye field, directional preference and contralateral bias are observed in pre-saccadic activity, but are not observed in post-saccadic activity (Bruce and Goldberg 1985). In the prefrontal cortex, both pre- and post-saccadic activities exhibit directional preference and contralateral bias, although the contralateral bias of pre-saccadic activity is stronger than post-saccadic activity (Fig. 5.24b) (Funahashi et al. 1991).

Thus, the prefrontal cortex contains a large number of saccade-related neurons, most of which exhibit post-saccadic activity. Although most of post-saccadic activity are initiated well after the termination of the saccadic eye movement, this activity is clearly related to saccadic eye movements, since this activity exhibits selectivity with respect to the saccade direction and directional tuning. Traditionally, post-saccadic activity is thought to be a corollary discharge from oculomotor centers.

However, post-saccadic activity observed in the prefrontal cortex has different feature from a corollary discharge, such that post-saccadic activity is context dependent. These results strongly suggest that post-saccadic activity in the prefrontal cortex plays important roles not in oculomotor functions but in cognitive functions.

5.3.4 *Contribution of Post-Saccadic Activity in Cognitive Functions*

5.3.4.1 Post-Trial Activity Observed in the Prefrontal Cortex

Post-saccadic activity is initiated after the initiation of saccadic eye movement. Since some neurons having post-saccadic activity start firing well after the termination of saccadic eye movement, post-saccadic activity can be classified as one of post-trial activities. Several post-trial activities have been reported in the prefrontal cortex. For example, Rosenkilde et al. (1981a) reported three types of post-trial activity (Fig. 5.25). These post-trial activities are linked with the presence or the absence of the reward. One type of post-trial activity (Type I cells) is linked with the reward delivery and observed after correct and reinforced responses regardless of kinds of the tasks. This activity is also observed after the delivery of free reward without any behavioral response. No activity is observed after unreinforced responses regardless of whether the monkey performed correct responses or not. Therefore, this activity is directly linked to the reward delivery. Another type of post-trial activity (Type II cells) is not observed after correct and reinforced responses, but is observed in trials with correct but unreinforced responses. Therefore, this type of activity is thought to represent the deviation from the reward expectation. The third type (Type III cells) is observed after every behavioral response regardless of the presence or the absence of the reward. This activity is thought to represent the termination of a trial.

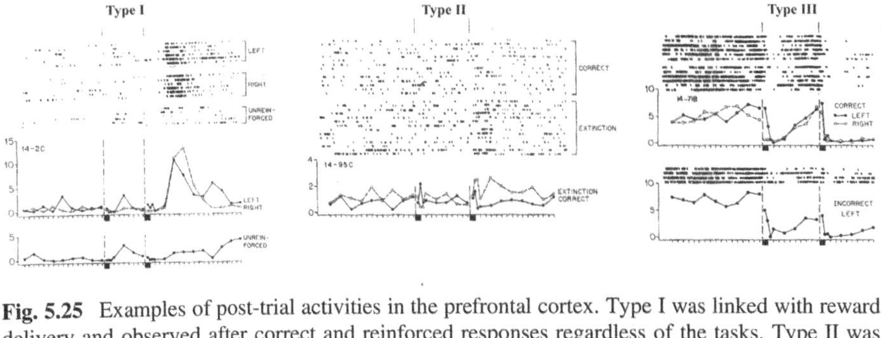

Fig. 5.25 Examples of post-trial activities in the prefrontal cortex. Type I was linked with reward delivery and observed after correct and reinforced responses regardless of the tasks. Type II was observed only in trials with correct but unreinforced responses (extinction trials). Type III (this example was inhibitory response) was observed after every response regardless of the presence or absence of the reward. From Rosenkilde et al. (1981a) with permission

Watanabe (1989) further characterized post-trial activity observed in the prefrontal cortex. He observed juice-related cells and end-of-trial-related cells which are similar as Type I and III cells described by Rosenkilde et al. (1981a). However, he observed several other types of neurons that exhibited post-trial activity. For example, reinforcement-related cells show post-trial activity only when the reward is given as a reinforcement of the correct response and do not show the activity when the reward is given as a free reward. Reinforcement-error-related cells show post-trial activity after both correct and error responses, but the magnitude of the activity is different between correct and error responses. Some reinforcement-related and reinforcement-error-related cells encode the correctness of the response since activity of these cells depends only on whether the response is correct or not and is independent of whether the reward is present or absent. In addition, two kinds of error-related cells are observed: one is related to error recognition and the other is related to encoding the absence of an expected reward.

Thus, prefrontal neurons exhibit several kinds of post-trial activity. Since the reward is delivered at the end of every correct trial, the activity linked with reward delivery can be observed in any study. Reward-related activity is observed when the ODR task is used to examine prefrontal activity (Ichihara-Takeda and Funahashi 2006). In the ODR task, the monkey is required to perform memory-guided saccades toward one of eight directions and the same amount of reward is given in every correct trial. Therefore, post-saccadic activity can be considered to be reward-related activity. However, a great majority of post-saccadic activity exhibit directional selectivity (Funahashi et al. 1991; Takeda and Funahashi 2002). Directional tuning should not be observed in reward-related activity. Therefore, post-saccadic activity observed in the ODR task is not reward-related activity.

Post-trial activity representing the end of the trial and the activity representing the correctness of the response are observed in the prefrontal cortex. Post-trial activity representing the end of the trial would play an important role to initiate the preparation for the next new trial. Similarly, post-trial activity representing the correctness of the response and the activity detecting an error would play an important role to monitor and adjust behavioral responses flexibly. Post-saccadic activity might also play similar roles as post-trial activity does.

5.3.4.2 Post-Saccadic Activity Has Function to Terminate Delay-Period Activity

A large number of prefrontal neurons exhibited post-saccadic activity while the monkey performed the ODR task. Post-saccadic activity is one of post-trial activities and post-trial activity plays important roles to monitor and adjust behavioral responses flexibly. Therefore, post-saccadic activity is also thought to play important roles similar as post-trial activity in the ODR task.

During the performance of the ODR task, a large number of prefrontal neurons exhibit delay-period activity, which is tonic, sustained activation during the delay period (Funahashi et al. 1989). Delay-period activity has several important features.

First, the duration of delay-period activity becomes either longer or shorter depending on the length of the delay period. Second, this activity is observed only when the monkey makes correct behavioral responses. Third, a great majority of delay-period activity exhibit directional preference, such that this activity is observed only when visual cues are presented at a particular area in the visual field. Since the preferred direction of this activity differs from neuron to neuron, neurons having directional delay-period activity are thought to have mnemonic receptive fields (memory fields) in the visual field. Fourth, a great majority of delay-period activity represent information regarding the direction of the visual cue (retrospective information), whereas a minority represents information regarding the direction of the saccade (prospective information) (Funahashi et al. 1993; Takeda and Funahashi 2002). Based on these observations, delay-period activity is considered to be a neural correlate of the mechanism for temporarily retaining necessary information in working memory (Goldman-Rakic 1987; Funahashi 2001; Fuster 2008).

While the monkey performs the ODR task, retaining spatial information of the visual cue during the delay period is essential to make a correct eye movement during the response period. Directional delay-period activity plays an essential role for retaining spatial information of the visual cue during the delay period at any given trial. However, retaining spatial information of current visual cue is no longer necessary after the monkey makes a saccadic eye movement during the response period, because the visual cue will present at different position in the next trial. The presence of delay-period activity after the termination of the trial would be counterproductive. Interestingly, the temporal pattern of delay-period activity shows that this activity is terminated immediately after the monkey makes saccadic eye movements. In addition, a comparison of the temporal patterns between delay-period activity and post-saccadic activity shows that the termination of delay-period activity corresponds to the beginning of post-saccadic activity. Post-saccadic activity is thought to be a feedback signal from oculomotor centers that are responsible for the initiation of purposive saccades. Therefore, post-saccadic activity may play a critical role to actively terminate delay-period activity after the execution of the eye movement during the response period.

Figure 5.26 shows population histograms of excitatory delay-period activity and post-saccadic activity both triggered at the initiation of the saccadic eye movement during the response period. As shown in Fig. 5.26, delay-period activity starts terminating rapidly when the saccadic eye movement is initiated (Fig. 5.26b), and at the same time, post-saccadic activity starts increasing (Fig. 5.26d). Thus, the termination of excitatory delay-period activity coincided with the initiation of post-saccadic activity based on population analyses of prefrontal activity (Fig. 5.26e) (Goldman-Rakic et al. 1990; Funahashi and Takeda 2002a, b). This result supports the idea that post-saccadic activity plays a critical role to terminate delay-period activity after the execution of the eye movement during the response period. The termination of delay-period activity by post-saccadic activity at the end of every trial could ensure to receive and retain new information for the next trial. This is an important mechanism for information processing in working memory. Thus, it can

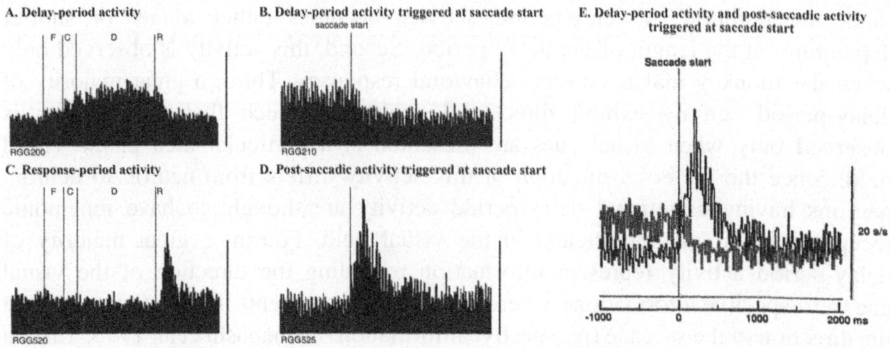

Fig. 5.26 (**a**) Population delay-period activity during a whole trial. (**b**) Population delay-period activity triggered at the saccade start. (**c**) Population response-period activity during a whole trial. (**d**) Population response-period activity triggered at the saccade start. (**e**) Interaction between delay-period activity and post-saccadic activity

be concluded that post-saccadic activity observed in many neurons in the prefrontal cortex significantly contributes to information processing in working memory.

5.3.4.3 Post-Saccadic Activity May Participate in Performance Monitoring

Post-trial activity representing the correctness of the response and detecting errors has been observed in the prefrontal cortex in monkeys as well as humans (see reviews by Mansouri et al. (2009) and Ridderinkhof et al. (2004)). This activity is thought to be important for monitoring behavioral responses and enabling the flexible modulation of response behaviors. The anterior cingulate cortex is known to contribute to monitor behavioral performance by detecting errors. For example, Carter et al. (1998) observed activation of the anterior cingulate area by fMRI while human subjects performed erroneous responses in the continuous performance task. However, the same area was also activated during correct responses when the response competition increased and errors were likely to occur. Similar results have also been reported by Botvinick et al. (1999) using a version of the flanker task. They observed anterior cingulate activation even in correct trials and this activation was greater in high conflict and incongruent trials than in low conflict and congruent trials. Thus, the anterior cingulate cortex not just detects errors but also participates in the online monitoring of behavioral performance.

Post-saccadic activity detecting errors and monitoring conflict among processes has been observed in the supplementary eye field. Stuphorn et al. (2000) analyzed activity of supplementary eye field neurons while the monkey performed a countermanding task and found several types of neurons that exhibited post-saccadic activity. One type was strongly activated after the initiation of the planned eye movement specifically when the monkey failed to cancel it. Although this activity was classified as post-saccadic, this activity was distinct from post-saccadic activity observed in the frontal eye field, because this activity was not observed in visually guided saccades and because this activation was observed in both ipsiversive and contraversive saccades. Based on these observations, they concluded that this activity was a neural signal indicating the occurrence of the error. A second type was also activated only when the planned eye movement was cancelled. However, this activation was not likely to be related to the cancelling of eye movement, because the activity occurred after the stop-signal reaction time, which is the timing after the initiation of the eye movement in non-cancelled stop-signal trials. When eye movements are cancelled in the countermanding task, gaze-shifting neurons (saccade-related neurons) and gaze-holding neurons need to be co-activated. However, this co-activation produces a conflict in oculomotor processing and the balance of activation between these two neuron groups is critical for determining whether the eye movement will be cancelled or not. Therefore, they suggested that the second type of activity is a signal indicating the presence of a conflict in processing. Thus, post-saccade-like activity is observed in the supplementary eye field while the monkey performed a countermanding task. Stuphorn et al. (2000) concluded that these activities represent performance monitoring and play as signals for detecting errors and indicating a conflict in processing.

Prefrontal participation in flexible behavioral control and behavioral adjustment to dissolve the conflict has been examined using the anti-saccade task and categorization tasks (Munoz and Everling 2004; Mansouri et al. 2009). Recently, Teichert et al. (2014) tried to understand post-saccadic activity of frontal eye field neurons in more cognitive functions such as reward expectation, decision uncertainty, or response conflict. If post-saccadic activity is related to these functions, the activity should differ between correct and error trials. To determine whether post-saccadic activity reflects choice errors, they examined error-related activity while the monkey performed a reward-biased speed-categorization task with delayed feedback. Since the monkey often made impulsive errors in this task, they could analyze a sufficient amount of internally generated error-related activity in the frontal eye field. As a result, they found that error-related activity is observed in a large fraction of frontal eye field neurons and that the magnitude of error-related activity is strongly correlated with the difficulty of the task and not with eye movements. Therefore, post-saccadic activity observed in the frontal eye field might be related to the evaluation of post-decision outcome. Since the evaluation of the preceding choice could play a role in optimizing the subsequent behavior, post-saccadic activity might play a role in monitoring performance.

5.3.5 Other Features of Prefrontal Saccade-Related Activity

Asaad et al. (2000) reported task-dependent modulation of saccade-related activity in prefrontal neurons. They used three types of working memory tasks (delayed-response task, delayed matching-to-sample task, and delayed conditional task). In these tasks, one sample object or a small light spot is briefly presented during the cue period and, after a brief delay (1.0 s), the monkey needs to make a saccade to an appropriate target position (right or left). The monkey needs to make a saccade to the position where the small light spot is presented in the delayed-response task, to the same object that is presented as the sample object in the delayed matching-to-sample task, and to the right or left position associated with the sample object in the delayed conditional task. They found that many prefrontal neurons exhibited task-dependent difference in saccade-related activity. Many neurons (41%, 125/305) exhibited direction-selective saccade-related activity whose magnitude was equal across three tasks. However, in 46% (139/305) neurons, the magnitude of this activity was modulated by the difference of tasks. They concluded that task-dependent difference of saccade-related activity is caused by the difference of behavioral rules. Although the monkey performed the same right or left saccade in all these tasks, visual stimuli presented during saccade performances were different among tasks. Therefore, task-dependent difference of saccade-related activity might be caused by the difference of visual stimuli presented during saccades.

Context-dependent pre-saccadic activity has been observed in the prefrontal cortex. Inoue and Mikami (2010) used a "serial probe reproduction task" and showed that 41% of prefrontal neurons having pre-saccadic activity exhibited context-dependent activity. In this task, the monkey is required to remember the rank order of presentation of two objects and select either the first rank object or the second rank object among three objects depending on the color signal. Among 56 neurons having pre-saccadic activity, 23 neurons exhibited saccade-related activity when the monkey made a saccade toward a particular direction, regardless of target objects. However, 9 neurons exhibited pre-saccadic activity only when the monkey made a saccade toward a particular target object, regardless of the directions of the saccades, and 17 exhibited pre-saccadic activity only when the monkey made a saccade toward a particular target object presented at a particular position. Thus, saccade-related activity in the prefrontal cortex is also affected by behavioral context.

References

Andersen RA, Asanuma C, Cowan WM (1985) Callosal and prefrontal associational projecting cell populations in area 7A of the macaque monkey: a study using retrogradely transported fluorescent dyes. J Comp Neurol 232:443–455. https://doi.org/10.1002/cne.902320403

Andersen RA, Asanuma C, Essick G, Siegel RM (1990a) Corticocortical connections of anatomically and physiologically defined subdivisions within the inferior parietal lobule. J Comp Neurol 296:65–113. https://doi.org/10.1002/cne.902960106

Andersen RA, Bracewell RM, Barash S, Gnadt JW, Fogassi L (1990b) Eye position effects on visual, memory, and saccade-related activity in areas LIP and 7a of macaque. J Neurosci 10: 1176–1196. https://doi.org/10.1523/jneurosci.10-04-01176.1990

Aou S, Oomura Y, Nishino H, Ono T, Yamabe K, Sikdar SK, Noda T, Inoue M (1983) Functional heterogeneity of single neuronal activity in the monkey dorsolateral prefrontal cortex. Brain Res 260:121–124. https://doi.org/10.1016/0006-8993(83)90770-9

Arcaro MJ, McMains SA, Singer BD, Kastner S (2009) Retinotopic organization of human ventral visual cortex. J Neurosci 29:10638–10652. https://doi.org/10.1523/jneurosci.2807-09.2009

Arcaro MJ, Pinsk MA, Li X, Kastner S (2011) Visuotopic organization of macaque posterior parietal cortex: a functional magnetic resonance imaging study. J Neurosci 31:2064–2078. https://doi.org/10.1523/jneurosci.3334-10.2011

Arikuni T, Sakai M, Hamada I, Kubota K (1980) Topographical projections from the prefrontal cortex to the post-arcuate area in the rhesus monkey, studies by retrograde axonal transport of horseradish peroxidase. Neurosci Lett 19:155–160. https://doi.org/10.1016/0304-3940(80)90187-1

Arikuni T, Watanabe K, Kubota K (1988) Connections of area 8 with area 6 in the brain of the macaque monkey. J Comp Neurol 277:21–40. https://doi.org/10.1002/cne.902770103

Asaad WF, Rainer G, Miller EK (2000) Task-specific neural activity in the primate prefrontal cortex. J Neurophysiol 84:451–459. https://doi.org/10.1152/jn.2000.84.1.451

Averbeck BB, Romanski LM (2006) Probabilistic encoding of vocalizations in macaque ventral lateral prefrontal cortex. J Neurosci 26:11023–11033. https://doi.org/10.1523/jneurosci.3466-06.2006

Azuma M, Suzuki H (1984) Properties and distribution of auditory neurons in the dorsolateral prefrontal cortex of the alert monkey. Brain Res 298:43–346. https://doi.org/10.1016/0006-8993(84)91434-3

Barbas H (1988) Anatomic organization of basoventral and mediodorsal visual recipient prefrontal regions in the rhesus monkey. J Comp Neurol 276:313–342. https://doi.org/10.1002/cne.902760302

Barbas H, Pandya DN (1987) Architecture and frontal cortical connections of the premotor cortex (area 6) in the rhesus monkey. J Comp Neurol 256:211–228. https://doi.org/10.1002/cne.902560203

Barbas H, Pandya DN (1989) Architecture and intrinsic connections of the prefrontal cortex in the rhesus monkey. J Comp Neurol 286:353–375. https://doi.org/10.1002/cne.902860306

Barbas H, Zikopoulos B, Timbie C (2011) Sensory pathways and emotional context for action in primate prefrontal cortex. Biol Psychiatry 69:1133–1139. https://doi.org/10.1016/j.biopsych.2010.08.008

Barone P, Joseph JP (1989) Prefrontal cortex and spatial sequencing in macaque monkey. Exp Brain Res 78:447–464. https://doi.org/10.1007/BF00230234

Bates JF, Goldman-Rakic PS (1993) Prefrontal connections of medial motor areas in the rhesus monkey. J Comp Neurol 336:211–228. https://doi.org/10.1002/cne.903360206

Baylis GC, Rolls ET, Leonard CM (1987) Functional subdivisions of the temporal lobe neocortex. J Neurosci 7:330–342. https://doi.org/10.1523/jneurosci.07-02-00330.1987

Bizzi E (1967) Discharge of frontal eye field neurons during eye movements in unanesthetized monkeys. Science 157:1588–1590. https://doi.org/10.1126/science.157.3796.1588

Bizzi E (1968) Discharge of frontal eye field neurons during saccadic and following eye movements in unanesthetized monkeys. Exp Brain Res 6:69–80. https://doi.org/10.1007/BF00235447

Bizzi E, Schiller PH (1970) Single unit activity in the frontal eye fields of unanesthetized monkeys during eye and head movement. Exp Brain Res 10:151–158. https://doi.org/10.1007/BF00234728

Boch RA, Goldberg ME (1989) Participation of prefrontal neurons in the preparation of visually guided eye movements in the rhesus monkey. J Neurophysiol 61:1064–1084. https://doi.org/10.1152/jn.1989.61.5.1064

Borra E, Gerbella M, Rozzi S, Luppino G (2011) Anatomical evidence for the involvement of the macaque ventrolateral prefrontal area 12r in controlling goal-directed actions. J Neurosci 31:12351–12363. https://doi.org/10.1523/jneurosci.1745-11.2011

Borra E, Ferroni CG, Gerbella M, Giorgetti V, Mangiaracina C, Rozzi S, Luppino G (2019) Rostro-caudal connectional heterogeneity of the dorsal part of the macaque prefrontal area 46. Cereb Cortex 29:485–504. https://doi.org/10.1093/cercor/bhx332

Botvinick M, Nystrom LE, Fissell K, Carter CS, Cohen JD (1999) Conflict monitoring versus selection-for-action in anterior cingulated cortex. Nature 402:179–181. https://doi.org/10.1038/46035

Brody CD, Hernandez A, Zainos A, Romo R (2003) Timing and neural encoding of somatosensory parametric working memory in macaque prefrontal cortex. Cereb Cortex 13:1196–1207. https://doi.org/10.1093/cercor/bhg100

Bruce CJ, Goldberg ME (1984) Physiology of the frontal eye fields. Trends Neurosci 7:436–441. https://doi.org/10.1016/S0166-2236(84)80149-6

Bruce CJ, Goldberg ME (1985) Primate frontal eye fields. I. Single neurons discharging before saccades. J Neurophysiol 53:603–635. https://doi.org/10.1152/jn.1985.53.3.603

Bruce CJ, Goldberg ME, Bushnell MC, Stanton GB (1985) Primate frontal eye fields. II. Physiological and anatomical correlates of electrically evoked eye movements. J Neurophysiol 54:714–734. https://doi.org/10.1152/jn.1985.54.3.714

Bruni S, Giorgetti V, Bonini L, Fogassi L (2015) Processing and integration of contextual information in monkey ventrolateral prefrontal neurons during selection and execution of goal-directed manipulative actions. J Neurosci 35:11877–11890. https://doi.org/10.1523/jneurosci.1938-15.2015

Bullier J, Schall JD, Morel A (1996) Functional streams in occipito-frontal connections in the monkey. Behav Brain Res 76:89–97. https://doi.org/10.1016/0166-4328(95)00182-4

Bushnell MC, Goldberg ME, Robinson DL (1981) Behavioral enhancement of visual responses in monkey cerebal cortex. I. Modulation in posterior parietal cortex related to selective visual attention. J Neurophysiol 46:755–772. https://doi.org/10.1152/jn.1981.46.4.755

Carmichael ST, Price JL (1995) Sensory and premotor connections of the orbital and medial prefrontal cortex of macaque monkeys. J Comp Neurol 363:642–664. https://doi.org/10.1002/cne.903630409

Carter CS, Braver TS, Barch DM, Botvinick MM, Noll D, Cohen JD (1998) Anterior cingulate cortex, error detection, and the online monitoring of performance. Science 280:747–749. https://doi.org/10.1126/science.280.5364.747

Cavada C, Goldman-Rakic PS (1989) Posterior parietal cortex in rhesus monkey. II. Evidence for segregated corticocortical networks linking sensory and limbic areas with the frontal lobe. J Comp Neurol 287:422–445. https://doi.org/10.1002/cne.902870403

Cipolloni PB, Pandya DN (1999) Cortical connections of the frontoparietal opercular areas in the rhesus monkey. J Comp Neurol 403:431–458. https://doi.org/10.1002/(SICI)1096-9861(19990125)403:4<431::AID-CNE2>3.0.CO;2-1

Cohen YE, Russ BE, Gifford GW III, Kiringoda R, MacLean KA (2004) Selectivity for the spatial and nonspatial attributes of auditory stimuli in the ventrolateral prefrontal cortex. J Neurosci 24:11307–11316. https://doi.org/10.1523/jneurosci.3935-04.2004

Crosby EC, Yoss RE, Henderson JW (1952) The mammalian midbrain and isthmus regions. Part II. The fiber connections. D. The pattern for eye movements on the frontal eye field and the discharge of specific portions of the field to and through midbrain levels. J Comp Neurol 97:357–383. https://doi.org/10.1002/cne.900970205

Deuel R, Mishkin M (1977) Limbic and prefrontal contributions to somesthetic learning in monkeys. Brain Res 132:521–535. https://doi.org/10.1016/0006-6993(77)90199-8

Engel SA, Glover GH, Wandell BA (1997) Retinotopic organization in human visual cortex and the spatial precision of functional MRI. Cereb Cortex 7:181–192. https://doi.org/10.1093/cercor/7.2.181

Evarts EV (1966) Pyramidal tract activity associated with a conditioned hand movement in the monkey. J Neurophysiol 29:1011–1027. https://doi.org/10.1152/jn.1966.29.6.1011

Evarts EV (1968) A technique for recording activity of subcortical neurons in moving animals. Electroenceph Clin Neurophysiol 24:83–86. https://doi.org/10.1016/0013-4694(68)90070-9

Felleman DJ, Van Essen DC (1991) Distributed hierarchical processing in the primate cerebral cortex. Cereb Cortex 1:1–47. https://doi.org/10.1093/cercor/1.1.1-a

Ferrier D (1875a) Experiments on the brain of monkeys. No. 1. Proc R Soc Lond 23:409–430. https://doi.org/10.1098/rspl.1874.0058
Ferrier D (1875b) The Croonian Lecture. Experiments on the brain of monkeys (second series). Phil Trans R Soc Lond 165:433–488. https://doi.org/10.1098/rstl.1875.0016
Fukushi T, Sawaguchi T (2005) Neural representation of response category and motor parameters in monkey prefrontal cortex. Exp Brain Res 164:472–483. https://doi.org/10.1007/s00221-005-2268-z
Funahashi S (1983) Responses of monkey prefrontal neurons during a visual tracking task reinforced by substantia innominata self- stimulation. Brain Res 276:267–276. https://doi.org/10.1016/0006-8993(83)90734-5
Funahashi S (2001) Neuronal mechanisms of executive control by the prefrontal cortex. Neurosci Res 39:147–165. https://doi.org/10.1016/S0168-0102(00)00224-8
Funahashi S, Kubota K (1994) Working memory and prefrontal cortex. Neurosci Res 21:1–11. https://doi.org/10.1016/0168-0102(94)90063-9
Funahashi S, Takeda K (2002a) Population vector analysis by primate prefrontal neuron activities. J Biol Physics 28:527–537. https://doi.org/10.1023/A:1020309916014
Funahashi S, Takeda K (2002b) Information processes in the primate prefrontal cortex in relation to working memory processes. Rev Neurosci 13:313–346. https://doi.org/10.1515/revneuro.2002.13.4.313
Funahashi S, Bruce CJ, Goldman-Rakic PS (1989) Mnemonic coding of visual space in the monkey's dorsolateral prefrontal cortex. J Neurophysiol 61:331–349. https://doi.org/10.1152/jn.1989.61.2.331
Funahashi S, Bruce CJ, Goldman-Rakic PS (1990) Visuospatial coding in primate prefrontal neurons revealed by oculomotor paradigms. J Neurophysio 63:814–831. https://doi.org/10.1152/jn.1990.63.4.814
Funahashi S, Bruce CJ, Goldman-Rakic PS (1991) Neuronal activity related to saccadic eye movements in the monkey's dorsolateral prefrontal cortex. J Neurophysiol 65:1464–1483. https://doi.org/10.1152/jn.1991.65.6.1464
Funahashi S, Goldman-Rakic CMV, PS. (1993) Prefrontal neuronal activity in rhesus monkeys performing a delayed anti-saccade task. Nature 365:753–756. https://doi.org/10.1038/365753a0
Fuster JM (1973) Unit activity in prefrontal cortex during delayed-response performance: neuronal correlates of transient memory. J Neurophysiol 36:61–78. https://doi.org/10.1152/jn.1973.36.1.61
Fuster JM (2008) The prefrontal cortex, 4th edn. Academic Press, New York, NY
Fuster JM, Bauer RH, Jervey JP (1982) Cellular discharge in the dorsolateral prefrontal cortex of the monkey in cognitive tasks. Exp Neurol 77:679–694. https://doi.org/10.1016/0014-4886(82)90238-2
Gardner JL, Merriam EP, Movshon JA, Heeger DJ (2008) Maps of visual space in human occipital cortex are retinotopic, not spatiotopic. J Neurosci 28:3988–3999. https://doi.org/10.1523/jneurosci.5476-07.2008
Gerbella M, Borra E, Tonelli S, Rozzi S, Luppino G (2013) Connectional heterogeneity of the ventral part of the macaque area 46. Cereb Cortex 23:967–987. https://doi.org/10.1093/cercor/bhs096
Ghosh S, Gattera R (1995) A comparison of the ipsilateral cortical projections to the dorsal and ventral subdivisions of the macaque premotor cortex. Somatosens Motor Res 12:359–378. https://doi.org/10.3109/08990229509093668
Goldberg ME, Bushnell MC (1981) Behavioral enhancement of visual responses in monkey cerebral cortex. II. Modulation in frontal eye fields specifically related to saccades. J Neurophysiol 46:773–787. https://doi.org/10.1152/jn.1981.46.4.773
Goldman-Rakic PS (1987) Circuitry of primate prefrontal cortex and regulation of behavior by representational memory. In: Plum F (ed) Handbook of physiology, section 1: the nervous system, Part 1: higher functions of the brain, vol V. American Physiological Society, Bethesda, MD, pp 373–417. https://doi.org/10.1002/cphy.cp010509

Goldman-Rakic PS, Funahashi S, Bruce CJ (1990) Neocortical memory circuits. Cold Spring Harbor Symp Quant Biol 55:1025–1038. https://doi.org/10.1101/SQB.1990.055.01.097
Hage SR, Nieder A (2013) Single neurons in monkey prefrontal cortex encode volitional initiation of vocalizations. Nat Commun 4:2409. https://doi.org/10.1038/ncomms3409
Hage SR, Nieder A (2015) Audio-vocal interaction in single neurons of the monkey ventrolateral prefrontal cortex. J Neurosci 35:7030–7040. https://doi.org/10.1523/jneurosci.2371-14.2015
Hagen MC, Zald DH, Thornton TA, Pardo JV (2002) Somatosensory processing in the human inferior prefrontal cortex. J Neurophysiol 88:1400–1406. https://doi.org/10.1152/jn.00940.2001
Hagler DJ Jr, Sereno MI (2006) Spatial maps in frontal and prefrontal cortex. Neuroimg 29:567–577
Haile TM, Bohon KS, Romero MC, Conway BR (2019) Visual stimulus-driven functional organization of macaque prefrontal cortex. Neuroimage 188:427–444. https://doi.org/10.1016/j.neuroimage.2018.11.060
Hasegawa R, Matsumoto M, Mikami A (2000) Search target selection in monkey prefrontal cortex. J Neurophysiol 84:1692–1696. https://doi.org/10.1152/jn.2000.84.3.1692
Hepp K, Henn V, Vilis T, Cohen B (1989) Brainstem regions related to saccade generation. In: Wurtz HW, Goldberg ME (eds) The neurobiology of saccadic eye movements. Elsevier, Amsterdam, pp 105–212
Hernández A, Salinas E, Garcia R, Romo R (1997) Discrimination in the sense of flutter: new psychophysical measurements in monkeys. J Neurosci 17:6391–6400. https://doi.org/10.1523/jneurosci.17-16-06391.1997
Hikosaka O, Wurtz RH (1983) Visual and oculomotor functions of monkey substantia nigra pars reticulata. III. Memory-contingent visual and saccade responses. J Neurophysiol 49:1268–1284. https://doi.org/10.1152/jn.1983.49.5.1268
Hikosaka O, Sakamoto M, Usui S (1989) Functional properties of monkey caudate neurons. III. Activities related to expectation of target and reward. J Neurophysiol 61:814–832. https://doi.org/10.1152/jn.1989.61.4.814
Holmes G (1938) The cerebral integration of the ocular movements. Br Med J 2:107–112. https://doi.org/10.1136/bmj.2.4045.107
Hoshi E, Shima K, Tanji J (1998) Task-dependent selectivity of movement-related neuronal activity in the primate prefrontal cortex. J Neurophysiol 80:3392–3397. https://doi.org/10.1152/jn.1998.80.6.3392
Hoshi E, Shima K, Tanji J (2000) Neuronal activity in the primate prefrontal cortex in the process of motor selection based on two behavioral rules. J Neurophysiol 83:2355–2373. https://doi.org/10.1152/jn.2000.83.4.2355
Huerta MF, Kaas JH (1990) Supplementary eye field as defined by intracortical microstimulation: connections in macaques. J Comp Neurol 293:299–330. https://doi.org/10.1002/cne.902930211
Hussar CR, Pasternak T (2009) Flexibility of sensory representations in prefrontal cortex depends on cell type. Neuron 64:730–743. https://doi.org/10.1016/j.neuron.2009.11.018
Hussar CR, Pasternak T (2013) Common riles guide comparisons of speed and direction of motion in the dorsolateral prefrontal cortex. J Neurosci 33:972–986. https://doi.org/10.1523/jneurosci.4075-12.2013
Hwang J, Romanski LM (2015) Prefrontal neuronal responses during audiovisual mnemonic processing. J Neurosci 35:960–971. https://doi.org/10.1523/jneurosci.1328-14.2015
Ichihara-Takeda S, Funahashi S (2006) Reward-period activity in primate dorsolateral prefrontal and orbitofrontal neurons is affected by reward schedules. J Cogn Neurosci 18:212–226. https://doi.org/10.1162/jocn.2006.18.2.212
Inoue M, Mikami A (2006) Prefrontal activity during serial probe reproduction task: encoding, mnemonic, and retrieval processes. J Neurophysiol 95:1008–1041. https://doi.org/10.1152/jn.00552.2005
Inoue M, Mikami A (2010) Feature to space conversion during target selection in the dorsolateral nad ventrolateral prefrontal cortex of monkeys. Eur J Neurosci 31:938–949. https://doi.org/10.1111/j.1460-9568.2010.07123.x

Ito S (1982) Prefrontal unit activity of macaque monkey during auditory and visual reaction time tasks. Brain Res 247:39–47. https://doi.org/10.1016/0006-8993(82)91025-3

Jacobson S, Trojanowski JQ (1977a) Prefrontal granular cortex of the rhesus monkey. I. Intrahemispheric cortical afferents. Brain Res 132:209–233. https://doi.org/10.1016/0006-8993(77)90417-6

Jacobson S, Trojanowski JQ (1977b) Prefrontal granular cortex of the rhesus monkey. II. Interhemispheric cortical afferents. Brain Res 132:235–245. https://doi.org/10.1016/0006-8993(77)90418-8

Jones EG, Powell TPS (1970) An anatomical study of converging sensory pathways within the cerebral cortex of the monkey. Brain 93:793–820. https://doi.org/10.1093/brain/93.4.793

Joseph JP, Barone P (1987) Prefrontal unit activity during a delayed oculomotor task in the monkey. Exp Brain Res 67:460–468. https://doi.org/10.1007/BF00247279

Kadohisa M, Kusunoki M, Petrov P, Sigala N, Buckley MJ, Gaffan D, Duncan J (2014) Spatial and temporal distribution of visual information coding in lateral prefrontal cortex. Eur J Neurosci 41: 89–96. https://doi.org/10.1111/ejn.12754

Kastner S, DeSimone K, Konen CS, Szczepanski SM, Weiner KS, Schneider KA (2007) Topographic maps in human frontal cortex revealed in memory-guided saccade and spatial working-memory tasks. J Neurophysiol 97:3494–3507. https://doi.org/10.1152/jn.00010.2007

Kawamura K, Naito J (1984) Corticocortical projections to the prefrontal cortex in the rhesus monkey investigated with horseradish peroxidase techniques. Neurosci Res 1:89–103. https://doi.org/10.1016/s0168-0102(84)80007-3

Kikuchi-Yorioka Y, Sawaguchi T (2000) Parallel visuospatial and audiospatial working memory processes in the monkey dorsolateral prefrontal cortex. Nat Neurosci 3:1075–1076. https://doi.org/10.1038/80581

Kojima S (1980) Prefrontal unit activity in the monkey: relation to visual stimuli and movements. Exp Neurol 69:110–123. https://doi.org/10.1016/0014-4886(80)90147-8

Kojima S, Goldman-Rakic PS (1984) Functional analysis of spatially discriminative neurons in prefrontal cortex of rhesus monkey. Brain Res 291:229–240. https://doi.org/10.1016/0006-8993(84)91255-1

Kojima S, Matsumura M, Kubota K (1981) Prefrontal neuron activity during delayed-response performance without imperative GO signals in the monkey. Exp Neurol 74:396–407. https://doi.org/10.1016/0014-4886(81)90178-3

Kolster H, Peeters R, Orban GA (2010) The retinotopic organization of the human middle temporal area MT/V5 and its cortical neighbors. J Neurosci 30:9801–9820. https://doi.org/10.1523/jneurosci.2069-10.2010

Komatsu H (1982) Prefrontal unit activity during a color discrimination task with go and no-go responses in the monkey. Brain Res 244:269–277. https://doi.org/10.1016/0006-8993(82)90085-3

Krieger HP, Wagman IH, Bender MB (1958) Changes in state of consciousness and patterns of eye movements. J Neurophysiol 21:224–230. https://doi.org/10.1152/jn.1958.21.3.224

Kubota K (1978) Neuron activity in the dorsolateral prefrontal cortex of the monkey and initiation of behavior. In: Ito M, Tsukahara N, Kubota K, Yagi K (eds) Integrative control functions of the brain, vol 1. Kohdansha-Elsevier, Tokyo, pp 407–417

Kubota K, Funahashi S (1982) Direction-specific activities of dorsolateral prefrontal and motor cortex pyramidal tract neurons during visual tracking. J Neurophysiol 47:362–376. https://doi.org/10.1152/jn.1982.47.3.362

Kubota K, Komatsu H (1985) Neuron activities of monkey prefrontal cortex during the learning of visual discrimination tasks with go/no-go performances. Neurosci Res 3:106–129. https://doi.org/10.1016/0168-0102(85)90025-2

Kubota K, Niki H (1971) Prefrontal cortical unit activity and delayed alternation performance in monkeys. J Neurophysiol 34:337–347. https://doi.org/10.1152/jn.1971.34.3.337

Kubota K, Iwamoto T, Suzuki H (1974) Visuokinetic activities of primate prefrontal neurons during delayed-response performance. J Neurophysiol 37:1197–1212. https://doi.org/10.1152/jn.1974.37.6.1197

Kubota K, Tonoike M, Mikam A (1980) Neuronal activity in the monkey dorsolateral prefrontal cortex during a discrimination task with delay. Brain Res 183:29–42. https://doi.org/10.1016/0006-8993(80)90117-1

Latto R, Cowey A (1971a) Visual field defects after frontal eye-field lesions in monkeys. Brain Res 30:1–24. https://doi.org/10.1016/0006-8993(71)90002-3

Latto R, Cowey A (1971b) Fixation changes after frontal eye-field lesions in monkeys. Brain Res 30:25–36. https://doi.org/10.1016/0006-8993(71)90003-5

Lauwereyns J, Sakagami M, Tsutsui K, Kobayashi S, Koizumi M, Hikosaka O (2001) Responses to task-irrelevant visual features by primate prefrontal neurons. J Neurophysiol 86:2001–2010. https://doi.org/10.1152/jn.2001.86.4.2001

Lu M-T, Preston JB, Strick PL (1994) Interconnections between the prefrontal cortex and the premotor areas in the frontal lobe. J Comp Neurol 341:375–392. https://doi.org/10.1002/cne.903410308

Luppino G, Rozzi S, Calzavara R, Matelli M (2003) Prefrontal and agranular cingulate projections to the dorsal premotor areas F2 and F7 in the macaque monkey. Eur J Neurosci 17:559–578. https://doi.org/10.1046/j.1460-9568.2003.02476.x

Lynch JC, Mountcastle VB, Talbot WH, Yin TCT (1977) Parietal lobe mechanisms for directed visual attention. J Neurophysiol 40:362–389. https://doi.org/10.1152/jn.1977.40.2.362

Mansouri FA, Tanaka K, Buckley MJ (2009) Conflict-induced behavioral adjustment: a clue to the executive functions of the prefrontal cortex. Nat Rev Neurosci 10:141–152. https://doi.org/10.1038/nrn2538

Medalla M, Barbas H (2014) Specialized prefrontal "auditory fields": organization of primate prefrontal-temporal pathways. Front Neurosci 8:77. https://doi.org/10.3389/fnins.2014.00077

Mikami A, Ito S, Kubota K (1982) Visual response properties of dorsolateral prefrontal neurons during visual fixation task. J Neurophysiol 47:593–605. https://doi.org/10.1152/jn.1982.47.4.593

Miller EK, Erickson CA, Desimone R (1996) Neural mechanisms of visual working memory in prefrontal cortex of the macaque. J Neurosci 16:5154–5167. https://doi.org/10.1523/jneurosci.16-16-05154.1996

Miyachi S, Lu X, Inoue S, Iwasaki T, Koike S, Nambu A, Takada M (2005) Organization of multisynaptic inputs from prefrontal cortex to primary motor cortex as revealed by retrograde transneuronal transport of rabies virus. J Neurosci 25:2547–2556. https://doi.org/10.1523/jneurosci.4186-04.2005

Mohler CW, Goldberg ME, Wurtz RH (1973) Visual receptive fields of frontal eye field neurons. Brain Res 61:385–389. https://doi.org/10.1016/0006-8993(73)90543-x

Moran J, Desimone R (1985) Selective attention gates visual processing in the extrastriate cortex. Science 229:782–784. https://doi.org/10.1126/science.4023713

Munoz DP, Everling S (2004) Look away: the anti-saccade task and the voluntary control of eye movement. Nat Rev Neurosci 5:218–228. https://doi.org/10.1038/nrn1345

Niki H (1974) Differential activity of prefrontal units during right and left delayed response trials. Brain Res 70:346–349. https://doi.org/10.1016/0006-8993(74)90324-2

Niki H, Watanabe M (1976) Prefrontal unit activity and delayed response: relation to cue location versus direction of response. Brain Res 105:79–88. https://doi.org/10.1016/0006-8993(76)90924-0

Ninokura Y, Mushiake H, Tanji J (2004) Integration of temporal order and object information in the monkey lateral prefrontal cortex. J Neurophysiol 91:555–560. https://doi.org/10.1152/jn.00694.2003

O'Rawe JF, Leung H-C (2020) Topographic mapping as a basic principle of functional organization for visual and prefrontal functional connectivity. eNeuro.0532-19:2019. https://doi.org/10.1523/eneuro.0532-19.2019

Pan X, Sawa K, Tsuda I, Tsukada M, Sakagami M (2008) Reward prediction based on stimulus categorization in primate lateral prefrontal cortex. Nat Neurosci 11:703–712. https://doi.org/10.1038/nn2128

Pandya DN, Dye P, Butters N (1971) Efferent cortico-cortical projections of the prefrontal cortex in the rhesus monkey. Brain Res 31:35–46. https://doi.org/10.1016/0006-8993(71)90632-9

Perret DI, Rolls ET, Caan W (1982) Visual neurones responsive to faces in the monkey temporal cortex. Exp Brain Res 47:329–342. https://doi.org/10.1007/BF00239352

Petrides M, Pandya DN (1984) Projections to the frontal cortex from the posterior parietal region in the rhesus monkey. J Comp Neurol 228:105–116. https://doi.org/10.1002/cne.902280110

Petrides M, Pandya DN (1994) Comparative architectonic analysis of the human and the macaque frontal cortex. In: Boller F, Spinnler H, Hendler JA (eds) Handbook of neuropsychology, The frontal lobes, vol 9, Sect. 12. Elsevier, Amsterdam, pp 17–58

Picard N, Strick PL (1996) Motor areas of the medial wall: a review of their location and functional activation. Cereb Cortex 6:342–353. https://doi.org/10.1093/cercor/6.3.342

Pigarev IN, Rizzolatt G, Scandolara C (1979) Neurons responding to visual stimuli in the frontal lobe of macaque monkeys. Neurosci Lett 12:207–212

Plakke B, Romanski LM (2014) Auditory connections and functions of prefrontal cortex. Front Neurosci 8:199. https://doi.org/10.3389/fnins.2014.00199

Plakke B, Diltz MD, Romanski LM (2013) Coding of vocalizations by single neurons in ventrolateral prefrontal cortex. Hearing Res 305:135–143. https://doi.org/10.1016/j.heares.2013.07.011

Preuss TM, Goldman-Rakic PS (1989) Connections of the ventral granular frontal cortex of macaques with perisylvian premotor and somatosensory areas: anatomical evidence for somatic representation in primate frontal association cortex. J Comp Neurol 282:293–316. https://doi.org/10.1002/cne.902820210

Quintana J, Yajeya J, Fuster JM (1988) Prefrontal representation of stimulus attributes during delay tasks. I. Unit activity in cross-temporal integration of sensory and sensory-motor information. Brain Res 474:211–221. https://doi.org/10.1016/0006-8993(88)90436-2

Rainer G, Asaad WF, Miller EK (1998a) Selective representation of relevant information by neurons in the primate prefrontal cortex. Nature 393:577–579. https://doi.org/10.1038/31235

Rainer G, Asaad WF, Miller EK (1998b) Memory fields of neurons in the primate prefrontal cortex. Proc Natl Acad Sci USA 95:15008–15013

Rao SC, Rainer G, Miller EK (1997) Integration of what and where in the primate prefrontal cortex. Science 276:821–824. https://doi.org/10.1126/science.276.5313.821

Raos V, Savaki HE (2017) The role of the prefrontal cortex in action perception. Cereb Cortex 27:4677–4690. https://doi.org/10.1093/cercor/bhw261

Reppas JB, Niyogi S, Dala AM, Sereno MI, Tootell RBH (1997) Representation of motion boundaries in retinotopic human visual cortical areas. Nature 388:175–179. https://doi.org/10.1038/40633

Ridderinkhof KR, van den Wildenberg WPM, Segalowitz SJ, Carter CS (2004) Neurocognitive mechanisms of cognitive control: the role of prefrontal cortex in action selection, response inhibition, performance monitoring, and reward-based learning. Brain Cogn 56:129–140. https://doi.org/10.1016/j.bandc.2004.09.016

Riley MR, Qi X-L, Constantinidis C (2017) Functional specialization of areas along the anterior-posterior axis of the primate prefrontal cortex. Cereb Cortex 27:3683–3697. https://doi.org/10.1093/cercor/bhw190

Rizzolatti G, Luppino G (2001) The cortical motor system. Neuron 31:889–901. https://doi.org/10.1016/S0896-6273(01)00423-8

Rizzolatti G, Scandolara C, Matelli M, Gentilucci M (1981) Afferent properties of periarcuate neurons in macaque monkeys. II. Visual responses. Behav Brain Res 2:147–163. https://doi.org/10.1016/0166-4328(81)90053-X

Rizzuto DS, Mamelak AN, Sutherling WW, Fineman I, Andersen RA (2005) Spatial selectivity in human ventrolateral prefrontal cortex. Nat Neurosci 8:415–417. https://doi.org/10.1038/nn1424

Robinson DA, Fuchs AF (1969) Eye movements evoked by stimulation of frontal eye fields. J Neurophysiol 32:637–648. https://doi.org/10.1152/jn.1969.32.5.637

Robinson DL, Goldberg ME, Stanton GB (1978) Parietal association cortex in the primate: sensory mechanisms and behavioral modulation. J Neurophysiol 41:910–932. https://doi.org/10.1152/jn.1978.41.4.910

Robinson DL, Petersen SE, Keys W (1986) Saccade-related and visual activities in the pulvinar nuclei of the behaving rhesus monkey. Exp Brain Res 62:625–634. https://doi.org/10.1007/bf00236042

Rocchi F, Oya H, Balezeau F, Billig AJ, Kocsis Z, Jenison RL, Hourski KV, Kovach CK, Steinschneider M, Kikuchi Y, Rhone AE, Dlouhy BJ, Kawasaki H, Adolphs R, Greenlee JDW, Griffiths TD, Howard MA III, Petkov CI (2021) Common fronto-temporal effective connectivity in humans and monkeys. Neuron 109:852–868.e8. https://doi.org/10.1016/j.neuron.2020.12.026

Rodman HR (1994) Development of inferior temporal cortex in the monkey. Cereb Cortex 4:484–498. https://doi.org/10.1093/cercor/4.5.484

Rolls ET, Baylis GC (1986) Size and contrast have only small effects on the responses to faces of neurons in the cortex of the superior temporal sulcus of the monkey. Ex Brain Res 65:38–48. https://doi.org/10.1007/BF00243828

Romanski LM (2007) Representation and integration of auditory and visual stimuli in the primate ventral lateral prefrontal cortex. Cereb Cortex 17:i61–i69. https://doi.org/10.1095/cercor/bhm099

Romanski LM (2012) Integration of faces and vocalizations in ventral prefrontal cortex: implications for the evolution of audiovisual speech. Proc Natl Acad Sci USA 109:10717–10724. https://doi.org/10.1073/pnas.1204335109

Romanski LM, Averbeck BB (2009) The primate cortical auditory system and neural representation of conspecific vocalizations. Annu Rev Neurosci 32:315–346. https://doi.org/10.1146/annurev.neuro.051508.135431

Romanski LM, Diehl MM (2011) Neuron responsive to face-view in the primate ventrolateral prefrontal cortex. Neuroscience 189:223–235. https://doi.org/10.1016/j.neuroscience.2011.05.014

Romanski LM, Goldman-Rakic PS (2002) An auditory domain in primate prefrontal cortex. Nat Neurosci 5:15–16. https://doi.org/10.1038/nn781

Romanski LM, Bates JF, Goldman-Rakic PS (1999a) Auditory belt and parabelt projections to the prefrontal cortex in the rhesus monkey. J Comp Neurol 403:141–157. https://doi.org/10.1002/(SICI)1096-9861(19990111)403:2<141::AID-CNE1>3.0.CO;2-V

Romanski LM, Tian B, Fritz J, Mishkin M, Goldman-Rakic PS, Rauschecker JP (1999b) Dual streams of auditory afferents target multiple domains in the primate prefrontal cortex. Nat Neurosci 2:1131–1136. https://doi.org/10.1038/16056

Romanski LM, Averbeck BB, Diltz M (2004) Neural representation of vocalizations in the primate ventrolateral prefrontal cortex. J Neurophysiol 93:734–747. https://doi.org/10.1152/jn.00675.2004

Romo R, Brody CD, Hernández A, Lemus L (1999) Neuronal correlates of parametric working memory in the prefrontal cortex. Nature 399:470–473. https://doi.org/10.1038/20939

Rosenkilde CE, Bauer RH, Fuster JM (1981a) Single cell activity in ventral prefrontal cortex of behaving monkeys. Brain Res 209:375–394. https://doi.org/10.1016/0006-8993(81)90160-8

Rosenkilde CE, Rosvold HE, Mishkin M (1981b) Time discrimination with positional responses after selective prefrontal lesions in monkeys. Brain Res 210:129–144. https://doi.org/10.1016/0006-8993(81)90890-8

Russo GS, Bruce CJ (1993) Effect of eye position within the orbit on electrically elicited saccadic eye movements: a comparison of the macaque monkey's frontal and supplementary eye fields. J Neurophysiol 69:800–818. https://doi.org/10.1152/jn.1993.69.3.800

Saga Y, Iba M, Tanji J, Hoshi E (2011) Development of multidimensional representations of task phases in the lateral prefrontal cortex. J Neurosci 31:10648–10665. https://doi.org/10.1523/jneurosci.0988-11.2011

Sakagami M, Niki H (1994a) Encoding of behavioral significance of visual stimuli by primate prefrontal neurons: relation to relevant task conditions. Exp Brain Res 97:423–436. https://doi.org/10.1007/BF00241536

Sakagami M, Niki H (1994b) Spatial selectivity of go/no-go neurons in monkey prefrontal cortex. Exp Brain Res 100:165–169. https://doi.org/10.1007/BF00227290

Sakagami M, Tsutsui K (1999) The hierarchical organization of decision making in the primate prefrontal cortex. Neurosci Res 34:79–89. https://doi.org/10.1016/S0168-0102(99)00038-3

Sakai M (1974) Prefrontal unit activity during visually guided lever pressing reaction in the monkey. Brain Res 81:297–309. https://doi.org/10.1016/0006-8993(74)90943-3

Sawaguchi T (1987) Properties of neuronal activity related to a visual reaction time task in the monkey prefrontal cortex. J Neurophysiol 58:1080–1099. https://doi.org/10.1152/jn.1987.58.5.1080

Sawaguchi T, Matsumura M, Kubota K (1989a) Delayed response deficits produced by local injection of bicuculline into the dorsolateral prefrontal cortex in Japanese macaque monkeys. Exp Brain Res 75:457–469. https://doi.org/10.1007/BF00249897

Sawaguchi T, Matsumura M, Kubota K (1989b) Depth distribution of neuronal activity related to a visual reaction time task in the monkey prefrontal cortex. J Neurophysiol 61:435–446. https://doi.org/10.1152/jn.1989.61.2.435

Ó Scalaidhe SP, Wilson FAW, Goldman-Rakic PS (1997) Areal segregation of face-processing neurons in prefrontal cortex. Science 278:1135–1138. https://doi.org/10.1126/science.278.5340.1135

Schiller PH, True SD, Conway JL (1980) Deficits in eye movements following frontal eye field and superior colliculus ablations. J Neurophysiol 44:1175–1189. https://doi.org/10.1152/jn.1980.44.6.1175

Schlag J, Schlag-Rey M (1987) Evidence for a supplementary eye field. J Neurophysiol 57:179–200. https://doi.org/10.1152/jn.1987.57.1.179

Schwartz ML, Goldman-Rakic PS (1984) Callosal and intrahemispheric connectivity of the prefrontal association cortex in rhesus monkey: relation between intraparietal and principal sulcal cortex. J Comp Neurol 226:403–420. https://doi.org/10.1002/cne.902260309

Schwartz EL, Desimone R, Albright TD, Gross CG (1983) Shape recognition and inferior temporal neurons. Proc Natl Acad Sci USA 80:5776–5778. https://doi.org/10.1073/pnas.80.18.5776

Schwartz AB, Kettner RE, Georgopoulos AP (1988) Primate motor cortex and free arm movements to visual targets in three-dimensional space. I. Relations between single cell discharge and direction of movement. J Neurosci 8:2913–2927. https://doi.org/10.1523/jneurosci.08-08-02913.1988

Seltzer B, Pandya DN (1989) Frontal lobe connections of the superior temporal sulcus in the rhesus monkey. J Comp Neurol 281:97–113. https://doi.org/10.1002/cne.902810108

Sereno MI, Pitzalis S, Martinez A (2001) Mapping of contralateral space in retinotopic coordinates by a parietal cortical area in humans. Science 294:1350–1354. https://doi.org/10.1126/science.1063695

Silver MA, Ress D, Heeger DJ (2005) Topographic maps of visual spatial attention in human parietal cortex. J Neurophysiol 94:1358–1371. https://doi.org/10.1152/jn.01316.2004

Simone L, Rozzi S, Bimbi M, Fogassi L (2015) Movement-related activity during goal-directed hand actions in the monkey ventrolateral prefrontal cortex. Eur J Neurosci 42:2882–2894. https://doi.org/10.1111/ejn.13040

Sparks DL, Mays LE (1980) Movement fields of saccade-related burst neurons in the monkey superior colliculus. Brain Res 190:39–50. https://doi.org/10.1016/0006-8993(80)91158-0

Stuphorn V, Taylor TL, Schall JD (2000) Performance monitoring by the supplementary eye field. Nature 408:857–860. https://doi.org/10.1038/35048576

Sugihara T, Diltz MD, Averbeck BB, Romanski LM (2006) Integration of auditory and visual communication information in the primate ventrolateral prefrontal cortex. J Neurosci 26:11138–11147. https://doi.org/10.1523/jneurosci.3650-06.2006

Suzuki H (1985) Distribution and organization of visual and auditory neurons in the monkey prefrontal cortex. Vision Res 25:465–469. https://doi.org/10.1016/0042-6989(85)90071-9

Suzuki H, Azuma M (1983) Topographic studies on visual neurons in the dorsolateral prefrontal cortex of the monkey. Exp Brain Res 53:47–58. https://doi.org/10.1007/BF00239397

Takada M, Nambu A, Hatanaka N, Tachibana Y, Muyachi S, Taira M, Inase M (2004) Organization of prefrontal outflow toward frontal motor-related areas in macaque monkeys. Eur J Neurosci 19:3328–3324. https://doi.org/10.1111/j.0953-816X.2004.03425.x

Takahara D, Inoue K, Hirata Y, Miyachi S, Nambu A, Takada M, Hoshi E (2012) Multisynaptic projections from the ventrolateral prefrontal cortex to the dorsal premotor cortex in macaques – anatomical substrate for conditional visuomotor behavior. Eur J Neurosci 36:3365–3375. https://doi.org/10.1111/j.1460-9568.2012.08251.x

Takeda K, Funahashi S (2002) Prefrontal task-related activity representing visual cue location or saccade direction in spatial working memory tasks. J Neurophysiol 87:567–588. https://doi.org/10.1152/jn.00249.2001

Tanila H, Carlson S, Linnankoski I, Lindroos F, Kahila H (1992) Functional properties of dorsolateral prefrontal cortical neurons in awake monkey. Behav Brain Res 47:169–180. https://doi.org/10.1016/s0166-4328(05)80123-8

Tanila H, Carlson S, Linnankoski I, Kahila H (1993) Regional distribution of functions in dorsolateral prefrontal cortex of the monkey. Behav Brain Res 53:63–71. https://doi.org/10.1016/s0166-4382(05)80266-9

Teichert T, Yu D, Ferrera VP (2014) Performance monitoring in monkey frontal eye field. J Neurosci 34:1657–1671. https://doi.org/10.1523/jneurosci.3694-13.2014

Teuber HL (1964) The riddle of frontal lobe function in man. In: Warren JM, Akert K (eds) The frontal granular cortex and behavior. McGraw-Hill, New York, NY, pp 410–441

Ungerleider LG, Mishkin M (1982) Two cortical visual systems. In: Engle DJ, Goodale MA, Mansfield RJ (eds) Analysis of visual behavior. MIT Press, Cambridge, MA, pp 549–586

Ungerleider LG, Gaffan D, Pelak VS (1989) Projections from inferior temporal cortex to prefrontal cortex via the uncinate fascicle in rhesus monkeys. Exp Brain Res 76:473–484. https://doi.org/10.1007/BF00248903

Vaadia E, Benson DA, Hien RD, Goldstein MH Jr (1986) Unit study of monkey frontal cortex: active localization of auditory and of visual stimuli. J Neurophysiol 56:934–952. https://doi.org/10.1153/jn.1986.56.4.934

Wagman IH, Krieger HP, Papatheodorou CA, Bender B (1961) Eye movements elicited by surface and depth stimulation of the frontal lobe of *Macaque mulatta*. J Comp Neurol 117:179–188. https://doi.org/10.1002/cne.901170204

Watanabe M (1981) Prefrontal unit activity during delayed conditional discriminations in the monkey. Brain Res 225:51–65. https://doi.org/10.1016/0006-8993(81)90317-6

Watanabe M (1986a) Prefrontal unit activity during delayed conditional go/no-go discrimination in the monkey. I. Relation to the stimulus. Brain Res 382:1–14. https://doi.org/10.1016/0006-8993(86)90104-6

Watanabe M (1986b) Prefrontal unit activity during delayed conditional go/no-go discrimination in the monkey. II. Relation to go and no-go responses. Brain Res 382:15–27. https://doi.org/10.1016/0006-8993(86)90105-8

Watanabe M (1989) The appropriateness of behavioral responses coded in post-trial activity of primate prefrontal units. Neurosci Lett 101:113–117. https://doi.org/10.1016/0304-3940(89)90450-3

Watanabe M (1990) Prefrontal unit activity during associative learning in the monkey. Exp Brain Res 80:96–309. https://doi.org/10.1007/BF00228157

Watanabe M (1992) Frontal units of the monkey coding the associative significance of visual and auditory stimuli. Exp Brain Res 89:233–247. https://doi.org/10.1007/BF00228241
Wilson FAW, Scalaidhe O, SP, Goldman-Rakic PS. (1993) Dissociation of object and spatial processing domains in primate prefrontal cortex. Science 260:1955–1958. https://doi.org/10.1126/science.8316836
Wurtz RH, Mohler CW (1976) Enhancement of visual responses in monkey striate cortex and frontal eye fields. J Neurophysiol 39:766–772. https://doi.org/10.1152/jn.1976.39.4.766
Yajeya J, Quintana J, Fuster JM (1988) Prefrontal representation of stimulus attributes during delay tasks. II. The role of behavioral significance. Brain Res 474:222–230. https://doi.org/10.1016/0006-8993(88)90473-4

Chapter 6
Cognitive Control

6.1 Dual-Task Interference, Interference Control, and Cognitive Control

In our everyday life, we often perform multiple tasks simultaneously. For example, while driving a car, we often listen to radio program or music. During cooking, we need to pay attention to a boiling pot while cutting vegetables. During discussion, we prepare answers and comments while listening to the opponent's questions. In an experimental condition, the subject is asked to perform two different tasks simultaneously. The conditions in which two independent tasks are performed simultaneously is called *dual-task condition*. Listening to music during driving a car on a narrow and crowded street, searching for a particular object among many others in a supermarket, or listening to a particular person's talk in conversation with friends are examples of dual-task conditions in our real life. Thus, dual-task conditions can be seen everywhere and anytime in our daily life.

To examine behavioral features and brain mechanisms of dual-task performances, a variety of behavioral tasks have been used in experimental conditions and *dual-task paradigm* is the general term for these behavioral tasks. As a dual-task paradigm, a wide range of combinations of two different tasks are used in experimental conditions, such as two speeded sensory-to-motor tasks, two perceptual tasks, two working memory tasks, cooking, or shopping.

The principle of the dual-task paradigm is that the subject performs two complex cognitive tasks (component tasks) simultaneously, each of which includes a distinct goal. Performing each component task correctly and efficiently is an important request to the subject in this paradigm. Despite such requests, the dual-task condition often results in poorer performance in one or both component tasks, compared to when each component task is performed independently. When the subjects perform two tasks in dual-task condition, they often exhibit decreased performance in either or both component tasks, since information processing for one task interferes with information processing for the other (Pashler 1994). The addition of a more

S. Funahashi, *Dorsolateral Prefrontal Cortex*, Brain Science,
https://doi.org/10.1007/978-981-19-7268-3_6

cognitively demanding secondary task strongly disrupts the performance of the primary task. For example, if we make a phone call during driving a car on a narrow and crowded street, we would have a high probability of getting into an accident. In this situation, we have to perform two tasks (driving a car and talking on a cell phone) simultaneously. Therefore, our attention is needed to be divided for these two tasks. However, our capacity of attention is not infinite but limited. If we pay more attention to talking on a cell phone because of receiving important information from the boss, the amount of attention necessary to focus on driving a car would become insufficient. Talking on a cell phone with the boss interferes with the performance of driving a car. Therefore, talking on a cell phone while driving along a city street causes four times increase in the risk of traffic accidents. Thus, the performance of one task (e.g., talking on a cell phone) often interferes with the performance of the other task (e.g., driving a car) in the dual-task condition. This interfering phenomenon observed in the dual-task condition is known as *dual-task interference* or *dual-task cost*.

On the other hand, we continuously perceive a variety of stimuli from internal and external environments. Among them, some stimuli are important information and need to be processed further, while the remaining are distractors or noise. Therefore, we need to select necessary stimuli among distractors to achieve a goal. Since distractors interfere with necessary operations in the brain, we continuously need to discriminate necessary information from distractors and inhibit to perceive distractors. This procedure is called *interference control* (Watanabe and Funahashi 2015a). Interference control is an ability to protect ongoing cognitive processing from internal or environmental distraction. In interference control, the subjects try to select and perceive necessary information and simultaneously try to inhibit to perceive unnecessary information from the environment. Therefore, necessary processes for interference control are similar to those for dual-task operation. Similar to the dual-task interference effect, the detection of necessary stimuli becomes more difficult for the subjects when the number of distractors increases, when distractors become more salient than necessary stimuli, or when distractors become more similar to necessary stimuli.

Dual-task interference is known as typical evidences of a strict limitation and a bottleneck in information processing capacity in the human brain. The cause of dual-task interference and the mechanism of interference control have been often explained using the concept of *cognitive resource*. The cognitive resource is thought to be neural collections or assemblies for task-general information processing in the brain and is shareable by concurrently operating tasks in a graded manner (Wickens 2002). An important feature of the cognitive resource is that the amount of available cognitive resources is limited, not infinite (Duncan 1980). Therefore, dual-task interference would occur if two tasks are accessing the same finite amount of the resource simultaneously and if the amount of the resource required to perform two tasks simultaneously is exceeding the total amount of available resource. The concept of capacity-limited cognitive resource has been frequently used to explain dual-task interference as well as the limitation observed in attention and working memory. Understanding neural mechanisms causing dual-task interference is an important issue. However, the exact nature of cognitive resource is not well known. In addition, the neural mechanism for flexible allocation of a limited capacity of the cognitive resource is also necessary to understand how multiple neural processes are operated concurrently and efficiently. However, this mechanism is also not known.

Although the mechanism of interference control has usually been analyzed and discussed from an inhibitory point of view (Friedman and Miyake 2004; Forstmann et al. 2008; Nee and Jonides 2008), this mechanism can be examined as a similar mechanism to that of dual-task interference. Heavy cognitive demand to process two tasks simultaneously causes dual-task interference, while heavy cognitive demand to process distractors requires interference control. Either a control mechanism to coordinate multiple information processing systems or a control mechanism to flexibly allocate the cognitive resource to each task is necessary for both dual-task performance and interference control.

6.2 Theories of Cognitive Resource

Cognitive resource is thought to be limited in capacity and sharable across different cognitive processes and resides somewhere in the brain. Theories in cognitive psychology suggested a close relationship between the cause of dual-task interference and the concept of cognitive resource (Moray 1967; Wickens 2002). These theories indicated that the cognitive resource corresponds to the brain's information-processing capacity which is shared by simultaneously operating multiple tasks cooperatively and competitively in a graded manner. The amount of available resource at a time defines the upper limit of the capacity that can be used for information processing. If the amount of the resource required to perform two cognitive tasks simultaneously exceeded the upper limit of the available resource, dual-task interference would occur. However, important questions regarding features of the cognitive resource are remained to be answered. These questions are whether the cognitive resource is composed of either a single modality-general (or process-general) resource (Kahneman 1973), multiple modality-specific (or process-specific) resources (Wickens 2002), or two independent resources controlled by each cerebral hemisphere (Friedman and Polson 1981).

The idea that dual-task interference is caused by capacity limitation of the cognitive resource was initially supported by studies using the P300 component of event-related potentials (Isreal et al. 1980; Wickens et al. 1983). P300 is usually recorded when the subject attends to the target stimulus during the task, in which the target stimulus is presented at low probability and embedded in frequently presenting non-target stimulus (an oddball situation). Therefore, P300 is thought to reflect the activity of the neural process related to attention to the target stimulus. The amplitude of P300 is considered to indicate the strength of attention to the target stimulus. Isreal et al. (1980) measured the amplitude of P300 recorded from parietal electrodes while the subject performed two tasks in dual-task condition, which consisted of an object-tracking task as a primary task and a counting task as a secondary task (Fig. 6.1). In the object-tracking task, a visual target jumped horizontally with a random distance at every 3 s and the subject is required to overlay a cursor on the jumping visual target by manipulating a joystick. Three levels of task difficulty are set in this task. In the easy condition, the target moves in the right-left sequence regularly and the subject needs to move the cursor with a joystick with constant displacement and velocity. In the intermediate condition, the target jumps toward an unpredictable direction. In the difficult condition, the target jumps toward an unpredictable direction and the relation between the joystick movement and the cursor movement is modified, such that the joystick movement accelerates the cursor movement. In the counting task, the subject needs to count the number of appearances of pre-specified auditory or visual signal. When the subject performed the object-tracking task in the single-task condition, an increase in the difficulty enhanced the amplitude of P300 associated with the horizontal jump of the visual target. However, when the subject performed two tasks in dual-task condition and when the object-tracking task became more difficult to perform, the magnitude of P300 associated with the appearance of the pre-specified signal of the counting task decreased (Fig. 6.1), suggesting the decrease of attention-related activity to the target signal. The decrease of P300 amplitude in dual-task condition might reflect a neural process that some amount of cognitive resource was taken from the counting task and was allocated for the object-tracking task to cope with the increase of its difficulty (Isreal et al. 1980; Wickens et al. 1983).

Similar changes in P300 amplitudes between single- and dual-task conditions have been reported in subsequent studies (Kok, 2001). Although these studies could not reveal the nature of the cognitive resource and neural mechanisms related to the limitation of the cognitive resource and its allocation depending on the demand of individual tasks, they provided evidence showing that, although the cognitive resource is a hypothetical concept in psychology, the cognitive resource can be depicted using neurobiological terms and that the dual-task paradigm is a useful tool to examine neural functions of the cognitive resource.

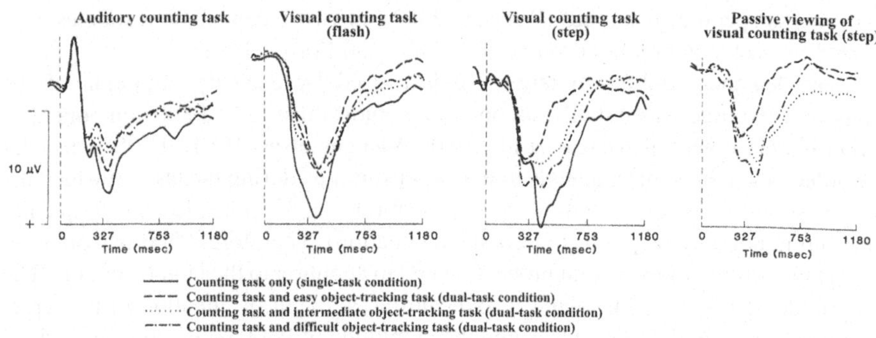

Fig. 6.1 ERP signals elicited by the presentation of auditory and visual stimuli used in the counting tasks. From Wickens et al. (1983) with permission

6.2.1 *Overlap Hypothesis: Modality-Specific Multiple Resource Model*

To describe a more specific relationship between dual-task performance and the cognitive resource, the *overlap hypothesis* has been proposed (Kinsbourne and Hicks 1978; Roland 1985). The overlap hypothesis proposes that the strength of dual-task interference can be determined by the functional distance on the cerebral cortex between two brain areas that are activated during each of the two component tasks. If a common brain region is activated when each of two tasks is performed independently (functional distance is short), stronger competition would happen between two tasks in dual-task condition, because each task tries to obtain enough amount of the neural resource from the common brain region in dual-task condition (stronger dual-task interference). Klingberg and Roland (1997) and Klingberg (1998) used positron emission tomography (PET) and examined whether the evidence supporting the overlap hypothesis exists or not (Fig. 6.2). They examined the size of commonly activated brain area in the cerebral cortex while two cognitive tasks were performed under two kinds of dual-task conditions. Auditory and visual go/no-go tasks were used in one dual-task condition, while auditory and visual working memory tasks were used in the other dual-task condition. According to the overlap hypothesis, if the commonly activated brain area is larger when tasks A and B are performed independently (larger overlap condition) and if the commonly

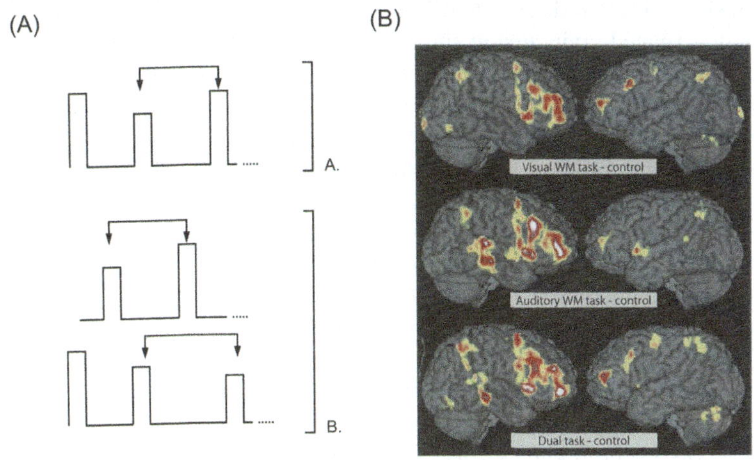

Fig. 6.2 (**a**) Schematic diagram of the two component tasks (top) and the dual task (bottom). Top: In the single working memory task condition, a brief tone or a circular visual stimulus was presented every 2–5 s. Participants were asked to compare the pitch or the brightness of the current stimulus to that of a preceding one. Bottom: In the dual-task condition, two working memory tasks were performed simultaneously. (**b**). Comparison of the increase in rCBF between the single-task and dual-task conditions showed no dual-task-specific activation throughout the brain including the prefrontal cortex. In the control condition, the participants passively viewed the stimuli. From Klingberg (1998) with permission

activated brain area is smaller when tasks C and D are performed independently (smaller overlap condition), stronger dual-task interference effect would be observed when tasks A and B are performed in dual-task condition compared with when tasks C and D are performed in dual-task condition. In fact, the results obtained by Klingberg and Roland (1997) and Klingberg (1998) confirmed this prediction. Behavioral results showed that larger interference effects in reaction times and percent correct rates were observed when visual and auditory working memory tasks were performed in dual-task condition compared with when visual and auditory go/no-go tasks were performed simultaneously. Neuroimaging studies using PET showed that commonly activated brain areas were observed in many cortical areas when each task was performed individually. However, the size of the commonly activated area was significantly larger when two working memory tasks were performed in dual-task condition compared with when two go/no-go tasks were performed (Fig. 6.2b). Thus, behavioral results and neuroimaging results agreed with the prediction based on the overlap hypothesis.

Similar results have been reported in other studies using fMRI. For example, Herath et al. (2001) used visual and somatosensory reaction time tasks in single-task and dual-task conditions and observed that activated area was increased in the right inferior frontal gyrus in dual-task condition. In particular, Nijboer et al. (2014) compared brain activation during three component tasks in single-task and dual-task conditions and showed that the size of a commonly activated brain area in single-task conditions negatively correlated with the degree of task performances in the dual-task condition. Further, Alavash et al. (2015) used a visuospatial task and a speech task in single-task and dual-task conditions and showed that the size of a commonly activated brain area in single-task condition significantly correlated with the decline of visuospatial accuracy in the visuospatial task in dual-task condition.

Wickens (2002) proposed modality-specific multiple resource model to explain the mechanism of dual-task interference. According to this model, when two tasks require the neural process of the same sensory modality to perform, stronger interference effect would be observed in dual-task condition. In fact, it has been shown that the cortical area whose activity decreases in one task (e.g., auditory task) does not overlay with the cortical area whose activity increases in the other task (e.g., visual task) (Klingberg and Roland 1997; Klingberg 1998). This indicates that no competition is present between neural processes of different modalities, suggesting that different neural resource is used for processing different modality information. Similar results have been obtained in fMRI studies by Mochizuki et al. (2007).

The overlap hypothesis predicts the presence of modality-specific multiple resource pools in the brain, because the competition between two tasks to obtain the necessary amount of the cognitive resource occurs predominantly within a common information processing channel. The demand for more neural resource for one particular information processing channel would have little effect on the state of other information processing channels. This notion has been supported by observations in some human psychophysical studies. For example, when the subject was required to attend two moving target objects that appeared in the visual field, the subject's performance improved when two target objects appeared in separate visual quadrants compared with when they appeared within the same quadrant (Carlson et al. 2007). A similar improvement in attention capacity was observed when two targets were presented in different hemifields compared with when they were presented in the same hemifield (Alvarez and Cavanagh 2005). The interference between concurrently attended two objects is predominantly restricted within the same visual quadrant or the same visual hemifield. In addition, behavioral evidences showed that, when the subject performed the tasks requiring working memory of items of different categories (faces, bodies, scenes, and objects), the subject's performance was superior when presented items were drawn from different categories (two faces and two scenes) compared with when items were drawn from one category (four faces) (Cohen et al. 2014). Thus, the interference effect occurred in dual-task condition is predominantly restricted within the same information processing channel. These observations support the presence of modality-specific multiple resource pools in the brain These results indicate that the cognitive resource is comprised of multiple reservoirs, which reside in functionally and spatially distinct brain regions.

6.2.2 Modality-General Single Resource Model

The modality-specific multiple resource model has been supported by several lines of evidences. However, an interdependent modulation of activity is also observed between two functionally and anatomically distant cortical areas during task performances in dual-task condition. Using fMRI, Just et al. (2001) showed that two distinct tasks, each of which activates spatially and functionally different cortical areas, interfered with each other behaviorally in dual-task condition (Fig. 6.3). Two tasks were an auditory language-comprehension task and a visual mental-rotation task (Fig. 6.3a). When these tasks were performed independently, the temporal cortex was activated in the language task, while the parietal cortex was activated in the mental-rotation task (Fig. 6.3b). If the resource competition occurs only within the same information processing channel (overlap hypothesis), the magnitude of activation in each brain area critical for each task in dual-task condition should be equal to the magnitude of brain activation during independent performance of each component task. However, if the modality-general cognitive resource is used to perform these two tasks, the magnitude of activation in the brain areas critical to perform two tasks should be less in dual-task condition than during independent performance of each component task, because dual-task interference is observed in dual-task condition. In fact, the results support the modality-general cognitive resource model. The magnitude of activation in the brain areas critical for two tasks was substantially less in dual-task condition than during independent performance of each component task (Fig. 6.3b right). In a subsequent study, Just et al. (2008) used a visuospatial driving task and an auditory language-comprehension task and found that these two tasks activated non-overlapping different brain areas when they were performed independently. However, simultaneous performance of

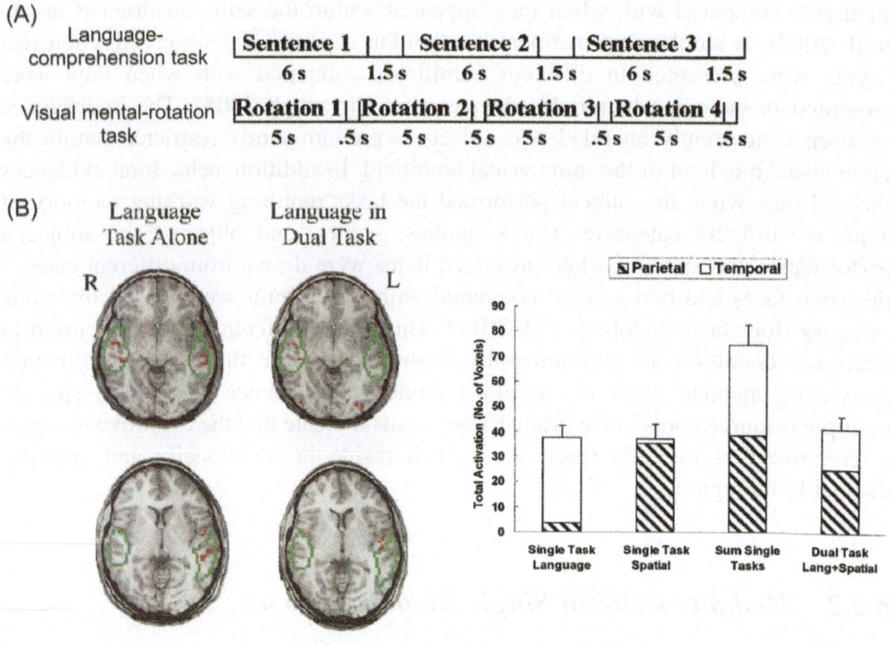

Fig. 6.3 (**a**) Schematic diagram of the timing of the dual-task condition. (**b**) (left). Comparison of brain activation in the temporal cortex ROIs (green border) in the language task between the single-task (left column) and dual-task conditions (right column). (**b**) (right). In the dual-task condition, the total number of activated voxels (rightmost bar) was significantly less than the sum of the two single-task conditions (second bar from the right). When the two-component tasks were performed separately, these two tasks elicited activations in largely non-overlapping brain areas: the language task activated the temporal lobe, and the mental-rotation task activated the parietal lobe (two leftmost bars). From Just et al. (2001) with permission

these two tasks caused a decrease in activation in the brain area associated with the performance of the visuospatial driving task. Similar results have been reported in other studies (Newman et al. 2007; Tombu et al. 2011). These results support the modality-specific multiple resource model.

Neuroimaging studies suggest that some mutual inhibitory mechanisms are present between anatomically and functionally distinct brain areas and that these inhibitory mechanisms come into play when the simultaneous performance of two tasks in dual-task condition activates these areas. At a given point in time, modality-general cognitive resource may be allocated at the whole brain level. When the capacity of the cognitive resource is about to exceed for performing two tasks simultaneously in dual-task condition, the magnitude of activation in two distinct brain areas each of which is recruited for performing either component task might be regulated via mutual inhibitory mechanisms present between these two areas. Therefore, these mutual inhibitory mechanisms would be a mechanism causing dual-task interference. This idea is supported by the results obtained using magneto-encephalography (MEG) studies by Marti et al. (2015).

The behavioral studies indicate that dual-task interference effect is observed not only between two similar tasks but also between two distinct tasks and that the interference effect is stronger between similar tasks than between different tasks. These results support both the modality-specific multiple resource model and the modality-general single resource model. In neuroimaging studies, some evidences support the presence of modality-specific multiple resource pools, while other evidences support the presence of modality-general single resource pool. Therefore, we should consider the cognitive resource not as a single neural component or assembly but as flexibly exchangeable component between single to multiple resources depending on a variety of situations.

6.3 Dual-Task Paradigm as a Tool to Examine Dual-Task Interference

The dual-task paradigm is the method that the subject is asked to perform two complex cognitive tasks simultaneously, each of which includes a distinct goal. When the subject performs two tasks in dual-task condition, he often exhibits decreased performance in either one or both tasks, since information processing for performing one task interferes with information processing for the other. Pairing with a more cognitively demanding secondary task strongly disrupts information processing for the primary task and causes impairment in the performance of the primary task. This interference effect is called dual-task interference. The cause of the dual-task interference has been explained with the concept that the cognitive resource to be able to operate a variety of cognitive activities is capacity-limited. In addition, this concept also includes the mechanism which allocates a limited capacity of the cognitive resource flexibly to information processing for each component task. This concept has also been used to explain limited ability in attention and working memory performances. Therefore, knowing the nature of the cognitive resource is an important issue. In addition, the neural mechanism for flexible allocation of a limited capacity of cognitive resource is also an important issue. The dual-task paradigm is a method to solve these questions.

In the dual-task paradigm, two relatively complex cognitive tasks are used. In one condition, two component tasks are initiated simultaneously triggered by two different external cues. In the other condition, a trial of the secondary task is inserted within a trial of the primary task. For example, a noun/verb judgment (secondary task) is inserted during the delay period of a verbal short-term memory task (primary task). A wide variety of pairs of behavioral tasks (e.g., speeded sensory-to-motor tasks, perceptual tasks, working memory tasks, or simulated tasks of cooking or shopping) have been used.

Although dual-task paradigm has been widely used to study neural mechanisms of dual-task interference and cognitive control in humans, this paradigm has not been well adopted for nonhuman animal studies. In experiments using monkeys, a delayed matching-to-sample (DMS) task has often been used as the primary task, and a variety of secondary tasks have been inserted during the delay period of the DMS task. These experiments have primarily relied on manual responses by pressing buttons attached to a screen (Moise 1970), moving a joystick (Washburn and Astur 1998; Smith et al. 2013), or pointing on a touch screen (Basile and Hampton 2013). Most of these studies did not focus on the psychological mechanism that is specifically related to the information processing for dual-task performance but used this paradigm as a tool to evaluate the functional similarity of short-term memory between humans and nonhuman animals. Therefore, the development of appropriate behavioral paradigms for animal experiments is desirable for a better understanding of psychological and neural mechanisms related to information processing for dual-task performance.

6.4 Prefrontal Cortex and Dual-Task Performance

6.4.1 Human Neuropsychological Studies

The simultaneous performance of two tasks in dual-task condition requires more mental effort than the performance of the individual task in single-task condition. To accommodate an increase in processing demand in dual-task condition, the activation of additional, spatially separate brain areas that are not activated in the performance in single-task condition would be necessary. According to this view, performance deficits in dual-task condition would arise from the limitation of information processing capacity or malfunctioning of information process in these brain areas that are devoted to the information processing specific for demands in dual-task condition. In fact, neuropsychological studies have shown that the prefrontal cortex plays specific roles in dual-task performance. Patients having lesions in the lateral prefrontal cortex have been shown to exhibit marked impairment in behavioral performances under multi-tasking conditions (Baddeley et al. 1997; Burgess 2000; Godbout et al. 2005; Frisch et al. 2012). However, these patients exhibited no impairment in the Wisconsin card sorting test and the verbal fluency test, which are well-known tests closely associated with prefrontal functions (Baddeley et al. 1997; Burgess 2000). These patients also exhibited no impairment in conventional psychological tests (e.g., perceptual, memory, or IQ tests) (Godbout et al. 2005; Frisch et al. 2012). These results suggest contributions of the prefrontal cortex specifically in behavioral performances under multi-tasking condition. Alzheimer's disease patients also exhibit clear behavioral impairment in dual-task conditions, while normal performance was maintained in single-task conditions (Baddeley et al. 1991). Alzheimer's disease patients impair the functions related to the central executive of working memory (Baddeley 1986) and the functions of the central executive are closely associated with prefrontal functions. Therefore, the behavioral impairment in dual-task conditions in Alzheimer's disease patients is thought to be caused by impairments of prefrontal functions. In addition, patients having traumatic brain injury in the frontal lobe exhibit behavioral impairments in dual-task condition (McDowell et al. 1997). Patients who had a traumatic brain injury and impaired working memory performance also exhibit behavioral impairments in dual-task condition (Park et al. 1999). Further, it has been shown that frontal patients exhibit behavioral impairments in multiple-task performances in daily life situations, such as cooking (Godbout et al. 2005; Frisch et al. 2012) and shopping (Shallice and Burgess 1991). However, these patients were known to exhibit very well behavioral scores on the tests of language, intelligence, and frontal lobe functions (Shallice and Burgess 1991; Godbout et al. 2005; Frisch et al. 2012). For example, in the "Multiple Errands Test" (Shallice and Burgess 1991), the participant is asked to perform several real-world tasks, such as buying specific items in different shops, collecting specific information, and arriving at a certain place at a certain time to meet the experimenter. This study found that frontal patients having lesions in the anterior lateral prefrontal cortex subtending Brodmann's areas 46 and 10 exhibited marked deficits in the simultaneous performance of multiple tasks, despite their relatively unaffected performance in conventional psychological tests such as memory or IQ tests and even in some executive tasks that are normally associated with prefrontal function, such as the Wisconsin card sorting test and the verbal fluency test. These results suggest that distinct neural components involved in task performances in dual-task condition are present in the prefrontal cortex. These results also suggest that the neural component involved in task performances in dual-task condition is related to the neural mechanism for working memory, especially to the central executive of the working memory model proposed by Baddeley (1986).

6.4.2 *Human fMRI Studies*

fMRI studies using human subjects also show an important contribution of the lateral prefrontal cortex to task performances in dual-task condition (D'Esposito et al. 1995; Klingberg 1998; Koechlin et al. 1999; Adcock et al. 2000; Derrfuss et al. 2004; Forstmann et al. 2008; Deprez et al. 2013). For example, D'Esposito et al. (1995) attempted to localize dual-task-specific brain areas by comparing brain activations under single-task and dual-task conditions. They used a semantic judgment task and a spatial rotation task (Fig. 6.4a). In the semantic judgment task, a series of words are presented aurally one by one at every 2 s and the subject is needed to respond to the words (e.g., onion) of a pre-defined target category (e.g., vegetable). In the spatial rotation task, three squares, each of which has one dot and one double line in a different arrangement, are simultaneously presented in a triangular position on the monitor. The subject is asked to indicate which of the two lower squares has a dot in the same location relative to the double line as the upper square has. Square stimuli are presented every 3 s. In the dual-task condition, the subject needs to perform these two tasks concurrently. Their fMRI study showed that, in the dual-task condition, a significant increase of activation compared to the single-task condition was observed only in the dorsolateral prefrontal cortex and the anterior cingulate cortex. These

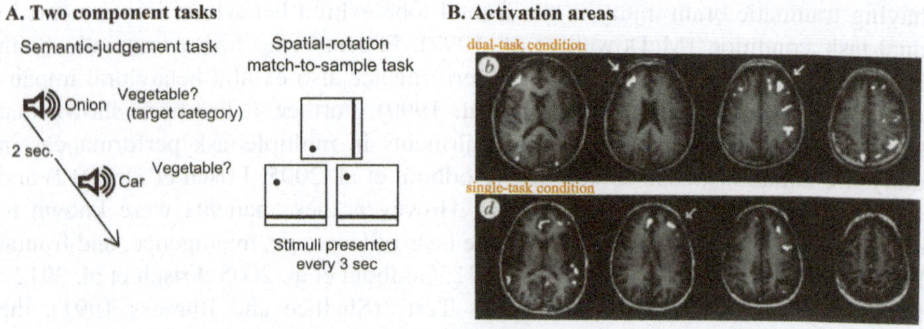

Fig. 6.4 (**a**) Schematic drawings of two-component tasks. (**b**). Comparison of activation areas between the dual-task condition (top) and the single-task condition (bottom). Arrows indicate activation areas in the prefrontal cortex during the dual-task condition. From D'Esposito et al. (1995) with permission

brain areas exhibited the same level of activation both in the single-task condition and in the resting control condition (Fig. 6.4b). However, significant activation was observed in the dorsolateral prefrontal cortex only during the dual-task condition and not during the single-task condition. Therefore, D'Esposito et al. (1995) suggested that significantly stronger activation in the dorsolateral prefrontal cortex reflects an additional requirement of the neural processing specific for dual-task performance. Thus, this study indicates the importance of the dorsolateral prefrontal cortex in dual-task performance and suggests that the dorsolateral prefrontal cortex is a brain area specific for dual-task performance.

Koechlin et al. (1999) also performed an fMRI study using a dual-task paradigm and showed the importance of the dorsolateral prefrontal cortex in dual-task performance. Their dual-task paradigm consists of the main task and the sub-task that is embedded in the main task. In their task, upper- and/or lower-case letters selected from the word "tablet" are presented in a pseudo-random sequence as visual stimuli under four conditions. The subject needs to hold the main goal in mind while he pays attention to the sub-goal (Fig. 6.5a). In the *control condition*, the subject is asked to judge whether two successively presented upper-case letters are also in immediate succession in the word "TABLET." In the *working memory condition*, lower-case letters (tablet) are irregularly inserted between upper-case letters (TABLET) during a trial of the control condition. Therefore, the insertion of lower-case letters causes a delay in behavioral responses required during the control task. In the *dual-task condition*, the subject is asked to respond to both upper- and lower-case letter sequences similar to the control condition. However, the subject is also asked to judge whether every first letter that makes a case change is the letter "T" or "t." In *task-switching condition*, which is another type of dual-task condition, the subject is asked to respond alternately to the letter "T" and the letter "t" exactly the same as in the working memory condition. Therefore, both working memory and switching attention are required in the task-switching condition. fMRI responses were compared among four task conditions. In the dual-task conditions, significant activation

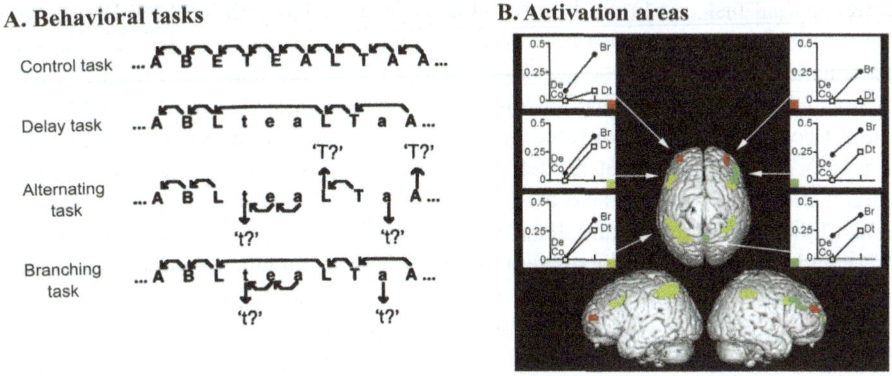

Fig. 6.5 (**a**) Schematic drawings of behavioral tasks. (**b**) Activation areas in the dual-task condition (branching task). Only the left and the right dorsal frontopolar areas were activated in the dual-task condition. From Koechlin et al. (1999) with permission

was observed selectively and bilaterally in the dorsolateral prefrontal cortex (areas 9 and 8). Koechlin et al. (1999) examined whether this activation was caused by an additive effect of the use of both working memory and switching attention or by an interaction of these two cognitive processes. They found that the activation caused by an additive effect was observed in the right anterior dorsolateral prefrontal cortex (areas 9/46 and 10) while the activation caused by an interaction was observed in the anterior prefrontal cortex (area 10) (Fig. 6.5b). Thus, these results also indicate the importance of the dorsolateral prefrontal cortex, especially the anterior prefrontal cortex (area 10), in dual-task performances. In addition, using similar behavioral tasks as Koechlin et al. (1999) used, Dreher et al. (2008) showed that not the total volume of the brain lesion in the prefrontal cortex but the size of the lesion in area 10 was correlated with error rates in dual-task performances. This result also supports the evidence that the anterior portion of the lateral prefrontal cortex plays important roles in dual-task performances.

Selective activation of the lateral prefrontal cortex in dual-task performances is further supported by fMRI studies. Braver and Bongiolatti (2002) used two tasks in dual-task condition. In one task, the subject monitors the presence of a specific concrete probe word (e.g., LIME), which occurred following a specific abstract cue word (e.g., FATE) (the main-goal task). In the other task, the subject monitors the presence of any concrete probe word immediately following any abstract cue word (the sub-goal task). In dual-task condition, the subject needs to perform the main-goal task during performing the sub-goal task. They found significant activation at the frontopolar prefrontal cortex (area 10) in dual-task condition and significant activation at the left dorsolateral prefrontal cortex in single-task conditions. Szameitat et al. (2002) used the visual and the auditory three-choice reaction tasks in single- and dual-task conditions and found dual-task interference effects in dual-task condition. By a cognitive subtraction method of fMRI signals, they found specific activation in the dorsolateral prefrontal cortex and the superior parietal cortex in dual-task condition. Ramsey et al. (2004) used a modified Sternberg's item recognition task as the primary task, in which the subject memorizes sets of consonants and decided whether subsequently presenting letters belong to this set or not. As the secondary task, a tone discrimination task was used, in which the subject counts brief tones differed in pitch from the baseline tone. They found that overlapped brain regions in the left dorsolateral prefrontal cortex and the anterior cingulate cortex were activated by these two tasks in dual-task condition. Thomsen et al. (2004) used binaural and dichotic listening tasks and observed that the dichotic presentation activates the dorsolateral prefrontal cortex and the cingulate cortex to a larger extent than the binaural presentation. Johnson and Zatorre (2006) compared activated brain areas when the subject attended to one sensory modality (e.g., hearing novel melody) and ignored another modality (e.g., viewing geometric shape) (selective attention condition) and when the subject attended to two sensory modalities simultaneously (divided attention condition). In the selective attention condition (single-task condition), relevant sensory cortices were activated but the dorsolateral prefrontal cortex was not activated. However, in the divided attention condition (dual-task condition), the left dorsolateral prefrontal cortex was activated and sensory cortices did not show an increase in activation and comparable activation to the passive presentation condition. Thus, these human fMRI studies also suggest that the dorsolateral prefrontal cortex participates in behavioral performance in dual-task condition.

6.4.3 Is the Prefrontal Cortex a Specific Area for Dual-Task Performance?

Neuroimaging and neuropsychological studies show that the lateral prefrontal cortex participates in behavioral performance in dual-task condition. However, using positron emission tomography (PET), Klingberg (1998) reported that the performance of two tasks in dual-task condition did not activate any specific brain area except the brain areas activated in single-task conditions. He measured regional cerebral blood flow while the subject performed three tasks (working memory tasks using auditory or visual stimuli and a control task) in single-task conditions and two tasks in dual-task condition (working memory tasks using auditory and visual stimuli). In the auditory working memory task, a tone is briefly presented every 2–5 s (Fig. 6.2a). The subject is asked to compare the current tone with the preceding tone and to press a button if the pitch of the current tone is lower than the preceding tone. In the visual working memory task, the subject is asked to compare the brightness of successively presented visual stimuli and to press a button if the current stimulus is brighter than the preceding one. In dual-task condition, the subject performs both working memory tasks simultaneously (Fig. 6.2a bottom). As a result, both working memory tasks commonly activated the dorsolateral prefrontal cortex and the anterior cingulate cortex in single-task condition. However, in dual-task condition, specific activation was not observed in these two cortical areas (Fig. 6.2b).

Adcock et al. (2000) performed an fMRI study and also reported the absence of dual-task-specific activation in the prefrontal cortex (Fig. 6.6). In their study, the subject performs tasks under two kinds of dual-task conditions (Fig. 6.6b). In the one dual-task condition, the subject simultaneously performs a semantic judgment task and a spatial matching task, which are similar to those used by D'Espositio et al. (1995). In the other dual-task condition, the subject simultaneously performs a semantic judgment task and a face matching task. When the subject performed tasks under two kinds of dual-task conditions, activated brain areas were different depending on which component tasks were included in dual-task condition (Fig. 6.6c). In addition, all brain areas activated in dual-task condition were also activated in single-task condition (Fig. 6.6c). Based on these results, Adcock et al. (2000) suggested that an executive process for dual-task performance is mediated by interactions among anatomically and functionally distinct neural systems each of which involves in the performance of each component task, and not by a specific neural system dedicated for dual-task performance.

Other studies also show the absence of dual-task-specific activation in the prefrontal cortex (Bunge et al. 2000; Jaeggi et al. 2003; Erickson et al. 2005; Newman et al. 2007; Salo et al. 2015). Bunge et al. (2000) asked the subject to perform a semantic reading task and a task requiring working memory of five words either in single-task condition or in dual-task condition. Behavioral performance in dual-task condition activated the prefrontal area to a greater extent than behavioral performance of either task in single-task condition. However, no new brain area was activated in dual-task condition beyond the brain areas activated in single-task condition. Jaeggi et al. (2003) used an n-back task at four levels of difficulty and auditory-verbal and visual-nonverbal tasks in single-task or dual-task conditions. Prefrontal activation increased continuously as a function of difficulty in both single- and dual-task conditions. An increase in prefrontal activation was observed in the dual-task condition even in the most difficult condition. Erickson et al. (2005) showed that dual-task condition activated the same brain areas (e.g., the inferior frontal and superior parietal areas) as in single-task condition. However, a greater magnitude of activation was observed in dual-task condition than in single-task condition. Salo et al. (2015) also found no specific activation of the prefrontal cortex in dual-task condition.

Some studies showed that the manipulation of task difficulty of a sensory perceptual task, in which complex cognitive control was not required, elicited significant activation throughout the dorsolateral prefrontal cortex (Crittenden and Duncan 2014; Jiang and Kanwisher 2003a, b). Activated areas in the dorsolateral prefrontal cortex included areas that had been considered to be specifically activated in dual-task condition (Braver and Bongiolatti 2002; D'Esposito et al. 1995; Koechlin et al. 1999; Sigman and Dehaene 2008). For example, Crittenden and Duncan (2014) used a simple visual oddball-search task, in which the subject is asked to indicate the position of the shortest of four vertical lines presented on a screen by pressing a corresponding key in a response box (baseline condition). The

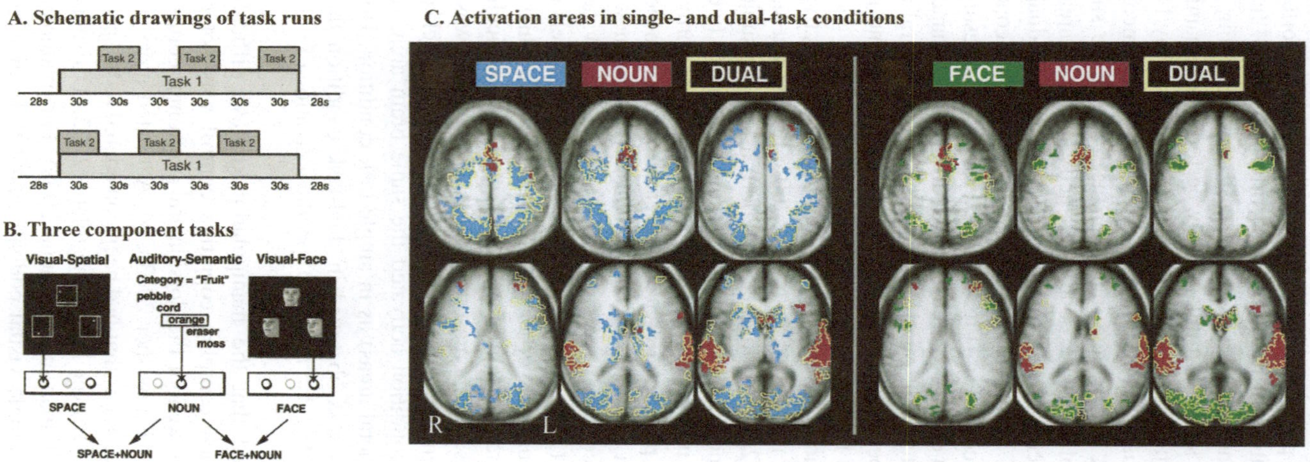

Fig. 6.6 (**a**) Schematic drawings of the experimental task runs. (**b**). Three-component tasks and sample stimuli for each task. (**c**) Activation areas during the dual-task condition (yellow outlines) and the single-task conditions (blue, SPACE; green, FACE; red, NOUN). From Adcock et al. (2000) with permission

difficulty of the task is manipulated in three ways: (1) by increasing the number of lines from four to eight (8 L condition), (2) by decreasing the difference in length among lines (FD condition), and (3) by changing the mapping between stimulus positions and response keys (MS condition). They showed that the greatest increase of activation was observed in the MS condition in the bilateral anterior dorsolateral prefrontal cortex. An increase in activation was also observed in the left rostrocaudal prefrontal cortex in the FD condition. Based on these results, they concluded that complex task demand, such as cognitive branching or comprehension of abstract task-rule, is not the only factor that recruits the anterior dorsolateral prefrontal cortex, which has been implicated as the locus necessary for information processing in dual-task condition (Burgess et al. 2007; D'Esposito et al. 1995; Koechlin et al. 1999; Koechlin and Summerfield 2007). Instead, they indicated that the recruitment of the anterior dorsolateral prefrontal cortex occurs even by a simple manipulation of task difficulty in sensory perception tasks. Recent fMRI studies suggest that requirements of dual-task-specific information processing mediate an increase in inter-regional synchronization of neural activity (Buchweitz et al. 2012; Mizuno et al. 2012). Buchweitz et al. (2012) compared the strength of inter-regional synchronization of brain activity between the simultaneous comprehension of two streams of dichotically presented spoken input (dual-task condition) and the comprehension of a single speech stream (single-task condition). Inter-regional synchronization of brain activity was measured as the correlation between average time courses of BOLD signal intensity in three cortical regions for language processing: the left frontal lobe (Broca's area), the left temporal lobe (Wernicke's area), and the right temporal lobe. As a result, inter-regional synchronization was significantly stronger in dual-task condition than in single-task condition. Left frontal activation changed between single-task and dual-task conditions, similar to that synchronization was increased between bilateral temporal areas. These results suggest that the increase of inter-regional synchronization in dual-task condition is related to the greater and more complex demand of information processing. Buchweitz et al. (2012) indicated in their review that, to search for brain regions that are specifically related to dual-task performance, we should focus on the coherent modulation of activity at the network level rather than on activity within individual brain areas.

Thus, although stronger prefrontal activation is commonly observed in dual-task condition compared to single-task condition, neuroimaging studies show that the prefrontal cortex is not a specific brain area supporting only dual-task performance. These studies suggest that distinct brain areas specifically involved in dual-task performance are not found in the brain. Rather, these studies suggest that dual-task performance relies largely on functional interactions among multiple brain areas each of that is activated by individual component tasks and that the prefrontal cortex is a part of this neural network.

6.4.4 Limitations of Neuroimaging Studies

Although many neuroimaging studies show the importance of the lateral prefrontal cortex in dual-task performance, it is not clear whether the lateral prefrontal cortex is a specific brain area supporting dual-task performance or not. On one hand, several neuroimaging studies showed that stronger activation of the lateral prefrontal cortex was observed only during task performances in dual-task condition and that in single-task conditions, the level of activation was similar to in the control condition. Therefore, the lateral prefrontal cortex was thought to be a specific brain area that supports task performance in dual-task condition. On the other hand, other neuroimaging studies showed that, although stronger and wider activation was observed in the lateral prefrontal cortex during task performance in dual-task condition, a similar degree of significant activation was also observed in single-task conditions. Therefore, although an additive effect in activation of the lateral prefrontal cortex was observed in dual-task condition, it was thought that the activation of the lateral prefrontal cortex was not specific for task performance in dual-task condition.

It is important to know whether or not the lateral prefrontal cortex is a specific brain area supporting task performance in dual-task condition. However, to perform two different cognitive tasks in dual-task condition, a limited capacity of cognitive resource is needed to allocate to each component task depending on the degree of its demand. Therefore, the most important questions are whether or not the lateral prefrontal cortex is a responsible brain area for allocating a limited capacity of cognitive resource to each component task with graded manner depending on the degree of its demand and what neural mechanism in the lateral prefrontal cortex supports these functions in dual-task condition.

Several difficulties are present in human neuroimaging studies to examine the neural mechanism and information processing supporting task performance in dual-task condition. First, coarse spatial and temporal resolution in neuroimaging studies makes it difficult to identify neural mechanisms participating in information processing specific for dual-task performance. For example, Jiang and Kanwisher (2003a, b) raised the difficulty to distinguish neural populations that underlie the central bottleneck in response selection from neural populations that underlie perceptual processing by fMRI. Since these two neural populations are intermixed in the brain, the limitation of spatial resolution in fMRI studies makes it difficult to distinguish these two populations. Similarly, if the same neural population participates in two or more functions simultaneously or in close temporal order, it is difficult to dissociate each of these functions into temporally distinct processing stages by fMRI studies because of the limitation of their temporal resolution.

Second, difference in behavioral tasks used in dual-task studies may cause inconsistent results. Any combination of two behavioral tasks can be used in dual-task studies. Therefore, cognitive processes required for behavioral tasks used in one study may be different from those in other studies. For example, the results obtained by D'Esposito et al. (1995) were different from those by Klingberg (1998). The difference in the results between these two studies might be caused by the difference in the behavioral tasks used. D'Esposito et al. (1995) used non-working memory tasks as two component tasks, while Klingberg (1998) used two working memory tasks. Since the lateral prefrontal cortex is known to participate in the working memory process, a similar degree of significant activation may be produced in the lateral prefrontal cortex during task performance both in single-task condition and in dual-task condition. However, when two non-working memory tasks are used, functional contributions of the lateral prefrontal cortex may be different between tasks and weaker compared with the situation when two working memory tasks are used. Therefore, the activation of the lateral prefrontal cortex only in dual-task condition suggests the importance of the lateral prefrontal cortex in information processing in dual-task condition of working memory tasks.

Third, individual differences in performance skills among subjects in dual-task condition may cause a difference in the degree of activation in different brain areas. It has been shown that extensive practice of task performance in dual-task condition reduces the strength of dual-task interference effect (Pashler et al. 1993; Passingham 1996; Van Selst et al. 1999; Ruthruff et al. 2001). Task-related activation measured by fMRI showed a significant difference between before and after the training of dual-task performance (Erickson et al. 2007; Dux et al. 2009). When the patterns of activation were compared between good and poor performers of the same tasks in dual-task condition, dual-task-specific activation was observed in the dorsolateral prefrontal cortex only in poor performers who exhibited stronger dual-task interference effects (Smith et al. 2001). In addition, when dual-task interference effect was not observed behaviorally (Bunge et al. 2000) or when its effect was negligible (Adcock et al. 2000), dual-task-specific activation was not observed in the prefrontal cortex.

As has been shown, several studies had shown the absence of dual-task-specific activation in the prefrontal cortex (Klingberg 1998; Adcock et al. 2000; Bunge et al. 2000; Jaeggi et al. 2003; Erickson et al. 2005; Newman et al. 2007; Salo et al. 2015). These studies concluded that no brain area was selectively activated in dual-task condition, because activated brain areas in dual-task condition are not spatially separated from those activated in single-task condition and are just extension of the brain areas activated in single-task condition (Buchweitz et al. 2012). However, even in these studies which showed that no brain area was specifically activated in dual-task condition, it was often observed that the total volume of significantly activated brain area was larger in dual-task condition than in single-task condition. Furthermore, it was also shown that the commonly activated areas in both component tasks in single-task condition were more strongly activated in dual-task condition. Therefore, it may be possible to explain the cause of the inconsistent observations and conclusions among previous neuroimaging studies as follows. Although the information processing for dual-task-specific demand depends on distinct neural populations which are different from those involved in information processing in single-task condition, these two different neural populations exist in interdigitated fashion within an overlapping brain region. Accordingly, the limitation in spatial and temporal resolution of neuroimaging techniques has made it difficult to distinguish these two neural populations. This explanation is supported by neurophysiological evidence in animal studies.

6.5 Search for Neural Mechanisms of Dual-Task Interference by Animal Studies

6.5.1 Advantages of Animal Studies

The use of animals could advance understanding neural mechanisms of dual-task interference, since a variety of physiological techniques are usable in higher temporal and spatial resolutions (Watanabe and Funahashi 2018). Activation and inactivation of a particular neuron group, a population of neurons, or a particular brain area by physiological, pharmacological, or optogenetic techniques make it possible to study causal relationships between specific neural components and behavioral outputs in dual-task condition. It is important to know which brain area contributes specifically to task performance in dual-task condition. However, from a physiological point of view, it is more important to know what is the neural basis of cognitive resource and how a limited capacity of cognitive resource is allocated to each component task depending on the demand of each task. For example, if each of two component tasks activates the same population of neurons within a certain brain area during single-task condition, we may directly examine the allocation of neural resource to each component task in dual-task condition by comparing the size of the neural population activated by each component task between single-task and dual-task conditions. Thus, the animal models are suitable to examine the neural mechanism of dual-task interference, neural correlates of cognitive resource, and the neural mechanism for flexible allocation of cognitive resource.

Although not many dual-task studies have been performed using animals, rats and monkeys have been shown to perform two behavioral tasks in dual-task condition (Watanabe and Funahashi 2018). Therefore, these animals can be used to examine the neural mechanism of dual-task interference. Improved methods for nonhuman animals have been proposed to train two tasks in dual-task condition (Moise 1970; Watanabe and Funahashi 2015b; Kleinman et al. 2016).

6.5.2 Behavioral Studies Using Rats

Rats have been shown to exhibit dual-task interference. For example, Cook and Brown (1985) used two 12-arm radial mazes (to-be-remembered primary maze and secondary maze). In single-task condition using the primary maze, the rat is placed on the center platform of the primary maze and the door to a single arm is opened, allowing the rat to get a reward at the end of the arm. After the rat got the reward and returned to the center, the door is closed and then the door to a second arm is opened. This procedure is repeated until the rat visit six arms (forced choice procedure). After a 30-min delay (the rat is in the individual cage), the rat is returned to the primary maze and all doors are opened to access all 12 arms. The rat is allowed to make choices of arms until the rat gets all remaining rewards or until 20 total choices are made. In single-task condition using the secondary maze, the rat is placed on the center platform of the secondary maze and all doors are open to access all 12 arms. The rat is allowed to either make 20 total choices or make choices until all 12 arms are visited. In dual-task condition, the rat first performs three or nine forced choices on the primary maze. Then, the rat has a 2-h interval period, during which the rat is asked to perform one or two trials on the secondary maze. After the interval period, the rat has a retention test on the primary maze, in which the rat is allowed free access to all arms but needs to access non-visited arms. Cook and Brown (1985) observed an interference effect in the retention test when the rat needed nine choices on the primary maze or when the secondary maze trial was introduced twice during the interval period. Thus, the spatial memory for the primary maze was impaired when the number of arms needed to visit in the primary maze was increased or when the rat needed to visit a substantial number of arms in the secondary maze. Roberts (1981) also used 8-arm radial mazes. In dual-task condition, the rat first performs four forced choices in the first maze, then has other forced choice experiences on one to three other mazes, and finally needs to take a free choice retention test in the first maze. Interference effect was observed only when the rat had forced choice maze experience on three mazes, but not observed when one or two mazes were introduced. Thus, in dual-task experiments using radial mazes, interference effect was observed when the number of interpolated maze trials increased (Roberts 1981; Cook and Brown 1985) or when either the primary maze or the secondary maze had a larger number of arms (Maki et al. 1979; Beatty and Shavalia 1980). Based on these results, it was concluded that interference effect is observed only when memory requirement is greatly overloaded by introducing interpolated mazes (Roberts 1981).

Timing behavior has also been used in rat experiments (Meck and Church 1984; Olton et al. 1988; Pang et al. 2001). These studies use a simultaneous temporal processing (STP) paradigm. The STP paradigm is a variant of the fixed-interval (FI) reinforcement schedule. This procedure includes successive presentation of two sensory stimuli (e.g., a light and a tone) and each stimulus is associated with a different FI reinforcement schedule (e.g., a light with 6-s FI schedule and a tone with 20-s FI schedule). Therefore, since the shorter FI schedule is embedded within the longer FI schedule, the rat is required to simultaneously measure two different intervals to obtain the reinforcement efficiently without a miss. Behavioral analyses indicated that the rat can measure both intervals simultaneously without any interference in the STP paradigm. Further, the rat can simultaneously measure two different intervals without interference even if two longer intervals (i.e., 10 s vs. 40 s, and 10 s vs. 50 s) are combined (Olton et al. 1988; Pang et al. 2001; Matsuo et al. 2017). However, the rat having lesions in the lateral portion of the frontal cortex exhibited impairment when two different FI schedules were

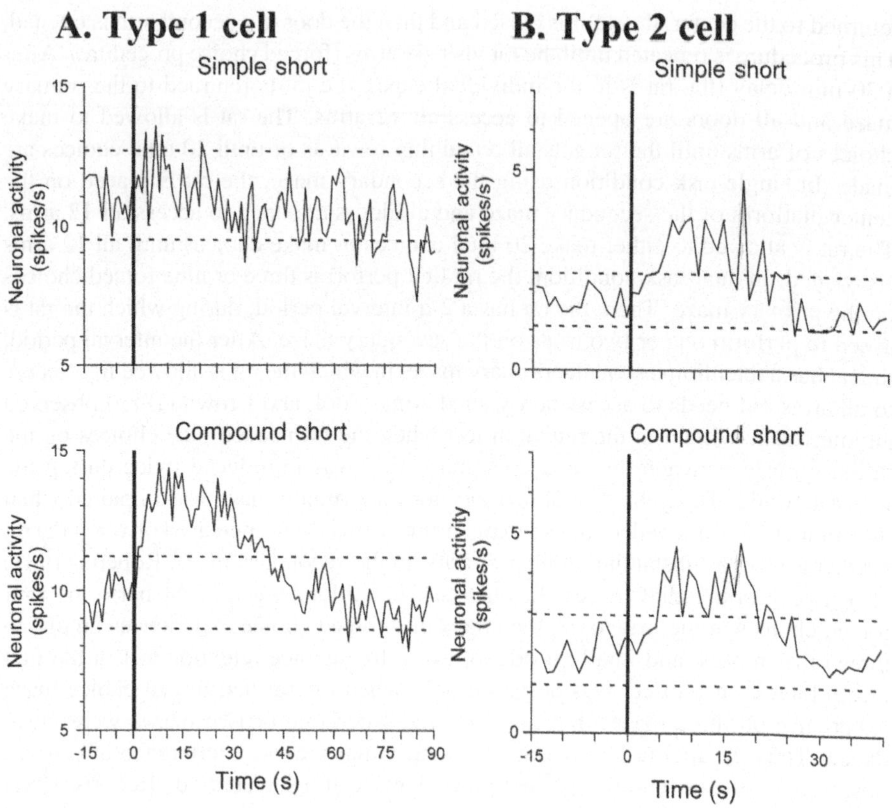

Fig. 6.7 Examples of prefrontal activity while the rat performed the STP task. (**a**) Type 1 cell which exhibited stronger excitatory activity only when two stimuli were presented simultaneously and exhibited no activity when either stimulus was presented. (**b**) Type 2 cell which exhibited excitatory activity when two stimuli were presented simultaneously and individually. From Pang et al. (2001) with permission

introduced simultaneously, but exhibited no impairment when a single FI schedule was introduced. (Olton et al. 1988). Pang et al. (2001) collected neural activity from the lateral agranular frontal cortex while the rat performed the STP task. This task uses two different stimuli (a light and a tone), each of which is associated with different FI schedule, and includes simple and compound trials. Simple trials correspond to single-task conditions and only either stimulus is presented, whereas compound trials correspond to dual-task conditions and both stimuli are presented simultaneously. They found that a large number of neurons (60%) exhibited stronger excitatory activity only when two stimuli were presented simultaneously and exhibited no activity when either stimulus was presented (Fig. 6.7a). Other neurons exhibited excitatory activity when two stimuli were presented simultaneously and individually (10%) (Fig. 6.7b), when one particular stimulus was presented alone or with the other stimulus (27%), and only when one particular stimulus was presented alone (3%). These results together with lesion studies suggest the importance of the prefrontal cortex in task performances under dual-task condition.

Thus, the rat can simultaneously perform two tasks each of which has a different goal. If behavioral conditions as well as sensory and motor parameters are appropriately selected and organized, the rat exhibits dual-task interference effect. In addition, lesion studies and neurophysiological studies suggest the importance of the frontal cortex to perform two tasks in dual-task condition. Recently, Kleinman et al. (2016) proposed a dual-task paradigm (concurrent-interval timing task) adopted from the STP paradigm for rhesus monkeys. In this task, the monkey is required to discriminate two different intervals (8-s FI and 16-s FI) simultaneously and report which FI schedule is currently active by making a saccade toward an appropriate visual target (e.g., a green circle and a blue circle associated to 8-s FI and 16-s FI, respectively). They indicated that the monkey was able to discriminate concurrently introduced two different intervals independently and accurately.

6.5.3 Behavioral Studies Using Monkeys

fMRI studies using human subjects have shown that the lateral prefrontal cortex participates in behavioral performance in dual-task condition, although precise roles of the lateral prefrontal cortex in dual-task performance are remained to understand. Since both humans and monkeys have shown to have similar characteristics in cognitive capacity limitations (Buschman et al. 2011; Elmore et al. 2011), it is expected that monkeys and humans have common neural mechanisms for the flexible allocation of a limited capacity of cognitive resources.

Moise (1970) first demonstrated dual-task interference effect in monkeys. He trained stump-tail monkeys and used a delayed matching-to-sample (DMS) task and a simple reaction time (RT) task as two component tasks. In the RT task, the subject is required to quickly touch a circular window when a white light is illuminated. In dual-task condition, the RT trial is repeatedly inserted during the delay period (<30 s) of the DMS task. He found that the correct rate of DMS performance decreased significantly as the number of inserted RT trials increased in dual-task condition. The decrease in DMS performance by the insert of RT trials during the delay period was consistently observed throughout weeks for testing. Because the monkey needs additional effort to perform RT trials inserted during the delay period of the DMS task, observed negative effect in DMS performances was thought to be caused by the same origin as dual-task interference is caused in human studies. This study suggested that a common processing resource is needed to perform simple sensory-motor actions for the RT task and memorize sensory information for the DMS task and that the strength of the interference effect depends on the degree of the demand of processing resource that each task requests. Thus, this study indicates that the information processing for the RT task and the short-term maintenance of information in the DMTS task need a common capacity-limited cognitive resource and that the degree of dual-task interference effect in one component task is determined by the demand of the cognitive resource to perform the other concurrent task. Thus, at any given moment, there is a limited amount of resource that can be allocated to simultaneously performing tasks and as more resource is allocated to one task, less resource is available for the other task.

When a limited amount of neural resource is allocated to two simultaneously performing tasks, if a larger amount of resource is allocated to the primary task, only a small amount of resource would be available for the secondary task and would not be enough to perform it correctly. Therefore, the performance of the secondary task would be deteriorated. The experiment by Washburn and Putney (1998) followed this logic. They used both humans and rhesus monkeys and examined whether or not the tradeoff of a common neural resource occurs between two component tasks in dual-task condition. They used a two-choice discrimination task as the primary task and a perceptual judgment task as the secondary task. In dual-task condition, the secondary task was inserted during the performance of the primary tasks. As a result, an increase in response times was observed in the secondary task in both species when the primary task became more demanding. This result indicated that the tradeoff of a common neural resource occurred between two component tasks in dual-task condition with similar manner in both humans and monkeys, suggesting that both humans and monkeys have common neural mechanisms for allocating cognitive resources depending on the demand of each task.

Basile and Hampton (2013) also observed the tradeoff of neural resource in dual-task condition in rhesus monkeys (Fig. 6.8). They used a delayed matching-to-sample (DMS) task as a primary task and three tasks (a motor-only task, an image perception task, and an image classification task) as secondary tasks (Fig. 6.8a). In dual-task condition, one of three secondary tasks or no secondary task was inserted during the delay period of the primary task. By adding three different secondary tasks one by one, they prepared four different dual-task conditions, each of which needed different levels of mental effort. Four dual-task conditions were (1) adding no secondary task (control condition), (2) adding a motor-only task, in which the monkey is required to touch a blue square presented at one of four screen corners, (3) adding an image perception task, in which the monkey is required to touch a complex image presented at a screen corner, and (4) adding an image classification task, in which the monkey is required to classify an image as flower, fish, bird, or person by touching the corresponding symbols presented at four screen corners. They also manipulated the difficulty of the primary task by changing the set size of images used. The small set condition included 4 images and the large set condition included 1400 images. In the small set condition, since the same four images are frequently and repetitively presented as either a target image or a non-target image across different trials, the familiarity of these four images increases. Therefore, a target image would be more difficult (more demanding) to distinguish from a non-target image during the test period because of the strong familiarity of these four images. In contrast, discrimination of the target image from distractor images would be easier (less demanding) in the large set condition than in the small set condition. Infrequent presentation of each image would make it easier to distinguish a target from a non-target by novelty of images during the test period of the DMS

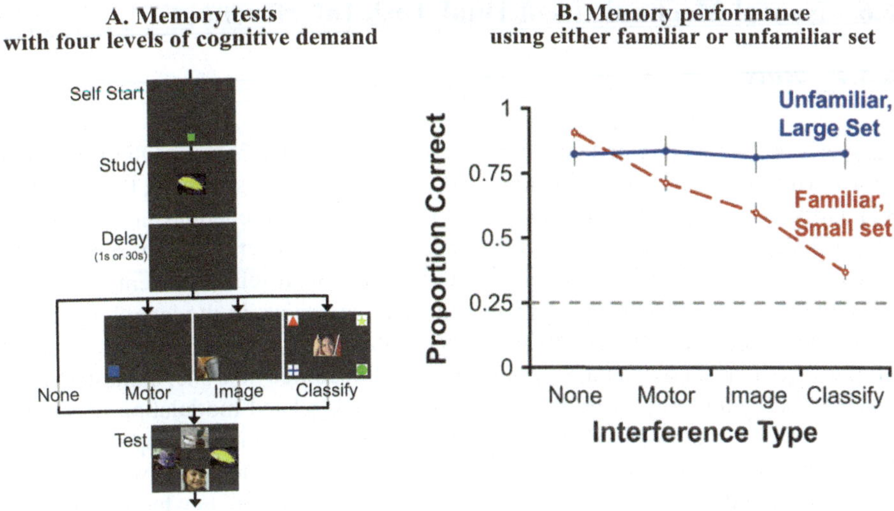

Fig. 6.8 (**a**) Temporal sequence of the memory tests which have four different levels of cognitive demand. (**b**) A comparison of memory performances when either small set of familiar stimuli or large set of unfamiliar stimuli were used. From Basile and Hampton (2013) with permission

task. In fact, the performance of the primary DMS task was impaired only when more effortful secondary task was coupled with the small set (more demanding) condition of the primary DMS task (Fig. 6.8b). This result indicates that the short-term maintenance of familiar information requires an active resource-demanding process similar to the human rehearsal process. This result also indicates that the additive effect of the magnitude in DMS impairments is similar to the dual-task interference effect in humans. Similar additive effects were also observed in rats (Roberts 1981; Cook and Brown 1985). The presence of interactions between difficulties (degree of required effort or demand) of both primary and secondary tasks and the strength of interference effects (degree of impairment) observed in the primary task supports the presence of the neural mechanism which allocates a limited capacity of cognitive resource depending on the demand of each component task.

Thus, these findings validate the effectiveness of the use of primates to investigate the neural mechanisms that underlie human dual-task interference effects. The similarity of dual-task interference effects between humans and animals suggests the presence of common cognitive processes related to dual-task interference. This could make it possible to apply a variety of neurobiological techniques to nonhuman primates to investigate the neural mechanisms related to dual-task interference effects.

6.6 Neural Mechanisms of Dual-Task Interference

6.6.1 Rats

Pang et al. (2001) performed single-unit recording from the lateral agranular frontal cortex while rats performed a simultaneous temporal processing (STP) task. The STP task is a variant of the fixed-interval (FI) reinforcement schedule and the shorter FI schedule is embedded within the longer FI schedule. The rat is required to simultaneously measure two different intervals to obtain reinforcement without a miss. Olton et al. (1988) had shown that rats having lesions in the lateral agranular frontal area impaired STP performances. Pang et al. (2001) reported that the activity of a large proportion of frontal neurons (30% of all recorded neurons) exhibited task-related modulation only in dual-task condition. Task-related modulation of single-neuron activity was observed during either short interval timing ("compound short" conditions) or long interval timing ('compound long' conditions) of two FI schedules. This modulation was not observed in single-task condition, in which only either FI schedule was introduced ("simple short" and "simple long" conditions). Task-related modulation in dual-task condition was triggered by the onset of the stimulus associated with the short FI schedule, which corresponded to the timing when simultaneous measuring of two different intervals was required. These observations suggested that activity of these neurons reflects some aspect of information processing related to dual-task performance. However, this activity continued even after the timing when simultaneous measuring of two different intervals was not required, suggesting a possibility that these neurons might not be directly involved in information processing for dual-task performance (Pang et al. 2001).

Neurons exhibited task-related modulation only during dual-task performance were distributed across the entire frontal cortex. However, there were other frontal neurons that exhibited task-related modulation either during both single-task and dual-task conditions, or during one of single-task conditions and dual-task condition, or during only one of single-task conditions. These neurons were also distributed across the entire frontal cortex. The coexistence of various task-related neurons within the frontal cortex suggests that, although information processing for dual-task performance may require distinct neural circuits comprised of neurons exhibited task-related modulation only during dual-task performance, these neural circuits may not be localized in spatially separate brain areas but maintain active interactions with neural circuits involved in information processing for single-task performance.

Psychopharmacological studies using the STP paradigm have suggested the importance of the cholinergic system for effective parallel processing of temporal information. Meck and Williams (1997) showed that prenatally choline-supplemented rats exhibited improvement in STP performance, while prenatally choline-deficient rats exhibited impairment of STP performance. Improvement of STP performance in choline-supplemented rats may presumably be caused by increasing the signal processing speed in brain regions related to attention and memory. Meck (1987) showed that vasopressin administration, which increases and maintains acetylcholine levels in the brain, also facilitated STP performance in rats. Further, lesions of the nucleus basalis magnocellularis, which contains cholinergic neurons that project to the entire neocortex, impaired STP performance.

The importance of the agranular frontal cortex and the cholinergic system has been reported in the STP performance, one type of dual-task paradigm, in rats. Further investigations using STP paradigms in rats may help to clarify whether or not rats are appropriate model animals to understand information processing underlying dual-task performance. Further investigation is also needed to examine whether or not the rat medial prefrontal cortex, which is proposed as a putative homolog of the primate lateral prefrontal cortex (Uylings et al. 2003; Seamans et al. 2008), and the hippocampus, which is known to participate in temporal information processing (Meck et al. 1984; Nakazono et al. 2015), contain neurons whose activity exhibits dual-task-specific modulation.

6.6.2 Monkeys

Single-neuron recordings from monkeys performing two tasks in dual-task condition have been started to reveal neural mechanisms responsible for dual-task interference (Lebedev et al. 2004; Messinger et al. 2009; Watanabe and Funahashi 2014).

As described before, a concept of cognitive resource has been proposed to explain the mechanism of dual-task interference. Cognitive resource corresponds to the neural resource in the brain for task-general information processing and is shareable across concurrently performing multiple tasks in a graded manner. Since the amount of available resource limits the amount of information that can be processed in the brain at a time, dual-task interference could occur if two tasks are accessing this finite resource and if the amount of resource necessary to perform these two tasks exceeds the total amount of resource that is currently available. This theory was supported by neuroimaging studies that had shown an overlap of activated areas in the lateral prefrontal cortex between dual-task condition and single-task condition (Klingberg 1998; Adcock et al. 2000; Bunge et al. 2000). In addition, the performance of two tasks in dual-task condition activated overlapped regions in the lateral prefrontal cortex in a greater degree than the performance of each task in single-task condition. These results suggested that the interference effect and the limitation of task performance in dual-task condition are caused by the competition between two concurrent tasks in order to obtain enough amount of resource for information processing and that the lateral prefrontal cortex is likely a region related to information processing for this competition. On the other hand, the limitation of information processing in the task-general processing system, such as the central executive of working memory model, has been proposed as the cause of dual-task interference (Baddeley 1986; D'Esposito et al. 1995; Baddeley et al. 1997). These studies consistently indicated that the lateral prefrontal cortex plays a key role in performing two tasks in dual-task condition and producing dual-task interference effect. However, neuronal mechanisms for these information processes in the lateral prefrontal cortex remain largely unknown because of the lack of neurophysiological investigations using appropriate animal models.

It is not known whether or not task performances in dual-task condition affect the activity of prefrontal neurons in a manner that could account for the behaviorally observed dual-task interference effect. It is not known whether or not the overlap of activity between two component tasks can be observed at the single-neuron level. This could show a possibility of the presence of competition between two tasks for obtaining enough amount of neural resource.

To address these issues at the level of single prefrontal neurons, nonhuman primates, especially macaque monkeys, would be appropriate subjects, because they have well-developed prefrontal cortex, because they exhibit similar characteristics in cognitive capacity limitations as humans do (Buschman et al. 2011; Elmore et al. 2011), and because they have an ability to perform complex behavioral tasks in dual-task condition (Moise 1970; Washburn and Putney 1998; Basile and Hampton 2013).

6.6.3 Researches by Wise and Colleagues

Wise and colleagues investigated neural mechanisms for parallel information processing in the lateral prefrontal cortex using dual-task-like behavioral conditions (Fig. 6.9A) (Lebedev et al. 2004; Messinger et al. 2009). In their task, the rhesus monkey is required to look at a fixation point presented at the center of a screen. After a 1.0–1.5 s initial fixation period, a visual cue is first presented at one of four initial positions (left, right, up or down from the center, remembered location) for 1.0–1.5 s. Then, the visual cue turns 90°, 180°, or 270° either clockwise or counterclockwise around the fixation point from the initial position to one of four final positions along a circular trajectory (attended location). After the visual cue stops turning around the fixation point, a variable delay period (1.0–2.5 s) is introduced. After the end of the delay period, the visual cue becomes brighter or dimmer for 150 ms and then disappears, which is the Go-signal to make a saccade and indicates the correct saccade target. If the visual cue dims, the monkey is required to make a saccade to the final cue position, while if the visual cue brightens, the monkey is required to make a saccade to the initial cue position. If the monkey makes a saccade to the correct target position, a new fixation point appears at the correct target position and the monkey needs to fixate it for 1.0–1.5 s to obtain a reward. In this task, brightness change of the visual cue occurs for a very short time (150 ms) at unpredictable timing (after a variable delay period of 1.0–2.5 s). The direction of the brightness change (brighter or dimmer) indicates the correct position for a saccade. In addition, the monkey is required to maintain gazing at the central fixation point during a whole trial except the response period. Therefore, the monkey needs to remember the initial position of the visual cue (remembered position) and, at the

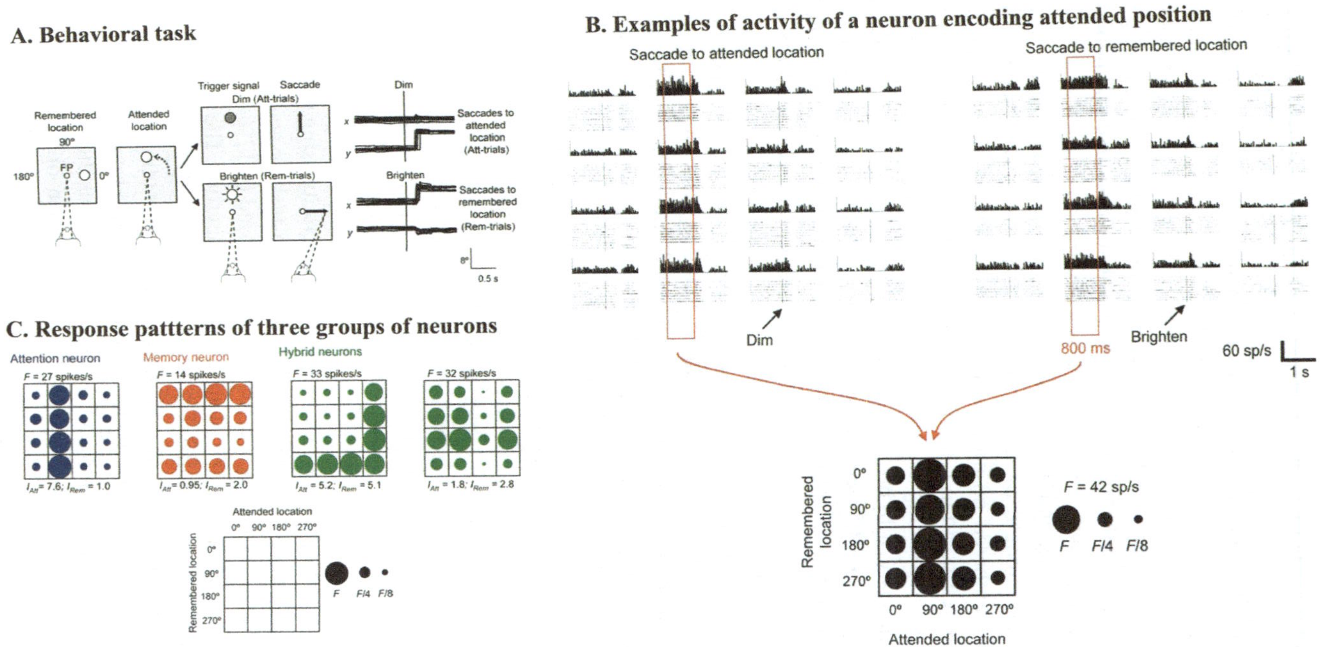

Fig. 6.9 (**a**) Behavioral task and examples of horizontal and vertical eye records during the task. (**b**) Examples of prefrontal activity representing attended position. Excitatory activity was observed only when the monkey attended to the 90° position, regardless of the direction of the saccade. (**c**) Examples of firing rate matrices of three groups of neurons. From Lebedev et al. (2004) with permission

same time, focus its attention to the visual cue at the final position (attended position) to detect its brief brightness change during the delay period.

The task used by Lebedev et al. (2004) and Messinger et al. (2009) differs from tasks that have been used in human and monkey dual-task studies, which are composed of two independent tasks each of which has a distinct goal. However, their task requests the monkey to engage in functionally distinct two information processing (attention and memory) simultaneously. Therefore, their task could allow to examine neural mechanisms for parallel and simultaneous processing of two functionally independent information streams, which are also essential neural mechanisms for information processing in dual-task condition.

Lebedev et al. (2004) found three kinds of task-related neurons in the prefrontal cortex based on the strength of spatial tuning of activity for either the attended position or the remembered position or both (Fig. 6.9c). Some neurons encoded either the remembered position ("memory" cells) or the attended position ("attention" cells) and the remaining neurons encoded both remembered and attended locations simultaneously ("hybrid" cells) (Fig. 6.9c). Attention cells occupied a large proportion (61% of 303 neurons) (Fig. 6.9b), while memory cells and hybrid cells occupied 16% and 23%, respectively. Hybrid cells exhibited several interesting features. For example, a large proportion (73%) of hybrid cells encoded the attended position and the remembered position independently and exhibited stronger spatial tuning than other cells. In decoding both remembered and attended positions at a given trial using population activity of these neurons, decoding accuracy by a population of hybrid cell activity was significantly higher than that by the same population size of memory and attention cell activity, suggesting that the activity of hybrid cells represents task-relevant information more efficiently than the activity of other cells. It has been shown that many prefrontal neurons encode more than one position (Funahashi et al. 1993b, 1997; Inoue and Funahashi 2002) or more than one object (Warden and Miller 2007, 2010), or a combination of spatial and non-spatial information (Rao et al. 1997) in single-task condition. However, the pattern to encode multiple information by hybrid cells seems to be different from that by neurons observed in previous studies. Lebedev et al. (2004) indicated that the activity of a single hybrid cell tuned the remembered location and the attended location independently.

Messinger et al. (2009) further examined prefrontal activity using the same task as was used by Lebedev et al. (2004) and showed that multi-tasking neurons, which are the same as hybrid cells by Lebedev et al. (2004), consisted of 17% of all prefrontal neurons recorded ($n = 957$) and 30% of all task-related neurons. The remaining task-related neurons were called specialized neurons, which correspond to attention cells and memory cells by Lebedev et al. (2004). Cortical distributions of multi-tasking neurons did not differ significantly from those of specialized neurons in the lateral prefrontal cortex. Both multi-tasking and specialized neurons did not form spatially separate clusters (Fig. 6.10). Similar results had also been obtained by Lebedev et al. (2004).

Independent representation of different information in single-neuron activity would be an economical and efficient method, because interference among different

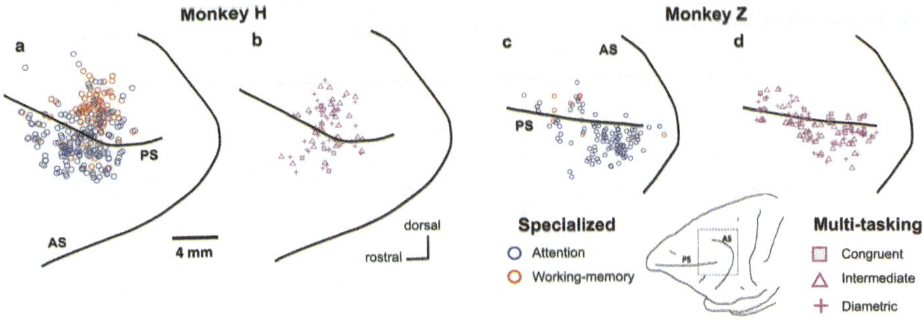

Fig. 6.10 Spatial distribution of three groups of neurons on the prefrontal cortex. Specialized attention and working-memory neurons correspond to attention neurons and memory neurons, respectively, in Fig. 6.9. Multi-tasking neurons correspond to hybrid neurons in Fig. 6.9. From Messinger et al. (2009) with permission.

information can be minimized. Therefore, dual-task specific information processing might be mediated by lateral prefrontal neurons having a hybrid cell-like activity that encodes information for two concurrent tasks independently. However, it is not known whether or not these hybrid cells in fact involve in information processing specific for behavioral performance in dual-task condition. If a neuron involves in information processing specific in dual-task condition, that neuron should exhibit task-relevant selective activation only when the simultaneous performance of multiple tasks is required. Therefore, it needs to be examined whether hybrid cells exhibit task-related activation specifically in dual-task condition or not.

6.6.4 *Researches by Watanabe and Funahashi*

Although Lebedev et al. (2004) and Messinger et al. (2009) obtained important results to understand prefrontal contributions in behavioral performances in dual-task condition and neural mechanisms related to the cause of dual-task interference effects, the task that they used differed from conventional tasks used in humans and monkeys for examining dual-task interference. Watanabe and Funahashi (2014) first developed behavioral tasks that can be used for primate neurophysiological studies and are comparable to conventional tasks used in human studies for examining dual-task interference.

Watanabe and Funahashi (2014) selected a visuospatial attention task and a visuospatial working memory task as two component tasks (Fig. 6.11). These two tasks are known to require intact functions of the lateral prefrontal cortex (Funahashi et al. 1993a; Rossi et al. 2007; Zanto et al. 2011). These two tasks are also known to recruit activation of many lateral prefrontal neurons (Funahashi et al. 1989; Constantinidis et al. 2001; Everling et al. 2002; Lebedev et al. 2004; Kadohisa et al., 2013). Therefore, when the monkey is required to perform these two tasks in

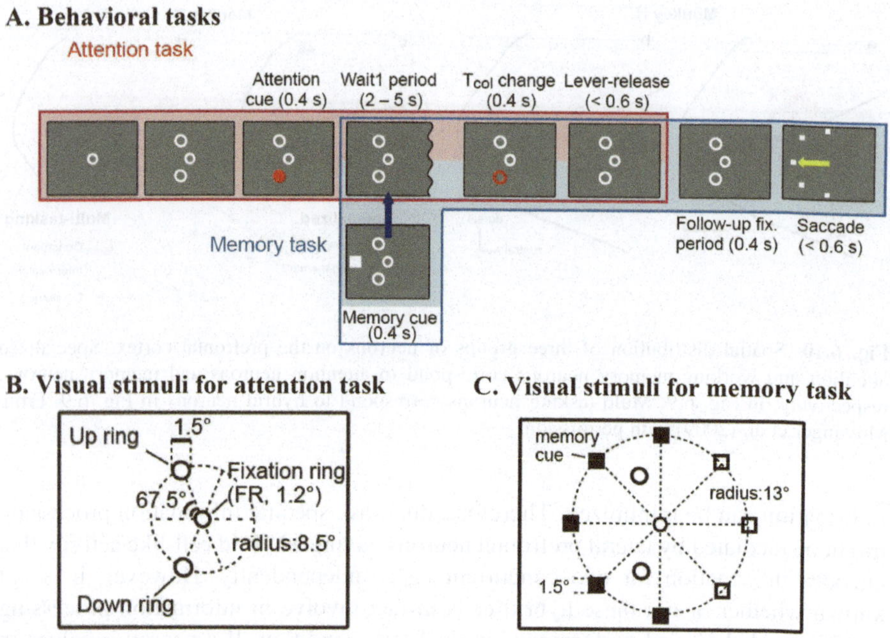

Fig. 6.11 (**a**) Schematic drawings of two-component tasks (attention task and memory task) in dual-task condition. The memory task was inserted during the wait 1 period of the attention task. (**b**) Spatial arrangement of three to-be-attended rings in the attention task. (**c**) Spatial arrangements of eight memory cues (filled and open squares) for the memory task and three to-be-attended rings (circles) for the attention task. From Watanabe and Funahashi (2014) with permission.

dual-task condition, simultaneous performance of these two tasks requires simultaneous processing of two independent information streams in the lateral prefrontal cortex. However, for simultaneous processing of these two information streams, a common population of lateral prefrontal neurons needs to be activated. Therefore, an interference effect would be expected to be observed when these two tasks are performed in dual-task condition.

6.6.4.1 Dual-Task Paradigm

Watanabe and Funahashi (2014) trained a spatial attention task and a spatial working memory task to two Japanese monkeys (Fig. 6.11). Each task requires different behavioral response; lever-release by the hand for the attention task and saccadic eye movement for the working memory task. In this experiment, the monkey is required to remember the position of a visual cue that is briefly presented in the peripheral visual field and execute a memory-guided saccade to that position after the end of the delay period (memory task). At the same time, the monkey is required to attend to a small ring presented on the monitor during the delay period and make a lever-release response when it detects a color change of the ring (attention task). To examine the load-dependency of dual-task interference effect, the difficulty of the attention task was parametrically manipulated by varying the position of a to-be-attended ring on the monitor.

In the spatial attention task (Fig. 6.11a), the lever-press by the monkey initiates a trial and a central fixation ring (FR) is presented simultaneously. The central fixation ring was also used as a fixation point for the memory task. The monkey is required to maintain fixation at the central fixation ring as long as the central fixation ring is presented. Two peripheral rings (Fig. 6.11b) are presented at both up and down positions1.0 s after the start of the fixation. Subsequently, a red-filled circle (attention cue) is presented for 0.4 s at one of the three rings (target ring), where the monkey needs to pay attention. Two to five seconds after the offset of the attention cue (Wait 1 period), the color of the target ring turns to red for 0.4 s, indicating the Go-signal for the monkey to release the lever within 0.6 s. Catch trials are unexpectedly added to confirm that the monkey actually pays attention to the target ring. In catch trials, the color of one non-target ring turns to red and the monkey needs to inhibit the lever-release response and keep pressing the lever until the color of the correct target ring changes.

The spatial working memory task is the same as an oculomotor delayed-response (ODR) task (Funahashi et al. 1989; Funahashi et al. 1993a) (Fig. 6.11a). The monkey is required to maintain fixation at the central fixation point (a small filled circle) as long as the fixation point is presented. After a brief fixation period, a visual cue (a small white-filled square) is presented for 0.4 s at one of 5 (in the right or the left hemifield) or 8 (in whole visual field) peripheral positions randomly (Fig. 6.11c). The monkey is required to remember the cue position during the subsequent delay period. At the end of the delay period, the fixation point is distinguished, which is the Go-signal for the monkey to make a saccade to the position where the visual cue was presented. If the monkey makes a correct saccade, it obtains a reward.

In dual-task condition, the attention task is always introduced first and the memory tasks introduced after the attention task. Similar to the attention task in single-task condition, the lever-press by the monkey initiates a trial and a central fixation point is presented simultaneously. The monkey is required to maintain fixation at the central fixation point as long as it is presented. One second after the start of the fixation, the central ring and two peripheral rings for the attention task are presented. Subsequently, a red-filled circle (attention cue) is presented for 0.4 s at one of the three rings (target ring) and then the Wait 1 period is introduced. During the Wait 1 period between the offset of the attention cue and the onset of the color change at the target ring (1.6–5.1 s), a memory cue is presented for 0.4 s at one of 5 peripheral locations at a random timing. The monkey needs to remember the location of the memory cue while performing the attention task. When the color of the target ring turns to red, the monkey needs to release the lever within 0.6 s to end the attention task. After the end of the attention task, all rings are distinguished and small place-holders are presented at 5 peripheral cue locations. The monkey needs to further maintain brief fixation (0.4 s) at the central fixation point and, after turning off the fixation point, the monkey is required to make a saccade within 0.6 s to the position where the memory cue was presented and to keep looking at it for additional 0.6 s.

The presentation of the memory cue is scheduled independently from the attention task and never occurs after the onset of the color change of the target ring. In addition, the memory cue is presented in two-thirds of attention trials, which makes these trials dual-task condition. The remaining one-third of attention trials are performed as single-task condition of the attention task. Dual-task condition is randomly introduced within single-task condition of the attention task. As a control condition, the monkey performs the memory task in single-task condition. To determine the spatial selectivity of neuron's activity, eight cue positions that covered both visual hemifields are used.

6.6.4.2 Behavioral Performance

Behavioral analysis of the attention task under single-task condition showed that the monkey exhibits different rates of performance and different response times (RTs) when the attention cue is presented at different rings (Fig. 6.12). In the attention task, the attention cue is presented randomly at one of three rings (Up, Down, and FR). Behavioral analysis revealed that significantly lower rate of correct performances and significantly longer RTs were observed when the attention cue was presented at either the Up or the Down ring than when the attention cue was presented at the central fixation ring (FR) (Fig. 6.12a, b), indicating that maintaining attention to the peripheral rings is more effortful than attention to the central fixation ring in single-task condition. Since the central fixation ring (FR) is used as the fixation point and since the monkey is required to maintain looking at the central fixation ring during single-task condition as well as dual-task condition, attention to the central fixation ring is easier to maintain than attention to the peripheral rings.

Behavioral analysis of the performance of the memory task revealed that the monkey exhibits significantly lower rates of correct performances and significantly longer RTs in dual-task condition compared with in single-task condition (Fig. 6.12c), indicating the presence of dual-task interference effect on the performance of the memory task. In addition, the strength of dual-task interference effect on the performance of the memory task increased proportionally as the difficulty of the attention task increased. Further, a greater magnitude of dual-task interference was observed in the performance of the memory task when the difficulty of the memory task increased by introducing longer delay intervals (Fig. 6.12d).

Thus, as predicted, behavioral studies showed dual-task interference effects on the performance of the memory task, such that the performance of the memory task is interfered by the concurrent performance of the attention task and that stronger interference effect is produced on the performance of the memory task as the increase of the difficulty of the attention task. The inverse relationship observed in two tasks (increasing difficulty of one task causes worsening performance in the other task) is comparable to the results observed in human dual-task studies (Pashler 1994; Strayer and Johnston 2001; Fougnie and Marois 2006). This demonstrates

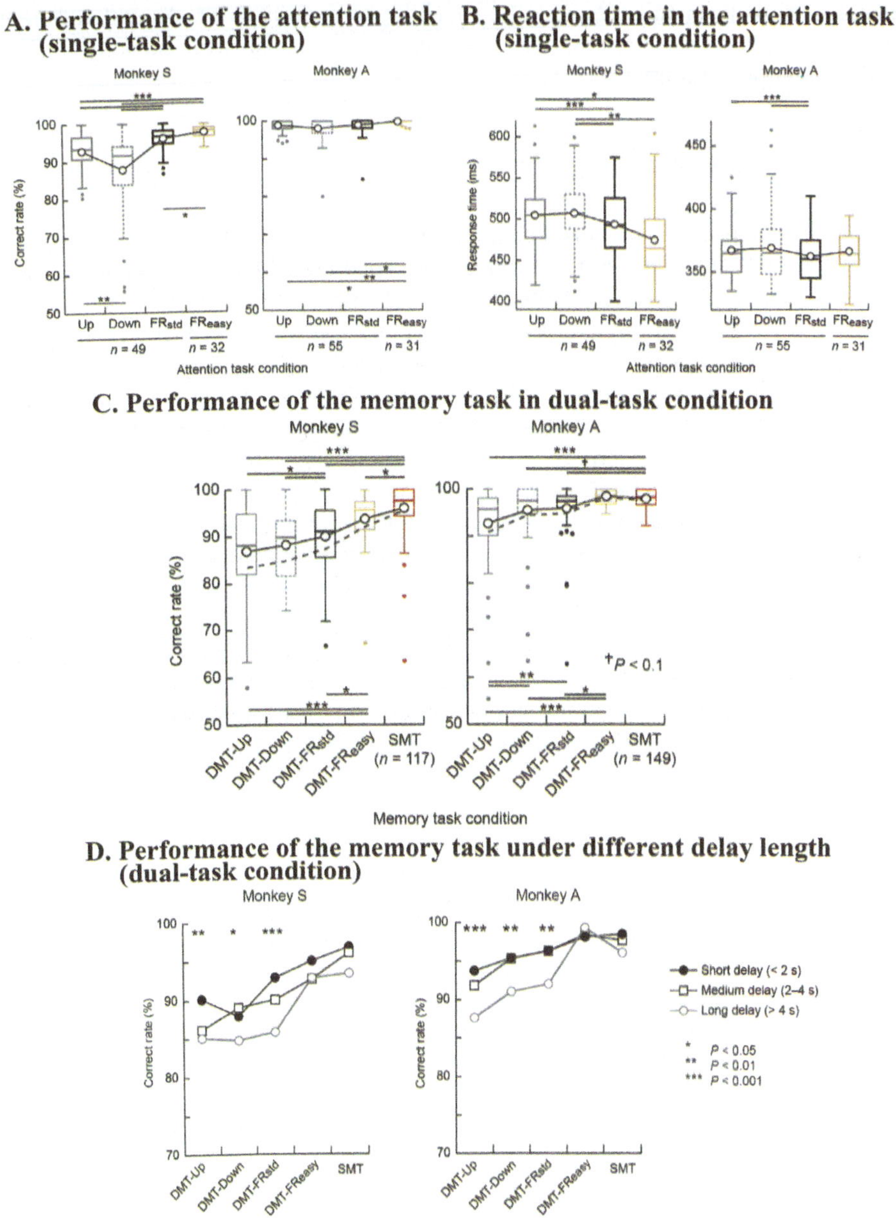

Fig. 6.12 (**a**) Behavioral performance of the attention task in single-task condition. Up, Down, and FP indicate positions of attention rings. (**b**) Reaction time in the attention task in single-task condition. (**c**) Behavioral performance of the memory task in dual-task condition. DMT-Up, for example, indicates dual-task condition of the memory task and the attention task with the attention ring at Up position. SMT indicates behavioral performance of the memory task in single-task condition. (**d**) Behavioral performance of the memory task in dual-task condition under three different lengths of the delay period. From Watanabe and Funahashi (2014) with permission

that, like humans, monkeys have sufficient ability to perform two tasks with different goals in dual-task condition and have human-like mechanisms for capacity-limited information processing.

6.6.4.3 Single-Neuron Activity

Behavioral analysis showed that concurrently performing attention task affected the performance of the memory task in dual-task condition. Many prefrontal neurons have been shown to exhibit spatially-selective cue- and/or delay-period activity during the ODR task in single-task condition (Funahashi et al. 1989, 1990). Spatially selective prefrontal activity is known to play essential roles in the performance of both spatial attention task and spatial working memory task in single-task condition. To investigate how the performance of these two tasks in dual-task condition affects prefrontal single-neuron activity, spatially-selective cue- and delay-period activity was compared between single-task condition and dual-task condition.

Analyses of prefrontal single-neuron activity showed that both tasks recruit activation of largely overlapped prefrontal neural populations in single-task condition. However, the performance of two tasks in dual-task condition affected prefrontal single-neuron activity related to the spatial working memory task by reducing both the magnitude of activity and the strength of spatial selectivity to a degree proportional to the difficulty of the concurrent attention task (Fig. 6.13). The strength of spatial selectivity of prefrontal activity observed in the spatial working memory task in single-task condition was remarkably attenuated in dual-task condition. The spatial selectivity of cue- and delay-period activity was both attenuated under dual-task condition (Fig. 6.13a, b). Especially, sustained delay-period activity that represents spatial memorandum in the spatial working memory task was significantly attenuated by the concurrent performance of the spatial attention task (Fig. 6.13b). The degree of attenuation of working memory-related activity became greater when the difficulty of the spatial attention task increased in dual-task condition. Population analysis of single-neuron activities also revealed that the difference in the magnitude of prefrontal activity between the best and the worst cue conditions of the spatial working memory task was remarkably reduced in dual-task condition.

To further confirm attenuation effect on prefrontal memory task-related activity, the strength of spatial selectivity of prefrontal activity for the memory cue location was quantified for each neuron using a proportion of explained variance (PEV) (Buschman et al. 2011). The PEV values in dual-task condition were significantly reduced relative to single-task condition in both the cue and delay periods of the working memory task (Fig. 6.14). These results indicate that a significant interference effect in dual-task condition can be observed at the level of prefrontal single-neuron activity as a significant attenuation of task-related activity necessary for performing either component task and that the degree of the interference effect in dual-task condition can be scaled by the degree of the attenuation effect observed in task-related activity for either component task, and that the degree of the attenuation

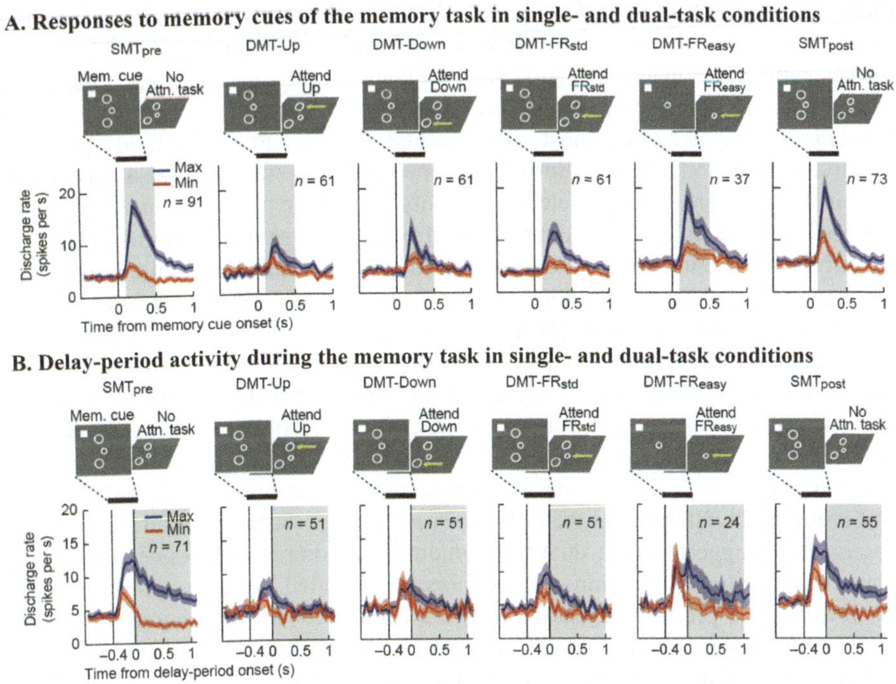

Fig. 6.13 (**a**) Population histograms showing responses to the memory cue presented at neurons' best (Max, blue) and worst (Min, red) positions under single- and dual-task conditions. (**b**) Population histograms of delay-period activity under single- and dual-task conditions when the memory cue was presented neurons' best (Max, blue) and worst (Min, red) positions. STM_{pre} and STM_{post} indicate activity during performance of the memory task in single-task condition before and after introduction of dual-task condition, respectively. DMT-Up, DMT-Down, and DMT-FR indicate that the memory task was introduced with the attention task with Up, Down, and FR rings, respectively, in dual-task conditions. From Watanabe and Funahashi (2014) with permission

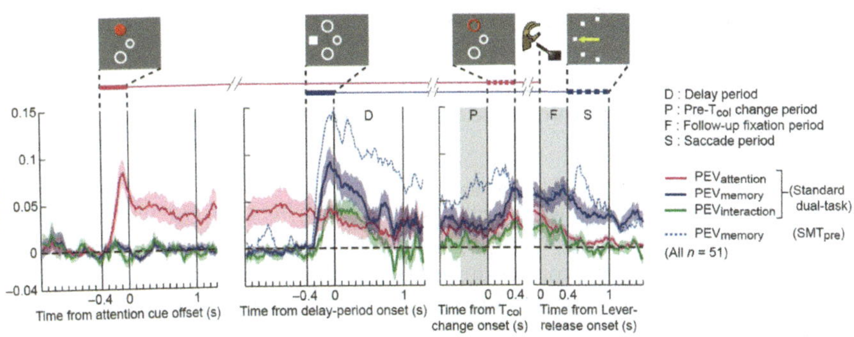

Fig. 6.14 Temporal changes of PEV (proportion of explained variance) values representing information regarding attention task (red), memory task (blue), and interaction of both tasks (green) in dual-task condition and information regarding memory task (blue dots) in single-task condition. From Watanabe and Funahashi (2014) with permission

effect observed in task-related activity for either component task is affected by the degree of the difficulty of the concurrently performing other tasks.

Watanabe and Funahashi (2014) showed that both memory and attention tasks used in this study recruited the activation of largely overlapped prefrontal neural populations. For example, most of prefrontal neurons that exhibited spatially-selective cue-period activity during the memory task in single-task condition exhibited significant spatially selective activity during the presentation of the attention cue in the attention task. Similarly, prefrontal neurons that exhibited spatially-selective delay-period activity during the memory task also exhibited sustained activation during the delay period during the attention task. These indicate that two component tasks in dual-task condition recruit a common population of prefrontal neurons. In addition, the size of the neural proportion that exhibited significant task-related modulation of activity during the cue period and the delay period was significantly different depending on the difficulty of the task in single-task condition, such that its size becomes larger in difficult condition but becomes smaller in easier condition. This indicates that more neurons are recruited for information processing as the difficulty of the task increases in single-task condition. Thus, these observations suggest that in dual-task condition, a competition between attention task-related neural population and memory task-related neural population takes place within the prefrontal cortex to acquire enough amount of neural population from a common neural pool for information processing by each task. The observations by single-neuron studies suggest that the limitation of the capacity of cognitive resource, at least that associated with spatial processing, corresponds to the limitation of the computational capacity of single prefrontal neurons.

These results support that the neural mechanism of dual-task interference resides in competitive and overloaded recruitment of neural population that participates in common information processing for two concurrently performing component tasks. This idea has been postulated in human neuroimaging studies (Klingberg 1998; Passingham 1996). These results also support that the psychological concept of cognitive resource can be implemented as a limited capacity for information processing in single prefrontal neurons.

6.6.4.4 Temporal Dynamics of Adaptive Allocation of Cognitive Resource

The temporal dynamics of the competitive interaction between two tasks in dual-task condition was examined by analyzing how neural signal encoding the information necessary for each task evolved during a trial by examining a partial ω^2 PEV measure (Buschman et al. 2011; Watanabe and Funahashi 2014). Analyzed neural signals included neural signals related to each of the attention task and the memory task and neural signals related to the interaction of two tasks. After the onset of the attention cue, neural signals related to the attention task exhibit sustained elevation and this elevation continues until the end of the attention task. On the other hand, the onset of the memory cue triggers the increase of neural signals related to both the memory task and the interaction of two tasks. The concurrent elevation of neural signals related to both tasks indicates that the neural resource in the prefrontal cortex is divided and allocated to each of two tasks for information processing depending on its demand. As is shown in Fig. 6.14, the strength of neural signals related to the memory task is significantly attenuated during the delay period in dual-task condition compared to the strength of neural signals for the memory task in single-task condition. This result indicates the reason why the performance of the memory test is deteriorated in behavioral experiments in dual-task condition and that dual-task interference effect is caused by competitive and overloaded recruitment of a limited capacity of the neural resource that requires for common information processing in concurrent performance of two component tasks.

Another important observation in this study is that a significant re-activation of neural signals related to the memory task was observed during the delay period after the end of the attention task (Watanabe and Funahashi 2014). The re-activation of neural signals related to the memory task occurred in concert with the end of the attention task, which corresponds to the timing of the change in the priority from the attention task to the memory task for information processing. Changing the priority from one task to the other indicates the evidence that the neural resource for information processing is flexibly reallocated from one component task to the other depending on the current processing demand of each task. The re-activation of neural signals related to the memory task also indicates that, even under the presence of the interference effect in dual-task condition caused by one component task, some neural mechanisms are operating in the prefrontal cortex to compensate for the interference effect produced by this task. The re-activation of neural signals related to the memory task would compensate for the loss of spatial information caused by the attention task in dual-task condition and support accurate behavioral responses for the memory task in dual-task condition (Watanabe and Funahashi 2015a).

A similar re-activation phenomenon has been reported by Miyazaki et al. (2013), who compared activity between lateral prefrontal neurons and dorsal premotor neurons while the monkey performed memory-guided bimanual actions (primary task) and visually-guided bimanual actions (secondary task) in dual-task condition. This experiment also showed that lateral prefrontal neurons play an important role to exert compensatory control by re-activating the information representation for the primary task after the end of the interference effect caused by the secondary task.

6.7 What Is Next?

Neurophysiological studies using animals provide important evidences for neural mechanisms causing dual-task interference effects, neural correlates of cognitive resource, and neural substrates of cognitive capacity limitations. Studies focused on neural mechanisms of dual-task interference using dual-task paradigms (Watanabe and Funahashi 2014, 2018) and those focused on neural mechanisms of cognitive capacity limitation using multiple-item working memory tasks (Buschman et al. 2011; Matsushima and Tanaka 2014) showed that, from neurobiological point of view, the cognitive resource can be understood as the total amount of neural resource in the prefrontal cortex which is able to represent task-relevant information. The neural resource in the prefrontal cortex is capacity-limited and sharable by multiple tasks for information processing. However, several important questions are remained to be answered.

Dual-task interference effect can be observed in either one of two component tasks or both tasks with different strengths of its effect in dual-task condition. One important question would be what factor determines which task becomes a looser (the task having stronger interference effect) in dual-task condition. One possible factor would be the difference in the difficulty between two component tasks. In some behavioral studies, a simple manual task (e.g., button press) was introduced as the secondary task during the delay period or the retention period of the delayed matching-to-sample task (Moise 1970; Washburn and Astur 1998; Smith et al. 2013; Basile and Hampton 2013). In this dual-task condition, dual-task interference effect was observed in the performance of the delayed matching-to-sample task and not observed in the performance of the manual task. Thus, dual-task interference effect might be observed in the component task that is more difficult to perform than the other component task. The cognitive resource is divided into information processing for two tasks depending on the demand of each task. The difficult task needs a larger amount of resource than the easier task. However, the available amount of cognitive resource is limited. Therefore, the difficult task would have more potential not to get enough amount of resource for information processing and get stronger interference effect than the easier task in dual task condition.

Another possible factor would be the order of two tasks in dual-task condition. Watanabe and Funahashi (2014) used the spatial attention task and the spatial working memory task and the degree of the difficulty was equivalent between two tasks. In dual-task condition, the spatial attention task started first and then the spatial working memory task was added during the delay period of the attention task. As a result, the performance of the memory task deteriorated and showed an interference effect. Prefrontal single-neuron activity related to the memory task was attenuated in dual-task condition, while activity related to the attention task was intact throughout its performance. Unfortunately, it is not known whether or not the same results can be also obtained when the order of these two tasks is reversed. The order of the component tasks introducing in dual-task condition may play a critical role to determine which task has dual-task interference effect.

Another important issue is to clarify whether the whole brain shares activation at the level of single-neuron activity and whether the general neural resource is present for the performance of multiple different tasks. For this purpose, it is necessary to increase the collection of behavioral tasks in animal neurophysiology. All of animal neurophysiology studies performed in dual-task condition have used a combination of two similar tasks that tax the same processing modality. Future studies should involve two tasks using different modalities and cognitive situations (e.g., an auditory DMTS task and a spatial delayed response task). To examine whether the whole brain shares activation at the level of single-neuron activity and whether the general resource is present or not, it is necessary to examine using a combination of dissimilar component tasks each of which activates non-overlapping brain regions in single-task condition.

Recently, Marcos et al. (2017) used a duration discrimination task and a distance discrimination task and examined whether cross-modal behavioral interference between the estimation of space and time could be explained by the magnitude of overlap of activation of prefrontal neurons. In the duration discrimination task, the monkey is required to report which of two visual stimuli (blue circle and red square) is presented longer than the other, while in the distance discrimination task, the monkey is required to report which of two stimuli is placed at greater distance from the fixation spot. The monkey performed each task separately. They found two independent neural populations participate in the estimation of distance and duration and concluded that cross-modal interference between time and space perception is not explained by prefrontal activity. This result, however, is based solely on the analysis of neural activity in single-task condition. The investigation regarding how these two independent neural populations react in dual-task condition would be a good method to test the overlap hypothesis.

The proposal that inter-regional synchronization of activity plays an important role to manage increasing processing demands in dual-task condition (Buchweitz et al. 2012) is directly testable by neurophysiological investigations using simultaneous recording of neural activity from multiple brain areas. A task-related increase in synchronization between the prefrontal cortex and the posterior brain areas has been observed in monkeys during the performance of a variety of cognitive tasks such as attention tasks (Buschman and Miller 2007), decision-making tasks (Pesaran et al. 2008), and working memory tasks (Salazar et al. 2012). Therefore, an examination of neural synchrony across multiple brain areas may provide important clues regarding the role of synchronization in information processing in dual-task condition.

As an experimental design, future studies should include comparisons of task-related neural activity of two component tasks in (1) single-task condition, in (2) easy dual-task condition without behavioral interference, and in (3) difficult dual-task condition with behavioral interference. The comparison of neural activity between the first and latter two conditions would enable to examine whether the neural circuitry specific for dual-task condition coexists with the neural circuits for single-task condition in an interdigitated manner. The comparison of neural activity between dual-task conditions with and without behavioral interference would allow to clarify whether the dual-task interference due to the increased task difficulty is attributed to (1) the competition for recruiting more neural resource and consequent overloading of the overlapping neural population (e.g., Klingberg and Roland 1997) or (2) the malfunctioning of neuronal processes specific in dual-task condition (e.g., Baddeley et al. 1997).

References

Adcock RA, Constable RT, Gore JC, Goldman-Rakic PS (2000) Functional neuroanatomy of executive processes involved in dual-task performance. Proc Natl Acad Sci USA 97:3567–3572. https://doi.org/10.1073/pnas.060588897

Alavash M, Hilgetag CC, Thiel CM, Gießing C (2015) Persistency and flexibility of complex brain networks underlie dual-task interference. Hum Brain Mapp 36:3542–3562. https://doi.org/10.1002/hbm.22861

Alvarez GA, Cavanagh P (2005) Independent resources for attentional tracking in the left and right visual hemifields. Psychol Sci 16:637–643. https://doi.org/10.1111/j.1467-9280.2005.01587.x

Baddeley A (1986) Working memory. Oxford University Press, Oxford, UK

Baddeley AD, Bressi S, Della Sala S, Logie R, Spinnler H (1991) The decline of working memory in Alzheimer's disease. Brain 114:2521–2542. https://doi.org/10.1093/brain/114.6.2521

Baddeley A, Della Sala S, Papagno C, Spinnler H (1997) Dual-task performance in dysexecutive and nondysexecutive patients with a frontal lesion. Neuropsychology 11:187–194. https://doi.org/10.1037/0894-4105.11.2.187

Basile BM, Hampton RR (2013) Dissociation of active working memory and passive recognition in rhesus monkeys. Cognition 126:391–396. https://doi.org/10.1016/j.cognition.2012.10.012

Beatty WW, Shavalia DA (1980) Rat spatial memory: resistance to retroactive interference at long retention intervals. Anim Learn Behav 8:550–552. https://doi.org/10.3758/bf03197768

Braver TS, Bongiolatti SR (2002) The role of frontopolar cortex in subgoal processing during working memory. Neuroimage 15:523–536. https://doi.org/10.1006/nimg.2001.1019

Buchweitz A, Keller TA, Meyler A, Just MA (2012) Brain activation for language dual-tasking: Listening to two people speak at the same time and a change in network timing. Hum Brain Mapp 33:1868–1882. https://doi.org/10.1002/hbm.21327

Bunge SA, Klingberg T, Jacobsen RB, Gabrieli JDE (2000) A resource model of the neural basis of executive working memory. Proc Natl Acad Sci USA 97:3573–3578

Burgess PW (2000) Strategy application disorder: the role of the frontal lobes in human multitasking. Psychol Res 63:279–288. https://doi.org/10.1007/s004269900006

Burgess PW, Dumontheil I, Gilbert SJ (2007) The gateway hypothesis of rostral prefrontal cortex (area 10) function. Trends Cogn Sci 11:290–298. https://doi.org/10.1016/j.tics.2007.05.004

Buschman TJ, Miller EK (2007) Top-down versus bottom-up control of attention in the prefrontal and posterior parietal cortices. Science 315:1860–1862. https://doi.org/10.1126/science.1138071

Buschman TJ, Siegel M, Roy JE, Miller EK (2011) Neuronal substrates of cognitive capacity limitations. Proc Natl Acad Sci USA 108:11252–11255. https://doi.org/10.1073/pnas.1104666108

Carlson TA, Alvarez GA, Cavanagh P (2007) Quadrantic deficit reveals anatomical constraints on selection. Proc Natl Acad Sci USA 104:13496–13500. https://doi.org/10.1073/pnas.0702685104

Cohen MA, Konkle T, Rhee Y, Nakayama K, Alvarez GA (2014) Processing multiple visual objects is limited by overlap in neural channels. Proc Natl Acad Sci USA 111:8955–8960. https://doi.org/10.1073/pnas.1317860111

Constantinidis C, Franowicz MN, Goldman-Rakic S (2001) The sensory nature of mnemonic representation in the primate prefrontal cortex. Nat Neurosci 4:311–316. https://doi.org/10.1038/85179

Cook RG, Brown MF (1985) Retroactive interference in rat radial maze performance: the role of point of delay interpolation and the similarity and amount of interpolated material. Anim Learn Behav 13:116–120. https://doi.org/10.3758/bf03199263

Crittenden BM, Duncan J (2014) Task difficulty manipulation reveals multiple demand activity but no frontal lobe hierarchy. Cereb Cortex 24:532–540. https://doi.org/10.1093/cercor/bhs333

Deprez S, Vandenbulcke M, Peeters R, Emsell L, Amant F, Sunaert S (2013) The functional neuroanatomy of multitasking: combining dual tasking with a short term memory task. Neuropsychologia 51:2251–2260. https://doi.org/10.1016/j.neuropsychologia.2013.07.024

Derrfuss J, Brass M, von Cramon DY (2004) Cognitive control in the posterior frontolateral cortex: evidence from common activations in task coordination, interference control, and working memory. Neuroimage 23:604–612. https://doi.org/10.1016/j.neuroimage.2004.06.007

D'Esposito M, Detre JA, Alsop DC, Shin RK, Atlas S, Grossman M (1995) The neural basis of the central executive system of working memory. Nature 378:279–281. https://doi.org/10.1038/378279a0

Dreher JC, Koechlin E, Tierney M, Grafman J (2008) Damage to the fronto-polar cortex is associated with impaired multitasking. PLoS One 3:e3227. https://doi.org/10.1371/journal.pone.0003227

Duncan J (1980) The demonstration of capacity limitation. Cogn Psychol 12:75–96. https://doi.org/10.1016/0010-0285(80)90004-3

Dux PE, Tombu MN, Harrison S, Rogers BP, Tong F, Marois R (2009) Training improves multitasking performance by increasing the speed of information processing in human prefrontal cortex. Neuron 63:127–138. https://doi.org/10.1016/j.neuron.2009.06.005

Elmore LC, Ma WJ, Magnotti JF, Leising KJ, Passaro AD, Katz JS, Wright AA (2011) Visual short-term memory compared in rhesus monkeys and humans. Curr Biol 21:975–979. https://doi.org/10.1016/j.cub.2011.04.031

Erickson KI, Colcombe SJ, Wadhwa R, Bherer L, Peterson MS, Scalf PE, Kramer AF (2005) Neural correlates of dual-task performance after minimizing task-preparation. Neuroimage 28:967–979. https://doi.org/10.1016/j.neuroimage.2005.06.047

Erickson KI, Colcombe SJ, Wadhwa R, Bherer L, Peterson MS, Scalf PE, Kim JS, Alvarado M, Kramer AF (2007) Training-induced functional activation changes in dual-task processing: an fMRI study. Cereb Cortex 17:192–204. https://doi.org/10.1093/cercor/bhj137

Everling S, Tinsley CJ, Gaffan D, Duncan J (2002) Filtering of neural signals by focused attention in the monkey prefrontal cortex. Nat Neurosci 5:671–676. https://doi.org/10.1038/nn874

Forstmann BU, van den Wildenberg WPM, Ridderinkhof KR (2008) Neural mechanisms, temporal dynamics, and individual differences in interference control. J Cogn Neurosci 20:1854–1865. https://doi.org/10.1162/jocn.2008.20122

Fougnie D, Marois R (2006) Distinct capacity limits for attention and working memory: evidence from attentive tracking and visual working memory. Psychol Sci 17:526–534. https://doi.org/10.1111/j.1467-9280.2006.01739.x

Friedman NP, Miyake A (2004) The relations among inhibition and interference control functions: a latent-variable analysis. J Exp Psychol Gen 133:101–135. https://doi.org/10.1037/0096-3445.133.1.101

Friedman A, Polson MC (1981) Hemispheres as independent resource systems: limited-capacity processing and cerebral specialization. J Exp Psychol Hum Percept Perform 7:1031–1058. https://doi.org/10.1037/0096-1523.7.5.1031

Frisch S, Förstl S, Legler A, Schöpe S, Goebel H (2012) The interleaving of actions in everyday life multitasking demands. J Neuropsychol 6:257–269. https://doi.org/10.1111/j.1748-6653.2012.02026.x

Funahashi S, Bruce CJ, Goldman-Rakic PS (1989) Mnemonic coding of visual space in the monkey's dorsolateral prefrontal cortex. J Neurophysiol 61:331–349. https://doi.org/10.1152/jn.1989.61.2.331

Funahashi S, Bruce CJ, Goldman-Rakic PS (1990) Visuospatial coding in primate prefrontal neurons revealed by oculomotor paradigms. J Neurophysio 63:814–831. https://doi.org/10.1152/jn.1990.63.4.814

Funahashi S, Bruce CJ, Goldman-Rakic PS (1993a) Dorsolateral prefrontal lesions and oculomotor delayed-response performance: evidence for mnemonic "scotomas.". J Neurosci 13:1479–1497. https://doi.org/10.1523/jneurosci.13-04-01479.1993

Funahashi S, Inoue M, Kubota K (1993b) Delay-related activity in the primate prefrontal cortex during sequential reaching tasks with delay. Neurosci Res 18:171–175. https://doi.org/10.1016/0168-0102(93)90019-m

Funahashi S, Inoue M, Kubota K (1997) Delay-period activity in the primate prefrontal cortex encoding multiple spatial positions and their order of presentation. Behav Brain Res 84:203–223. https://doi.org/10.1016/S0166-4328(96)00151-9

Godbout L, Grenier MC, Braun CMJ, Gagnon S (2005) Cognitive structure of executive deficits in patients with frontal lesions performing activities of daily living. Brain Injury 19:337–348. https://doi.org/10.1080/02699050400005093

Herath P, Klingberg T, Young J, Amunts K, Roland P (2001) Neural correlates of dual task interference can be dissociated from those of divided attention: an fMRI study. Cereb Cortex 11:796–805. https://doi.org/10.1093/cercor/11.9.796

Inoue M, Funahashi S (2002) Prefrontal delay-period activity is affected by visual cues presented outside the memory field. NeuroReport 13:2097–2101

Isreal JB, Chesney GL, Wickens CD, Donchin E (1980) P300 and tracking difficulty: evidence for multiple resources in dual-task performance. Psychophysiol 17:259–273. https://doi.org/10.1111/j.1469-8986.1980.tb00146.x

Jaeggi SM, Seewer R, Nirkko AC, Eckstein S, G, Groner R, Gutbrod K. (2003) Does excessive memory load attenuate activation in the prefrontal cortex? Load-dependent processing in single and dual tasks: functional magnetic resonance imaging study. Neuroimage 19:210–225. https://doi.org/10.1016/s1053-8119(03)00098-3

Jiang Y, Kanwisher N (2003a) Common neural substrates for response selection across modalities and mapping paradigms. J Cogn Neurosci 15:1080–1094. https://doi.org/10.1162/089892903322598067

Jiang Y, Kanwisher N (2003b) Common neural mechanisms for response selection and perceptual processing. J Cogn Neurosci 15:1095–1110. https://doi.org/10.1162/089892903322598076

Johnson JA, Zatorre RJ (2006) Neural substrates for dividing and focusing attention between simultaneous auditory and visual events. Neuroimage 31:1673–1681. https://doi.org/10.1016/j.neuroimage.2006.02.026

Just MA, Carpenter PA, Keller TA, Emery L, Zajac H, Thulborn KR (2001) Interdependence of nonoverlapping cortical systems in dual cognitive tasks. Neuroimage 14:417–426. https://doi.org/10.1006/nimg.2001.0826

Just MA, Keller TA, Cynkar J (2008) A decrease in brain activation associated with driving when listening to someone speak. Brain Res 1205:70–80. https://doi.org/10.1016/j.brainres.2007.12.075

Kadohisa M, Petrov P, Stokes M, Sigala N, Buckley M, Gaffan D, Kusunoki M, Duncan J (2013) Dynamic construction of a coherent attentional state in a prefrontal cell population. Neuron 80:235–246. https://doi.org/10.1016/j.neuron.2013.07.041

Kahneman D (1973) Attention and effort. Prentice Hall, Englewood Cliffs, NJ

Kinsbourne M, Hicks RE (1978) Functional cerebral space: a model for overflow, transfer, and interference effects in human performance. In: Requin J (ed) Attention and performance VII. Erlbaum, New Jersey, pp 342–362

Kleinman MR, Sohn H, Lee D (2016) A two-stage model of concurrent interval timing in monkeys. J Neurophysiol 116:1068–1081. https://doi.org/10.1152/jn.00375.2016

Klingberg T (1998) Concurrent performance of two working memory tasks: potential mechanisms of interference. Cereb Cortex 8:593–601. https://doi.org/10.1093/cercor/8.7.593

Klingberg T, Roland PE (1997) Interference between two concurrent tasks is associated with activation of overlapping fields in the cortex. Cogn Brain Res 6:1–8. https://doi.org/10.1016/s0926-6410(97)00010-4

Koechlin E, Summerfield C (2007) An information theoretical approach to prefrontal executive function. Trends Cogn Sci 11:229–235. https://doi.org/10.1016/j.tics.2007.04.005

Koechlin E, Basso G, Pietrini P, Panzer S, Grafman J (1999) The role of the anterior prefrontal cortex in human cognition. Nature 399:148–151. https://doi.org/10.1038/20178

Kok A (2001) On the utility of P3 amplitude as a measure of processing capacity. Psychophysiology 38:557–577

Lebedev MA, Messinger A, Kralik JD, Wise SP (2004) Representation of attended versus remembered locations in prefrontal cortex. PLoS Biol 2:e365. https://doi.org/10.1371/journal.pbio.0020365

Maki WS, Brokofsky S, Berg B (1979) Spatial memory in rats: resistance to retroactive interference. Anim Learn Behav 7:25–30. https://doi.org/10.3758/bf03209652

Marcos E, Tsujimoto S, Genovesio A (2017) Independent coding of absolute duration and distance magnitudes in the prefrontal cortex. J Neurophysiol 117:195–203. https://doi.org/10.1152/jn.00245.2016

Marti S, King J-R, Dehaene S (2015) Time-resolved decoding of two processing chains during dual-task interference. Neuron 88:1297–1307. https://doi.org/10.1016/j.neuron.2015.10.040

Matsuo K, Kaiju T, Nakazono T, Watanabe K, Suzuki T (2017) A behavioral paradigm to study rats' dual-task performance under head-direction and body-location tracking. Electron Commun Jpn 100:3–14

Matsushima A, Tanaka M (2014) Different neuronal computations of spatial working memory for multiple locations within versus across visual hemifields. J Neurosci 34:5621–5626. https://doi.org/10.1523/JNEUROSCI.0295-14.2014

McDowell S, Whyte J, D'Esposito M (1997) Working memory impairments in traumatic brain injury: evidence from a dual-task paradigm. Neuropsychologia 35:1341–1353. https://doi.org/10.1016/s0028-3932(97)00082-1

Meck WH (1987) Vasopressin metabolite neuropeptide facilitates simultaneous temporal processing. Behav Brain Res 23:147–157. https://doi.org/10.1016/0166-4328(87)90051-9

Meck WH, Church RM, Olton DS (1984) Hippocampus, time, and memory. Behav Neurosci 98:3–22. https://doi.org/10.1037/0735-7044.98.1.3

Meck WH, Williams CL (1997) Simultaneous temporal processing is sensitive to prenatal choline availability in mature and aged rats. Neuroreport 8:3045–3051. https://doi.org/10.1097/00001756-199709290-00009

Meck WH, Church RM, Olton DS. (1984) Hippocampus, time, and memory. Behav Neurosci 98:3-22. https://doi.org/10.1037/0735-7044.98.1.3

Messinger A, Lebedev MA, Kralik JD, Wise SP (2009) Multitasking of attention and memory functions in the primate prefrontal cortex. J Neurosci 29:5640–5653. https://doi.org/10.1523/jneurosci.3857-08.2009

Miyazaki A, Nakajima T, Shima K, Mushiake H (2013) Neuronal activity in the prefrontal cortex during performance of a dual task consisting of a main- and an interrupting-task. In: Advances in cognitive neurodynamics (III), pp 795–801. https://doi.org/10.1007/978-94-007-4792-0_104

Mizuno K, Tanaka M, Tanabe HC, Sadato N, Watanabe Y (2012) The neural substrates associated with attentional resources and difficulty of concurrent processing of the two verbal tasks. Neuropsychologia 50:1998–2009. https://doi.org/10.1016/j.neuropsychologia.2012.04.025

Mochizuki H, Tashiro M, Gyoba J, Suzuki M, Okamura N, Itoh M, Yanai K (2007) Brain activity associated with dual-task management differs depending on the combinations of response modalities. Brain Res 1172:82–92. https://doi.org/10.1016/j.brainres.2007.07.046

Moise SL (1970) Short-term retention in Macaca speciosa following interpolated activity during delayed matching from sample. J Comp Physiol Psychol 73:506–514. https://doi.org/10.1037/h0030228

Moray N (1967) Where is capacity limited? A survey and a model. Acta Psychol 27:84–92. https://doi.org/10.1016/0001-6918(67)90048-0

Nakazono T, Sano T, Takahashi S, Sakurai Y (2015) Theta oscillation and neuronal activity in rat hippocampus are involved in temporal discrimination of time in seconds. Front Syst Neurosci 9:95. https://doi.org/10.3389/fnsys.2015.00095

Nee DE, Jonides J (2008) Dissociable interference-control processes in perception and memory. Psychol Sci 19:490–500. https://doi.org/10.2307/40064784

Newman SD, Keller TA, Just MA (2007) Volitional control of attention and brain activation in dual task performance. Hum Brain Mapp 28:109–117. https://doi.org/10.1002/hbm.20257

Nijboer M, Borst J, van Rijn H, Taatgen N (2014) Single-task fMRI overlap predicts concurrent multitasking interference. Neuroimage 100:60–74. https://doi.org/10.1016/j.neuroimage.2914.05.082

Olton DS, Wenk GL, Church RM, Meck WH (1988) Attention and the frontal cortex as examined by simultaneous temporal processing. Neuropsychologia 26:307–318. https://doi.org/10.1016/0028-3932(88)90083-8

Pang KCH, Yoder RM, Olton DS (2001) Neurons in the lateral agranular frontal cortex have divided attention correlates in a simultaneous temporal processing task. Neuroscience 103:615–628. https://doi.org/10.1016/S0306-4522(01)00018-5

Park NW, Moscovitch M, Robertson IH (1999) Divided attention impairments after traumatic brain injury. Neuropsychologia 37:1119–1133. https://doi.org/10.1016/S0028-3932(99)00034-2

Pashler H (1994) Dual-task interference in simple tasks: data and theory. Psychol Bull 116:220–244. https://doi.org/10.1037/0033-2909.116.2.220

Pashler H, Carrier M, Hoffman J (1993) Saccadic eye movements and dual-task interference. Quar J Exp Psychol Sect A 46:51–82. https://doi.org/10.1080/14640749308401067

Passingham RE (1996) Attention to action. Phil Trans R Soc Lond B 351:1473–1479. https://doi.org/10.1098/rstb.1996.0132

Pesaran B, Nelson MJ, Andersen RA (2008) Free choice activates a decision circuit between frontal and parietal cortex. Nature 453:406–409. https://doi.org/10.1038/nature06849

Ramsey NF, Jansma JM, Jager G, Van Raalten T, Kahn RS (2004) Neurophysiological factors in human information processing capacity. Brain 127:517–525. https://doi.org/10.1093/brain/awh060

Rao SC, Rainer G, Miller EK (1997) Integration of what and where in the primate prefrontal cortex. Science 276:821–824. https://doi.org/10.1126/science.276.5313.821

Roberts WA (1981) Retroactive inhibition in rat spatial memory. Anim Learn Behav 9:566–574

Roland PE (1985) Cortical organization of voluntary behavior in man. Hum Neurobiol 4:155–167

Rossi AF, Bichot NP, Desimone R, Ungerleider LG (2007) Top-down attentional deficits in macaques with lesions of lateral prefrontal cortex. J Neurosci 27:11306–11314. https://doi.org/10.1523/JNEUROSCI.2939-07.2007

Ruthruff E, Johnston JC, Van Selst M (2001) Why practice reduces dual-task interference. J Exp Psychol Hum Percep Perform 27:3–21. https://doi.org/10.1037/0096-1523.27.1.3

Salazar RF, Dotson NM, Bressler SL, Gray CM (2012) Content-specific fronto-parietal synchronization during visual working memory. Science 338:1097–1100. https://doi.org/10.1126/science.1224000

Salo E, Rinne T, Salonen O, Alho K (2015) Brain activations during bimodal dual tasks depend on the nature and combination of component tasks. Front Hum Neurosci 9:102

Seamans JK, Lapish CC, Durstewitz D (2008) Comparing the prefrontal cortex of rats and primates: Insights from electrophysiology. Neurotoxity Res 14:249–262. https://doi.org/10.1007/BF03033814

Shallice T, Burgess PW (1991) Deficits in strategy application following frontal lobe damage in man. Brain 114:727–741. https://doi.org/10.1093/brain/114.2.727

Sigman M, Dehaene S (2008) Brain mechanisms of serial and parallel processing during dual-task performance. J Neurosci 28:7585–7598. https://doi.org/10.1523/jneurosci.0948-08.2008

Smith EE, Geva A, Jonides J, Miller A, Reuter-Lorenz P, Koeppe RA (2001) The neural basis of task-switching in working memory: effects of performance and aging. Proc Natl Acad Sci USA 98:2095–2100. https://doi.org/10.1073/pnas.98.4.2095

Smith JD, Coutinho MV, Church BA, Beran MJ (2013) Executive-attentional uncertainty responses by rhesus macaques (*Macaca mulatta*). J Exp Psychol Gen 142:458–475. https://doi.org/10.1037/a0029601

Strayer DL, Johnston WA (2001) Driven to distraction: dual-task studies of simulated driving and conversing on a cellular telephone. Psychol Sci 12:462–466. https://doi.org/10.1111/1467-9280.00386

Szameitat A, Schubert T, Müller K, Von Cramon D (2002) Localization of executive functions in dual-task performance with fMRI. J Cogn Neurosci 14:1184–1199. https://doi.org/10.1162/089892902760807195

Thomsen T, Rimol LM, Ersland L, Hugdahl K (2004) Dichotic listening reveals functional specificity in prefrontal cortex: an fMRI study. Neuroimage 21:211–218. https://doi.org/10.1016/j.neuroimage.2003.08.039

Tombu MN, Asplund CL, Dux E, Godwin D, Martin JW, Marois R (2011) A unified attentional bottleneck in the human brain. Proc Natl Aca Sci USA 108:13426–13431. https://doi.org/10.1073/pnas.1103583108

Uylings HB, Groenewegen HJ, Kolb B (2003) Do rats have a prefrontal cortex? Behav Brain Res 146:3–17. https://doi.org/10.1016/j.bbr.2003.09.028

Van Selst M, Ruthruff E, Johnston JC (1999) Can practice eliminate the psychological refractory period effect? J Exp Psychol Hum Percept Perform 25:1268–1283

Warden MR, Miller EK (2007) The representation of multiple objects in prefrontal neuronal delay activity. Cereb Cortex 17:i41–i50. https://doi.org/10.1093/cercor/bhm070

Warden MR, Miller EK (2010) Task-dependent changes in short-term memory in the prefrontal cortex. J Neurosci 30:15801–15810. https://doi.org/10.1523/jneurosci.1569-10.2010

Washburn DA, Astur RS (1998) Nonverbal working memory of humans and monkeys: rehearsal in the sketchpad? Mem Cogn 26:277–286. https://doi.org/10.3758/bf03201139

Washburn DA, Putney RT (1998) Stimulus movement and the intensity of attention. Psychol Record 48:555–570. https://doi.org/10.1007/bf03395289

Watanabe K, Funahashi S (2014) Neural mechanisms of dual-task interference and cognitive capacity limitation in the prefrontal cortex. Nat Neurosci 17:601–611. https://doi.org/10.1038/nn.3667

Watanabe K, Funahashi S (2015a) Primate models of interference control. Curr Opin Behav Sci 1: 9–16. https://doi.org/10.1016/j.cobeha.2014.07.004

Watanabe K, Funahashi S (2015b) A dual-task paradigm for behavioral and neurobiological studies in nonhuman primates. J Neurosci Methods 246:1–12. https://doi.org/10.1016/j.neumeth.2015.03.006

Watanabe K, Funahashi S (2018) Toward an understanding of the neural mechanisms underlying dual-task performance: contribution of comparative approeches using animal models. Neurosci Biobehav Rev 84:12–28. https://doi.org/10.1016/j.neubiorev.2017.08.008

Wickens CD (2002) Multiple resources and performance prediction. Theor Issues Ergon 3:159–177. https://doi.org/10.1080/14639220210123806

Wickens C, Kramer A, Vanasse L, Donchin E (1983) Performance of concurrent tasks: a psycho-physiological analysis of the reciprocity of information-processing resources. Science 221: 1080–1082. https://doi.org/10.1126/science.6879207

Zanto TP, Rubens MT, Thangavel A, Gazzaley A (2011) Causal role of the prefrontal cortex in top-down modulation of visual processing and working memory. Nat Neurosci 14:656–661. https://doi.org/10.1038/nn.2773

Chapter 7
Decision-Making

7.1 Decision-Making

In our daily life, we make any decision every moment. An appropriate and flexible decision is essential to acquire better and appropriate solutions against various and unpredictable problems and is also important to survive in the natural and social environment around us. When we make a decision, this decision is executed often based on some specific information. For example, when we visit a convenience store and try to get a drink, we may select a particular drink based on the information acquired before (e.g., information obtained from a talk with a friend or a TV advertisement) or based on our own preference or by chance. We often make a decision to select an item based on specific information acquired from a TV advertisement, for example, but we do not know whether we prefer it or not. This type of decision-making can be called memory-based decision-making. On the other hand, we make a decision to select an item based on our own preference (like or dislike) or experience (had high or low value). This type of decision-making can be called value-based decision-making. Further, we sometime select an item which we know nothing about our preference for it and its actual value or need to select one item among same items, or select an item by chance. This type of decision-making can be called free-choice decision-making, because we can use neither our memory, preference, nor experience to select an item. In the experimental situation, perceptual decision-making, decision-making in competitive condition, and decision-making in conflict conditions have also been used to examine neural mechanisms of decision-making. The prefrontal cortex has been shown to contribute to these decision-making processes significantly.

S. Funahashi, *Dorsolateral Prefrontal Cortex*, Brain Science,
https://doi.org/10.1007/978-981-19-7268-3_7

7.1.1 Memory-Based Decision-Making

Prefrontal contribution to memory-based decision-making is well known. We often select a particular item based on our memory acquired from an advertisement and TV commercials or based on the recommendation from a friend. In animal behavioral studies to examine functions of the prefrontal cortex, behavioral paradigms based on memory-based decision-making have been frequently used. Typical behavioral tasks using memory-based decision-making are the tasks including the delay period, such as the delayed-response task and the delayed matching-to-sample task. In these tasks, a particular sensory cue is presented briefly during the cue period. The subject is required to remember it during the subsequent delay period and use it to make a correct choice behavior during the response period. Therefore, the subject needs to use memorized information to make a correct behavioral response. Therefore, these tasks are tasks requiring memory-based decision-making. The feature of these tasks is that the information needed to remember is shown at the beginning of each trial and changes at every trial.

Neurophysiological studies using monkeys show that prefrontal neurons exhibit delay-period activity while the monkey performs behavioral tasks that need memory-based decision-making. Delay-period activity is tonic sustained activation during the delay period. This activity exhibits selectivity to the visual cues, such that delay-period activity is observed only when the visual cue is presented at a particular position or when the visual cue has a particular physical feature. In the tasks requiring memory-based decision-making, holding information regarding the visual cue during the delay period is essential to make a correct choice during the response period. Delay-period activity has been shown to be observed only when the subject made a correct response, showing that information representing by delay-period activity is the information necessary to make a correct choice in any given trial. Therefore, delay-period activity is thought to be a neural correlate of the mechanism for temporarily holding information necessary to make a correct choice, which corresponds to the temporary storage mechanism for working memory. The most of delay-period activity observed in the prefrontal cortex has been shown to represent retrospective information, such as the information of the visual cue, while a significant proportion of delay-period activity has also been shown to represent prospective information, such as the direction of the motor response. Both types of delay-period activity play essential roles in the working memory process and memory-based decision-making for these behavioral tasks. Therefore, working memory is an essential mechanism for memory-based decision-making.

7.1.2 Value-Based Decision-Making

Another type of decision-making is *value-based decision-making* (Rangel et al. 2008). When we try to select one option from multiple known alternatives, we usually select an option that we know is more satisfying, more pleasurable, or more valuable than others. The option which is more satisfying, more pleasurable, or more valuable is thought to have a higher value as a reward (reward value) compared with other options. Reward value is a subjective value based on the strength of satisfaction, pleasure, or worth that the subject feels about each option. Each option is associated with a certain amount of reward value. For example, if option A has a higher reward value than other options for the subject, he would choose option A more often than other options. Thus, in value-based decision-making, which option would be selected is based on the comparison of reward values among options, and the option having higher reward value would be selected.

Reward value of each option is created by the subject's learning and repeated experience of the choice between that option and others. For example, if the selection of option A always causes more pleasurable experience for the subject than option B, the reward value of option A becomes higher than option B. Therefore, when the subject encounters the selection between options A and B, he tends to select option A. An accumulation of the reward value by choosing a particular option is called reward history and the experience of the choice of a particular option is called choice history. The reward history and choice history associated with a particular option play important roles to construct the subjective reward value of the option. Subjective reward value assigned to each option plays an important role in value-based decision-making and, by comparing reward values among options, the option with higher reward value is selected in this decision process. Therefore, reward values associated with options and created by reward and choice histories not only affect current selection but also will affect selections in the future significantly.

Learning mechanism to acquire the reward value of an option is explained using a reinforcement learning model. In this model, each option is associated with an expected reward value, which is a relative value compared with other options. When the subject needs to select one option from multiple options, the subject compares expected reward values among these options and makes decision to choose the option that is thought to have the highest expected reward value. The expected value associated with each option changes systematically by the choice history. For example, if a positive outcome is obtained after the selection of a particular option, its expected reward value increases. However, if a negative outcome is obtained by the selection of the option, its expected reward value decreases. The increase or decrease of the expected reward value is accumulated as the reward history of that option. When the subject needs to make a decision to select one option among others, the cumulative effect of the reward value for each option influences the subject's decision. For example, suppose that the subject tries to select either option A or B. When the selection of option A is expected to produce more satisfying and pleasurable results than the selection of option B, option A would be selected. If the selection of option A causes an expected positive result, its reward value increases and option A would be selected in future selection. However, if the selection of option A causes unexpected negative result, its reward value decreases and option A would not be selected but option B would be selected in future selection between options A and B. Thus, in the reinforcement learning model, the expected reward value of each option is updated after every choice and either increases or decreases based on the result of each choice. In value-based decision-making, accumulated reward history associated with each option plays an important role and the option with higher reward value would be selected more often. Value-based decision-making is a typical form of decision-making that is performed in our daily life.

7.1.3 Free-Choice Decision-Making

Free-choice decision-making is also an important process for decision-making. As was considered in value-based decision-making, when we try to select one option from others of similar category, we evaluate the expected reward value of each option and select the one having higher reward value. However, when we encounter a vending machine as shown in Fig. 7.1, which bottle will we choose? This vending machine dispenses the same bottles of mineral water. All bottles are the same appearance, include the same quality and quantity of mineral water, and are selling at the same price. Whichever bottle we select, the outcome will be exactly the same and therefore, the expected reward value is the same. This is a typical situation of *free-choice decision-making*.

On the other hand, when we encounter a situation that we need to choose one option from unique unknown options for the first time, we also select one option without knowing which option has higher reward value for us. Decision-making in

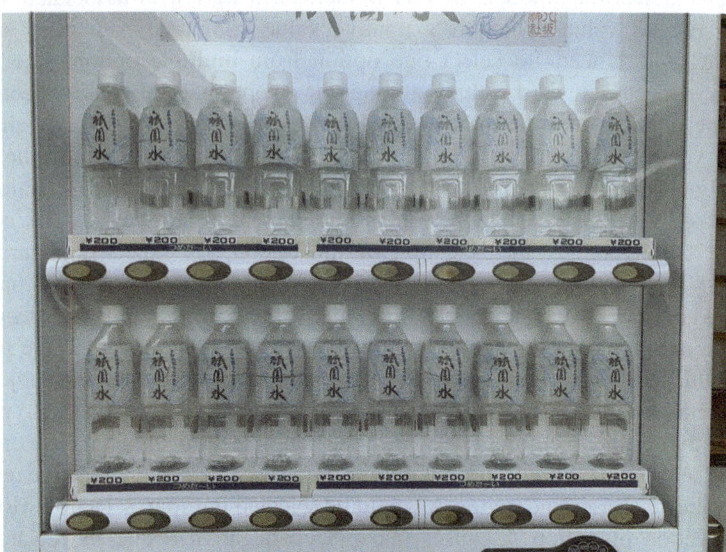

Fig. 7.1 A vending machine selling bottles of mineral water

this situation can be also considered to be free-choice decision-making. By repeating selections, we eventually realize which option is better or worse to select, accumulate this information and create the reward value of each option, and use this value for decision-making in future. Thus, free-choice decision-making is the first method to select one option from unknown options for the first time. At the same time, free-choice decision-making is the first step to constructing the reward history and the choice history of that option, which are necessary to construct the reward value of the selected option for value-based decision-making. Therefore, it is important to understand the neural mechanisms of free-choice decision-making in order to know how one option is selected from multiple alternatives having unknown reward values and how reward history and choice history are created. Although not many studies have examined neural mechanisms for free-choice decision-making, several studies have indicated that the prefrontal cortex contributes significantly to free-choice decision-making.

7.2 Prefrontal Contribution to Free-Choice Decision-Making: Human Imaging Studies

Human neuroimaging studies have shown that the prefrontal cortex plays significant roles in spontaneous or self-generated behavior and internally-driven decision-making (Frith et al. 1991; Hyder et al. 1997; Lau et al. 2004; Haynes et al. 2007; Soon et al. 2008). Spontaneous or self-generated actions are internally driven and not generated by external stimuli. Frith et al. (1991) used PET (positron emission tomography) and examined activated brain areas in relation to task performances while subjects performed two types of behavioral tasks. In routine tasks, each response was specified by an external stimulus, while in novel tasks, each response was selected by the subject's willed action. They found an increase in regional blood flow in the dorsolateral prefrontal cortex and the anterior cingulate cortex when the subjects performed novel tasks under both a speaking-a-word condition and lifting-a-finger condition. Hyder et al. (1997) repeated the study of Frith et al. (1991) using fMRI and confirmed the bilateral activation of the dorsolateral prefrontal cortex while subjects performed a willed action task (lifting-a-finger). They observed only left dorsolateral prefrontal activation in the verbal task (speaking-a-word). Thus, although modality difference in activated brain areas was observed, these results indicated that the dorsolateral prefrontal cortex plays an important role in self-generated willed actions.

On the other hand, Haynes et al. (2007) used fMRI and examined whether the activity of the prefrontal cortex encoded the subject's current intention or not. In their task, human subjects were required to first select either addition or subtraction of two numbers by themselves and then provide the answer. In the task (Fig. 7.2a), when the cue word "select" was presented on the monitor (selection period), the subject was first required to choose either addition or subtraction of the numbers by

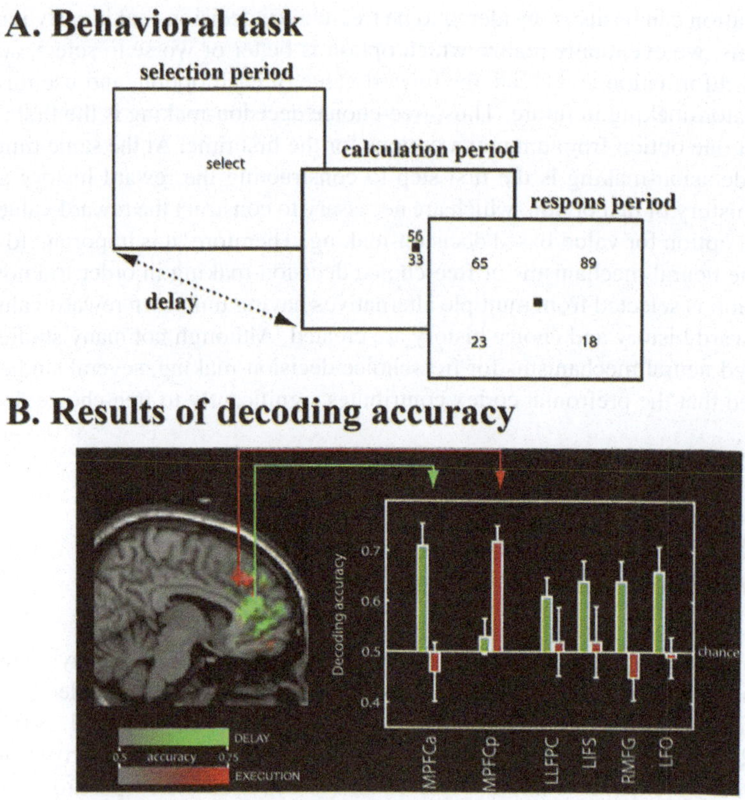

Fig. 7.2 (**a**) Behavioral task. (**b**) Brain regions encoding the subject's intention during the delay period (green) and the response period (red). Decoding accuracy was used to indicate which information these brain regions represented during two periods. MPFCa, anterior medial prefrontal cortex; MPFCp, posterior medial prefrontal cortex; LLFPC, left lateral frontopolar cortex; LIFS, left inferior frontal sulcus; RMFG, right middle frontal gyrus; LFO, left frontal operculum. From Haynes et al. (2007) with permission

himself. After a variable delay (2.7–10.8 s), two numbers were presented on the monitor (calculation period). Shortly after the presentation of two numbers, a response screen that included four numbers (one was the answer of addition, one was the answer of subtraction, and the remaining two were incorrect numbers) was presented (response period). While the response screen was presented, the subject was required to select a correct number based on his choice of calculation. The variable delay between the presentation of the cue word and the presentation of two numbers made the appearance of the two numbers unpredictable. This procedure forced the subject to maintain and prepare the chosen calculation during this period. Using this task, Haynes et al. (2007) obtained spatial patterns of fMRI signals recorded from a limited area within a local brain region, decoded information represented in these fMRI signals, and calculated decoding accuracies (how accurately MRI signals can predict the subject's decision regarding the calculation

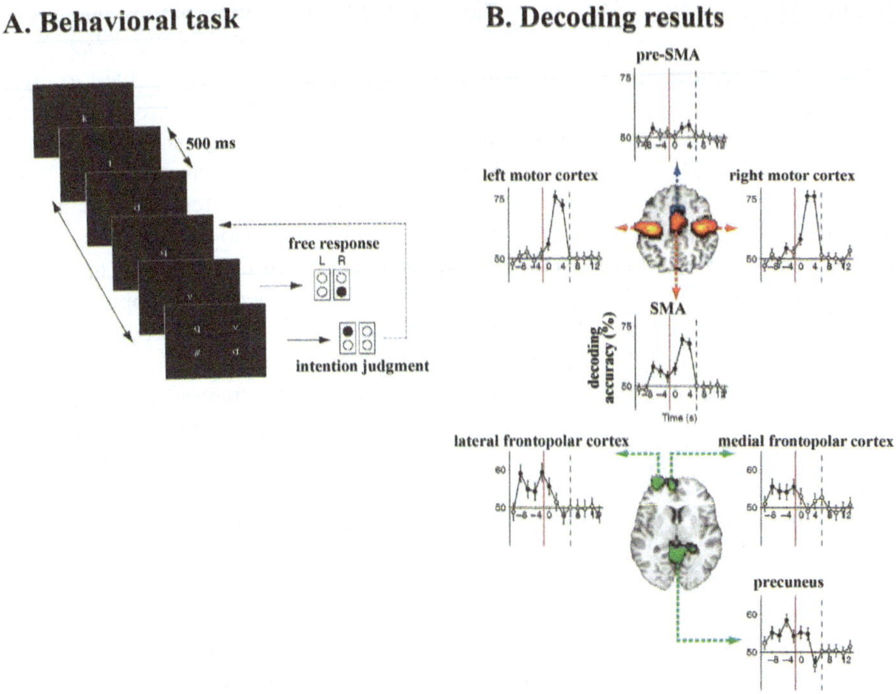

Fig. 7.3 (**a**) Behavioral task. (**b**) Temporal changes of decoding accuracy along a trial. The vertical red line indicates the earliest time when the subjects became aware of their choice. The dashed line indicates the onset of the next trial. Green areas and red areas show where decision outcome is decoded before and after the decision is made, respectively. From Soon et al. (2008) with permission

method) across several brain regions (Fig. 7.2b). As a result, during the delay period, high decoding accuracy was found in the anterior medial frontal cortex and the lateral prefrontal cortex. On the other hand, during the response period, high decoding accuracy was observed in the posterior medial frontal cortex (presumably the supplementary motor area). Since the subjects were not instructed which calculation they needed to select and since the four numbers were presented in random arrangement on the response screen, the lack of an explicit instruction prevented the subjects from preparing behavioral responses during the delay period. Therefore, the results were thought to indicate that the activity of the prefrontal cortex reflects the subject's own mental state or intention and that this mental state or intention determines the subject's subsequent choice or action in free-choice conditions.

Similarly, Soon et al. (2008) showed that the prefrontal cortex plays an important role in decision-making under free-choice conditions. They used a free-paced motor-decision task (Fig. 7.3a). In this task, the subjects were asked to freely decide to press one of two buttons by the left or right index finger. The subjects needed to gaze at the center of the screen, where alphabetical letters were presented one by one at every

500 ms. At any moment, when the subject wanted to press either button, they could freely decide to press that button (free response). At the same time, they were required to remember the letter presented on the screen when they decided to press the button. After they pressed either button, a response mapping screen with four letters was presented (intention judgment). During the response period, they were asked to indicate when they decided to press by selecting the button with corresponding letter. Soon et al. (2008) examined the temporal change of decoding accuracy, which indicates how accurately information regarding which button would be pressed was decoded from local patterns of fMRI signals in various brain regions (Fig. 7.3b). They found that the lateral and medial frontopolar cortex (area 10) and the medial parietal cortex (cortical area from the precuneus to the posterior cingulate cortex) exhibited high decoding accuracy before a conscious decision (when the selected letter had been presented on the screen) (lower figures of Fig. 7.3b), while the primary motor cortex and the supplementary motor area (SMA) exhibited high decoding accuracy in the execution phase of the button press (upper figures of Fig. 7.3b). Another study, in which the subjects needed to decide to press either left or right button when cued by an external trigger, showed that the frontopolar cortex was the first cortical area where the actual decision was made. Thus, these results also indicate that the prefrontal cortex, especially the frontopolar cortex, participates in free-choice decision-making.

Lau et al. (2004) also showed that the dorsolateral prefrontal cortex contributed to decision-making in free-choice conditions. They compared the magnitude of activation among three choice conditions. In the FREE condition, human subjects were required to select one of four targets presented on the monitor randomly and move the cursor to the selected target. In the SPECIFIED condition, the subjects were required to move the cursor to the target with the same features as the cursor. In the ROUTINE condition, the subjects were required to move the cursor to the highlighted target. By comparing fMRI signals among these three conditions, they concluded that the pre-supplementary motor area was closely associated with the free-choice of responses, because this area was activated only in the FREE condition. However, since the dorsolateral prefrontal cortex was activated in both the FREE and SPECIFIED conditions, the dorsolateral prefrontal cortex was also thought to be the brain area associated with the free-choice of responses.

Recently, Thiery et al. (2020) analyzed intracerebral electroencephalography recorded from surgical epilepsy patients who performed delayed oculomotor decision tasks in free-choice and instructed-choice conditions. Patients were asked to perform either rightward or leftward saccade from the central fixation point either freely or by instruction. They found distinct differences in the temporal dynamics of high-gamma (60–140 Hz) neural activities in frontal and parietal areas between free-choice and instructed-choice conditions. In free-choice trials, delayed and longer-lasting high-gamma activities were observed during the delay period. Based on the results, they indicated that sustained high-frequency gamma activities observed in the frontoparietal cortex drove neural process for free-choice. These results also support the prefrontal contribution in free-choice decision-making.

7.3 Prefrontal Contribution to Free-Choice Decision-Making: Animal Physiological Studies

As indicated by Soon et al. (2008), the prefrontal cortex, especially the frontopolar cortex and the dorsolateral prefrontal cortex, implicitly make decision well before the execution of an action. Therefore, the prefrontal cortex is thought to be a leading brain area for making spontaneous and self-generated behaviors and internally driven decision-making. Thus, the prefrontal cortex is thought to significantly contribute to decision-making in free-choice conditions, under which no external signal specifies a particular choice. Although few neurophysiological studies using animals have been performed to investigate neural mechanisms of decision-making in free-choice conditions in the prefrontal cortex (Watanabe et al. 2006; Watanabe and Funahashi 2007; Mochizuki and Funahashi 2014, 2016; Marcos and Genovesio 2016), these studies have also indicated that the prefrontal cortex contributes to free-choice decision-making.

On the other hand, the supplementary motor area and the pre-supplementary motor area have been shown to participate in internally driven motor actions (Mushiake et al. 1991; Halsband et al. 1994; Cunnington et al. 2002; Nachev et al. 2008). The cingulate motor area (Shima and Tanji 1998) and the dorsal anterior cingulate cortex (Walton et al. 2004) have also been shown to participate in behavioral responses based on the subject's own decision. Therefore, these areas are also thought to participate in free-choice decision-making. Recently, Sendhilnathan et al. (2021) compared neural activities of frontal eye field neurons between goal-directed saccade conditions and non-goal-directed saccade conditions, which are similar conditions to free choice of saccade targets. They found some differences in saccade-related activities between two conditions, such that the variability of spike discharge across trials was smaller in goal-directed condition and that saccade latencies were significantly longer in goal-directed condition, suggesting that neural activities are different between different conditions of decision-making even though the same selection is achieved. The neural mechanism for free-choice decision-making might be different from the mechanism for externally driven decision-making.

7.3.1 Neural Correlates of Free-Choice Decision-Making in the Prefrontal Cortex

Neurophysiological studies using monkeys have investigated the neural mechanism of free-choice decision-making in the prefrontal cortex and cortical three eye fields. Coe et al. (2002) used a free-choice delayed saccade task and examined single-neuron activity in the frontal eye field, the supplemental eye field, and the lateral intraparietal cortex. In their task, while monkeys maintained fixation at a fixation point, two visual targets (one located within the neuron's receptive field and the

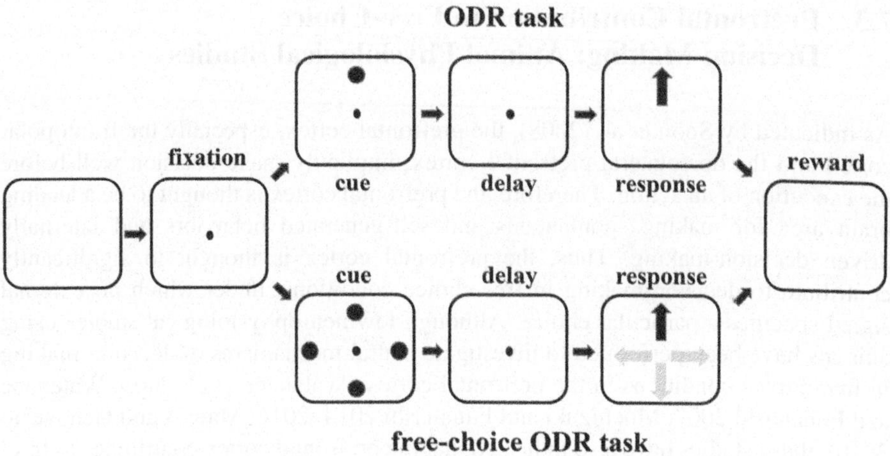

Fig. 7.4 Schematic drawings of the ODR task and a free-choice ODR task

other located outside of the receptive field) were presented simultaneously. After the fixation point was turned off, monkeys were required to freely choose either target and make a saccade to the selected target. They found not only an enhanced response to the presentation of the visual target but also stronger activation before the visual target presentation (*anticipatory bias*) in all three eye fields, when monkeys chose the target located within the receptive field. More neurons in the supplementary eye field exhibited stronger anticipatory bias and earlier start of activation than neurons in the frontal eye field and the lateral intraparietal cortex. Therefore, they concluded that the supplementary eye field plays more important roles in internally driven decision-making task using eye movements. The supplementary motor area and the pre-supplementary motor area have been shown to participate in internally driven skeleto-motor actions. This study showed the importance of internally driven decision-making in the supplementary eye field for eye movement control and the importance of anticipatory bias in internally driven decision-making.

7.3.2 Study by Watanabe et al. (2006)

Watanabe et al. (2006) examined neural mechanisms for free-choice decision-making in the prefrontal cortex. They examined single-neuron activities while monkeys performed an oculomotor delayed-response (ODR) task and a free-choice version of the ODR task (free-choice ODR task) (Fig. 7.4). In the free-choice ODR task, four identical visual cues were presented simultaneously during a 0.5-s cue period, and monkeys were required to choose one visual cue and make a saccade to the selected cue position after a 3-s delay period. The locations of the four visual cues were fixed during the experiment. Monkeys received the same amount of the same reward regardless of which visual cue they selected as a saccade target. They could freely choose any one of the four visual cues by themselves.

In the experiment, Watanabe et al. (2006) first tried to estimate when monkeys decided which visual cue would be used as a saccade target by comparing monkeys' behavioral performance between the ODR and the free-choice ODR tasks. To estimate the decision timing, saccade reaction times were compared between two tasks. In the ODR task, the saccade target was informed externally by the presentation of a single visual cue during the cue period (Fig. 7.4). Therefore, the monkey could prepare the saccade toward the target position in advance, probably during the delay period. However, in the free-choice ODR task, the monkey was able to decide the saccade target any time during the cue, the delay, or the response period. If the monkey decided the saccade direction during the response period, saccade preparation would start after the Go-signal presentation (offset of the fixation point). Therefore, reaction times would become longer in the free-choice ODR task than in the ODR task. However, if the monkey made the decision before the response period in the free-choice ODR task, it would have enough time to prepare the saccade similar to the ODR task. Therefore, reaction times would be similar between two tasks. In fact, reaction times in the free-choice ODR task were not significantly different from those in the ODR task. This result indicated that in the free-choice ODR task, the monkey decided saccade directions before the Go-signal presentation. Since the monkey probably prepared the saccade direction during the delay period, it is expected that the activity during the delay period exhibited selectivity to the saccade direction in the free-choice ODR task.

The analysis of single-neuron activity confirmed the result obtained by the comparison of saccade reaction times between two tasks. Three typical task-related activities have been observed in prefrontal neurons while monkeys performed the ODR task. These are cue-, delay-, and response-period activities, and majority of these activities exhibited directional selectivity. Since directional selectivity of these activities is an essential determinant regarding where the monkey made a saccade during the response period in the ODR task, characteristics of directional selectivity of these task-related activities would also be an important factor to determine the saccade direction in the free-choice ODR task. Therefore, directional selectivity of three task-related activities was compared between the ODR task and the free-choice ODR task (Fig. 7.5) (Watanabe and Funahashi 2007). As a result, in neurons having directional cue-period activity in the ODR task (Fig. 7.5-1a), directional selectivity was diminished in the free-choice ODR task (Fig. 7.5-1b). In addition, these neurons did not exhibit directional selectivity in delay-period activity in both tasks. These results suggest that neurons having directional cue-period activity in the ODR task did not contribute to the decision-making of the saccade target selection in the free-choice ODR task. Prefrontal neurons having response-period activity exhibited significant activation whenever monkeys made saccades to the neurons' response fields, regardless of which task monkeys performed (Fig. 7.5-2). These neurons did not exhibit directional selectivity in delay-period activity. Therefore, these results suggest that neurons having directional response-period activity in the ODR task do

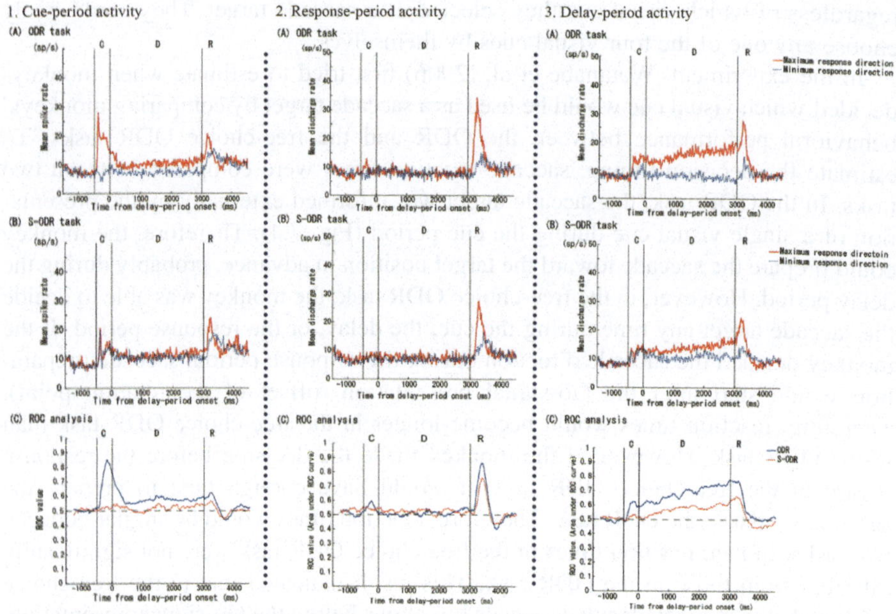

Fig. 7.5 Temporal patterns of neural activities of neurons having cue-period activity, those having delay-period activity, and those having delay-period activity during the ODR task (**a**) and the free-choice ODR (S-ODR) task (**b**). Red lines in figures (**a**) and (**b**) indicate temporal patterns of neural activities during correct trials in which the visual cue was presented at the position where neurons exhibited the maximum or the best response (best trials). Similarly, blue lines in figures (**a**) and (**b**) indicate temporal patterns of neural activities during correct trials in which the visual cue was presented at the position where neurons exhibited the minimum or the worst response (worst trials). The figures (**c**) show temporal changes of ROC values along a trial. ROC values indicate the difference of neural activities between best trials and worst trials for the ODR task (blue) and the free-choice ODR (S-ODR) task (red). C, D, R in each figure indicate the cue, delay, response period, respectively. From Watanabe and Funahashi (2006) with permission

not contribute to the decision-making of the saccade target selection but contributed to the saccade performance in the free-choice ODR task.

On the other hand, prefrontal neurons having directional delay-period activity in the ODR task (Fig. 7.5-3a) exhibited the same directional preference in the free-choice ODR task (Fig. 7.5-3b). Delay-period activity of these neurons exhibited a gradual increase toward the response period in both tasks. Although the degree of gradually increasing activation was weaker in the free-choice ODR task compared with the ODR task, the difference in delay-period activity between preferred and non-preferred directions was evident in the ROC analysis in the free-choice ODR task (Fig. 7.5-3c). These results indicate that the directional selectivity of delay-period activity plays an important role in determining the saccade direction in the free-choice ODR task. Further, the gradual development of delay-period activity in the free-choice ODR task suggests that the decision of the saccade direction is made during the delay period and that neural operations necessary for decision-making, such as accumulating sensory information, integrating sensory, motivational, and emotional information, and transforming those into motor information, are performed during the delay period (Watanabe and Funahashi 2007).

Thus, spatially selective delay-period activity (Watanabe et al. 2006; Watanabe and Funahashi 2007) and preparatory bias (Coe et al. 2002) play important roles in the animal's decision for the saccade direction in free-choice saccade tasks. However, in these studies, a fixed set of spatial targets were repeatedly presented as saccade targets in a block of free-choice trials. Therefore, monkeys often exhibited a strong positional bias, such that monkeys chose the same target repeatedly for several consecutive trials, because monkeys received the same reward regardless of which of two or four targets they chose. Coe et al. (2002) presented two visual targets at the same locations throughout the entire recording period of a given neuron. To prevent the monkey's selection of the same target repeatedly, they introduced specific reward schedules, such that, when the monkey selected the same target repeatedly, the amount of reward increased in the first few trials, became a peak value, and then began to decrease in subsequent trials. As a result, the monkey chose the same target for several consecutive trials until the amount of reward became a peak, then changed the target to obtain a full amount of reward. Therefore, Coe et al. (2002) considered that anticipatory bias toward the specific target would be added on neural processes of voluntary attention or motor preparation, such that the monkey had decided the saccade direction and allocated their attention to a specific target by the time when the visual targets appeared. Similarly, Watanabe et al. (2006) and Watanabe and Funahashi (2007) used a fixed four target positions throughout the experiment. They set the maximum number to permit repetitive selection of the same target. Monkeys could receive a reward every trial until they selected the same target for four consecutive trials, but would not receive a reward when they chose the same target more than five times. This procedure forced monkeys to choose another target after repetitive selection of the same target. However, since monkeys maintained the positional bias for several consecutive trials, this experimental procedure would not be ideal to examine the neural mechanism of decision-making in the free-choice condition.

7.3.3 Studies by Mochizuki and Funahashi (2014, 2016)

To examine the neural mechanism of decision-making in the free-choice condition more accurately without the effects of reward history, choice history, and anticipatory bias, Mochizuki and Funahashi (2014, 2016) developed a free-choice saccade task, in which monkeys were required to choose either of two targets freely by a memory-guided saccade (Fig. 7.6). In the free choice task (FCT), after a 1-s fixation period, eight potential saccade targets were presented as small "+" signs on the monitor (pre-cue period) and remained on it until the end of the response period. After a 1-s pre-cue period, visual cues were presented at two positions which were randomly selected from eight potential target positions (cue period). Then, the delay period was introduced. At the end of the delay period, the fixation point was

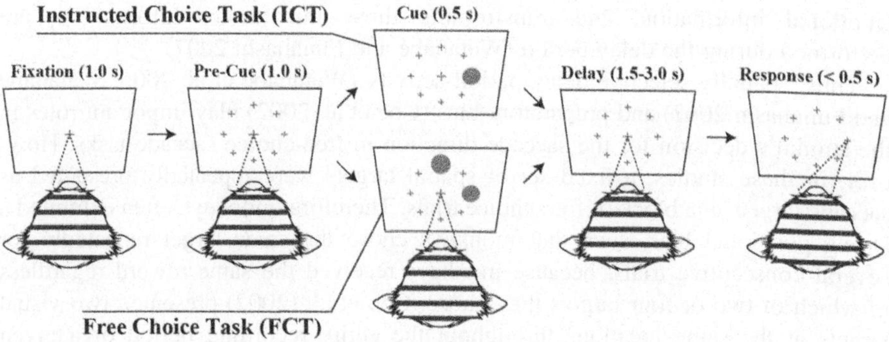

Fig. 7.6 Behavioral tasks used by Mochizuki and Funahashi (2014, 2016)

extinguished and monkeys were required to make a saccade to either of two cue positions to obtain the reward (response period). Monkeys selected either cue position freely by themselves. Since the positions of two visual cues changed from trial to trial, monkeys could not predict where visual cues would be presented in any given trial. In addition, monkeys obtained the same reward regardless of which visual cue they chose. Therefore, these procedures could eliminate the monkey's behavioral bias such as choosing the same target repeatedly. In addition, to compare prefrontal activity between free-choice conditions and forced-choice conditions, an instructed choice task (ICT) was introduced as a control task (Fig. 7.6). The ICT task was the same as the ODR task. In the ICT task, only one visual cue was presented during the cue period and the monkey was required to make a saccade to the cue position after the delay period.

7.3.4 Choice-Predictive Activity in the Prefrontal Cortex

Using these two tasks, Mochizuki and Funahashi (2016) examined activity related to decision-making in free-choice conditions in the dorsolateral prefrontal cortex. They found that, in 65% of examined prefrontal neurons, cue-period activity was significantly enhanced when the monkeys chose the visual cue presented within the neuron's receptive field (T_{in} condition) as a saccade target during the response period. However, when they chose the visual cue presented outside of the receptive field either in the ipsilateral visual field (T_{ipsi}), in the contralateral visual field (T_{contra}), or opposite to the T_{in} position (T_{opp}) as a saccade target, an enhancement of cue-period activity was not observed, even though unselected visual cue was presented within the neuron's receptive field during the cue period (Fig. 7.7a).

As is seen in Fig. 7.7a, Mochizuki and Funahashi (2016) found gradually increasing activation just before the visual cue presentation in the FCT task (pre-cue activity). The development of pre-cue activity began several 100 ms before the presentation of the visual cue only in the FCT trials in which monkeys eventually

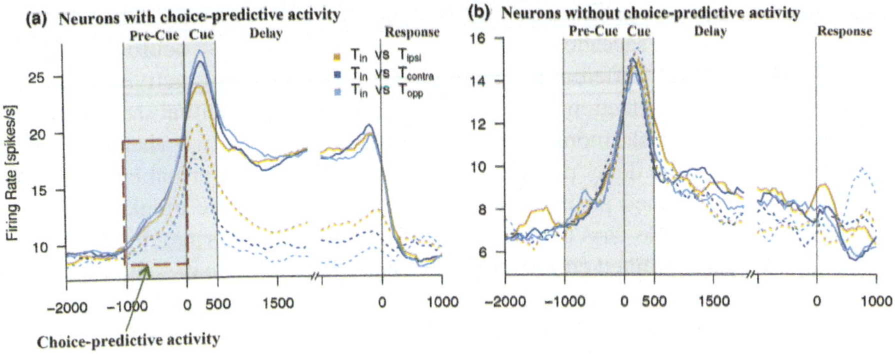

Fig. 7.7 (**a**) Activities of prefrontal neurons with choice-predictive activity. Significant choice-predictive activity and delay-period activity were observed when the monkey selected the neuron's preferred direction (solid lines), regardless of wherever the remaining cue was presented (dotted lines). (**b**) Activities of prefrontal neurons without choice-predictive activity. The same activity was observed regardless of whether the monkey selected the neuron's preferred direction (solid lines) or non-preferred directions (dotted lines). T_{in} indicates the monkey's selection of the neuron's preferred direction (solid lines) and T_{ipsi}, T_{contra}, and T_{opp} indicate the monkey's selection of the neuron's non-preferred directions (dotted lines)

chose the visual cue presented within the neuron's receptive field (T_{in}) as a saccade target. The appearance of pre-cue activation seemed to be related to monkey's eventual choice of T_{in} as a saccade target during the response period. This suggests that the activity that appeared just before the cue presentation predicts the animal's eventual choice of the saccade direction after the delay period. Therefore, this pre-cue activity was named *choice-predictive activity* (Mochizuki and Funahashi 2016).

Prefrontal neurons exhibited choice-predictive activity not only showed differential magnitude of cue-period activity between the selection of T_{in} and other targets but also showed significantly higher discharge rates during the delay period when monkeys chose the T_{in} as an eventual saccade target compared with when monkeys chose other targets (Fig. 7.7a). Differential magnitude of delay-period activity between T_{in} and other conditions (T_{ipsi}, T_{contra}, and T_{opp}) was maintained until the end of the delay period. Directional selectivity of delay-period activity in the ODR task is thought to be a neural mechanism for temporary storage of the spatial information of the visual cue. Therefore, delay-period activity observed in neurons having choice-predictive activity is thought to maintain spatial information of T_{in} during the delay period and this activity is utilized to perform a saccade toward T_{in} during the response period.

On the other hand, not all prefrontal neurons having cue-period activity exhibited choice-predictive activity. Neurons that did not exhibit choice-predictive activity but exhibited significant cue-period activity were present in the prefrontal cortex (35% of examined neurons) (Fig. 7.7b). These neurons exhibited significant cue-period activity whenever the visual cue was presented within the neuron's visual receptive field, regardless of the position where paired visual cue was presented (Fig. 7.7b).

These neurons did not exhibit enhancement of cue-period activity when T_{in} was selected as an eventual saccade target. In addition, although these neurons exhibited delay-period activity, differential magnitude of delay-period activity was not observed between the selection of T_{in} and other targets as an eventual saccade target.

Thus, in the FCT task, monkeys could freely select one of two visual cues as a saccade target after the delay period. Random selection of two visual cue positions from eight predetermined positions in every trial suppressed the monkey's behavioral bias in the saccade target selection. While monkeys performed the FCT task, prefrontal neurons exhibited choice-predictive activity before the presentation of the visual cue. Choice-predictive activity was observed only when the visual cue presented within the neuron's receptive field was selected as a saccade target after the delay period. Neurons having choice-predictive activity exhibited enhanced cue-period activity and directional delay-period activity encoding information regarding the visual cue presented within the neuron's receptive field. Thus, the presence or the absence of choice-predictive activity seems to play an important role to determine whether or not the visual cue presented within the neuron's receptive field is selected as a target for the behavioral response in the free-choice saccade task.

7.3.5 *Anticipatory Bias and Pre-Stimulus Activity*

Coe et al. (2002) had been reported *anticipatory bias* in prefrontal activity in a free-choice saccade task. Anticipatory bias was an enhancement of anticipatory activity during the visual period and was observed only when the monkey made a saccade toward the target presented in the neuron's receptive field. They considered that anticipatory bias represents a neural process of voluntary attention or motor preparation toward a specific saccade target. Both anticipatory bias and choice-predictive activity can be observed before the presentation of the visual cue. However, choice-predictive activity is not the same as anticipatory bias. Coe et al. (2002) presented the visual cues at the same two positions repeatedly during a block of trials. However, in Mochizuki and Funahashi (2016), since two positions of the visual cues were selected randomly from eight predetermined positions in every trial, it was impossible for the monkeys to predict where two visual cues would be presented in any given trial. Therefore, choice-predictive activity would not be a result of voluntary attention or motor preparation toward a specific saccade target.

Activity similar to choice-predictive activity was reported in prefrontal neurons by Marcos and Genovesio (2016). They used a strategy task, in which the monkey needed to perform a visually-guided saccade to one of three visual targets in either a repeat-stay strategy (make a saccade to the same target as the one selected in the preceding trial) or a change-shift strategy (make a saccade to a different target from the one selected in the preceding trial). The monkey needed to choose either strategy depending on whether the same imperative stimulus was presented repeatedly (take repeat-stay strategy) or not (take change-shift strategy). Two strategies were introduced randomly in any given trial. When the strategy shifted from the repeat-stay strategy to the change-shift strategy, in the first trial, the monkey was required to select one of the remaining two visual targets without knowing which visual target was associated with a reward. Therefore, they considered that the selection at the first trial of the change-shift strategy would be similar to the selection in the free-choice condition. In the first trial of the change-shift strategy, they observed a significantly different magnitude of prefrontal activity during approximately 200 ms before the presentation of the imperative stimulus when the monkey eventually selected the visual target in the neuron's receptive field. They called this activity *pre-stimulus activity*. Pre-stimulus activity was observed in similar situation as choice-predictive activity was observed and exhibited similar features as choice-predictive activity exhibited.

Thus, when monkeys are asked to select one visual cue from multiple cues as the target for a saccade in a free-choice saccade task, choice-predictive activity or pre-stimulus activity can be observed in prefrontal visual neurons several 100 ms before the presentation of the visual cue only when the visual cue presented within the neuron's receptive field becomes the target of the saccadic eye movement after the delay period. The appearance of choice-predictive activity is associated with the enhancement of visual responses, the presentation of sustained directional delay-period activity, and the monkey's eventual choice of the visual cue as the saccade target in the free-choice saccade tasks. Therefore, choice-predictive activity and pre-stimulus activity are thought to play an important role in decision-making for the selection of the saccade target in the free-choice saccade tasks.

7.3.6 What Causes Choice-Predictive Activity?

Choice-predictive activity was observed several 100 ms before the presentation of the visual cue in the free-choice saccade task. During the pre-cue period, only a fixation point was presented on the monitor and the monkey only kept looking at the fixation point quietly. No external stimulus except the fixation point was presented to the monkey and the monkey exhibited no overt movement. Since the position of the visual cue was randomly selected from multiple predetermined positions in every trial, the monkey could not predict where the visual cue would be presented in any given trial. Therefore, it is difficult to consider that choice-predictive activity is caused by some external stimulus or body movement, or voluntary attention to a particular stimulus or position, or preparation of a movement.

Most prefrontal neurons have spontaneous activity of 1–10 spikes/s (Fuster 1973; Quintana et al. 1988). Spontaneous activity is irregular and arrhythmic in nature. Such irregularity and arrhythmicity of spontaneous activity in individual neurons may cause variability in their responses and outputs. Fluctuation of spontaneous activity is thought to reflect the fluctuation of the neuron's active state, such that transient increase or decrease of spontaneous discharge rates reflects transient increase or decrease of the neuron's active state, respectively. Therefore, irregular and arrhythmic fluctuations of spontaneous activity of prefrontal neurons are considered to be a source of choice-predictive activity. Choice-predictive activity might be caused by a transient increase of the neuron's spontaneous activity that occurred just before the presentation of visual cues.

If this is the case, the following mechanism might explain the effect of choice-predictive activity on subsequent neural processing. Transient increase of spontaneous activity occurs in neurons having cue-period activity just before the visual cue presentation. This transient activation causes additive effect to the response to the visual cue presentation and produces an enhancement of cue-period activity in these neurons. Prefrontal neurons having cue-period activity often exhibit delay-period activity with similar directional selectivity or send spatial information to the neurons having only delay-period activity. Since most of prefrontal neurons having cue-period activity exhibit directional selectivity, the enhancement of cue-period activity occurred in prefrontal neurons having a particular directional selectivity produces a biasing effect in directional selectivity in subsequent information processing, especially during the delay period and this biasing effect determines the direction of the saccade during the response period.

A biasing effect by the spontaneous fluctuation of baseline activity on the subsequent choice has been reported (Platt and Glimcher 1999; Shadlen and Newsome 2001). For example, Shadlen and Newsome (2001) examined activity of lateral intraparietal (LIP) neurons while monkeys performed a perceptual decision-making task using random-dot motion. They showed that, when the monkey was difficult to discriminate the direction of overall motion direction because of very low coherence in motion directions among dots, LIP neurons often exhibited higher discharge rates just before the onset of motion stimuli in the trials in which the monkey eventually chose the neuron's preferred motion direction. They suggested that this higher discharge rate just before the onset of the motion stimuli was caused by the spontaneous fluctuation of baseline activity and that the transient activation of baseline activity biased the result of subsequent competition among LIP neurons, each of which represents different directions of visual motion. As a result of this competition, the preferred motion direction of winning neurons becomes the final decision. Thus, when the discrimination between stimuli is difficult because of the ambiguity of these stimuli, spontaneous fluctuation of baseline activity biases the decision of which stimulus is detected. By a theoretical approach using a network model with integrate-and-fire neurons, Rolls and Deco (2011) showed that biasing effect can be produced by fluctuation of noise inputs to the network and that this biasing effect affects the output of the network. These studies indicate that a small change of the activation level of the network produced by the fluctuation of spontaneous neural activity affects the competition among networks, each of which participates in different information processing, and biases the selection of the network.

Recently, Waschke et al. (2021) discussed the importance of variability in neural activity, local field potentials, EEG/MEG powers, and BOLD signals during task performances and indicated that neural variability represents a key and undervalued dimension to understanding brain–behavior relationships. They also suggested that inter-individual differences in behavioral strategies might be reflected in the momentary level of neural variability. Therefore, individual differences in free-choice decision-making may be explained by inter-individual variability of neural activity in the momentary level.

7.3.7 Functional Relations Between Choice-Predictive Activity and Delay-Period Activity

Prefrontal neurons showing choice-predictive activity not only exhibited enhanced directional cue-period activity but also exhibited directional delay-period activity in the free-choice saccade task (Fig. 7.7a). Directional selectivity was the same between cue-period activity and delay-period activity. However, neurons that did not show choice-predictive activity exhibited neither directional cue-period activity nor directional delay-period activity in the free-choice saccade task (Fig. 7.7b). These observations suggest that some common cellular mechanism causes both choice-predictive activity, enhanced cue-period activity, and delay-period activity in the same prefrontal neuron.

Mochizuki and Funahashi (2016) indicated that prefrontal neurons showing choice-predictive activity had higher sustainability of their firing patterns. The degree of sustainability in firing patterns was predicted by Pearson's correlation coefficient between the lengths of successive inter-spike intervals obtained during the fixation period (control period). A stronger serial correlation of inter-spike intervals indicates that, once the neuron becomes an active state, this state tends to be maintained for a while. Higher sustainability in neural firings is observed in prefrontal neurons exhibiting choice-predictive activity. Therefore, a transient increase of spontaneous discharge by some noise inputs causes these neurons in active state and this state is maintained for a while. If the active state of the neuron is produced just before the cue presentation and if this active state is maintained until sometime after the start of the cue period, choice-predictive activity and an enhancement of cue-period activity would be observed.

Prefrontal neurons having choice-predictive activity also exhibit delay-period activity. Delay-period activity is tonic sustained activation during the delay period. Choice-predictive activity is related to the neuron's higher sustainable firing pattern. These observations suggest that choice-predictive activity and delay-period activity are caused by a common cellular mechanism that is able to maintain an active state for some time when an excitatory input is received. Cortical neurons receive excitatory and inhibitory synaptic inputs from other cortical neurons and form local recurrent networks. Local recurrent networks are balanced between recurrent excitation and inhibition, which allows to generate sustained excitation or inhibition during a certain period. Shu et al. (2003) used slice preparations of the ferret prefrontal cortex and showed that cortical local circuits were in fact operated through a proportional balance between excitation and inhibition generated by local recurrent connections. These circuits generated self-sustained activity which was turned on and off by excitatory and inhibitory synaptic inputs. Theoretical studies suggested that tonic sustained activation like delay-period activity is maintained by excitatory synaptic reverberation in a local recurrent circuit and that the stability of sustained activation could be achieved if reverberation is mediated by NMDA receptors at recurrent synapses (Wang 2001). Therefore, having both self-sustained mechanism in local recurrent networks and excitatory synaptic reverberation mediated by NMDA receptors may be important features of prefrontal neurons that exhibit sustained delay-period activity and choice-predictive activity.

By these intrinsic cellular mechanisms, a transient increase of spontaneous discharge triggers to produce activated state in a local neural circuit and an active state is maintained stably in this circuit for some time (e.g., 100–200 ms). If this active state is produced just before the visual cue presentation and is maintained for some time until the early phase of the cue period, the active state could enhance subsequent responses to the visual cue. An enhanced effect observed in cue-period activity would be more obvious when the visual cue is presented in the neuron's receptive field. An enhancement of cue-period activity would cause further maintenance of an active state during the delay period and could generate tonic sustained delay-period activity. Thus, the neuron's higher sustainable firing pattern might be related to the facts that a group of prefrontal neurons exhibit choice-predictive activity, enhanced cue-period activity, and sustained delay-period activity with the same directional selectivity as cue-period activity in the free-choice saccade task with delay.

7.4 Neural Mechanism for Free-Choice Decision-Making in the Prefrontal Cortex

A possible mechanism for free-choice decision-making can be explained based on the features of choice-predictive activity (Mochizuki and Funahashi 2016) and pre-stimulus activity (Marcos and Genovesio 2016) and their possible source (Fig. 7.8). As described before, prefrontal neurons seem to have distinctive features in firing patterns. First, every prefrontal neuron exhibits irregularly fluctuating spontaneous firing. This irregular fluctuation of spontaneous firing can be observed at any time in any given trial during the task. Second, some prefrontal neurons have intrinsic mechanisms that produce a specific sustained firing pattern. In these prefrontal neurons, an active state would be maintained for several 100 ms by self-sustained mechanism of local recurrent connections once an excitatory input is received. Therefore, these neurons tend to exhibit tonic sustained firing patterns after a transient increase in spontaneous firing rates. These are important features of prefrontal neurons that exhibit tonic sustained activation during the delay period.

While monkeys perform the ODR tasks, most prefrontal neurons exhibit directionally selective activity during the cue, delay, and response periods (Funahashi et al. 1989, 1990, 1991; Funahashi 2014, 2015, 2017a, b). During the

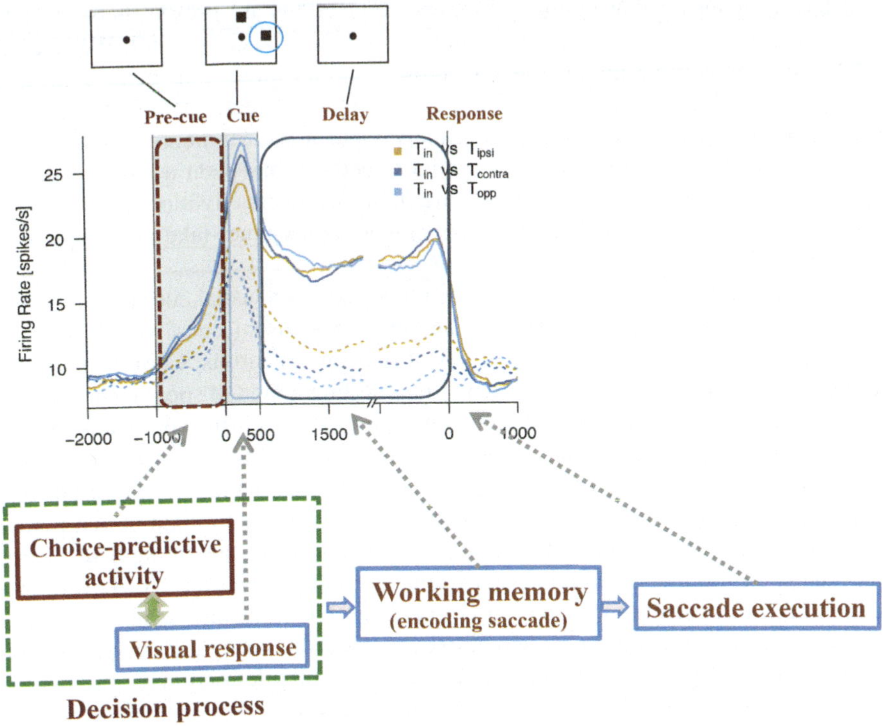

Fig. 7.8 A diagram explaining how choice-predictive activity could contribute to decision-making in free-choice conditions

pre-cue fixation period in the free-choice ODR task, monkeys can see no external stimulus except the central fixation point. Therefore, prefrontal neurons are in the baseline state during the pre-cue fixation period across trials. However, every prefrontal neuron has irregular and arrhythmic spontaneous firings and transient increase or decrease of spontaneous firings constantly occurs at any given time during the pre-cue fixation period. These transient and irregular changes of spontaneous firings reflect an increase or decrease in baseline active state in a prefrontal neuron. If transient increase of spontaneous firings happens to occur just before the visual cue presentation in a prefrontal neuron having directional cue-period activity, this transient increase of spontaneous firings increases the baseline state in this neuron, and activated baseline state is maintained by intrinsic mechanisms for a while. If this active state is maintained until sometime after the start of the cue period, the response to the visual cue would be enhanced. Prefrontal neurons having cue-period activity are known to have visual receptive fields. In the free-choice ODR task, two visual cues are presented simultaneously and either of two visual cues is presented within the neuron's receptive field. Therefore, the neuron would respond more quickly and more strongly to the visual cue presented in the neuron's visual receptive field if the neuron is in active state.

It has been shown that mutual inhibitory connections are present between prefrontal neurons exhibiting different spatial selectivity (Rao et al. 1999; Wang et al. 2004a, b). Competitive interactions are present among prefrontal neurons having different spatial selectivity. Therefore, competitive interactions among neurons having different spatial selectivity eventually yield a winner neuron or a winner group of neurons having a particular spatial selectivity based on a winner-take-all principle. Prefrontal neurons that exhibit faster and stronger activation in response to the presentation of the visual cue become winners in a winner-take-all competition (Compte et al. 2000; Wang 2008).

As shown in Fig. 7.8, in the free-choice ODR task, two visual cues are presented simultaneously at different positions. Therefore, two visual cues would generate cue-period activity in two different groups of prefrontal neurons. Choice-predictive activity is a transient activation caused by irregular fluctuation of spontaneous firings of prefrontal neurons. If irregular fluctuation of spontaneous firings produces choice-predictive activity in one group of prefrontal neurons and if the choice-predictive activity causes an enhancement of cue-period activity only in this group of prefrontal neurons, enhanced cue-period activity produces a bias in the strength of cue-period activity between two groups of neurons. This bias affects the strength of neural representation for directional information among groups of neurons. As a result of a winner-take-all competition, one group of prefrontal neurons that exhibit enhanced cue-period activity exhibit faster and stronger activation during the cue period and become winners of this competition. And, the directional information represented by these neurons becomes the final output for the direction of the saccadic eye movement. Thus, choice-predictive activity is an important neural activity that affects the decision of the final output in free-choice decision-making. Transient activation occurred in spontaneous activity just before the cue presentation and a winner-take-all principle in competitive interactions among neuron groups with different spatial selectivity would be two main mechanisms to understand how a decision is made in the decision-making of the free-choice ODR task.

7.5 Prefrontal Contribution to Other Types of Decision-Making

Neural mechanisms for other types of decision-making than free-choice decision-making have also been examined in the prefrontal cortex. Using monkeys, several behavioral paradigms have been used to examine prefrontal involvement in decision-making. These behavioral paradigms include perceptual decision-making, value-based decision-making, decision-making under competitive conditions, and decision-making in conflict conditions.

7.5.1 *Perceptual Decision-Making*

A variety of behavioral paradigms have been used to examine neural mechanisms for perceptual decision-making. In one task, subjects are asked to report the direction of the visual motion when they see random dot motions in different levels of motion coherences of dots (e.g., Shadlen and Newsome 2001). In other tasks, subjects need to identify whether a picture shown is a dog or a cat when they see a picture distorted in various degrees (e.g., Freedman et al. 2001, 2002). Or, subjects need to report whether the frequency of the current vibration stimulus is the same or not as that of the preceding stimulus (e.g., Romo et al. 2002; Romo and Salinas 2003). In these behavioral paradigms, a stimulus with various degrees of ambiguity is presented to the subject and the subject is asked to identify the stimulus itself (e.g., dog or cat) or a feature of the stimulus (e.g., motion direction). In the most difficult condition to discriminate or identify the stimulus, since a stimulus with the highest degree of ambiguity is presented to the subject (e.g., completely random motion), it is impossible for the subject to identify the stimulus or its feature. Therefore, decision-making in this condition is similar to decision-making in free-choice conditions.

The perceptual decision-making paradigm has been widely used to examine neural mechanisms of decision-making. Single-neuron recording studies have been performed in the lateral intraparietal cortex (areas LIP) while monkeys performed motion discrimination tasks with a saccade response under various coherent motion conditions (Shadlen and Newsome 2001; Mazurek et al. 2003; Huk and Shadlen 2005; Hanks et al. 2006; Churchland et al. 2008; Kiani and Shadlen 2009; Shadlen and Kiani 2013). The neural mechanisms of perceptual decision-making in the motion discrimination task have also been examined in human neuroimaging studies using fMRI (Kayser et al. 2010). Theoretical considerations based on the results obtained from single-neuron studies (Gold and Shadlen 2007; Furman and Wang 2008; Wang 2008) and a comparison of perceptual decision-making among rodents, monkeys, and humans (Hanks and Summerfield 2017) have also been performed.

Neural mechanisms of perceptual decision-making have also been examined in the monkey prefrontal cortex using the motion discrimination task with a saccade response. Kim and Shadlen (1999) used the perceptual decision task with a two-alternative forced-choice motion discrimination. In this task, monkeys were required to discriminate the direction of random dot motions presented on a monitor and report the perceived direction using a saccadic eye movement toward either of two visual targets. While monkeys looked at the fixation point, two targets (one was within the neuron's response field and the other was outside the response field) were presented. After 200–300 ms, randomly moving dots appeared for 1 s at the outside of the neuron's response field (motion viewing period). The dot motion was directed toward either of the two targets. Motion strength (motion coherence) was varied from trial to trial by systematically changing the percentage of coherently moving dots. After the delay period of 0.5–1.5 s, the fixation point was extinguished and monkeys were required to indicate the direction of the dot motion by making a saccade toward one of the targets. Kim and Shadlen (1999) recorded single-neuron activity from the posterior portion of the principal sulcus and the frontal eye field. When monkeys performed a memory-guided saccade task which was introduced as a control task, neurons exhibited differential activity during the motion viewing period and the delay period, such that stronger activation was observed when impending saccades were directed toward the response field, while weaker activation was observed when impending saccades were directed away from the response field. When monkeys performed the motion discrimination task, significant difference in activity was not observed during the motion viewing period, but observed during the delay period. The magnitude of prefrontal activity during the delay period was affected not only by the direction of the impending saccade but also by the sensory information which determined the direction of the saccade. The magnitude of prefrontal activity during the delay period was larger in trials in which motion strength was larger (e.g., 50% in motion coherence) than in trials in which motion strength was small (e.g., 0% in motion coherence). Kim and Shadlen (1999) found that the activity of most neurons (86%, 76/88) reliably predicted the monkey's choice of the saccade direction during the motion-viewing period or the subsequent delay period. They also found that 60% (53/88) of neurons predicted the monkey's decision during the motion viewing period and reflected combined information of motion directions and saccade directions. In addition, the monkey's decision can be predicted by prefrontal activity occurred 100–200 ms after the onset of random dot motion (predictive activity). Further, these predictive activities were observed even in the most difficult conditions (0% in motion coherence) and the magnitude of predictive activity increased as the motion strength increased (behaviorally easier condition).

Ding and Gold (2012) examined single-neuron activity in the frontal eye field while monkeys performed perceptual decision-making using a motion discrimination task with a saccade response. They observed predictive activity in the frontal eye field with similar features as those of prefrontal activity. Predictive activity has also been observed in the parietal cortex (area LIP) while monkeys performed the same behavioral task (Shadlen and Newsome 2001). In area LIP, predictive activity began early in the motion-viewing period. This activity was observed even in the most difficult condition (e.g., 0% in motion coherence) and the timing and magnitude of the activity were affected by the strength of motion coherence (Shadlen and Newsome 2001).

In perceptual decision-making tasks, subjects are shown randomly moving dots for a short period of time and asked the direction of moving dots by selecting one of two targets by eye movements. One target is usually set in the neuron's response field and the other is placed outside the response field. And, the dot motion is directed toward one of two targets. Neurons in area LIP have been known to exhibit directional selectivity to visual motion stimuli. When different percentages of dots moving toward different directions coherently are presented to the subjects, different proportions of LIP neurons, each of which has specific direction selectivity to visual motion, are activated simultaneously. Predictive activity can be observed in one population of LIP neurons even in the most difficult condition to identify the motion direction and the magnitude of predictive activity increases as the motion strength increases. Therefore, a final input to LIP neurons was determined by comparing inputs from different populations of sensory neurons with different direction selectivity. Thus, in the prefrontal cortex and the parietal cortex, a decision regarding which direction of visual motion the subject perceived would be made by comparing the magnitude differences of input activity from two groups of sensory neurons (e.g., MT neurons), one of which is selective to one direction of visual motion and another of which is selective to another direction of visual motion (Kim and Shadlen 1999; Gold and Shadlen 2000; Shadlen and Newsome 2001).

The integration and accumulation of input activity to prefrontal and parietal neurons are also important components for perceptual decision-making (Gold and Shadlen 2000, 2007; Mazurek et al. 2003; Huk and Shadlen 2005). Shadlen and Kiani (2013) used bounded evidence accumulation to explain neural mechanisms of perceptual decision-making in the LIP. The area MT plays essential role in visual motion perception. Each MT neuron exhibits selectivity to the visual motion direction and has a preferred direction. Firing rates of MT neurons are proportional to the motion strength toward their preferred direction. Firing rates of MT neurons in a given condition can be assessed by an average of spontaneous firing rates and the firing rates proportional to the strength of visual motion toward their preferred direction. LIP neurons receive the outputs of MT neurons and exhibit dynamic changes in firing rates 100–200 ms after the onset of the motion viewing period. Since this dynamic firing change in LIP neurons are proportional to the motion strength of a particular motion direction and predicts the direction of the forthcoming saccade, this firing change is considered as a decision-related dynamic change. Since LIP neurons receive outputs from MT neurons having different preferred motion directions, LIP neurons accumulate information from two groups of MT neurons, each of which has different preferred visual motion direction, and the average firing rates of LIP neurons approximate the integration of the difference between the averaged firing rates of two groups of MT neurons. If firing rates of LIP neurons resulted from the integration of this difference reach a certain boundary or threshold, it leads to a decision.

Predictive activity observed even in the most difficult condition predicted the monkey's behavioral response (saccade direction). In addition, this activity was observed 100–200 ms after the presentation of visual motion stimuli in both prefrontal and LIP neurons. The most difficult condition in the perceptual decision-making paradigm is the condition impossible to discriminate or identify the stimulus or the stimulus feature and is similar condition as free-choice decision-making. Therefore, the predictive activity observed in the most difficult condition of the perceptual decision-making tasks might be the same activity as choice-predictive activity, pre-stimulus activity, or anticipatory bias observed in prefrontal neurons during free-choice decision-making tasks. Predictive activity observed in the perceptual decision-making tasks may also correspond to a transient activation caused by irregular fluctuation of spontaneous discharges. Similar to the mechanism to make a decision in the free-choice ODR task, the decision of the saccade direction in the most difficult condition of the perceptual decision-making tasks would be determined by the winner-take-all competition between two groups of neurons each of which encodes different saccade directions.

7.5.2 Value-Based Decision-Making

The paradigm using value-based decision-making has also been used to examine neural mechanisms of decision-making in the prefrontal cortex. Decision-making is an evaluation process for making a particular choice from a set of alternatives. A particular choice usually predicts a particular outcome or expects a value associated with the choice. The outcome associated with the choice is sometimes beneficial and rewarding for the subject, but is sometimes risky and costly. Therefore, the neural mechanism of decision-making is closely related to the neural processes to evaluate how much benefits or risks would be taken by the choice. The link between a particular choice and a value of its outcome is formed by the subject's learning and experience. The mechanism to form the link between a choice and a value of its outcome can be described using a reinforcement learning model (Lee et al. 2012). When subjects make decision in any given situation, they first assess and evaluate the relative value of each option, and then select the option having the largest value.

For example, the monkey was asked to perform a two-choice visually-guided saccade task, in which the monkey was required to select either of two targets to get a reward. Selection of one target resulted in reward delivery in a high probability, while the selection of the other target resulted in either no reward delivery or reward delivery in a low probability. In this reward condition, the monkey evaluated the reward value of each target based on the choice history and the reward history and would decide to continuously select the target associated with high probability of reward delivery (Platt and Glimcher 1999; Dorris and Glimcher 2004; Sugrue et al. 2004, 2005). These studies also showed that the monkey's choice of the target would change systematically depending on the probability of reward delivery assigned to each target, such that the monkey tended to choose the target associated with higher probability of reward delivery (higher reward value) and maintained this selection until the contingency between the target choice and the reward value changed. On the other hand, when the reward probability for the choice of each option changed randomly from trial to trial, monkeys tended to select recently rewarded option more often than when the reward probability for each option was fixed (Massi et al. 2018). Thus, behavioral observations indicate that monkeys use a value-based decision-making strategy to perform this task.

Activities of parietal neurons have been analyzed using the two-choice visually-guided saccade task to understand neural mechanisms of value-based decision-making. These studies indicated that the activities of parietal neurons were affected by the reward probability associated with a particular saccade response. When monkeys were able to choose one of two alternative responses freely, both the monkey's choice of the response and the activation of parietal neurons during the task performance were correlated with the relative amount of expected outcome associated with each response (Platt and Glimcher 1999; Sugrue et al. 2004). On the other hand, Dorris and Glimcher (2004) indicated that the activity of parietal neurons was correlated with the subject's relative subjective desirability associated with a particular action, regardless of the specific combination of reward magnitude, reward probability, and response probability associated with each action.

The prefrontal contribution to value-based decision-making has also been extensively examined in human neuroimaging studies (Krawczyk 2002; Walton et al. 2004; Kennerley et al. 2006; Gläscher et al. 2009; Rushworth et al. 2011). These studies indicate that several regions of the prefrontal cortex play distinct and important roles in reward-based decision-making. These regions include the ventromedial prefrontal cortex, the medial orbitofrontal cortex, the lateral orbitofrontal cortex, the anterior cingulate cortex, and the lateral anterior prefrontal cortex (Rushworth et al. 2011).

Animal studies also showed prefrontal participation in value-based decision-making. Kennerley et al. (2011) indicated that, although the prefrontal cortex participates in value computation since many neurons represent the value of the chosen option (chosen value), anterior cingulate neurons and orbitofrontal neurons encode different types of chosen values, such that orbitofrontal neurons evaluate current choices whereas anterior cingulate neurons encode choice prediction and prediction error. Similarly, Rich and Wallis (2016) showed that the orbitofrontal cortex computes a value for each option. Further, Rudebeck et al. (2013) showed that the orbitofrontal cortex has a role to update a value of the option. Since the orbitofrontal cortex has been known to participate in specialized roles of reward-guided behaviors (Kringelbach 2005), these results well agree with the notion that the orbitofrontal cortex participates in value-based decision-making (Wallis 2012).

Dorsolateral prefrontal neurons also contribute to value-based decision-making. Donahue and Lee (2015) indicated that task-related activity in the dorsolateral prefrontal cortex was affected by reward outcomes so as to favor the choices that maximize reward value. Assad et al. (2017) showed that the activity of dorsolateral prefrontal neurons represented information relevant for reward credit assignment and information regarding outcomes at the time of feedback while monkeys performed a trial-and-error learning task. In this task, monkeys were required to find a correct object among four by a trial-and-error manner and keep selecting the correct object for 40–50 trials once they found it. Bartolo and Averbeck (2020) also showed that a population of dorsolateral prefrontal neurons accurately encoded the preferential choice based on the reward probabilities during the reversal learning task in which a reward was delivered stochastically.

Further, Lin et al. (2020) examined how reward value was accumulated for decision-making and which brain areas participated in this process by comparing neuron activity between the orbitofrontal cortex and the dorsolateral prefrontal cortex. In their decision-making task, monkeys were required to choose a target between two by eye movements. Reward probability of the target choice was determined by the combination of sequentially presented four visual stimuli. They found that orbitofrontal neurons encoded the reward probability of each stimulus, responded transiently to that stimulus, and did not encode the combined probability of four stimuli. On the other hand, dorsolateral prefrontal neurons encoded the combined reward probability of four stimuli and exhibited reward-probability-related activation just before eye movements, suggesting that not orbitofrontal neurons but dorsolateral prefrontal neurons participate in the accumulation of the reward value for decision-making and the computation of the transformation from reward information to action generation. Thus, although the prefrontal cortex participates in value-based decision-making, the contribution to this decision-making seems to be different among different sectors of the prefrontal cortex.

7.5.3 Decision-Making Under Competitive Conditions

Decision-making is a process of making a particular choice from a set of alternatives. If a particular choice always predicts a specific outcome or a certain size of value, the decision to select the option associated with the best outcome is optimal. Therefore, the best decision in a particular environment will depend on the subject's own choice history or reward history. However, in a natural environment, since we do not always have choice history and reward history for each option, we cannot make a decision based only on these histories.

In competitive condition, our decisions are often affected by the decisions of others. A game is a typical situation of decision-making in a competitive condition. A game is usually played by multiple players and a payoff table specifies the amount of reward or penalty for each player based on the decisions made by all players. To win the competitive game, each player needs to find an optimal strategy by constantly changing the choice strategy. However, Nash (1950) showed that a competitive game with multiple players has at least one equilibrium condition in which no players can obtain any benefit by changing their choice strategies individually. This equilibrium condition is called the Nash equilibrium. Barraclough et al. (2004) asked two rhesus monkeys to perform an oculomotor free-choice task and play a game analogous to matching pennies against a computer. In their task, monkeys needed to select either the right or the left visual target presented on the monitor by a saccadic eye movement. Monkeys were rewarded when they selected the same target as that selected by the computer. Three task conditions were used. In condition 0, the computer selected the target randomly with an equal probability (the reward rate was 0.5) regardless of the monkey's choice patterns. Since the reward rate was fixed at 0.5, monkeys exhibited spatial bias in their target selection (e.g., more frequent selection of the right target). In condition 1, the computer analyzed the monkey's choice history but not the reward history, while in condition 2, the computer analyzed both the monkey's choice and reward histories. In both conditions 1 and 2, the computer chose the target randomly if no systematic bias was found in the monkey's choice behavior. Therefore, if a reward rate was near 0.5, the result indicates that the monkey's performance was optimal. In these two conditions, the monkey's selection of the right target was closer to 0.5, which agreed with the Nash equilibrium in the matching penny game (Barraclough et al. 2004). The probability of the target choice was not affected by the choice in the preceding trial. However, the monkey's choice was influenced by the computer's choice in the preceding trial. Barraclough et al. (2004) applied a reinforcement learning algorithm to explain behavioral results obtained by two monkeys and concluded that both monkeys approximated the optimal decision strategy using a reinforcement learning algorithm.

Barraclough et al. (2004) then examined the activity of dorsolateral prefrontal neurons while monkeys performed this oculomotor free-choice task. The activity during the fore-period (monkeys looked at a fixation point) and sometimes during the delay period (two targets were presented while monkeys maintained fixation at the fixation point) were affected by the monkey's choice in the preceding trial. In addition, the signals related to the combination of the monkey's previous choice and its outcome were processed differentially in prefrontal neurons depending on the type of decision-making by the monkey. Since monkeys flexibly changed their strategy based on the choice and reward histories and the computer's choice strategy, these results support the notion that the prefrontal cortex plays a key role to optimize the strategy for decision-making during a competitive game. Further, Seo et al. (2007) examined prefrontal activity while monkeys performed a similar penny-matching game and found neural signals that would influence the monkeys' forthcoming choice behavior. Thus, the results obtained by Barraclough et al. (2004) and Seo et al. (2007) indicate that the prefrontal cortex plays an important role in neural mechanisms of decision-making and that neural signals that affect forthcoming decision-making play a role to optimize the strategy of decision-making under the competitive conditions. However, in these studies, monkeys played against the computer. Therefore, the monkeys might not fully understand that the game was competitive. Recently, prefrontal single-neuron activity has been examined in social and real competitive conditions of two monkeys (Yoshida et al. 2012; Chang et al. 2013; Haroush and Williams 2015). These studies indicated that different sectors of the prefrontal cortex participate in different functions in decision-making under social and competitive conditions. Further studies may need to understand prefrontal contributions to decision-making under competitive conditions.

7.5.4 *Decision-Making in Conflict Conditions*

When a subject is exposed to a high-conflict condition immediately preceding a trial, the task-relevant information is enhanced while the task-irrelevant information is suppressed. Therefore, the detrimental effect on the performance produced by the conflict is reduced. This reduction is often called behavioral adaptation and produces an advantageous effect in decision-making in the conflict condition. Mansouri et al. (2007) showed that monkeys having dorsolateral prefrontal lesions exhibited impairments in behavioral adaptation when they performed an analogous task of Wisconsin card sorting test (WCST). Mansouri et al. (2015) also showed that the dorsolateral prefrontal cortex maintains information regarding the happening of the conflict in working memory and indicated that this information was used for optimal decision-making in a dynamic situation like WCST. Similarly, Boschin et al. (2017a, b) applied repeated transcranial magnetic stimulations (rTMS) to the dorsolateral prefrontal cortex to interfere its activity while human subjects performed a conflict version of the WCST analog. They showed that the dorsolateral prefrontal cortex is a fundamental structure to optimize conflict-induced behavioral adaptation by maintaining conflict-history information online across trials.

If you are informed that you would have obtained a better reward if you selected another option after selecting one option, what strategy do you take in subsequent trials? It has been shown that humans and animals change their strategy by simulating the hypothetical outcome that could have resulted from alternative unchosen options. Abe and Lee (2011) used a rock-paper-scissors task with a specific payoff matrix and asked to perform this task between a computer and a monkey. They examined behavioral effects of actual and hypothetical outcomes and neural coding of these outcomes by prefrontal neurons. All monkeys tended to choose the same option again after getting a better outcome (actual outcome). However, when monkeys obtained worse outcome or lost the reward by choosing one option, they tended to choose the other option that would have obtained a better outcome (hypothetical outcome). They analyzed the activity of both dorsolateral prefrontal and orbitofrontal neurons and found that, although both groups of neurons encoded signals related to both actual and hypothetical outcomes, dorsolateral prefrontal neurons encode hypothetical outcomes related to specific actions, while orbitofrontal neurons play roles to monitor both actual and hypothetical outcomes.

7.6 Conclusions

Decision-making is an important executive function that the prefrontal cortex involves. Although the prefrontal cortex participates in various types of decision-making, understanding the neural mechanism related to free-choice decision-making is important to understand basic neural mechanisms for various types of decision-making, because free-choice decision-making corresponds to the very first phase of various types of decision-making. Recent neuroimaging studies have shown that the prefrontal cortex plays an important role in free-choice decision-making. Neurophysiological studies using monkeys showed that prefrontal neurons exhibit transient activation just before the presentation of the imperative cue. This transient activation was called choice-predictive activity (Mochizuki and Funahashi 2016) or prestimulus activity (Marcos and Genovesio 2016). Studies using the free-choice ODR task suggest that choice-predictive activity is produced by a transient increase of spontaneous firings occurred just before the cue presentation, that this transient increase of activity enhances the response to the cue presentation, and that the enhancement of the cue response biases the strength of the neuron's selectivity to the particular option. Further, this biasing effect affects the directional selectivity of neurons having sustained delay-period activity, and eventually, one particular direction is determined by mutual inhibitory interactions among neurons having different directional selectivity and by a winner-take-all principle. These results indicate that the biasing effect caused by transient activation (e.g., choice-predictive activity) is an important neural component of free-choice decision-making and that this biasing effect not only influences the animal's decision regarding the upcoming choice but also links with working memory mechanisms in the prefrontal cortex. Recently, Katz et al. (2016) indicated that decision-related activity observed in LIP neurons is not essential in the final outcome of perceptual decision-making and may reflect some secondary processes. Therefore, further studies may need to confirm that the biasing effect by transient activation in prefrontal neurons having directional activity is causally related to free-choice decision-making.

References

Abe H, Lee D (2011) Distributed coding of actual and hypothetical outcomes in the orbital and dorsolateral prefrontal cortex. Neuron 70:731–741. https://doi.org/10.1016/j.neuron.2011.03.026

Assad WF, Lauto PM, Perge JA, Eskandar EN (2017) Prefrontal neurons encode a solution to the credir-assignment problem. J Neurosci 37:6995–7007. https://doi.org/10.1523/jneurosci.3311-16.2017

Barraclough DJ, Conroy MI, Lee D (2004) Prefrontal cortex and decision making in a mixed-strategy game. Nat Neurosci 7:404–410. https://doi.org/10.1038/nn.1209

Bartolo R, Averbeck BB (2020) Prefrontal cortex predicts state switches during reversal learning. Neuron 106:1044–1054.e4. https://doi.org/10.1016/j.neuron.2020.03.024

Boschin EA, Brkic MM, Simons JS, Buckley MJ (2017a) Distinct roles for the anterior cingulate and dorsolateral prefrontal cortices during conflict between abstract rules. Cereb Cortex 27:34–45. https://doi.org/10.1093/cercor/bhw350

Boschin EA, Mars RB, Buckley MJ (2017b) Transcranial magnetic stimulation to dorsolateral prefrontal cortex affects conflict-induced behavioural adaptation in a Wisconsin Card Sorting Test analogue. Neuropsychologia 94:36–43. https://doi.org/10.1016/j.neuropsychologia.2016.11.015

Chang SWC, Gariépy J-F, Platt ML (2013) Neuronal reference frames for social decisions in primate frontal cortex. Nat Neurosci 16:243–250. https://doi.org/10.1038/nn.3287

Churchland AK, Kiani R, Shadlen MN (2008) Decision-making with multiple alternatives. Nat Neurosci 11:693–702. https://doi.org/10.1038/nn.2123

Coe B, Tomihara K, Matsuzawa M, Hikosaka O (2002) Visual and anticipatory bias in three cortical eye fields of the monkey during an adaptive decision-making task. J Neurosci 22:5081–5090. https://doi.org/10.1523/jneurosci.22-12-0581.2002

Compte A, Brunel N, Goldman-Rakic PS, Wang X-J (2000) Synaptic mechanisms and network dynamics underlying spatial working memory in a cortical network model. Cereb Cortex 10: 910–923. https://doi.org/10.1093/cercor/10.9.910

Cunnington R, Windischberger C, Deecke L, Moser E (2002) The preparation and execution of self-initiated and externally-triggered movement: a study of event-related fMRI. Neuroimage 15:373–385. https://doi.org/10.1006/nimg.2001.0976

Ding L, Gold JI (2012) Neural correlates of perceptual decision making before, during, and after decision commitment in monkey frontal eye field. Cereb Cortex 22:1052–1067. https://doi.org/10.1093/cercor/bhr178

Donahue CH, Lee D (2015) Dynamic routing of task-relevant signals for decision making in dorsolateral prefrontal cortex. Nat Neurosci 18:295–301. https://doi.org/10.1038/nn.3918

Dorris MC, Glimcher PW (2004) Activity in posterior parietal cortex is correlated with the relative subjective desirability of action. Neuron 44:365–378. https://doi.org/10.1016/j.neuron.2004.09.009

Freedman DJ, Riesenhuber M, Poggio T, Miller EK (2001) Categorical representation of visual stimuli in the primate prefrontal cortex. Science 291:312–316. https://doi.org/10.1126/science.291.5502.312

Freedman DJ, Riesenhuber M, Poggio T, Miller EK (2002) Visual categorization and the primate prefrontal cortex: neurophysiology and behavior. J Neurophysiol 88:929–941. https://doi.org/10.1152/jn.00040.2002

Frith CD, Friston K, Liddle PF, Fracowiak RSJ (1991) Willed action and the prefrontal cortex in man: a study with PET. Proc R Soc Lon B 244:241–246. https://doi.org/10.1098/rspb.1991.0077

Funahashi S (2014) Saccade-related activity in the prefrontal cortex: its role in eye movement control and cognitive functions. Front Integr Neurosci 8:54. https://doi.org/10.3389/fnint.2014.00054

Funahashi S (2015) Functions of delay-period activity in the prefrontal cortex and mnemonic scotomas revisited. Front Syst Neurosci 9:2. https://doi.org/10.3389/fnsys.2015.00002

Funahashi S (2017a) Working memory in the prefrontal cortex. Brain Sci 7:49. https://doi.org/10.3390/brainsci7050049

Funahashi S (2017b) Prefrontal contribution to decision-making under free-choice conditions. Front Neurosci 11:431. https://doi.org/10.3389/fnins.2017.00431

Funahashi S, Bruce CJ, Goldman-Rakic PS (1989) Mnemonic coding of visual space in the monkey's dorsolateral prefrontal cortex. J Neurophysiol 61:331–349. https://doi.org/10.1152/jn.1989.61.2.331

Funahashi S, Bruce CJ, Goldman-Rakic PS (1990) Visuospatial coding in primate prefrontal neurons revealed by oculomotor paradigms. J Neurophysio 63:814–831. https://doi.org/10.1152/jn.1990.63.4.814

Funahashi S, Bruce CJ, Goldman-Rakic PS (1991) Neuronal activity related to saccadic eye movements in the monkey's dorsolateral prefrontal cortex. J Neurophysiol 65:1464–1483. https://doi.org/10.1152/jn.1991.65.6.1464

Furman M, Wang X-J (2008) Similarity effect and optimal control of multiple-choice decision making. Neuron 60:1153–1168. https://doi.org/10.1016/j.neuron.2008.12.003

Fuster JM (1973) Unit activity in prefrontal cortex during delayed-response performance: neuronal correlates of transient memory. J Neurophysiol 36:61–78. https://doi.org/10.1152/jn.1973.36.1.61

Gläscher J, Hampton AN, O'Doherty JP (2009) Determining a role for ventromedial prefrontal cortex in encoding action-based value signals during reward-related decision making. Cereb Cortex 19:483–495. https://doi.org/10.1093/cercor/bhn098

Gold JI, Shadlen MN (2000) Representation of a perceptual decision in developing oculomotor commands. Nature 404:390–394. https://doi.org/10.1038/35006062

Gold JI, Shadlen MN (2007) The neural basis of decision making. Annu Rev Neurosci 30:535–574. https://doi.org/10.1146/annurev.neuro.29.051605.113038

Halsband U, Matsuzaka Y, Tanji J (1994) Neuronal activity in the primate supplementary, pre-supplementary and premotor cortex during externally and internally instructed sequential movements. Neurosci Res 20:149–155. https://doi.org/10.1016/0168-0102(94)90032-9

Hanks TD, Summerfield C (2017) Perceptual decision making in rodents, monkeys, and humans. Neuron 93:15–31. https://doi.org/10.1016/j.neuron.2016.12.003

Hanks TD, Ditterich J, Shadlen MN (2006) Microstimulation of macaque area LIP affects decision-making in a motion discrimination task. Nat Neurosci 9:682–689. https://doi.org/10.1038/nn1683

Haroush K, Williams ZM (2015) Neuronal prediction of opponent's behavior during cooperative social interchange in primates. Cell 160:1233–1245. https://doi.org/10.1016/j.cell.2015.01.045

Haynes J-D, Sakai K, Rees G, Gilbert S, Frith C, Passingham RE (2007) Reading hidden intentions in the human brain. Curr Biol 17:323–328. https://doi.org/10.1016/j.cub.2006.11.072

Huk AC, Shadlen MN (2005) Neural activity in macaque parietal cortex reflects temporal integration of visual motion signals during perceptual decision making. J Neurosci 25:10420–10436. https://doi.org/10.1523/JNEUROSCI.4684-04.2005

Hyder F, Phelps EA, Wiggins CJ, Labar KS, Blamire AM, Shulman RG (1997) "Willed action": a functional MRI study of the human prefrontal cortex during a sensorimotor task. Proc Nat Acad Sci. USA 94:6989–6994. https://doi.org/10.1073/pnas.94.13.6989

Katz LN, Yates JL, Pillow JW, Huk AC (2016) Dissociated functional significance of decision-related activity in the primate dorsal stream. Nature 535:285–288. https://doi.org/10.1038/nature18617

Kayser AS, Buchsbaum BR, Erickson DT, D'Esposito M (2010) The functional anatomy of a perceptual decision in the human brain. J Neurophysiol 103:1179–1194. https://doi.org/10.1152/jn.00364.2009

Kennerley SW, Walton ME, Behrens TEJ, Buckley MJ, Rushworth MFS (2006) Optimal decision making and anterior cingulate cortex. Nat Neurosci 9:940–947. https://doi.org/10.1038/nn1724

Kennerley SW, Behrens TEJ, Wallis JD (2011) Double dissociation of value computations in orbitofrontal and anterior cingulate neurons. Nat Neurosci 14:1581–1589. https://doi.org/10.1038/nn.2961

Kiani R, Shadlen MN (2009) Representation of confidence associated with a decision by neurons in the parietal cortex. Science 324:759–764. https://doi.org/10.1126/science.1169405

Kim J-N, Shadlen MN (1999) Neural correlates of a decision in the dorsolateral prefrontal cortex of the macaque. Nat Neurosci 2:176–185. https://doi.org/10.1038/5739

Krawczyk DC (2002) Contributions of the prefrontal cortex to the neural basis of human decision making. Neurosci Biobehav Rev 26:631–664. https://doi.org/10.1016/S0149-7634(02)00021-0

Kringelbach ML (2005) The human orbitofrontal cortex: linking reward to hedonic experience. Nat Rev Neurosci 6:691–702. https://doi.org/10.1038/nrn1747

Lau HC, Rogers RD, Ramnani N, Passingham RE (2004) Willed action and attention to the selection of action. Neuroimage 21:1407–1415. https://doi.org/10.1016/j.neuroimage.2003.10.034

Lee D, Seo H, Jung MW (2012) Neural basis of reinforcement learning and decision making. Annu Rev Neurosci 35:287–308. https://doi.org/10.1146/annurev-neuro-062111-150512

Lin Z, Nie C, Zhang Y, Chen Y, Yang T (2020) Evidence accumulation for value computation in the prefrontal cortex during decision making. Proc Natl Acad Sci USA 117:30728–30737. https://doi.org/10.1073/pnas.2019077117

Mansouri FA, Buckley MJ, Tanaka K (2007) Mnemonic function of the dorsolateral prefrontal cortex in conflict-induced behavioral adjustment. Science 318:987–990. https://doi.org/10.1126/science.1146384

Mansouri F, Rosa MGP, Atapour N (2015) Working memory in the service of executive control functions. Front Syst Neurosci 9:166. https://doi.org/10.3389/fnsys.2015.00166

Marcos E, Genovesio A (2016) Determining monkey free choice long before the choice is made: the principal role of prefrontal neurons involved in both decision and motor processes. Front Neural Circuits 10:75. https://doi.org/10.3389/fncir.2016.00075

Massi B, Donahue CH, Lee D (2018) Volatility facilitates value updating in the prefrontal cortex. Neuron 99:598–608.e4. https://doi.org/10.1016/j.neuron.2018.06.033

Mazurek ME, Roitman JD, Ditterich J, Shadlen MN (2003) A role for neural integrators in perceptual decision making. Cereb Cortex 13:1257–1269. https://doi.org/10.1093/cercor/bhg097

Mochizuki K, Funahashi S (2014) Opponent history effect of preceding decision and action in the free choice of saccade direction. J Neurophysiol 112:923–932. https://doi.org/10.1152/jn.00846.2013

Mochizuki K, Funahashi S (2016) Prefrontal spatial working memory network predicts animal's decision making in a free choice saccade task. J Neurophysiol 115:127–142. https://doi.org/10.1152/jn.00255.2015
Mushiake H, Inase M, Tanji J (1991) Neuronal activity in the primate premotor, supplementary, and precentral motor cortex during visually guided and internally determined sequential movements. J Neurophysiol 66:705–718. https://doi.org/10.1152/jn.1991.66.3.705
Nachev P, Kennard C, Husain M (2008) Functional role of the supplementary and pre-supplementary motor areas. Nat Rev Neurosci 9:856–869. https://doi.org/10.1038/nrn2478
Nash JF (1950) Equilibrium points in n-person games. Proc Natl Acad Sci USA 36:48–49. https://doi.org/10.1073/pnas.36.1.48
Platt ML, Glimcher PW (1999) Neural correlates of decision variables in parietal cortex. Nature 400:233–238. https://doi.org/10.1038/22268
Quintana J, Yajeya J, Fuster JM (1988) Prefrontal representation of stimulus attributes during delay tasks. I. Unit activity in cross-temporal integration of sensory and sensory-motor information. Brain Res 474:211–221. https://doi.org/10.1016/0006-8993(88)90436-2
Rangel A, Camerer C, Montague PR (2008) A framework for studing the neurobiology of value-based decision making. Nat Rev Neurosci 9:545–556. https://doi.org/10.1038/nrn2357
Rao SG, Williams GV, Goldman-Rakic PS (1999) Isodirectional tuning of adjacent interneurons and pyramidal cells during working memory: evidence for microcolumnar organization in PFC. J Neurophysiol 81:1903–1916. https://doi.org/10.1152/jn.1999.81.4.1903
Rich EL, Wallis JD (2016) Decoding subjective decisions from orbitofrontal cortex. Nat Neurosci 19:973–980. https://doi.org/10.1038/nn.4320
Rolls ET, Deco G (2011) Prediction of decisions from noise in the brain before the evidence is provided. Front Neurosci 5:33. https://doi.org/10.3389/fnins.2011.00033
Romo R, Salinas E (2003) Flutter discrimination: neural codes, perception, memory and decision making. Nat Rev Neurosci 4:203–218. https://doi.org/10.1038/nrn1058
Romo R, Hernández A, Zainos A, Lemus L, Brody CD (2002) Neuronal correlates of decision-making in secondary somatosensory cortex. Nat Neurosci 6:1217–1225. https://doi.org/10.1038/nn950
Rudebeck PH, Saunders RC, Prescott AT, Chau LS, Murray EA (2013) Prefrontal mechanisms of behavioral flexibility, emotion regulation and value updating. Nat Neurosci 16:1140–1145. https://doi.org/10.1038/nn.3440
Rushworth MFS, Noonan MP, Boorman ED, Walton ME, Behrens TE (2011) Frontal cortex and reward-guided learning and decision-making. Neuron 70:1054–1069. https://doi.org/10.1016/j.neuron.2011.05.014
Sendhilnathan N, Basu D, Goldberg ME, Schall JD, Murthy A (2021) Neural correlates of goal-directed and non-goal-directed movements. Proc Natl Acad Sci USA 118:e2006372118. https://doi.org/10.1073/pnas.2006372118
Seo H, Barraclough DJ, Lee D (2007) Dynamic signals related to choices and outcomes in the dorsolateral prefrontal cortex. Cereb Cortex 17:i110–i117. https://doi.org/10.1093/cercor/bhm064
Shadlen MN, Kiani R (2013) Decision making as a window on cognition. Neuron 80:791–806. https://doi.org/10.1016/j.neuron.2013.10.047
Shadlen MN, Newsome WT (2001) Neural basis of a perceptual decision in the parietal cortex (area LIP) of the rhesus monkey. J Neurophysiol 86:1916–1936. https://doi.org/10.1152/jn.2001.86.4.1916
Shima K, Tanji J (1998) Role for cingulate motor area cells in voluntary movement selection based on reward. Science 282:1335–1338. https://doi.org/10.1126/science.282.5329.1335
Shu Y, Hasenstaub A, McCormick DA (2003) Turning on and off recurrent balanced cortical activity. Nature 423:288–293. https://doi.org/10.1038/nature01616
Soon CS, Brass M, Heinze H-J, Haynes J-D (2008) Unconscious determinants of free decisions in the human brain. Nat Neurosci 11:543–545. https://doi.org/10.1038/nn2112

Sugrue LP, Corrado GS, Newsome WT (2004) Matching behavior and the representation of value in the parietal cortex. Science 304:1782–1787. https://doi.org/10.1126/science.1094765

Sugrue LP, Corrado GS, Newsome WT (2005) Choosing the greater of two goods: neural currencies for valuation and decision making. Nat Rev Neurosci 6:363–375. https://doi.org/10.1038/nrn1666

Thiery T, Saive A-L, Combrisson E, Dehgan A, Bastin J, Kahane P, Berthoz A, Lachaux J-P, Jerbi K (2020) Decoding the neural dynamics of free choice in humans. PLoS Biol 18:e3000864. https://doi.org/10.1371/journal.pbio.3000864

Wallis JD (2012) Cross-species studies of orbitofrontal cortex and value-based decision-making. Nat Neurosci 15:13–19. https://doi.org/10.1038/nn.2956

Walton ME, Devlin JT, Rushworth MFS (2004) Interactions between decision making and performance monitoring within prefrontal cortex. Nat Neurosci 7:1259–1265. https://doi.org/10.1038/nn1339

Wang XJ (2001) Synaptic reverberation underlying mnemonic persistent activity. Trends Neurosci 24:455–463. https://doi.org/10.1016/S0166-2236(00)01868-3

Wang X-J (2008) Decision making in recurrent neuronal circuits. Neuron 60:215–234. https://doi.org/10.1016/j.neuron.2008.09.034

Wang M, Vijayraghavan S, Goldman-Rakic PS (2004a) Selective D2 receptor actions on the functional circuitry of working memory. Science 303:853–856. https://doi.org/10.1126/science.1091162

Wang X-J, Tegner J, Constantinidis C, Goldman-Rakic PS (2004b) Division of labor among distinct subtypes of inhibitory neurons in a cortical microcircuit of working memory. Proc Natl Acad Sci USA 101:1368–1373. https://doi.org/10.1073/pnas.0305337101

Waschke L, Kloosterman NA, Obleser J, Garrett DD (2021) Behavior needs neural variability. Neuron 109:761–766. https://doi.org/10.1016/j.neuron.2021.01.023

Watanabe K, Funahashi S (2007) Prefrontal delay-period activity reflects the decision process of a saccade direction during a free-choice ODR task. Cereb Cortex 17:i88–i100. https://doi.org/10.1093/cercor/bhm102

Watanabe K, Igaki S, Funahashi S (2006) Contributions of prefrontal cue-, delay-, and response-period activity to the decision process of saccade direction in a free-choice ODR task. Neural Networks 19:1203–1222. https://doi.org/10.1016/j.neunet.2006.05.033

Yoshida K, Saito N, Iriki A, Isoda M (2012) Social error monitoring in macaque frontal cortex. Nat Neurosci 15:1307–1312. https://doi.org/10.1038/nn.3180

Chapter 8
Metacognition

8.1 Metacognition

I know what I am doing now and what I want to do. I can continuously monitor whether I am doing correctly or not. I know what I know and what I do not know. I can monitor what I remember and what I do not remember. If someone asks me "Do you know where is the capital of Japan?" I would say "Yes, I know. That is Tokyo." If someone asks me "Do you know where Charlie Brown is living?" I would say "I don't know. Who is Charlie Brown?" When we are asked these questions, we may answer in these ways, because we are able to correctly monitor our knowledge and discriminate between what we know and what we do not know. On the other hand, when we feel uncertainty about our answer, we suspend responses and will seek information further. If someone asks me "Do you know the name of the city where Egas Moniz was born?" I would say "I knew it, but I can't recall it now. If you give me some hint, I may recall it."

Thus, we can consciously monitor whether we have certain information in our memory and knowledge or not. The degree of whether we can recall correct information from our memory or not and the degree of how strongly we feel confident regarding recalled information is different depending on the questions and the situations. Thus, the human mind has an executive function to monitor its memory and knowledge and control its behavior or decision based on the result of monitoring. This sophisticated mental ability to monitor and control perception, cognition, and memory is called "*metacognition*" (Flavell 1979; Nelson 1996). Metacognition is explained as "thinking about thinking," "cognition about cognition," or "knowing about knowing." If we notice that we have more trouble in learning mathematics than biology, it can be said that we are engaging in metacognition.

Metacognitive ability is an important human capacity and allows us to perform adaptive and flexible responses to various problems. Metacognition also plays an important role in oral communication of information, oral persuasion, oral

S. Funahashi, *Dorsolateral Prefrontal Cortex*, Brain Science,
https://doi.org/10.1007/978-981-19-7268-3_8

comprehension, reading comprehension, writing, language acquisition, attention, memory, problem solving, social cognition, and, various types of self-control and self-instruction (Flavell 1979). Thus, metacognition is strongly linked to declarative consciousness and language (Nelson 1996). Since metacognition is based on the subject's introspection, the metacognitive ability can be estimated only using language. Therefore, metacognition is strongly linked to the human mind and all humans are thought to have this capacity. An important function of metacognition is to monitor our own cognitive processes. This function also links to the human's capacity to understand and estimate others' minds and can be used to predict others' behavior (Carruthers 2009). Metacognitive ability would be a cognitive basis to understand others' minds. Therefore, understanding metacognitive ability is an important subject to understanding human mind and consciousness (Flavell 1979; Nelson 1996).

Metacognition has been thought to be a unique human capacity. However, recent behavioral studies have shown that non-human animals including monkeys and birds share functional parallels with human's metacognition (Smith 2009; Terrace and Son 2009; Hampton 2010). Therefore, behavioral or neurobiological studies of metacognition using non-human animals are useful new methods to examine and understand neural mechanisms of metacognition and the nature of human mind.

8.2 Metamemory

Metacognition is explained as "cognition about cognition" or "knowing about knowing." Metacognitive abilities include knowledge about how to obtain particular information, when and where and how to use particular information or strategy for solving problems, and judgment or estimation of how strongly confident own memory is. Among metacognitive abilities, *metamemory* is an important ability. Metamemory is the knowledge about one's own memory and is a higher cognitive function that enables us to monitor and control our own memory processes. For theoretical frameworks of metamemory, Nelson and Narens (1990) proposed the following three abstract principles. (1) The cognitive processes are split into two specifically interrelated levels; the meta-level and the object-level (Fig. 8.1, left). (2) The meta-level contains a dynamical model (e.g., mental simulation) of the object-level. (3) There are two dominant relations, called *control* and *monitoring*, which are defined in terms of the direction of the information flow between the meta-level and the object-level. *Control* is the operation that the meta-level modifies information processing in the object-level. Modification includes the initiation, the maintenance, and the termination of information processing. *Control* does not produce any information from the object-level. *Monitoring* is the operation that the meta-level receives information from the object-level. The information from the object-level changes the state of the model of the object-level placed in the meta-level. *Monitoring* and *control* are two important processes for metamemory.

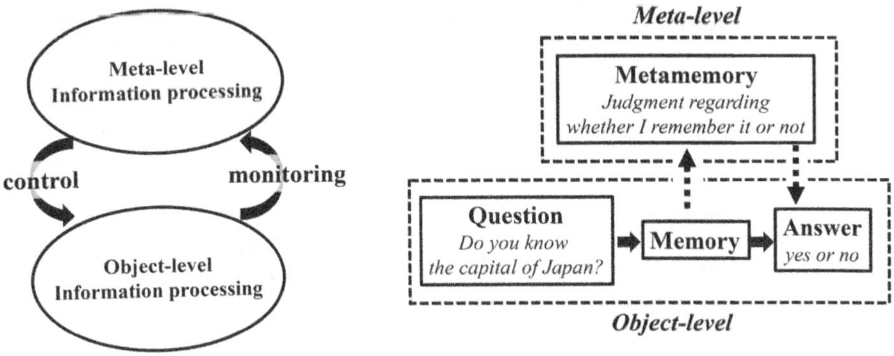

Fig. 8.1 Models of metacognition. Modified based on Nelson and Narens (1990)

Two levels of processing (meta-level processing and object-level processing), which is the model of metacognition proposed by Nelson and Narens (1990), explain functions of metamemory (Fig. 8.1, right). If someone asks you a question ("Do you know the capital of Japan?"), you will access long-term memory, retrieve relevant information from the memory, and make an answer ("Yes, I know"). This straight forward information processing is called *object-level processing* (Fig. 8.1). However, before you access long-term memory, you may ask yourself whether you have information relevant to the question in your memory or not (*monitoring*), and make a judgment based on the information regarding whether you can retrieve a correct answer or not and the degree of confidence about the answer (*control*). Then, you make an answer ("Yes" or "No") to the question based on your judgment. This process includes both *object-level* and *meta-level processing*. In metamemory, the meta-level monitors object-level memory process and verifies whether a specific or relevant information is in memory or not and, at the same time, controls the output of the object-level based on the degree of confidence and certainty. Thus, when we try to understand a neural system supporting metamemory, it is important to know how meta-level neural system monitors information processing in the object-level memory systems and how meta-level neural system makes a judgment and controls the output from the object-level neural system.

8.3 Metamemory: Human Studies

The subject's judgment of whether particular information is remembered or not and the degree of the subject's confidence about the judgment are important variables for examining mechanisms of metamemory. However, these are based on subjective processes. Metamemory is a subjective process and metamemory capacity can only be examined by introspective reports using language. Therefore, metamemory has been considered to be a unique human capacity and can be examined only by human subjects.

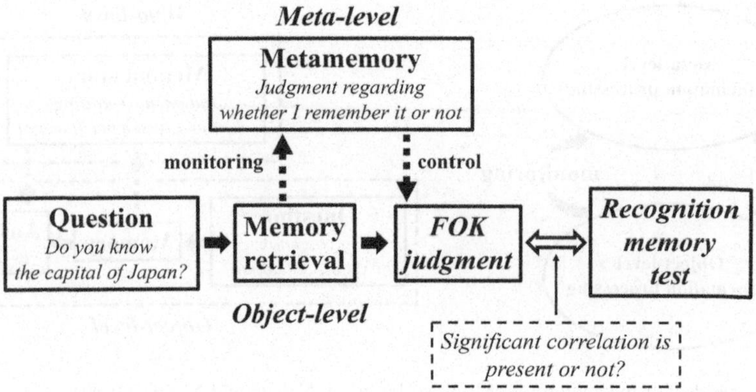

Fig. 8.2 Metamemory study paradigm

In psychological studies, the paradigm using *feeling-of-knowing (FOK) judgment* has been used (Fig. 8.2) to examine whether humans have metamemory capacity or not. The usefulness of feeling-of-knowing measurement in metamemory study was first shown by Hart (1965). In his study, the accuracy of feeling-of-knowing was examined by college students. The students first took memory tests, in which they were asked to recall correct answers to general questions (e.g., Who developed the nonsense syllable in studies of learning?"). After the memory test, for the questions that the subjects could not recall correct answers, they were asked to predict whether they could or could not recall correct answers to these questions (feeling-of-knowing) when some likely but incorrect choices were provided as hints. The subjects' predictions were validated by subsequent recognition tests (e.g., choose from a. Ebbinghaus, b. Hull, c. Pavlov, d. Wundt). As a result, the performance of the recognition tests was better for the questions eliciting stronger feeling-of-knowing than for the questions that did not elicit feeling-of-knowing. This result showed that the strength of feeling-of-knowing would be a useful measure of metamemory capacity in human subjects.

In the paradigm using FOK judgment, human subjects are asked to answer several questions about general knowledge. When the subject cannot make a correct answer to a question, he is requested to report the strength of feeling-of-knowing about the correct answer for that question. Feeling-of-knowing is the subjective feeling to one's own memory. Suppose you are asked a general question. In one case, you may feel that the correct answer is at the tip of the tongue (high FOK). In another case, you may feel that you can retrieve the correct answer if some hint is given (medium FOK). In the other case, you may feel that, although you knew the correct answer, you cannot recall it now even if a hint is given (low FOK). And, in other cases, you know that you do not know the correct answer (pure guess). These are examples of different levels of FOK judgment. The subject is asked his subjective scale of FOK judgment for each question (e.g., 1, high FOK; 2, medium FOK; 3, low FOK; 4, pure guess). To prove whether the level of the subject's FOK judgment to each question is relevant or not, the subject is asked to take the recognition memory test using the same questions after finishing the memory test with FOK judgment. In the recognition memory test, the subject is required to select a correct answer among multiple choices (e.g., 5 or 7 choices) to each question. Using the level of the subject's FOK judgment and the score of the recognition memory test for each question, correlations between the levels of FOK judgment and the scores of the recognition memory test are calculated for each subject. If the subject has metamemory capacity, it is expected that the subject shows high correct scores in the recognition memory test to the questions in that the subject shows a high level of FOK judgment. Therefore, significant positive correlation could be observed between the strength of FOK judgment and the score of the recognition memory test. If significant positive correlation is observed, it can be concluded that the subject correctly monitors his own memory and therefore has metamemory capacity. Using this paradigm, it has been examined whether human subjects have metamemory capacity or not. Since significant positive correlation was observed between the strength of FOK judgment and the score of the recognition memory test in normal human subjects, it has been concluded that humans can monitor their own memory accurately and have metamemory capacity.

8.4 Prefrontal Contribution to Metamemory and Metacognition

8.4.1 Human Neuropsychological Studies

Metacognition is unique human capacity and is thought to be closely related to human mind and consciousness. Therefore, metacognition must be one of the important subjects to examine human mind and consciousness (Grimaldi et al. 2015). Metamemory is a higher cognitive function and enables to monitor and control our own memory processes. Therefore, it is important to examine how metamemory is achieved, what neural mechanism supports metamemory, and which brain area contributes to metamemory.

Metamemory can be thought to be one kind of memory function. However, it was not known whether or not metamemory is operated by the same mechanism as that of memory. Amnesic patients exhibit severe impairments on memory tests including recall and recognition of information. Therefore, it was expected that amnesic patients exhibit severe impairment in metamemory. Since amnesic patients lack conscious knowledge about what information is in their memory, they might exhibit difficulty in metamemory tests similar to difficulty in conventional memory tests. Shimamura and Squire (1986) examined the accuracy of FOK judgment in eight patients with Korsakoff's syndrome, eight patients with electroconvulsive therapy, and four amnesic patients. All subjects were asked to take memory tests including a set of 100 general questions (e.g., What is the name of the ship on which Charles Darwin made his scientific voyage?). After the test phase, each subject was asked to make FOK judgment to each of the nonrecalled questions on a 4-point scale (1, high FOK; 2, medium FOK; 3, low FOK; 4, pure guess). Then, the subject took a seven-alternative forced-choice recognition test for each of nonrecalled questions and asked to select a correct answer. Correlations between FOK judgments and scores of the recognition test showed that only patients with Korsakoff's syndrome exhibited impairment in FOK judgment (no or low correlations between the rates of FOK judgments and the scores of the recognition memory test). Amnesic patients exhibited accurate FOK judgments similar to normal subjects. These results indicate that amnesia does not cause an impairment of metamemory and that the medial temporal memory system does not contribute to metamemory function.

Next, Janowsky et al. (1989) examined the accuracy of FOK judgment using 7 patients with frontal lobe lesions, 6 patients with Korsakoff's syndromes, and 5 patients with non-Korsakoff's amnesia. Each subject was first presented with 24 sentences (e.g., "At the museum we saw some ancient relics made of clay") and, after a delay period (1–3 days), the subject was asked to recall missing words of each sentence (e.g., "At the museum we saw some ancient relics made of."). After the recall phase, each subject was asked to make FOK judgment, such that, for non-recalled words, the subject was asked to rate the strength of the FOK on a 4-point scale (high, medium, low, pure guess). In the recognition phase, the subject was asked to perform a seven-alternative forced-choice recognition test for 24 sentences. Comparison between the score of FOK judgment and the score of the recognition memory test showed that patients with frontal lobe lesions and patients with Korsakoff's syndrome exhibited poor performance of FOK judgment. However, non-Korsakoff amnesic patients exhibited no impairment in FOK judgment. These results indicate that the prefrontal cortex, not the medial temporal lobe including the hippocampus, participates in metamemory function. The impairment of metamemory in patients with Korsakoff's syndrome is thought to be caused by prefrontal damage accompanied by Korsakoff's syndrome.

Prefrontal contribution to FOK judgment was further confirmed by Souchay et al. (2000). They asked normal old (60–98 years) and young (20–32 years) adult subjects to perform a pair-association task with FOK judgment. The subject was first asked to learn 36 cue-target word pairs and then asked to recall a target word when a cue word was presented. The judgment of FOK was the selection of either a "yes" or "no" prediction. If the subject thought that he knows the correct target word, he should make a "yes" prediction, while if the subject thought that he cannot recall the target word, he should make a "no" prediction. After this procedure, the subject was given a recognition test. Compared with young subjects, old subjects performed FOK judgment at chance level. Age difference in performance on both the Wisconsin card sorting test and the verbal fluency test was also observed; younger subjects performed better than older subjects. Performance scores of the Wisconsin card sorting test and the verbal fluency test were closely correlated with the accuracy of FOK judgment in old subjects, suggesting that age-related decline in the accuracy of FOK judgment is related to the decline of the executive function of the prefrontal cortex. Thus, these results further support prefrontal contribution to FOK judgment.

Schnyer et al. (2004) showed that patients with frontal lobe lesions exhibited lower scores of FOK judgment compared with normal subjects. Fourteen patients with frontal damage and 18 control subjects participated in this experiment. The subject was first asked to read and remember 18 sentences randomly selected from 24 sentences and presented one at a time. After a 20-min delay, the subject was shown each sentence with the final word missing and was asked to recall the missing word. When the subject made an answer or felt unable to recall the missing word, he was asked to make a confidence rating to his answer using a six-point scale (1, pure guess; 2, very uncertain; 3, somewhat uncertain; 4, somewhat certain; 5, very certain; 6, extremely certain) (confidence judgment or retrospective method). After the subject recalled the missing word or made confidence rating, he was again asked to make a FOK judgment to each of incorrectly recalled missing words using a four-point scale (1, pure guess; 2, low FOK; 3, medium FOK; 4, high FOK) (FOK judgment or prospective method). During the FOK judgment period, the subject was shown a sample sentence with six alternative choices and told to judge the likelihood. After finishing FOK judgment, the subject was asked to perform a recognition task to select a correct word from six alternatives. Although frontal patients exhibited impairment in recalling the correct word, they were able to make accurate confidence judgment. However, they exhibited impairment in FOK judgment. Damaged areas of frontal patients who exhibited impairment in FOK judgment were found as an overlapped region in the right medial prefrontal cortex. These results suggest that functional dissociation is present between the confidence judgment of recalled information and the memory monitoring (FOK judgment) and that the right medial prefrontal cortex participates in memory monitoring.

Pannu et al. (2005) showed evidences that patients with frontal lobe damage exhibit impairment in metamemory and that the retrospective confidence judgment is dissociated from the prospective FOK judgment. They asked frontal patients and normal control subjects to perform a face identification task using famous persons' faces. In the confidence judgment condition, a face was first presented to the subject, and then the subject was asked the question "Who is this person?" and needed to select the correct name from eight choices of possible names. After the subject made a choice of the name, he was asked to make a confidence rating to his choice using a 6-point scale. In the FOK judgment condition, the subject was shown a face and asked to recall the name of the person. If the subject could not recall the name or answered an incorrect name, he first took FOK judgment using a 6-point scale and then took a recognition test with eight choices. Patients with frontal lobe damage showed worse performance than controls in the confidence judgment but showed less worse performance in the FOK judgment. This result was somewhat different from the result obtained by Schnyer et al. (2004). The difference might be caused by the differences in the items used for two different studies and the differences in size and location of frontal damages among patients. However, both studies show the importance of the prefrontal cortex in metamemory and suggest the presence of functional segregation between retrospective confidence judgment process and prospective FOK judgment process within the prefrontal cortex.

Pannu and Kaszniak (2005) overviewed evidences related to metamemory obtained by neuropsychological studies and concluded that the frontal lobe critically involves as the central component of metamemory, especially for both monitoring and control processes. The studies using patients with local brain damage could make following conclusions. (1) Strong correlation is present between the degree of impairment of frontal lobe function and metamemory accuracy, especially accuracy of FOK judgment. (2) Metamemory ability is severely impaired by frontal lobe dysfunction and poor memory. However, the evidence that amnesic patients exhibit normal metamemory ability indicates that the poor memory itself and functions of the medial temporal lobe are not related to metamemory ability. (3) Subject's performance level varies in different metamemory tasks and the underlying processes that these tasks measure are different. Therefore, the difference in the brain areas responsible for impairments in metamemory functions among different studies might be caused by the differences in underlying information processing for the tasks used. (4) There have been two methods to examine metamemory ability; measuring the strength of the confidence about the decision (confidence judgment or retrospective method) and measuring the strength of FOK about the answer (FOK judgment or prospective method). Some prefrontal patients exhibited different degrees of impairments between these two measures.

8.4.2 Human Brain Imaging Studies

Human brain imaging studies also indicate prefrontal contribution to metamemory functions. Kikyo et al. (2002) showed that the strength of the FOK judgment positively correlated with the magnitude of activation in several prefrontal areas. In their study (Fig. 8.3, left), 15 human subjects were first required to answer 204 general questions (e.g., "Who first conquered Mt. Everest?") while brain activation was measured in the fMRI scanner. Whether the subject knew the answer to each question or not was reported by pressing either button (recall trial). After finishing all recall trials in the scanner, the subject was asked to write his answer to each question outside the scanner. If the subject did not know the correct answer to a particular question, he was asked to make the FOK judgment about his answer using a three-point scale (1, pure guess; 2, low FOK; 3, high FOK) (FOK trial). Then, the subject took an identification test for all FOK trials, in which the question and its correct answer were presented to the subject and he was asked to report whether he could identify the answer as correct. Event-related fMRI analysis revealed that the degree of prefrontal activation was significantly correlated with the increase of the FOK rating. Activated regions included the left dorsolateral portion, the left anterior portion, bilateral inferior portion, and the medial portion of the prefrontal cortex (Fig. 8.3, right). Among these prefrontal areas, the bilateral inferior frontal gyrus was activated in correlation with the increase of the FOK rating but was not activated in a successful recall, suggesting that these prefrontal areas contribute to metamemory processes.

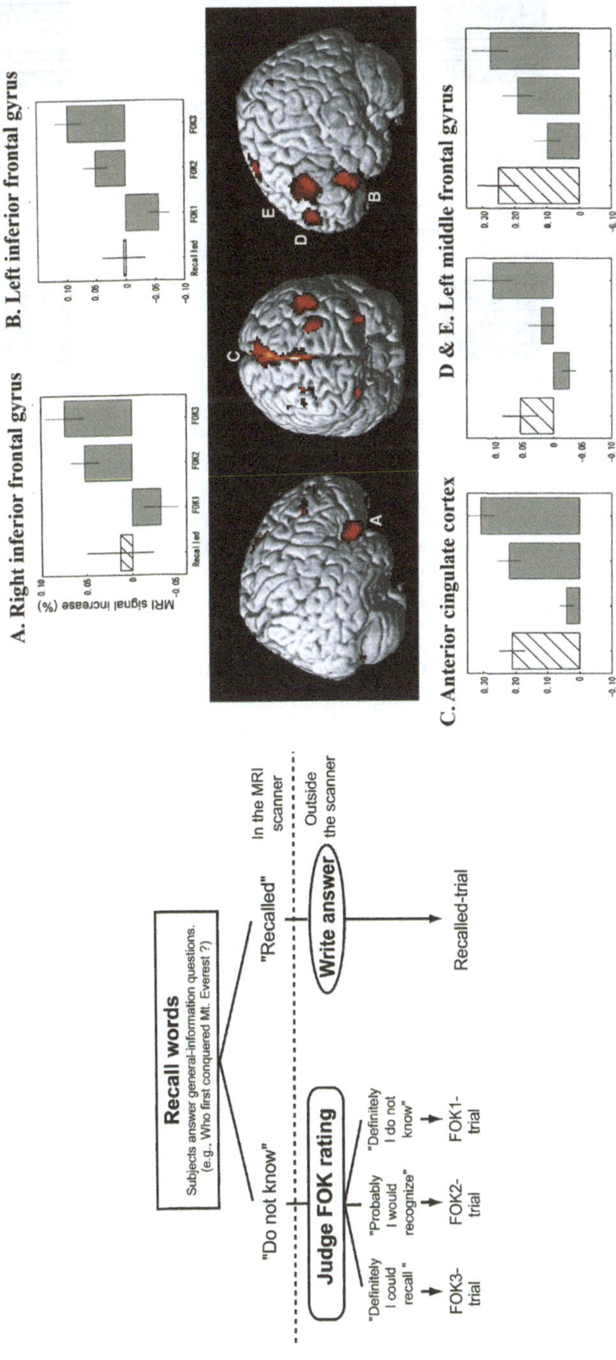

Fig. 8.3 *Left*: Experimental procedures. *Right*: Five prefrontal areas whose activation was correlated with the strength of FOK ratings. From Kikyo et al. (2002) with permission

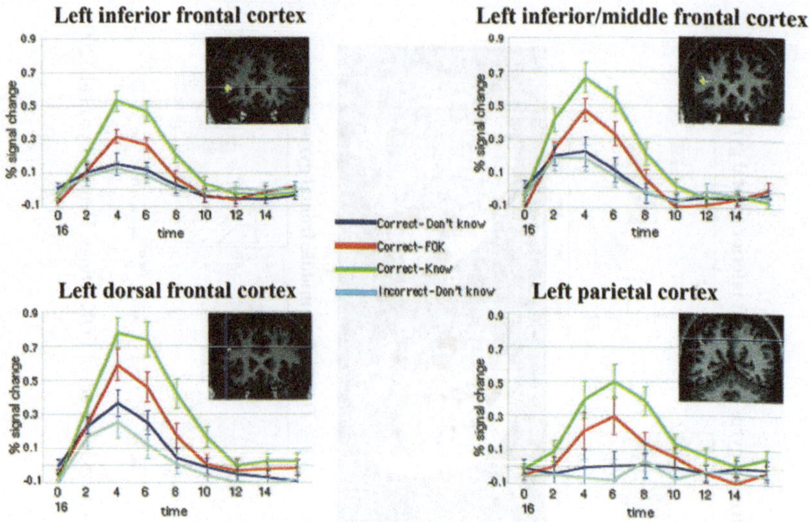

Fig. 8.4 Percent signal changes observed in three prefrontal areas and one parietal area under four conditions during performances of metamemory tasks using FOK judgments. From Maril et al. (2003) with permission

Maril et al. (2003) also showed prefrontal contribution to FOK judgment by event-related fMRI study. During the initial encoding scan, the subject was presented with 180 word-pairs and asked to form a mental image that included the two concepts represented by each word in each pair. For each trial, the subject was asked to indicate the quality of that image by selecting one of three keys (1, forming good detailed image; 2, forming less good image; 3, forming poor image). The encoding scan was followed by the cued-recall scan. One word of each word-pair was presented and the subject needed to recall the paired word. For each trial, the subject was asked to press one of three keys depending on the retrieved outcome ("know," "FOK," and "don't know"). After these scans, the subject was given a recognition test. As a result, the magnitude of activation of frontal and left parietal cortices was related to "FOK" selection rather than "know" or "don't know" selections (Fig. 8.4). The left mid-lateral prefrontal cortex exhibited greater activation in "FOK" and "know" selections than "don't know" selection, and the left middle temporal lobe exhibited greater activation in "know" selection than "FOK" and "don't know" selections. Thus, these results also indicate prefrontal contribution to the FOK judgment.

Further, Schnyer et al. (2005) used an episodic FOK paradigm, the same task as they used before, and tried to confirm the results that the medial prefrontal cortex participates in memory monitoring (Schnyer et al. 2004) by fMRI. Results obtained from fMRI studies showed that the ventromedial prefrontal cortex participates in accurate FOK judgment and that the degree of activation of the ventromedial prefrontal cortex is modulated by the rating of the FOK judgment.

Chua et al. (2009) used a face-name associative memory test and examined which brain areas participate in prospective and retrospective monitoring in metamemory by fMRI. In this study, the subject was presented with 270 faces with fictional first names one by one before the scanning. During the scanning, a face was presented and the subject was asked whether they "know" the first name, or had "high" FOK or "low" FOK about the name (prospective monitoring). After the FOK judgment, the subject was tested whether he recognized the name or not by selecting the correct name from the lists, and then the subject was asked how much he was confident about his selection by choosing either "knew," "high confident," or "low confident" (retrospective monitoring). Both prospective and retrospective monitoring produced greater activation in the medial prefrontal cortex and the medial and lateral parietal regions, and mild activation in the lateral inferior and dorsal medial prefrontal cortex.

Further, Rounis et al. (2010) applied theta-burst transcranial magnetic stimulation to the bilateral prefrontal cortex and showed impairment of metacognitive ability in humans. Chua and Ahmed (2016) used a metamemory task with the FOK judgment and examined dorsolateral prefrontal contribution to memory monitoring by high-definition transcranial direct current stimulation. They found that the stimulation at the dorsolateral prefrontal cortex improved FOK judgment (memory monitoring ability) compared to the stimulation at the control site, where any improvement in memory performance was not observed.

Fleming et al. (2010) examined the ability to introspect self-performance. Since accurate introspection requires discrimination between correct decisions from incorrect decisions, metacognitive ability is required in accurate introspection. They showed that introspective ability is correlated with the gray matter volume in the anterior prefrontal cortex. In addition, Fleming et al. (2012) showed that the right rostrolateral prefrontal cortex exhibited greater activation during self-report condition compared to a control condition. They also showed that the activity in this region correlated with the strength of the confidence reported by the subject and that the strength of the relationship between the activity in this region and the confidence reported by the subject predicted metacognitive ability of individual subjects. Thus, these results suggest the contribution of the rostral prefrontal cortex to metacognitive functions. The importance of the rostral prefrontal cortex in metacognitive ability has also been observed by Yokoyama et al. (2010) and Baird et al. (2013). Recently, Miyamoto et al. (2018) performed fMRI studies using monkeys and showed that the bilateral frontopolar cortices (area 10) were activated in metacognitive evaluation of non-experienced events and that reversible inactivation of area 10 selectively impaired the confidence judgment of non-experienced events. Thus, the frontopolar cortex (area 10) seems to play an important role for metacognitive confidence judgment. Contribution of the anterior prefrontal cortex to metacognitive functions is also shown by Miyamoto et al. (2021) in human fMRI studies. They considered two types of probabilities in decision-making; external probability and internal probability. The external probability is the probability of the outcome irrespective of the subject's effort and the result is determined by a given environment. On the other hand, the internal probability is determined by the subject's ability to make a correct choice among options and the decision in this condition is based on the subject's metacognitive ability. In their task, the subject needed to make a decision regarding the direction of the random dot motion. Their task consisted of two sub-tasks (main task and follow-up task). In the main task, which was introduced first, two random dot motion stimuli were presented simultaneously and the subject was asked to select either stimulus. Each of the two stimuli had different number of dots and was modulated by different motion coherence. Then, the follow-up task was introduced, in which the same stimulus as the one that the subject selected in the main task was presented and the subject was asked to answer the direction of the random dot motion. In this task condition, which of two stimuli the subject would select in the main task depends on the subject's ability how well to make correct decision in the follow-up task, which is based on internal probability and corresponds to the metacognitive judgment. On the other hand, the follow-up task is based on external probability and corresponds to the perceptual decision. fMRI studies using this task revealed that the anterior lateral prefrontal cortex is specialized for metacognitive judgment. And, the disruption of the anterior lateral prefrontal cortex by the TMS caused an alteration of the metacognitive judgment. Thus, the anterior portion of the prefrontal cortex seems to be an important brain area for metacognition. The importance of the anterior prefrontal region in metacognition is also observed in human TMS study (Shekhar and Rahnev 2018).

Thus, neuropsychological studies using human patients having local brain damage and neuroimaging studies using fMRI indicate that the prefrontal cortex participates in correct metacognitive judgment. Although activation areas in the prefrontal cortex somewhat vary across studies, the difference in information used for task performances and the difference in underlying information processing for different metamemory tasks might cause the difference in activation areas across studies. Difference in used metamemory paradigms (confidence judgment or FOK judgment) might also cause the difference in activation areas. Morales et al. (2018) examined whether metacognition relies on a domain-general resource, which is applied to different tasks, or on a domain-specific resource, which applied to a specific task. They used both a perceptual task and a memory task and found that the anterior prefrontal cortex predicted confidence levels in a domain-specific manner, while a widespread network in the frontal cortex predicted confidence and accuracy in a domain-general manner. However, correct metacognitive judgment and confidence judgment are both based on correct monitoring of own memory. Thus, both neuropsychological and neuroimaging studies show that the prefrontal cortex, maybe the anterior prefrontal area, is an important brain structure for metamemory.

8.5 Metamemory Studies Using Non-Human Animals

Since metamemory is subjective function, metamemory ability is thought to be examined only by introspective reports using language. It is thought that metamemory is a unique human capacity and that there is no appropriate method to examine this capacity by non-human animals. Therefore, it has not been examined whether non-human animals can also exhibit similar capacity as human's metamemory. However, recent behavioral studies accumulate evidences that non-human animals including rhesus monkeys share functional parallels with human's metacognition (Hampton 2009, 2010; Templer and Hampton 2012; Basile et al. 2015). There is firm evidence that monkeys can make confidence judgments about their performance and memory on various cognitive tasks. Monkey's ability to make confidence judgments about their own performance and memory shows clear evidences that monkeys have the capacity for metacognition and metamemory. It is important to examine how similar metamemory in monkeys is as metamemory in humans and whether the neural mechanism of monkey's metamemory is analogous to that of human's metamemory or not. Since evidences that rhesus monkeys share functional parallels with human's metacognition have been accumulated, behavioral or neurobiological studies of metamemory using monkeys would become a powerful new tool to understand the nature of human mind.

8.5.1 Behavioral Paradigms for Animal Studies

Two different methods have been used to examine the capacity of metamemory; prospective and retrospective paradigms. In *prospective paradigms*, the subject is first asked to make confidence judgments about its memory of prior event (e.g., whether the subject remembers the cue or not) (correspond to FOK judgment in human studies), and then make a decision (e.g., whether the subject takes a memory test or declines it) based on its confidence judgment. In *retrospective paradigms*, the subject first makes a decision (e.g., the subject takes a test and answers the question), and then make confidence judgment about its response (e.g., the subject answers whether the response was correct or not). Whether or not and how strong positive correlation is present between the subject's accuracy on the memory test and the degree of its confidence are used as a measure of the subject's capacity of metamemory.

8.5.1.1 Prospective Paradigm

In the prospective paradigm, three basic principles are needed to be considered. *First*, the subject has to take a difficult memory test. The difficulty of the memory test would be controlled by increasing the duration of the memory delay, increasing the number or the complexity of items that the subject needs to remember, or increasing the number or the characteristics of distracters. Introducing the difficult memory test guarantees that the animal subject often feels uncertain to make a correct response. *Second*, the behavioral paradigm includes an *escape option* (an option that the subject can decline to take a memory test). The memory paradigm used to examine metamemory includes two conditions; *forced test condition* and *free test condition*. In the forced test condition, the subject needs to take a memory test without an escape option. However, in the free test condition, the subject has a chance to select whether it takes a memory test or decline to take a test (*escape option*) by itself. If the correct answer is uncertain in the current trial, the subject can use an escape option to handle this trial. Therefore, the prospective paradigm includes the following three types of trials. *Chosen test trials* are trials that the subject selects to take a memory test by itself. *Chosen escape trials* are trials that the subject selects to decline a memory test by itself. And, *forced test trials* are trials that the subject is forced to take a memory test. *Third*, the subject receives different amounts of reward depending on its selection of trials and its behavioral performance. When the subject makes a correct response in the chosen test trial and the forced test trial, it will receive a full amount of the most favorable reward. However, when the subject makes an error in the chosen test trial and the forced test trial, it will receive no reward and have long time-out penalty. When the subject selects an escape option, it will receive the least favorable reward or a small amount of reward in 30–50% chance. Thus, the escape option leads to a favorable result compared with a failure on the memory test (no reward), while the correct response on the test yields the most favorable result.

The basic idea underlying the prospective paradigm is that, if the subject has metamemory capacity, it could effectively use the escape option to decline the memory test when the correct answer is uncertain and to obtain reward effectively. If the subject is able to monitor its own memory state, it can be predicted that the subject more often selects the escape option in more difficult trials. Therefore, the ratio that the subject selects the escape option would increase in proportion with the increase of task difficulty. In addition, if the subject effectively uses the escape option and selectively escapes from the memory test in difficult trials, it can improve the accuracy of the performance of the memory test in the free choice condition, compared to when it is required to take the same test in the forced choice condition. Thus, it can be concluded that non-human animals have metamemory capacity if the ratio of correct performances is higher in the chosen test condition than in the forced test condition and if the ratio of the selection of escape options is higher in difficult trials than in easy trials.

8.5.1.2 Experiments by Smith et al.

Based on this idea, Smith et al. (1997) examined whether animals escaped some trials adaptively when the correct choice was uncertain, like humans do. Human subjects and rhesus monkeys were asked to perform visual discrimination tasks.

Three visual stimuli ("Box" shape, a letter "S," and "Star" shape) were presented on the screen and the number of illuminated pixels presented within the "Box" changed from 450 to 2950 pixels. If the number of illuminated pixels in the "Box" was 2950 pixels ("dense condition"), the subject's correct response was to move the cursor to the "Box," while if the number of illuminated pixels was between 450 and 2949 pixels ("sparse condition"), the subject's correct response was to move the cursor to the letter "S." If the subject was unable to decide whether illuminated pixels were dense or sparse, it could move the cursor to the "Star" (escape option). The subject always received reward when it selected the "Box" and the "S," but received a small amount of delayed reward when they selected the "Star." As a result, both humans and monkeys selectively chose the escape option in which the dense-sparse discrimination was difficult. This result indicated that humans and monkeys exhibited similar cognitive ability to monitor uncertainty in sensory perception, suggesting that monkeys have metacognitive capacity similar to humans.

Smith et al. (1998) further compared the capacity to monitor own memory (metamemory ability) between humans and rhesus monkeys using the same behavioral tasks. They used a serial probe recognition task, which is able to examine different difficulty in retrieving particular information from memory across its serial position. Human studies had shown that humans exhibited stronger serial position effects in this task, such as primacy and recency effects. In the experiment using monkeys, a list of pictures (2, 4, or 6 pictures) was first presented one by one to the monkey. Then, a probe picture, a large letter "D," and a "Star" shape were presented simultaneously on the screen and the monkey was asked to make a decision whether the probe picture was in the list or not. If the monkey thought that the probe picture was in the list, it needed to move the cursor to the picture. If the monkey thought that the probe picture was not in the list, it needed to move the cursor to the letter "D." When the monkey selected the "Star" (escape option), the screen was cleared, then either "D" (if the probe was not in the list) or the probe picture (if the probe picture was in the list) was presented, and the monkey was required to move the cursor to either stimulus to get a reward. Similar to humans, monkeys exhibited a strong recency effect. In addition, monkeys used escape options less in shorter-list conditions but more in longer-list conditions. And also, monkeys used escape options less when the probe picture appeared in later serial position, but more when it appeared in an early serial position. Thus, similar to humans, monkeys used escape options selectively when the information retrieved from memory became uncertain. The pattern of selecting escape options mirrored the pattern of difficulty in retrieving information from memory across its serial position. These results also indicate that rhesus monkeys share functional parallels with human's metamemory abilities.

8.5.1.3 Experiments by Hampton

One of the typical behavioral experiments for examining a monkey's ability of metamemory using the prospective paradigm is the experiment performed by Hampton (2001). He used two rhesus monkeys who performed a delayed matching-to-

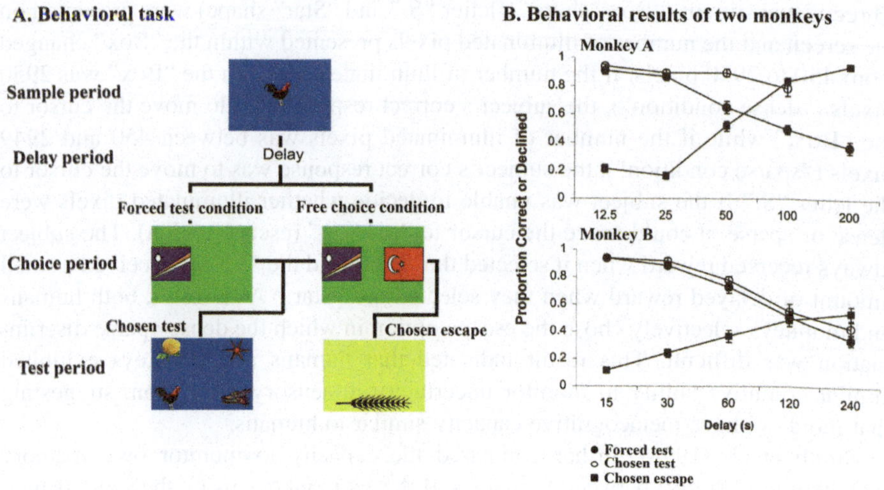

Fig. 8.5 (a) Schematic drawing of the behavioral paradigm used for monkeys. (b) Behavioral results of two monkeys. From Hampton (2001) with permission

sample task (Fig. 8.5a). In this task, after the inter-trial interval, a sample stimulus (colored clip-art image) was presented on the monitor. The sample stimulus was selected randomly from a set of 4 images. The monkey was required to touch the sample stimulus three times during the sample period. Then, the monitor screen was cleared and the delay period was introduced. The task difficulty was controlled by changing the length of the delay period from 12.5 s to 240 s. During the choice period, either *free choice condition* or *forced test condition* was introduced in two-third or one-third of trials, respectively. In the *free choice condition*, two visual stimuli were presented simultaneously on the left and the right of the monitor. If the monkey touched the left-side stimulus, it needed to take a memory test (*chosen test trial*). In the test period of the test trial, the sample stimulus and three distracters were presented randomly at four corners on the screen. If the monkey selected the sample stimulus correctly by touch, it would receive a favorable reward (e.g., peanut). However, if the monkey made an error, no reward was given and a 15-s timeout period was introduced. If the monkey touched the right-side stimulus during the choice period, it would escape the memory test (*chosen escape trial*). In the test period of the escape trial, one image that differed from the sample stimulus was presented on the monitor. If the monkey touched this image, it would get the least favorable reward (e.g., primate pellet). In the choice period of the *forced test condition*, only one visual image was presented on the monitor, and the monkey needed to touch this image. In the test period, similar to the chosen test trial, the sample stimulus and three distractors were presented randomly at four corners on the screen. If the monkey selected the sample stimulus correctly by touch, it would receive a favorable reward (e.g., peanut). However, if the monkey made an error, no reward was given and a 15-s timeout period was introduced. The free choice condition and the forced test condition were intermixed randomly during daily sessions so that the monkey could not predict the trial type at any given trial.

In this experiment, Hampton compared proportions of correct performance between forced test trials and chosen test trials of the free choice condition and found that the proportion of correct performances was higher during chosen test trials of the free choice condition than the forced test condition, especially in more difficult conditions (longer delay conditions), although monkeys' correct performances declined proportionally with the increase of the delay length (Fig. 8.5b). He also examined the proportions of chosen escape trials across different levels of task difficulty and found that the ratio to select the escape option increased in proportion with the increase of task difficulty (increase of delay length) in both monkeys (Fig. 8.5b). These results show that monkeys can improve the accuracy of their performance effectively using the escape option in the memory test, especially in difficult trials of the free choice condition, compared with the forced test condition, in which they needed to perform the same task without the escape option.

However, another explanation is also possible, such that events that occurred during the delay period (e.g., noises, self-grooming, or changes in motivation) might results in escaping the memory test. To rule out this possibility, Hampton (2001) unpredictably introduced trials having no sample stimulus in normal experimental sessions. If the absence of the memory trace causes the escape of the memory test, the monkeys would select the escape option in trials having no sample stimulus, similar to trials in which they forgot the sample stimulus. However, if the decision to escape the memory test is caused by some environmental or behavioral event, the probability that the monkeys select the escape option would be a value between the value obtained under no-sample trials and the value obtained under normal trials. Hampton (2001) found that monkeys selected escape options with a significantly higher probability in no-sample trials than in normal free choice trials, supporting the notion that monkeys' selection of the escape option is caused by the loss of memory. These results indicate that monkeys can effectively use the escape option to escape the memory test when the correct answer is uncertain. These results also indicate that monkeys can improve their accuracy of memory test performance and increase the chance to get the reward using the escape option effectively in uncertain trials.

Hampton et al. (2004) further examined whether or not rhesus monkeys have the metamemory capacity to access the contents of short-term memory. They used four tubes mounted on a tray, one of which contained a reward. The tray was placed at one of four levels; the lowest was near the bottom of the cage and the highest was near the monkey's eye level. Initially, the tray was set at the monkey's eye level and the monkey was trained to look down the length of each tube to see whether the reward was placed at the far end of the tube or not. By this training, the monkey spontaneously looked down the length of the tubes to search which tube had a reward. After this training, the monkey was asked to obtain a reward from the tube either in the *seen condition* or in the *unseen condition*. In the seen condition, the monkey could see which tube the reward was baited by the experimenter, while in the unseen condition, the monkey could not know which tube contained the reward. During the choice period after the delay period, the monkey had a chance to look down each tube to determine whether the reward was hidden or not. If the monkey knew which tube had a reward (seen condition), it would choose a correct tube without looking down other tubes. However, in the unseen condition, the monkey needed to make the effort to search for the reward by looking down the tubes one by one. Therefore, if the difference was present in looking-down behavior between the seen condition and the unseen condition, it could be concluded that the monkey can discriminate between knowing and not knowing. In fact, the monkeys chose a baited tube in shorter latencies without looking down at other tubes in the seen condition, while they took more time to find a baited tube in the unseen condition. Thus, these results show clear evidence that monkeys have a capacity to monitor their memory and its certainty. Therefore, it can be concluded that monkeys have similar metamemory capacity as humans have.

The results obtained by Hampton (2001), Hampton et al. (2004) were further supported by the results by Templer and Hampton (2012) and by Basile et al. (2015).

8.5.1.4 Experiments by Fujita

Fujita (2009) also used the prospective paradigm and reported behavioral studies of metamemory using tufted capuchin monkeys (*Cebus apella*). The basic idea of his behavioral paradigm is the same as Hampton's (2001). He used a delayed matching-to-shape task (Fig. 8.6b). Nine visual stimuli constructed by the combination of three different shapes (circle, triangle, and plus) and three different texture patterns (Fig. 8.6a) were used as sample stimuli and comparison stimuli. In this task, the monkey needed to remember both the shape and the texture of the sample stimulus during the delay period and make a choice response based on either the shape or the

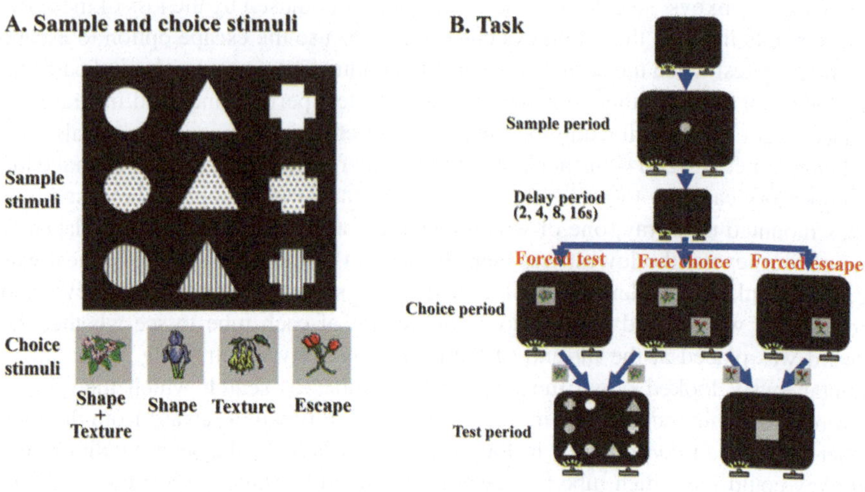

Fig. 8.6 (**a**) Sample and choice stimuli used for the study. (**b**) Schematic drawing of the behavioral paradigm used for capuchin monkeys. From Fujita (2009) with permission

texture or both of the comparison stimulus. When the monkey depressed a lever and kept it down for 1 s, a sample stimulus was presented at a center of the touch panel, and the monkey was required to touch the sample stimulus five times to extinct the stimulus (sample period) and then, the delay period was introduced for 2–32 s. To examine whether the monkey showed metamemory ability or not, the choice period was introduced at the end of the delay period. During the choice period, one or two choice stimuli (Fig. 8.6a) were presented, which indicated the trial type. The choice stimulus with blue flowers indicated that the trial is a *shape test trial*, and the monkey needed to select a comparison stimulus with the same shape as the sample stimulus during the test period. The choice stimulus with yellow flowers indicated that the trial is a *texture test trial*, and the monkey needed to select a comparison stimulus with the same texture as the sample stimulus. The choice stimulus with pink flowers indicated that the trial is a *shape and texture test trial*, and the monkey needed to select a comparison stimulus exactly the same as the sample stimulus. On the other hand, the choice stimulus with red flowers indicated that the trial is an *escape trial*, and the monkey could escape the memory test by selecting this stimulus (Fig. 8.6). In the choice period, one of three conditions (*free choice condition*, *forced test condition*, and *forced escape condition*) was introduced. In the *free choice condition*, two choice stimuli were presented and the monkey could select either stimulus to take or escape a test. In the *forced test condition* and the *forced escape condition*, only one choice stimulus was presented and the monkey had to either take a memory test or escape a test depending on the stimulus. These three conditions were introduced at the ratios of 60%, 30%, and 10% of trials in a daily session, respectively. During the test period, 9 comparison stimuli (Fig. 8.6a) were presented at 9 positions randomly selected from 10 positions, and the monkey was required to touch one corresponding stimulus depending on the trial type to get a reward.

Fujita (2009) compared proportions of the correct performance between trials in the forced test condition and chosen test trials in the free choice condition and proportions of chosen escape trials in the free choice condition among different levels of task difficulty. The ratio of selecting escape options increased in proportion with the increase of task difficulty in both capuchin monkeys, suggesting that capuchin monkeys are able to monitor the strength of their memory trace. In addition, he occasionally introduced trials with no sample stimulus and found that these monkeys selected escape options more frequently in no-sample-stimulus trials than in regular trials of the free choice condition. These results indicate that capuchin monkeys can effectively use escape options to improve their accuracy in a memory test and effectively obtain reward. Thus, these results show evidence that capuchin monkeys also have metamemory capacity.

8.5.2 *Retrospective Paradigms*

In prospective metamemory paradigms, the subjects first make a confidence judgment about their memory and then make a decision based on their judgment.

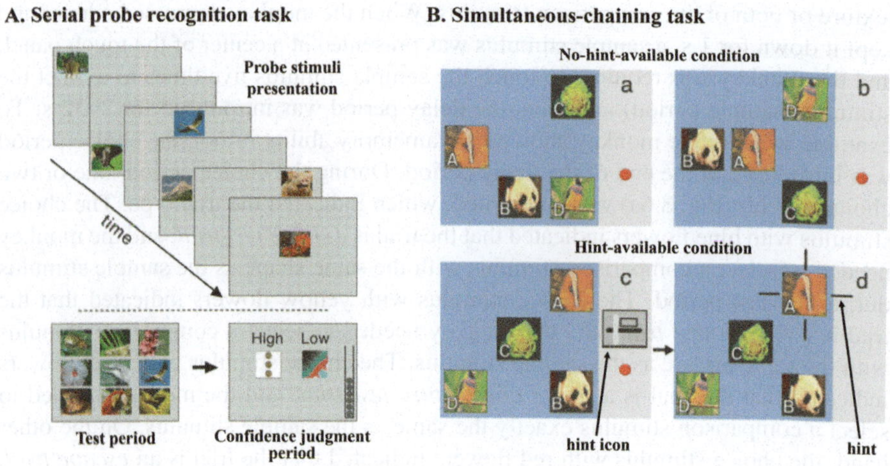

Fig. 8.7 Behavioral tasks used by Kornell et al. (2007). (**a**) Serial probe recognition task. (**b**) Simultaneous-chaining task. From Kornell et al. (2007) with permission

However, in retrospective metamemory paradigms, the subjects first make a decision and then make a confidence judgment about their decision. Similar to the prospective paradigm, the subjects first have to take a difficult memory test. Introducing the difficult memory test guarantees that animal subjects often feel uncertain to make a correct answer. After the subjects make an answer, they are asked the degree of their confidence about the accuracy of their answer.

For example, Kornell et al. (2007) trained monkeys to perform a serial probe recognition task (Fig. 8.7a). In this task, six probe pictures were presented one by one with inter-stimulus interval of 2 s. Following the presentation of the last probe picture and a 0.5-s delay period, nine test pictures were presented (test period). These nine test pictures included one of six probe pictures (familiar picture) and eight distractors (unfamiliar picture) and the monkey was required to identify the familiar picture. After the monkey made a decision, two confidence icons (one for high confidence and one for low confidence) were presented and the monkey was asked to select either icon. If the monkey selected a high-confidence icon and made a correct choice, it would get a favorable reward. However, if the monkey selected a high-confidence icon and made an incorrect choice, it could not get any reward. If the monkey selected a low confidence icon, it would get a less favorable reward, regardless of whether it had made a correct choice or not. In fact, when two rhesus monkeys selected a high confidence icon, they showed highly accurate selection in this task, indicating that monkeys can accurately monitor their memory.

Further, Kornell et al. (2007) asked monkeys to perform a simultaneous-chaining task either in the hint-available condition or in the no-hint-available condition (Fig. 8.7b). In this task, four pictures were presented on a monitor and their positions varied randomly from trial to trial. The monkey's task was to touch these pictures in a particular order, regardless of their spatial positions. In the no-hint-available

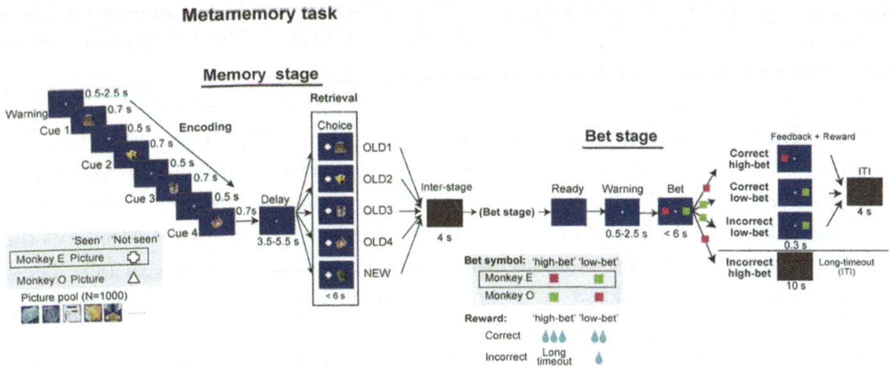

Fig. 8.8 Retrospective metamemory task used by Miyamoto et al. (2017). Modified from Miyamoto et al. (2017) with permission

condition, the monkey needed to learn the order for touching the pictures by trial and error. In the hint-available condition, a hint icon and a red circle were presented on the right side of the monitor. If the monkey touched the red circle, it needed to perform the task without a hint and would get a favorable reward if it made a correct response. If the monkey touched the hint icon, four blinking lines (hint) were presented around the correct picture that it needed to select and it would get a less favorable reward after finishing the trial. Kornell et al. (2007) found that monkeys could learn to request a hint when they faced a condition that they had to solve the task by trial and error, which is a similar behavior as humans exhibit in the same situation. These results show that monkeys can not only make an appropriate judgment about the accuracy of their responses based on the confidence of their memory but also can request the hint when they do not know the correct response. Thus, these studies also demonstrate that monkeys exhibit a similar ability of metacognition as humans have.

Miyamoto et al. (2017) also used a retrospective metamemory task and showed by pharmacological intervention that the dorsolateral prefrontal cortex (areas 9 and 46d) contributes to metamemory for temporally remote items, while the premotor area (area 6) contributes to metamemory for more recent items. In their task, the monkeys first performed a visual recognition memory task (memory stage) and then made a confidence judgment (bet stage) (Fig. 8.8). In the memory stage, four cue pictures selected from 1000 natural or artificial objects appeared serially one by one during the encoding period and, after the delay period, two choice stimuli (one test picture and one cross or triangle symbol) were presented simultaneously during the retrieval period. If the test picture was in the list of the cue pictures, the monkey was required to move a joystick in the direction of the test picture. If the test picture was not on the list, the monkey was required to move the joystick in the direction of the symbol. After the monkey made a behavioral response, the task moved to the bet stage without reward. In the bet stage, two bet targets (e.g., a pink target for high bet and a green target for low bet) appeared and the monkey was required to report its bet by moving the joystick toward either target. If the monkey made a correct answer in the memory stage and selected a high-bet target, it obtained the maximum amount of reward. If the monkey made an incorrect answer in the memory stage and selected a high-bet target, it received no reward and a 10-s time-out penalty. They obtained three main behavioral results. (1) Monkeys placed high bet frequently after a successful performance in the memory stage. (2) Significant primacy and recency effects in serial presentation of cue pictures were observed in the high-bet condition but not observed in the low-bet condition. (3) Monkeys did not use response latencies as behavioral cues for making decisions of bet. These results indicate that monkeys actually used metamemory ability to evaluate the confidence of their memory.

Then, Miyamoto et al. (2017) injected muscimol ($GABA_A$ receptor agonist) into the dorsolateral prefrontal cortex bilaterally and examined its effect on confidence judgment in the metamemory task. As a result, bilateral injections of muscimol in the dorsolateral prefrontal cortex produced significant impairment in metamemory capacity for temporally remote items (e.g., cue 1 and cue 2) than for recent items (e.g., cue 3 and cue 4) (see Fig. 8.8). On the contrary, bilateral injections of muscimol into the premotor cortex produced impairment in metamemory capacity for recent items. Moscimol injections into both areas produced no impairment in the performance of the recognition memory task (memory stage). Kwok et al. (2019) reported similar results. They showed that lesions to the dorsolateral prefrontal cortex impaired spatial recognition performance, while lesions to the superior dorsolateral prefrontal cortex impaired in the performance of a spatial metamemory task.

Although both prospective and retrospective metamemory paradigms demonstrated that the monkeys exhibit metamemory ability, behavioral tasks based on the retrospective paradigm have some advantages over those based on the prospective paradigm. Retrospective metamemory task includes the memory stage and the bet stage. A variety of memory tests can be used in the memory stage across different experiments. However, the experimental situation would be the same in the bet stage across different experiments. Therefore, it may be easier to construct or transfer to a new metamemory task from one metamemory task that was trained for a qualitatively different memory task (Kornell et al. 2007; Terrace and Son 2009). Therefore, the retrospective paradigm would be an important tool to examine the metamemory ability of monkeys under a variety of behavioral conditions.

8.6 Examination of Neural Mechanisms of Metamemory

Neuropsychological studies using human patients having local brain damages and functional brain imaging studies using normal human subjects have revealed that the prefrontal cortex plays an important role in metamemory ability. However, neural mechanisms supporting metamemory ability in the prefrontal cortex are not yet known. In order to understand neural mechanisms supporting metamemory ability in the prefrontal cortex, we first need to develop an appropriate behavioral task that requires metamemory to perform and that the animal subjects can learn and perform. Since there are firm evidences that rhesus monkeys exhibit metamemory ability, macaque monkeys would be appropriate subjects for examining neural mechanisms for metamemory. In addition, a large amount of neurophysiological data has been accumulated for neural mechanisms of prefrontal functions using monkeys. Therefore, using these available resources, it would be possible to examine and understand neural mechanisms related to metamemory in the prefrontal cortex.

8.6.1 Experiment by Tanaka and Funahashi (2012)

Tanaka and Funahashi (2012) first developed a behavioral task based on the prospective paradigm by modifying an oculomotor delayed-response (ODR) task and confirmed that monkeys exhibited behavioral evidences of metamemory ability. Then, they performed neurophysiological experiments using this task to examine neural mechanisms related to metamemory in the prefrontal cortex. The reasons why they used an ODR task as a basic behavioral task are as follows. (1) The ODR task is a spatial working memory task and the subject needs to maintain spatial information of the visual cue during the delay period. (2) The ODR task is easy to manipulate the difficulty of the task by changing the length of the delay period or introducing different numbers of distractors during the delay period. And importantly, (3) a variety of neurophysiological data have been accumulated using the ODR task and help to understand prefrontal functions. They used the prospective metamemory paradigm, because this paradigm has been frequently used in behavioral studies to prove that monkeys exhibit metamemory ability and because prefrontal activation has been observed in relation to the FOK judgment in human neuroimaging studies.

Figure 8.9 shows a diagram of the metamemory task used for this study. Japanese monkeys sat in a primate chair and faced a TV monitor, on which necessary visual stimuli were presented. The monkey's head was immobilized and eye movements were monitored by a magnetic search coil technique. After a 4-s inter-trial interval, a fixation point was presented at the center of the monitor. If the monkey maintained fixation at the fixation point for 1 s, a visual cue was presented for 0.5 s (cue period) at one of eight fixed positions around the fixation point. The monkey was required to maintain fixation at the fixation point during the cue period and a subsequent delay period. During the delay period, various numbers (0, 3, or 7) of distractors were presented one by one around the fixation point for 0.2 s with a 0.3-s inter-stimulus interval. The difficulty of each trial was manipulated by presenting different numbers of distracters and introducing different lengths of the delay period. At the end of the delay period, the fixation point was extinguished and two colored squares (green, red, or blue) appeared at the upper and lower positions of the fixation point (choice period). The monkey was required to select one of the colored squares by a saccade and maintain fixation at the selected square for 1 s. This task included three test conditions (*free-choice*, *forced-test*, and *forced-escape* conditions) and the

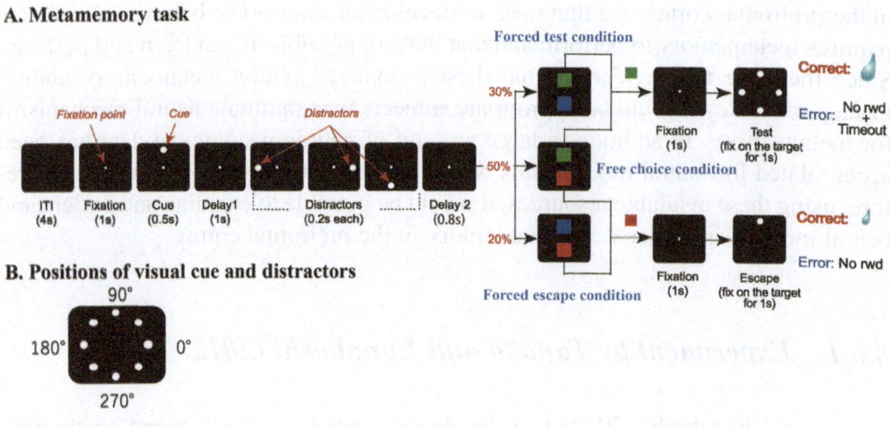

Fig. 8.9 (**a**) A metamemory task modified from the ODR task. (**b**) Positions of the visual cues and distractors

combination of the colored squares indicated the test condition. In the *free-choice condition*, both green and red squares were presented. When the monkey selected the green square (*chosen-test trial*), the monkey was required to take a matching-to-place test after a 1-s fixation period. In this test, visual stimuli were presented at the positions where the visual cue and distracters had been presented during the cue and the delay periods, and the monkey needed to make a saccade to the position where the visual cue had been presented. If the monkey made a correct saccade, it would get a larger and favorable reward. However, if the monkey made an error, it would not get any reward and get a 20-s time-out penalty. When the monkey selected the red square (*chosen-escape trial*), only one visual stimulus was presented at the cue position and the monkey was required to make a visually guided saccade to that position. If the monkey made a correct saccade, it would get a small and less favorable reward in only 30–40% of trials. In the *forced-test condition*, both green and blue squares were presented. When the monkey selected the green square, the monkey was required to take a matching-to-place test, same as the *chosen-test trial* in the *free-choice condition*. When the monkey selected the blue square (*trap trial*), the task was immediately terminated without reward and a 30-s time-out penalty was introduced. In the *forced-escape condition*, both red and blue squares were presented. When the monkey selected the red square, the monkey was required to perform a visually guided saccade same as the *chosen-escape trial* in the free-choice condition. When the monkey selected the blue square (*trap trial*), the task was terminated immediately without reward and a 30-s time-out penalty was introduced.

First, Tanaka and Funahashi (2012) used four Japanese monkeys and tried to confirm that the current task was appropriate for examining metamemory ability in monkeys and that monkeys used metamemory to perform this task. The basic idea underlying the prospective paradigm is that, if monkeys have metamemory capacity, they effectively use an escape option to decline the memory test when the correct answer is uncertain. Therefore, it is expected that the ratio of selecting the escape

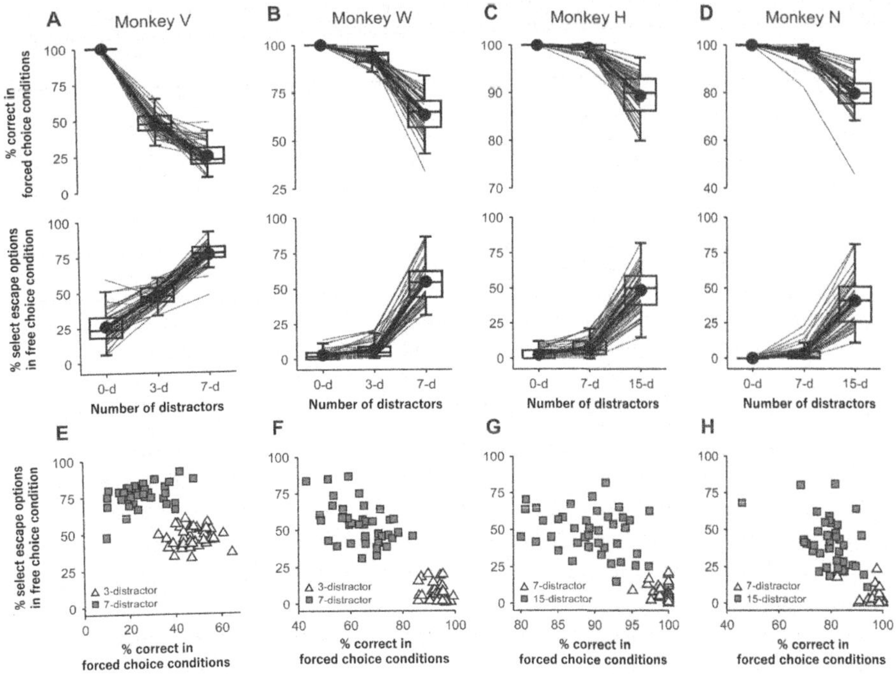

Fig. 8.10 Behavioral results obtained from four monkeys while they performed the metamemory task modified from the ODR task. From Tanaka and Funahashi (2012) with permission

option increases in proportion to the increase in task difficulty. In addition, if the monkeys can effectively use the escape option in uncertain trials, it is expected that the accuracy of the memory task is improved in chosen-test trials of the free-choice condition, compared to that in the forced-test condition. Therefore, if the ratio of the correct performance is higher in chosen-test trials than in forced-test trials and if the ratio of selecting the escape option is higher in difficult trials than in easy trials, it can be concluded that the behavioral task is appropriate to examine metamemory ability in monkeys and that monkeys use metamemory to perform this task.

The proportion of the correct performance in the forced-choice condition decreased for all four monkeys as task difficulty (number of distractors) increased (Fig. 8.10a–d). In addition, all monkeys selected the escape option in the free-choice condition more often as task difficulty increased (Fig. 8.10a–d). Negative correlations were observed between the proportion of the correct performance in the forced-choice condition and the proportion of selecting the escape option in the free-choice condition (Fig. 8.10e–h). These results indicate that monkeys experienced the uncertainty of their memory more frequently in more difficult trials of this task. A comparison of the memory performance between chosen-test trials of the free-choice condition and trials of the forced-choice condition revealed that the memory performance was improved in the free-choice condition relative to the forced-choice condition (Fig. 8.11a–d). Although monkey V did not show a statistically significant

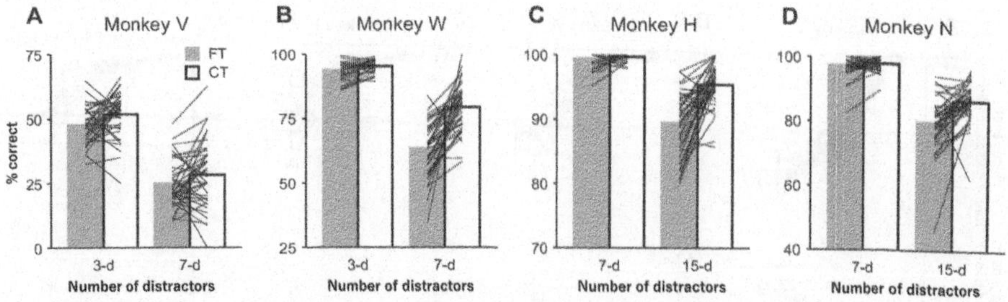

Fig. 8.11 Behavioral results obtained from four monkeys while they performed the metamemory task modified from the ODR task. FT: trials in the forced-choice condition. CT: chosen-test trials in the free-choice condition. From Tanaka and Funahashi (2012) with permission

difference in the memory performance between two conditions even in difficult trials, the other three monkeys exhibited a significantly higher proportion of the correct performance in chosen-test trials of the free-choice condition than in the forced-choice condition, especially in more difficult trials. Better performance in chosen-test trials of the free-choice condition compared with trials of the forced-choice condition were maintained for a rather long period, especially in more difficult task condition (Fig. 8.12).

Thus, behavioral results show that the ratio of the correct performance is higher when monkeys take the memory test in the free-choice condition than in the forced-choice condition and that the ratio of selecting the escape option is higher in more difficult trials than easier trials. Therefore, it is concluded that the behavioral task used in this study is appropriate as a prospective metamemory task and that monkeys used metamemory to perform this task.

Then, Tanaka and Funahashi (2016) recorded single-neuron activity from the dorsolateral prefrontal cortex while monkeys performed this metamemory task (see Fig. 8.9a). The memory task used in this study is the task similar to the ODR task. Using the ODR task, a variety of prefrontal activities have been reported, including cue-, delay-, and response-period activity with or without directional selectivity (see Chap. 3). When the monkey performs this metamemory task under the free-choice condition, the decision of whether it takes the memory test or escapes the test would depend on whether correct spatial information is maintained during the delay period or not. It has been shown that directional delay-period activity is truncated or is not observed on error trials. Therefore, whether a significant delay-period activity is maintained during the delay period or not and whether directional selectivity of delay-period activity is maintained during the delay period or not must be important determinants of whether the monkey takes the memory test or escapes it. Or, the strength of directional selectivity of delay-period activity could also be a determinant of whether the monkey takes the memory test or escapes it.

Figure 8.13a, c show population histograms of prefrontal activity while two monkeys (W and H) performed the metamemory version of the ODR task. Thick lines indicate activity during trials in which the spatial cue was presented at neurons' preferred position, while thin lines indicate activity during trials in which the spatial

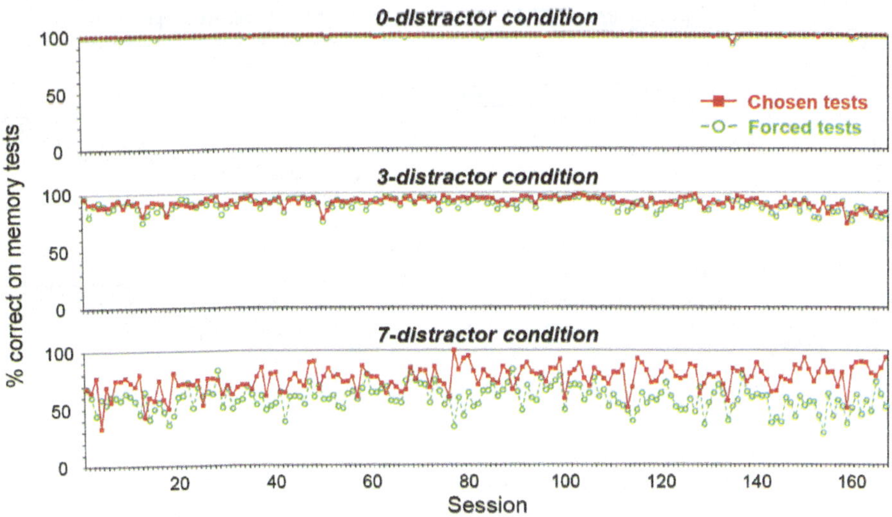

Fig. 8.12 Consistent difference in behavioral performances between chosen-test trials of the free-choice condition (Chosen tests) and trials of the forced-choice condition (Forced tests) was observed in the most difficult condition (7-distractor condition) for a long time

cue was presented at the opposite of neurons' preferred position. Figure 8.13b, d indicate temporal changes of ROC values, which indicate the degree of similarity in magnitude of activity between preferred and non-preferred cue conditions. ROC values were obtained by mean discharge rates of a 200-ms time window moving at a 20-ms step from 1.0 s before the onset of the visual cue until 6.0 s or 12.0 s after the offset of the visual cue. Black, red, and blue lines indicate activity during chosen-test trials in the forced-test condition, chosen-test trials in the free-choice condition, and chosen-escape trials in the free-choice condition, respectively.

In monkey W, when the monkey selected to take the memory test in the forced-test and the free-choice conditions, a significantly larger magnitude of delay-period activity was observed during trials of preferred cue condition than trials of non-preferred cue condition (see black and red lines in Fig. 8.13a). Increased ROC values were maintained during the delay period in both forced-test and free-choice conditions (see black and red lines in Fig. 8.13b), indicating that directional selectivity of delay-period activity is maintained during the delay period. However, when the monkey selected the escape option, the difference in the magnitude of delay-period activity between preferred and non-preferred cue conditions was gradually diminished with the progress of the delay period (see blue lines in Fig. 8.13a). Increased ROC values gradually decreased toward a 0.5 value during the delay period (see blue lines in Fig. 8.13b), indicating that directional selectivity of delay-period activity gradually diminishes with the progress of the delay period.

Similarly, in monkey H, when the monkey selected to take the memory test in the forced-test and the free-choice conditions, population histograms show that the

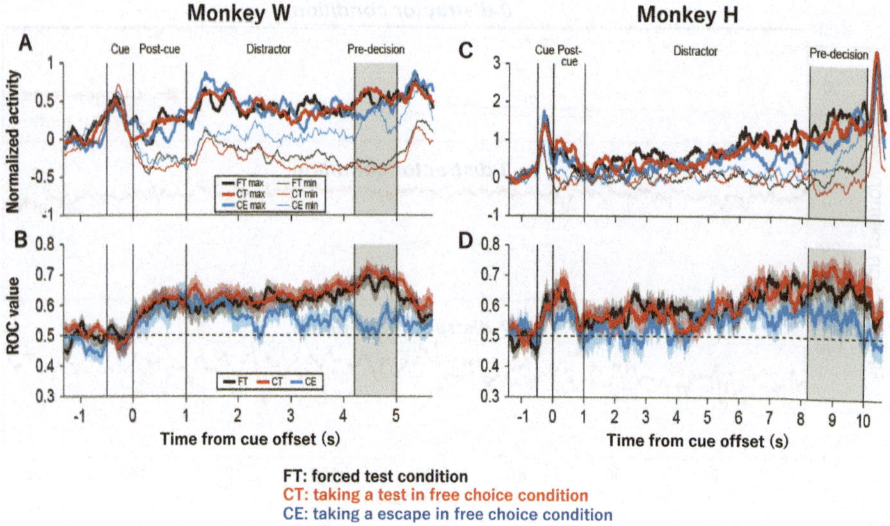

Fig. 8.13 Population histograms (**a** and **c**) of prefrontal activity and temporal changes of ROC values (**b** and **d**) when monkeys performed trials of the forced-test condition (FT, black), chosen-test trials of the free-choice condition (CT, red), and chosen-escape trials of the free-choice condition (CE, blue). Thick lines and thin lines in A and C indicate activities when the cue was presented at the preferred position and at the non-preferred position, respectively

magnitude of delay-period activity in the preferred cue condition was gradually developed with the progress of the delay period and differentiated from that in the non-preferred cue condition (see black and red lines in Fig. 8.13c). ROC values also gradually increased with the progress of the delay period (see black and red lines in Fig. 8.13d), However, when the monkey selected the escape option in the free-choice condition, the difference in the magnitude of delay-period activity between preferred and non-preferred cue conditions became weak and ROC values became smaller than chosen-test trials (see blue lines in Fig. 8.13c, d).

These results suggest that the magnitude difference of delay-period activity between preferred and non-preferred cue conditions is one factor to determine whether the monkey takes a memory test or not in the metamemory version of the ODR task. These results also suggest that directional selectivity of delay-period activity is another factor to make a decision whether taking a test or escaping it and that the monkey selects the escape option when prefrontal neurons lose direction selectivity of delay-period activity, which is caused by increasing activation in non-preferred cue conditions.

The relation between the reduction of direction selectivity of delay-period activity and the increase of the selection of the escape option is further supported by comparing mean discharge rates during the delay 2 period (see Fig. 8.9) among chosen-test trials in the forced-test condition, chosen-test trials in the free-choice condition, and chosen-escape trials in the free-choice condition (Fig. 8.14a, b). The delay 2 period is the last 0.8 s of the delay period. Since the delay 2 period is followed by the choice period, the activity during the delay 2 period is thought to play an important role in the decision whether the monkey takes a test or not during the choice period. When the monkey took the escape option, the difference in the discharge rate of delay-period activity and ROC values between preferred and non-preferred cue conditions were diminished or reduced (Fig. 8.14a, b). A similar reduction in the discharge rate of delay-period activity and ROC values were observed in error trials in the forced-test condition (Fig. 8.14c).

Thus, these results indicate that monkeys' selection of the escape option is closely linked with the loss of direction selectivity of delay-period activity in prefrontal neurons. The loss of direction selectivity of delay-period activity was not caused by the decrease of discharge rates in preferred cue conditions but caused by the increase of discharge rates in non-preferred cue conditions. In error trials, although discharge rates in preferred cue conditions decreased, discharge rates in non-preferred cue conditions increased, causing the loss of direction selectivity of delay-period activity. When prefrontal neurons lose direction selectivity of delay-period activity and exhibit omni-directional selectivity, this causes the loss of spatial information regarding the cue position. Maintaining spatial information regarding the cue position during the delay period is essential to perform the correct response during the response period. Therefore, the monkey would tend to select the escape option if it notices or feels the loss of spatial information of the visual cue. Thus, whether prefrontal neurons having directional delay-period activity can maintain direction selectivity along a trial or not would be a critical factor to decide whether to take a memory test or not in the metamemory version of the ODR task.

8.6.2 *Experiments by Middlebrooks and Sommer (2011, 2012)*

Tanaka and Funahashi (2012, 2016) used a metamemory task based on the prospective paradigm. On the other hand, Middlebrooks and Sommer (2011, 2012) used a metamemory task based on the retrospective paradigm.

In the retrospective paradigm, the monkey first takes a memory test under several different levels of difficulty, and, after making a response, the monkey places a bet to indicate the degree of the confidence of its response by selecting a visual stimulus that corresponds to the degree of its confidence. Whether the monkey receives a reward or not is based on whether the bet is appropriate or not. If a correct response is followed by a high bet or if an incorrect response is followed by a low bet, the monkey receives a large amount of preferable reward. The idea based on this paradigm is that, if the monkey is able to monitor its memory and decision, it could advantageously bet on its own decision.

In the betting task that Middlebrooks and Sommer (2011) used (Fig. 8.15a), the monkey needed to maintain fixation at a fixation point during a trial. A visual target was presented at one of four positions 500–800 ms after the start of the fixation.

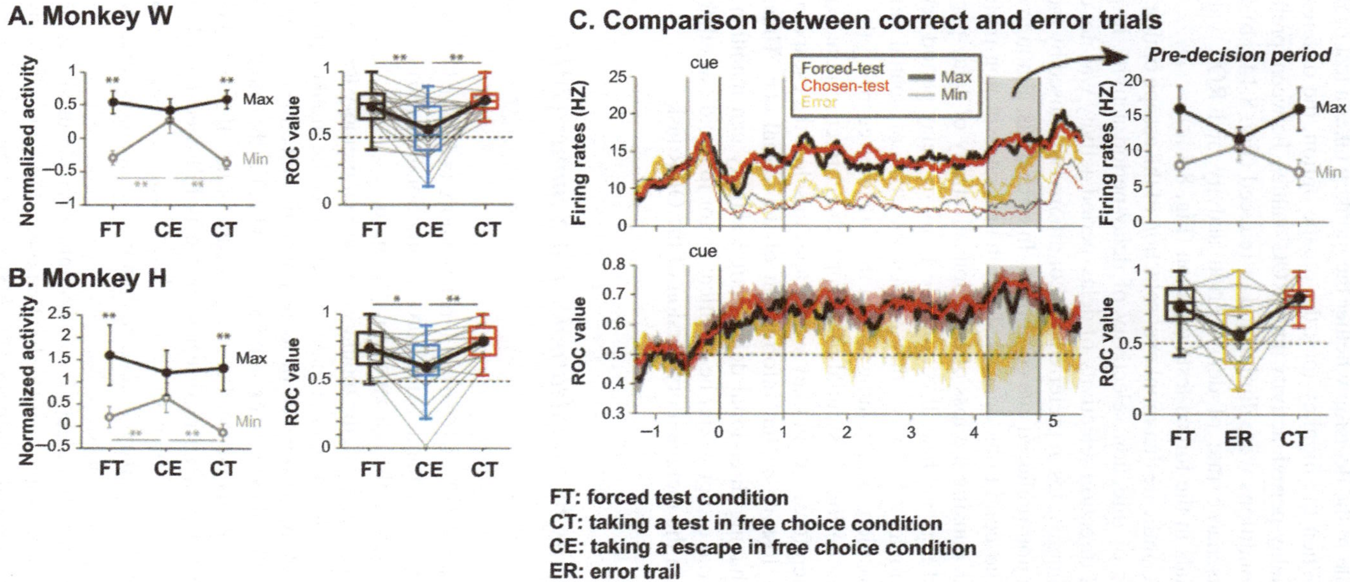

Fig. 8.14 (**a, b**) Difference of discharge rates and ROC values between preferred and non-preferred cue conditions was diminished or reduced during the delay 2 period when monkeys took escape options. (**c**) Comparison of activities among correct trials in the forced choice condition and the free choice condition and error trials in the forced choice condition

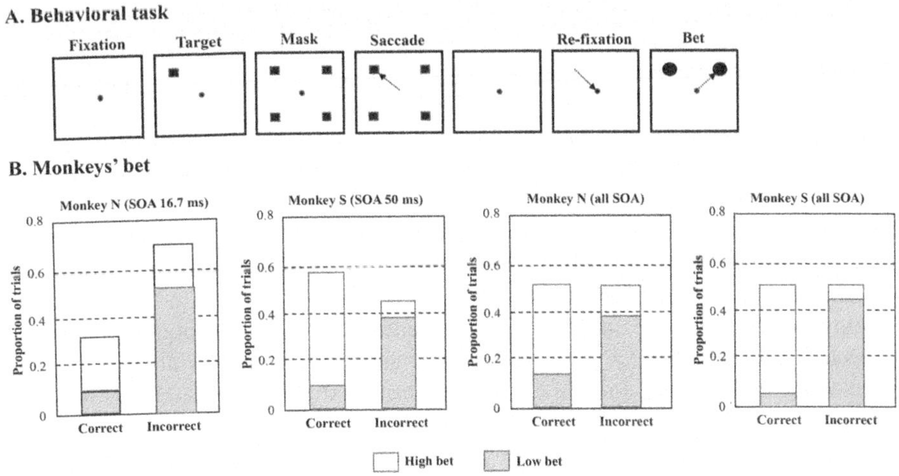

Fig. 8.15 (**a**) Schematic drawing of the temporal sequence of events in the betting task. (**b**) Behavioral results of the betting task obtained from two monkeys. SOA: stimulus onset asynchrony. From Middlebrooks and Sommer (2011) with permission

Then, mask stimuli were presented at all four positions. The interval between the target presentation and the mask presentation (stimulus onset asynchrony, SOA) was randomly selected from 16.7, 33, 50, and 66.7 ms. After the monkey was maintained fixation for an additional 500–1000 ms while mask stimuli remained visible, the fixation point was extinguished and the monkey was asked to make a saccade to the position where the visual target had been presented within 1.0 s (decision period). The fixation point reappeared after the monkey made a saccade, and the monkey needed to fixate it for 500–800 ms. Then, two bet targets (a red high bet target and a green low bet target) were presented. The monkey needed to report the bet by making a saccade to one of the bet targets (bet period). If the monkey made a saccade to the correct visual target and selected the high bet target, it would get a larger amount of preferable reward. If the monkey made a saccade to an incorrect visual target and selected the high bet target, it would receive no reward and a 5-s time-out penalty. If the monkey made a saccade to a correct visual target and selected the low bet target, or if it made a saccade to an incorrect visual target and selected the low bet target, it would receive a small amount of reward.

They obtained the following results. (1) The proportion of selecting the high bet target increased with the increase of SOA in parallel with selecting the correct visual target. (2) The monkey selected the low bet target more frequently after it made a saccade to an incorrect visual target, while it selected the high bet target more often after it made a saccade to the correct visual target (Fig. 8.15b). These results showed that, to obtain the maximum amount of reward, the monkey monitored its decision and used this information to bet advantageously, indicating that the monkey performed this task using metacognitive ability.

Then, Middlebrooks and Sommer (2012) performed single-neuron recordings at the dorsolateral prefrontal cortex, the frontal eye field, and the supplementary eye field while the monkeys performed the betting task (Fig. 8.15a). To examine whether neural activity correlates with metacognitive monitoring or not, they compared neural activity between the trials in which the monkey made saccades to the correct visual target (C) and selected the high bet targets (H) (CH trials) and the trials in which the monkey made saccades to the correct visual target (C) and selected the low bet targets (L) (CL trials). In CH and CL trials, visual stimuli and saccade directions were equivalent. They focused on neural activity during the time epoch from the decision period until the bet period.

Figure 8.16a shows examples of single-neuron activity and population activity collected from the frontal eye field, the dorsolateral prefrontal cortex, and the supplementary eye field. Frontal eye field neurons exhibited no difference between CH and CL trials during examined time epoch in single-neuron activity as well as population activity (upper figures in Fig. 8.16a). Dorsolateral prefrontal neurons exhibited small difference between CH and CL trials (more active in CH trials than in CL trials) in single-neuron activity but exhibited no difference between CH and CL trials in population activity (middle figures in Fig. 8.16a). On the other hand, supplementary eye field neurons exhibited a significant difference between CH and CL trials in both single-neuron activity and population activity (lower figures in Fig. 8.16a). These neurons exhibited more activation in CH trials than in CL trials. As shown in Fig. 8.16b, differential CH-CL activity can be seen before the decision period in supplementary eye field neurons. Some of these neurons exhibited more activity in CH trials than in CL trials (upper figure in Fig. 8.16b), but some other neurons exhibited more activity in CL trials than in CH trials (middle figure in Fig. 8.16b). These results indicate that the supplementary eye field exclusively participates in the performance of the betting task, although some neurons in the dorsolateral prefrontal cortex also participate in this task.

Putative metacognitive activity (differential CH-CL activity) started before the monkey made a saccade toward the visual target during the decision period and this activity was maintained until the monkey made a saccade to the bet target. This temporal pattern suggests that the neural process for monitoring a decision occurs almost simultaneously when the monkey makes the decision where to make a saccade.

8.6.3 Experiments by Miyamoto et al. (2017, 2018)

Miyamoto et al. (2017) used a retrospective metamemory task and showed by fMRI that the dorsolateral prefrontal cortex (areas 9 and 46d) contributes to metamemory for temporally remote items, while the premotor cortex (area 6) contributes to metamemory for more recent items. In their task, the monkeys first performed a visual recognition memory task (memory stage) and then made a confidence judgment (bet stage) (Fig. 8.8). They examined cortical areas involved in metamemory

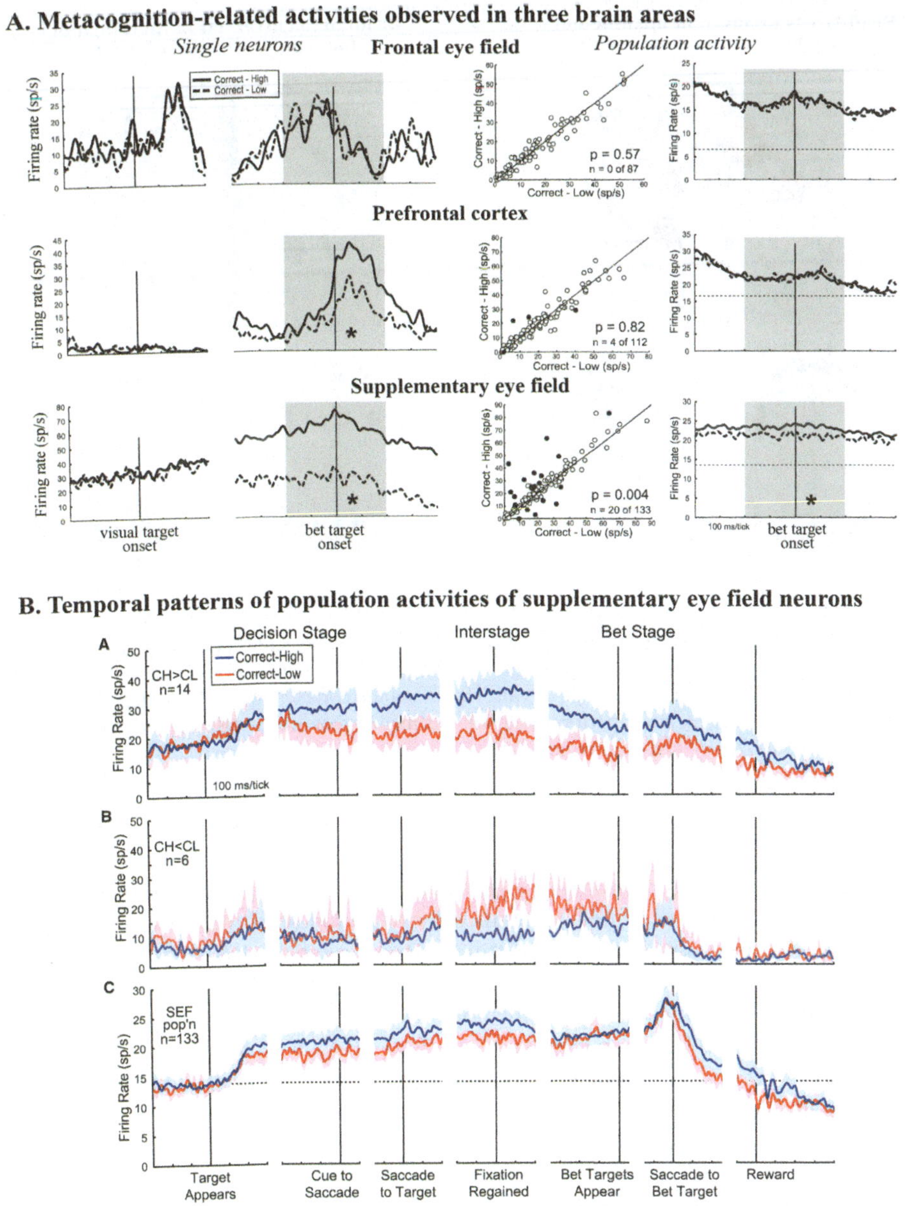

Fig. 8.16 (**a**) Metacognition-related activity observed in three brain areas. Frontal eye field neurons exhibit no difference, dorsolateral prefrontal neurons exhibit a small difference, but supplementary eye field neurons exhibit a significant difference between CH and CL trials. (**b**) Different temporal patterns of population activities of supplementary eye field neurons during CH and CL trials. From Middlebrooks and Sommer (2012) with permission

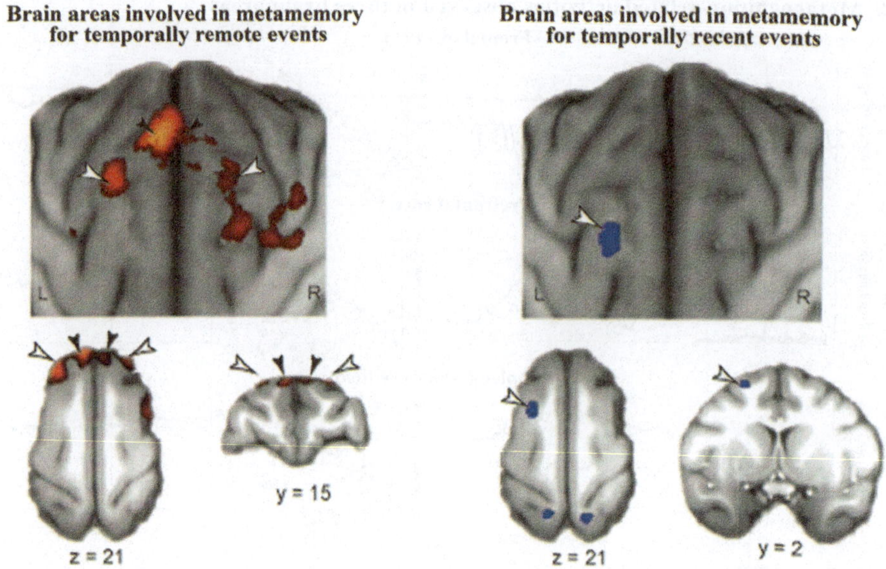

Fig. 8.17 Activated prefrontal areas while monkeys performed a metamemory task shown in Fig. 8.8. *Left*: The dorsolateral prefrontal cortex around the posterior portion of the superior principal sulcal area exhibited a significant correlation with metamemory performance for the memory of temporally remote items. *Right*: The premotor cortex (area 6) exhibited significant correlation with metamemory performance for the memory of temporally recent items. From Miyamoto et al. (2017) with permission

processes by comparing brain activity between high-bet and low-bet conditions using functional mapping of the monkey's whole brain. They found that the activation of the dorsolateral prefrontal cortex around the posterior portion of the superior principal sulcal area exhibited a significant correlation with metamemory performance for the memory of temporally remote items (Fig. 8.17, left). On the contrary, the activation of the premotor cortex (area 6) exhibited a significant correlation with metamemory performance for the memory of temporally recent items (Fig. 8.17, right). As described before, bilateral muscimol ($GABA_A$ receptor agonist) injections into the dorsolateral prefrontal cortex produced significant impairment in metamnemonic ability for temporally remote items, while bilateral muscimol injections into the premotor cortex produced impairment in metamnemonic ability for recent items. These results suggest that separate metamemory processes supervise recognition for remote and recent memory, respectively.

Metacognition is related to awareness of our own cognition, such as knowing what we know or knowing what we do not know. To be confident that the item that we see now is actually a never-seen item, an introspective search of our own memory is required. Miyamoto et al. (2018) searched neural substrates of the metacognitive judgment on non-experienced events and found that only bilateral frontopolar cortices (area 10) were recruited for the metacognitive judgment of non-experienced event by fMRI while the monkeys performed the metamemory task (see Fig. 8.8). Unlike the dorsolateral prefrontal cortex (area 9 and 46d) which is recruited for metacognitive judgment of experienced event (Miyamoto et al. 2017), the frontopolar cortex is not recruited for metacognitive judgment of experienced event. Muscimol injections to the frontopolar cortex produced impairment in metacognitive judgment of non-experienced event, but not impaired metacognitive judgment of experienced event. Thus, although both the dorsolateral prefrontal cortex and the frontopolar cortex participate in metacognitive judgment, the frontopolar cortex involves in different aspects of metacognitive judgment from that of the dorsolateral prefrontal cortex.

8.6.4 Metacognition and Confidence Judgment

Two behavioral paradigms (prospective and retrospective) are present to examine neural mechanisms of metacognition. In tasks using the prospective paradigm, the subjects are asked to select whether they take or decline a cognitive test before answering the question. On the contrary, in tasks using the retrospective paradigm, after the subjects make an answer of a cognitive test, they are asked the degree of confidence (strong or weak) about their answer. In either paradigm, the subjects are required to monitor their own knowledge and memory and, based on the result of monitoring, they decide to take or decline the test or they make a judgment regarding the degree of their confidence about the decision.

The retrospective paradigm has recently been used to examine the neural representation of the confidence or the certainty associated with perceptual decision-making. For example, Kiani and Shadlen (2009) used a perceptual decision-making task, in which the monkeys need to answer the direction of the visual motion stimulus by eye movements, and showed that parietal neurons represent the degree of certainty about monkeys' decision in a motion-direction detection task. In their task, while the monkey looks at a fixation point, two visual targets (direction targets) are presented around the fixation point, one in the neuron's receptive field (T_{in}) and the other at the opposite side (T_{opp}). Then, the motion stimulus is presented around the fixation point for 100–900 ms, and then a delay of 1200–1800 ms is introduced. The motion stimulus consists of 100 small dots, each of which is moving randomly. The degree of coherence of dots' movements (0–50%) is used to indicate the strength of the motion stimulus. During the delay period, on one-half of the trials, a third visual target (sure target) is presented. After the delay, the fixation point is extinguished and the monkey is allowed to make a saccade to one of two or three visual targets. If the monkey answers the correct motion direction by selecting the corresponding visual target by an eye movement, it would get a reward. If the monkey selects the sure target, it would get a smaller amount of reward. If the monkey makes an error, it would not get reward.

Monkeys' behavioral performance showed that the frequency of choosing the sure target increased when the motion strength (% coherence) became weaker or when the duration of motion stimulus presentation became shorter, indicating that the frequency of choosing the sure target represents monkeys' strength of confidence about their decision. Lateral intraparietal neurons exhibited differential activity between when the monkey correctly selected the direction target presented within the neuron's receptive field (T_{in}) and when it correctly selected the direction target presented in the outside of the receptive field (T_{opp}), indicating that these neurons participate in decision of motion direction. At the same time, when the monkey selected the sure target, these neurons' activities were an intermediate level between T_{in} and T_{opp} conditions and gradually reached to the T_{opp} condition, suggesting that the intermediate level of activation represents a low state of certainty. Therefore, these results indicate that the activity of parietal neurons that participate in perceptual decision-making of the motion direction also carries information regarding the degree of certainty about the decision.

Komura et al. (2013) used a perceptual categorization task similar to the task used by Kiani and Shadlen (2009) and showed that neurons in the pulvinar represent the degree of certainty about the decision. In their task, the monkey needs to press either the left or right key depending on the detected direction (upward or downward) of moving colored (red and green) dots, whose color to be attended is initially assigned by the color (red or green) of a fixation point. In the metacognitive condition, the third key (escape option) is added. While the monkey performs the categorization task, pulvinar neurons did not exhibit differential activation depending on the difference in perceived motion directions. However, when the monkey chose the escape option, pulvinar activity decreased. Based on the result, they suggested that the level of pulvinar activity represents the degree of confidence and that the decrease of pulvinar activity associated with the escape option represents less confidence about the monkeys' decision for the perceptual categorization.

Thus, neural correlates of decision confidence have been observed in the supplementary eye field (Middlebrooks and Sommer 2012), the lateral intraparietal cortex (Kiani and Shadlen 2009), and the pulvinar (Komura et al. 2013) in monkeys, and the orbitofrontal cortex in rats (Kepecs et al. 2008), while subjects performed perceptual decision-making tasks with the escape option. These studies showed that the ratio of the subject's selection of the escape option correlated with the strength of uncertainty of its decision and that the subject's choice of the escape option was associated with the decrease of activity of neurons that exhibited stimulus-feature or movement-direction selective activity, compared with the activity observed when the subject made a correct choice. Thus, in the tasks based on the retrospective paradigm, the change of the magnitude of stimulus-feature or movement-direction selective activity seems to play an important role in metacognitive judgment.

8.7 Conclusions

Metacognition is an important and unique human capacity. Metacognition is explained as "cognition about cognition" or "knowing about knowing." Since metacognition is based on the subject's introspection, the metacognitive ability has been considered to be estimated only by language and link to the human mind. Metacognition is the function to monitor our own cognitive processes, and this function is thought to link to the human's capacity to understand and estimate other humans' minds. Therefore, understanding metacognitive ability is an important subject to understanding human mind. Metamemory is a function to monitor own memory and an important function of metacognition. Although metamemory was considered one kind of memory function, studies by human patients with focal brain damages revealed that impairments of metamemory functions were not observed in amnesic patients having damages in the medial temporal lobe but observed in patients having damages in the prefrontal cortex. Brain imaging studies also revealed prefrontal activation while human subjects performed metamemory tasks. These indicate prefrontal contribution of metamemory functions in humans.

On the other hand, it has been thought that metacognitive ability can only be examined by introspective reports using language. Therefore, metacognition or metamemory can only be examined by human subjects, not by animals. However, recently, animal psychologists developed several behavioral tasks and paradigms in which they could examine whether or not non-human animals exhibit similar capacity of metacognition as humans exhibit and revealed evidences that non-human animals, especially monkeys, share functional parallels with human's metacognition. These findings triggered to start neurophysiological studies to examine neural mechanisms related to metacognition.

References

Baird B, Smallwood J, Gorgolewski KJ, Margulies DS (2013) Medial and lateral networks in anterior prefrontal cortex support metacognitive ability for memory and perception. J Neurosci 33:16657–16665. https://doi.org/10.1523/jneurosci.0786-13.2013

Basile BM, Schroeder GR, Brown EK, Templer VL, Hampton RR (2015) Evaluation of seven hypotheses for metamemory performance in rhesus monkeys. J Exp Psychol: Gen 144:85–102. https://doi.org/10.1037/xge0000031

Carruthers P (2009) How we know our own minds: the relationship between mindreading and metacognition. Behav Brain Sci 32:121–182. https://doi.org/10.1017/S0140525X09000545

Chua EF, Ahmed R (2016) Electrical stimulation of the dorsolateral prefrontal cortex improves memory monitoring. Neuropsychologia 85:74–79. https://doi.org/10.1016/j.neuropsychologia.2016.03.008

Chua EF, Schacter DL, Sperling RA (2009) Neural correlates of metamemory: a comparison of feeling-of-knowing and retrospective confidence judgments. J Cogn Neurosci 21:1751–1765. https://doi.org/10.1162/jocn.2009.21123

Flavell JH (1979) Metacognition and cognitive monitoring: a new area of cognitive-developmental inquiry. Am Psychol 34:906–911. https://doi.org/10.1037/0003-066x.34.10.906

Fleming SM, Weil RS, Nagy Z, Dolan RJ, Rees G (2010) Relating introspective accuracy to individual differences in brain structure. Science 329:1541–1543. https://doi.org/10.1126/science.1191883

Fleming SM, Huijgen J, Dolan RJ (2012) Prefrontal contribution to metacognition in perceptual decision making. J Neurosci 32:6117–6125. https://doi.org/10.1523/JNEUROSCI.6489-11.2012

Fujita K (2009) Metamemory in tufted capuchin monkeys (Cebus apella). Animal Cogn 12:575–585. https://doi.org/10.1007/s10071-009-0217-0

Grimaldi P, Lau H, Basso MA (2015) There are things that we know that we know, and there are things that we do not know we do not know: confidence in decision-making. Neurosci Biobehav Rev 55:88–97. https://doi.org/10.1016/j.neubiorev.2015.04.006

Hampton RR (2001) Rhesus monkeys know when they remember. Proc Natl Acad Sci USA 98:5359–5362. https://doi.org/10.1073/pnas.071600998

Hampton RR (2009) Multiple demonstrations of metacognitionin nonhumans: converging evidence or multiple mechanisms? Comp Cogn Behav Rev 4:17–28. https://doi.org/10.3819/ccbr.2009.40002

Hampton RR (2010) Metacognition and metamemory in non-human animals. Encyclopedia of animal behavior. Elsevier, Amsterdam, pp 443–448. https://doi.org/10.1016/B978-0-12-809633-8.01089-X

Hampton RR, Zivin A, Murray EA (2004) Rhesus monkeys (*Macaca mulatta*) discriminate between knowing and not knowing and collect information as needed before acting. Anim Cogn 7:239–246. https://doi.org/10.1007/s10071-004-0215-1

Hart JT (1965) Memory and the feeling-of-knowing experience. J Educ Psychol 56:208–216. https://doi.org/10.1037/h0022263

Janowsky JS, Shimamura AP, Squire LR (1989) Memory and metamemory: Comparisons between patients with frontal lobe lesions and amnesiac patients. Psychobiol 17:3–11. https://doi.org10.3758/BF03337811

Kepecs A, Uchida N, Zariwala H, Mainen ZF (2008) Neural correlates, computation and behavioural impact of decision confidence. Nature 455:227–231. https://doi.org/10.1038/nature07200

Kiani R, Shadlen MN (2009) Representation of confidence associated with a decision by neurons in the parietal cortex. Science 324:759–764. https://doi.org/10.1126/science.1169405

Kikyo H, Ohki K, Miyashita Y (2002) Neural correlates for feeling-of-knowing: an fMRI parametric analysis. Neuron 36:177–186. https://doi.org/10.1016/S0896-6273(02)00939-X

Komura Y, Nikkuni A, Hirashima N, Uetake T, Miyamoto A (2013) Responses of pulvinar neurons reflect a subject's confidence in visual categorization. Nat Neurosci 16:749–755. https://doi.org/10.1038/nn.3393

Kornell N, Son LK, Terrace HS (2007) Transfer of metacognitive skills and hint seeking in monkeys. Psychol Sci 18:64–71. https://doi.org/10.1111/j.1467-9280.2007.01850.x

Kwok SC, Cai Y, Buckley MJ (2019) Mnemonic introspection in macaques is dependent on superior dorsolateral prefrontal cortex but not orbitofrontal cortex. J Neurosci 39:5922–5934. https://doi.org/10.1523/jneurosci.0330-19.2019

Maril A, Simons JS, Mitchell JP, Schwartz BL, Schacter DL (2003) Feeling-of-knowing in episodic memory: an event-related fMRI study. Neuroimage 18:827–836. https://doi.org/10.1016/S1053-8119(03)00014-4

Middlebrooks PG, Sommer MA (2011) Metacognition in monkeys during an oculomotor task. J Exp Psychol Lear Mem Cogn 37:325–337. https://doi.org/10.1037/a0021611

Middlebrooks PG, Sommer MA (2012) Neuronal correlates of metacognition in primate frontal cortex. Neuron 75:517–530. https://doi.org/10.1016/j.neuron.2012.05.028

Miyamoto K, Osada T, Setsuie R, Takeda M, Tamura K, Adachi Y, Miyashita Y (2017) Causal neural network of metamemory for retrospection in primates. Science 355:188–193. https://doi.org/10.1126/science.aal0162

Miyamoto K, Setsuie R, Osada T, Miyashita Y (2018) Reversible silencing of the frontopolar cortex selectively impairs metacognitive judgment on non-experience in primates. Neuron 97:980–989.e6. https://doi.org/10.1016/j.neuron.2017.12.040

Miyamoto K, Trudel N, Kamermans K, Lim MC, Lazari A, Verhagen L, Wittmann MK, Ruchworth MFS (2021) Identification and disruption of a neural mechanism for accumulating prospective metacognitive information prior to decision-making. Neuron 109:1396–1408.e7. https://doi.org/10.1016/j.neuron.2021.02.024

Morales J, Lau H, Fleming SM (2018) Domain-general and domain-specific patterns of activity supporting metacognition in human prefrontal cortex. J Neurosci 38:3534–3546. https://doi.org/10.1523/jneurosci.2360-17.2018

Nelson TO (1996) Consciousness and metacognition. Am Psychol 51:102–116. https://doi.org/10.1037/0003-066x.51.2.102

Nelson TO, Narens L (1990) Metamemory: a theoretical framework and new findings. In: Bower G (ed) The psychology of learning and motivation: advances in research and theory. Academic Press, New York, pp 125–173. https://doi.org/10.1016/s0079-7421(08)60053-5

Pannu JK, Kaszniak AW (2005) Metamemory experiments in neurological populations: a review. Neuropsychol Rev 15:105–130. https://doi.org/10.1007/s11065-005-7091-6

Pannu JK, Kaszniak AW, Rapcsak SZ (2005) Metamemory for faces following frontal lobe damage. J Internat Neuropsychol Soc 11:668–676. https://doi.org/10.1017/S1355617705050873

Rounis E, Maniscalco B, Rothwell JC, Passingham RE, Lau H (2010) Theta-burst transcranial magnetic stimulation to the prefrontal cortex impairs metacognitive visual awareness. Cogn Neurosci 1:165–175. https://doi.org/10.1080/17588921003632529

Schnyer DM, Verfaellie M, Alexander M, LaFleche G, Nicholls L, Kaszniak AW (2004) A role for right medial prefrontal cortex in accurate feeling-of-knowing judgments: evidence from patients with lesions to frontal cortex. Neuropsychologia 42:957–966. https://doi.org/10.1016/j.neuropsychologia.2003.11.020

Schnyer DM, Nicholls L, Verfaellie M (2005) The role of VMPC in metamemorial judgments of content retrievability. J Cogn Neurosci 17:832–846. https://doi.org/10.1162/0898929053747694

Shekhar M, Rahnev D (2018) Distinguishing the roles of dorsolateral and anterior PFC in visual metacognition. J Neurosci 38:5078–5087. https://doi.org/10.1523/jneurosci.3484-17.2018

Shimamura AP, Squire LR (1986) Memory and metamemory: a study of the feeling-og-knowing phenomenon in amnesic patients. J Exp Psychol Learn Mem Cogn 12:452–460. https://doi.org/10.1037/0278-7393.12.3.452

Smith JD (2009) The study of animal metacognition. Trends Cogn Sci 13:389–396. https://doi.org/10.1016/j.tics.2009.06.009

Smith JD, Shields WE, Schull J, Washburn DA (1997) The uncertain response in humans and animals. Cognition 62:75–97. https://doi.org/10.1016/S0010-0277(96)00726-3

Smith JD, Shields WE, Allendoerfer KR, Washburn DA (1998) Memory monitoring by animals and humans. J Exp Psychol Gen 127:227–250. https://doi.org/10.1037/0096-3445.127.3.227

Souchay CS, Isingrini M, Espagnet L (2000) Aging, episodic memory feeling-of-knowing, and frontal functioning. Neuropsychology 14:299–309. https://doi.org/10.1037/0894-4105.14.2.199

Tanaka A, Funahashi S (2012) Macaque monkeys exhibit behavioral signs of metamemory in an oculomotor working memory task. Behav Brain Res 233:256–270. https://doi.org/10.1016/j.bbr.2012.05.013

Tanaka A, Funahashi S (2016) Persistent activity of prefrontal neurons as a source of confidence in working memory. In: 2016 Neuroscience Meeting Planner, Program No. 550.01, Online. Society for Neuroscience, San Diego, CA

Templer VL, Hampton RR (2012) Rhesus monkeys (*Macaca mulatta*) show robust evidence for memory awareness across multiple generalization tests. Anim Cogn 15:409–419. https://doi.org/10.1007/s10071-011-0468-4

Terrace HS, Son LK (2009) Comparative metacognition. Curr Opin Neurobiol 19:67–74. https://doi.org/10.1016/j.conb.2009.06.004

Yokoyama O, Miura N, Watanabe J, Takemoto A, Uchida S, Sugiura M, Horie K, Sato S, Kawashima R, Nakamura K (2010) Right frontopolar cortex activity correlates with reliability of retrospective rating of confidence in short-term recognition memory performance. Neurosci Res 68:199–206. https://doi.org/10.1016/j.neures.2010.07.2014

Chapter 9
Top-Down Control

9.1 Evidences of Top-Down Control by the Prefrontal Cortex

Top-down control or top-down modulation can be observed when we focus our attention on task-relevant stimulus and ignore task-irrelevant distractors or when we try to retrieve task-specific information from long-term memory. Top-down signals often enhance task-relevant activity and suppress task-irrelevant activity. By fMRI and event-related potential (ERP) studies, Gazzaley et al. (2005) showed that both the magnitude of neural activity and the speed of neural processing are modulated by top-down control signals in visual information processing and that both enhancement and suppression are affected by the difference of the instructions for task performance. It has been widely accepted that top-down control is not an intrinsic function within a particular cortical area but relies on long-range functional interactions between the cortical area and the control area.

The prefrontal cortex has been considered an important area that contributes to top-down control functions. For example, Barceló et al. (2000) showed that prefrontal activity modulates visual information processing in extra-striate neurons. They recorded event-related potentials (ERPs) while patients having damages in the dorsolateral prefrontal cortex performed a visual detection task and found the reduction of ERPs in the extra-striate cortex of the damaged hemisphere. In addition, abnormal ERPs were accompanied by the behavioral deficit in detection capacity when the stimulus was presented in the hemifield contralateral to the damaged hemisphere. Merchant et al. (2011) used a categorization task, in which the monkey is required to categorize the height of a horizontal bar presented in the rectangular frame into either high or low compared to the position of the frame's midline. They found that prefrontal neurons exhibited differential responses based on the difference of the categories (high or low) earlier than posterior parietal neurons, such that prefrontal neurons exhibited differential responses depending on the difference of the categories after the presentation of the sample bar and continued to discriminate

S. Funahashi, *Dorsolateral Prefrontal Cortex*, Brain Science,
https://doi.org/10.1007/978-981-19-7268-3_9

two categories throughout the delay and the choice periods, while posterior parietal neurons exhibited differential responses depending the difference of the categories after the choice period was started. This result suggests that the top-down categorical signal originated in the prefrontal cortex and is provided to the posterior cortices when it becomes necessary for making a decision.

Prefrontal outputs are shown to act as executive signals to control parietal functions. Crowe et al. (2013) collected neural activity from prefrontal and posterior parietal neurons simultaneously while the monkey performed a rule-based spatial categorization task. In this task, a sample stimulus is presented somewhere on the monitor and, after the delay, the monkey is asked to categorize whether the position of the sample stimulus is on the right, the left, the up, or the down based on the boundary presented on the screen. The boundary is presented as either a horizontal line or a vertical line. Therefore, the same position of the sample stimulus is categorized as either right or up, for example, depending on the orientation of the boundary. Although both prefrontal and parietal neurons encoded signals reflecting rule-dependent spatial categories, these signals were transmitted from prefrontal neurons to parietal neurons. Signal transmission was directional from prefrontal neurons to parietal neurons, occurred at a restricted time scale, and modulated as a function of the behavior. These results indicate that prefrontal outputs act as top-down executive signals to control cognitive processes in the posterior parietal cortex.

Interactions between the prefrontal cortex and the posterior visual areas have been reported. Mendoza-Halliday and Martinez-Trujillo (2017) examined the activity of lateral prefrontal neurons while the monkey performed a delayed matching-to-motion direction task and showed that direction-selective neurons were accumulated within the posterior end of the principal sulcus (area 8 Av), where directly projects to areas MT and MST. This suggests that prefrontal neurons provide top-down control signals to areas MT and MST to modulate their operations. Merrikhi et al. (2017) first identify frontal eye field (FEF) neurons projecting to area MST and then examined the effects of working memory-related activity of FEF neurons on the activity of MST neurons while the monkey performed an oculomotor delayed-response task. Persistent activity of FEF neurons during the delay period did not evoke spiking responses in MST neurons. However, in MST neurons, the gain of the visual response increased, the size of the visual receptive field expanded, the position of the visual receptive field shifted toward the remembered position, and the stimulus information represented by a population of MST neurons improved. These results indicate an enhancement of stimulus representation in MST neurons by the persistent working memory-related activity of FEF neurons. These results also indicate that working memory-related top-down signals originated in the FEF modulate bottom-up visual information processing in area MST. Working memory-related top-down modulation of visual information processing is also observed in the modulation (increase) of power of the local field potential in visual areas (e.g., MT) (Bahmani et al. 2018).

Top-down control in tactile discrimination by the prefrontal cortex has also been reported. Gogulski et al. (2015) first determined anatomical connections between the prefrontal cortex and the primary somatosensory cortex by probabilistic tractography and then applied single-pulse transcranial magnetic stimulation to the connected area in the middle frontal gyrus. Magnetic stimulation impaired the ability to discriminate between single and twin tactile pulses and the impairment occurred within a restricted time window and skin area. Thus, prefrontal cortical activity is shown to modulate activity in the somatosensory cortex and affect ability for tactile discrimination.

Functional connectivity between two brain areas is shown to be modulated by top-down control signals provided from the prefrontal cortex. Hwang et al. (2019) tracked time-varying and task-evoked changes in functional connectivity obtained by fMRI studies between the primary visual cortex and the fusiform face area and between the primary visual cortex and the parahippocampus place area, while the subject performed different tasks (a localizer task, a categorization task, and a 1-back task) using the same visual stimuli. They found a selective increase of functional connectivity during the processing of task-relevant information and strengthening of this connectivity by additional task demands. An analysis to identify candidate brain regions that contribute to increasing functional connectivity between these areas revealed that frontoparietal regions provide top-down biasing signals to modulate task-specific functional connectivity between these two areas.

The evidence of top-down control by the prefrontal cortex has been shown by experiments using local field potentials and oscillations. Chan et al. (2015) examined the effects of bilateral deactivation of the dorsolateral prefrontal cortex on local field potentials in the superior colliculus while the monkey performed pro- and anti-saccade tasks. The power of event-related local field potentials was related to the difference in behavioral performances (correct or error) and the difference in the tasks. Deactivation of the dorsolateral prefrontal cortex reduced high beta (20–30 Hz) and high gamma (60–150 Hz) powers in the superior colliculus during the preparatory period in both tasks. Spike rates in the superior colliculus were positively correlated with the gamma power and this correlation was reduced by deactivation of the prefrontal cortex. Although it had been shown that the deactivation of the dorsolateral prefrontal cortex induced modulation of task-related activity in the superior colliculus, these results suggest that the dorsolateral prefrontal cortex provides top-down control signals to the superior colliculus by neural oscillations.

Thus, evidences of the top-down modulation of information processing by the prefrontal cortex have been accumulated. These studies indicate that the prefrontal cortex is the source of the top-down control signals and that the top-down control signals from the prefrontal cortex affect not only on the bottom-up sensory information processing but also a variety of information processing operating in the cortical and the subcortical structures by modulating the neural activity or functional connectivity between different brain areas.

9.2 Top-Down Control and Contribution of the Prefrontal Cortex

9.2.1 Top-Down Interactions between the Prefrontal Cortex and Other Cortical and Subcortical Areas

The presence of extensive anatomical connections between the prefrontal cortex and other cortical and subcortical areas is anatomical and functional bases for top-down control mechanisms by the prefrontal cortex. In fact, the prefrontal cortex has reciprocal anatomical connections with the posterior association cortices. These reciprocal connections support functionally tight interactions between the prefrontal cortex and posterior association cortices. Neurophysiological studies and human neuroimaging studies have demonstrated that activation or inactivation of the prefrontal cortex modulates the neural activity of the posterior cortical areas. These evidences suggest that the prefrontal cortex generates some specific neural signals to send to the posterior association cortices through efferent anatomical connections and that these neural signals modulate information processing in these cortical areas.

9.2.1.1 Reversible Inactivation Studies

The evidence of functional interactions between the prefrontal cortex and the posterior association cortices has been demonstrated by the effects of reversible inactivation of the prefrontal cortex using cooling or drug injections (e.g., muscimol). When the temperature of the local brain tissue becomes lower than 20 °C, synaptic communications among neurons within this tissue cease, causing functional impairment in this brain area. Differing from brain lesions, cooling of the local brain area produces no damage in nervous tissue and no reorganization of brain functions, but produces only temporary and reversible impairment of neural functions that the cooled brain area participates in. Therefore, cooling of the local brain area has been used as one of the useful methods to examine the functions of a particular brain area (see Brooks (1983) in detail).

Fuster et al. (1985) applied the cooling method to understand functional interactions between the prefrontal cortex and the inferior temporal cortex. They prepared two groups of monkeys and examined single-neuron activity while monkeys performed a delayed matching-to-sample task. One group had cooling probes over the dorsolateral prefrontal cortex and the other group had cooling probes over the inferior temporal cortex. Cooling of either the prefrontal cortex or the inferior temporal cortex produced modulation of both spontaneous activity and responses to sample stimuli and also caused a reduction of stimulus selectivity during the sample period and the delay period in neurons of either intact cortex. Further, cooling of either cortical area produced behavioral impairment of this task. Therefore, they concluded that mutual reciprocal functional interaction is present between the prefrontal cortex and the inferior temporal cortex and that the interaction between

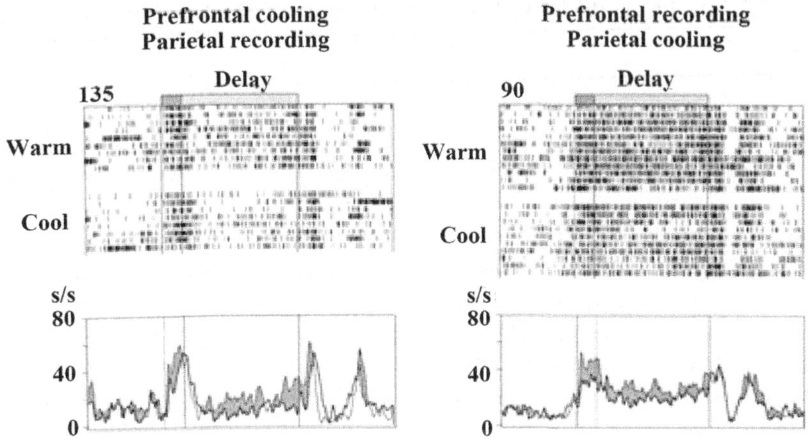

Fig. 9.1 Effects of local cooling on prefrontal and parietal activities while monkeys performed the ODR task. Shaded areas in histograms indicate activities during warm condition, while open areas indicate activities during cooling. From Chafee and Goldman-Rakic (2000) with permission

two cortical areas plays an important role to perform cognitive tasks, such as a delayed matching-to-sample task.

Quintana et al. (1989) further examined the effects of parietal cooling on prefrontal activity while monkeys performed delayed matching-to-sample or delayed matching-to-position tasks. Bilateral parietal cooling not only produced significant changes in spontaneous and task-related activity in prefrontal neurons but also produced behavioral changes such as mis-reaching, slow and inaccurate eye movements, and longer choice reaction times. However, behavioral performances of these tasks in terms of correct responses were not impaired by bilateral parietal cooling, suggesting that the performance of these delay tasks requires intact function of the prefrontal cortex. These results indicate that, although bilateral parietal cooling affects bottom-up sensory information transmission to the prefrontal cortex, top-down information transmission from the prefrontal cortex is maintained. Thus, these results demonstrate the importance of top-down information transmission from the prefrontal cortex to the posterior cortices for performing cognitive tasks.

Chafee and Goldman-Rakic (2000) also examined the effects of local cooling of either the prefrontal or the parietal cortex on task-related activity of these brain areas while monkeys performed an oculomotor version of the delayed-response (ODR) task. Similar percentages of prefrontal and parietal neurons exhibited either an increase or decrease in firing rates by parietal and prefrontal cooling, respectively. The magnitude and the direction selectivity of cue-, delay-, and response-period activity were also affected by either prefrontal or parietal cooling (Fig. 9.1). Similarity of characteristics of task-related activity between two brain areas during ODR performances (Chafee and Goldman-Rakic 1998) and symmetrical cooling effects indicate that prefrontal and parietal neurons symmetrically exchange neural signals during every epoch of the task. However, the shift of the best direction of task-related activity was larger in parietal cooling than in prefrontal cooling. The mean firing rate of task-related activity was modulated more in parietal neurons during prefrontal cooling than in prefrontal neurons during parietal cooling. These results suggest that parietal cooling affects more on bottom-up sensory information transmission to the prefrontal cortex, while the prefrontal cortex participates in top-down modulation of parietal activity.

Thus, cooling studies demonstrated the presence of functional interactions between the prefrontal cortex and the posterior association cortices. Since mutual reciprocal anatomical connections are present between the prefrontal cortex and the parietal and the inferior temporal cortex, cooling either cortex produces similar and largely symmetrical effects in spontaneous activity and task-related activity in the counterpart cortex. However, cooling the posterior association cortices tends to modulate the magnitude of sensory responses in prefrontal neurons, while cooling the prefrontal cortex tends to modulate the magnitude of both spontaneous activity and task-related activity in neurons of the posterior association cortices, suggesting that neural outputs from the prefrontal cortex control information processing in the posterior association cortices by modulating (increasing or decreasing) spontaneous as well as task-related activities.

9.2.1.2 TMS Studies

Top-down functional modulation of the information processing in the posterior association cortices has also been demonstrated by transcranial magnetic stimulation (TMS) at the prefrontal cortex in humans. The presence of functional interactions has been known between the prefrontal cortex and the extra-striate and the inferior temporal cortices (Tomita et al. 1999; Moore and Armstrong 2003; Monosov et al. 2011). Lee and D'Esposito (2012) examined how top-down signals generated in the prefrontal cortex modulate visual information processing in the extra-striate cortex. Human subjects performed a modified n-back task using images of faces and scenes. The subject was asked to selectively attend to images of a relevant category (either faces, scenes, or both). At the beginning of each trial block, the subject was informed which category of images he needed to attend and remember. Functions of the prefrontal cortex were first disrupted by applying theta burst TMS, and then the subject performed the task in the fMRI scanner. Disruption of prefrontal functions by TMS decreased selectivity of extra-striate responses to the stimuli of attended category. The decrease of response selectivity in the extra-striate areas was in parallel with the decrease in n-back task performance. These results suggest that the difference in the functional state of the prefrontal cortex affects the functional state of the posterior cortices through top-down efferent pathways originated in the prefrontal cortex and that the change of the functional state in these cortical areas causes the change in behavioral performance.

Similarly, Morishima et al. (2009) applied a single-pulse TMS over the prefrontal cortex and examined the strength of the top-down signal by measuring how much the

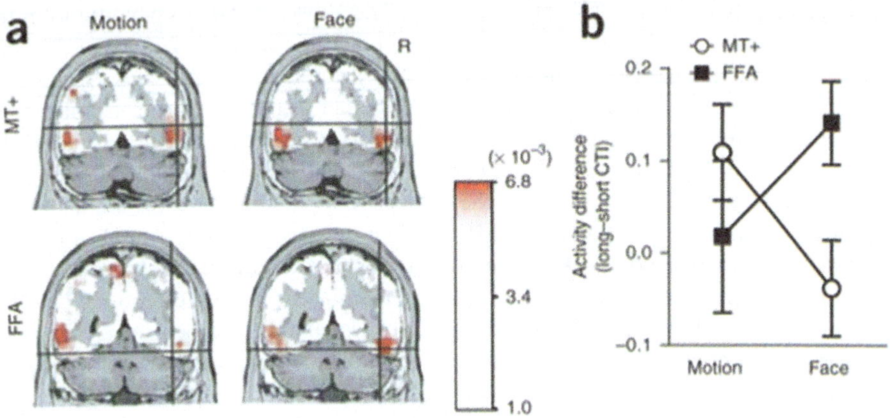

Fig. 9.2 (**a**) Mean TMS-induced current source density at 20–40 ms after TMS. (**b**) Difference in the current source density between long and short CTI trials in MT and FFA. From Morishima et al. (2009) with permission

TMS pulse increases the efficacy of impulse transmission. Human subjects are asked to perform a visual discrimination task of the motion direction of vertical gratings or the gender of the face image. The visual target consists of vertical gratings moving rightward or leftward, which are superimposed on an image of male or female face. An instruction cue is presented before the presentation of the visual target and indicates the subject which stimulus feature (motion direction or face gender) is needed to be attended. A single-pulse TMS is applied 134 ms after the onset of the visual target over the scalp position of FC2 electrode (correspond to the frontal eye field). The magnitude of event-related potential (ERP) produced by a single-pulse TMS at the frontal cortex was recorded in several posterior visual areas and compared among these areas. They found that the magnitude of recorded ERP was different among posterior visual areas and different between two stimulus features needed to attend. For example, when the subject performed the motion-direction discrimination task, larger TMS-induced ERP was observed in the middle temporal (MT) area (Fig. 9.2b). However, when the subject performed the face-gender discrimination task, larger TMS-induced ERP was observed in the fusiform face area (Fig. 9.2b). These results indicate that frontal eye field neurons send impulses to different posterior visual areas depending on the visual feature that the subject needed to attend. In addition, the amount of impulse transmission from the frontal area was associated with the level of attention and the level of the performance of the task. These results clearly demonstrate that prefrontal neurons send signals to the posterior visual areas and that these signals control the excitability of the posterior visual areas depending on the task demand.

Zanto et al. (2011) also examined the effects of repetitive TMS (rTMS) applied at the inferior frontal junction while human subjects performed a delayed recognition task with selective attention to a stimulus. In this task, four cue stimuli are presented one by one for 800 ms (cue period), two of which consist of moving color dots, and the remaining two consist of stable color dots. The subject needs to pay attention to either color or motion direction of dots and remember this stimulus feature of the cue stimuli with a black bar below the stimulus. After the delay period (4 or 8 s), a probe stimulus is presented and the subject is asked whether or not the probe stimulus has the same stimulus feature (color or motion direction) as the cue stimuli with a black bar. The subject first received 1-Hz rTMS at the right inferior frontal junction for 10 min. Immediately after the rTMS, the subject was asked to perform the task and an EEG recording was started. rTMS applied at the inferior frontal junction decreased the accuracy of working memory in the task condition requiring memory of color, but not in the task condition requiring memory of motion direction. The P1 amplitude of the ERP recorded from the posterior visual areas can be used as a neural marker of top-down modulation occurring in information processing for visual features. rTMS applied in the prefrontal cortex decreased the P1 amplitude. In addition, the subject who had stronger functional connectivity between the prefrontal cortex and the posterior visual areas exhibited greater disruptive effects on both working memory performance and the P1 amplitude by rTMS. Based on these results, they suggested that the prefrontal cortex mediates top-down modulation of information processing in the posterior visual area and that top-down modulation mediated by the prefrontal cortex contributes to information processing for attention control and memory performance. Top-down modulation mediated by the prefrontal cortex for information processing in visual areas has also been shown by other studies using TMS (Taylor et al. 2007; Ruff et al. 2008; Miller et al. 2011).

9.2.2 Top-Down Control of Attention by the Prefrontal Cortex

If we unexpectedly hear a crash sound from the outside of the room, our attention will automatically orient toward the source of the crash sound. This kind of orienting behavior is reflexive and automatic. Usually, peripheral sensory stimulus triggers this orienting behavior and we reflexively pay attention to that sensory stimulus. This type of attention is called bottom-up attention. On the other hand, when we need to get off the bus at a particular bus stop, we have to pay attention to the announcement of the particular name of the bus stop. In this case, we voluntarily and consciously keep paying attention to particular sensory information. This type of attention is called top-down attention. The bottom-up attention is managed by peripheral sensory inputs, while the top-down attention is managed by the central nervous system and controls information processing in the sensory cortices. Evidences show that the prefrontal cortex participates in top-down control of attention.

9.2.2.1 Top-Down Control of Attention by the Frontal Eye Field

It has been known that visual responses of frontal eye field (FEF) neurons are enhanced when a visual stimulus is used as the target of the saccadic eye movement

(Goldberg and Bushnell 1981; Bruce and Goldberg 1985; Kodaka et al. 1997). An enhancement of the visual response in FEF neurons is thought to be the evidence that the FEF participates in selective spatial attention (Bushnell et al. 1981). Therefore, the FEF is known to be the source of top-down control signals for selective attention to a particular visual stimulus (Thompson and Bichot 2005). Therefore, physiological studies have been performed in the FEF to examine neural mechanisms for top-down control of attention.

Subthreshold stimulation at the FEF does not produce eye movements, but consistently produces a facilitatory effect on target selection of the saccade and reduction of the saccade latency (Schiller and Tehovnik 2001). This suggests that saccade-related mechanisms of the FEF provide a neural basis of spatial attention. To examine whether the activity of the FEF modulates the performance of visual attention or not, Moore and Fallah (2004) examined the effects of the micro-stimulation applied at the FEF on luminance discrimination performances in monkeys. In their task, the monkey is required to maintain fixation at a fixation point and detect the luminance change of the peripheral visual target while ignoring a flashing distractor. The luminance change of the target was preceded by the FEF stimulation at the current strength below the level that evoked eye movements. They found that the monkey could detect a smaller luminance change of the target if the luminance change was preceded by the FEF stimulation. The increase in the sensitivity of the luminance change was found only when the target stimulus was presented within visual fields of FEF neurons that were placed at the stimulation site. The magnitude of increased sensitivity depended on the temporal asynchrony between the onset of the FEF stimulation and the target presentation. When a longer interval of this temporal asynchrony was introduced, no or weak effect was observed in the sensitivity of the luminance change by the FEF stimulation. Thus, these results indicate that micro-stimulations in the FEF produce a transient increase of the sensitivity to detect changes in visual stimulus features, and suggest that the stimulation in the FEF transiently improves covert spatial attention.

Although Moore and Fallah (2004) showed a contribution of the FEF to covert spatial attention, how the FEF contribures to covert spatial attention was not clear. Focusing attention on a visual stimulus was known to enhance visual responses in the striate and the extra-striate cortices (Moran and Desimone 1985; Motter 1993). The FEF projects to the extra-striate cortex (Ninomiya et al. 2012) and is known to participate in the target selection for eye movements. Therefore, Moore and Armstrong (2003) considered that, since micro-stimulations in the FEF facilitate saccade initiation and target selection, the stimulation in the FEF might induce attention-like modulation in extra-striate visual responses. Based on this assumption, they examined the effects of subthreshold FEF stimulations on visual responses observed in extra-striate neurons while the monkey fixated at a central fixation point. They put a recording electrode into the visual area V4 and mapped the visual receptive field of an isolated V4 neuron (Fig. 9.3 left). At the same time, they put a stimulation electrode into a particular site of the FEF and determined the strength of the threshold current to evoke saccadic eye movements (Fig. 9.3 left). The stimulation site of the FEF was selected at the place where the end point of the evoked saccade

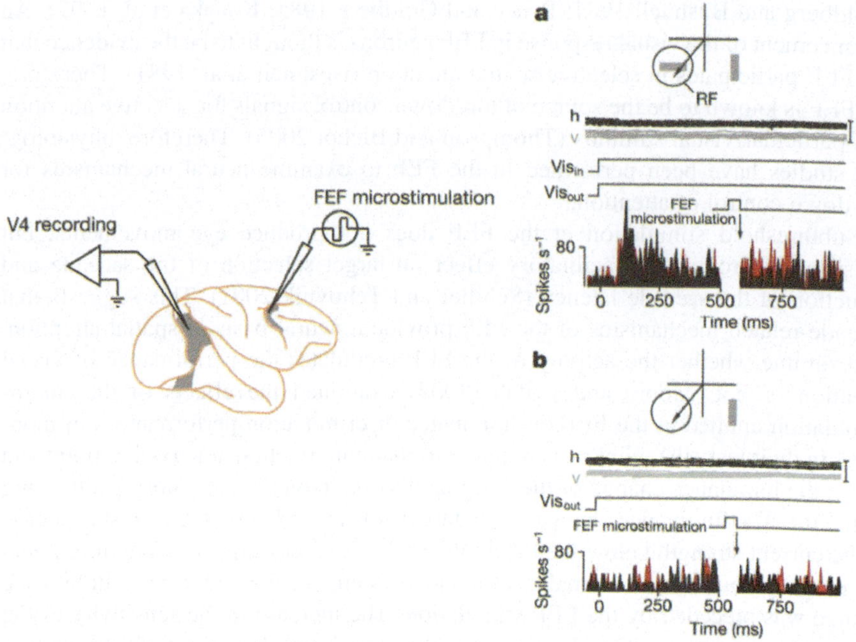

Fig. 9.3 Enhanced visual responses were observed in a V4 neuron by micro-stimulation at the FEF. *Left:* Experimental situation. *Right* (**a**) Enhanced response (red) was observed when visual stimulus was presented within the receptive field of the V4 neuron and the response field of the stimulation site of the FEF. (**b**) Enhanced response was not observed when visual stimulus was presented within the response field of the stimulation site of the FEF but outside of the receptive field of the V4 neuron. From Moore and Armstrong (2003) with permission

(response field) was either within or outside the visual receptive field of the V4 neuron. The subthreshold stimulation of the FEF (e.g., 50-ms trains of 20 μA current pulse) was applied 200–500 ms after the presentation of the visual stimulus in order to focus only on the effect of FEF stimulations on visual responses of the V4 neuron. They found that visual responses of V4 neurons were enhanced by subthreshold stimulations at the FEF if the position of the receptive field of the V4 neuron matched to the response field at the stimulation site of the FEF (Fig. 9.3 right). However, visual responses of V4 neurons were suppressed if the position of the receptive field of the V4 neuron was different from the response field at the stimulation site of the FEF. The magnitude of the enhancement depended on the effectiveness of the visual stimulus presented within the receptive field. Significant enhancement was observed only when the visual stimulus with preferred orientation, for example, was presented within the receptive field of the V4 neuron (Fig. 9.3 right). In addition, the magnitude of the enhancement of visual responses depended on the presence or the absence of the distractor. The largest enhancement was observed when the preferred stimulus was presented within the receptive field of the V4 neuron simultaneously with a distractor presented outside the receptive field. Thus, FEF stimulations modulate the magnitude of visual responses in extra-striate neurons and the degree of the enhancement of visual responses depends on the condition of visual stimulation. The phenomenon that the modulation of visual responses of extra-striate neurons by the FEF stimulation depended on the condition of visual stimulation seems to correspond to the top-down control of information processing in sensory areas by efferent outputs from the FEF. The outputs from the FEF are also thought to act as a mechanism for focusing attention on a specific stimulus presented in a particular visual field for a particular purpose (e.g., saccade). Similarly, attention-related top-down oscillatory coupling in local field potentials has been observed between the FEF and V4. Gregoriou et al. (2009) found that the attention to a visual stimulus causes enhanced oscillatory coupling at gamma frequencies between the FEF and V4 and that the coupling was initiated by the FEF. Thus, these results indicate that the FEF would be a source of top-down control signals to modulate information processing in extra-striate visual areas.

The results indicate that the FEF contributes to top-down control of attention. However, the FEF is well known to participate in eye movement control. A question was whether FEF neurons participating in eye movement control also contribute to control of spatial attention or are separate from FEF neurons participating in control of spatial attention. Schafer and Moore (2011) examined whether activation of FEF neurons was primarily associated with visual attention or oculomotor preparation. They used a method for an operant control of neural activity, in which the monkey was required to alternately increase (up trials) and decrease (down trials) firing rates of FEF neurons while it maintained fixation. Operant increase or decrease of FEF activity had no effect on saccade preparation (e.g., no difference in ratios and latencies of saccades toward the response field). However, operant control of increase or decrease of FEF activity produced a spatially selective effect on visual search performances, such that the proportion of trials on which the monkey failed to detect the target appeared in the response field was significantly increased during down trials. This study showed that voluntary increase or decrease of FEF activity modulates the performance of the visual search. Visual search is known to require top-down control of selective attention. Therefore, these results demonstrate specific association between voluntarily driven FEF activation and top-down control of selective attention. These results also demonstrate that neurons related to top-down control of selective attention are different from neurons related to eye movement control in the FEF.

Moore and Armstrong (2003) and Schafer and Moore (2011) indicate the contribution of the FEF to covert spatial attention. Thompson et al. (2005) used a pop-out visual search task without eye movements and further examined how the FEF contributes to covert spatial attention. In this task, the monkey was required to maintain fixation at the central fixation point and make a response by turning the lever to the right or the left. While the monkey looked at the fixation point, a search array, which consisted of six or eight stimuli, was presented and one of the stimuli was different in color or shape from the others. In the location search condition, iso-luminant green and red disks were used for the search array and the monkey was required to turn the lever to the same direction as a color singleton was present. In the identity search condition, Landolt "C" and "O" were used for the search array and the monkey was required to turn the lever to the same direction as the gap in the "C" stimulus. They also used a memory-guided saccade task to identify whether a recorded FEF neuron was a visually responsive neuron or saccade-related neuron. They found that the majority of visually responsive neurons exhibited significantly greater responses (enhancement of responses) when the target stimulus was in the receptive field than when the distractor was in the receptive field. On the other hand, saccade-related neurons exhibited a significant decrease in activity during the presentation of the search array and the decrease in activity was maintained until the search array was removed. These results show that the neural signal of spatial attention is enhanced activity of visually responsive neurons in the FEF and that, when the attention is allocated to searching the visual target, saccade-related neurons are inhibited, indicating that neural signals related to spatial attention are independent of saccade command signals. Thompson et al. (2005) proposed that the spatially selective activity of visually responsive neurons in the FEF corresponds to the "mental spotlight" of spatial attention.

The effect of FEF activity on spatial attention was also examined by reversible inactivation of the FEF using multiple microinjections of muscimol (Wardak et al. 2006). They used a visually-guided saccade task and a covert visual search task without eye movement. In the covert visual search task, similar to Thompson et al. (2005), while the monkey maintains fixation at a fixation point, up to three visual search displays are presented for 200 ms in succession separated by 1 s blank interval. One search display contains a predefined visual target with 2, 4, or 8 distractors. If the monkey presses the lever within 900 ms after the appearance of the search display that contains the visual target, it would get a reward. If the target is not included in the search display, the monkey is required to refrain from the response and maintain fixation until the search display containing the visual target appears. Inactivation of the FEF by local injections of muscimol produced spatially selective impairment in visually-guided saccade performance and the decrease of the ability to detect the visual target among distractors (e.g., an increase of the detection time of the visual target in a spatially selective manner) in covert visual search performance. The impairment of the visual search task by local inactivation of the FEF was neither caused by impairment of oculomotor functions nor caused by impairment of pure visual functions, indicating an involvement of the FEF in spatial attention function. Bollimunta et al. (2018) also obtained similar results by local injections of muscimol into the FEF while monkeys performed a covert attention task involving visual change detection. Thus, neural mechanisms related to spatial attention are independent of neural mechanisms related to oculomotor functions in the FEF.

The FEF participates not only in control of spatial attention but also in feature-based attention. Zhou and Desimone (2011) compared the contribution to feature-based attention between the FEF and area V4 while the monkey performed a visual search task with free saccades. In this task, a search array with 20 visual stimuli (colored shapes) including the search target is presented to the monkey and the monkey is required to find the target by free eye movements. Visual responses of neurons in both areas were enhanced by the attention to the visual feature of the target, not by the attention to the spatial position of the target. The latency of visual responses and the latency of attention effect on visual responses showed that area V4 provides bottom-up sensory information about stimulus features to the FEF while the FEF provides top-down bias in information processing of the target feature to area V4. Thus, top-down attentional bias provided by the FEF modulates information processing for visual features in area V4. Anderson et al. (2011) showed that anatomical projections from the FEF terminate all cortical layers of area V4 and mostly make excitatory synaptic contacts on pyramidal neurons in area V4. Top-down biasing signals generated in the FEF would modulate activity of V4 neurons through these anatomical projections.

Bressler et al. (2008) used Granger causality effects measured by BOLD signals while human subjects performed a visual spatial attention task and showed that the top-down control signals of the FEF and the intraparietal sulcus (IPS) area modulate the activity of occipital visual areas. They showed that top-down levels of Granger causality from the FEF and the IPS to occipital visual areas were significantly greater than bottom-up levels and cortex-wide levels of Granger causality effects. In addition, Granger causality effect was significantly greater from the FEF to the IPS than from the IPS to the FEF. These results indicate that the FEF and the IPS modulate the activity of the occipital visual areas and that the FEF modulates the activity of the IPS in relation to visual attention processes. Thus, the FEF plays an essential role in selective visual attention and participates in the control of attention using top-down command signals.

Although the FEF is an important brain area for voluntary eye movement control, behavioral and neurophysiological results indicate that the FEF is an important source of top-down control signals for visual selective attention and that neurons having attention-related activity is independent from neurons having eye-movement-related activity in the FEF. It has been shown that the FEF sends mainly excitatory outputs to neurons in the posterior cortical areas and modulates the activity of these neurons (e.g., enhancement of visual responses). Thus, neural signals output from the FEF act as top-down control signals and play important roles in covert visual attention.

9.2.2.2 Top-Down Control of Attention by the Dorsolateral Prefrontal Cortex

The FEF plays an important role in top-down control of attention. However, behavioral and neuropsychological studies using monkeys show that the prefrontal cortex also participates in top-down control of attention.

Rossi et al. (2007) made unilateral lesions in the primate prefrontal cortex in combination with transection of the corpus callosum and the anterior commissure. Therefore, top-down control from the prefrontal cortex to the visual areas was not effective in the ipsilateral hemisphere to the lesion but was effective in the contralateral hemisphere. The monkey was asked to look at a fixation point, whose color indicated the color of the target grating, and to discriminate the orientation of the target grating from the orientations of distractors presented in either visual hemifield without eye movements. The monkey needed to respond by a bar release. When the color of the target grating was kept the same for many trials, the monkey's discrimination performance was normal even if the target grating and distractor gratings were presented in the affected visual hemifield. However, the monkey's discrimination performance was severely impaired when the color of the target grating was switched frequently across trials. The monkey exhibited no impairment in a pop-out task, which required bottom-up control of attention. These results showed that, although the prefrontal cortex plays an essential role in top-down control of attention, this cortical area also plays a critical role in flexible switching of attended target on the basis of task demand. Rossi et al. (2009) used the same task as was used in monkey experiments and showed the importance of the prefrontal cortex in top-down attention control by human neuroimaging studies.

Neurophysiological studies have also shown prefrontal participation in top-down control of attention. For example, Everling et al. (2002) used a focused attention task, in which the monkey was required to monitor a stream of three visual objects (e.g., hamburger, bear, and fish) and wait for the presentation of the single target object (e.g., fish) at an attended location. To examine visual information represented by prefrontal neurons, all objects were presented at a single position in either right or left visual field (unilateral condition). To examine the effect of spatial attention, objects were simultaneously presented in both visual fields (bilateral condition). The visual field that the monkey needed to attend was informed by the visual cue presented in either side before the object presentation. In both conditions, the monkey needed to keep fixation at a fixation point until the appearance of the target object (e.g., fish) in the attended visual field. In the unilateral condition, prefrontal neurons discriminated the target object from non-target objects (e.g., significant response was observed only to the target object) and the magnitude of the visual response was enhanced when the target object was presented in the preferred position. In the bilateral condition, the visual response to the target object was enhanced only when it was presented in the attended visual field. Neither visual responses to the non-target object nor their enhancement were observed when the non-target object was presented in either attended or unattended visual field. These results indicate that prefrontal neurons discriminate between the target object and the non-target object and that the discriminability between the target object and the non-target object is weakened when the target object is presented in the unattended visual field. Based on these results, Everling et al. (2002) considered that the prefrontal cortex plays a role in the active filtering of visual stimulus presented in the unattended position. Although it is not clear how prefrontal neurons actively filter necessary information from unnecessary information, this may be achieved by top-down modulation of activity in the brain areas for visual information processing by the neural outputs from the prefrontal cortex.

Buschman and Miller (2007) examined neural activity in both the prefrontal cortex and the posterior parietal cortex and showed that top-down and bottom-up signals for attention arise from the prefrontal cortex and the posterior parietal cortex,

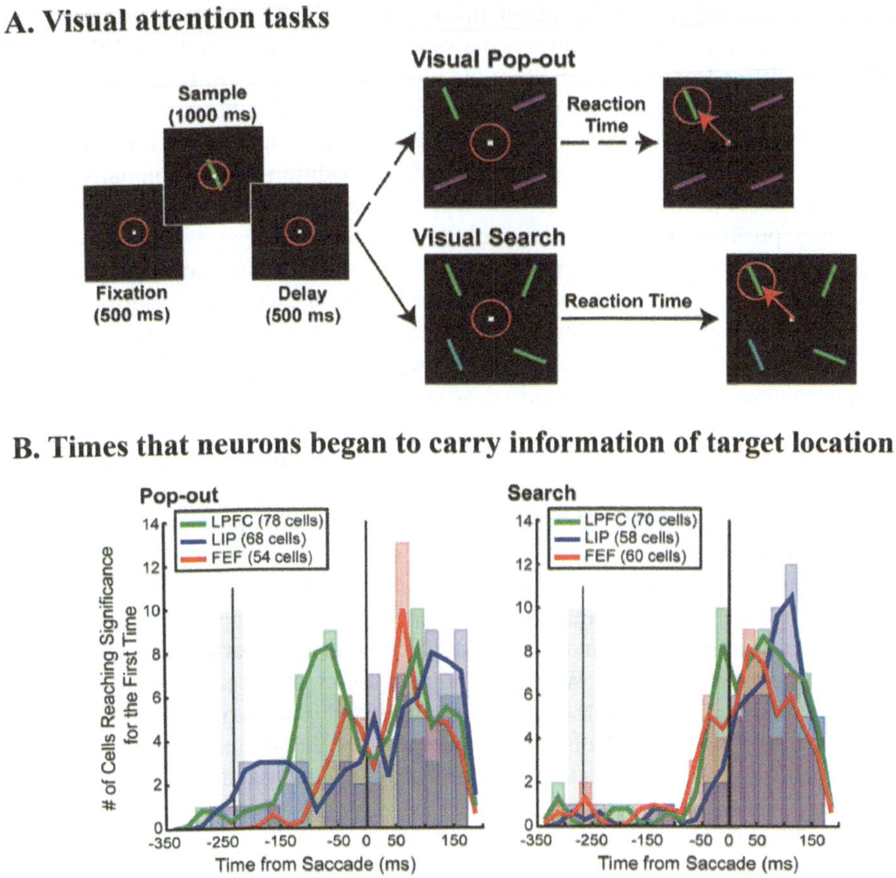

Fig. 9.4 (**a**) Two conditions of visual attention tasks. In the visual pop-out condition, bottom-up attention works, while in the visual search condition, top-down attention is required. (**b**) Comparisons of response timings of neuron activities in three brain areas between two conditions. In the pop-out condition, activation of LIP neurons was faster than LPFC neurons, while in the visual search condition, activation of LPFC neurons was faster than LIP neurons. From Buschman and Miller (2007) with permission

respectively. In this study, the monkey performed visual attention tasks, in which the monkey is required to find a visual target under either a visual "pop-out" (bottom-up) condition or a visual search (top-down) condition (Fig. 9.4a). The monkey is required to keep fixation at a fixation point throughout a trial. After a brief fixation period (500 ms), a target stimulus (a colored bar with a particular orientation) is presented for 1000 ms over the fixation point, and then a brief delay period (500 ms) is introduced. After the delay period, the target stimulus and three distractors are presented simultaneously around the fixation point and the monkey is required to make a saccade to the target stimulus. In the visual pop-out condition, the target stimulus is different from distractors either in color or in orientation, so that the salience of the target stimulus automatically draws the monkey's attention to the target stimulus (bottom-up attention condition). In the visual search condition, each distractor differs from the target stimulus independently in both color and orientation, so that the feature of the target stimulus matches some of the distractors in either color or orientation. Therefore, the monkey needs to use remembered information of the target stimulus to find it (top-down attention condition). They compared distributions of times at which neurons first began to carry significant information about the target position relative to the saccade initiation between two cortical areas (Fig. 9.4b). They found that prefrontal activity reflected the target position first during the visual search condition (Search in Fig. 9.4b), whereas posterior parietal activity reflected the target position first during the visual pop-out condition (Pop-out in Fig. 9.4b). To find the target stimulus in the visual search condition, the top-down spatial attention mechanism is required, whereas in the visual pop-out condition, the bottom-up spatial attention mechanism is activated. Therefore, these results indicate that top-down spatial attention signals arise from the prefrontal cortex and are transmitted to the parietal cortex, while bottom-up spatial attention signals arise from the sensory cortices and are transmitted to the prefrontal cortex through the parietal cortex.

Prefrontal participation in the top-down control of attention has also been demonstrated by human brain imaging studies. For example, Hopfinger et al. (2000) used a cued spatial attention task and examined which brain areas participate in top-down control of spatial attention. While the subject looks at a central fixation point, attention-directing cue, which consists of yellow and blue arrows, is presented for 500 ms around the fixation point. At the beginning of the session, the subject is told that which side is needed to attend is indicated by the color and the direction of the arrow of the attention-directing cue and that this information is maintained across the session. After the delay interval (1000 or 8160 ms), two black and white checkerboards are presented simultaneously at both sides of the fixation point. The checkerboard containing some gray checks is the target checkerboard and is presented in either side of the fixation point. The subject is required to discriminate whether the target checkerboard is presented at the attended side or not by pressing a button. Event-related fMRI analysis showed that the superior frontal, inferior parietal, and superior temporal cortices were activated only when the attention-directing cue was presented, not when the target stimulus was presented. In order to perform this task, the subject needs to maintain attention during the delay period toward the side that the attention-directing cue indicates. Therefore, top-down control of spatial attention is required. The present result suggests that the brain areas activated during the presentation of the attention-directing cue participate in top-down spatial attention. They suggested that these brain areas consist of a network for voluntary control of attention and that this attention-control network biases neural activity in multiple visual areas and enables particular visual areas to perform selective sensory processing for relevant visual target.

Visual search paradigm has been widely used to examine neural mechanisms of covert visual attention and most studies have focused on space-based attention mechanisms. However, in our real life, we use both space-based and feature-based attention mechanisms in visual search. Egner et al. (2008) examined using fMRI how human brain represents and integrates concurrent spatial- and feature-based attention in visual search and which brain areas participate in this process. They used a cued visual search task. In this task, while the subject maintains fixation at the fixation point, a cue stimulus is presented, which consists of a central square and two flanking triangles. The cue contains two kinds of information; feature-based information (either the letter "B" for blue or the letter "R" for red is presented at the center of the square) and space-based information (the "filling in" flanking triangle is presented on the right or the left). The cue informs the subject which side is needed to attend (space-based search information) or which color is needed to attend (feature-based search information). Then, a search array is presented, which consists of four colored diamonds; red and blue diamonds on both sides of the fixation point. On every trial, a small fragment is removed from either the top or the bottom corner of one (target diamond) of four diamonds. The subject needs to identify the target diamond and indicate which corner is missing by pressing a button. They analyzed only cue-related BOLD responses during the cue period in order to exclude the effects of bottom-up sensory signals, response behavior, and other processes for actual visual search. As a result, the increase of space-based search information and the increase of feature-based search information were both associated with enhanced bilateral activation in the dorsolateral prefrontal cortex and the posterior parietal cortex. In addition, space-based search information and feature-based search information were represented additively and shared in frontal and parietal regions during the cue period. Based on these results, they suggested that representation of space-based and feature-based search information are integrated in the prefrontal and the posterior parietal areas and that these brain areas are the sources of top-down biasing signals for both space-based and feature-based attention.

Thus, similar as the FEF, the lateral prefrontal cortex participates in the control of top-down visual attention both for space-based and feature-based information processing. Neurophysiological studies and human brain imaging studies indicate that the prefrontal cortex is the source of the top-down control signal which modulates information processing operating in the posterior cortices, such as selecting and discriminating relevant stimuli from irrelevant distractors.

9.2.3 Top-Down Control of Object Recognition by the Prefrontal Cortex

We can easily recognize and identify a particular group of objects (e.g., glass) despite their differences in shape, size, color, or view. In the early visual areas (e.g., V1), neurons extract lower-level local features of the object (e.g., color, shape, texture). Therefore, since different objects usually have a common local feature, the same neuron often responds to different groups of objects and different objects activate widely overlapped populations of neurons in the early visual areas.

Therefore, when decoding of object identity is performed by a population of activities in the early visual areas, decoding accuracy would be low (Romanski and Chafee 2021). On the other hand, in the higher levels of the visual system (e.g., temporal association areas), neurons respond to complex visual features (e.g., combination of a particular color and a particular shape). Therefore, these neurons respond to a specific set of objects and different objects activate different populations of neurons. Therefore, when decoding of object identity is performed in the higher levels of the visual system, decoding accuracy becomes high (Romanski and Chafee 2021). The inferior temporal cortex is known to play important roles in object recognition, because neurons in the inferior temporal cortex respond to "moderately complex" object features rather than lower-level local features of the object (Tanaka 1996). Therefore, object recognition is thought to be achieved by feedforward information processing through the bottom-up visual pathway from V1 to the inferior temporal cortex.

However, Kar and DiCarlo (2021) showed that neural mechanisms in the ventrolateral prefrontal cortex provide top-down neural signals to the inferior temporal cortex and that these top-down neural signals play important roles in object recognition. They recorded multiple neuron activities from the inferior temporal cortex by multiple electrodes during object discrimination performances. While the monkey maintains fixation at a fixation point, a sample image is presented for 100 ms, and after a delay of 100 ms, standard images of the target object (correct choice) and the distractor object are presented simultaneously. Sample images are made by modifying the position, the size, and the background of standard images (changing difficulty). Kar et al. (2019) showed that the monkey took longer time to recognize when object images difficult to identify were presented. They calculated the object solution time which was obtained from population of inferior temporal activity and was defined as the time when the decoding accuracy reached the level of behavioral recognition accuracy for each object image. They found that the object solution times for the images difficult to recognize were about 30 ms later than those for the images easier to recognize. This result suggests that, although feedforward information processing plays an important role to recognize the images easy to identify, some late-coming inputs to the inferior temporal cortex play important roles to recognize the images difficult to identify. Kar and DiCarlo (2021) further examined the source of late-coming inputs to the inferior temporal cortex and showed that inactivation of the ventrolateral prefrontal cortex by muscimol injection selectively impaired object discrimination performance and degraded the encoding accuracy of object identity obtained by a population of inferior temporal activity only when the presented sample image was difficult to identify. This result indicates that the ventrolateral prefrontal cortex is the source of late-coming inputs to the inferior temporal cortex. This result also indicates that easily identified images are recognized by the feedforward visual process from V1 to the inferior temporal cortex, while for the recognition of more difficult images, top-down inputs to the inferior temporal cortex from the ventrolateral prefrontal cortex are required. Thus, these results indicate that top-down neural signals originating in the ventrolateral prefrontal cortex play important roles in object recognition.

Figure-ground segregation is also an important function for object recognition. It has been shown that the figure enhances neural responses, while the background suppresses responses. This phenomenon is known as figure-ground modulation and plays important roles in the awareness of an object (identifying and paying attention to the object). Huang et al. (2020) showed that awareness-dependent figure-background segregation was not derived through intracortical neural interactions within the primary visual cortex but was derived by top-down feedback signals from the frontal eye field and the dorsolateral prefrontal cortex to the visual cortex.

When the subject performs object recognition, object identification, or figure-ground segregation, feedback neural inputs to the inferior temporal cortex are shown to play important roles to identify an object and differentiate the figure from the ground. Studies show that these feedback neural inputs are generated in the lateral prefrontal cortex and provided to the inferior temporal cortex. Thus, in addition to the control of attention, top-down neural signals generated in the lateral prefrontal cortex play important roles in object recognition and for figure-background segregation.

9.2.4 Top-Down Modulation of Long-Term Memory Process by the Prefrontal Cortex

Although working memory has been paid attention to as an important function of the prefrontal cortex, it is known that the prefrontal cortex also participates in the long-term memory process and that top-down control signals from the prefrontal cortex play an important role in the long-term memory process, especially in the process for encoding and retrieving information. The importance of top-down control signals from the prefrontal cortex in long-term memory process is supported by neuropsychological and neuroimaging studies (Nolde et al. 1998; Buckner et al. 1999; Blumenfeld and Ranganath 2007). Patients having prefrontal cortical damages exhibit impairment in the strategic utilization of memory. Their deficits are observed in the domains such as free recall, proactive interference, temporal-order memory, and source memory (Fuster 2008), indicating that impairments caused by prefrontal damages are not only observed in encoding, storage, and retrieval of information but also observed in the efficient and appropriate use of stored information depending on the conditions or the demands. The prefrontal cortex is thought to play a role to achieve the efficient use of information stored in long-term memory by sending top-down control signals to appropriate memory-related brain areas. Neuroimaging studies have also revealed prefrontal participation in long-term memory and regional differences in neural processes related to long-term memory in the prefrontal cortex. For example, the ventrolateral prefrontal region participates in memory encoding (Brewer et al. 1998; Wagner et al. 1998b), whereas the anterior prefrontal region participates in memory retrieval (Buckner and Koutsaal 1998; Cabeza and Nyberg 2000). It has been shown that the degree of activity in the ventrolateral prefrontal region during encoding predicts the probability of the successful retrieval of memorized materials (Rugg et al. 1996; Brewer et al. 1998; Wagner et al. 1998a; Kao et al. 2005; Kluen et al. 2019). This result suggests that the prefrontal cortex manages efficient and selective encoding and retrieval of information depending on the situations by top-down control signals. Hemispheric difference in prefrontal contribution to long-term memory has also been reported. Rossi et al. (2001) examined how the prefrontal cortex participates in episodic memory by transient functional interference using repetitive TMS applied to either the left or right prefrontal cortex. They found that the right prefrontal cortex is important for retrieving memorized items, whereas the left prefrontal cortex is important for encoding items in memory. Buckner and Petersen (1996) summarized hemispheric difference in the prefrontal role in memory retrieval. Both PET and fMRI studies have shown that the left inferior prefrontal area is active during a wide range of tasks that require the subject to retrieve words or information about words from semantic memory, whereas the right anterior and posterior prefrontal areas are active during tasks that require the subject to retrieve information from specific episodes (episodic memory). Importantly, the prefrontal cortex is not the brain area where information is stored as long-term memory. These results indicate that the prefrontal cortex participates in both storing information in long-term memory and retrieving appropriate information from long-term memory and suggest that top-down control signals generated in the prefrontal cortex play important roles in both encoding and retrieving processes in long-term memory.

Several neuroimaging studies have shown interactions between working memory and long-term memory in the prefrontal cortex. For example, Braver et al. (2001) used the n-back task as a working memory task and item encoding and retrieval tasks as long-term memory tasks and examined the regional difference in the participation of the prefrontal cortex in these tasks. They found that the dorsolateral prefrontal cortex mainly participates in working memory process, whereas the ventrolateral prefrontal cortex participates in encoding and retrieving processes of long-term memory. They also found that the anterior prefrontal region participates in both working memory and long-term memory. On the other hand, Blumenfeld and Ranganath (2006) used two working memory conditions (rehearse condition and reorder condition), in that the subject is required to retain triplets of words during a delay, and examined prefrontal activity using fMRI. Behavioral analysis showed that the reorder condition enhanced long-term memory by strengthening inter-item association. fMRI analysis showed that dorsolateral prefrontal activity during the reorder condition was predictive for subsequent long-term memory performance. Therefore, they concluded that the dorsolateral prefrontal cortex contributes to long-term memory formation by strengthening associations among items that are organized in working memory (Blumenfeld and Ranganath 2006, 2007).

The prefrontal cortex is thought to provide top-down control signals to control encoding and retrieving processes of long-term memory. In order to achieve this control, functional connectivity between the prefrontal cortex and the posterior cortices, especially the temporal cortex, is inevitable (Buckner et al. 1999; Simons and Spiers 2003; Miyashita 2004). In fact, strong reciprocal anatomical connections are present between the prefrontal cortex and the temporal cortex (Petrides and Pandya 1999, 2001). Top-down control signals would be sent from the prefrontal cortex to the temporal cortex through these anatomical connections. Recently, Ninomiya et al. (2012) used retrograde trans-synaptic tracing methods with rabies virus and double injections of two different dyes into areas MT and V4 and showed segregated projections from the prefrontal cortex to areas MT and V4 in macaque monkeys, such that area V4 primarily receives inputs from the prefrontal area 46, while area MT receives inputs from both area 46 and the FEF through distinct FEF and parietal neurons. These results suggest that these segregated pathways from the prefrontal cortex carry functionally different top-down signals to different temporal areas.

In addition to anatomical studies, neurophysiological studies have also shown top-down influence by the prefrontal cortex in long-term memory processes. Hasegawa et al. (1998a, b) used posterior-split-brain monkeys and showed that the retrieval of specific information stored as long-term memory in the temporal cortex is under the executive control of the prefrontal cortex. They used a visual paired association task, which requires the memory of the association of two particular visual stimuli. In intact monkeys, information regarding the association of two visual stimuli acquired by training of the paired association task is stored in the temporal cortex. This information is transferred between two hemispheres via the posterior corpus callosum and is not usually transferred via the anterior corpus callosum. They used posterior-split-brain monkeys that have transections of occipito-temporal visual commissural fibers. In these monkeys, when one visual stimulus of a stimulus pair is introduced to one hemisphere, this information will not be transferred to the other hemisphere via the intact anterior callosum. Therefore, it is expected that the posterior-split-brain monkey exhibits behavioral deficits in performing the visual paired association task, when one visual cue is introduced to one hemisphere and when the paired associate of the visual cue is introduced to the opposite hemisphere. However, behavioral studies showed that posterior-split-brain monkeys exhibited normal correct performances when the paired associate of the visual cue was introduced to the hemisphere opposite to the hemisphere where the visual cue was introduced. It has been known that the anterior corpus callosum is an important region for bilateral communication of the prefrontal cortex. In posterior-split-brain monkeys, the information of the visual cue is thought to be transferred to the opposite hemisphere by bilateral communication of the prefrontal cortex via the anterior corpus callosum. This bilateral communication of the prefrontal cortex via the anterior corpus callosum could allow retrieving the information regarding the paired associate of that visual cue. Therefore, they concluded that the retrieval of specific information stored in long-term memory is under the executive control of the prefrontal cortex.

Tomita et al. (1999) also used posterior-split-brain monkeys and the visual paired association task and showed that a large number of inferior temporal neurons receive top-down signals from the prefrontal cortex. The visual paired association task has two conditions (bottom-up condition and top-down condition). A pair of the cue and the choice stimulus are presented in either the right or the left visual field. In the bottom-up condition, the cue stimulus and the choice stimulus are presented in the same visual field, whereas in the top-down condition, the choice stimulus is presented in the visual field opposite to the visual field where the cue stimulus is presented. Single-neuron activity was recorded from one hemisphere of the inferior temporal cortex. In the bottom-up condition, since the cue stimulus is presented on the same side of the recording hemisphere, bottom-up visual signals from the visual cortex are present in the recording hemisphere. In the top-down condition, since the cue stimulus is presented on the opposite of the recording hemisphere, bottom-up visual signals from the visual cortex are absent in the recording hemisphere. In fact, the response latency to the presentation of the cue stimulus was faster in the bottom-up condition than in the top-down condition, suggesting that, in the top-down condition, the information regarding the visual cue reaches the inferior temporal cortex from the prefrontal cortex as top-down signals. Then, they added the split of the anterior corpus callosum in the posterior-split-brain monkeys. Significant responses to the cue stimulus were observed and the selectivity to the cue stimulus was significant in the bottom-up condition, whereas responses to the visual cue were diminished and the selectivity of the visual cue was not observed in the top-down condition. These results indicate that the top-down signal related to the visual cue from the prefrontal cortex is important to retrieve information regarding the paired associate of the visual cue in the paired association task. These results also indicate that top-down signals from the prefrontal cortex play important roles to retrieve information from long-term memory.

Thus, although the prefrontal cortex does not participate in long-term memory per se, the prefrontal cortex plays important roles in both encoding information in long-term memory and retrieving appropriate information from long-term memory. The prefrontal cortex plays roles to achieve the efficient use of information stored in long-term memory depending on the situations and demands by sending top-down control signals to appropriate brain areas.

9.3 Search for Top-Down Signals in the Prefrontal Cortex

A top-down control signal is an important component for understanding how the prefrontal cortex performs executive control and how the prefrontal cortex supervises and manages information processing in other cortical and subcortical areas. However, although there are evidences showing that the prefrontal cortex controls information processing in other cortical and subcortical areas by top-down signals, it is not known which activity is a neural correlate of the top-down signal and how this activity controls activities in other brain areas.

9.3.1 *Neural Activity Related to Long-Term Memory in the Inferior Temporal Cortex*

The prefrontal cortex is known to participate in retrieving information from long-term memory by sending top-down signals to the posterior cortices. Neural activity related to the retrieval of information from long-term memory has been reported by Sakai and Miyashita (1991). They used a visual paired association task with 12 arbitrarily constructed pairs of fractal images and recorded single-neuron activity from the IT cortex (Fig. 9.5a). In this task, the monkey is required to first remember these 12 arbitral pairs of fractal images during training and, after finishing this training, the monkey is required to correctly recall the paired image during test trials. Therefore, long-term memory of these 12 pairs of images is required to perform this task correctly. In their task, while the monkey looks at a central fixation point, one fractal image is presented for 1 s over the fixation point as a visual cue (cue period). After a 4-s delay period, two fractal images are presented simultaneously (choice period); one is the paired associate of the visual cue and one is an image selected from a different pair. If the monkey touches the paired associate of the visual cue, it will get a reward.

Sakai and Miyashita (1991) found two kinds of task-related activity (pair-coding activity and pair-recall activity) in the inferior temporal (IT) cortex while the monkey performed the visual paired association task. One group of IT neurons exhibited stimulus-selective excitatory responses during the cue period (Fig. 9.5b). Some IT neurons exhibited broader stimulus selectivity and responded to more than three images during the cue period. However, other IT neurons exhibited the strongest and the second-strongest responses to a particular pair of images during the cue period (Fig. 9.5c) and two images of this pair had no apparent geometrical similarity (Fig. 9.5a). These results indicate that cue-period activity of latter IT neurons represents information of a particular pair of images. Therefore, they named this activity *pair-coding activity*. The characteristics of pair-coding activity suggest that the storage of information in memory is achieved by such a method that a single neuron encodes two different images as a pair by association.

On the other hand, other groups of IT neurons exhibited stimulus-selective excitatory activity during the delay period. Some IT neurons exhibited a strong excitatory response during the cue period but exhibited no significant activity during the delay period when a particular image was presented as the visual cue. However, these neurons exhibited a weak response during the cue period but exhibited the highest tonic activity or gradually increasing activity during the delay period when the paired associate of this visual cue was presented (Fig. 9.5d). Since tonic or gradually increasing delay-period activity was observed only when a few limited pairs of images were presented during the cue period and since the paired associate of a particular visual cue elicited the highest delay-period activity, they named this delay-period activity *pair-recall activity*. Interestingly, pair-recall activity is not only stimulus selective but also closely coupled with the paired associate of the visual cue which is not presented during the cue period in a given trial but is necessary to

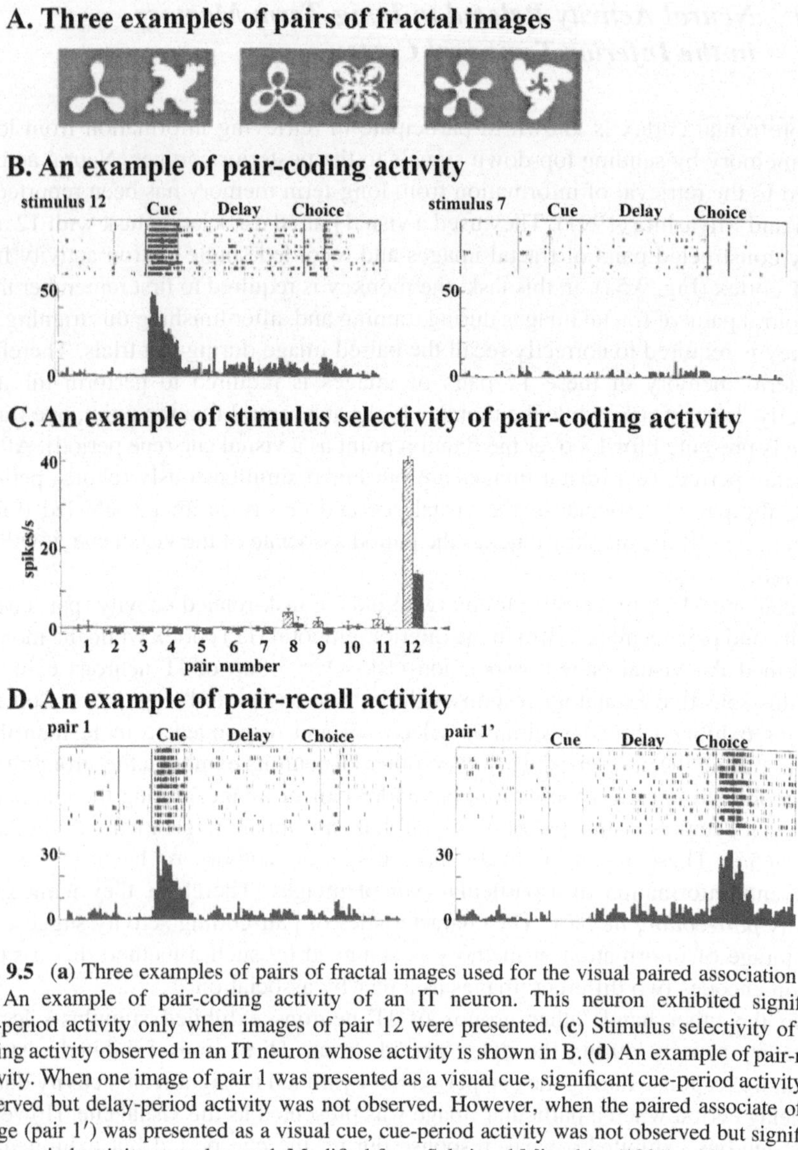

Fig. 9.5 (**a**) Three examples of pairs of fractal images used for the visual paired association task. (**b**) An example of pair-coding activity of an IT neuron. This neuron exhibited significant cue-period activity only when images of pair 12 were presented. (**c**) Stimulus selectivity of pair-coding activity observed in an IT neuron whose activity is shown in B. (**d**) An example of pair-recall activity. When one image of pair 1 was presented as a visual cue, significant cue-period activity was observed but delay-period activity was not observed. However, when the paired associate of that image (pair 1′) was presented as a visual cue, cue-period activity was not observed but significant delay-period activity was observed. Modified from Sakai and Miyashita (1991) with permission

retrieve from memory to correctly select during the choice period. Therefore, IT neurons having pair-recall activity would participate significantly in retrieving necessary information from long-term memory.

Although the IT cortex participates in the long-term storage of information, it has been shown that the retrieval of specific information from long-term memory is under the executive control of the prefrontal cortex (Miyashita and Hayashi 2000). In addition, Tomita et al. (1999) showed that a large number of IT neurons received top-down visual inputs from the prefrontal cortex as well as bottom-up visual inputs from the visual cortex. These results strongly support the notion that the prefrontal cortex provides top-down control signals to the IT cortex and that these top-down control signals play important roles in retrieving specific information from long-term memory. Pair-recall activity reported by Sakai and Miyashita (1991) is tonic or gradually increasing delay-period activity observed only when a few limited pairs of images were presented during the cue period. This pair-recall activity is closely related to retrieving information of the visual cue from the memory to correctly select the visual cue during the choice period. Since it has been shown that the retrieval of specific information from long-term memory is under the executive control of the prefrontal cortex, it is expected that top-down signals generated in the prefrontal cortex are related to producing pair-recall activity during the delay period in the IT cortex.

9.3.2 Prefrontal Activity During a Visual Paired Association Task

The paired association task requires the subject to remember arbitrarily constructed pairs of words or symbols or images by association, store these pairs in long-term memory, and retrieve a particular paired information from the memory to perform a correct response. When the subject performs the visual paired association task, the subject is required to retrieve a specific visual image associated with the sample image from long-term memory. Therefore, the paired association task would be an appropriate task to examine neural mechanisms for retrieving specific information from long-term memory. In the IT cortex, Sakai and Miyashita (1991) found pair-recall activity while the monkey performed the visual paired association task. Pair-recall activity is thought to be related to retrieving the information regarding the paired associate of the visual cue from long-term memory to perform correct selection during the choice period. Although it is expected that top-down signals generated in the prefrontal cortex produce pair-recall activity in the IT cortex, it has not yet been known what kind of neural activity is observed in prefrontal neurons while the monkey performs the visual pair association task and whether or not some task-related activity observed in prefrontal neurons corresponds to top-down signal to produce pair-recall activity in the IT cortex.

Andreau and Funahashi (2011) analyzed prefrontal single-neuron activity while the monkey performed a visual paired association task with Go/No-Go responses using 12 pairs of visual stimuli (Fig. 9.6). In this task, when the monkey presses a lever, a fixation point appears at the center of the monitor. After a 1-s fixation period,

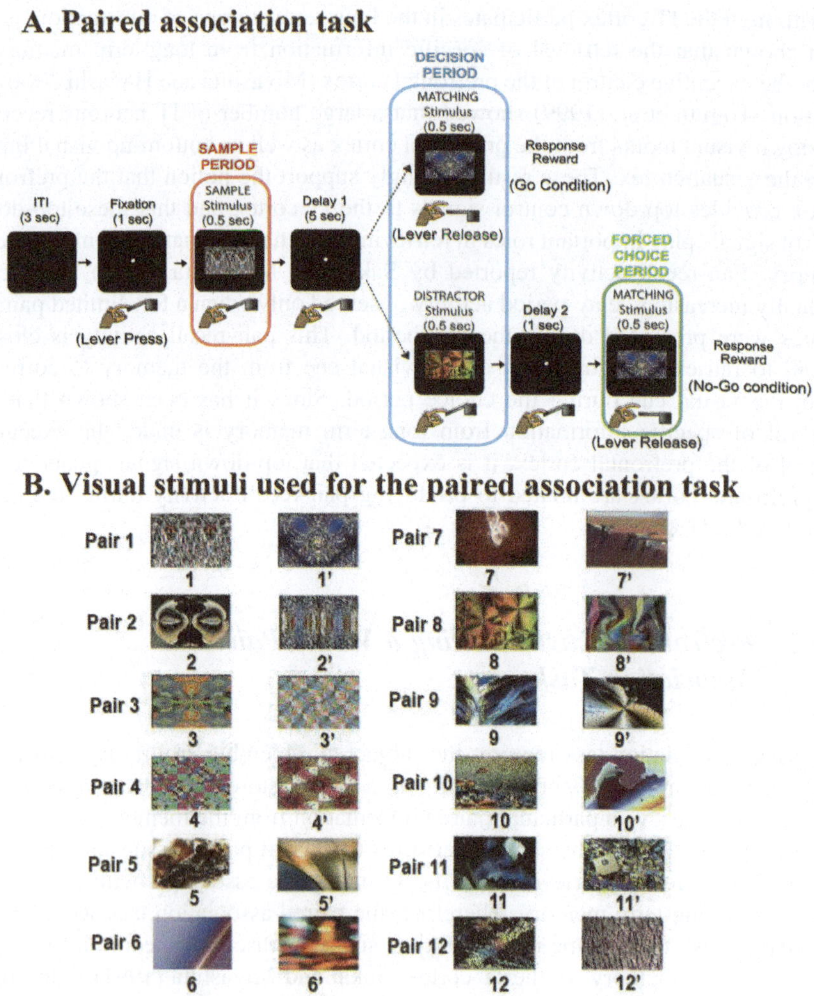

Fig. 9.6 A visual paired association task used for prefrontal studies by Andreau and Funahashi (2011). (**a**) Schematic drawing of temporal sequence of task events in the paired association task. (**b**) Twelve pairs of visual stimuli used for this task

the sample stimulus is presented over the fixation point for 0.5 s (sample period). The sample stimulus is selected randomly from 24 visual stimuli (Fig. 9.6b). At the end of the sample period, the fixation point reappears and a delay period is introduced for 5 s (delay 1 period). The monkey is required to keep looking at the fixation point during this period. At the end of the delay 1 period, another visual stimulus is presented over the fixation point (decision period). This stimulus is either the paired associate of the sample stimulus (matching stimulus) or one of the remaining 22 stimuli (distractor stimulus). If the matching stimulus is presented, the monkey is required to release the lever within 0.5 s to receive a reward (Go condition). If a distractor stimulus is presented during the choice period, the monkey is required to keep pressing the lever for an additional 1 s (delay 2 period) and release the lever within 0.5 s when the matching stimulus is presented (forced choice period) (No-Go condition). The monkey is required to discriminate the matching stimulus from the distractor stimulus during the decision period based on the sample stimulus.

9.3.2.1 Prefrontal Activity During the Sample Period

Andreau and Funahashi (2011) collected activities of 217 neurons from the lateral prefrontal cortex. Among them, 74 neurons showed statistically significant sample-period activity. A large number of prefrontal neurons having sample-period activity exhibited selectivity to sample stimuli (Fig. 9.7a). The response magnitude of sample-period activity was different depending on which stimulus was presented as the sample stimulus. Among prefrontal neurons exhibiting stimulus selectivity, some neurons exhibited significant excitatory responses when either stimulus of a particular pair was presented as a sample stimulus (Fig. 9.7b). Similar to a neuron shown in Fig. 9.7b, the neurons having stimulus selectivity exhibited the largest response to a particular sample stimulus and the second largest response to the paired associate of that stimulus. This feature of stimulus selectivity is similar to pair-coding activity observed in IT neurons. Among prefrontal neurons exhibiting stimulus selectivity, 32% showed similar pair-coding activity. In this task, the same visual stimulus is presented as either the sample stimulus, the matching stimulus, or the distractor. Behavioral significance of the same visual stimulus is different among these three conditions. However, as shown in Fig. 9.7a, the magnitude of the response, the response latency, and the stimulus selectivity were the same among three conditions. These results indicate that these prefrontal neurons exhibit pair-coding activity even if the behavioral significance of the stimulus is different among three conditions.

Sakai and Miyashita (1991) calculated pair indices (PIs) for IT neurons having pair-coding activity to describe the strength of pair selectivity. To compare the strength of pair selectivity between prefrontal neurons and IT neurons, PIs were calculated in prefrontal neurons using the same method as Sakai and Miyashita (1991) had used. The result showed that more prefrontal neurons had positive PIs than IT neurons, suggesting that the pair selectivity is stronger in prefrontal neurons than IT neurons (Andreau and Funahashi 2011). Further, Naya et al. (2003) calculated pair coding indices (PCIs) to quantify the strength of pair selectivity using sample-period activity. They compared these indices between IT neurons (area TE) and entorhinal neurons (area 36) and found that the mean PCI of entorhinal neurons ($PCI = 0.51$) was larger than that of IT neurons ($PCI = 0.14$), suggesting that visual information from IT neurons selectively converges on entorhinal neurons. Pair coding indices were also calculated using sample-period activity of prefrontal

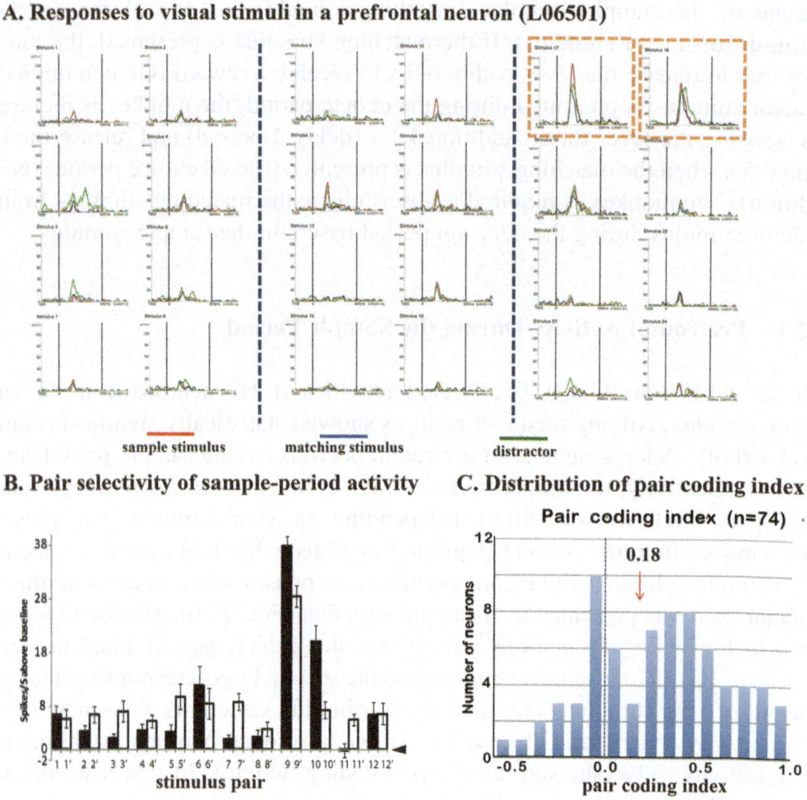

Fig. 9.7 (**a**) Examples of responses to visual stimuli in a prefrontal neuron (L06501). Red, blue, and green lines indicate responses to the same stimulus presented as a sample stimulus, a matching stimulus, and a distractor, respectively. (**b**) Selectivity of sample-period activity in a prefrontal neuron (L06501). This neuron exhibited significant sample-period activity only when pair 9 was presented. (**c**) Distribution of pair coding index for 74 prefrontal neurons

neurons (Fig. 9.7c). The mean PCI of prefrontal neurons was 0.18, indicating that the strength of pair selectivity was comparable between prefrontal neurons and IT neurons, but was weaker than entorhinal neurons.

As has shown before, Tomita et al. (1999) showed that a large number of IT neurons received top-down signals from the prefrontal cortex as well as bottom-up signals from the visual cortex and that the onset latency of sample-period activity under the top-down condition (178 ms) was significantly longer than that under the bottom-up condition (73 ms). If the onset latency of sample-period activity in prefrontal neurons was significantly longer than that in IT neurons under the bottom-up condition and significantly shorter than that in IT neurons under the top-down condition, this result strongly supports the notion that sample-period activity in the prefrontal cortex acts as top-down signal to activate IT neurons and that top-down signals from the prefrontal cortex could play essential roles in retrieving information from long-term memory in the paired association task. In fact, the mean onset latency of sample-period activity was 144 ms in prefrontal neurons (Andreau and Funahashi 2011), which was longer than the onset latency observed in IT neurons under the bottom-up condition (73 ms), but shorter than the onset latency under the top-down condition (178 ms).

Thus, visual responses of prefrontal neurons observed after training of the visual paired association task exhibit stimulus selectivity and stimulus-pair selectivity with similar strength as IT neurons exhibit. Pair-selectivity of prefrontal neurons was observed not only when each paired visual stimuli was presented as either the sample stimulus or the matching stimulus but also when the stimulus was presented as the distractor. These results indicate that pair-selective activity observed in prefrontal neurons has the same feature as pair-coding activity in IT neurons has. Importantly, the mean onset latency of sample-period activity in prefrontal neurons was longer than the onset latency observed in IT neurons under the bottom-up condition but shorter than the onset latency under the top-down condition. These results indicate that pair-selective sample-period activity acts as top-down signal which is generated in the prefrontal cortex and provided to the IT cortex in order to retrieve specific information assigned by the sample stimulus from long-term memory. Thus, the prefrontal cortex plays significant roles in retrieving information from long-term memory by sending top-down signals to the posterior brain areas.

9.3.2.2 Prefrontal Activity During the Delay Period

Delay-period activity is also observed in prefrontal neurons while the monkey performed the visual paired association task. As is observed during the delayed-response task, various temporal patterns of delay-period activity can be observed including tonic sustained type, gradually increasing type, and gradually decreasing type. Delay-period activity observed in the paired association task exhibits stimulus selectivity. An example shown in Fig. 9.8a exhibited a gradually increasing type of delay-period activity when pair 10 (consisting of stimuli 19 and 20) was presented as the sample stimulus. This neuron exhibited significant delay-period activity only in the trials, in which stimulus 20 was presented as the sample stimulus and the monkey was required to retrieve stimulus 19 to make a correct response during the decision period. This neuron exhibited stronger excitatory visual responses when stimulus 19 was presented during the sample period of trial 19 (index 19) and during the decision period of trial 20 (index 20), whereas weak visual responses were observed when stimulus 20 was presented during the sample period of trial 20 (index 20) and during the decision period of trial 19 (index 19) (Fig. 9.8b). The same patterns of delay-period activity and visual responses were observed in both Go and No-Go conditions, because retrieving the information of the paired associate of the sample stimulus is necessary during the delay period in both conditions. This pattern of delay-period activity is similar to the pattern of pair-recall activity observed in IT

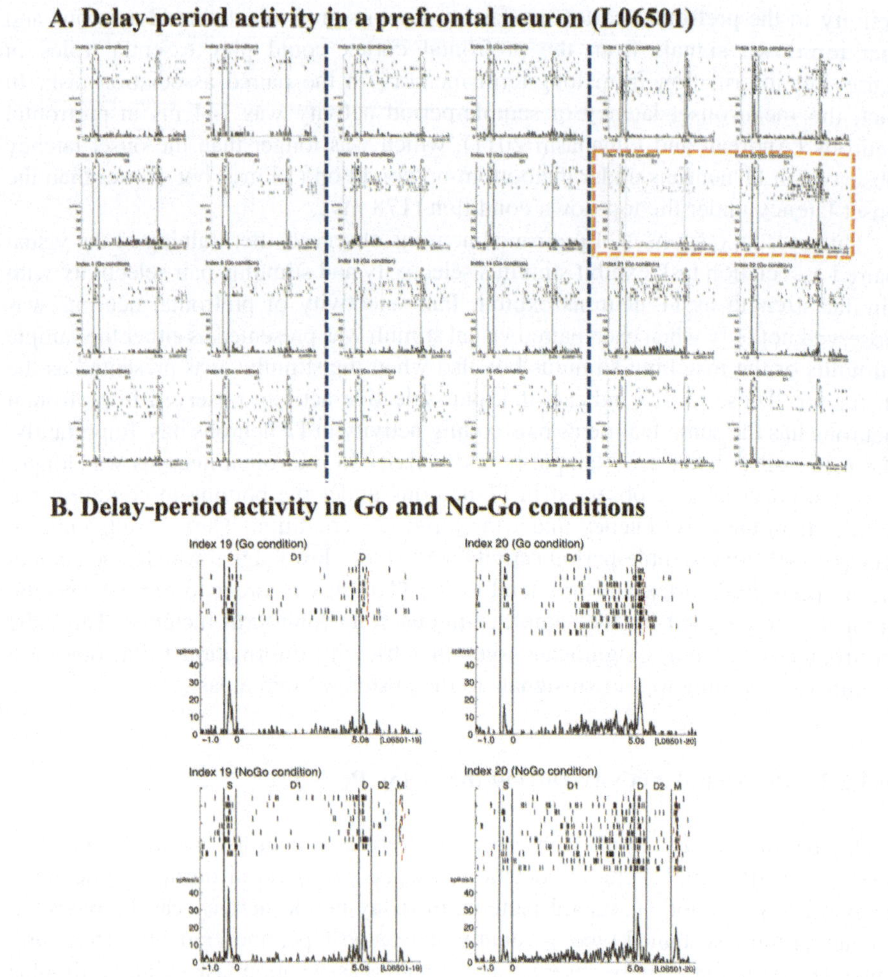

Fig. 9.8 (**a**) Examples of delay-period activity in a prefrontal neuron (L06501) during the visual paired association task. (**b**) Pair-recall-type delay-period activity was observed when stimulus pair 10 was presented as the sample stimulus

neurons (Fig. 9.5d). Pair-recall activity is thought to be related to retrieving the visual information from the memory to correctly select the visual stimulus during the choice period. Therefore, delay-period activity similar as pair-recall activity in prefrontal neurons is also thought to contribute to retrieve the information of the paired associate of the sample stimulus from long-term memory. It is thought that retrieving specific visual information from long-term memory is under the executive

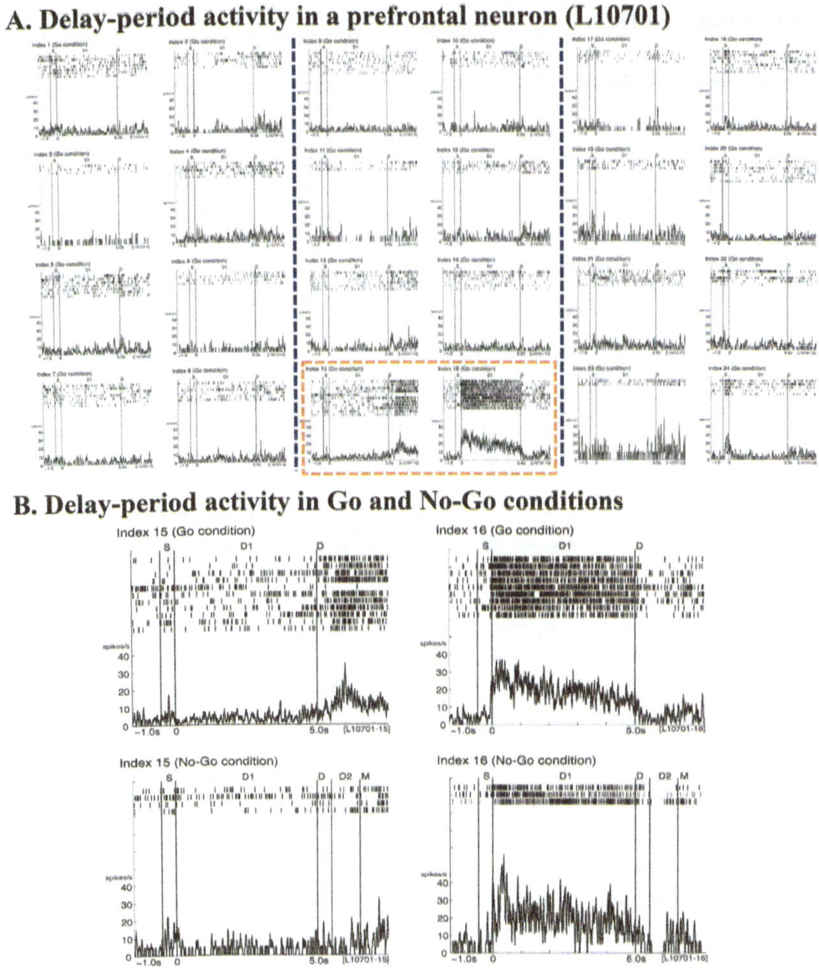

Fig. 9.9 (**a**) Examples of tonic sustained delay-period activity in a prefrontal neuron (L10701) during the visual paired association task. (**b**) Tonic sustained delay-period activity was observed only when stimulus pair 16 was presented during the sample period in both Go and No-Go conditions

control of the prefrontal cortex. Therefore, delay-period activity similar to pair-recall activity might act as top-down signals and produce pair-recall activity in IT neurons.

Prefrontal neurons exhibited not only delay-period activity having similar characteristics of pair-recall activity in IT neurons but also tonic sustained delay-period activity (Fig. 9.9). The prefrontal neuron shown in Fig. 9.9 showed stimulus selectivity and exhibited delay-period activity only when stimulus 16 was presented

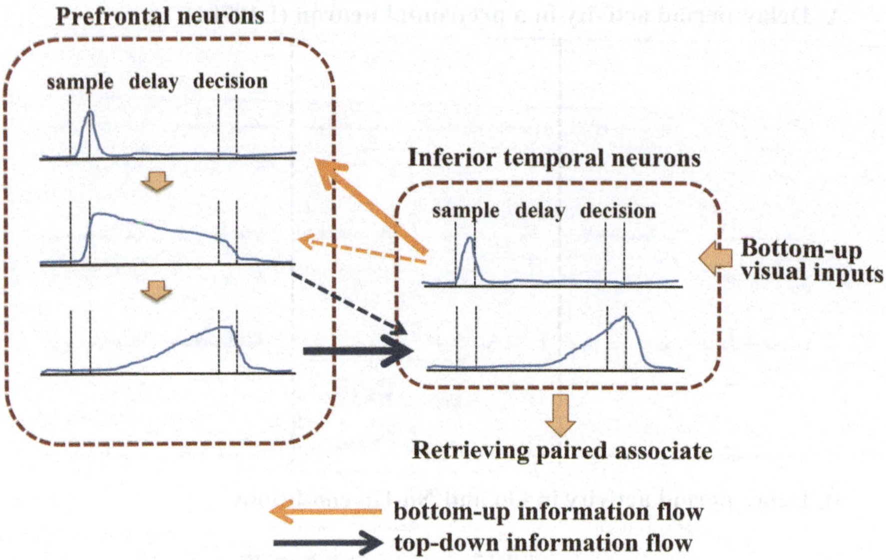

Fig. 9.10 A possible information flow between the prefrontal cortex and the inferior temporal cortex when the information of the paired associate of the sample stimulus is retrieved from the memory during the visual paired association task

during the sample period in both Go and No-Go conditions (Fig. 9.9b). Stimulus-selective tonic sustained delay-period activity allows to maintain the information regarding the sample stimulus along the delay period and is able to provide this information to objective brain areas to help to retrieve the information regarding the paired associate of the sample stimulus. Therefore, stimulus-selective tonic sustained delay-period activity may also act as top-down control signals to retrieve particular information such as the information of the paired associate (stimulus 15) of the sample stimulus (stimulus 16) from long-term memory.

Based on several different task-related activities observed in prefrontal neurons and IT neurons during the visual paired association task, a possible information flow can be considered within the prefrontal cortex and between the prefrontal cortex and the IT cortex (Fig. 9.10). Long-term training of the visual paired association task constructs specific associative response patterns for selective pairs of visual stimuli (pair-coding activity, pair-recall activity, stimulus-selective delay-period activity) in prefrontal neurons and IT neurons. When a particular visual stimulus is presented during the sample period, the information of the sample stimulus is provided first to IT neurons, produces stimulus-selective responses in IT neurons, then is transferred to prefrontal neurons through bottom-up visual pathways, and produces sample-period activity in prefrontal neurons. As described in Chap. 10, there are extensive functional interactions among prefrontal neurons exhibiting different task-related activities and different stimulus selectivity and information flows are present from the neurons having cue-period activity to the neurons having response-period activity through the neurons having delay-period activity in the prefrontal cortex. Therefore, neurons having sample-period activity activate neurons having tonic sustained delay-period activity. Then, neurons having sample-period activity and neurons having tonic sustained delay-period activity may activate neurons having pair-coding-type delay-period activity in the prefrontal cortex. The outputs of prefrontal neurons having tonic sustained delay-period activity and neurons having pair-coding-type delay-period activity are sent to the IT cortex through top-down information flow pathways and may act as triggers to initiate pair-recall activity in IT neurons. Thus, bottom-up visual inputs to the prefrontal cortex produce activation of specific groups of prefrontal neurons, this activation produces specific patterns of activity in prefrontal neurons, and these activities are provided to the objective brain areas and act as top-down control signals for retrieving necessary information from long-term memory.

9.4 Conclusions

Executive functions are products of coordinated operation of various neural systems and are essential for achieving a particular goal in a flexible manner. The prefrontal cortex is an important brain area for achieving executive functions. To accomplish coordinated operations of multiple brain areas, the prefrontal cortex needs to monitor the activities of other cortical and subcortical areas through bottom-up signals and to control their activities by sending top-down control signals. Although studies show evidences that the prefrontal cortex outputs top-down control signals to the posterior cortices and controls their information processing, neural correlate of top-down control signal is not well known. Top-down control signals from the prefrontal cortex are shown to contribute to retrieving specific information from long-term memory. Sakai and Miyashita (1991) examined activity of IT neurons while the monkey performed a paired association task, and found pair-coding activity and pair-recall activity. Pair-coding activity suggests how two different stimuli are stored in a single IT neuron, while pair-recall activity suggests how specific information is retrieved from memory by other information associated with that information. Since pair-recall activity is thought to be under top-down control by the prefrontal cortex, activities of prefrontal neurons were examined using the paired association task and found pair-coding-like sample-period activity and pair-recall-like delay-period activity in prefrontal neurons. The mean latency of visual responses in prefrontal neurons was longer than bottom-up visual responses but faster than top-down responses in IT neurons. Some prefrontal neurons exhibit tonic sustained delay-period activity only when a particular visual stimulus is presented during the sample period. These results suggest that pair-coding-like sample-period activity, pair-recall-like delay-period activity, and tonic sustained delay-period activity in prefrontal neurons are neural correlates of top-down signals and control activities of IT cortex to retrieve necessary information.

References

Anderson JC, Kennedy H, Martin KAC (2011) Pathways of attention: synaptic relationships of frontal eye field to V4, lateral intraparietal cortex, and area 46 in macaque monkey. J Neurosci 31:10872–10881. https://doi.org/10.1523/jneurosci.0622-11.2011

Andreau JM, Funahashi S (2011) Primate prefrontal neurons encode the association of paired visual stimuli during the pair-association task. Brain Cogn 76:58–69. https://doi.org/10.1016/j.bandc.2011.03.001

Bahmani Z, Daliri MR, Merrikhi Y, Clark K, Noudoost B (2018) Working memory enhances cortical representations via spatially specific coordination of spike times. Neuron 97:967–979.e6. https://doi.org/10.1016/j.neuron.2018.01.012

Barceló F, Suwazono S, Knight RT (2000) Prefrontal modulation of visual processing in humans. Nat Neurosci 3:399–403. https://doi.org/10.1038/73975

Blumenfeld RS, Ranganath C (2006) Dorsolateral prefrontal cortex promotes long-term memory formation through its role in working memory organization. J Neurosci 26:916–925. https://doi.org/10.1523/jneurosci.2353-05.2006

Blumenfeld RS, Ranganath C (2007) Prefrontal cortex and long-term memory encoding: an integrative review of findings from neuropsychology and neuroimaging. Neuroscientist 13: 280–291. https://doi.org/10.1177/1073858407299290

Bollimunta A, Bogadhi AR, Krauzlis RJ (2018) Comparing frontal eye field and superior colliculus contributions to covert spatial attention. Nat Comm 9:533. https://doi.org/10.1038/s41467-018-06042-2

Braver TS, Barch DM, Kelley WM, Buckner RL, Cohen NJ, Miezin FM, Snyder AZ, Ollinger JM, Akbudak E, Conturo TE, Petersen SE (2001) Direct comparison of prefrontal cortex regions engaged by working and long-term memory tasks. Neuroimage 14:48–59. https://doi.org/10.1006/nimg.2001.0791

Bressler SL, Tang W, Sylvester CM, Shulman GL, Corbetta M (2008) Top-down control of human visual cortex by frontal and parietal cortex in anticipatory visual spatial attention. J Neurosci 28: 10056–10061. https://doi.org/10.1523/jneurosci.1776-08.2008

Brewer JB, Zhao Z, Desmond JE, Glover GH, Gabrieli JDE (1998) Making memories: activity that predicts how well visual experience will be remembered. Science 281:1185–1187. https://doi.org/10.1126/science.281.5380.1185

Brooks VB (1983) Study of brain function by local, reversible cooling. Rev Physiol Biochem Pharmacol 95:1–109. https://doi.org/10.1007/bfb0034097

Bruce CJ, Goldberg ME (1985) Primate frontal eye fields. I. Single neurons discharging before saccades. J Neurophysiol 53:603–635. https://doi.org/10.1152/jn.1985.53.3.603

Buckner RL, Koutsaal W (1998) Functional neuroimaging studies of encoding, priming and explicit memory retrieval. Proc Natl Acad Sci USA 95:891–898. https://doi.org/10.1073/pnas.95.3.891

Buckner RL, Petersen SE (1996) What does neuroimaging tell us about the role of prefrontal cortex in memory retrieval? Seminars Neurosci 8:47–55. https://doi.org/10.1006/smns.1996.0007

Buckner RL, Kelley WM, Petersen SE (1999) Frontal cortex contributes to human memory formation. Nat Neurosci 2:311–314. https://doi.org/10.1038/7221

Buschman TJ, Miller EK (2007) Top-down versus bottom-up control of attention in the prefrontal and posterior parietal cortices. Science 315:1860–1862. https://doi.org/10.1126/science.1138071

Bushnell MC, Goldberg ME, Robinson DL (1981) Behavioral enhancement of visual responses in monkey cerebal cortex. I. Modulation in posterior parietal cortex related to selective visual attention. J Neurophysiol 46:755–772. https://doi.org/10.1152/jn.1981.46.4.755

Cabeza R, Nyberg L (2000) Imaging cognition II: an empirical review of 275 PET and fMRI studies. J Cogn Neurosci 12:1–47. https://doi.org/10.1162/08989290051137585

Chafee MV, Goldman-Rakic PS (1998) Matching patterns of activity in primate prefrontal area 8a and parietal area 7ip neurons during a spatial working memory task. J Neurophysiol 79:2919–2940. https://doi.org/10.1152/jn.1998.79.6.2919

Chafee MV, Goldman-Rakic PS (2000) Inactivation of parietal and prefrontal cortex reveals interdependence of neural activity during memory-guided saccades. J Neurophysiol 83:1550–1566. https://doi.org/10.1152/jn.2000.83.3.1550

Chan JL, Koval MJ, Womelsdorf T, Lomber SG, Everling S (2015) Dorsolateral prefrontal cortex deactivation in monkeys reduces preparatory beta and gamma power in the superior colliculus. Cereb Cortex 25:4704–4714. https://doi.org/10.1093/cercor/bhu154

Crowe DA, Goodwin SJ, Blackman RK, Sakellaridi S, Sponheim SR, McDonald AW III, Chafee MV (2013) Nat Neurosci 16:1484–1491. https://doi.org/10.1038/nn.3509

Egner T, Monti JMP, Trittschuh EH, Wieneke CA, Hirsch J, Mesulam M-M (2008) Neural integration of top-down spatial and feature-based information in visual search. J Neurosci 28: 6141–6151. https://doi.org/10.1523/jneurosci.1262-08.2008

Everling S, Tinsley CJ, Gaffan D, Duncan J (2002) Filtering of neural signals by focused attention in the monkey prefrontal cortex. Nat Neurosci 5:671–676. https://doi.org/10.1038/nn874

Fuster JM (2008) The prefrontal cortex, 4th edn. Academic Press, New York, NY

Fuster JM, Bauer RH, Jervey JP (1985) Functional interactions between inferotemporal and prefrontal cortex in a cognitive task. Brain Res 220:299–307. https://doi.org/10.1016/0006-8993(85)90689-4

Gazzaley A, Cooney JW, McEvoy K, Knight RT, D'Esposito M (2005) Top-down enhancement and suppression of the magnitude and speed of neural activity. J Cogn Neurosci 17:507–517. https://doi.org/10.1162/0898929053279522

Gogulski J, Boldt R, Savolainen P, Guzmán-López J, Carlson S, Pertovaara A (2015) A segregated neural pathway for prefrontal top-down control of tactile discrimination. Cereb Cortex 25:161–166. https://doi.org/10.1093/cercor/bht211

Goldberg ME, Bushnell MC (1981) Behavioral enhancement of visual responses in monkey cerebral cortex. II. Modulation in frontal eye fields specifically related to saccades. J Neurophysiol 46:773–787. https://doi.org/10.1152/jn.1981.46.4.773

Gregoriou GG, Gotts SJ, Zhou H, Desimone R (2009) High-frequency, long-range coupling between prefrontal and visual cortex during attention. Science 324:1207–1210. https://doi.org/10.1126/science.1171402

Hasegawa I, Fukushima T, Ihara T, Miyashita Y (1998a) Callosal window between prefrontal cortices: cognitive interaction to retrieve long-term memory. Science 281:814–818. https://doi.org/10.1126/science.281.5378.814

Hasegawa R, Sawaguchi T, Kubota K (1998b) Monkey prefrontal neuronal activity coding the forthcoming saccade in an oculomotor delayed matching-to-sample task. J Neurophysiol 79: 322–333. https://doi.org/10.1152/jn.1998.79.1.322

Hopfinger JB, Buonocore MH, Mangum GR (2000) The neural mechanisms of top-down attentional control. Nat Neurosci 3:284–291. https://doi.org/10.1038/72999

Huang L, Wang L, Shen W, Li M, Wang S, Wang X, Ungerleider LG, Zhang X (2020) A source for awareness-dependent figure-ground segregation in human prefrontal cortex. Proc Natl Acad Sci USA 117:30836–30847. https://doi.org/10.1073/pnas.1922832117

Hwang K, Shine JM, D'Esposito M (2019) Frontoparietal activity interacts with task-evoked changes in functional connectivity. Cereb Cortex 29:802–813. https://doi.org/10.1093/cercor/bhy011

Kao Y-C, Davis ES, Gabrieli JDE (2005) Neural correlates of actual and predicted memory formation. Nat Neurosci 8:1776–1783. https://doi.org/10.1038/nn1595

Kar K, DiCarlo JJ (2021) Fast recurrent processing via ventrolateral prefrontal cortex is needed by the primate ventral stream for robust core visual object recognition. Neuron 109:164–176.e5. https://doi.org/10.1016/j.neuron.2020.09.035

Kar K, Kubilius J, Schmidt K, Issa EB, DiCarlo JJ (2019) Evidence that recurrent circuits are critical to the ventral stream's execution of core object recognition behavior. Nat Neurosci 22: 974–983. https://doi.org/10.1038/s41593-019-0392-5

Kluen LM, Dandolo LC, Jocham G, Schwabe L (2019) Dorsolateral prefrontal cortex enables updating of established memories. Cereb Cortex 29:4154–4168. https://doi.org/10.1093/cercor/bhy298

Kodaka Y, Mikami A, Kubota K (1997) Neuronal activity in the frontal eye field of the monkey is modulated while attention is focused on to a stimulus in the peripheral visual field, irrespective of eye movement. Neurosci Res 28:291–298. https://doi.org/10.1016/S0168-0102(97)00055-2

Lee TG, D'Esposito M (2012) The dynamic nature of top-down signals originating from prefrontal cortex: a combined fMRI-TMS study. J Neurosci 32:15458–15466. https://doi.org/10.1523/JNEUROSCI.0627-12.2012

Mendoza-Halliday D, Martinez-Trujillo JC (2017) Neuronal population coding of perceived and memorized visual features in the lateral prefrontal cortex. Nat Commun 8:15471. https://doi.org/10.1038/ncomms15471

Merchant H, Crowe DA, Robertson MS, Fortes AF, Georgopoulos AP (2011) Top-down spatial categorization signal from prefrontal to posterior parietal cortex in the primate. Front Syst Neurosci 5:69. https://doi.org/10.3389/fnsys.2011.00069

Merrikhi Y, Clark K, Albarran E, Parsa M, Zirnsak M, Moore T, Noudoost B (2017) Spatial working memory alters the efficacy of input to visual cortex. Nat Commun 8:15041. https://doi.org/10.1038/ncommn15041

Miller BT, Vytlacil J, Fegen D, Pradhan S, D'Esposito M (2011) The prefrontal cortex modulates category selectivity in human extrastriate cortex. J Cogn Neurosci 23:1–10. https://doi.org/10.1162/jocn.2010.21516

Miyashita Y (2004) Cognitive memory: cellular and network machineries and their top-down control. Science 306:435–440. https://doi.org/10.1126/science.1101864

Miyashita Y, Hayashi T (2000) Neural representation of visual objects: encoding and top-down activation. Curr Opin Neurobiol 10:187–194. https://doi.org/10.1016/S0959-4388(00)00071-4

Monosov IE, Sheinberg DL, Thompson KG (2011) The effects of prefrontal cortex inactivation on object responses of single neurons in the inferotemporal cortex during visual search. J Neurosci 31:15956–15961. https://doi.org/10.1523/jneurosci.2995-11.2011

Moore T, Armstrong KM (2003) Selective gating of visual signals by microstimulation of frontal cortex. Nature 421:370–373. https://doi.org/10.1038/nature01341

Moore T, Fallah M (2004) Microstimulation of the frontal eye field and its effects on covert spatial attention. J Neurophysiol 91:152–162. https://doi.org/10.1152/jn.00741.2002

Moran J, Desimone R (1985) Selective attention gates visual processing in the extrastriate cortex. Science 229:782–784. https://doi.org/10.1126/science.4023713

Morishima Y, Akaishi R, Yamada Y, Okuda J, Toma K, Sakai K (2009) Task-spacific signal transmission from prefrontal cortex in visual selective attention. Nat Neurosci 12:85–91. https://doi.org/10.1038/nn.2237

Motter BC (1993) Focal attention produces spatially selective processing in visual cortical areas V1, V2, and V4 in the presence of competing stimuli. J Neurophysiol 70:909–919. https://doi.org/10.1152/jn.1993.70.3.909

Naya Y, Yoshida M, Miyashita Y (2003) Forward processing of long-term associative memory in monkey inferotemporal cortex. J Neurosci 23:2861–2871. https://doi.org/10.1523/jneurosci.23-07-02861.2003

Ninomiya T, Sawamura H, Inoue K-I, Takada M (2012) Segregated pathways carrying frontally derived top-down signals to visual areas MT and V4 in macaques. J Neurosci 32:6851–6858. https://doi.org/10.1523/jneurosci.6295-11.2012

Nolde SF, Johnson MK, Raye CL (1998) The role of prefrontal cortex during tests of episodic memory. Trends Cogn Sci 2:99–406. https://doi.org/10.1016/S1364-6613(98)01233-9

Petrides M, Pandya DN (1999) Dorsolateral prefrontal cortex: comparative cytoarchitectonic analysis in the human and the macaque brain and corticocortical connection patterns. Eur J Neurosci 11:1011–1036. https://doi.org/10.1046/j.1460-9568.1999.00518.x

Petrides M, Pandya DN (2001) Comparative cytoarchitectonic analysis of the human and the macaque ventrolateral prefrontal cortex and corticocortical connection patterns in the monkey. Eur J Neurosci 16:291–310. https://doi.org/10.1002/cne.21048

Quintana J, Fuster JM, Yajeya J (1989) Effects of cooling parietal cortex on prefrontal units in delay tasks. Brain Res 503:100–110. https://doi.org/10.1016/0006-8993(89)91709-5

Romanski LM, Chafee MV (2021) A view from the top: prefrontal control of object recognition. Neuron 109:6–8. https://doi.org/10.1016/j.neuron.2020.12.014

Rossi S, Cappa SF, Babiloni C, Pasqualetti P, Miniussi C, Carducci F, Babiloni F, Rossini PM (2001) Prefrontal cortex in long-term memory: an "interference" approach using magnetic stimulation. Nat Neurosci 4:948–952. https://doi.org/10.1038/nn0901-948

Rossi AF, Bichot NP, Desimone R, Ungerleider LG (2007) Top-down attentional deficits in macaques with lesions of lateral prefrontal cortex. J Neurosci 27:11306–11314. https://doi.org/10.1523/JNEUROSCI.2939-07.2007

Rossi AF, Pessoa L, Desimone R, Ungerleider LG (2009) The prefrontal cortex and the executive control of attention. Exp Brain Res 192:489–497. https://doi.org/10.1007/s00221-008-1642-z

Ruff CC, Bestmann S, Blankenburg F, Bjoertomt O, Josephs O, Weiskopf N, Deichmann R, Driver J (2008) Distinct causal influences of parietal versus frontal areas on human visual cortex: evidence from concurrent TMS-fMRI. Cereb Cortex 18:817–827. https://doi.org/10.1093/cercor/bhm128

Rugg MD, Fletcher PC, Frith CD, Frackowiak RSJ, Dolan RJ (1996) Differential activation of the prefrontal cortex in successful and unsuccessful memory retrieval. Brain 119:2073–2083. https://doi.org/10.1093/brain/119.6.2073

Sakai K, Miyashita Y (1991) Neural organization for the long-term memory of paired associates. Nature 354:152–155. https://doi.org/10.1038/354152a0

Schafer RJ, Moore T (2011) Selective attention from voluntary control of neurons in prefrontal cortex. Science 332:1568–1571. https://doi.org/10.1126/science.1199892

Schiller PH, Tehovnik EJ (2001) Look and see: how the brain moves your eye about. In: Casanova C, Ptito M (eds) Progress in brain research, vol 134. Elsevier Science BV, pp 127–142. https://doi.org/10.1016/S0079-6123(01)34010-4

Simons JS, Spiers HJ (2003) Prefrontal and medial temporal lobe interactions in long-term memory. Nat Rev Neurosci 4:637–648. https://doi.org/10.1038/nrn1178

Tanaka K (1996) Inferotemporal cortex and object vision. Annu Rev Neurosci 19:109–139. https://doi.org/10.1146/annurev.ne.19.030196.000545

Taylor PCJ, Nobre AC, Rushworth MFS (2007) FEF TMS affects visual cortical activity. Cereb Cortex 17:391–399. https://doi.org/10.1093/cercor/bhj156

Thompson KG, Bichot NP (2005) A visual salience map in the primate frontal eye field. Progr Brain Res 147:249–262. https://doi.org/10.1016/S0079-6123(04)47019-8

Thompson KG, Biscoe KL, Sato TR (2005) Neuronal basis of covert spatial attention in the frontal eye field. J Neurosci 25:9479–9487. https://doi.org/10.1523/jneurosci.0741-05.2005

Tomita H, Ohbayashi M, Nakahara K, Hasegawa I, Miyashita Y (1999) Top-down signal from prefrontal cortex in executive control of memory retrieval. Nature 401:699–703. https://doi.org/10.1038/44372

Wagner AD, Desmond JE, Glover GH, Gabrieli JDE (1998a) Prefrontal cortex and recognition memory. Functional-MRI evidence for context-dependent retrieval processes. Brain 121:1985–2002. https://doi.org/10.1093/brain/121.10.1985

Wagner AD, Schacter DL, Rotte M, Koutstaal W, Maril A, Dale AM, Rosen BR, Buckner RL (1998b) Building memories: remembering and forgetting of verbal experiences as predicted by brain activity. Science 281:1188–1191. https://doi.org/10.1126/science.281.5380.1188

Wardak C, Ibos G, Duhamel J-R, Olivier E (2006) Contribution of the monkey frontal eye field to covert visual attention. J Neurosci 26:4228–4235. https://doi.org/10.1523/jneurosci.3336-05.2006

Zanto TP, Rubens MT, Thangavel A, Gazzaley A (2011) Causal role of the prefrontal cortex in top-down modulation of visual processing and working memory. Nat Neurosci 14:656–661. https://doi.org/10.1038/nn.2773

Zhou H, Desimone R (2011) Feature-based attention in the frontal eye field and area V4 during visual search. Neuron 70:1205–1217. https://doi.org/10.1016/j.neuron.2011.04.032

Chapter 10
Prefrontal Cortex and Working Memory

10.1 Baddeley's Working Memory Model

Working memory has been described as a system that includes mechanisms for temporarily maintaining information as well as manipulating and processing information. Baddeley and Logie (1999) described working memory as a mechanism to "allow humans to comprehend and mentally represent their immediate environment, to retain information about their immediate past experience, to support the acquisition of new knowledge, to solve problems, and to formulate, relate, and act on current goals." Kieras et al. (1999) described working memory as a mechanism to "encompass the entire ensemble of temporary stored codes, knowledge representations, and procedures whereby information is maintained, updated, and applied for performing perceptual-motor and cognitive tasks." Further, Miyake and Shah (1999) described working memory as "mechanisms or processes that are involved in the control, regulation, and active maintenance of task-relevant information in the service of complex cognition, including novel as well as familiar, skilled tasks." Thus, working memory is an important concept to understand mechanisms to perform higher cognitive functions such as thinking, planning, reasoning, decision-making, and language comprehension.

Several models of working memory have been proposed (e.g., see Miyake and Shah (1999)). The most influential model of working memory is the one proposed by Baddeley and Hitch (1974) and Baddeley (1986), which included one master component (central executive) and two slave components (visuospatial sketchpad and phonological loop). Then, Baddeley (2000) proposed his revised model of working memory, which includes one master component (central executive) and three slave components (visuospatial sketchpad, phonological loop, and episodic buffer). The phonological loop is a system for speech perception and language comprehension and includes mechanisms for temporary maintenance of speech-based information by sub-vocal rehearsal. The visuospatial sketchpad is a system for processing visuospatial information as well as information that cannot be

S. Funahashi, *Dorsolateral Prefrontal Cortex*, Brain Science,
https://doi.org/10.1007/978-981-19-7268-3_10

processed by language. The visuospatial sketchpad includes mechanisms for temporarily maintaining information of visual images. The episodic buffer is a temporary storage buffer with a limited capacity to integrate information arising from a variety of sources including long-term memory and stored as "chunks." The central executive supervises the operation of these slave systems to achieve an appropriate goal. The central executive manages a limited capacity of memory resource and divides and re-organizes memory resource depending on the demand from each slave component so that each slave component can accomplish the most effective operation under the current condition. Further, the central executive selects appropriate control processes or strategies for performing a current task accurately. Thus, the central executive supervises the operation of the slave components by monitoring and controlling the operations of these components.

The model of working memory proposed by Baddeley and Hitch (1974) and Baddeley (2000) is an abstract model and based on the results obtained by psychological experiments. They proposed three slave components each of which is considered to temporarily maintain different modalities or types of information. Three slave components to process different modality information and one executive component to manage a limited capacity of memory resource for slave components must be important and necessary to accurately interpret the results of psychological experiments. As Baddeley himself stated (Baddeley 1986), in general, no particular brain region or neural system is assigned to each of the four components of the working memory model. However, these components must be supported by certain and distinct neural systems in the brain.

10.2 Physiologically Plausible Model of Working Memory

Working memory is defined as a system that includes neural mechanisms for temporarily maintaining, as well as manipulating and processing information. From a physiological point of view, neural mechanisms for temporarily maintaining and processing information are the basic and essential mechanisms for any information processing in the brain and, therefore, must be ubiquitous in the brain. The neural mechanism for temporarily maintaining information must be the basic and common neural mechanism for three slave components. The only difference among the three components is that different information is managed in each slave component. In order to understand the neural mechanism of working memory, it would be suitable to construct a physiologically plausible model of working memory, which is constructed by physiologically identifiable neural components. The working memory model based on the results of physiological experiments would be more helpful to understand how working memory is achieved in the brain and how information is temporarily stored and processed in the brain. It has been shown that working memory is an important function to understand functions of the prefrontal cortex. A physiologically plausible model of working memory could help to understand how the prefrontal cortex participates in working memory and what neural mechanisms in the prefrontal cortex support working memory processes. In addition, the central executive is considered to be the control mechanism for executive functions. However, it is not clear how the central executive manages executive functions and monitors and controls other systems. The prefrontal cortex has been considered to play a role as the central executive (Miyake and Shah 1999). Physiologically plausible model of working memory could help to understand how the prefrontal cortex manages executive control and monitors and controls information processing of other brain areas.

10.3 Working Memory-Related Activities in the Prefrontal Cortex

Goldman-Rakic (1987) has proposed that working memory is an important concept to understand prefrontal functions, especially the functions of the dorsolateral prefrontal cortex, in both humans and animals. Since then, the dorsolateral prefrontal cortex has been known to be an important brain area to understand neural mechanisms of working memory. The importance of this cortical area in working memory has been supported by a variety of experiments including lesion studies (see reviews by Goldman-Rakic 1987; Petrides 1994; Fuster 1997), neurophysiological studies using non-human primates (see reviews by Funahashi and Kubota 1994; Fuster 1997; Goldman-Rakic 1998; Funahashi and Takeda 2002a), and brain imaging studies using human subjects (see Stuss and Knight 2002). Close relationships between working memory and functions of the dorsolateral prefrontal cortex indicate that neurophysiological investigations in this cortical area could provide important clues to understanding common neural mechanisms for working memory, for example, how active maintenance of information is achieved and how maintained information is processed.

In fact, neurophysiological studies have shown that many neurons in the dorsolateral prefrontal cortex exhibit specific working memory-related activity, which is tonic sustained activation during the delay period (delay-period activity), while monkeys perform working memory tasks. Delay-period activity is maintained during the whole delay period and is prolonged or shortened depending on the length of the delay period. Delay-period activity is observed only when the subject performed correct responses, indicating that delay-period activity maintaining during the delay period is essential for making a correct response during the response period. Importantly, delay-period activity is observed only when the visual cue is presented in particular positions of the visual field (spatial selectivity) or only when a particular physical feature of the visual cue is needed for the response (feature selectivity), indicating that delay-period activity represents and holds a particular information during the delay period. Therefore, delay-period activity observed in working memory tasks is thought to be a neural correlate of the mechanism for temporarily active maintenance of information (Goldman-Rakic 1987; Fuster 1997; Funahashi 2001; Funahashi and Takeda 2002b).

While the subject performs the delayed-response task, a variety of task-related activities have been observed. Among these, cue-period activity, delay-period activity, and response-period activity are included. In the delayed-response task, the subject is required to perform a correct behavioral response during the response period based on the sensory cue that was presented during the cue period. Therefore, sensory information is needed to be transformed into motor information during the delay period. It has been shown that delay-period activity representing information regarding either the visual cue (retrospective information) or the response behavior (prospective information) is present in the prefrontal cortex, although delay-period activity representing retrospective information is the majority (Takeda and Funahashi 2002). This suggests that information transformation occurs during the delay period in the prefrontal cortex. In fact, the population vector analysis (Takeda and Funahashi 2004) shows that the transformation from sensory information to motor information occurs in the prefrontal cortex during the delay period and that the process for information transformation in the prefrontal cortex can be seen as a temporal change of the directions of population vectors which represent directional information represented by a population of prefrontal neural activities. Cross-correlation analysis of simultaneously recorded pairs of neural activities shows functional interactions among prefrontal neurons having either same task-related activity or different task-related activities (Funahashi and Inoue 2000; Constantinidis et al. 2001). Therefore, functional interactions among prefrontal neurons having same or different task-related activities are thought to play essential roles for temporal change of the information represented by a population of prefrontal neural activity. In addition, functional interactions among neurons having different selectivity are also thought to play important roles in information processing during the delay period.

10.4 A Process-Based Model of Working Memory

Based on the neurophysiological findings in the prefrontal cortex, we propose a process-based model of working memory to understand how information processing of working memory is achieved in the prefrontal cortex (Funahashi 2001; Funahashi 2007). As shown in Fig. 10.1, this model consists of four neural processes: a *selection process*, a *temporary information-storage process*, an *output process*, and *modulatory signals*. The prefrontal cortex receives a variety of information including not only sensory and motor information but also motivational and emotional information as well as information stored in long-term memory. Although the prefrontal cortex receives a variety of information from other cortical and subcortical areas, only specific information is needed to temporarily store and process for achieving a current goal. Therefore, a neural mechanism must select specific and relevant information for achieving a current goal among a variety of information.

This mechanism is the *selection process* and is thought to be an essential component of working memory process. To select necessary information among others, discrimination of relevant information from irrelevant information is inevitable. Therefore, the *selection process* not only functions to input selected information into the *temporary information-storage process* but also functions to discriminate between relevant and irrelevant information. The selection process may be operated within the prefrontal cortex or outside of the prefrontal cortex including sensory cortices. The top-down control signals generated in the prefrontal cortex must play important roles to discriminate relevant information from irrelevant information.

The *temporary information-storage process* is another essential component for working memory process. The relevant information selected in the *selection process* is sent to the *temporary information-storage process*. This neural process can temporarily but actively hold selected relevant information. The *temporary information-storage process* consists of neural mechanisms each of which can hold different kinds of information similar as different prefrontal neurons having delay-period activity exhibit selectivity to different stimulus features. According to the results obtained from neurophysiological experiments, prefrontal neurons or a population of neurons having delay-period activity correspond to neural mechanisms for the temporary information-storage process. The *temporary information-storage process* not only participates in temporary active holding of information but also

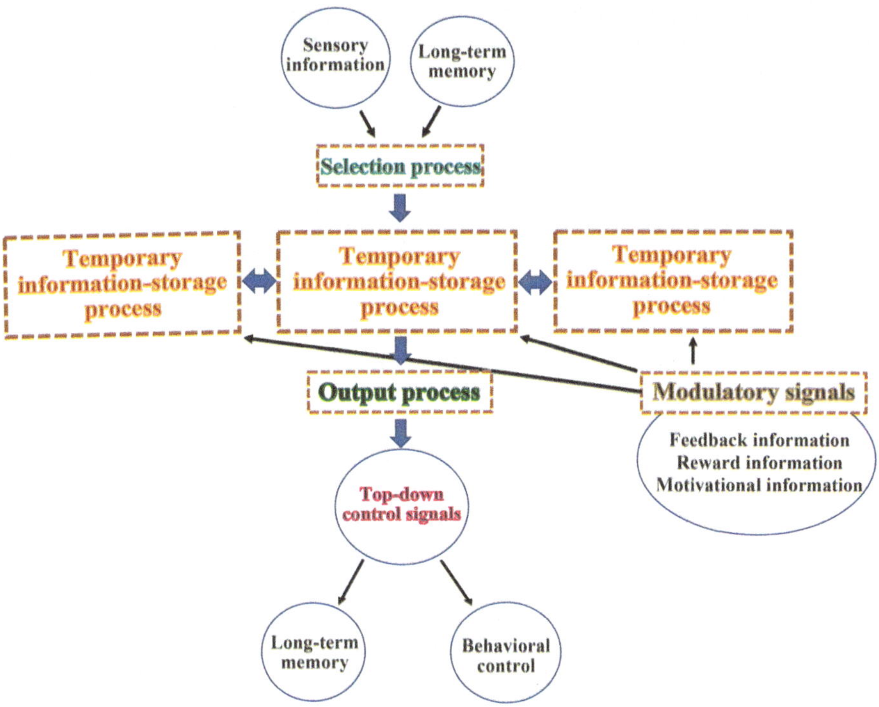

Fig. 10.1 A process-based model of working memory

participates in processing information. The information stored in neural mechanisms of the *temporary information-storage process* would be processed or modified by functional interactions among neural mechanisms holding different information, replaced by new information received from the *selection process*, or updated by the result of information processing. Therefore, functional interactions among neural mechanisms of the *temporary information-storage process*, each of which actively maintains different information, the inputs from the *selection process*, and the inputs from the *modulatory signals* would be the basic mechanism for dynamic and flexible information processing to prepare appropriate information to achieve a goal.

Further, in order for the prefrontal cortex to manage and participate in such higher cognitive functions as thinking, reasoning, decision making, and language comprehension, the prefrontal cortex not only receives a variety of information from internal as well as external environments but also selects relevant information. To receive relevant information at a given moment, the prefrontal cortex needs to monitor the results of the outputs by receiving feedback information from brain areas related to behavioral controls and long-term memory and, at the same time, the prefrontal cortex needs to control information processing of these brain areas by sending appropriate outputs to these areas. The *output process* sends appropriate information to the brain areas where the information stored in the *temporary information-storage process* is utilized for their operations. In addition, the *output process* sends top-down control signals to control the operations of other cortical and subcortical areas for achieving a specific goal. Target areas where the *output process* sends information, for example, include the motor areas to behave adaptively and flexibly and the areas related to long-term memory to retrieve appropriate information.

The *modulatory signal* includes feedback signals from motor areas, motivational or emotional information from the limbic areas (Barbas 1992), and modulatory signals by catecholaminergic or monoaminergic inputs (Sawaguchi and Goldman-Rakic 1994; Williams and Goldman-Rakic 1995; Arnsten 1998; Sawaguchi 1998; Wang et al. 2004a, b), which modulate information processing of prefrontal neurons related to the *temporary information-storage process*. The *modulatory signal* is also important for the prefrontal cortex to monitor information process operating in other brain areas. The modulatory signals are important for the flexible and adaptive operation of temporary information-storage processes and flexible modulation of functional interactions among temporary information-storage processes.

We hypothesize dynamic and flexible modulation of interactions among different processes as well as among temporary information-storage processes as the process for information processing. The strength of the functional interaction among neurons or a population of neurons has been shown to be modulated dynamically and flexibly depending on the behavioral context of the task or external and internal states. Dynamic and flexible modulation of the strength of functional interactions among neurons has been observed in prefrontal neurons (Vaadia et al. 1995; Funahashi 2001; Tsujimoto et al. 2008). In addition, modulatory signals including reward-related information and neurochemical inputs to the prefrontal cortex have also been known to play a significant role to modulate activity related to the temporary information-storage process. Therefore, information processing in working memory process can be considered as the alteration of the information represented by a neuron or a population of neurons caused by dynamic functional interactions among neurons each of which encodes different information as well as by the modulatory inputs.

10.5 Temporary Information-Storage Process

Delay-period activity, most of which is tonic sustained activation during an imposed delay period, is considered to be a neural correlate of the temporary information-storage process in the prefrontal cortex (Goldman-Rakic 1987; Funahashi and Kubota 1994; Fuster 1997; Miller 2000; Funahashi 2001; Funahashi and Takeda 2002a; Funahashi 2015, 2017a, b). Neurophysiological studies have revealed that many neurons in the dorsolateral prefrontal cortex exhibit delay-period activity while monkeys performed spatial working memory tasks such as the delayed-response task (Fuster 1973; Niki 1974; Niki and Watanabe 1976a; Kojima and Goldman-Rakic 1982; Funahashi et al. 1989). Delay-period activity has several important features, supporting that this activity is a neural correlate of the mechanisms for temporarily maintaining information during the delay period. First, the duration of delay-period activity is prolonged or shortened depending on the length of the delay period (Fuster 1973; Kojima and Goldman-Rakic 1982; Funahashi et al. 1989). Second, this activity is observed only when monkeys perform correct behavioral responses (Fuster 1973; Funahashi et al. 1989). If the monkey made an error, delay-period activity is truncated or not present. Third, a great majority of delay-period activity exhibit directional selectivity (directional delay-period activity) (Funahashi et al. 1989), such that delay-period activity is observed only when a visual cue is presented at a particular area in the visual field. Fourth, many prefrontal neurons exhibit directional delay-period activity and the preferred direction of this activity differs from neuron to neuron. Based on these observations, it has been proposed that neurons having directional delay-period activity have mnemonic receptive fields (memory fields) in the visual field (Funahashi et al. 1989; Rainer et al. 1998a, b). Fifth, using delayed pro- and anti-saccade tasks, it had been shown that the great majority (about 70%) of directional delay-period activity represent information regarding the position of the visual cue, whereas the minority represent information regarding the direction of the saccade (Funahashi et al. 1993a, b). Similar results have been observed by Niki and Watanabe (1976a, b) using a manual delayed-response task and a conditional position task and by Takeda and Funahashi (2002) using conventional and rotatory versions of the oculomotor delayed-response (ODR) task. Thus, delay-period activity represents either retrospective sensory information or prospective motor information, although the majority of delay-period activity represents retrospective sensory information in prefrontal neurons.

Delay-period activity of prefrontal neurons maintains not only visuo-spatial information but also non-spatial visual information. Experiments using non-spatial working memory tasks (e.g., delayed matching-to-sample tasks and delayed conditional tasks) have revealed that delay-period activity represents non-spatial information including faces (Wilson et al. 1993; O'Scalaidhe et al. 1997), object shapes, patterns, and colors (Sakagami and Niki 1994; Miller et al. 1996; Rao et al. 1997; Asaad et al. 1998; Rainer et al. 1999; Freedman et al. 2002).

Further, Romo et al. (1999) showed that the magnitude of delay-period activity varied as a monotonic function of the base stimulus frequency in a somatosensory discrimination task, in which the monkey was required to discriminate the difference of frequency between two mechanical vibrations (the base stimulus and the test stimulus) applied to the fingertips. Based on the result, Romo et al. (1999) concluded that the monotonic encoding of the stimulus frequency by the magnitude of delay-period activity is a fundamental method to represent the quality of one-dimensional sensory stimulus, such as vibration frequency, in working memory.

In addition, White and Wise (1999) showed behavioral rule-dependent activity in the prefrontal cortex. They showed that task-related activity of prefrontal neurons exhibited a statistically significant difference in its magnitude depending on the difference of the rules (e.g., either a conditional rule or a spatial rule), indicating that prefrontal neurons can retain the difference of the rules between behavioral tasks by modulating the magnitude of task-related activity. A similar rule-dependent difference of prefrontal activity has also been reported by Wallis et al. (2001). Further, it has been shown that delay-period activity represents reward information (Watanabe 1996; Hikosaka and Watanabe 2000; Kobayashi et al. 2002; Wallis and Miller 2003).

Thus, these results indicate that delay-period activity of prefrontal neurons maintains a wide variety of information which includes visuospatial information, non-spatial visual features, quality and quantity differences in one stimulus modality, motor outputs, and task rules. Regional difference for the difference in information processed has been proposed within the lateral prefrontal cortex. For example, Goldman-Rakic proposed that the mid-dorsolateral sector of the prefrontal cortex mainly participates in visuospatial information processing, whereas the mid-ventrolateral sector mainly participates in information processing of non-spatial visual features (Wilson et al. 1993; ÓScalaidhe et al. 1997; Goldman-Rakic 1998). Hoshi (2006) also suggested that at least two distinct subregions (dorsal sector and ventral sector) can be identified within the lateral prefrontal cortex based on the region-specific neural activity. However, it has also been shown that neurons having various task-related activity and neurons having task-related activity representing various spatial and non-spatial features are widely distributed within the lateral prefrontal cortex with substantial overlaps (Carlson et al. 1997; Quintana and Fuster 1999; Rainer et al. 1999; Sakagami and Tsutsui 1999). In addition, a number of neurons exhibit task-related activity in both spatial and non-spatial working memory conditions (Rao et al. 1997). Thus, neurons in the prefrontal cortex encode and maintain a variety of information although information encoding and maintaining is different from neuron to neuron. Neurons encoding and maintaining different information are intermingled within the lateral prefrontal cortex.

The temporary information-storage process is an essential component for working memory. In the prefrontal cortex, neurons having delay-period activity are principal elements for the temporary information-storage process. Delay-period activity of different prefrontal neurons represents different sensory features and different behavioral information. Therefore, although a variety of information is temporarily maintained and processed in the prefrontal cortex, different information is maintained as delay-period activity by different neurons or different groups of neurons in the prefrontal cortex.

10.6 Information Processing

Information processing in working memory can be described as an alteration of the information maintained in the temporary information-storage process caused by dynamic functional interactions among this process and other processes. Neurophysiological studies have revealed the presence of extensive functional interactions among prefrontal neurons having delay-period activity and interactions among prefrontal neurons having delay-period activity and those having other task-related activities.

10.6.1 Temporal Change of Delay-Period Activity Along the Delay Period

Several studies have shown that the information represented by prefrontal activity changes with the progress of the task. For example, using a delayed paired association task, Rainer et al. (1999) showed that, although prefrontal activity primarily represented characteristics of the sample stimuli (sensory-related coding) during the early phase of the delay period, this activity began to represent characteristics of anticipated targets (prospective coding) toward the end of the delay period. Similarly, using a spatial delayed matching-to-sample task, Sawaguchi and Yamane (1999) showed that spatial information was tuned broadly by delay-period activity in the early phase of the delay period. However, the proportion of neurons exhibiting sharper spatial tuning and high spatial discriminability increased in the later phase of the delay period. Further, Asaad et al. (1998) showed that neural activity conveyed the direction of the impending eye movement progressively earlier along successive trials while the monkey performed arbitrary cue-response association tasks. Quintana and Fuster (1999) observed prefrontal neurons attuned to the cue color and neurons attuned to response directions while the monkey performed working memory tasks using color cues. They found that the discharge of neurons attuned to the cue color gradually diminished during the delay period, whereas the discharge of neurons attuned to response directions gradually increased. These results indicate that the modulation of the temporal pattern of neural discharges reflects the change of the information represented by the neuron's activity and that the information represented by the temporal pattern of the firing of a neuron or an assembly of neurons alters gradually along the progress of the trial.

10.6.2 Delay-Period Activity Representing Composite Information

One evidence suggesting that functional interactions among prefrontal neurons having delay-period activity play important roles for processing and integrating information is shown by experiments using delayed-response tasks with sequential responses (Funahashi et al. 1997; Inoue and Funahashi 2002). These studies showed that delay-period activity represents composite information when the monkey performed delayed-response tasks combined with sequential hand-reaching behavior toward two different levers (Funahashi et al. 1997) or with sequential saccades to two different targets (Barone and Joseph 1989; Inoue and Funahashi 2002). Delay-period activity representing composite information includes delay-period activity representing a pair of different spatial positions, delay-period activity representing the temporal order of multiple cue presentation, and delay-period activity representing a pair of spatial positions of the visual cues as well as the temporal order of their presentation (Barone and Joseph 1989; Funahashi et al. 1997; Inoue and Funahashi 2002) (see Fig. 3.24). However, when the monkey performed a conventional delayed-response task in which a single spatial cue was presented in each trial, most of these neurons exhibited delay-period activity representing one spatial position and showed simple directional selectivity (Funahashi et al. 1997; Inoue and Funahashi 2002) (see Fig. 3.24). Therefore, these results suggest that each neuron encodes different spatial information when simple delayed-response task is introduced. However, when a complex delayed-response task, which includes multiple cue presentations and requires retention of the temporal order of cue presentations, is introduced, some neurons exhibit delay-period activity representing composite information by functional interactions with neighboring neurons whose delay-period activity represents different spatial information.

10.6.3 Interactions Among Neurons Having Different Task-Related Activity

Other example of functional interactions is observed between neurons having delay-period activity and neurons having post-saccadic activity. Many prefrontal neurons exhibit saccade-related activity when saccadic eye movements are used as behavioral responses (Boch and Goldberg 1989; Funahashi et al. 1991; Takeda and Funahashi 2002). However, most of saccade-related activities are post-saccadic (Funahashi et al. 1991; Takeda and Funahashi 2002). Post-saccadic activity is initiated after the initiation of the saccadic eye movement. Therefore, post-saccadic activity is not directly related to the initiation as well as the control of eye movement. However, the most of post-saccadic activity in the prefrontal cortex exhibits directional selectivity, such that post-saccadic activity is observed only when the saccade is made toward a particular direction. Therefore, post-saccadic activity is thought to be a feedback signal from oculomotor centers which directly participate in the initiation and the control of the eye movement (Funahashi et al. 1991). On the other hand, delay-period activity usually terminates soon after the saccade is made during the response period regardless of whether the response is correct or wrong. A comparison of temporal profiles between post-saccadic activity and delay-period activity revealed that the termination of delay-period activity coincided with the initiation of post-saccadic activity (see Fig. 3.30). This suggests that post-saccadic activity controls the timing of the termination of delay-period activity (Goldman-Rakic et al. 1990; Funahashi and Kubota 1994; Funahashi and Takeda 2002a). Once the response behavior is performed in the delayed-response task, the information maintained in the temporary information-storage process is no longer necessary, because the position of the visual cue is selected randomly from trial to trial. Therefore, erasing unnecessary information from the temporary information-storage process is an important action of working memory process. These observations indicate that inputs from neurons having post-saccadic activity terminate delay-period activity of neurons having delay-period activity, causing the erase of the information stored in the temporary information-storage process. The feedback inputs from motor centers serve as a modulatory signal of task-related activity. Functional interactions between the modulatory signals and the temporary information-storage process act for erasing unnecessary information stored in the temporary information-storage process by terminating delay-period activity. This is an important process for information processing in working memory.

10.6.4 Information Transformation During the Delay Period

Takeda and Funahashi (2002) examined prefrontal activity using two kinds of oculomotor delayed-response tasks (ODR and R-ODR tasks). In the ODR task, the monkey was required to make a saccade to the direction where the visual cue was presented, whereas in the R-ODR task, the monkey was required to make a saccade 90° clockwise from the direction where the visual cue was presented. They showed that the majority of delay-period activity represented the direction of the visual cue and that the minority of delay-period activity and the majority of response-period activity represented the direction of the saccade. These results suggest that the transformation from visuospatial information to motor information is carried out in the prefrontal cortex during the delay period. Since both visuospatial information and motor information can be depicted as direction selectivity of task-related activity, transformation from visual information to motor information can be demonstrated as the temporal change of the preferred direction of task-related activity along trials of ODR and R-ODR tasks. Takeda and Funahashi (2004) used population vectors to demonstrate the temporal change of the preferred direction of a population of prefrontal activity with the progress of a trial of the ODR task and the R-ODR task. An analysis using population vectors showed that population vectors were mostly directed toward one direction (e.g., 180° direction) during the delay period in the ODR task (see Fig. 3.27). Since the direction of the visual cue and the direction of the saccade were the same in the ODR task, this result indicates that the same directional information is maintained along the delay period. However, in the R-ODR task, the visual cue was presented in one direction (e.g., 180° direction) and the direction of the correct saccade was the 90° clockward direction (e.g., 90° direction) from the cue direction. Population vectors were directed toward one direction (e.g., 180° direction) at the beginning of the delay period. However, population vectors began to rotate in the middle of the delay period, continued to rotate slowly from one direction (e.g., 180° direction) to 90° clockward direction from that direction (e.g., 90° direction) during the late half of the delay period, and finally directed toward the new direction (e.g., 90° direction) at the response period (see Fig. 3.27). This result indicates that the information represented by a population of prefrontal activity changes during the delay period. The change of the directions of population vectors during the delay period clearly indicates neural process corresponding to the transformation from sensory information to motor information.

Fuster (1997) proposed that the mediation of cross-temporal contingency is an important function of the prefrontal cortex. He stated that the delay period is the period for the cross-temporal bridging of sensory-to-motor information transformation, which is a dynamic process for the internal transfer of information as well as a process of cross-temporal matching. The present result supports the notion that the prefrontal cortex plays a significant role in mediating cross-temporal contingency. This result also suggests that delay-period activity observed in the prefrontal cortex contributes significantly to the dynamic process related to internal information transfer.

Thus, the information temporarily stored in the prefrontal cortex as delay-period activity is shown to alter or transform from one kind to another kind with the progress of the trial depending on the demand of the task. Therefore, information processing in working memory is thought to be an alteration of the information maintained in temporary information-storage processes. This alteration is caused by functional interactions among neurons having a different task-related activity which represents different information.

10.7 Interactions Among Prefrontal Neurons

While the monkey performs the ODR task, we can observe not only prefrontal neurons exhibiting either cue-, delay-, or response-period activity but also neurons exhibiting combinations of these task-related activities. Neurons having task-related activity have specific functions to perform the task. For example, neurons having cue-period activity encode visuo-spatial information regarding where the visual cue is presented, while neurons having delay-period activity encode the information regarding either where the visual cue was presented or which direction the eye movement is going to move. Mendoza-Halliday and Martinez-Trujillo (2017) analyzed prefrontal activity while the monkey performed a working memory task, in which the monkey is required to memorize the motion direction of cue stimuli and respond by releasing a button depending on whether the motion directions are same or different between the cue and the sample stimulus. They found that prefrontal neurons exhibit a wide variety of combinations of coding strength in perceived and memorized motion directions and showed that population activity including the combination of diverse activity contains information to effectively decode perceived and memorized motion directions. The presence of neurons exhibiting a wide variety of task-related activities and their combinations is important for encoding, maintaining, and decoding perceived and memorized information effectively.

10.7.1 Information Flow During a Trial

In addition to the presence of neurons exhibiting a wide variety of task-related activities in the prefrontal cortex, Takeda and Funahashi (2007) showed the presence of the information flow among these neurons to support the performance of the task. For example, the information flow is present from neurons having cue-period activity to neurons having saccade-related activity through neurons having delay-period activity in the prefrontal cortex. While the monkey performs an oculomotor version of the delayed-response task (the ODR task), task-related activity including cue- (C), delay- (D), or response-period (R) activity or their combinations (CD, CR, DR, CDR) are observed in prefrontal neurons. In addition, by comparing direction selectivity of the same task-related activity of the same prefrontal neuron between the ODR task and the R-ODR task, it is possible to determine whether task-related activity represents the direction of the visual cue (*cue*) or the direction of the saccade (*sac*) (Takeda and Funahashi 2002). Using this method, Takeda and Funahashi (2007) characterized each prefrontal neuron based on which task-related activity is exhibited and what information (*cue* or *sac*) each task-related activity represents. As a result, prefrontal neurons were classified into 9 groups (C, D*cue*, D*sac*, CD*cue*, D*cue*R*cue*, D*sac*R*sac*, D*cue*R*sac*, CD*cue*R*cue* and CD*cue*R*sac*) (Fig. 10.2). In CD*cue*, CD*cue*R*cue*, and CD*cue*R*sac* groups, direction selectivity was the same between cue- and delay-period activity and both cue- and delay-period activity represented the direction of the visual cue, suggesting that the directional selectivity of delay-period activity is affected by the directional selectivity of cue-period activity in these groups of neurons. In D*cue*R*cue*, CD*cue*R*cue*, and D*sac*R*sac* groups, direction selectivity was also the same between delay- and response-period activity, suggesting that the directional selectivity of delay-period activity affects the directional selectivity of response-period activity in these groups of neurons. A comparison of temporal profiles of delay-period activity among these groups of

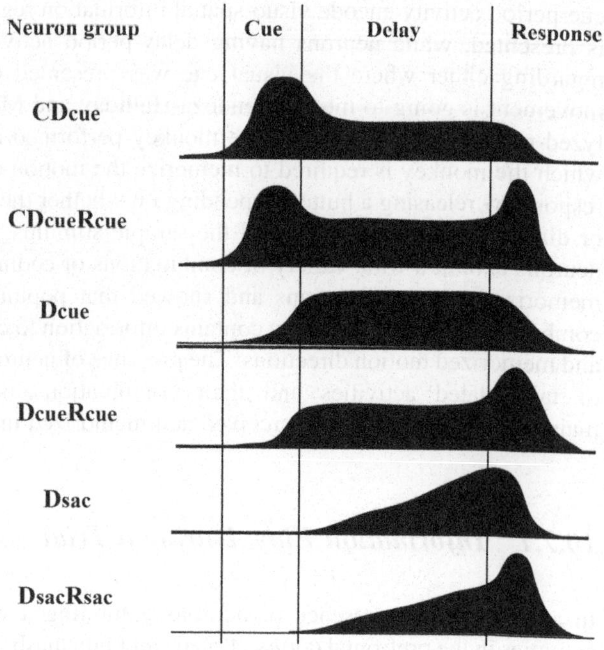

Fig. 10.2 Prefrontal neurons classified based on the combination of task-related activity and the information represented by the activity. From Takeda and Funahashi (2007) with permission

neurons (Fig. 10.2) suggests that cue-period activity contributes to initiating delay-period activity of CD*cue*, CD*cue*R*cue*, D*cue*, and D*cue*R*cue* groups of neurons and that delay-period activity of these groups of neurons maintain information regarding the direction of the visual cue. In addition, a comparison of temporal profiles of D*sac* and D*sac*R*sac* groups of neurons suggests that the directional selectivity of delay-period activity of D*sac* and D*sac*R*sac* groups affects the directional selectivity of saccade-related activity during the response period. Therefore, while the monkey performs the ODR tasks, the visuospatial information flows from neurons having cue-period activity to neurons having response-period activity through neurons having delay-period activity in the prefrontal cortex. Observed information flow must be performed by functional interactions among prefrontal neurons having different task-related activity with similar directional or stimulus selectivity. Information processing for transforming visual information into oculomotor information might be performed through this information flow.

A comparison of the directional selectivity of delay-period activity between the ODR and R-ODR tasks reveals that delay-period activity represents either visual information or saccade information in prefrontal neurons. Delay-period activity representing prospective motor information should be generated during the delay period based on the visual information. However, delay-period activity representing both visual and saccade information was not found by this comparison and it is not clear how prospective motor information is generated in the prefrontal cortex during the delay period. As seen in Fig. 10.2, delay-period activity representing prospective motor information exhibits a gradual increase of activity along the delay period, suggesting gradual recruitment of activation by excitatory inputs from other brain areas. Funahashi (2013) suggested the mediodorsal (MD) nucleus of the thalamus as a candidate brain structure that provides prospective motor information to the prefrontal cortex. In the MD, a majority of neurons having delay-period activity encode saccade information (Watanabe and Funahashi 2004a, b, 2012). Since neurons in the MD project to the prefrontal cortex, these MD neurons might be the source that provides prospective saccade information to the prefrontal cortex. However, other brain structures including the frontal eye field, the supplementary eye field, the posterior parietal cortex may also be candidates for providing prospective motor information to the prefrontal cortex. We need further experiments to elucidate how delay-period activity representing prospective saccade information is generated in the prefrontal cortex during the delay period.

10.7.2 Dynamic Modulation of Functional Interactions Among Neurons

To understand the neural mechanism of information processing through information flow from sensory-related neurons to motor-related neurons, dynamic, flexible, and adaptive modulation of functional interactions among neurons encoding different information must be a key process. In the prefrontal cortex, Wilson et al. (1994) showed that types of responses (excitatory or inhibitory) of adjacent pyramidal and non-pyramidal neurons were often inverted, such that a pyramidal neuron exhibited excitatory response while adjacent non-pyramidal neuron exhibited inhibitory response in the same condition. In addition, they showed that the timing of excitatory and inhibitory responses of these neurons appeared to be phased. This result suggests the presence of functional interaction between adjacent pyramidal and non-pyramidal neurons. Rao et al. (1999) used the cross-correlation analysis to examine functional interactions between pyramidal and non-pyramidal neurons in the prefrontal cortex and found inhibitory interactions between pyramidal neurons and adjacent non-pyramidal interneurons. Funahashi and Inoue (2000) also applied the cross-correlation analysis to simultaneously isolated pairs of prefrontal neurons during ODR performances and examined functional interactions between task-related prefrontal neurons (Fig. 10.3). Although excitatory interactions (either having a central peak or a displaced peak in the cross-correlogram) were observed among neurons having same task-related activity and among neurons having different task-related activity, the strength and the proportion of excitatory interactions were larger among neurons having same task-related activity. In addition, when a pair of neurons exhibited different task-related activities, these neurons tended to show excitatory interactions if the best directions of task-related activities were similar. Excitatory interactions were frequently observed between two neurons both having delay-period activity. A pair of neurons both exhibited delay-period activity and similar best directions showed significant excitatory interactions. These

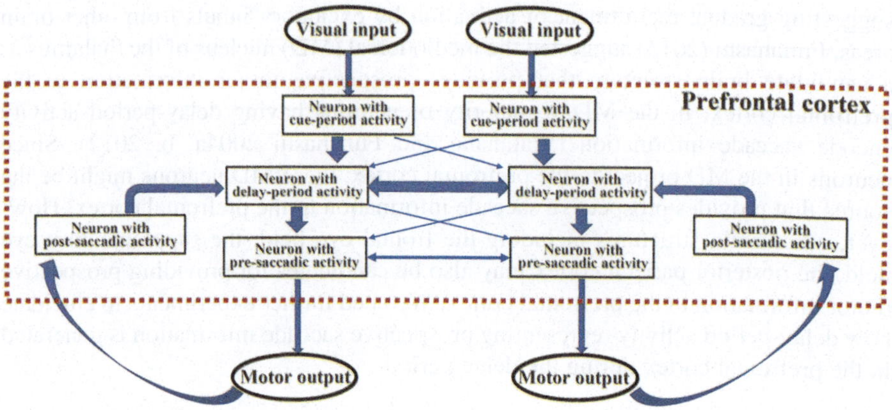

Fig. 10.3 Information flow among task-related neurons in the prefrontal cortex during ODR task performance

results indicate the presence of a wide range of functional interactions among prefrontal neurons having different task-related activity and different visuospatial selectivity. These results indicate that these functional interactions among neurons encoding different information are the fundamental mechanism for information processing in working memory.

Each prefrontal neuron changes the magnitude of task-related activity depending on the difference of the trial condition, the temporal context of the trial, and the trial event. Therefore, the strength of the functional interaction among neurons would change depending on the trial condition or the temporal context of the trial. In fact, the height of the peak of the cross-correlogram changed depending on the difference of the position where the visual cue was presented in the ODR task (Funahashi 2001). The modulation of the peak height in the cross-correlogram was observed in most of the neuron pairs examined in prefrontal neurons. These results suggest that, although functional interaction is present among neurons exhibiting same or different task-related activity, the strength of functional interaction changes dynamically depending on the trial condition, the temporal context of the trial, or the trial event. Thus, dynamic and flexible change of functional interaction among prefrontal neurons depending on the different conditions would be a fundamental mechanism for information processing in working memory.

10.8 Process-Based Model of Working Memory Can Be Applied to Many Information Processes

Working memory is an essential neural system for a variety of information processing which is operating not only in the prefrontal cortex but also in other cortical areas. Delay-period activity in the prefrontal cortex is known to represent not only visuo-spatial information but also non-spatial information including faces, object shapes, patterns, or colors, tactile information, behavioral rules, and reward information. Delay-period activity is a neural correlate of the mechanism for temporarily maintaining information, which is a neural component of the temporary information-storage process of the working memory model. Since the temporary information-storage process can maintain a variety of information, the process-based model of working memory can be applied to explain neural mechanisms for a variety of information processing including the phonological loop, the visuospatial sketchpad, and the episodic buffer of Baddeley's working memory model. Tonic sustained delay-period activity has been observed not only in the prefrontal cortex but also in the frontal eye field (Funahashi et al. 1989; Lawrence et al. 2005; Roesch and Olson 2005), the inferior temporal cortex (Fuster and Jervey 1982; Miyashita and Chang 1988; Fuster 1990; Miller et al. 1993; Naya et al. 1996), the posterior parietal cortex (Murata et al. 1996; Chafee and Goldman-Rakic 1998; Pesaran et al. 2002; Zhang and Barash 2004; Sereno and Amador 2006), and the premotor cortex (Weinrich and Wise 1982; Kurata and Wise 1988; Crammond and Kalaska 2000; Ohbayashi et al. 2003). In these cortical areas, not only temporary maintenance of information but also processing information are essential operations. Therefore, the process-based model of working memory can not only be applied to understand functions of the prefrontal cortex but also applied to understand functions of other cortical and subcortical areas where a variety of information is processed.

10.9 Hierarchical Model of Working Memory

The process-based model of working memory is originally proposed to explain functions of the prefrontal cortex. The prefrontal cortex participates not in processing for few particular kinds of information but in processing a variety of information for various purposes. Therefore, working memory functioning in the prefrontal cortex can be described as general-purpose working memory (Fig. 10.4). On the other hand, the primary visual cortex processes visual information and does not process other sensory modalities. The primary motor cortex processes skeletomotor information and does not process oculomotor information. Thus, although a wide variety of information is processed in the brain, each brain area, in general, participates in information processing for some specific purposes and contributes to achieving specific functions. In order to accomplish a specific purpose and achieve specific functions, neural mechanisms for working memory are also important neural components in this brain area and contribute to information processing for the purpose. Therefore, working memory functioning in this brain area can be described as purpose-specific working memory (Fig. 10.4). Basic neural operations are the same in both general-purpose and purpose-specific working memories and the difference between these two working memory systems is whether working memory system is used either for general information processing or for processing of specific information. Therefore, both working memory systems are

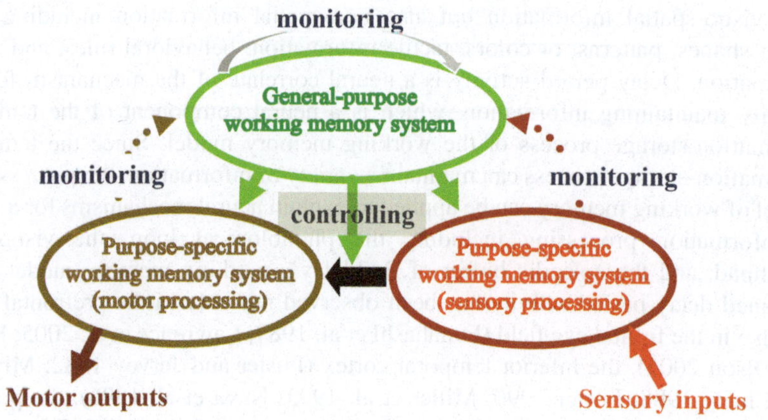

Fig. 10.4 A hierarchical model of working memory systems

thought to consist of four neural processes of the working memory, shown in Fig. 10.1.

Purpose-specific working memory system participates in information processing for specific purposes and functions, such as sensory information processing, linguistic information processing, or skeletomotor or oculomotor information processing. Purpose-specific working memory system can be found in a cortical area where a particular sensory or motor information processing is performed.

On the other hand, general-purpose working memory is a neural mechanism for maintaining and processing a variety of information which include not only information of multiple sensory and motor modalities but also information for the behavioral context. However, the most important function of the general-purpose working memory would be monitoring and controlling purpose-specific working memory systems by receiving and sending neural signals, respectively. Neural signals receiving from purpose-specific working memory systems and monitoring their performance can be called bottom-up signals, while neural signals sending to purpose-specific working memory systems and controlling their performance can be called top-down signals. Using top-down and bottom-up signals, the general-purpose working memory system can regulate the operations of purpose-specific working memory systems. Therefore, a hierarchical relation is present between the general-purpose working memory system and purpose-specific working memory systems. Top-down control signals from the general-purpose working memory system regulate information processing in purpose-specific working memory systems as well as information flow between purpose-specific working memory systems. Using both top-down and bottom-up signals flexibly and appropriately, the general-purpose working memory system leads information processing of purpose-specific working memory systems to the necessary and right direction to achieve a specific goal. The functions and proposed operations of the general-purpose working memory system correspond to the functions and operations that the prefrontal cortex participates. Therefore, prefrontal functions can be explained using this hierarchical model of working memory systems.

Cohen and others (Cohen and Servan-Schreiber 1992; Cohen et al. 1996; Miller and Cohen 2001) explained prefrontal executive functions by computational architectures with the connectionist framework. They hypothesized that the prefrontal cortex represents the context information, which corresponds to the goal representation in the computational architecture of the production system. Context information can be defined as the information necessary to mediate an appropriate behavioral response, and includes a set of task instructions, a specific prior stimulus, or the result of processing a sequence of prior stimuli (Cohen et al. 1996). To accomplish a particular goal, some brain region needs to maintain an internal representation of the goal, to suppress unnecessary behaviors, and to maintain temporally coordinate series of behaviors. Cohen et al. (1996) considered the prefrontal cortex as this cortical region. In this regard, prefrontal executive functions that Cohen and others proposed are similar to the functions of the general-purpose working memory system. The general-purpose working memory system in the prefrontal cortex not only maintains context information and the goal but also uses both top-down and bottom-up signals to control operations of purpose-specific working memory systems in sensory and motor areas, which enable to suppress unnecessary behavior and maintain coordinate series of behaviors to accomplish the goal. Thus, the hierarchical model of working memory systems is useful to explain and understand the functions of the prefrontal cortex.

10.10 Dynamic Interactions Between the Prefrontal Cortex and Other Cortical Areas

10.10.1 Anatomical Evidences

To understand how the prefrontal cortex acts as the general-purpose working memory system, it is important to consider dynamic and flexible functional interactions between the prefrontal cortex and other cortical and subcortical areas. Cortico-cortical and cortico-subcortical anatomical connections with the prefrontal cortex have been examined extensively [see reviews by Petrides (1994), Fuster (1997), Petrides and Pandya (2002)]. The prefrontal cortex has reciprocal connections with the posterior parietal cortex, the inferior temporal cortex, the superior temporal polysensory areas, the anterior cingulate cortex, the retrosplenial cortex, and the parahippocampal gyrus. The prefrontal cortex also has strong reciprocal connections with the mediodorsal nucleus of the thalamus. In addition, the prefrontal cortex has connections with the frontal eye field, the pre-supplementary motor area, the premotor cortex, and the caudate nucleus. Morris et al. (1999) reported that the mid-dorsolateral prefrontal cortex (areas 46 and 9) projects to the retrosplenial area 30 and the posterior presubiculum, and suggested that this system could be the anatomical substrate of functional interaction between the prefrontal cortex and the hippocampus. Thus, the prefrontal cortex has anatomical connections with various cortical and subcortical areas including sensory and motor association areas and the limbic brain areas. Most of these connections are reciprocal. These anatomical results indicate that the prefrontal cortex can receive various modalities of information from sensory and motor areas, emotional and motivational information, and information stored in long-term memory. Anatomical results also indicate that the prefrontal cortex can send appropriate information and signals to various brain areas and regulate information processing operated in various brain regions. Reciprocal anatomical connections between the prefrontal cortex and various brain areas are the anatomical basis of dynamic and flexible interactions between the general-purpose working memory system and purpose-specific working memory systems.

10.10.2 Non-Invasive Brain Imaging Studies

Non-invasive brain imaging studies have also revealed dynamic and flexible functional interactions between the prefrontal cortex and other cortical areas. N-back tasks have often been used as working memory tasks. Bilateral prefrontal cortices have been shown to be activated in visuospatial n-back tasks (Cohen et al. 1997; Smith and Jonides 1997, 1998). The magnitude of the activation in the prefrontal cortex depended on the memory load required to perform the task (Carlson et al. 1998; Callicott et al. 1999). However, working memory tasks like n-back tasks are composed of numerous elementary neural processes. Based on the analysis using evoked potentials, McEvoy et al. (1998) showed that, although some of these neural processes were affected by the type of information (verbal or spatial), the neural processes associated with frontal and parietal slow waves were affected not by the type of information, but by the amount of information being retained. McEvoy et al. (1998) suggested that working memory is operating by dynamic cortical networks of information-specific processes (e.g., modality-specific working memory system) and non-specific, capacity-limited, higher-order processes (e.g., general-purpose working memory system).

Dynamic cortical networks including the prefrontal cortex have been shown directly by correlation analyses in functional brain imaging studies as well as by covariance and correlation analyses of event-related potentials. Using a working memory task of faces, McIntosh et al. (1996) examined functional interactions among different cortical regions by an anatomy-based covariance structural equation model. As a result, they found that different functional interactions among different cortical areas were observed in different task conditions (e.g., different delay lengths) and that a top-down interaction from the prefrontal cortex to the temporal and occipital cortices was observed in a long delay condition. McIntosh et al. (1999) also showed that only aware subjects exhibited learning-related modulations of brain activation in the left prefrontal cortex during an associative learning task and that a correlation analysis of this modulation revealed significant functional connectivity among bilateral prefrontal cortices, the sensory association cortices, and the cerebellum. Gazzaley et al. (2004) used a multivariate analysis method of fMRI signals to examine functional connectivity among different brain areas during a face recognition task with delay. They used activity of the fusiform face area (FFA) as a seed and generated whole-brain correlation maps during the delay period. Brain areas exhibited significant correlations with the FFA including the dorsolateral and ventrolateral prefrontal cortex, the premotor cortex, the intraparietal sulcus, and the occipitotemporal cortex. These results indicate that coordinated functional interaction among widely distributed brain regions is necessary to actively maintain perceptual information as sustained activation in working memory.

Evidences for homotopic interaction or co-activation of the bilateral prefrontal cortices have been reported (McGuire et al. 1991a, b; Klingberg et al. 1997). Klingberg et al. (1997) reported the increase in bilateral co-activation of the prefrontal cortex and the inferior parietal cortex in association with the increase in task difficulty. They also found that the inferior parietal cortex was often coactivated with the prefrontal cortex in working memory tasks and tasks involving planning. Further, Huettel et al. (2005) showed that the degree of uncertainty on decision-making influenced the magnitude of activation of both prefrontal and parietal cortices. Collette et al. (2005) showed activation of both the prefrontal and the inferior parietal cortex in dual-task performances. Thus, all of these results indicate the importance of functional interactions as well as co-activation between the prefrontal cortex and other cortical and subcortical areas when subjects perform cognitive tasks. These results also indicate that functional interactions and co-activation between the prefrontal cortex and other cortical and subcortical areas are dynamically and flexibly modulated depending on the context of the cognitive task.

10.10.3 Top-Down Modulation

Evidences of top-down modulatory effects by the prefrontal cortex have also been shown. Using visual associative memory tasks, Hasegawa et al. (1998a, b) showed that retrieval of visual images from long-term memory was under the executive control of the prefrontal cortex by the experiment using partial split-brain monkeys. In addition, by neurophysiological experiments using monkeys having resection of the corpus callosum, Tomita et al. (1999) showed that top-down signals from the prefrontal cortex regulated retrieval of associative memory in inferior temporal neurons. These observations indicate that the prefrontal cortex can control information processing in other cortical areas by sending top-down signals (Miyashita 2004).

Lumer and Rees (1999) showed by fMRI that covariation of activation was observed in multiple extrastriate areas, the parietal cortex, and the prefrontal cortex in bistable viewing conditions. Coordinated activation of these areas was not linked to either sensory or motor events, but rather reflected perceptual events or internal changes of perception. Therefore, they concluded that functional interactions between visual areas and the prefrontal cortex are important for conscious vision. On the other hand, either enhancement or suppression of the visual response was observed in inferior temporal neurons when two visual stimuli (one preferred and the other non-preferred) were presented simultaneously in the neuron's receptive field. Chelazzi et al. (1998) proposed that this phenomenon can be explained by a "biased competition" model of attention. Based on this model, simultaneously presented objects in the visual receptive field would compete with others for representation within the cortex. They suggested that this competition could be biased by "top-down" feedback signals from brain structures involved in working memory.

Based on the results by Hasegawa et al. (1998a, b), Tomita et al. (1999), and Lumer and Rees (1999), it would be reasonable to think that top-down signals from the prefrontal cortex bias the magnitude of the activity in the visual system. These results indicate the presence and importance of top-down modulatory effects by the prefrontal cortex to perform cognitive tasks. These results also suggest that top-down modulatory effects by the prefrontal cortex affect a wide variety of cognitive activities including sensory perception, directed attention, episodic memory encoding and retrieval through dynamic and flexible functional interactions. This could be the typical function of the prefrontal cortex as the general-purpose working memory system.

10.11 Conclusions

A general-purpose working memory system is proposed to explain how the prefrontal cortex can perform executive control. The general-purpose working memory system can (1) represent and process various kinds of information for a variety of purposes, (2) access various neural systems (e.g., purpose-specific working memory systems), and (3) control their operations and monitor their information processing by sending top-down control signals and receiving bottom-up signals, respectively. To represent and process information in both general-purpose and purpose-specific working memory systems, both working memory systems are composed of four neural processes: selection process, temporary information-storage process, output process, and modulatory signals. Extensive functional interactions among temporary information-storage processes are important components for information processing. Dynamic and flexible interactions among these four processes are also essential features of the working memory system and important for information processing. Dynamic and flexible interactions between the general-purpose and purpose-specific working memory systems through top-down and bottom-up signals are essential mechanisms to achieve the specific goal. However, it is not known how the information relevant for a particular purpose is selected among a variety of information, how the general-purpose working memory system creates top-down signals to control a particular operation in a particular brain area, and how the general-purpose working memory system can manage a variety of functions efficiently and in a context-dependent manner. Therefore, we need further experiments to understand the neural mechanisms of working memory system and how the prefrontal cortex accomplishes a variety of functions.

References

Arnsten AFT (1998) Catecholamine modulation of prefrontal cortical cognitive function. Trends Cogn Sci 2:436–447. https://doi.org/10.1016/S1364-6613(98)01240-6

Asaad WF, Rainer G, Miller EK (1998) Neural activity in the primate prefrontal cortex during associative learning. Neuron 21:1399–1407. https://doi.org/10.1016/s0896-6273(00)80658-3

Baddeley A (1986) Working memory. Oxford University Press, Oxford, UK

Baddeley A (2000) The episodic buffer: a new component of working memory? Trends Cogn Sci 4: 417–423. https://doi.org/10.1016/S1364-6613(00)01538-2

Baddeley A, Hitch GJ (1974) Working memory. In: Bower GH (ed) The psychology of learning and motivation, vol 8. Academic Press, New York, pp 47–89

Baddeley A, Logie RH (1999) Working memory: the multiple-component model. In: Miyake A, Shah P (eds) Models of working memory: mechanisms of active maintenance and executive control. Cambridge University Press, Cambridge, UK, pp 28–61

Barbas H (1992) Architecture and cortical connections of the prefrontal cortex in the rhesus monkey. In: Chauvel AV et al (eds) Advances in neurology, vol 57. Raven Press, New York, pp 91–115

Barone P, Joseph JP (1989) Prefrontal cortex and spatial sequencing in macaque monkey. Exp Brain Res 78:447–464. https://doi.org/10.1007/BF00230234

Boch RA, Goldberg ME (1989) Participation of prefrontal neurons in the preparation of visually guided eye movements in the rhesus monkey. J Neurophysiol 61:1064–1084. https://doi.org/10.1152/jn.1989.61.5.1064

Callicott JH, Mattay VS, Bertolino A, Finn K, Coppola R, Frank JA, Goldberg TE, Weinberger DR (1999) Physiological characteristics of capacity constraints in working memory as revealed by functional MRI. Cereb Cortex 9:20–26. https://doi.org/10.1093/cercor/9.1.20

Carlson S, Rämä P, Tanila H, Linnankoski I, Mansikka H (1997) Dissociation of mnemonic coding and other functional neuronal processing in the monkey prefrontal cortex. J Neurophysiol 77: 761–774. https://doi.org/10.1152/jn.1997.77.2.761

Carlson S, Martinkauppi S, Rämä P, Salli E, Korvenoja A, Aronen HJ (1998) Distribution of cortical activation during visuospatial n-back tasks as revealed by functional magnetic resonance imaging. Cereb Cortex 8:743–752. https://doi.org/10.1093/cercor/8.8.743

Chafee MV, Goldman-Rakic PS (1998) Matching patterns of activity in primate prefrontal area 8a and parietal area 7ip neurons during a spatial working memory task. J Neurophysiol 79:2919–2940. https://doi.org/10.1152/jn.1998.79.6.2919

Chelazzi L, Duncan J, Miller EK, Desimone R (1998) Responses of neurons in inferior temporal cortex during memory-guided visual search. J Neurophysiol 80:2918–2940. https://doi.org/10.1152/jn.1998.80.6.2918

Cohen JD, Servan-Schreiber D (1992) Context, cortex, and dopamine: a connectionist approach to behavior and biology in schizophrenia. Psychol Rev 99:45–77. https://doi.org/10.1037/0033-295X.99.1.45

Cohen JD, Braver TS, O'Reilly RC (1996) A computational approach to prefrontal cortex, cognitive control and schizophrenia: recent developments and current challenges. Phil Trans R Soc Lond B 351:1515–1527. https://doi.org/10.1098/rstb.1996.0138

Cohen JD, Peristein WM, Braver TS, Nystrom LE, Noll DC, Jonides J, Smith EE (1997) Temporal dynamics of brain activation during a working memory task. Nature 386:604–607. https://doi.org/10.1038/386604a0

Collette F, Olivier L, Van der Linden M, Laureys S, Delfiore G, Luxen A, Salmon E (2005) Involvement of both prefrontal and inferior parietal cortex in dual-task performance. Cogn Brain Res 24:237–251. https://doi.org/10.1016/j.cogbrainres.2005.01.023

Constantinidis C, Franowicz MN, Goldman-Rakic PS (2001) Coding specificity in cortical microcircuits: a multiple-electrode analysis of primate prefrontal cortex. Neurosci 21:3646–3655. https://doi.org/10.1523/jneurosci.21-10-03646.2001

Crammond DJ, Kalaska JF (2000) Prior information in motor and premotor cortex: activity during the delay period and effect on pre-movement activity. J Neurophysiol 84:986–1005. https://doi.org/10.1152/jn.2000.84.2.986

Freedman DJ, Riesenhuber M, Poggio T, Miller EK (2002) Visual categorization and the primate prefrontal cortex: neurophysiology and behavior. J Neurophysiol 88:929–941. https://doi.org/10.1152/jn.00040.2002

Funahashi S (2001) Neuronal mechanisms of executive control by the prefrontal cortex. Neurosci Res 39:147–165. https://doi.org/10.1016/S0168-0102(00)00224-8

Funahashi S (2007) The prefrontal cortex as a model system to understand representation and processing of information. In: Funahashi S (ed) Representation and brain. Springer Verlag, Tokyo, pp 311–336

Funahashi S (2013) Thalamic mediodorsal nucleus and its participation in spatial working memory processes: comparison with the prefrontal cortex. Front Syst Neurosci 7:36. https://doi.org/10.3389/fnsys.2013.00036

Funahashi S (2015) Functions of delay-period activity in the prefrontal cortex and mnemonic scotomas revisited. Front Syst Neurosci 9:2. https://doi.org/10.3389/fnsys.2015.00002

Funahashi S (2017a) Working memory in the prefrontal cortex. Brain Sci 7:49. https://doi.org/10.3390/brainsci7050049

Funahashi S (2017b) Prefrontal contribution to decision-making under free-choice conditions. Front Neurosci 11:431. https://doi.org/10.3389/fnins.2017.00431

Funahashi S, Inoue M (2000) Neuronal interactions related to working memory processes in the primate prefrontal cortex revealed by cross-correlation analysis. Cereb Cortex 10:535–551. https://doi.org/10.1093/cercor/10.6.535

Funahashi S, Kubota K (1994) Working memory and prefrontal cortex. Neurosci Res 21:1–11. https://doi.org/10.1016/0168-0102(94)90063-9

Funahashi S, Takeda K (2002a) Population vector analysis by primate prefrontal neuron activities. J Biol Physics 28:527–537. https://doi.org/10.1023/A:1020309916014

Funahashi S, Takeda K (2002b) Information processes in the primate prefrontal cortex in relation to working memory processes. Rev Neurosci 13:313–346. https://doi.org/10.1515/revneuro.2002.13.4.313

Funahashi S, Bruce CJ, Goldman-Rakic PS (1989) Mnemonic coding of visual space in the monkey's dorsolateral prefrontal cortex. J Neurophysiol 61:331–349. https://doi.org/10.1152/jn.1989.61.2.331

Funahashi S, Bruce CJ, Goldman-Rakic PS (1991) Neuronal activity related to saccadic eye movements in the monkey's dorsolateral prefrontal cortex. J Neurophysiol 65:1464–1483. https://doi.org/10.1152/jn.1991.65.6.1464

Funahashi S, Bruce CJ, Goldman-Rakic PS (1993a) Dorsolateral prefrontal lesions and oculomotor delayed-response performance: evidence for mnemonic "scotomas.". J Neurosci 13:1479–1497. https://doi.org/10.1523/jneurosci.13-04-01479.1993

Funahashi S, Goldman-Rakic CMV, PS. (1993b) Prefrontal neuronal activity in rhesus monkeys performing a delayed anti-saccade task. Nature 365:753–756. https://doi.org/10.1038/365753a0

Funahashi S, Inoue M, Kubota K (1997) Delay-period activity in the primate prefrontal cortex encoding multiple spatial positions and their order of presentation. Behav Brain Res 84:203–223. https://doi.org/10.1016/S0166-4328(96)00151-9

Fuster JM (1973) Unit activity in prefrontal cortex during delayed-response performance: neuronal correlates of transient memory. J Neurophysiol 36:61–78. https://doi.org/10.1152/jn.1973.36.1.61

Fuster JM (1990) Inferotemporal units in selective visual attention and short-term memory. J Neurophysiol 64:681–697. https://doi.org/10.1152/jn.1990.64.3.681
Fuster JM (1997) The prefrontal cortex: anatomy, physiology, and neuropsychology of the frontal lobe, 3rd edn. Lippincott-Raven, Philadelphia, PA
Fuster JM, Jervey JP (1982) Neuronal firing in the inferotemporal cortex of the monkey in a visual memory task. J Neurosci 2:361–375. https://doi.org/10.1523/jneurosci.02-03-00361.1982
Gazzaley A, Rissman J, D'Esposito M (2004) Functional connectivity during working memory maintenance. Cogn Affect Behav Neurosci 4:580–599. https://doi.org/10.3758/CABN.4.4.580
Goldman-Rakic PS (1987) Circuitry of primate prefrontal cortex and regulation of behavior by representational memory. In: Plum F (ed) Handbook of physiology, section 1: the nervous system, Part 1: higher functions of the brain, vol V. American Physiological Society, Bethesda, MD, pp 373–417. https://doi.org/10.1002/cphy.cp010509
Goldman-Rakic PS (1998) The prefrontal landscape: implications of functional architecture for understanding human mentation and the central executive. In: Roberts AC, Robbins TW, Weiskrantz L (eds) The prefrontal cortex: executive and cognitive functions. Oxford University Press, Oxford, UK, pp 87–102
Goldman-Rakic PS, Funahashi S, Bruce CJ (1990) Neocortical memory circuits. Cold Spring Harbor Symp Quant Biol 55:1025–1038. https://doi.org/10.1101/SQB.1990.055.01.097
Hasegawa I, Fukushima T, Ihara T, Miyashita Y (1998a) Callosal window between prefrontal cortices: cognitive interaction to retrieve long-term memory. Science 281:814–818. https://doi.org/10.1126/science.281.5378.814
Hasegawa R, Sawaguchi T, Kubota K (1998b) Monkey prefrontal neuronal activity coding the forthcoming saccade in an oculomotor delayed matching-to-sample task. J Neurophysiol 79:322–333. https://doi.org/10.1152/jn.1998.79.1.322
Hikosaka K, Watanabe M (2000) Delay activity of orbital and lateral prefrontal neurons of the monkey varying with different rewards. Cereb Cortex 10:263–271. https://doi.org/10.1093/cercor/10.3.263
Hoshi E (2006) Functional specialization within the dorsolateral prefrontal cortex: a review of anatomical and physiological studies of non-human primates. Neurosci Res 54:73–84. https://doi.org/10.1016/j.neures.2005.10.013
Huettel SA, Song AW, McCarthy G (2005) Decisions under uncertainty: probabilistic context influences activation of prefrontal and parietal cortices. J Neurosci 25:3304–3311. https://doi.org/10.1523/jneurosci.5070-04.2005
Inoue M, Funahashi S (2002) Prefrontal delay-period activity is affected by visual cues presented outside the memory field. NeuroReport 13:2097–2101
Kieras DE, Meyer DE, Mueller S, Seymour T (1999) Insights into working memory from the perspective of the EPIC architecture for modeling skilled perceptual-motor and cognitive human performance. In: Miyake A, Shah P (eds) Models of working memory: mechanisms of active maintenance and executive control. Cambridge University Press, Cambridge, UK, pp 183–223
Klingberg T, O'Sullivan BT, Roland PE (1997) Bilateral activation of fronto-parietal networks by incrementing demand in a working memory task. Cereb Cortex 7:465–471. https://doi.org/10.1093/cercor/7.5.465
Kobayashi S, Lauwereyns J, Koizumi M, Sakagami M, Hikosaka O (2002) Influence of reward expectation on visuospatial processing in macaque lateral prefrontal cortex. J Neurophysiol 87:1488–1498. https://doi.org/10.1152/jn.00472.2001
Kojima S, Goldman-Rakic PS (1982) Delay-related activity of prefrontal neurons in rhesus monkeys performing delayed response. Brain Res 248:43–49. https://doi.org/10.1016/0006-8993(82)91145-3
Kurata K, Wise SP (1988) Premotor cortex of rhesus monkeys: set-related activity during two conditional motor tasks. Exp Brain Res 69:327–343. https://doi.org/10.1007/BF00247578
Lawrence BM, White RL III, Snyder LH (2005) Delay-period activity in visual, visuomovement, and movement neurons in the frontal eye field. J Neurophysiol 94:1498–1508. https://doi.org/10.1152/jn.00214.2005

Lumer ED, Rees G (1999) Covariation of activity in visual and prefrontal cortex associated with subjective visual perception. Proc Natl Acad Sci USA 96:1669–1673. https://doi.org/10.1073/pnas.96.4.1669

McEvoy LK, Smith ME, Gevins A (1998) Dynamic cortical networks of verbal and spatial working memory: effects of memory load and task practice. Cereb Cortex 8:563–574. https://doi.org/10.1093/cercor/8.7.563

McGuire PK, Bates JF, Goldman-Rakic PS (1991a) Interhemispheric integration: I. Symmetry and convergence of the corticocortical connections of the left and the right principal sulcus (PS) and the left and the right supplementary motor area (SMA) in the rhesus monkey. Cereb Cortex 1: 390–407. https://doi.org/10.1093/cercor/1.5.390

McGuire PK, Bates JF, Goldman-Rakic PS (1991b) Interhemispheric integration: II. Symmetry and convergence of the corticostriatal projections of the left and the right principal sulcus (PS) and the left and the right supplementary motor area (SMA) of the rhesus monkey. Cereb Cortex 1: 408–417. https://doi.org/10.1093/cercor/1.5.408

McIntosh AR, Grady CL, Haxby JV, Ungerleider LG, Horwitz B (1996) Changes in limbic and prefrontal functional interactions in a working memory task for faces. Cereb Cortex 6:571–584. https://doi.org/10.1093/cercor/6.4.571

McIntosh AR, Rajah MN, Lobaugh NJ (1999) Interactions of prefrontal cortex in relation to awareness in sensory learning. Science 284:1531–1533. https://doi.org/10.1126/science.284.5419.1531

Mendoza-Halliday D, Martinez-Trujillo JC (2017) Neuronal population coding of perceived and memorized visual features in the lateral prefrontal cortex. Nat Commun 8:15471. https://doi.org/10.1038/ncomms15471

Miller EK (2000) The prefrontal cortex and cognitive control. Nat Rev Neurosci 1:59–65. https://doi.org/10.1038/35036228

Miller EK, Cohen JD (2001) An integrative theory of prefrontal cortex function. Annu Rev Neurosci 24:167–202. https://doi.org/10.1146/annurev.neuro.24.1.167

Miller EK, Li L, Desimone R (1993) Activity of neurons in anterior inferior temporal cortex during a short-term memory task. J Neurosci 13:1460–1478. https://doi.org/10.1523/jneurosci.13-04-01460.1993

Miller EK, Erickson CA, Desimone R (1996) Neural mechanisms of visual working memory in prefrontal cortex of the macaque. J Neurosci 16:5154–5167. https://doi.org/10.1523/jneurosci.16-16-05154.1996

Miyake A, Shah P (1999) Toward unified theories of working memory: emerging general consensus, unresolved theoretical issues, and future research directions. In: Miyake A, Shah P (eds) Models of working memory: mechanisms of active maintenance and executive control. Cambridge University Press, Cambridge, UK, pp 442–481

Miyashita Y (2004) Cognitive memory: cellular and network machineries and their top-down control. Science 306:435–440. https://doi.org/10.1126/science.1101864

Miyashita Y, Chang HS (1988) Neuronal correlate of pictorial short-term memory in the primate temporal cortex. Nature 331:68–70. https://doi.org/10.1038/331068a0

Morris R, Pandya DN, Petrides M (1999) Fiber system linking the mid-dorsolateral frontal cortex with the retrosplenial/presubicular region in the rhesus monkey. J Comp Neurol 407:183–192. https://doi.org/10.1002/(SICI)1096-9861(19990503)407:2<183::AID-CNE3>3.0.CO;2-N

Murata A, Gallese V, Kaseda M, Sakata H (1996) Parietal neurons related to memory-guided hand manipulation. J Neurophysiol 75:2180–2186. https://doi.org/10.1152/jn.1996.75.5.2180

Naya Y, Sakai K, Miyashita Y (1996) Activity of primate inferotemporal neurons related to a sought target in pair-association task. Proc Natl Acad Sci USA 93:2664–2669. https://doi.org/10.1073/pnas.93.7.2664

Niki H (1974) Differential activity of prefrontal units during right and left delayed response trials. Brain Res 70:346–349. https://doi.org/10.1016/0006-8993(74)90324-2

Niki H, Watanabe M (1976a) Prefrontal unit activity and delayed response: relation to cue location versus direction of response. Brain Res 105:79–88. https://doi.org/10.1016/0006-8993(76)90924-0

Niki H, Watanabe M (1976b) Cingulate unit activity and delayed response. Brain Res 110:381–386. https://doi.org/10.1016/0006-8993(76)90412-1

Ohbayashi M, Ohki K, Miyashita Y (2003) Conversion of working memory to motor sequence in the monkey premotor cortex. Science 301:233–236. https://doi.org/10.1126/science.1084884

ÓScalaidhe SP, Wilson FAW, Goldman-Rakic PS (1997) Areal segregation of face-processing neurons in prefrontal cortex. Science 278:1135–1138. https://doi.org/10.1126/science.278.5340.1135

Pesaran B, Pezaris JS, Sahani M, Mitra PP, Andersen RA (2002) Temporal structure in neuronal activity during working memory in macaque parietal cortex. Nat Neurosci 5:805–811. https://doi.org/10.1038/nn890

Petrides M (1994) Frontal lobes and working memory: evidence from investigations of the effects of cortical excisions in nonhuman primates. In: Boller F, Spinnler H, Hendler JA (eds) Handbook of neuropsychology, vol 9. Elsevier, Amsterdam, pp 17–58

Petrides M, Pandya DN (2002) Association pathways of the prefrontal cortex and functional observations. In: Stuss DF, Knight RT (eds) Principles of frontal lobe function. Oxford University Press, New York, pp 31–50

Quintana J, Fuster JM (1999) From perception to action: temporal integrative functions of prefrontal and parietal neurons. Cereb Cortex 9:213–221. https://doi.org/10.1093/cercor/9.3.213

Rainer G, Asaad WF, Miller EK (1998a) Selective representation of relevant information by neurons in the primate prefrontal cortex. Nature 393:577–579. https://doi.org/10.1038/31235

Rainer G, Asaad WF, Miller EK (1998b) Memory fields of neurons in the primate prefrontal cortex. Proc Natl Acad Sci USA 95:15008–15013

Rainer G, Rao SC, Miller EK (1999) Prospective coding for objects in primate prefrontal cortex. J Neurosci 19:5493–5505. https://doi.org/10.1523/jneurosci.19-13-05493.1999

Rao SC, Rainer G, Miller EK (1997) Integration of what and where in the primate prefrontal cortex. Science 276:821–824. https://doi.org/10.1126/science.276.5313.821

Rao SG, Williams GV, Goldman-Rakic PS (1999) Isodirectional tuning of adjacent interneurons and pyramidal cells during working memory: evidence for microcolumnar organization in PFC. J Neurophysiol 81:1903–1916. https://doi.org/10.1152/jn.1999.81.4.1903

Roesch MR, Olson CR (2005) Neuronal activity dependent on anticipated and elapsed delay in macaque prefrontal cortex, frontal and supplementary eye fields, and premotor cortex. J Neurophysiol 94:1469–1497. https://doi.org/10.1152/jn.00064-2005

Romo R, Brody CD, Hernández A, Lemus L (1999) Neuronal correlates of parametric working memory in the prefrontal cortex. Nature 399:470–473. https://doi.org/10.1038/20939

Sakagami M, Niki H (1994) Encoding of behavioral significance of visual stimuli by primate prefrontal neurons: relation to relevant task conditions. Exp Brain Res 97:423–436. https://doi.org/10.1007/BF00241536

Sakagami M, Tsutsui K (1999) The hierarchical organization of decision making in the primate prefrontal cortex. Neurosci Res 34:79–89. https://doi.org/10.1016/S0168-0102(99)00038-3

Sawaguchi T (1998) Attenuation of delay-period activity of monkey prefrontal neurons by an alpha2-adrenergic antagonist during an oculomotor delayed-response task. J Neurophysiol 80:2200–2205. https://doi.org/10.1152/jn.1998.80.4.2200

Sawaguchi T, Goldman-Rakic PS (1994) The role of D1-dopamine receptor in working memory: local injections of dopamine antagonists into the prefrontal cortex of rhesus monkeys performing an oculomotor delayed-response task. J Neurophysiol 71:515–528. https://doi.org/10.1152/jn.1994.71.2.5152

Sawaguchi T, Yamane I (1999) Properties of delay-period neuronal activity in the monkey dorsolateral prefrontal cortex during a spatial delayed matching-to-sample task. J Neurophysiol 82:2070–2080. https://doi.org/10.1152/1999.82.5.2070

Sereno AB, Amador SC (2006) Attention and memory-related responses of neurons in the lateral intraparietal area during spatial and shape-delayed match-to-sample tasks. J Neurophysiol 95: 1078–1098. https://doi.org/10.1152/jn.00491.2005
Smith EE, Jonides J (1997) Working memory: a review from neuroimaging. Cogn Psychol 33:5–42. https://doi.org/10.1006/cogp.1997.0658
Smith EE, Jonides J (1998) Neuroimaging analyses of human working memory. Proc Natl Acad Sci USA 95:12061–12068. https://doi.org/10.1073/pnas.95.20.12061
Stuss DT, Knight RT (2002) Principles of frontal lobe function. Oxford University Press, New York, NY
Takeda K, Funahashi S (2002) Prefrontal task-related activity representing visual cue location or saccade direction in spatial working memory tasks. J Neurophysiol 87:567–588. https://doi.org/10.1152/jn.00249.2001
Takeda K, Funahashi S (2004) Population vector analysis of primate prefrontal activity during spatial working memory. Cereb Cortex 14:1328–1339. https://doi.org/10.1093/cercor/bhh093
Takeda K, Funahashi S (2007) Relationship between prefrontal task-related activity and information flow during spatial working memory performances. Cortex 43:38–52. https://doi.org/10.1016/S0010-9452(08)70444-1
Tomita H, Ohbayashi M, Nakahara K, Hasegawa I, Miyashita Y (1999) Top-down signal from prefrontal cortex in executive control of memory retrieval. Nature 401:699–703. https://doi.org/10.1038/44372
Tsujimoto S, Genovesio A, Wise SP (2008) Transient neuronal correlations underlying foal selection and maintenance in prefrontal cortex. Cereb Cortex 18:2748–2761. https://doi.org/10.1093/cercor/bhn033
Vaadia E, Haalman I, Abeles M, Bergman H, Prut Y, Slovin H, Aertsen A (1995) Dynamics of neuronal interactions in monkey cortex in relation to behavioral events. Nature 373:515–518. https://doi.org/10.1038/373515a0
Wallis JD, Miller EK (2003) Neuronal activity in primate dorsolateral and orbital prefrontal cortex during performance of a reward preference task. Eur J Neurosci 18:2069–2081. https://doi.org/10.1046/j.1460-9568.2003.02922.x
Wallis JD, Anderson KC, Miller EK (2001) Single neurons in prefrontal cortex encode abstract rules. Nature 411:953–956. https://doi.org/10.1038/35082081
Wang M, Vijayraghavan S, Goldman-Rakic PS (2004a) Selective D2 receptor actions on the functional circuitry of working memory. Science 303:853–856. https://doi.org/10.1126/science.1091162
Wang X-J, Tegner J, Constantinidis C, Goldman-Rakic PS (2004b) Division of labor among distinct subtypes of inhibitory neurons in a cortical microcircuit of working memory. Proc Natl Acad Sci USA 101:1368–1373. https://doi.org/10.1073/pnas.0305337101
Watanabe M (1996) Reward expectancy in primate prefrontal neurons. Nature 382:629–632. https://doi.org/10.1038/382629a0
Watanabe Y, Funahashi S (2004a) Neuronal activity throughout the primate mediodorsal nucleus of the thalamus during oculomotor delayed-response. I. Cue-, delay-, and response-period activity. J Neurophysiol 92:1738–1755. https://doi.org/10.1152/jn.00994.2003
Watanabe Y, Funahashi S (2004b) Neuronal activity throughout the primate mediodorsal nucleus of the thalamus during oculomotor delayed-response. II. Activity encoding visual versus motor signal. J Neurophysiol 92:1756–1769. https://doi.org/10.1152/jn.00995.2003
Watanabe Y, Funahashi S (2012) Thalamic mediodorsal nucleus and working memory. Neurosci Biobehav Rev 36:134–142. https://doi.org/10.1016/j.neubiorev.2011.05.003
Weinrich M, Wise SP (1982) The premotor cortex of the monkey. J Neurosci 2:1329–1345. https://doi.org/10.1523/jneurosci.02-09-01329.1982
White IM, Wise SP (1999) Rule-dependent neuronal activity in the prefrontal cortex. Exp Brain Res 126:315–335. https://doi.org/10.1007/s002210050740

Williams GV, Goldman-Rakic PS (1995) Modulation of memory fields by dopamine D1 receptors in prefrontal cortex. Nature 376:572–575. https://doi.org/10.1038/376572a0

Wilson FAW, Scalaidhe O, SP, Goldman-Rakic PS. (1993) Dissociation of object and spatial processing domains in primate prefrontal cortex. Science 260:1955–1958. https://doi.org/10.1126/science.8316836

Wilson FAW, O'Scalaidhe SP, Goldman-Rakic PS (1994) Functional synergism between putative gamma-aminobutyrate-containing neurons and pyramidal neurons in prefrontal cortex. Proc Natl Acad Sci USA 91:4009–4013. https://doi.org/10.1073/pnas.91.9.4009

Zhang M, Barash S (2004) Persistent LIP activity in memory antisaccades: working memory for a sensorimotor transformation. J Neurophysiol 91:1424–1441. https://doi.org/10.1152/jn.00504.2003

The manufacturer's authorised representative in the EU is Springer Nature Customer Service Centre GmbH, Europaplatz 3, 69115 Heidelberg, Germany. If you have any concerns regarding our products, please contact ProductSafety@springernature.com

Printed and bound by CPI Group (UK) Ltd, Croydon, CR0 4YY
30/09/2026
02225245-0019